COURSE OF THEORETICAL PHYSICS

Volume 3

QUANTUM MECHANICS

Non-relativistic Theory

Third edition, revised and enlarged

Other Titles in the Course of Theoretical Physics

Vol. 1. MECHANICS*

Vol. 2. THE CLASSICAL THEORY OF FIELDS

Vol. 4. RELATIVISTIC QUANTUM THEORY (published in two parts)

Vol. 5. STATISTICAL PHYSICS

Vol. 6. FLUID MECHANICS

Vol. 7. THEORY OF ELASTICITY

Vol. 8. ELECTRODYNAMICS OF CONTINUOUS MEDIA

Vol. 9. PHYSICAL KINETICS

Also of interest:

A Shorter Course of Theoretical Physics
(*Based on the Course of Theoretical Physics*)

Vol. 1. Mechanics and Electrodynamics

Vol. 2. Quantum Mechanics

* Latest English edition published 1976

QUANTUM MECHANICS

NON-RELATIVISTIC THEORY

by

L. D. LANDAU AND E. M. LIFSHITZ
INSTITUTE OF PHYSICAL PROBLEMS, U.S.S.R. ACADEMY OF SCIENCES

Volume 3 of *Course of Theoretical Physics*

Third edition, revised and enlarged with the assistance of L. P. PITAEVSKII

Translated from the Russian by
J. B. SYKES AND J. S. BELL

PERGAMON PRESS

OXFORD . NEW YORK . TORONTO . SYDNEY
PARIS . FRANKFURT

U.K.	Pergamon Press Ltd., Headington Hill Hall, Oxford OX3 0BW, England
U.S.A.	Pergamon Press Inc., Maxwell House, Fairview Park, Elmsford, New York 10523, U.S.A.
CANADA	Pergamon of Canada Ltd., 75 The East Mall, Toronto, Ontario, Canada
AUSTRALIA	Pergamon Press (Aust.) Pty. Ltd., 19a Boundary Street, Rushcutters Bay, N.S.W. 2011, Australia
FRANCE	Pergamon Press SARL, 24 rue des Ecoles, 75240 Paris, Cedex 05, France
FEDERAL REPUBLIC OF GERMANY	Pergamon Press GmbH, 6242 Kronberg-Taunus, Pferdstrasse 1, Federal Republic of Germany

First published in English in 1958

2nd Impression 1959

3rd Impression 1962

2nd Revised edition 1965

3rd Revised edition 1977

Translated from the third edition of *Kvantovaya mekhanika: nerelyativistskaya teoriya*, Izdatel'stvo "Nauka" Moscow, 1974

Library of Congress Cataloging in Publication Data

Landau, Lev Davidovich, 1908–1968.
Quantum mechanics.

(Their course of theoretical physics; v. 3)
Translation of Kvantovaia mekhanika.
Includes bibliographical references and index.
1. Quantum theory. I. Lifshitz, Evgenii Mikhailovich, joint author. II. Title.
QC174.12.L3513 1976 530.1'2 76-18223
ISBN 0-08-020940-8 (Hardcover)
ISBN 0-08-019012-X (Flexicover)

Printed by A. Wheaton & Co. Ltd., Exeter

CONTENTS

	Page
From the Preface to the first English edition	xi
Preface to the second English edition	xii
Preface to the third Russian edition	xiii
Notation	xiv

I. THE BASIC CONCEPTS OF QUANTUM MECHANICS

§1.	The uncertainty principle	1
§2.	The principle of superposition	6
§3.	Operators	8
§4.	Addition and multiplication of operators	13
§5.	The continuous spectrum	15
§6.	The passage to the limiting case of classical mechanics	19
§7.	The wave function and measurements	21

II. ENERGY AND MOMENTUM

§8.	The Hamiltonian operator	25
§9.	The differentiation of operators with respect to time	26
§10.	Stationary states	27
§11.	Matrices	30
§12.	Transformation of matrices	35
§13.	The Heisenberg representation of operators	37
§14.	The density matrix	38
§15.	Momentum	41
§16.	Uncertainty relations	45

III. SCHRÖDINGER'S EQUATION

§17.	Schrödinger's equation	50
§18.	The fundamental properties of Schrödinger's equation	53
§19.	The current density	55
§20.	The variational principle	58
§21.	General properties of motion in one dimension	60
§22.	The potential well	63
§23.	The linear oscillator	67
§24.	Motion in a homogeneous field	74
§25.	The transmission coefficient	76

IV. ANGULAR MOMENTUM

§26.	Angular momentum	82
§27.	Eigenvalues of the angular momentum	86
§28.	Eigenfunctions of the angular momentum	89

Page

§29. Matrix elements of vectors 92
§30. Parity of a state 96
§31. Addition of angular momenta 99

V. MOTION IN A CENTRALLY SYMMETRIC FIELD

§32. Motion in a centrally symmetric field 102
§33. Spherical waves 105
§34. Resolution of a plane wave 112
§35. Fall of a particle to the centre 114
§36. Motion in a Coulomb field (spherical polar coordinates) 117
§37. Motion in a Coulomb field (parabolic coordinates) 128

VI. PERTURBATION THEORY

§38. Perturbations independent of time 133
§39. The secular equation 138
§40. Perturbations depending on time 142
§41. Transitions under a perturbation acting for a finite time 146
§42. Transitions under the action of a periodic perturbation 151
§43. Transitions in the continuous spectrum 154
§44. The uncertainty relation for energy 157
§45. Potential energy as a perturbation 159

VII. THE QUASI-CLASSICAL CASE

§46. The wave function in the quasi-classical case 164
§47. Boundary conditions in the quasi-classical case 167
§48. Bohr and Sommerfeld's quantization rule 170
§49. Quasi-classical motion in a centrally symmetric field 175
§50. Penetration through a potential barrier 178
§51. Calculation of the quasi-classical matrix elements 185
§52. The transition probability in the quasi-classical case 189
§53. Transitions under the action of adiabatic perturbations 194

VIII. SPIN

§54. Spin 197
§55. The spin operator 201
§56. Spinors 204
§57. The wave functions of particles with arbitrary spin 208
§58. The operator of finite rotations 213
§59. Partial polarization of particles 219
§60. Time reversal and Kramers' theorem 221

IX. IDENTITY OF PARTICLES

§61. The principle of indistinguishability of similar particles 225
§62. Exchange interaction 228

Page

§63. Symmetry with respect to interchange 232
§64. Second quantization. The case of Bose statistics 239
§65. Second quantization. The case of Fermi statistics 245

X. THE ATOM

§66. Atomic energy levels 249
§67. Electron states in the atom 250
§68. Hydrogen-like energy levels 254
§69. The self-consistent field 255
§70. The Thomas–Fermi equation 259
§71. Wave functions of the outer electrons near the nucleus 264
§72. Fine structure of atomic levels 265
§73. The Mendeleev periodic system 269
§74. X-ray terms 277
§75. Multipole moments 279
§76. An atom in an electric field 282
§77. A hydrogen atom in an electric field 287

XI. THE DIATOMIC MOLECULE

§78. Electron terms in the diatomic molecule 298
§79. The intersection of electron terms 300
§80. The relation between molecular and atomic terms 303
§81. Valency 307
§82. Vibrational and rotational structures of singlet terms in the diatomic molecule 314
§83. Multiplet terms. Case *a* 319
§84. Multiplet terms. Case *b* 323
§85. Multiplet terms. Cases *c* and *d* 327
§86. Symmetry of molecular terms 329
§87. Matrix elements for the diatomic molecule 332
§88. Λ-doubling 336
§89. The interaction of atoms at large distances 339
§90. Pre-dissociation 342

XII. THE THEORY OF SYMMETRY

§91. Symmetry transformations 354
§92. Transformation groups 357
§93. Point groups 360
§94. Representations of groups 368
§95. Irreducible representations of point groups 376
§96. Irreducible representations and the classification of terms 380

Page

§97. Selection rules for matrix elements 383
§98. Continuous groups 387
§99. Two-valued representations of finite point groups 391

XIII. POLYATOMIC MOLECULES

§100. The classification of molecular vibrations 396
§101. Vibrational energy levels 403
§102. Stability of symmetrical configurations of the molecule 405
§103. Quantization of the rotation of a top 410
§104. The interaction between the vibrations and the rotation of the molecule 419
§105. The classification of molecular terms 423

XIV. ADDITION OF ANGULAR MOMENTA

§106. 3*j*-symbols 431
§107. Matrix elements of tensors 439
§108. 6*j*-symbols 442
§109. Matrix elements for addition of angular momenta 448
§110. Matrix elements for axially symmetric systems 450

XV. MOTION IN A MAGNETIC FIELD

§111. Schrödinger's equation in a magnetic field 453
§112. Motion in a uniform magnetic field 456
§113. An atom in a magnetic field 461
§114. Spin in a variable magnetic field 468
§115. The current density in a magnetic field 470

XVI. NUCLEAR STRUCTURE

§116. Isotopic invariance 472
§117. Nuclear forces 476
§118. The shell model 480
§119. Non-spherical nuclei 489
§120. Isotopic shift 494
§121. Hyperfine structure of atomic levels 496
§122. Hyperfine structure of molecular levels 499

XVII. ELASTIC COLLISIONS

§123. The general theory of scattering 502
§124. An investigation of the general formula 505
§125. The unitary condition for scattering 508
§126. Born's formula 512
§127. The quasi-classical case 518

Page

§128. Analytical properties of the scattering amplitude 523
§129. The dispersion relation 529
§130. The scattering amplitude in the momentum representation 532
§131. Scattering at high energies 535
§132. The scattering of slow particles 542
§133. Resonance scattering at low energies 548
§134. Resonance at a quasi-discrete level 555
§135. Rutherford's formula 560
§136. The system of wave functions of the continuous spectrum 563
§137. Collisions of like particles 567
§138. Resonance scattering of charged particles 570
§139. Elastic collisions between fast electrons and atoms 575
§140. Scattering with spin–orbit interaction 579
§141. Regge poles 585

XVIII. INELASTIC COLLISIONS

§142. Elastic scattering in the presence of inelastic processes 591
§143. Inelastic scattering of slow particles 597
§144. The scattering matrix in the presence of reactions 599
§145. Breit and Wigner's formulae 603
§146. Interaction in the final state in reactions 611
§147. Behaviour of cross-sections near the reaction threshold 614
§148. Inelastic collisions between fast electrons and atoms 620
§149. The effective retardation 629
§150. Inelastic collisions between heavy particles and atoms 633
§151. Scattering of neutrons 636
§152. Inelastic scattering at high energies 640

MATHEMATICAL APPENDICES

§a. Hermite polynomials 647
§b. The Airy function 650
§c. Legendre polynomials 652
§d. The confluent hypergeometric function 655
§e. The hypergeometric function 659
§f. The calculation of integrals containing confluent hypergeometric functions 662

Index 667

FROM THE PREFACE TO THE FIRST ENGLISH EDITION

THE present book is one of the series on *Theoretical Physics*, in which we endeavour to give an up-to-date account of various departments of that science. The complete series will contain the following nine volumes:

1. *Mechanics*. 2. *The classical theory of fields*. 3. *Quantum mechanics (non-relativistic theory)*. 4. *Relativistic quantum theory*. 5. *Statistical physics*. 6. *Fluid mechanics*. 7. *Theory of elasticity*. 8. *Electrodynamics of continuous media*. 9. *Physical kinetics*.

Of these, volumes 4 and 9 remain to be written.

The scope of modern theoretical physics is very wide, and we have, of course, made no attempt to discuss in these books all that is now included in the subject. One of the principles which guided our choice of material was not to deal with those topics which could not properly be expounded without at the same time giving a detailed account of the existing experimental results. For this reason the greater part of nuclear physics, for example, lies outside the scope of these books. Another principle of selection was not to discuss very complicated applications of the theory. Both these criteria are, of course, to some extent subjective.

We have tried to deal as fully as possible with those topics that are included. For this reason we do not, as a rule, give references to the original papers, but simply name their authors. We give bibliographical references only to work which contains matters not fully expounded by us, which by their complexity lie "on the borderline" as regards selection or rejection. We have tried also to indicate sources of material which might be of use for reference. Even with these limitations, however, the bibliography given makes no pretence of being exhaustive.

We attempt to discuss general topics in such a way that the physical significance of the theory is exhibited as clearly as possible, and then to build up the mathematical formalism. In doing so, we do not aim at "mathematical rigour" of exposition, which in theoretical physics often amounts to self-deception.

The present volume is devoted to non-relativistic quantum mechanics. By "relativistic theory" we here mean, in the widest sense, the theory of all quantum phenomena which significantly depend on the velocity of light. The volume on this subject (volume 4) will therefore contain not only Dirac's relativistic theory and what is now known as quantum electrodynamics, but also the whole of the quantum theory of radiation.

Institute of Physical Problems — L. D. LANDAU
USSR Academy of Sciences — E. M. LIFSHITZ

August 1956

PREFACE TO THE SECOND ENGLISH EDITION

For this second edition the book has been considerably revised and enlarged, but the general plan and style remain as before. Every chapter has been revised. In particular, extensive changes have been made in the sections dealing with the theory of the addition of angular momenta and with collision theory. A new chapter on nuclear structure has been added; in accordance with the general plan of the course, the subjects in question are discussed only to the extent that is proper without an accompanying detailed analysis of the experimental results.

We should like to express our thanks to all our many colleagues whose comments have been utilized in the revision of the book. Numerous comments were received from V. L. Ginzburg and Ya. A. Smorodinskiĭ. We are especially grateful to L. P. Pitaevskiĭ for the great help which he has given in checking the formulae and the problems.

Our sincere thanks are due to Dr. Sykes and Dr. Bell, who not only translated excellently both the first and the second edition of the book, but also made a number of useful comments and assisted in the detection of various misprints in the first edition.

Finally, we are grateful to the Pergamon Press, which always acceded to our requests during the production of the book.

L. D. Landau
E. M. Lifshitz

October 1964

PREFACE TO THE THIRD RUSSIAN EDITION

THE previous edition of this volume was the last book on which I worked together with my teacher L. D. Landau. The revision and expansion that we then carried out was very considerable, and affected every chapter.

For the third edition, naturally, much less revision was needed. Nevertheless, a fair amount of new material has been added, including some more problems, and relating both to recent research and to earlier results that have now become of greater significance.

Landau's astonishing grasp of theoretical physics often enabled him to dispense with any consultation of original papers: he was able to derive results by methods of his own choice. This may have been the reason why our book did not contain certain necessary references to other authors. In the present edition, I have tried to supply them as far as possible. I have also added references to the work of Landau himself where we describe results or methods that are due to him personally and have not been published elsewhere.

As when dealing with the revision of other volumes in the *Course of Theoretical Physics*, I have had the assistance of numerous colleagues who informed me either of deficiencies in the treatment given previously, or of new material that should be added. Many useful suggestions incorporated in this book have come from A. M. Brodskii, G. F. Drukarev, I. G. Kaplan, V. P. Krainov, I. B. Levinson, P. E. Nemirovskii, V. L. Pokrovskii, I. I. Sobel'man, and I. S. Shapiro. My sincere thanks are due to all of these.

The whole of the work on revising this volume has been done in close collaboration with L. P. Pitaevskii. In him I have had the good fortune to find a fellow-worker who has passed likewise through the school of Landau and is inspired by the same ideals in the service of science.

E. M. LIFSHITZ

Moscow
November 1973

NOTATION

Operators are denoted by a circumflex: $\hat{f}$

$\mathrm{d}V$ volume element in coordinate space

$\mathrm{d}q$ element in configuration space

d^3p element in momentum space

$f_{nm} = f^n_m = \langle n|f|m\rangle$ matrix elements of the quantity f (see definition in §11)

$\omega_{nm} = (E_n - E_m)/\hbar$ transition frequency

$\{\hat{f}, \hat{g}\} = \hat{f}\hat{g} - \hat{g}\hat{f}$ commutator of two operators

$\hat{H}$ Hamiltonian

δ_l phase shifts of wave functions

Atomic and Coulomb units (see beginning of §36)

Vector and tensor indices are denoted by Latin letters i, k, l

e_{ikl} antisymmetric unit tensor (see §26)

References to other volumes in the *Course of Theoretical Physics*:
Mechanics = Vol. 1 (*Mechanics*, third English edition, 1976).
Fields = Vol. 2 (*The Classical Theory of Fields*, fourth English edition, 1975).
RQT or *Relativistic Quantum Theory* = Vol. 4 (*Relativistic Quantum Theory*, first English edition, Part 1, 1971; Part 2, 1974).
All are published by Pergamon Press.

CHAPTER I

THE BASIC CONCEPTS OF QUANTUM MECHANICS

§1. The uncertainty principle

WHEN we attempt to apply classical mechanics and electrodynamics to explain atomic phenomena, they lead to results which are in obvious conflict with experiment. This is very clearly seen from the contradiction obtained on applying ordinary electrodynamics to a model of an atom in which the electrons move round the nucleus in classical orbits. During such motion, as in any accelerated motion of charges, the electrons would have to emit electromagnetic waves continually. By this emission, the electrons would lose their energy, and this would eventually cause them to fall into the nucleus. Thus, according to classical electrodynamics, the atom would be unstable, which does not at all agree with reality.

This marked contradiction between theory and experiment indicates that the construction of a theory applicable to atomic phenomena—that is, phenomena occurring in particles of very small mass at very small distances—demands a fundamental modification of the basic physical concepts and laws.

As a starting-point for an investigation of these modifications, it is convenient to take the experimentally observed phenomenon known as *electron diffraction*.† It is found that, when a homogeneous beam of electrons passes through a crystal, the emergent beam exhibits a pattern of alternate maxima and minima of intensity, wholly similar to the diffraction pattern observed in the diffraction of electromagnetic waves. Thus, under certain conditions, the behaviour of material particles—in this case, the electrons—displays features belonging to wave processes.

How markedly this phenomenon contradicts the usual ideas of motion is best seen from the following imaginary experiment, an idealization of the experiment of electron diffraction by a crystal. Let us imagine a screen impermeable to electrons, in which two slits are cut. On observing the passage of a beam of electrons‡ through one of the slits, the other being covered, we obtain, on a continuous screen placed behind the slit, some pattern of intensity distribution; in the same way, by uncovering the second

† The phenomenon of electron diffraction was in fact discovered after quantum mechanics was invented. In our discussion, however, we shall not adhere to the historical sequence of development of the theory, but shall endeavour to construct it in such a way that the connection between the basic principles of quantum mechanics and the experimentally observed phenomena is most clearly shown.

‡ The beam is supposed so rarefied that the interaction of the particles in it plays no part.

slit and covering the first, we obtain another pattern. On observing the passage of the beam through both slits, we should expect, on the basis of ordinary classical ideas, a pattern which is a simple superposition of the other two: each electron, moving in its path, passes through one of the slits and has no effect on the electrons passing through the other slit. The phenomenon of electron diffraction shows, however, that in reality we obtain a diffraction pattern which, owing to interference, does not at all correspond to the sum of the patterns given by each slit separately. It is clear that this result can in no way be reconciled with the idea that electrons move in paths.

Thus the mechanics which governs atomic phenomena—*quantum mechanics* or *wave mechanics*—must be based on ideas of motion which are fundamentally different from those of classical mechanics. In quantum mechanics there is no such concept as the path of a particle. This forms the content of what is called the *uncertainty principle*, one of the fundamental principles of quantum mechanics, discovered by W. Heisenberg in 1927.†

In that it rejects the ordinary ideas of classical mechanics, the uncertainty principle might be said to be negative in content. Of course, this principle in itself does not suffice as a basis on which to construct a new mechanics of particles. Such a theory must naturally be founded on some positive assertions, which we shall discuss below (§2). However, in order to formulate these assertions, we must first ascertain the statement of the problems which confront quantum mechanics. To do so, we first examine the special nature of the interrelation between quantum mechanics and classical mechanics. A more general theory can usually be formulated in a logically complete manner, independently of a less general theory which forms a limiting case of it. Thus, relativistic mechanics can be constructed on the basis of its own fundamental principles, without any reference to Newtonian mechanics. It is in principle impossible, however, to formulate the basic concepts of quantum mechanics without using classical mechanics. The fact that an electron‡ has no definite path means that it has also, in itself, no other dynamical characteristics.‖ Hence it is clear that, for a system composed only of quantum objects, it would be entirely impossible to construct any logically independent mechanics. The possibility of a quantitative description of the motion of an electron requires the presence also of physical objects which obey classical mechanics to a sufficient degree of accuracy. If an electron interacts with such a "classical object", the state of the latter is, generally speaking, altered. The nature and magnitude of this change depend on the state of the electron, and therefore may serve to characterize it quantitatively.

In this connection the "classical object" is usually called *apparatus*, and

---

† It is of interest to note that the complete mathematical formalism of quantum mechanics was constructed by W. Heisenberg and E. Schrödinger in 1925–6, before the discovery of the uncertainty principle, which revealed the physical content of this formalism.

‡ In this and the following sections we shall, for brevity, speak of "an electron", meaning in general any object of a quantum nature, i.e. a particle or system of particles obeying quantum mechanics and not classical mechanics.

‖ We refer to quantities which characterize the motion of the electron, and not to those, such as the charge and the mass, which relate to it as a particle; these are parameters.

its interaction with the electron is spoken of as *measurement*. However, it must be emphasized that we are here not discussing a process of measurement in which the physicist-observer takes part. By *measurement*, in quantum mechanics, we understand any process of interaction between classical and quantum objects, occurring apart from and independently of any observer. The importance of the concept of measurement in quantum mechanics was elucidated by N. Bohr.

We have defined "apparatus" as a physical object which is governed, with sufficient accuracy, by classical mechanics. Such, for instance, is a body of large enough mass. However, it must not be supposed that apparatus is necessarily macroscopic. Under certain conditions, the part of apparatus may also be taken by an object which is microscopic, since the idea of "with sufficient accuracy" depends on the actual problem proposed. Thus, the motion of an electron in a Wilson chamber is observed by means of the cloudy track which it leaves, and the thickness of this is large compared with atomic dimensions; when the path is determined with such low accuracy, the electron is an entirely classical object.

Thus quantum mechanics occupies a very unusual place among physical theories: it contains classical mechanics as a limiting case, yet at the same time it requires this limiting case for its own formulation.

We may now formulate the problem of quantum mechanics. A typical problem consists in predicting the result of a subsequent measurement from the known results of previous measurements. Moreover, we shall see later that, in comparison with classical mechanics, quantum mechanics, generally speaking, restricts the range of values which can be taken by various physical quantities (for example, energy): that is, the values which can be obtained as a result of measuring the quantity concerned. The methods of quantum mechanics must enable us to determine these admissible values.

The measuring process has in quantum mechanics a very important property: it always affects the electron subjected to it, and it is in principle impossible to make its effect arbitrarily small, for a given accuracy of measurement. The more exact the measurement, the stronger the effect exerted by it, and only in measurements of very low accuracy can the effect on the measured object be small. This property of measurements is logically related to the fact that the dynamical characteristics of the electron appear only as a result of the measurement itself. It is clear that, if the effect of the measuring process on the object of it could be made arbitrarily small, this would mean that the measured quantity has in itself a definite value independent of the measurement.

Among the various kinds of measurement, the measurement of the coordinates of the electron plays a fundamental part. Within the limits of applicability of quantum mechanics, a measurement of the coordinates of an electron can always be performed† with any desired accuracy.

---

† Once again we emphasize that, in speaking of "performing a measurement", we refer to the interaction of an electron with a classical "apparatus", which in no way presupposes the presence of an external observer.

Let us suppose that, at definite time intervals Δt, successive measurements of the coordinates of an electron are made. The results will not in general lie on a smooth curve. On the contrary, the more accurately the measurements are made, the more discontinuous and disorderly will be the variation of their results, in accordance with the non-existence of a path of the electron. A fairly smooth path is obtained only if the coordinates of the electron are measured with a low degree of accuracy, as for instance from the condensation of vapour droplets in a Wilson chamber.

If now, leaving the accuracy of the measurements unchanged, we diminish the intervals Δt between measurements, then adjacent measurements, of course, give neighbouring values of the coordinates. However, the results of a series of successive measurements, though they lie in a small region of space, will be distributed in this region in a wholly irregular manner, lying on no smooth curve. In particular, as Δt tends to zero, the results of adjacent measurements by no means tend to lie on one straight line.

This circumstance shows that, in quantum mechanics, there is no such concept as the velocity of a particle in the classical sense of the word, i.e. the limit to which the difference of the coordinates at two instants, divided by the interval Δt between these instants, tends as Δt tends to zero. However, we shall see later that in quantum mechanics, nevertheless, a reasonable definition of the velocity of a particle at a given instant can be constructed, and this velocity passes into the classical velocity as we pass to classical mechanics. But whereas in classical mechanics a particle has definite coordinates and velocity at any given instant, in quantum mechanics the situation is entirely different. If, as a result of measurement, the electron is found to have definite coordinates, then it has no definite velocity whatever. Conversely, if the electron has a definite velocity, it cannot have a definite position in space. For the simultaneous existence of the coordinates and velocity would mean the existence of a definite path, which the electron has not. Thus, in quantum mechanics, the coordinates and velocity of an electron are quantities which cannot be simultaneously measured exactly, i.e. they cannot simultaneously have definite values. We may say that the coordinates and velocity of the electron are quantities which do not exist simultaneously. In what follows we shall derive the quantitative relation which determines the possibility of an inexact measurement of the coordinates and velocity at the same instant.

A complete description of the state of a physical system in classical mechanics is effected by stating all its coordinates and velocities at a given instant; with these initial data, the equations of motion completely determine the behaviour of the system at all subsequent instants. In quantum mechanics such a description is in principle impossible, since the coordinates and the corresponding velocities cannot exist simultaneously. Thus a description of the state of a quantum system is effected by means of a smaller number of quantities than in classical mechanics, i.e. it is less detailed than a classical description.

A very important consequence follows from this regarding the nature of the predictions made in quantum mechanics. Whereas a classical description suffices to predict the future motion of a mechanical system with complete accuracy, the less detailed description given in quantum mechanics evidently cannot be enough to do this. This means that, even if an electron is in a state described in the most complete manner possible in quantum mechanics, its behaviour at subsequent instants is still in principle uncertain. Hence quantum mechanics cannot make completely definite predictions concerning the future behaviour of the electron. For a given initial state of the electron, a subsequent measurement can give various results. The problem in quantum mechanics consists in determining the probability of obtaining various results on performing this measurement. It is understood, of course, that in some cases the probability of a given result of measurement may be equal to unity, i.e. certainty, so that the result of that measurement is unique.

All measuring processes in quantum mechanics may be divided into two classes. In one, which contains the majority of measurements, we find those which do not, in any state of the system, lead with certainty to a unique result. The other class contains measurements such that for every possible result of measurement there is a state in which the measurement leads with certainty to that result. These latter measurements, which may be called *predictable*, play an important part in quantum mechanics. The quantitative characteristics of a state which are determined by such measurements are what are called *physical quantities* in quantum mechanics. If in some state a measurement gives with certainty a unique result, we shall say that in this state the corresponding physical quantity has a definite value. In future we shall always understand the expression "physical quantity" in the sense given here.

We shall often find in what follows that by no means every set of physical quantities in quantum mechanics can be measured simultaneously, i.e. can all have definite values at the same time. We have already mentioned one example, namely the velocity and coordinates of an electron. An important part is played in quantum mechanics by sets of physical quantities having the following property: these quantities can be measured simultaneously, but if they simultaneously have definite values, no other physical quantity (not being a function of these) can have a definite value in that state. We shall speak of such sets of physical quantities as *complete sets*.

Any description of the state of an electron arises as a result of some measurement. We shall now formulate the meaning of a *complete description* of a state in quantum mechanics. Completely described states occur as a result of the simultaneous measurement of a complete set of physical quantities. From the results of such a measurement we can, in particular, determine the probability of various results of any subsequent measurement, regardless of the history of the electron prior to the first measurement.

From now on (except in §14) we shall understand by the states of a quantum system just these completely described states.

§2. The principle of superposition

The radical change in the physical concepts of motion in quantum mechanics as compared with classical mechanics demands, of course, an equally radical change in the mathematical formalism of the theory. We must therefore consider first of all the way in which states are described in quantum mechanics.

We shall denote by q the set of coordinates of a quantum system, and by dq the product of the differentials of these coordinates. This dq is called an element of volume in the *configuration space* of the system; for one particle, dq coincides with an element of volume dV in ordinary space.

The basis of the mathematical formalism of quantum mechanics lies in the proposition that the state of a system can be described by a definite (in general complex) function $\Psi(q)$ of the coordinates. The square of the modulus of this function determines the probability distribution of the values of the coordinates: $|\Psi|^2 dq$ is the probability that a measurement performed on the system will find the values of the coordinates to be in the element dq of configuration space. The function Ψ is called the *wave function* of the system.†

A knowledge of the wave function allows us, in principle, to calculate the probability of the various results of any measurement (not necessarily of the coordinates) also. All these probabilities are determined by expressions bilinear in Ψ and Ψ^*. The most general form of such an expression is

$$\iint \Psi(q)\,\Psi^*(q')\,\phi(q, q')\,dq\,dq', \tag{2.1}$$

where the function $\phi(q, q')$ depends on the nature and the result of the measurement, and the integration is extended over all configuration space. The probability $\Psi\Psi^*$ of various values of the coordinates is itself an expression of this type.‡

The state of the system, and with it the wave function, in general varies with time. In this sense the wave function can be regarded as a function of time also. If the wave function is known at some initial instant, then, from the very meaning of the concept of complete description of a state, it is in principle determined at every succeeding instant. The actual dependence of the wave function on time is determined by equations which will be derived later.

The sum of the probabilities of all possible values of the coordinates of the system must, by definition, be equal to unity. It is therefore necessary that the result of integrating $|\Psi|^2$ over all configuration space should be equal to unity:

$$\int |\Psi|^2\,dq = 1. \tag{2.2}$$

† It was first introduced into quantum mechanics by Schrödinger in 1926.

‡ It is obtained from (2.1) when $\phi(q, q') = \delta(q-q_0)\,\delta(q'-q_0)$, where δ denotes the *delta function*, defined in §5 below; q_0 denotes the value of the coordinates whose probability is required.

This equation is what is called the *normalization condition* for wave functions. If the integral of $|\Psi|^2$ converges, then by choosing an appropriate constant coefficient the function Ψ can always be, as we say, *normalized*. However, we shall see later that the integral of $|\Psi|^2$ may diverge, and then Ψ cannot be normalized by the condition (2.2). In such cases $|\Psi|^2$ does not, of course, determine the absolute values of the probability of the coordinates, but the ratio of the values of $|\Psi|^2$ at two different points of configuration space determines the relative probability of the corresponding values of the coordinates.

Since all quantities calculated by means of the wave function, and having a direct physical meaning, are of the form (2.1), in which Ψ appears multiplied by Ψ^*, it is clear that the normalized wave function is determined only to within a constant *phase factor* of the form $e^{i\alpha}$ (where α is any real number). This indeterminacy is in principle irremovable; it is, however, unimportant, since it has no effect upon any physical results.

The positive content of quantum mechanics is founded on a series of propositions concerning the properties of the wave function. These are as follows.

Suppose that, in a state with wave function $\Psi_1(q)$, some measurement leads with certainty to a definite result (result 1), while in a state with $\Psi_2(q)$ it leads to result 2. Then it is assumed that every linear combination of Ψ_1 and Ψ_2, i.e. every function of the form $c_1\Psi_1+c_2\Psi_2$ (where c_1 and c_2 are constants), gives a state in which that measurement leads to either result 1 or result 2. Moreover, we can assert that, if we know the time dependence of the states, which for the one case is given by the function $\Psi_1(q, t)$, and for the other by $\Psi_2(q, t)$, then any linear combination also gives a possible dependence of a state on time. These propositions constitute what is called the *principle of superposition of states*, the chief positive principle of quantum mechanics. In particular, it follows from this principle that all equations satisfied by wave functions must be linear in Ψ.

Let us consider a system composed of two parts, and suppose that the state of this system is given in such a way that each of its parts is completely described.† Then we can say that the probabilities of the coordinates q_1 of the first part are independent of the probabilities of the coordinates q_2 of the second part, and therefore the probability distribution for the whole system should be equal to the product of the probabilities of its parts. This means that the wave function $\Psi_{12}(q_1, q_2)$ of the system can be represented in the form of a product of the wave functions $\Psi_1(q_1)$ and $\Psi_2(q_2)$ of its parts:

$$\Psi_{12}(q_1, q_2) = \Psi_1(q_1)\Psi_2(q_2). \tag{2.3}$$

If the two parts do not interact, then this relation between the wave function of the system and those of its parts will be maintained at future instants also,

† This, of course, means that the state of the whole system is completely described also. However, we emphasize that the converse statement is by no means true: a complete description of the state of the whole system does not in general completely determine the states of its individual parts (see also §14).

i.e. we can write

$$\Psi_{12}(q_1, q_2, t) = \Psi_1(q_1, t)\,\Psi_2(q_2, t). \tag{2.4}$$

§3. Operators

Let us consider some physical quantity f which characterizes the state of a quantum system. Strictly, we should speak in the following discussion not of one quantity, but of a complete set of them at the same time. However, the discussion is not essentially changed by this, and for brevity and simplicity we shall work below in terms of only one physical quantity.

The values which a given physical quantity can take are called in quantum mechanics its *eigenvalues*, and the set of these is referred to as the *spectrum of eigenvalues* of the given quantity. In classical mechanics, generally speaking, quantities run through a continuous series of values. In quantum mechanics also there are physical quantities (for instance, the coordinates) whose eigenvalues occupy a continuous range; in such cases we speak of a *continuous spectrum* of eigenvalues. As well as such quantities, however, there exist in quantum mechanics others whose eigenvalues form some discrete set; in such cases we speak of a *discrete spectrum.*

We shall suppose for simplicity that the quantity f considered here has a discrete spectrum; the case of a continuous spectrum will be discussed in §5. The eigenvalues of the quantity f are denoted by f_n, where the suffix n takes the values 0, 1, 2, 3, ... We also denote the wave function of the system, in the state where the quantity f has the value f_n, by Ψ_n. The wave functions Ψ_n are called the *eigenfunctions* of the given physical quantity f. Each of these functions is supposed normalized, so that

$$\int |\Psi_n|^2 \, dq = 1. \tag{3.1}$$

If the system is in some arbitrary state with wave function Ψ, a measurement of the quantity f carried out on it will give as a result one of the eigenvalues f_n. In accordance with the principle of superposition, we can assert that the wave function Ψ must be a linear combination of those eigenfunctions Ψ_n which correspond to the values f_n that can be obtained, with probability different from zero, when a measurement is made on the system and it is in the state considered. Hence, in the general case of an arbitrary state, the function Ψ can be represented in the form of a series

$$\Psi = \Sigma\, a_n \Psi_n, \tag{3.2}$$

where the summation extends over all n, and the a_n are some constant coefficients.

Thus we reach the conclusion that any wave function can be, as we say, expanded in terms of the eigenfunctions of any physical quantity. A set of functions in terms of which such an expansion can be made is called a *complete* (or *closed*) *set.*

The expansion (3.2) makes it possible to determine the probability of finding (i.e. the probability of getting the corresponding result on measurement), in a system in a state with wave function Ψ, any given value f_n of the quantity f. For, according to what was said in the previous section, these probabilities must be determined by some expressions bilinear in Ψ and Ψ^*, and therefore must be bilinear in a_n and a_n^*. Furthermore, these expressions must, of course, be positive. Finally, the probability of the value f_n must become unity if the system is in a state with wave function $\Psi = \Psi_n$, and must become zero if there is no term containing Ψ_n in the expansion (3.2) of the wave function Ψ. The only essentially positive quantity satisfying these conditions is the square of the modulus of the coefficient a_n. Thus we reach the result that the squared modulus $|a_n|^2$ of each coefficient in the expansion (3.2) determines the probability of the corresponding value f_n of the quantity f in the state with wave function Ψ. The sum of the probabilities of all possible values f_n must be equal to unity; in other words, the relation

$$\sum_n |a_n|^2 = 1 \tag{3.3}$$

must hold.

If the function Ψ were not normalized, then the relation (3.3) would not hold either. The sum $\Sigma\, |a_n|^2$ would then be given by some expression bilinear in Ψ and Ψ^*, and becoming unity when Ψ was normalized. Only the integral $\int \Psi\Psi^*\, dq$ is such an expression. Thus the equation

$$\sum_n a_n a_n^* = \int \Psi\Psi^*\, dq \tag{3.4}$$

must hold.

On the other hand, multiplying by Ψ the expansion $\Psi^* = \Sigma\, a_n^* \Psi_n^*$ of the function Ψ^* (the complex conjugate of Ψ), and integrating, we obtain

$$\int \Psi\Psi^*\, dq = \sum_n a_n^* \int \Psi_n^* \Psi\, dq.$$

Comparing this with (3.4), we have

$$\sum_n a_n a_n^* = \sum_n a_n^* \int \Psi_n^* \Psi\, dq,$$

from which we derive the following formula determining the coefficients a_n in the expansion of the function Ψ in terms of the eigenfunctions Ψ_n:

$$a_n = \int \Psi\Psi_n^*\, dq. \tag{3.5}$$

If we substitute here from (3.2), we obtain

$$a_n = \sum_m a_m \int \Psi_m \Psi_n^*\, dq,$$

from which it is evident that the eigenfunctions must satisfy the conditions

$$\int \Psi_m \Psi_n^*\, dq = \delta_{nm}, \tag{3.6}$$

where $\delta_{nm} = 1$ for $n = m$ and $\delta_{nm} = 0$ for $n \neq m$. The fact that the integrals of the products $\Psi_m\Psi_n^*$ with $m \neq n$ vanish is called the *orthogonality* of the functions Ψ_n. Thus the set of eigenfunctions Ψ_n forms a complete set of normalized and orthogonal (or, for brevity, *orthonormal*) functions.

We shall now introduce the concept of the *mean value* $\bar{f}$ of the quantity f in the given state. In accordance with the usual definition of mean values, we define $\bar{f}$ as the sum of all the eigenvalues f_n of the given quantity, each multiplied by the corresponding probability $|a_n|^2$. Thus

$$\bar{f} = \sum_n f_n |a_n|^2. \tag{3.7}$$

We shall write $\bar{f}$ in the form of an expression which does not contain the coefficients a_n in the expansion of the function Ψ, but this function itself. Since the products $a_n a_n^*$ appear in (3.7), it is clear that the required expression must be bilinear in Ψ and Ψ^*. We introduce a mathematical operator, which we denote† by $\hat{f}$ and define as follows. Let $(\hat{f}\Psi)$ denote the result of the operator $\hat{f}$ acting on the function Ψ. We define $\hat{f}$ in such a way that the integral of the product of $(\hat{f}\Psi)$ and the complex conjugate function Ψ^* is equal to the mean value $\bar{f}$:

$$\bar{f} = \int \Psi^* (\hat{f}\Psi) \, dq. \tag{3.8}$$

It is easily seen that, in the general case, the operator $\hat{f}$ is a linear‡ integral operator. For, using the expression (3.5) for a_n, we can rewrite the definition (3.7) of the mean value in the form

$$\bar{f} = \sum_n f_n a_n a_n^* = \int \Psi^* (\sum_n a_n f_n \Psi_n) \, dq.$$

Comparing this with (3.8), we see that the result of the operator $\hat{f}$ acting on the function Ψ has the form

$$(\hat{f}\Psi) = \sum_n a_n f_n \Psi_n. \tag{3.9}$$

If we substitute here the expression (3.5) for a_n, we find that $\hat{f}$ is an integral operator of the form

$$(\hat{f}\Psi) = \int K(q, q') \Psi(q') \, dq', \tag{3.10}$$

where the function $K(q, q')$ (called the *kernel* of the operator) is

$$K(q, q') = \sum_n f_n \Psi_n^*(q') \Psi_n(q). \tag{3.11}$$

Thus, for every physical quantity in quantum mechanics, there is a definite corresponding linear operator.

It is seen from (3.9) that, if the function Ψ is one of the eigenfunctions Ψ_n

---

† By convention, we shall always denote operators by letters with circumflexes.

‡ An operator is said to be *linear* if it has the properties

$$\hat{f}(\Psi_1 + \Psi_2) = \hat{f}\Psi_1 + \hat{f}\Psi_2 \text{ and } \hat{f}(a\Psi) = a\hat{f}\Psi,$$

where Ψ_1 and Ψ_2 are arbitrary functions and a is an arbitrary constant.

(so that all the a_n except one are zero), then, when the operator $\hat{f}$ acts on it, this function is simply multiplied by the corresponding eigenvalue f_n:

$$\hat{f}\Psi_n = f_n\Psi_n. \tag{3.12}$$

(In what follows we shall always omit the parentheses in the expression $(\hat{f}\Psi)$, where this cannot cause any misunderstanding; the operator is taken to act on the expression which follows it.) Thus we can say that the eigenfunctions of the given physical quantity f are the solutions of the equation

$$\hat{f}\Psi = f\Psi,$$

where f is a constant, and the eigenvalues are the values of this constant for which the above equation has solutions satisfying the required conditions. As we shall see below, the form of the operators for various physical quantities can be determined from direct physical considerations, and then the above property of the operators enables us to find the eigenfunctions and eigenvalues by solving the equations $\hat{f}\Psi = f\Psi$.

Both the eigenvalues of a real physical quantity and its mean value in every state are real. This imposes a restriction on the corresponding operators. Equating the expression (3.8) to its complex conjugate, we obtain the relation

$$\int \Psi^*(\hat{f}\Psi)\,dq = \int \Psi(\hat{f}^*\Psi^*)\,dq, \tag{3.13}$$

where $\hat{f}^*$ denotes the operator which is the complex conjugate of $\hat{f}$.† This relation does not hold in general for an arbitrary linear operator, so that it is a restriction on the form of the operator $\hat{f}$. For an arbitrary operator $\hat{f}$ we can find what is called the *transposed operator* $\tilde{\hat{f}}$, defined in such a way that

$$\int \Phi(\hat{f}\Psi)\,dq = \int \Psi(\tilde{\hat{f}}\Phi)\,dq, \tag{3.14}$$

where Ψ and Φ are two different functions. If we take, as the function Φ, the function Ψ^* which is the complex conjugate of Ψ, then a comparison with (3.13) shows that we must have

$$\tilde{\hat{f}} = \hat{f}^*. \tag{3.15}$$

Operators satisfying this condition are said to be *Hermitian*.‡ Thus the operators corresponding, in the mathematical formalism of quantum mechanics, to real physical quantities must be Hermitian.

We can formally consider complex physical quantities also, i.e. those whose eigenvalues are complex. Let f be such a quantity. Then we can introduce its complex conjugate quantity f^*, whose eigenvalues are the complex conjugates of those of f. We denote by $\hat{f}^+$ the operator corresponding to the quantity f^*. It is called the *Hermitian conjugate* of the operator $\hat{f}$ and,

---

† By definition, if for the operator $\hat{f}$ we have $\hat{f}\psi = \phi$, then the complex conjugate operator $\hat{f}^*$ is that for which we have $\hat{f}^*\psi^* = \phi^*$.

‡ For a linear integral operator of the form (3.10), the Hermitian condition means that the kernel of the operator must be such that $K(q, q') = K^*(q', q)$.

in general, will be different from the definition of the operator $\hat{f}^*$: the mean value of the quantity f^* in a state Ψ is

$$\overline{f^*} = \int \Psi^* \hat{f}^+ \Psi \, dq.$$

We also have

$$(\overline{f})^* = [\int \Psi^* \hat{f} \Psi \, dq]^* = \int \Psi \hat{f}^* \Psi^* \, dq = \int \Psi^* \tilde{\hat{f}}^* \Psi \, dq.$$

Equating these two expressions gives

$$\hat{f}^+ = \tilde{\hat{f}}^*, \tag{3.16}$$

from which it is clear that $\hat{f}^+$ is in general not the same as $\hat{f}^*$.

The condition (3.15) can now be written

$$\hat{f} = \hat{f}^+, \tag{3.17}$$

i.e. the operator of a real physical quantity is the same as its Hermitian conjugate (Hermitian operators are also called *self-conjugate*).

We shall show how the orthogonality of the eigenfunctions of an Hermitian operator corresponding to different eigenvalues can be directly proved. Let f_n and f_m be two different eigenvalues of the real quantity f, and Ψ_n, Ψ_m the corresponding eigenfunctions:

$$\hat{f}\Psi_n = f_n\Psi_n, \quad \hat{f}\Psi_m = f_m\Psi_m.$$

Multiplying both sides of the first of these equations by Ψ_m^*, and both sides of the complex conjugate of the second by Ψ_n, and subtracting corresponding terms, we find

$$\Psi_m^* \hat{f} \Psi_n - \Psi_n \hat{f}^* \Psi_m^* = (f_n - f_m)\Psi_n \Psi_m^*.$$

We integrate both sides of this equation over q. Since $\hat{f}^* = \tilde{\hat{f}}$, by (3.14) the integral on the left-hand side of the equation is zero, so that we have

$$(f_n - f_m) \int \Psi_n \Psi_m^* \, dq = 0,$$

whence, since $f_n \neq f_m$, we obtain the required orthogonality property of the functions Ψ_n and Ψ_m.

We have spoken here of only one physical quantity f, whereas, as we said at the beginning of this section, we should have spoken of a complete set of simultaneously measurable physical quantities. We should then have found that to each of these quantities $f, g, \ldots$ there corresponds its operator

$\hat{f}, \hat{g}, \ldots$. The eigenfunctions Ψ_n then correspond to states in which all the quantities concerned have definite values, i.e. they correspond to definite sets of eigenvalues $f_n, g_n, \ldots$, and are simultaneous solutions of the system of equations

$$\hat{f}\Psi = f\Psi, \quad \hat{g}\Psi = g\Psi, \ldots .$$

§4. Addition and multiplication of operators

If $\hat{f}$ and $\hat{g}$ are the operators corresponding to two physical quantities f and g, the sum $f+g$ has a corresponding operator $\hat{f}+\hat{g}$. However, the significance of adding different physical quantities in quantum mechanics depends considerably on whether the quantities are or are not simultaneously measurable. If f and g are simultaneously measurable, the operators $\hat{f}$ and $\hat{g}$ have common eigenfunctions, which are also eigenfunctions of the operator $\hat{f}+\hat{g}$, and the eigenvalues of the latter operator are equal to the sums f_n+g_n. But if f and g cannot simultaneously take definite values, their sum $f+g$ has a more restricted significance. We can assert only that the mean value of this quantity in any state is equal to the sum of the mean values of the separate quantities:

$$\overline{f+g} = \bar{f}+\bar{g}. \tag{4.1}$$

The eigenvalues and eigenfunctions of the operator $\hat{f}+\hat{g}$ will not, in general, now bear any relation to those of the quantities f and g. It is evident that, if the operators $\hat{f}$ and $\hat{g}$ are Hermitian, the operator $\hat{f}+\hat{g}$ will be so too, so that its eigenvalues are real and are equal to those of the new quantity $f+g$ thus defined.

The following theorem should be noted. Let f_0 and g_0 be the smallest eigenvalues of the quantities f and g, and $(f+g)_0$ that of the quantity $f+g$. Then

$$(f+g)_0 \geqslant f_0+g_0. \tag{4.2}$$

The equality holds if f and g can be measured simultaneously. The proof follows from the obvious fact that the mean value of a quantity is always greater than or equal to its least eigenvalue. In a state in which the quantity $f+g$ has the value $(f+g)_0$ we have $\overline{f+g} = (f+g)_0$, and since, on the other hand, $\overline{f+g} = \bar{f}+\bar{g} \geqslant f_0+g_0$, we arrive at the inequality (4.2).

Next, let f and g once more be quantities that can be measured simultaneously. Besides their sum, we can also introduce the concept of their *product* as being a quantity whose eigenvalues are equal to the products of those of the quantities f and g. It is easy to see that, to this quantity, there corresponds an operator whose effect consists of the successive action on the function of first one and then the other operator. Such an operator is represented mathematically by the product of the operators $\hat{f}$ and $\hat{g}$. For, if Ψ_n are the eigenfunctions common to the operators $\hat{f}$ and $\hat{g}$, we have

$$\hat{f}\hat{g}\Psi_n = \hat{f}(\hat{g}\Psi_n) = \hat{f}g_n\Psi_n = g_n\hat{f}\Psi_n = g_n f_n \Psi_n$$

(the symbol $\hat{f}\hat{g}$ denotes an operator whose effect on a function Ψ consists of the successive action first of the operator $\hat{g}$ on the function Ψ and then of the operator $\hat{f}$ on the function $\hat{g}\Psi$). We could equally well take the operator $\hat{g}\hat{f}$ instead of $\hat{f}\hat{g}$, the former differing from the latter in the order of its factors. It is obvious that the result of the action of either of these operators on the functions Ψ_n will be the same. Since, however, every wave function Ψ can be represented as a linear combination of the functions Ψ_n, it follows that the result of the action of the operators $\hat{f}\hat{g}$ and $\hat{g}\hat{f}$ on an arbitrary function will also be the same. This fact can be written in the form of the symbolic equation $\hat{f}\hat{g} = \hat{g}\hat{f}$ or

$$\hat{f}\hat{g} - \hat{g}\hat{f} = 0. \tag{4.3}$$

Two such operators $\hat{f}$ and $\hat{g}$ are said to be *commutative*, or to *commute* with each other. Thus we arrive at the important result: if two quantities f and g can simultaneously take definite values, then their operators commute with each other.

The converse theorem can also be proved (§11): if the operators $\hat{f}$ and $\hat{g}$ commute, then all their eigenfunctions can be taken common to both; physically, this means that the corresponding physical quantities can be measured simultaneously. Thus the commutability of the operators is a necessary and sufficient condition for the physical quantities to be simultaneously measurable.

A particular case of the product of operators is an operator raised to some power. From the above discussion we can deduce that the eigenvalues of an operator $\hat{f}^p$ (where p is an integer) are equal to the pth powers of the eigenvalues of the operator $\hat{f}$. Any function $\phi(\hat{f})$ of an operator can be defined as an operator whose eigenvalues are equal to the same function $\phi(f)$ of the eigenvalues of the operator $\hat{f}$. If the function $\phi(f)$ can be expanded as a Taylor series, this expresses the effect of the operator $\phi(\hat{f})$ in terms of those of various powers $\hat{f}^p$.

In particular, the operator $\hat{f}^{-1}$ is called the *inverse* of the operator $\hat{f}$. It is evident that the successive action of the operators $\hat{f}$ and $\hat{f}^{-1}$ on any function leaves the latter unchanged, i.e. $\hat{f}\hat{f}^{-1} = \hat{f}^{-1}\hat{f} = 1$.

If the quantities f and g cannot be measured simultaneously, the concept of their product does not have the same direct meaning. This appears in the fact that the operator $\hat{f}\hat{g}$ is not Hermitian in this case, and hence cannot correspond to any real physical quantity. For, by the definition of the transpose of an operator we can write

$$\int \Psi \hat{f}\hat{g}\Phi \,\mathrm{d}q = \int \Psi \hat{f}(\hat{g}\Phi) \,\mathrm{d}q = \int (\hat{g}\Phi)(\tilde{\hat{f}}\Psi) \,\mathrm{d}q.$$

Here the operator $\tilde{\hat{f}}$ acts only on the function Ψ, and the operator $\hat{g}$ on Φ, so that the integrand is a simple product of two functions $\hat{g}\Phi$ and $\tilde{\hat{f}}\Psi$. Again using the definition of the transpose of an operator, we can write

$$\int \Psi \hat{f}\hat{g}\Phi \, dq = \int (\tilde{\hat{f}}\Psi)(\hat{g}\Phi) \, dq = \int \Phi \tilde{\hat{g}}\tilde{\hat{f}}\Psi \, dq.$$

Thus we obtain an integral in which the functions Ψ and Φ have changed places as compared with the original one. In other words, the operator $\tilde{\hat{g}}\tilde{\hat{f}}$ is the transpose of $\hat{f}\hat{g}$, and we can write

$$\widetilde{\hat{f}\hat{g}} = \tilde{\hat{g}}\tilde{\hat{f}}, \tag{4.4}$$

i.e. the transpose of the product $\hat{f}\hat{g}$ is the product of the transposes of the factors written in the opposite order. Taking the complex conjugate of both sides of equation (4.4), we have

$$(\hat{f}\hat{g})^+ = \hat{g}^+\hat{f}^+. \tag{4.5}$$

If each of the operators $\hat{f}$ and $\hat{g}$ is Hermitian, then $(\hat{f}\hat{g})^+ = \hat{g}\hat{f}$. It follows from this that the operator $\hat{f}\hat{g}$ is Hermitian if and only if the factors $\hat{f}$ and $\hat{g}$ commute.

We note that, from the products $\hat{f}\hat{g}$ and $\hat{g}\hat{f}$ of two non-commuting Hermitian operators, we can form an Hermitian operator, the *symmetrized product*

$$\tfrac{1}{2}(\hat{f}\hat{g}+\hat{g}\hat{f}). \tag{4.6}$$

It is easy to see that the difference $\hat{f}\hat{g}-\hat{g}\hat{f}$ is an *anti-Hermitian* operator (i.e. one for which the transpose is equal to the complex conjugate taken with the opposite sign). It can be made Hermitian by multiplying by i; thus

$$i(\hat{f}\hat{g}-\hat{g}\hat{f}) \tag{4.7}$$

is again an Hermitian operator.

In what follows we shall sometimes use for brevity the notation

$$\{\hat{f},\hat{g}\} = \hat{f}\hat{g}-\hat{g}\hat{f}, \tag{4.8}$$

called the *commutator* of these operators. It is easily seen that

$$\{\hat{f}\hat{g}, \hat{h}\} = \{\hat{f}, \hat{h}\}\hat{g}+\hat{f}\{\hat{g}, \hat{h}\}. \tag{4.9}$$

We notice that, if $\{\hat{f}, \hat{h}\} = 0$ and $\{\hat{g}, \hat{h}\} = 0$, it does not in general follow that $\hat{f}$ and $\hat{g}$ commute.

§5. The continuous spectrum

All the relations given in §§3 and 4, describing the properties of the eigenfunctions of a discrete spectrum, can be generalized without difficulty to the case of a continuous spectrum of eigenvalues.

Let f be a physical quantity having a continuous spectrum. We shall denote its eigenvalues by the same letter f simply, and the corresponding eigenfunctions by Ψ_f. Just as an arbitrary wave function Ψ can be expanded in a series (3.2) of eigenfunctions of a quantity having a discrete spectrum, it can also be expanded (this time as an integral) in terms of the complete set of eigenfunctions of a quantity with a continuous spectrum. This expansion has the form

$$\Psi(q) = \int a_f \Psi_f(q)\, df, \tag{5.1}$$

where the integration is extended over the whole range of values that can be taken by the quantity f.

The subject of the normalization of the eigenfunctions of a continuous spectrum is more complex than in the case of a discrete spectrum. The requirement that the integral of the squared modulus of the function should be equal to unity cannot here be satisfied, as we shall see below. Instead, we try to normalize the functions Ψ_f in such a way that $|a_f|^2\, df$ is the probability that the physical quantity concerned, in the state described by the wave function Ψ, has a value between f and $f+df$. Since the sum of the probabilities of all possible values of f must be equal to unity, we have

$$\int |a_f|^2\, df = 1 \tag{5.2}$$

(similarly to the relation (3.3) for a discrete spectrum).

Proceeding in exactly the same way as in the derivation of formula (3.5), and using the same arguments, we can write, firstly,

$$\int \Psi\Psi^*\, dq = \int |a_f|^2\, df$$

and, secondly,

$$\int \Psi\Psi^*\, dq = \int\int a_f^* \Psi_f^* \Psi\, df dq.$$

By comparing these two expressions we find the formula which determines the expansion coefficients,

$$a_f = \int \Psi(q)\Psi_f^*(q)\, dq, \tag{5.3}$$

in exact analogy to (3.5).

To derive the normalization condition, we now substitute (5.1) in (5.3), and obtain

$$a_f = \int a_{f'} \left(\int \Psi_{f'}\Psi_f^*\, dq\right) df'.$$

This relation must hold for arbitrary a_f, and therefore must be satisfied identically. For this to be so, it is necessary that, first of all, the coefficient of $a_{f'}$ in the integrand (i.e. the integral $\int \Psi_{f'}\Psi_f^*\, dq$) should be zero for all $f' \neq f$. For $f' = f$, this coefficient must become infinite (otherwise the integral over f' would vanish). Thus the integral $\int \Psi_{f'}\Psi_f^*\, dq$ is a function of the difference $f'-f$, which becomes zero for values of the argument different from zero and is infinite when the argument is zero. We denote this function by $\delta(f'-f)$:

$$\int \Psi_{f'}\Psi_f^*\, dq = \delta(f'-f). \tag{5.4}$$

The manner in which the function $\delta(f'-f)$ becomes infinite for $f'-f = 0$ is determined by the fact that we must have

$$\int \delta(f'-f)\, a_{f'} \mathrm{d}f' = a_f.$$

It is clear that, for this to be so, we must have

$$\int \delta(f'-f)\, \mathrm{d}f' = 1.$$

The function thus defined is called a *delta function*, and was first used in theoretical physics by P. A. M. Dirac. We shall write out once more the formulae which define it. They are

$$\delta(x) = 0 \text{ for } x \neq 0, \quad \delta(0) = \infty, \tag{5.5}$$

while

$$\int_{-\infty}^{\infty} \delta(x)\, \mathrm{d}x = 1. \tag{5.6}$$

We can take as limits of integration any numbers such that $x = 0$ lies between them. If $f(x)$ is some function continuous at $x = 0$, then

$$\int_{-\infty}^{\infty} \delta(x) f(x)\, \mathrm{d}x = f(0). \tag{5.7}$$

This formula can be written in the more general form

$$\int \delta(x-a) f(x)\, \mathrm{d}x = f(a), \tag{5.8}$$

where the range of integration includes the point $x = a$, and $f(x)$ is continuous at $x = a$. It is also evident that

$$\delta(-x) = \delta(x), \tag{5.9}$$

i.e. the delta function is even. Finally, writing

$$\int_{-\infty}^{\infty} \delta(\alpha x)\, \mathrm{d}x = \int_{-\infty}^{\infty} \delta(y) \frac{\mathrm{d}y}{|\alpha|} = \frac{1}{|\alpha|},$$

we can deduce that

$$\delta(\alpha x) = (1/|\alpha|)\, \delta(x), \tag{5.10}$$

where α is any constant.

The formula (5.4) gives the normalization rule for the eigenfunctions of a continuous spectrum; it replaces the condition (3.6) for a discrete spectrum. We see that the functions Ψ_f and $\Psi_{f'}$ with $f \neq f'$ are, as before, orthogonal. However, the integrals of the squared moduli $|\Psi_f|^2$ of the functions diverge for a continuous spectrum.

The functions $\Psi_f(q)$ satisfy still another relation similar to (5.4). To derive this, we substitute (5.3) in (5.1), which gives

$$\Psi(q) = \int \Psi(q') \left(\int \Psi_f^*(q') \Psi_f(q)\, \mathrm{d}f \right) \mathrm{d}q',$$

whence we can at once deduce that we must have

$$\int \Psi_f^*(q') \Psi_f(q)\, \mathrm{d}f = \delta(q'-q). \tag{5.11}$$

There is, of course, an analogous relation for a discrete spectrum:

$$\sum_n \Psi_n^*(q')\Psi_n(q) = \delta(q'-q). \tag{5.12}$$

Comparing the pair of formulae (5.1), (5.4) with the pair (5.3), (5.11), we see that, on the one hand, the function $\Psi(q)$ can be expanded in terms of the functions $\Psi_f(q)$ with expansion coefficients a_f and, on the other hand, formula (5.3) represents an entirely analogous expansion of the function $a_f \equiv a(f)$ in terms of the functions $\Psi_f^*(q)$, while the $\Psi(q)$ play the part of expansion coefficients. The function $a(f)$, like $\Psi(q)$, completely determines the state of the system; it is sometimes called a wave function *in the f representation* (while the function $\Psi(q)$ is called a wave function in the q representation). Just as $|\Psi(q)|^2$ determines the probability for the system to have coordinates lying in a given interval dq, so $|a(f)|^2$ determines the probability for the values of the quantity f to lie in a given interval df. On the one hand, the functions $\Psi_f(q)$ are the eigenfunctions of the quantity f in the q representation; on the other hand, their complex conjugates are the eigenfunctions of the coordinate q in the f representation.

Let $\phi(f)$ be some function of the quantity f, such that ϕ and f are related in a one-to-one manner. Each of the functions $\Psi_f(q)$ can then be regarded as an eigenfunction of the quantity ϕ. Here, however, the normalization of these functions must be changed: the eigenfunctions $\Psi_\phi(q)$ of the quantity ϕ must be normalized by the condition

$$\int \Psi_{\phi(f')}\Psi_{\phi(f)}^* \, \mathrm{d}q = \delta[\phi(f')-\phi(f)],$$

whereas the functions $\Psi_{f'}$ are normalized by the condition (5.4). The argument of the delta function becomes zero only for $f' = f$. As f' approaches f, we have $\phi(f')-\phi(f) = [\mathrm{d}\phi(f)/\mathrm{d}f] \,.\, (f'-f)$. By (5.10) we can therefore write†

$$\delta[\phi(f')-\phi(f)] = \frac{1}{|\mathrm{d}\phi(f)/\mathrm{d}f|}\delta(f'-f). \tag{5.13}$$

Comparing this with (5.4), we see that the functions Ψ_ϕ and Ψ_f are related by

$$\Psi_{\phi(f)} = \frac{1}{\sqrt{|\mathrm{d}\phi(f)/\mathrm{d}f|}}\Psi_f. \tag{5.14}$$

There are also physical quantities which in one range of values have a discrete spectrum, and in another a continuous spectrum. For the eigenfunctions of such a quantity all the relations derived in this and the previous sections are, of course, true. It need only be noted that the complete set of functions is formed by combining the eigenfunctions of both spectra.

---

† In general, if $\phi(x)$ is some one-valued function (the inverse function need not be one-valued), we have

$$\delta[\phi(x)] = \sum_i \frac{1}{|\phi'(\alpha_i)|}\delta(x-\alpha_i), \tag{5.13a}$$

where α_i are the roots of the equation $\phi(x) = 0$.

Hence the expansion of an arbitrary wave function in terms of the eigenfunctions of such a quantity has the form

$$\Psi(q) = \sum_n a_n\Psi_n(q) + \int a_f\Psi_f(q)\,df, \tag{5.15}$$

where the sum is taken over the discrete spectrum and the integral over the whole continuous spectrum.

The coordinate q itself is an example of a quantity having a continuous spectrum. It is easy to see that the operator corresponding to it is simply multiplication by q. For, since the probability of the various values of the coordinate is determined by the square $|\Psi(q)|^2$, the mean value of the coordinate is

$$\bar{q} = \int q|\Psi|^2\,dq \equiv \int \Psi^* q\Psi\,dq.$$

Comparison of this with the definition (3.8) of an operator shows that†

$$\hat{q} = q. \tag{5.16}$$

The eigenfunctions of this operator must be determined, according to the usual rule, by the equation $q\Psi_{q_0} = q_0\Psi_{q_0}$, where q_0 temporarily denotes the actual values of the coordinate as distinct from the variable q. Since this equation can be satisfied either by $\Psi_{q_0} = 0$ or by $q = q_0$, it is clear that the eigenfunctions which satisfy the normalization condition are‡

$$\Psi_{q_0} = \delta(q-q_0). \tag{5.17}$$

§6. The passage to the limiting case of classical mechanics

Quantum mechanics contains classical mechanics in the form of a certain limiting case. The question arises as to how this passage to the limit is made.

In quantum mechanics an electron is described by a wave function which determines the various values of its coordinates; of this function we so far know only that it is the solution of a certain linear partial differential equation. In classical mechanics, on the other hand, an electron is regarded as a material particle, moving in a path which is completely determined by the equations of motion. There is an interrelation, somewhat similar to that between quantum and classical mechanics, in electrodynamics between wave optics

† In future we shall always, for simplicity, write operators which amount to multiplication by some quantity in the form of that quantity itself.

‡ The expansion coefficients for an arbitrary function Ψ in terms of these eigenfunctions are

$$a_{q_0} = \int \Psi(q)\delta(q-q_0)\,dq = \Psi(q_0).$$

The probability that the value of the coordinate lies in a given interval dq_0 is

$$|a_{q_0}|^2\,dq_0 = |\Psi(q_0)|^2\,dq_0,$$

as it should be.

and geometrical optics. In wave optics, the electromagnetic waves are described by the electric and magnetic field vectors, which satisfy a definite system of linear differential equations, namely Maxwell's equations. In geometrical optics, however, the propagation of light along definite paths, or rays, is considered. Such an analogy enables us to see that the passage from quantum mechanics to the limit of classical mechanics occurs similarly to the passage from wave optics to geometrical optics.

Let us recall how this latter transition is made mathematically (see *Fields*, §53). Let u be any of the field components in the electromagnetic wave. It can be written in the form $u = ae^{i\phi}$ (with a and ϕ real), where a is called the *amplitude* and ϕ the *phase* of the wave (called in geometrical optics the *eikonal*). The limiting case of geometrical optics corresponds to small wavelengths; this is expressed mathematically by saying that ϕ varies by a large amount over short distances; this means, in particular, that it can be supposed large in absolute value.

Similarly, we start from the hypothesis that, to the limiting case of classical mechanics, there correspond in quantum mechanics wave functions of the form $\Psi = ae^{i\phi}$, where a is a slowly varying function and ϕ takes large values. As is well known, the path of a particle can be determined in mechanics by means of the variational principle, according to which what is called the *action* S of a mechanical system must take its least possible value (the *principle of least action*). In geometrical optics the path of the rays is determined by what is called *Fermat's principle*, according to which the *optical path length* of the ray, i.e. the difference between its phases at the beginning and end of the path, must take its least (or greatest) possible value.

On the basis of this analogy, we can assert that the phase ϕ of the wave function, in the limiting (classical) case, must be proportional to the mechanical action S of the physical system considered, i.e. we must have $S = \text{constant} \times \phi$. The constant of proportionality is called *Planck's constant*† and is denoted by $\hbar$. It has the dimensions of action (since ϕ is dimensionless) and has the value

$$\hbar = 1{\cdot}054 \times 10^{-27} \text{ erg sec.}$$

Thus, the wave function of an "almost classical" (or, as we say, *quasi-classical*) physical system has the form

$$\Psi = ae^{iS/\hbar}. \tag{6.1}$$

Planck's constant $\hbar$ plays a fundamental part in all quantum phenomena. Its relative value (compared with other quantities of the same dimensions) determines the "extent of quantization" of a given physical system. The transition from quantum mechanics to classical mechanics, corresponding to large phase, can be formally described as a passage to the limit $\hbar \to 0$ (just

† It was introduced into physics by M. Planck in 1900. The constant $\hbar$, which we use everywhere in this book, is, strictly speaking, Planck's constant divided by 2π; this is Dirac's notation.

as the transition from wave optics to geometrical optics corresponds to a passage to the limit of zero wavelength, $\lambda \to 0$).

We have ascertained the limiting form of the wave function, but the question still remains how it is related to classical motion in a path. In general, the motion described by the wave function does not tend to motion in a definite path. Its connection with classical motion is that, if at some initial instant the wave function, and with it the probability distribution of the coordinates, is given, then at subsequent instants this distribution will change according to the laws of classical mechanics (for a more detailed discussion of this, see the end of §17).

In order to obtain motion in a definite path, we must start from a wave function of a particular form, which is perceptibly different from zero only in a very small region of space (what is called a *wave packet*); the dimensions of this region must tend to zero with $\hbar$. Then we can say that, in the quasi-classical case, the wave packet will move in space along a classical path of a particle.

Finally, quantum-mechanical operators must reduce, in the limit, simply to multiplication by the corresponding physical quantity.

§7. The wave function and measurements

Let us again return to the process of measurement, whose properties have been qualitatively discussed in §1; we shall show how these properties are related to the mathematical formalism of quantum mechanics.

We consider a system consisting of two parts: a classical apparatus and an electron (regarded as a quantum object). The process of measurement consists in these two parts' coming into interaction with each other, as a result of which the apparatus passes from its initial state into some other; from this change of state we draw conclusions concerning the state of the electron. The states of the apparatus are distinguished by the values of some physical quantity (or quantities) characterizing it—the "readings of the apparatus". We conventionally denote this quantity by g, and its eigenvalues by g_n; these take in general, in accordance with the classical nature of the apparatus, a continuous range of values, but we shall—merely in order to simplify the subsequent formulae—suppose the spectrum discrete. The states of the apparatus are described by means of quasi-classical wave functions, which we shall denote by $\Phi_n(\xi)$, where the suffix n corresponds to the "reading" g_n of the apparatus, and ξ denotes the set of its coordinates. The classical nature of the apparatus appears in the fact that, at any given instant, we can say with certainty that it is in one of the known states Φ_n with some definite value of the quantity g; for a quantum system such an assertion would, of course, be unjustified.

Let $\Phi_0(\xi)$ be the wave function of the initial state of the apparatus (before the measurement), and $\Psi(q)$ some arbitrary normalized initial wave function of the electron (q denoting its coordinates). These functions describe the state of the apparatus and of the electron independently, and therefore the initial wave function of the whole system is the product

$$\Psi(q)\Phi_0(\xi). \tag{7.1}$$

Next, the apparatus and the electron interact with each other. Applying the equations of quantum mechanics, we can in principle follow the change of the wave function of the system with time. After the measuring process it may not, of course, be a product of functions of ξ and q. Expanding the wave function in terms of the eigenfunctions Φ_n of the apparatus (which form a complete set of functions), we obtain a sum of the form

$$\sum_n A_n(q)\Phi_n(\xi), \tag{7.2}$$

where the $A_n(q)$ are some functions of q.

The classical nature of the apparatus, and the double role of classical mechanics as both the limiting case and the foundation of quantum mechanics, now make their appearance. As has been said above, the classical nature of the apparatus means that, at any instant, the quantity g (the "reading of the apparatus") has some definite value. This enables us to say that the state of the system apparatus + electron after the measurement will in actual fact be described, not by the entire sum (7.2), but by only the one term which corresponds to the "reading" g_n of the apparatus,

$$A_n(q)\Phi_n(\xi). \tag{7.3}$$

It follows from this that $A_n(q)$ is proportional to the wave function of the electron after the measurement. It is not the wave function itself, as is seen from the fact that the function $A_n(q)$ is not normalized. It contains both information concerning the properties of the resulting state of the electron and the probability (determined by the initial state of the system) of the occurrence of the nth "reading" of the apparatus.

Since the equations of quantum mechanics are linear, the relation between $A_n(q)$ and the initial wave function of the electron $\Psi(q)$ is in general given by some linear integral operator:

$$A_n(q) = \int K_n(q, q')\Psi(q')\,dq', \tag{7.4}$$

with a kernel $K_n(q, q')$ which characterizes the measurement process concerned.

We shall suppose that the measurement concerned is such that it gives a complete description of the state of the electron. In other words (see §1), in the resulting state the probabilities of all the quantities must be independent of the previous state of the electron (before the measurement). Mathematically, this means that the form of the functions $A_n(q)$ must be determined by the measuring process itself, and does not depend on the initial wave function $\Psi(q)$ of the electron. Thus the A_n must have the form

$$A_n(q) = a_n\phi_n(q), \tag{7.5}$$

where the ϕ_n are definite functions, which we suppose normalized, and only the constants a_n depend on $\Psi(q)$. In the integral relation (7.4) this corresponds to a kernel $K_n(q, q')$ which is a product of a function of q and a function of q':

$$K_n(q, q') = \phi_n(q)\Psi_n^*(q'); \tag{7.6}$$

then the linear relation between the constants a_n and the function $\Psi(q)$ is

$$a_n = \int \Psi(q)\Psi_n^*(q)\, \mathrm{d}q, \tag{7.7}$$

where the $\Psi_n(q)$ are certain functions depending on the process of measurement.

The functions $\phi_n(q)$ are the normalized wave functions of the electron after measurement. Thus we see how the mathematical formalism of the theory reflects the possibility of finding by measurement a state of the electron described by a definite wave function.

If the measurement is made on an electron with a given wave function $\Psi(q)$, the constants a_n have a simple physical meaning: in accordance with the usual rules, $|a_n|^2$ is the probability that the measurement will give the nth result. The sum of the probabilities of all results is equal to unity:

$$\sum_n |a_n|^2 = 1. \tag{7.8}$$

In order that equations (7.7) and (7.8) should hold for an arbitrary normalized function $\Psi(q)$, it is necessary (cf. §3) that an arbitrary function $\Psi(q)$ can be expanded in terms of the functions $\Psi_n(q)$. This means that the functions $\Psi_n(q)$ form a complete set of normalized and orthogonal functions.

If the initial wave function of the electron coincides with one of the functions $\Psi_n(q)$, then the corresponding constant a_n is evidently equal to unity, while all the others are zero. In other words, a measurement made on an electron in the state $\Psi_n(q)$ gives with certainty the nth result.

All these properties of the functions $\Psi_n(q)$ show that they are the eigenfunctions of some physical quantity (denoted by f) which characterizes the electron, and the measurement concerned can be spoken of as a measurement of this quantity.

It is very important to notice that the functions $\Psi_n(q)$ do not, in general, coincide with the functions $\phi_n(q)$; the latter are in general not even mutually orthogonal, and do not form a set of eigenfunctions of any operator. This expresses the fact that the results of measurements in quantum mechanics cannot be reproduced. If the electron was in a state $\Psi_n(q)$, then a measurement of the quantity f carried out on it leads with certainty to the value f_n. After the measurement, however, the electron is in a state $\phi_n(q)$ different from its initial one, and in this state the quantity f does not in general take any definite value. Hence, on carrying out a second measurement on the electron immediately after the first, we should obtain for f a value which did not agree with that obtained from the first measurement.† To predict (in the sense of calculating probabilities) the result of the second measurement from the known result of the first, we must take from the first measurement the wave function $\phi_n(q)$ of the state in which it resulted, and from the second measurement the wave function $\Psi_n(q)$ of the state whose probability is required. This means that from the equations of quantum mechanics we determine the wave function $\phi_n(q, t)$ which, at the instant when the first measurement is made, is equal to $\phi_n(q)$; the probability of the mth result of the second measurement, made at time t, is then given by the squared modulus of the integral $\int \phi_n(q, t)\Psi_m^*(q)\, dq$.

We see that the measuring process in quantum mechanics has a "two-faced" character: it plays different parts with respect to the past and future of the electron. With respect to the past, it "verifies" the probabilities of the various possible results predicted from the state brought about by the previous measurement. With respect to the future, it brings about a new state (see also §44). Thus the very nature of the process of measurement involves a far-reaching principle of irreversibility.

This irreversibility is of fundamental significance. We shall see later (at the end of §18) that the basic equations of quantum mechanics are in themselves symmetrical with respect to a change in the sign of the time; here quantum mechanics does not differ from classical mechanics. The irreversibility of the process of measurement, however, causes the two directions of time to be physically non-equivalent, i.e. creates a difference between the future and the past.

---

† There is, however, an important exception to the statement that results of measurements cannot be reproduced: the one quantity the result of whose measurement can be exactly reproduced is the coordinate. Two measurements of the coordinates of an electron, made at a sufficiently small interval of time, must give neighbouring values; if this were not so, it would mean that the electron had an infinite velocity. Mathematically, this is related to the fact that the coordinate commutes with the operator of the interaction energy between the electron and the apparatus, since this energy is (in non-relativistic theory) a function of the coordinates only.

CHAPTER II

ENERGY AND MOMENTUM

§8. The Hamiltonian operator

THE wave function Ψ completely determines the state of a physical system in quantum mechanics. This means that, if this function is given at some instant, not only are all the properties of the system at that instant described, but its behaviour at all subsequent instants is determined (only, of course, to the degree of completeness which is generally admissible in quantum mechanics). The mathematical expression of this fact is that the value of the derivative $\partial\Psi/\partial t$ of the wave function with respect to time at any given instant must be determined by the value of the function itself at that instant, and, by the principle of superposition, the relation between them must be linear. In the most general form we can write

$$i\hbar\, \partial\Psi/\partial t = \hat{H}\Psi, \tag{8.1}$$

where $\hat{H}$ is some linear operator; the factor $i\hbar$ is introduced here for a reason that will become apparent.

Since the integral $\int\Psi^*\Psi\, \mathrm{d}q$ is a constant independent of time, we have

$$\frac{\mathrm{d}}{\mathrm{d}t}\int|\Psi|^2\, \mathrm{d}q = \int\frac{\partial\Psi^*}{\partial t}\Psi\, \mathrm{d}q + \int\Psi^*\frac{\partial\Psi}{\partial t}\, \mathrm{d}q = 0.$$

Substituting here (8.1) and using in the first integral the definition of the transpose of an operator, we can write (omitting the common factor $i/\hbar$)

$$\int\Psi\hat{H}^*\Psi^*\, \mathrm{d}q - \int\Psi^*\hat{H}\Psi\, \mathrm{d}q = \int\Psi^*\tilde{\hat{H}}^*\Psi\, \mathrm{d}q - \int\Psi^*\hat{H}\Psi\, \mathrm{d}q$$
$$= \int\Psi^*(\tilde{\hat{H}}^* - \hat{H})\Psi\, \mathrm{d}q = 0.$$

Since this equation must hold for an arbitrary function Ψ, it follows that we must have identically $\hat{H}^+ = \hat{H}$; the operator $\hat{H}$ is therefore Hermitian. Let us find the physical quantity to which it corresponds. To do this, we use the limiting expression (6.1) for the wave function and write

$$\frac{\partial\Psi}{\partial t} = \frac{i}{\hbar}\frac{\partial S}{\partial t}\Psi;$$

the slowly varying amplitude a need not be differentiated. Comparing this equation with the definition (8.1), we see that, in the limiting case, the operator $\hat{H}$ reduces to simply multiplying by $-\partial S/\partial t$. This means that

$-\partial S/\partial t$ is the physical quantity into which the Hermitian operator $\hat{H}$ passes.

The derivative $-\partial S/\partial t$ is just Hamilton's function H for a mechanical system. Thus the operator $\hat{H}$ is what corresponds in quantum mechanics to Hamilton's function; this operator is called the *Hamiltonian operator* or, more briefly, the *Hamiltonian* of the system. If the form of the Hamiltonian is known, equation (8.1) determines the wave functions of the physical system concerned. This fundamental equation of quantum mechanics is called the *wave equation.*

§9. The differentiation of operators with respect to time

The concept of the derivative of a physical quantity with respect to time cannot be defined in quantum mechanics in the same way as in classical mechanics. For the definition of the derivative in classical mechanics involves the consideration of the values of the quantity at two neighbouring but distinct instants of time. In quantum mechanics, however, a quantity which at some instant has a definite value does not in general have definite values at subsequent instants; this was discussed in detail in §1.

Hence the idea of the derivative with respect to time must be differently defined in quantum mechanics. It is natural to define the *derivative* $\dot{f}$ of a quantity f as the quantity whose mean value is equal to the derivative, with respect to time, of the mean value $\bar{f}$. Thus we have the definition

$$\bar{\dot{f}} = \dot{\bar{f}}. \tag{9.1}$$

Starting from this definition, it is easy to obtain an expression for the quantum-mechanical operator $\hat{\dot{f}}$ corresponding to the quantity $\dot{f}$:

$$\bar{\dot{f}} = \dot{\bar{f}} = \frac{\mathrm{d}}{\mathrm{d}t}\int \Psi^*\hat{f}\Psi\,\mathrm{d}q = \int \Psi^*\frac{\partial \hat{f}}{\partial t}\Psi\,\mathrm{d}q + \int \frac{\partial \Psi^*}{\partial t}\hat{f}\Psi\,\mathrm{d}q + \int \Psi^*\hat{f}\frac{\partial \Psi}{\partial t}\,\mathrm{d}q.$$

Here $\partial\hat{f}/\partial t$ is the operator obtained by differentiating the operator $\hat{f}$ with respect to time; $\hat{f}$ may depend on the time as a parameter. Substituting for $\partial\Psi/\partial t$, $\partial\Psi^*/\partial t$ their expressions according to (8.1), we obtain

$$\bar{\dot{f}} = \int \Psi^*\frac{\partial \hat{f}}{\partial t}\Psi\,\mathrm{d}q + \frac{i}{\hbar}\int (\hat{H}^*\Psi^*)\hat{f}\Psi\,\mathrm{d}q - \frac{i}{\hbar}\int \Psi^*\hat{f}(\hat{H}\Psi)\,\mathrm{d}q.$$

Since the operator $\hat{H}$ is Hermitian, we have

$$\int (\hat{H}^*\Psi^*)(\hat{f}\Psi)\,\mathrm{d}q = \int \Psi^*\hat{H}\hat{f}\Psi\,\mathrm{d}q;$$

thus

$$\bar{\dot{f}} = \int \Psi^*\left(\frac{\partial \hat{f}}{\partial t} + \frac{i}{\hbar}\hat{H}\hat{f} - \frac{i}{\hbar}\hat{f}\hat{H}\right)\Psi\,\mathrm{d}q.$$

Since, on the other hand, we must have, by the definition of mean values, $\bar{\dot{f}} = \int \Psi^*\hat{\dot{f}}\Psi\,\mathrm{d}q$, it is seen that the expression in parentheses in the integrand is the required operator $\hat{\dot{f}}$:†

$$\hat{\dot{f}} = \frac{\partial \hat{f}}{\partial t} + \frac{i}{\hbar}(\hat{H}\hat{f} - \hat{f}\hat{H}). \tag{9.2}$$

If the operator $\hat{f}$ is independent of time, $\hat{\dot{f}}$ reduces, apart from a constant factor, to the commutator of the operator $\hat{f}$ and the Hamiltonian.

A very important class of physical quantities is formed by those whose operators do not depend explicitly on time, and also commute with the Hamiltonian, so that $\hat{\dot{f}} = 0$. Such quantities are said to be *conserved*. For these $\dot{\bar{f}} = \bar{\dot{f}} = 0$, that is, $\bar{f}$ is constant. In other words, the mean value of the quantity remains constant in time. We can also assert that, if in a given state the quantity f has a definite value (i.e. the wave function is an eigenfunction of the operator $\hat{f}$), then it will have a definite value (the same one) at subsequent instants also.

§10. Stationary states

The Hamiltonian of a closed system (and of a system in a *constant* external field) cannot contain the time explicitly. This follows from the fact that, for such a system, all times are equivalent. Since, on the other hand, any operator of course commutes with itself, we reach the conclusion that Hamilton's function is conserved for systems which are not in a varying external field. As is well known, a Hamilton's function which is conserved is called the *energy*. The law of conservation of energy in quantum mechanics signifies that, if in a given state the energy has a definite value, this value remains constant in time.

† In classical mechanics we have for the total derivative, with respect to time, of a quantity f which is a function of the generalized coordinates q_i and momenta p_i of the system

$$\frac{\mathrm{d}f}{\mathrm{d}t} = \frac{\partial f}{\partial t} + \sum_i \left(\frac{\partial f}{\partial q_i}\dot{q}_i + \frac{\partial f}{\partial p_i}\dot{p}_i\right).$$

Substituting, in accordance with Hamilton's equations, $\dot{q}_i = \partial H/p_i$ and $\dot{p}_i = -\partial H/\partial q_i$, we obtain

$$\mathrm{d}f/\mathrm{d}t = \partial f/\partial t + [H, f],$$

where

$$[H, f] \equiv \sum_i \left(\frac{\partial f}{\partial q_i}\frac{\partial H}{\partial p_i} - \frac{\partial f}{\partial p_i}\frac{\partial H}{\partial q_i}\right)$$

is what is called the *Poisson bracket* for the quantities f and H (see *Mechanics*, §42). On comparing with the expression (9.2), we see that, as we pass to the limit of classical mechanics, the operator $i(\hat{H}\hat{f} - \hat{f}\hat{H})$ reduces in the first approximation to zero, as it should, and in the second approximation (with respect to $\hbar$) to the quantity $\hbar[H, f]$. This result is true also for any two quantities f and g; the operator $i(\hat{f}\hat{g} - \hat{g}\hat{f})$ tends in the limit to the quantity $\hbar[f, g]$, where $[f, g]$ is the Poisson bracket

$$[f, g] \equiv \sum_i \left(\frac{\partial g}{\partial q_i}\frac{\partial f}{\partial p_i} - \frac{\partial g}{\partial p_i}\frac{\partial f}{\partial q_i}\right).$$

This follows from the fact that we can always formally imagine a system whose Hamiltonian is $\hat{g}$.

States in which the energy has definite values are called *stationary states* of a system. They are described by wave functions Ψ_n which are the eigenfunctions of the Hamiltonian operator, i.e. which satisfy the equation $\hat{H}\Psi_n = E_n\Psi_n$, where E_n are the eigenvalues of the energy. Correspondingly, the wave equation (8.1) for the function Ψ_n,

$$i\hbar\, \partial\Psi_n/\partial t = \hat{H}\Psi_n = E_n\Psi_n$$

can be integrated at once with respect to time and gives

$$\Psi_n = e^{-(i/\hbar)E_n t}\psi_n(q), \tag{10.1}$$

where ψ_n is a function of the coordinates only. This determines the relation between the wave functions of stationary states and the time.

We shall denote by the small letter ψ the wave functions of stationary states without the time factor. These functions, and also the eigenvalues of the energy, are determined by the equation

$$\hat{H}\psi = E\psi. \tag{10.2}$$

The stationary state with the smallest possible value of the energy is called the *normal* or *ground state* of the system.

The expansion of an arbitrary wave function Ψ in terms of the wave functions of stationary states has the form

$$\Psi = \sum_n a_n e^{-(i/\hbar)E_n t}\psi_n(q). \tag{10.3}$$

The squared moduli $|a_n|^2$ of the expansion coefficients, as usual, determine the probabilities of various values of the energy of the system.

The probability distribution for the coordinates in a stationary state is determined by the squared modulus $|\Psi_n|^2 = |\psi_n|^2$; we see that it is independent of time. The same is true of the mean values

$$\bar{f} = \int \Psi_n^* \hat{f} \Psi_n \, dq = \int \psi_n^* \hat{f} \psi_n \, dq$$

of any physical quantity f (whose operator does not depend explicitly on the time).

As has been said, the operator of any quantity that is conserved commutes with the Hamiltonian. This means that any physical quantity that is conserved can be measured simultaneously with the energy.

Among the various stationary states, there may be some which correspond to the same value of the energy (the same *energy level* of the system), but differ in the values of some other physical quantities. Such energy levels, to which several different stationary states correspond, are said to be *degenerate*. Physically, the possibility that degenerate levels can exist is related to the fact that the energy does not in general form by itself a complete set of physical quantities.

If there are two conserved physical quantities f and g whose operators do not commute, then the energy levels of the system are in general degenerate. For, let ψ be the wave function of a stationary state in which, besides the energy, the quantity f also has a definite value. Then we can say that the function $\hat{g}\psi$ does not coincide (apart from a constant factor) with ψ; if it did, this would mean that the quantity g also had a definite value, which is impossible, since f and g cannot be measured simultaneously. On the other hand, the function $\hat{g}\psi$ is an eigenfunction of the Hamiltonian, corresponding to the same value E of the energy as ψ:

$$\hat{H}(\hat{g}\psi) = \hat{g}\hat{H}\psi = E(\hat{g}\psi).$$

Thus we see that the energy E corresponds to more than one eigenfunction, i.e. the energy level is degenerate.

It is clear that any linear combination of wave functions corresponding to the same degenerate energy level is also an eigenfunction for that value of the energy. In other words, the choice of eigenfunctions of a degenerate energy level is not unique. Arbitrarily selected eigenfunctions of a degenerate energy level are not, in general, orthogonal. By a proper choice of linear combinations of them, however, we can always obtain a set of orthogonal (and normalized) eigenfunctions (and this can be done in infinitely many ways; for the number of independent coefficients in a linear transformation of n functions is n^2, while the number of normalization and orthogonality conditions for n functions is $\frac{1}{2}n(n+1)$, i.e. less than n^2).

These statements concerning the eigenfunctions of a degenerate energy level relate, of course, not only to eigenfunctions of the energy, but also to those of any operator. Only those functions are automatically orthogonal which correspond to different eigenvalues of the operator concerned; functions which correspond to the same degenerate eigenvalue are not in general orthogonal.

If the Hamiltonian of the system is the sum of two (or more) parts, $\hat{H} = \hat{H}_1 + \hat{H}_2$, one of which contains only the coordinates q_1 and the other only the coordinates q_2, then the eigenfunctions of the operator $\hat{H}$ can be written down as products of the eigenfunctions of the operators $\hat{H}_1$ and $\hat{H}_2$, and the eigenvalues of the energy are equal to the sums of the eigenvalues of these operators.

The spectrum of eigenvalues of the energy may be either discrete or continuous. A stationary state of a discrete spectrum always corresponds to a finite motion of the system, i.e. one in which neither the system nor any part of it moves off to infinity. For, with eigenfunctions of a discrete spectrum, the integral $\int |\Psi|^2 \, dq$, taken over all space, is finite. This certainly means that the squared modulus $|\Psi|^2$ decreases quite rapidly, becoming zero at infinity. In other words, the probability of infinite values of the coordinates is zero; that is, the system executes a finite motion, and is said to be in a *bound state*.

For wave functions of a continuous spectrum, the integral $\int |\Psi|^2 \, dq$ diverges. Here the squared modulus $|\Psi|^2$ of the wave function does not directly determine the probability of the various values of the coordinates, and must be regarded only as a quantity proportional to this probability. The divergence of the integral $\int |\Psi|^2 \, dq$ is always due to the fact that $|\Psi|^2$ does not become zero at infinity (or becomes zero insufficiently rapidly). Hence we can say that the integral $\int |\Psi|^2 \,dq$, taken over the region of space outside any arbitrarily large but finite closed surface, will always diverge. This means that, in the state considered, the system (or some part of it) is at infinity. For a wave function which is a superposition of the wave functions of various stationary states of a continuous spectrum, the integral $\int |\Psi|^2 \,dq$ may converge, so that the system lies in a finite region of space. However, in the course of time, this region moves unrestrictedly, and eventually the system moves off to infinity. This can be seen as follows. Any superposition of wave functions of a continuous spectrum has the form

$$\Psi = \int a_E e^{-(i/\hbar)Et}\psi_E(q)\,dE.$$

The squared modulus of Ψ can be written in the form of a double integral:

$$|\Psi|^2 = \iint a_E a_{E'}^* e^{(i/\hbar)(E'-E)t}\psi_E(q)\psi_{E'}^*(q)\,dE dE'.$$

If we average this expression over some time interval T, and then let T tend to infinity, the mean values of the oscillating factors $e^{(i/\hbar)(E'-E)t}$, and therefore the whole integral, tend to zero in the limit. Thus the mean value, with respect to time, of the probability of finding the system at any given point of configuration space tends to zero. This is possible only if the motion takes place throughout infinite space.† Thus the stationary states of a continuous spectrum correspond to an infinite motion of the system.

§11. Matrices

We shall suppose for convenience that the system considered has a discrete energy spectrum; all the relations obtained below can be generalized at once to the case of a continuous spectrum. Let $\Psi = \Sigma a_n \Psi_n$ be the expansion of an arbitrary wave function in terms of the wave functions Ψ_n of the stationary states. If we substitute this expansion in the definition (3.8) of the mean value of some quantity f, we obtain

$$\bar{f} = \sum_n \sum_m a_n^* a_m f_{nm}(t), \tag{11.1}$$

where $f_{nm}(t)$ denotes the integral

$$f_{nm}(t) = \int \Psi_n^* \hat{f} \Psi_m \,dq. \tag{11.2}$$

The set of quantities $f_{nm}(t)$ with all possible n and m is called the *matrix* of the

† Note that, for a function Ψ which is a superposition of functions of a discrete spectrum, we should have

$$\overline{|\Psi|^2} = \sum_{n}\sum_{m} \overline{a_n a_m^* e^{(i/\hbar)(E_m-E_n)t}\psi_n\psi_m^*} = \sum_n |a_n\psi_n(q)|^2,$$

i.e. the probability density remains finite on averaging over time.

quantity f, and each of the $f_{nm}(t)$ is called the *matrix element* corresponding to the *transition* from state m to state n.†

The dependence of the matrix elements $f_{nm}(t)$ on time is determined (if the operator $\hat{f}$ does not contain the time explicitly) by the dependence of the functions Ψ_n on time. Substituting for them the expressions (10.1), we find that

$$f_{nm}(t) = f_{nm}e^{i\omega_{nm}t}, \tag{11.3}$$

where

$$\omega_{nm} = (E_n - E_m)/\hbar \tag{11.4}$$

is what is called the *transition frequency* between the states n and m, and the quantities

$$f_{nm} = \int \psi_n^* \hat{f} \psi_m \, dq \tag{11.5}$$

form the matrix of the quantity f which is independent of time, and which is commonly used.‡

The matrix elements of the derivative $\dot{f}$ are obtained by differentiating the matrix elements of the quantity f with respect to time; this follows directly from the fact that

$$\bar{\dot{f}} = \dot{\bar{f}} = \sum_{mn}\sum a_n^* a_m \dot{f}_{nm}(t). \tag{11.6}$$

From (11.3) we thus have for the matrix elements of $\dot{f}$

$$\dot{f}_{nm}(t) = i\omega_{nm} f_{nm}(t) \tag{11.7}$$

or (cancelling the time factor $e^{i\omega_{nm}t}$ from both sides) for the matrix elements independent of time

$$(\dot{f})_{nm} = i\omega_{nm} f_{nm} = (i/\hbar)(E_n - E_m) f_{nm}. \tag{11.8}$$

To simplify the notation in the formulae, we shall derive all our relations below for the matrix elements independent of time; exactly similar relations hold for the matrices which depend on the time.

For the matrix elements of the complex conjugate f^* of the quantity f we obtain, taking into account the definition of the Hermitian conjugate operator,

$$(f^*)_{nm} = \int \psi_n^* \hat{f}^+ \psi_m \, dq = \int \psi_n^* \tilde{\hat{f}}^* \psi_m \, dq = \int \psi_m \hat{f}^* \psi_n^* \, dq$$

or

$$(f^*)_{nm} = (f_{mn})^*. \tag{11.9}$$

For real physical quantities, which are the only ones we usually consider,

---

† The matrix representation of physical quantities was introduced by Heisenberg in 1925, before Schrödinger's discovery of the wave equation. "Matrix mechanics" was later developed by M. Born, W. Heisenberg and P. Jordan.

‡ Because of the indeterminacy of the phase factor in normalized wave functions (see §2), the matrix elements f_{nm} (and $f_{nm}(t)$) also are determined only to within a factor of the form $e^{i(\alpha_m - \alpha_n)}$. Here again this indeterminacy has no effect on any physical results.

we consequently have

$$f_{nm} = f_{mn}^* \tag{11.10}$$

(f_{mn}^* stands for $(f_{mn})^*$). Such matrices, like the corresponding operators, are said to be *Hermitian*.

Matrix elements with $n = m$ are called *diagonal elements*. These are independent of time, and (11.10) shows that they are real. The element f_{nn} is the mean value of the quantity f in the state ψ_n.

It is not difficult to obtain the "multiplication rule" for matrices. To do so, we first observe that the formula

$$\hat{f}\psi_n = \sum_m f_{mn}\psi_m \tag{11.11}$$

holds. This is simply the expansion of the function $\hat{f}\psi_n$ in terms of the functions ψ_m, the coefficients being determined in accordance with the general formula (3.5). Remembering this formula, let us write down the result of the product of two operators acting on the function ψ_n:

$$\hat{f}\hat{g}\psi_n = \hat{f}(\hat{g}\psi_n) = \hat{f}\sum_k g_{kn}\psi_k = \sum_k g_{kn}\hat{f}\psi_k = \sum_{k,m} g_{kn}f_{mk}\psi_m.$$

Since, on the other hand, we must have

$$\hat{f}\hat{g}\psi_n = \sum_m (fg)_{mn}\psi_m,$$

we arrive at the result that the matrix elements of the product fg are determined by the formula

$$(fg)_{mn} = \sum_k f_{mk}g_{kn}. \tag{11.12}$$

This rule is the same as that used in mathematics for the multiplication of matrices: the rows of the first matrix in the product are multiplied by the columns of the second matrix.

If the matrix is given, then so is the operator itself. In particular, if the matrix is given, it is in principle possible to determine the eigenvalues of the physical quantity concerned and the corresponding eigenfunctions.

We shall now consider the values of all quantities at some definite instant, and expand an arbitrary wave function Ψ (at that instant) in terms of the eigenfunctions of the Hamiltonian, i.e. of the wave functions ψ_m of the stationary states (these wave functions are independent of time).

$$\Psi = \sum_m c_m\psi_m, \tag{11.13}$$

where the expansion coefficients are denoted by c_m. We substitute this expansion in the equation $\hat{f}\Psi = f\Psi$ which determines the eigenvalues and eigenfunctions of the quantity f. We have

$$\sum_m c_m(\hat{f}\psi_m) = f\sum_m c_m\psi_m.$$

We multiply both sides of this equation by ψ_n^* and integrate over q. Each of the integrals $\int \psi_n^*\hat{f}\psi_m\, dq$ on the left-hand side of the equation is the corresponding matrix element f_{nm}. On the right-hand side, all the integrals $\int \psi_n^*\psi_m\, dq$ with $m \neq n$ vanish by virtue of the orthogonality of the functions

ψ_m, and $\int \psi_n{}^*\psi_n \, dq = 1$ by virtue of their normalization.† Thus

$$\sum_m f_{nm}c_m = fc_n, \tag{11.14}$$

or

$$\sum_m (f_{nm} - f\delta_{nm})c_m = 0,$$

where $\delta_{nm} = 0$ for $m \neq n$ and $= 1$ for $m = n$.

Thus we have obtained a system of homogeneous algebraic equations of the first degree (with the c_m as unknowns). As is well known, such a system has solutions different from zero only if the determinant formed by the coefficients in the equations vanishes, i.e. only if

$$|f_{nm} - f\delta_{nm}| = 0. \tag{11.15}$$

The roots of this equation (in which f is regarded as the unknown) are the possible values of the quantity f. The set of values c_m satisfying the equations (11.14) when f is equal to any of these values determines the corresponding eigenfunction.

If, in the definition (11.5) of the matrix elements of the quantity f, we take as ψ_n the eigenfunctions of this quantity, then from the equation $\hat{f}\psi_n = f_n\psi_n$ we have

$$f_{nm} = \int \psi_n{}^*\hat{f}\psi_m \, dq = f_m \int \psi_n{}^*\psi_m \, dq.$$

By virtue of the orthogonality and normalization of the functions ψ_m, this gives $f_{nm} = 0$ for $n \neq m$ and $f_{mm} = f_m$. Thus only the diagonal matrix elements are different from zero, and each of these is equal to the corresponding eigenvalue of the quantity f. A matrix with only these elements different from zero is said to be put in *diagonal form*. In particular, in the usual representation, with the wave functions of the stationary states as the functions ψ_n, the energy matrix is diagonal (and so are the matrices of all other physical quantities having definite values in the stationary states). In general, the matrix of a quantity f, defined with respect to the eigenfunctions of some operator $\hat{g}$, is said to be the matrix of *f in a representation in which g is diagonal*. We shall always, except where the subject is specially mentioned, understand in future by the matrix of a physical quantity its matrix in the usual representation, in which the energy is diagonal. Everything that has been said above regarding the dependence of matrices on time refers, of course, only to this usual representation.‡

---

† In accordance with the general rule (§5), the set of coefficients c_n in the expansion (11.13) can be considered as the wave function in the "energy representation" (the variable being the suffix n that gives the number of the energy eigenvalue). The matrix f_{nm} here acts as the operator $\hat{f}$ in this representation, the action of which on the wave function is given by the left-hand side of (11.14). The formula $\bar{f} = \sum\sum c_n^*(f_{nm}c_m)$ then corresponds to the general expression for the mean value of a quantity in terms of its operator and the wave function of the state concerned.

‡ Bearing in mind the diagonality of the energy matrix, it is easy to see that equation (11.8) is the operator relation (9.2) written in matrix form.

By means of the matrix representation of operators we can prove the theorem mentioned in §4: if two operators commute with each other, they have their entire sets of eigenfunctions in common. Let $\hat{f}$ and $\hat{g}$ be two such operators. From $\hat{f}\hat{g} = \hat{g}\hat{f}$ and the matrix multiplication rule (11.12), it follows that

$$\sum_k f_{mk}g_{kn} = \sum_k g_{mk}f_{kn}.$$

If we take the eigenfunctions of the operator $\hat{f}$ as the set of functions ψ_n with respect to which the matrix elements are calculated, we shall have $f_{mk} = 0$ for $m \neq k$, so that the above equation reduces to $f_{mm}g_{mn} = g_{mn}f_{nn}$, or

$$g_{mn}(f_m - f_n) = 0.$$

If all the eigenvalues f_n of the quantity f are different, then for all $m \neq n$ we have $f_m - f_n \neq 0$, so that we must have $g_{mn} = 0$. Thus the matrix g_{mn} is also diagonal, i.e. the functions ψ_n are eigenfunctions of the physical quantity g also. If, among the values f_n, there are some which are equal (i.e. if there are eigenvalues to which several different eigenfunctions correspond), then the matrix elements g_{mn} corresponding to each such group of functions ψ_n are, in general, different from zero. However, linear combinations of the functions ψ_n which correspond to a single eigenvalue of the quantity f are evidently also eigenfunctions of f; one can always choose these combinations in such a way that the corresponding non-diagonal matrix elements g_{mn} are zero, and thus, in this case also, we obtain a set of functions which are simultaneously the eigenfunctions of the operators $\hat{f}$ and $\hat{g}$.

The following formula is useful in applications:

$$(\partial \hat{H}/\partial \lambda)_{nn} = \partial E_n/\partial \lambda, \tag{11.16}$$

where λ is a parameter on which the Hamiltonian $\hat{H}$ (and therefore the energy eigenvalues E_n) depends. It is proved as follows. Differentiating the equation $(\hat{H} - E_n)\psi_n = 0$ with respect to λ and then multiplying on the left by ψ_n^*, we obtain

$$\psi_n^*(\hat{H} - E_n)\frac{\partial \psi_n}{\partial \lambda} = \psi_n^*\left(\frac{\partial E_n}{\partial \lambda} - \frac{\partial \hat{H}}{\partial \lambda}\right)\psi_n.$$

On integration with respect to q, the left-hand side gives zero, since

$$\int \psi_n^*(\hat{H} - E_n)\frac{\partial \psi_n}{\partial \lambda}\,dq = \int \frac{\partial \psi_n}{\partial \lambda}(\hat{H} - E_n)^*\psi_n^*\,dq,$$

the operator $\hat{H}$ being Hermitian. The right-hand side gives the required equation.

A widely used notation (introduced by Dirac) in recent literature is that which denotes the matrix elements f_{nm} by†

$$\langle n|\, f\, |m\rangle. \tag{11.17}$$

This symbol is written so that it may be regarded as "consisting" of the quantity f and the symbols $|m\rangle$ and $\langle n|$ which respectively stand for the initial and final states as such (independently of the representation of the wave functions of the states). With the same symbols we can construct notations for the expansion coefficients of wave functions: if there is a complete set of wave functions corresponding to the states $|n_1\rangle$, $|n_2\rangle$, ... , the coefficients in the expansion in terms of these of the wave function of a state $|m\rangle$ are denoted by

$$\langle n_i|m\rangle = \int \psi_{n_i}{}^*\psi_m\, dq. \tag{11.18}$$

§12. **Transformation of matrices**

The matrix elements of a given physical quantity can be defined with respect to various sets of wave functions, for example the wave functions of stationary states described by various sets of physical quantities, or the wave functions of stationary states of the same system in various external fields. The problem therefore arises of the transformation of matrices from one representation to another.

Let $\psi_n(q)$ and $\psi_n'(q)$ $(n = 1, 2, ...)$ be two complete sets of orthonormal functions, related by some linear transformation:

$$\psi_n' = \sum_m S_{mn}\psi_m, \tag{12.1}$$

which is simply an expansion of the function ψ_n' in terms of the complete set of functions ψ_n. This transformation may be conventionally written in the operator form

$$\psi_n' = \hat{S}\psi_n. \tag{12.2}$$

The operator $\hat{S}$ must satisfy a certain condition in order that the functions ψ_n' should be orthonormal if the functions ψ_n are. Substituting (12.2) in the condition

$$\int \psi_m'^*\psi_n'\, dq = \delta_{mn},$$

and using the definition of the transposed operator (3.14), we have

$$\int (\hat{S}\psi_n)\hat{S}^*\psi_m{}^*\, dq = \int \psi_m{}^*\tilde{\hat{S}}{}^*\hat{S}\psi_n\, dq = \delta_{mn}.$$

If these equations hold for all m and n, we must have $\tilde{\hat{S}}{}^*\hat{S} = 1$, or

$$\tilde{\hat{S}}{}^* \equiv \hat{S}^+ = \hat{S}^{-1}, \tag{12.3}$$

† Both notations are used in the present book. The form (11.17) is especially convenient when each suffix has to be written as several letters.

i.e. the inverse operator is equal to the Hermitian conjugate operator. Operators having this property are said to be *unitary*. Owing to this property, the transformation $\psi_n = \hat{S}^{-1}\psi_n'$ inverse to (12.1) is given by

$$\psi_n = \sum_m S_{nm}^*\psi_m'. \tag{12.4}$$

Writing the equations $\hat{S}^+S = 1$ and $\hat{S}\hat{S}^+ = 1$ in matrix form, we obtain the following forms of the unitarity condition:

$$\sum_l S_{lm}^* S_{ln} = \delta_{mn}, \tag{12.5}$$

$$\sum_l S_{ml}^* S_{nl} = \delta_{mn}. \tag{12.6}$$

Let us now consider some physical quantity f and write down its matrix elements in the "new" representation, i.e. with respect to the functions ψ_n'. These are given by the integrals

$$\begin{aligned}\int \psi_m'^*\hat{f}\psi_n'\,dq &= \int (\hat{S}^*\psi_m^*)(\hat{f}\hat{S}\psi_n)\,dq\\ &= \int \psi_m^*\tilde{\hat{S}}^*\hat{f}\hat{S}\psi_n\,dq\\ &= \int \psi_m^*\hat{S}^{-1}\hat{f}\hat{S}\psi_n\,dq.\end{aligned}$$

Hence we see that the matrix of the operator $\hat{f}$ in the new representation is equal to the matrix of the operator

$$\hat{f}' = \hat{S}^{-1}\hat{f}\hat{S} \tag{12.7}$$

in the old representation.†

The sum of the diagonal elements of a matrix is called the *trace* or *spur*‡ of the matrix and denoted by tr f:

$$\mathrm{tr}\, f = \sum_n f_{nn}. \tag{12.8}$$

It may be noted first of all that the trace of a product of two matrices is independent of the order of multiplication:

$$\mathrm{tr}\,(fg) = \mathrm{tr}\,(gf), \tag{12.9}$$

† If $\{\hat{f}, \hat{g}\} = -i\hbar\hat{c}$ is the commutation rule for two operators $\hat{f}$ and $\hat{g}$, the transformation (12.7) gives $\{\hat{f}', \hat{g}'\} = -i\hbar\hat{c}'$, i.e. the rule is unchanged. We have shown in the footnote in §9 that $\hat{c}$ is the quantum analogue of the classical Poisson bracket $[f, g]$. In classical mechanics, however, the Poisson brackets are invariant under canonical transformations of the variables (generalized coordinates and momenta); see *Mechanics*, §45. In this sense we can say that unitary transformations in quantum mechanics play a role analogous to that of canonical transformations in classical mechanics.

‡ From the German word *Spur*. The notation sp f is also used. The trace can be defined, of course, only if the sum over n is convergent.

since the rule of matrix multiplication gives

$$\operatorname{tr}(fg) = \sum_n \sum_k f_{nk} g_{kn} = \sum_k \sum_n g_{kn} f_{nk} = \operatorname{tr}(gf).$$

Similarly we can easily see that, for a product of several matrices, the trace is unaffected by a cyclic permutation of the factors; for example,

$$\operatorname{tr}(fgh) = \operatorname{tr}(hfg) = \operatorname{tr}(ghf). \tag{12.10}$$

An important property of the trace is that it does not depend on the choice of the set of functions with respect to which the matrix elements are defined, since

$$(\operatorname{tr} f)' = \operatorname{tr}(S^{-1}fS) = \operatorname{tr}(SS^{-1}f) = \operatorname{tr} f. \tag{12.11}$$

A unitary transformation leaves unchanged the sum of the squared moduli of the functions that are transformed: from (12.6) we have

$$\sum_i |\psi_i'|^2 = \sum_{k,l,i} S_{ki}\psi_k S_{li}^*\psi_l^* = \sum_{k,l} \psi_k\psi_l^*\delta_{kl} = \sum_k |\psi_k|^2. \tag{12.12}$$

Any unitary operator may be written as

$$\hat{S} = e^{i\hat{R}}, \tag{12.13}$$

where $\hat{R}$ is an Hermitian operator: since $\hat{R}^+ = \hat{R}$, we have

$$\hat{S}^+ = e^{-i\hat{R}^+} = e^{-i\hat{R}} = \hat{S}^{-1}.$$

The expansion

$$f' = \hat{S}^{-1} f \hat{S} = f + \{f, i\hat{R}\} + \tfrac{1}{2}\{\{f, i\hat{R}\}, i\hat{R}\} + \ldots \tag{12.14}$$

is easily verified by a direct expansion of the factors $\exp(\pm i\hat{R})$ in powers of $\hat{R}$. This expansion may be useful when $\hat{R}$ is proportional to a small parameter, so that (12.14) becomes an expansion in powers of the parameter.

§13. The Heisenberg representation of operators

In the mathematical formalism of quantum mechanics described here, the operators corresponding to various physical quantities act on functions of the coordinates and do not usually depend explicitly on time. The time dependence of the mean values of physical quantities is due only to the time dependence of the wave function of the state, according to the formula

$$\bar{f}(t) = \int \Psi^*(q, t)\hat{f}\Psi(q, t)\, dq. \tag{13.1}$$

The quantum-mechanical treatment can, however, be formulated also in a somewhat different but equivalent form, in which the time dependence is transferred from the wave functions to the operators. Although we shall not use this *Heisenberg representation* (as opposed to the *Schrödinger representation*) of operators in the present volume, a statement of it is given here with a view to applications in the relativistic theory.

We define the operator (which is unitary; see (12.13))

$$\hat{S} = \exp[-(i/\hbar)\hat{H}t], \tag{13.2}$$

where $\hat{H}$ is the Hamiltonian of the system. By definition, its eigenfunctions are the same as those of the operator $\hat{H}$, i.e. the stationary-state wave functions $\psi_n(q)$, where

$$\hat{S}\psi_n(q) = e^{-(i/\hbar)E_n t}\psi_n(q). \tag{13.3}$$

Hence it follows that the expansion (10.3) of an arbitrary wave function Ψ in terms of the stationary-state wave functions can be written in the operator form

$$\Psi(q, t) = \hat{S}\Psi(q, 0), \tag{13.4}$$

i.e. the effect of the operator $\hat{S}$ is to convert the wave function of the system at some initial instant into the wave function at an arbitrary instant.

Defining, as in (12.7), the time-dependent operator

$$\hat{f}(t) = \hat{S}^{-1}\hat{f}\hat{S}, \tag{13.5}$$

we have

$$\bar{f}(t) = \int \Psi^*(q, 0)\hat{f}(t)\Psi(q, 0)\, dq, \tag{13.6}$$

and thus obtain the formula (3.8) for the mean value of the quantity f in a form in which the time dependence is entirely transferred to the operator (for our definition of an operator rests on formula (3.8)).

It is evident that the matrix elements of the operator (13.5) with respect to the stationary-state wave functions are the same as the time-dependent matrix elements $f_{nm}(t)$ defined by formula (11.3).

Finally, differentiating the expression (13.5) with respect to time (assuming that the operators $\hat{f}$ and $\hat{H}$ do not themselves involve t), we obtain

$$\frac{\partial}{\partial t}\hat{f}(t) = \frac{i}{\hbar}[\hat{H}\hat{f}(t) - \hat{f}(t)\hat{H}], \tag{13.7}$$

which is similar in form to (9.2) but has a somewhat different significance: the expression (9.2) defines the operator $\hat{\dot{f}}$ corresponding to the physical quantity $\dot{f}$, while the left-hand side of equation (13.7) is the time derivative of the operator of the quantity f itself.

§14. The density matrix

The description of a system by means of a wave function is the most complete description possible in quantum mechanics, in the sense indicated at the end of §1.

States that do not allow such a description are encountered if we consider a system that is part of a larger closed system. We suppose that the closed system as a whole is in some state described by the wave function $\Psi(q, x)$, where x denotes the set of coordinates of the system considered, and q the remaining coordinates of the closed system. This function in general does not fall into a product of functions of x and of q alone, so that the system does not have its own wave function.†

Let f be some physical quantity pertaining to the system considered. Its operator therefore acts only on the coordinates x, and not on q. The mean value of this quantity in the state considered is

$$\bar{f} = \iint \Psi^*(q, x)\hat{f}\Psi(q, x)\, \mathrm{d}q\mathrm{d}x. \tag{14.1}$$

We introduce the function $\rho(x, x')$ defined by

$$\rho(x, x') = \int \Psi(q, x)\Psi^*(q, x')\, \mathrm{d}q, \tag{14.2}$$

where the integration is extended only over the coordinates q; this function is called the *density matrix* of the system. From the definition (14.2) it is evident that the function is "Hermitian":

$$\rho^*(x, x') = \rho(x', x). \tag{14.3}$$

The "diagonal elements" of the density matrix

$$\rho(x, x) = \int |\Psi(q, x)|^2\, \mathrm{d}q$$

determine the probability distribution for the coordinates of the system.

Using the density matrix, we can write the mean value f in the form

$$\bar{f} = \int [\hat{f}\rho(x, x')]_{x'=x}\, \mathrm{d}x. \tag{14.4}$$

Here $\hat{f}$ acts only on the variables x in the function $\rho(x, x')$; after calculating the result of its action, we put $x' = x$. We see that, if we know the density matrix, we can calculate the mean value of any quantity characterizing the system. It follows from this that, by means of $\rho(x, x')$, we can also determine the probabilities of various values of the physical quantities in the system. Thus the state of a system which does not have a wave function can be described by means of a density matrix. This does not contain the coordinates q which do not belong to the system concerned, though, of course, it depends essentially on the state of the closed system as a whole.

The description by means of the density matrix is the most general form of quantum-mechanical description of the system. The description by means of the wave function, on the other hand, is a particular case of this, corresponding to a density matrix of the form $\rho(x, x') = \Psi(x)\Psi^*(x')$. The following important difference exists between this particular case and the general one.‡ For a state having a wave function there is always a complete

† In order that $\Psi(q, x)$ should (at a given instant) fall into such a product, the measurement as a result of which this state was brought about must completely describe the system considered and the remainder of the closed system separately. In order that $\Psi(q, x)$ should continue to have this form at subsequent instants, it is necessary in addition that these parts of the closed system should not interact (see §2). Neither of these conditions is now assumed.

‡ States having a wave function are called "pure" states, as distinct from "mixed" states, which are described by a density matrix.

set of measuring processes such that they lead with certainty to definite results (mathematically, this means that Ψ is an eigenfunction of some operator). For states having only a density matrix, on the other hand, there is no complete set of measuring processes whose result can be uniquely predicted.

Let us now suppose that the system is closed, or became so at some instant. Then we can derive an equation giving the change in the density matrix with time, similar to the wave equation for the Ψ function. The derivation can be simplified by noticing that the required linear differential equation for $\rho(x, x', t)$ must be satisfied in the particular case where the system has a wave function, i.e.

$$\rho(x, x', t) = \Psi(x, t)\Psi^*(x', t).$$

Differentiating with respect to time and using the wave equation (8.1), we have

$$\begin{aligned} i\hbar\frac{\partial\rho}{\partial t} &= i\hbar\Psi^*(x', t)\frac{\partial\Psi(x, t)}{\partial t}+i\hbar\Psi(x, t)\frac{\partial\Psi^*(x', t)}{\partial t} \\ &= \Psi^*(x', t)\hat{H}\Psi(x, t)-\Psi(x, t)\hat{H}'^*\Psi^*(x', t), \end{aligned}$$

where $\hat{H}$ is the Hamiltonian of the system, acting on a function of x, and $\hat{H}'$ is the same operator acting on a function of x'. The functions $\Psi^*(x', t)$ and $\Psi(x, t)$ can obviously be placed behind the respective operators $\hat{H}$ and $\hat{H}'$, and we thus obtain the required equation:

$$i\hbar\, \partial\rho(x, x', t)/\partial t = (\hat{H}-\hat{H}'^*)\rho(x, x', t). \tag{14.5}$$

Let $\Psi_n(x, t)$ be the wave functions of the stationary states of the system, i.e. the eigenfunctions of its Hamiltonian. We expand the density matrix in terms of these functions; the expansion consists of a double series in the form

$$\begin{aligned} \rho(x, x', t) &= \sum_m\sum_n a_{mn}\Psi_n^*(x', t)\Psi_m(x, t) \\ &= \sum_m\sum_n a_{mn}\psi_n^*(x')\psi_m(x)e^{(i/\hbar)(E_n-E_m)t}. \end{aligned} \tag{14.6}$$

For the density matrix, this expansion plays a part analogous to that of the expansion (10.3) for wave functions. Instead of the set of coefficients a_n, we have here the double set of coefficients a_{mn}. These clearly have the property of being "Hermitian", like the density matrix itself:

$$a_{nm}^* = a_{mn}. \tag{14.7}$$

For the mean value of some quantity f we have, substituting (14.6) in (14.4),

$$\bar{f} = \sum_m\sum_n a_{mn}\int \Psi_n^*(x, t)\hat{f}\Psi_m(x, t)\, dx,$$

or

$$\bar{f} = \sum_m\sum_n a_{mn}f_{nm}(t) = \sum_m\sum_n a_{mn}f_{nm}e^{(i/\hbar)(E_n-E_m)t}, \tag{14.8}$$

where f_{nm} are the matrix elements of the quantity f. This expression is similar to formula (11.1).†

The quantities a_{mn} must satisfy certain inequalities. The "diagonal elements" $\rho(x, x)$ of the density matrix, which determine the probability distribution for the coordinates, must obviously be positive quantities. It therefore follows from the expression (14.6) (with $x' = x$) that the quadratic form

$$\sum_n \sum_m a_{mn} \xi_n^* \xi_m$$

constructed with the coefficients a_{mn} (where the ξ_n are arbitrary complex quantities) must be positive. This places certain conditions, known from the theory of quadratic forms, on the quantities a_{nm}. In particular, all the "diagonal" quantities must clearly be positive:

$$a_{nn} \geqslant 0, \tag{14.9}$$

and any three quantities a_{nn}, a_{mm} and a_{mn} must satisfy the inequality

$$a_{nn} a_{mm} \geqslant |a_{mn}|^2. \tag{14.10}$$

To the "pure" case, where the density matrix reduces to a product of functions, there evidently corresponds a matrix a_{mn} of the form

$$a_{mn} = a_m a_n^*. \tag{14.11}$$

We shall indicate a simple criterion which enables us to decide, from the form of the matrix a_{mn}, whether we are concerned with a "pure" or a "mixed" state. In the pure case we have

$$\begin{aligned}(a^2)_{mn} &= \sum_k a_{mk} a_{kn} \\ &= \sum_k a_k^* a_m a_n^* a_k \\ &= a_m a_n^* \sum_k |a_k|^2 \\ &= a_m a_n^*,\end{aligned}$$

or

$$(a^2)_{mn} = a_{mn}, \tag{14.12}$$

i.e. the density matrix is equal to its own square.

§15. Momentum

Let us consider a closed system of particles not in an external field. Since all positions in space of such a system as a whole are equivalent, we can say, in particular, that the Hamiltonian of the system does not vary when the system undergoes a parallel displacement over any distance. It is sufficient that this condition should be fulfilled for an arbitrary small displacement.

An infinitely small parallel displacement over a distance $\delta\mathbf{r}$ signifies a transformation under which the radius vectors $\mathbf{r}_a$ of all the particles (a being the number of the particle) receive the same increment $\delta\mathbf{r} : \mathbf{r}_a \to \mathbf{r}_a + \delta\mathbf{r}$. An

† The quantities a_{mn} form the density matrix in the energy representation. The description of the states of a system by means of this matrix was introduced independently by L. Landau and F. Bloch in 1927.

arbitrary function $\psi(\mathbf{r}_1, \mathbf{r}_2, \ldots)$ of the coordinates of the particles, under such a transformation, becomes the function

$$\psi(\mathbf{r}_1+\delta\mathbf{r}, \mathbf{r}_2+\delta\mathbf{r}, \ldots) = \psi(\mathbf{r}_1, \mathbf{r}_2, \ldots)+\delta\mathbf{r} \cdot \sum_a \nabla_a\psi$$
$$= (1+\delta\mathbf{r} \cdot \sum_a \nabla_a)\psi(\mathbf{r}_1, \mathbf{r}_2, \ldots)$$

(∇_a denotes the operator of differentiation with respect to $\mathbf{r}_a$). The expression

$$1+\delta\mathbf{r} \cdot \sum_a \nabla_a$$

is the operator of an infinitely small displacement, which converts the function $\psi(\mathbf{r}_1, \mathbf{r}_2, \ldots)$ into the function

$$\psi(\mathbf{r}_1+\delta\mathbf{r}, \mathbf{r}_2+\delta\mathbf{r}, \ldots).$$

The statement that some transformation does not change the Hamiltonian means that, if we make this transformation on the function $\hat{H}\psi$, the result is the same as if we make it only on the function ψ and then apply the operator $\hat{H}$. Mathematically, this can be written as follows. Let $\hat{O}$ be the operator which effects the transformation in question. Then we have $\hat{O}(\hat{H}\psi) = \hat{H}(\hat{O}\psi)$, whence

$$\hat{O}\hat{H}-\hat{H}\hat{O} = 0,$$

i.e. the Hamiltonian must commute with the operator $\hat{O}$.

In the case considered, the operator $\hat{O}$ is the operator of an infinitely small displacement. Since the unit operator (the operator of multiplying by unity) commutes, of course, with any operator, and the constant factor $\delta\mathbf{r}$ can be taken in front of $\hat{H}$, the condition $\hat{O}\hat{H}-\hat{H}\hat{O} = 0$ reduces here to

$$(\sum_a \nabla_a)\hat{H}-\hat{H}(\sum_a \nabla_a) = 0. \tag{15.1}$$

As we know, the commutability of an operator (not containing the time explicitly) with $\hat{H}$ means that the physical quantity corresponding to that operator is conserved. The quantity whose conservation for a closed system follows from the homogeneity of space is the *momentum* of the system (cf. *Mechanics*, §7). Thus the relation (15.1) expresses the law of conservation of momentum in quantum mechanics; the operator $\sum \nabla_a$ must correspond, apart from a constant factor, to the total momentum of the system, and each term ∇_a of the sum to the momentum of an individual particle.

The coefficient of proportionality between the operator $\hat{\mathbf{p}}$ of the momentum of a particle and the operator ∇ can be determined by means of the passage to the limit of classical mechanics, and is $-i\hbar$:

$$\hat{\mathbf{p}} = -i\hbar\nabla, \tag{15.2}$$

or, in components,

$$\hat{p}_x = -i\hbar\partial/\partial x, \quad \hat{p}_y = -i\hbar\partial/\partial y, \quad \hat{p}_z = -i\hbar\partial/\partial z.$$

Using the limiting expression (6.1) for the wave function, we have

$$\hat{\mathbf{p}}\Psi = -i\hbar(i/\hbar)\Psi\nabla S = \Psi\nabla S,$$

i.e. in the classical approximation the effect of the operator $\hat{\mathbf{p}}$ reduces to multiplication by ∇S. The gradient ∇S of the action is the classical momentum $\mathbf{p}$ of the particle (see *Mechanics*, §43).

It is easy to see that the operator (15.2) is Hermitian, as it should be. For, with arbitrary functions $\psi(x)$ and $\phi(x)$ which vanish at infinity, we have

$$\int \phi\hat{p}_x\psi \,\mathrm{d}x = -i\hbar \int \phi\frac{\partial\psi}{\partial x}\,\mathrm{d}x = i\hbar \int \psi\frac{\partial\phi}{\partial x}\,\mathrm{d}x = \int \psi\hat{p}_x{}^*\phi \,\mathrm{d}x,$$

and this is the condition that the operator should be Hermitian.

Since the result of differentiating functions with respect to two different variables is independent of the order of differentiation, it is clear that the operators of the three components of momentum commute with one another:

$$\hat{p}_x\hat{p}_y - \hat{p}_y\hat{p}_x = 0, \quad \hat{p}_x\hat{p}_z - \hat{p}_z\hat{p}_x = 0, \quad \hat{p}_y\hat{p}_z - \hat{p}_z\hat{p}_y = 0. \tag{15.3}$$

This means that all three components of the momentum of a particle can simultaneously have definite values.

Let us find the eigenfunctions and eigenvalues of the momentum operators. They are determined by the vector equation

$$-i\hbar\nabla\psi = \mathbf{p}\psi. \tag{15.4}$$

The solutions are of the form

$$\psi = Ce^{(i/\hbar)\mathbf{p}.\mathbf{r}}, \tag{15.5}$$

where C is a constant. If all three components of the momentum are given simultaneously, we see that this completely determines the wave function of the particle. In other words, the quantities p_x, p_y, p_z form one of the possible complete sets of physical quantities for a particle. Their eigenvalues form a continuous spectrum extending from $-\infty$ to $+\infty$.

According to the rule (5.4) for normalizing the eigenfunctions of a continuous spectrum, the integral $\int \psi^*_{\mathbf{p}'}\psi_{\mathbf{p}}\,\mathrm{d}V$ taken over all space ($\mathrm{d}V = \mathrm{d}x\,\mathrm{d}y\,\mathrm{d}z$) must be equal to the delta function $\delta(\mathbf{p}'-\mathbf{p})$.† However, for reasons that will become clear from subsequent applications, it is more natural to normalize the eigenfunctions of the particle momentum by the delta function of the momentum difference divided by $2\pi\hbar$:

$$\int \psi_{\mathbf{p}'}{}^*\psi_{\mathbf{p}}\,\mathrm{d}V = \delta\left(\frac{\mathbf{p}'-\mathbf{p}}{2\pi\hbar}\right),$$

or, equivalently,

$$\int \psi_{\mathbf{p}'}{}^*\psi_{\mathbf{p}}\,\mathrm{d}V = (2\pi\hbar)^3\,\delta(\mathbf{p}'-\mathbf{p}) \tag{15.6}$$

(since each of the three factors in the three-dimensional delta function is $\delta[(p'_x-p_x)/2\pi\hbar] = 2\pi\hbar\,\delta(p'_x-p_x)$, and so on).

† The three-dimensional function $\delta(\mathbf{a})$ of a vector $\mathbf{a}$ is defined as a product of delta functions of the components of the vector $\mathbf{a}$: $\delta(\mathbf{a}) = \delta(a_x)\delta(a_y)\delta(a_z)$.

The integration is effected by means of the formula†

$$\frac{1}{2\pi}\int_{-\infty}^{\infty} e^{i\alpha\xi}\,\mathrm{d}\xi = \delta(\alpha). \tag{15.7}$$

This shows that the constant in (15.5) is equal to unity if the normalization is according to (15.6):‡

$$\psi_{\mathbf{p}} = e^{(i/\hbar)\mathbf{p}\cdot\mathbf{r}}. \tag{15.8}$$

The expansion of an arbitrary wave function $\psi(\mathbf{r})$ of a particle in terms of the eigenfunctions $\psi_{\mathbf{p}}$ of its momentum operator is simply the expansion as a Fourier integral:

$$\psi(\mathbf{r}) = \int a(\mathbf{p})\psi_p(\mathbf{r})\frac{\mathrm{d}^3p}{(2\pi\hbar)^3} = \int a(\mathbf{p})e^{(i/\hbar)\mathbf{p}\cdot\mathbf{r}}\frac{\mathrm{d}^3p}{(2\pi\hbar)^3} \tag{15.9}$$

(where $\mathrm{d}^3p = \mathrm{d}p_x\mathrm{d}p_y\mathrm{d}p_z$). The expansion coefficients $a(\mathbf{p})$ are, according to formula (5.3),

$$a(\mathbf{p}) = \int \psi(\mathbf{r})\psi_p{}^*(\mathbf{r})\,\mathrm{d}V = \int \psi(\mathbf{r})e^{-(i/\hbar)\mathbf{p}\cdot\mathbf{r}}\,\mathrm{d}V. \tag{15.10}$$

The function $a(\mathbf{p})$ can be regarded (see §5) as the wave function of the particle in the "momentum representation"; $|a(\mathbf{p})|^2\, d^3p/(2\pi\hbar)^3$ is the probability that the momentum has a value in the interval d^3p.

Just as the operator $\hat{\mathbf{p}}$ corresponds to the momentum, determining its eigenfunctions in the coordinate representation, we can introduce the operator $\hat{\mathbf{r}}$ of the coordinates of the particle in the momentum representation. It must be defined so that the mean value of the coordinates can be represented in the form

$$\bar{\mathbf{r}} = \int a^*(\mathbf{p})\hat{\mathbf{r}}a(\mathbf{p})\frac{\mathrm{d}^3p}{(2\pi\hbar)^3}. \tag{15.11}$$

On the other hand, this mean value is determined from the wave function $\psi(\mathbf{r})$ by

$$\bar{\mathbf{r}} = \int \psi^*\mathbf{r}\psi\,\mathrm{d}V.$$

† The conventional meaning of this formula is that the function on the left-hand side has the property (5.8) of the delta function. Substituting $\delta(x-a)$ in the form (15.7), we obtain from (5.8) the well-known Fourier integral formula

$$f(a) = \int\int_{-\infty}^{\infty} f(x)e^{i\xi(x-a)}\,\mathrm{d}x\,\mathrm{d}\xi/2\pi.$$

‡ Note that with this normalization the probability density $|\psi|^2 = 1$, i.e. the function is normalized to "one particle per unit volume". This agreement of normalizations is, of course, no accident; see the last footnote to §48.

Substituting $\psi(\mathbf{r})$ in the form (15.9) we have (integrating by parts)

$$\mathbf{r}\psi(\mathbf{r}) = (2\pi\hbar)^{-3} \int \mathbf{r}a(\mathbf{p})e^{(i/\hbar)\mathbf{p}.\mathbf{r}}\,\mathrm{d}^3p$$
$$= (2\pi\hbar)^{-3} \int i\hbar e^{(i/\hbar)\mathbf{p}.\mathbf{r}}[\partial a(\mathbf{p})/\partial\mathbf{p}]\,\mathrm{d}^3p.$$

Using this expression and (15.10), we find

$$\bar{\mathbf{r}} = (2\pi\hbar)^{-3} \int\int \psi^*(\mathbf{r})i\hbar[\partial a(\mathbf{p})/\partial\mathbf{p}]e^{(i/\hbar)\mathbf{p}.\mathbf{r}}\,\mathrm{d}^3p\,\mathrm{d}V$$
$$= \int i\hbar a^*(\mathbf{p})[\partial a(\mathbf{p})/\partial\mathbf{p}]\frac{\mathrm{d}^3p}{(2\pi\hbar)^3}.$$

Comparing with (15.11), we see that the radius vector operator in the momentum representation is

$$\hat{\mathbf{r}} = i\hbar\partial/\partial\mathbf{p}. \tag{15.12}$$

The momentum operator in this representation reduces simply to multiplication by **p**.

Finally, we shall express in terms of $\hat{\mathbf{p}}$ the operator of a parallel displacement in space over any finite (not only infinitesimal) distance **a**. By the definition of this operator ($\hat{T}_{\mathbf{a}}$) we must have

$$\hat{T}_{\mathbf{a}}\psi(\mathbf{r}) = \psi(\mathbf{r}+\mathbf{a}).$$

Expanding the function $\psi(\mathbf{r}+\mathbf{a})$ in a Taylor series, we have

$$\psi(\mathbf{r}+\mathbf{a}) = \psi(\mathbf{r})+\mathbf{a}\,.\,\partial\psi(\mathbf{r})/\partial\mathbf{r}+\ \dots,$$

or, introducing the operator $\hat{\mathbf{p}} = -i\hbar\nabla$,

$$\psi(\mathbf{r}+\mathbf{a}) = \left[1+\frac{i}{\hbar}\mathbf{a}\,.\,\hat{\mathbf{p}}+\frac{1}{2}\left(\frac{i}{\hbar}\mathbf{a}\,.\,\hat{\mathbf{p}}\right)^2+\ \dots\right]\psi(\mathbf{r}).$$

The expression in brackets is the operator

$$\hat{T}_{\mathbf{a}} = e^{(i/\hbar)\mathbf{a}\,.\,\hat{\mathbf{p}}}. \tag{15.13}$$

This is the required *operator of the finite displacement*.

§16. Uncertainty relations

Let us derive the rules for commutation between momentum and coordinate operators. Since the result of successively differentiating with respect to one of the variables x, y, z and multiplying by another of them does not depend on the order of these operations, we have

$$\hat{p}_x y - y\hat{p}_x = 0, \quad \hat{p}_x z - z\hat{p}_x = 0, \tag{16.1}$$

and similarly for $\hat{p}_y$, $\hat{p}_z$.

To derive the commutation rule for $\hat{p}_x$ and x, we write

$$(\hat{p}_x x - x\hat{p}_x)\psi = -i\hbar\, \partial(x\psi)/\partial x + i\hbar x\, \partial\psi/\partial x$$
$$= -i\hbar\psi.$$

We see that the result of the action of the operator $\hat{p}_x x - x\hat{p}_x$ reduces to multiplication by $-i\hbar$; the same is true, of course, of the commutation of $\hat{p}_y$ with y and $\hat{p}_z$ with z. Thus we have†

$$\hat{p}_x x - x\hat{p}_x = -i\hbar, \quad \hat{p}_y y - y\hat{p}_y = -i\hbar, \quad \hat{p}_z z - z\hat{p}_z = -i\hbar. \tag{16.2}$$

All the relations (16.1) and (16.2) can be written jointly in the form

$$\hat{p}_i x_k - x_k \hat{p}_i = -i\hbar\delta_{ik} \qquad (i, k = x, y, z). \tag{16.3}$$

Before going on to examine the physical significance of these relations and their consequences, we shall set down two formulae which will be useful later. Let $f(\mathbf{r})$ be some function of the coordinates. Then

$$\hat{\mathbf{p}}f(\mathbf{r}) - f(\mathbf{r})\hat{\mathbf{p}} = -i\hbar\nabla f. \tag{16.4}$$

For

$$(\hat{\mathbf{p}}f - f\hat{\mathbf{p}})\psi = -i\hbar[\nabla(f\psi) - f\nabla\psi] = -i\hbar\psi\nabla f.$$

A similar relation holds for the commutator of $\mathbf{r}$ with a function of the momentum operator:

$$f(\hat{\mathbf{p}})\mathbf{r} - \mathbf{r}f(\hat{\mathbf{p}}) = -i\hbar\partial f/\partial\mathbf{p}. \tag{16.5}$$

It can be derived in the same way as (16.4) if we calculate in the momentum representation, using the expression (15.12) for the coordinate operators.

The relations (16.1) and (16.2) show that the coordinate of a particle along one of the axes can have a definite value at the same time as the components of the momentum along the other two axes; the coordinate and momentum component along the same axis, however, cannot exist simultaneously. In particular, the particle cannot be at a definite point in space and at the same time have a definite momentum $\mathbf{p}$.

Let us suppose that the particle is in some finite region of space, whose dimensions along the three axes are (of the order of magnitude of) Δx, Δy, Δz. Also, let the mean value of the momentum of the particle be $\mathbf{p}_0$. Mathematically, this means that the wave function has the form $\psi = u(\mathbf{r})e^{(i/\hbar)\mathbf{p}_0\cdot\mathbf{r}}$, where $u(\mathbf{r})$ is a function which differs considerably from zero only in the region of space concerned. We expand the function ψ in terms of the eigenfunctions of the momentum operator (i.e. as a Fourier integral). The coefficients $a(\mathbf{p})$ in this expansion are determined by the integrals (15.10) of functions of the form $u(\mathbf{r})e^{(i/\hbar)(\mathbf{p}_0-\mathbf{p})\cdot\mathbf{r}}$. If this integral is to differ considerably from zero, the periods of the oscillatory factor $e^{(i/\hbar)(\mathbf{p}_0-\mathbf{p})\cdot\mathbf{r}}$ must not be small in comparison with the dimensions Δx, Δy, Δz of the region in which the function $u(\mathbf{r})$ is different from zero. This means that $a(\mathbf{p})$ will be con-

† These relations, discovered in matrix form by Heisenberg in 1925, formed the genesis of quantum mechanics.

siderably different from zero only for values of $\mathbf{p}$ such that $(1/\hbar)(p_{0x}-p_x)\Delta x \lesssim 1$, etc. Since $|a(\mathbf{p})|^2$ determines the probability of the various values of the momentum, the ranges of values of p_x, p_y, p_z in which $a(\mathbf{p})$ differs from zero are just those in which the components of the momentum of the particle may be found, in the state considered. Denoting these ranges by $\Delta p_x, \Delta p_y, \Delta p_z$, we thus have

$$\Delta p_x \Delta x \sim \hbar, \quad \Delta p_y \Delta y \sim \hbar, \quad \Delta p_z \Delta z \sim \hbar. \tag{16.6}$$

These relations, known as the *uncertainty relations*, were obtained by Heisenberg in 1927.

We see that, the greater the accuracy with which the coordinate of the particle is known (i.e. the less Δx), the greater the uncertainty Δp_x in the component of the momentum along the same axis, and *vice versa*. In particular, if the particle is at some completely definite point in space ($\Delta x = \Delta y = \Delta z = 0$), then $\Delta p_x = \Delta p_y = \Delta p_z = \infty$. This means that all values of the momentum are equally probable. Conversely, if the particle has a completely definite momentum $\mathbf{p}$, then all positions of it in space are equally probable (this is seen directly from the wave function (15.8), whose squared modulus is quite independent of the coordinates).

If the uncertainties of the coordinates and momenta are specified by the standard deviations

$$\delta x = \sqrt{[\overline{(x-\bar{x})^2}]}, \; \delta p_x = \sqrt{[\overline{(p_x-\bar{p}_x)^2}]},$$

we can specify exactly the least possible value of their product (H. Weyl). Let us consider the one-dimensional case of a wave packet with wave function $\psi(x)$ depending on only one coordinate, and assume for simplicity that the mean values of x and p_x in this state are zero. We consider the obvious inequality

$$\int_{-\infty}^{\infty} \left| \alpha x \psi + \frac{d\psi}{dx} \right|^2 dx \geqslant 0,$$

where α is an arbitrary real constant. On calculating this integral, noticing that

$$\int x^2 |\psi|^2 \, dx = (\delta x)^2,$$

$$\int \left(x \frac{d\psi^*}{dx} \psi + x\psi^* \frac{d\psi}{dx} \right) dx = \int x \frac{d|\psi|^2}{dx} \, dx = -\int |\psi|^2 \, dx = -1,$$

$$\int \frac{d\psi^*}{dx} \frac{d\psi}{dx} \, dx = -\int \psi^* \frac{d^2\psi}{dx^2} \, dx = \frac{1}{\hbar^2} \int \psi^* \hat{p}_x^2 \psi \, dx = \frac{1}{\hbar^2} (\delta p_x)^2,$$

we obtain

$$\alpha^2 (\delta x)^2 - \alpha + (1/\hbar^2)(\delta p_x)^2 \geqslant 0.$$

If this quadratic (in α) trinomial is positive for all α, its discriminant must be negative, which gives the inequality

$$\delta x\, \delta p_x \geqslant \tfrac{1}{2}\hbar. \tag{16.7}$$

The least possible value of the product is $\frac{1}{2}\hbar$, and occurs for wave packets with wave functions of the form

$$\psi = \frac{1}{(2\pi)^{1/4}\sqrt{(\delta x)}} \exp\left(\frac{i}{\hbar} p_0 x - \frac{x^2}{4(\delta x)^2}\right), \tag{16.8}$$

where p_0 and δx are constants. The probabilities of the various values of the coordinates in such a state are

$$|\psi|^2 = \frac{1}{\sqrt{(2\pi)}\,.\,\delta x} \exp\left(-\frac{x^2}{2(\delta x)^2}\right),$$

and thus have a Gaussian distribution about the origin (the mean value $\bar{x} = 0$) with standard deviation δx. The wave function in the momentum representation is

$$a(p_x) = \frac{1}{\sqrt{(2\pi\hbar)}} \int_{-\infty}^{\infty} \psi(x) e^{-(i/\hbar)p_x x}\, dx.$$

Calculation of the integral gives

$$a(p_x) = \text{constant} \times \exp\left[-\frac{(\delta x)^2 (p_x - p_0)^2}{\hbar^2}\right].$$

The distribution of probabilities of values of the momentum, $|a(p_x)|^2$, is also Gaussian about the mean value $\overline{p_x} = p_0$, with standard deviation $\delta p_x = \hbar/2\delta x$, so that the product $\delta p_x \delta x$ is indeed $\frac{1}{2}\hbar$.

Finally, we shall derive another useful relation. Let f and g be two physical quantities whose operators obey the commutation rule

$$\hat{f}\hat{g} - \hat{g}\hat{f} = -i\hbar\hat{c}, \tag{16.9}$$

where $\hat{c}$ is the operator of some physical quantity c. On the right-hand side of the equation the factor $\hbar$ is introduced in accordance with the fact that in the classical limit (i.e. as $\hbar \to 0$) all operators of physical quantities reduce to multiplication by these quantities and commute with one another. Thus, in the "quasi-classical" case, we can, to a first approximation, regard the right-hand side of equation (16.9) as being zero. In the next approximation, the operator $\hat{c}$ can be replaced by the operator of simple multiplication by the quantity c. We then have

$$\hat{f}\hat{g} - \hat{g}\hat{f} = -i\hbar c.$$

This equation is exactly analogous to the relation $\hat{p}_x x - x\hat{p}_x = -i\hbar$, the only difference being that, instead of the constant $\hbar$, we have† the quantity $\hbar c$. We can therefore conclude, by analogy with the relation $\Delta x \Delta p_x \sim \hbar$, that in the quasi-classical case there is an uncertainty relation

$$\Delta f \Delta g \sim \hbar c \tag{16.10}$$

for the quantities f and g.

In particular, if one of these quantities is the energy ($\hat{f} \equiv \hat{H}$) and the operator ($\hat{g}$) of the other does not depend explicitly on the time, then by (9.2) $c = \dot{g}$, and the uncertainty relation in the quasi-classical case is

$$\Delta E \Delta g \sim \hbar \dot{g}. \tag{16.11}$$

† The classical quantity c is the Poisson bracket of the quantities f and g; see the footnote in §9.

CHAPTER III

SCHRÖDINGER'S EQUATION

§17. Schrödinger's equation

THE form of the wave equation of a physical system is determined by its Hamiltonian, which is therefore of fundamental significance in the whole mathematical formalism of quantum mechanics.

The form of the Hamiltonian for a free particle is established by the general requirements imposed by the homogeneity and isotropy of space and by Galileo's relativity principle. In classical mechanics, these requirements lead to a quadratic dependence of the energy of the particle on its momentum: $E = p^2/2m$, where the constant m is called the mass of the particle (see *Mechanics*, §4). In quantum mechanics, the same requirements lead to a corresponding relation for the energy and momentum eigenvalues, these quantities being conserved and simultaneously measurable (for a free particle).

If the relation $E = p^2/2m$ holds for every eigenvalue of the energy and momentum, the same relation must hold for their operators also:

$$\hat{H} = (1/2m)(\hat{p}_x^2+\hat{p}_y^2+\hat{p}_z^2). \tag{17.1}$$

Substituting here from (15.2), we obtain the Hamiltonian of a freely moving particle in the form

$$\hat{H} = -(\hbar^2/2m)\triangle, \tag{17.2}$$

where $\triangle = \partial^2/\partial x^2+\partial^2/\partial y^2+\partial^2/\partial z^2$ is the Laplacian operator.

The Hamiltonian of a system of non-interacting particles is equal to the sum of the Hamiltonians of the separate particles:

$$\hat{H} = -\tfrac{1}{2}\hbar^2 \sum_a (1/m_a)\triangle_a \tag{17.3}$$

(the suffix a is the number of the particle; $\triangle_a$ is the Laplacian operator in which the differentiation is with respect to the coordinates of the ath particle).

In classical (non-relativistic) mechanics, the interaction of particles is described by an additive term in the Hamiltonian, the potential energy of the interaction $U(\mathbf{r}_1, \mathbf{r}_2, \ldots)$, which is a function of the coordinates of the particles. By adding a similar function to the Hamiltonian of the system, the interaction of particles can be represented in quantum mechanics:†

$$\hat{H} = -\tfrac{1}{2}\hbar^2 \sum_a \triangle_a/m_a+U(\mathbf{r}_1, \mathbf{r}_2, \ldots). \tag{17.4}$$

† This statement is, of course, not a logical consequence of the basic principles of quantum mechanics, and is to be regarded as a deduction from experiment.

The first term can be regarded as the operator of the kinetic energy and the second as that of the potential energy. In particular, the Hamiltonian for a single particle in an external field is

$$\hat{H} = \hat{p}^2/2m + U(x, y, z) = -(\hbar^2/2m)\triangle + U(x, y, z), \tag{17.5}$$

where $U(x, y, z)$ is the potential energy of the particle in the external field.

Substituting the expressions (17.2) to (17.5) in the general equation (8.1), we obtain the wave equations for the corresponding systems. We shall write out here the wave equation for a particle in an external field:

$$i\hbar\, \partial\Psi/\partial t = -(\hbar^2/2m)\triangle\Psi + U(x, y, z)\Psi. \tag{17.6}$$

The equation (10.2), which determines the stationary states, takes the form

$$(\hbar^2/2m)\triangle\psi + [E - U(x, y, z)]\psi = 0. \tag{17.7}$$

The equations (17.6) and (17.7) were obtained by Schrödinger in 1926 and are called *Schrödinger's equations*.

For a free particle, equation (17.7) has the form

$$(\hbar^2/2m)\triangle\psi + E\psi = 0. \tag{17.8}$$

This equation has solutions finite in all space for any positive value of the energy E. For states with definite directions of motion, these solutions are eigenfunctions of the momentum operator, with $E = p^2/2m$. The complete (time-dependent) wave functions of such stationary states are

$$\Psi = \text{constant} \times e^{-(i/\hbar)Et + (i/\hbar)\mathbf{p}\cdot\mathbf{r}} \tag{17.9}$$

Each such function, a *plane wave*, describes a state in which the particle has a definite energy E and momentum $\mathbf{p}$. The angular frequency of this wave is $E/\hbar$ and its wave vector $\mathbf{k} = \mathbf{p}/\hbar$; the corresponding wavelength $2\pi\hbar/p$ is called the *de Broglie wavelength* of the particle.†

The energy spectrum of a freely moving particle is thus found to be continuous, extending from zero to $+\infty$. Each of these eigenvalues (except $E = 0$) is degenerate, and the degeneracy is infinite. For there corresponds to every value of E, different from zero, an infinite number of eigenfunctions (17.9), differing in the direction of the vector $\mathbf{p}$, which has a constant absolute magnitude.

Let us enquire how the passage to the limit of classical mechanics occurs in Schrödinger's equation, considering for simplicity only a single particle in an external field. Substituting in Schrödinger's equation (17.6) the limiting expression (6.1) for the wave function, $\Psi = ae^{(i/\hbar)S}$, we obtain, on performing the differentiation,

$$a\frac{\partial S}{\partial t} - i\hbar\frac{\partial a}{\partial t} + \frac{a}{2m}(\nabla S)^2 - \frac{i\hbar}{2m}a\triangle S - \frac{i\hbar}{m}\nabla S \cdot \nabla a - \frac{\hbar^2}{2m}\triangle a + Ua = 0.$$

† The idea of a wave related to a particle was first introduced by L. de Broglie in 1924.

In this equation there are purely real and purely imaginary terms (we recall that S and a are real); equating each separately to zero, we obtain two equations

$$\frac{\partial S}{\partial t}+\frac{1}{2m}(\nabla S)^2+U-\frac{\hbar^2}{2ma}\Delta a = 0,$$

$$\frac{\partial a}{\partial t}+\frac{a}{2m}\Delta S+\frac{1}{m}\nabla S\cdot\nabla a = 0.$$

Neglecting the term containing $\hbar^2$ in the first of these equations, we obtain

$$\frac{\partial S}{\partial t}+\frac{1}{2m}(\nabla S)^2+U = 0, \tag{17.10}$$

that is; the classical Hamilton–Jacobi equation for the action S of a particle, as it should be. We see, incidentally, that, as $\hbar \to 0$, classical mechanics is valid as far as quantities of the first (and not only the zero) order in $\hbar$ inclusive.

The second equation obtained above, on multiplication by $2a$, can be rewritten in the form

$$\frac{\partial a^2}{\partial t}+\operatorname{div}\left(a^2\frac{\nabla S}{m}\right) = 0. \tag{17.11}$$

This equation has an obvious physical meaning: a^2 is the probability density for finding the particle at some point in space ($|\Psi|^2 = a^2$); $\nabla S/m = \mathbf{p}/m$ is the classical velocity $\mathbf{v}$ of the particle. Hence equation (17.11) is simply the equation of continuity, which shows that the probability density "moves" according to the laws of classical mechanics with the classical velocity $\mathbf{v}$ at every point.

PROBLEM

Find the transformation law for the wave function in a Galilean transformation.

SOLUTION. Let us apply the transformation to the wave function for free motion of a particle (a plane wave). Since any function Ψ can be expanded in plane waves, this will also give the transformation law for any wave function.

The plane waves in the frames of reference K and K' (K' moving with velocity $\mathbf{V}$ relative to K) are

$$\Psi(\mathbf{r}, t) = \text{constant} \times e^{(i/\hbar)(\mathbf{p}\cdot\mathbf{r}-Et)},$$

$$\Psi'(\mathbf{r}', t) = \text{constant} \times e^{(i/\hbar)(\mathbf{p}'\cdot\mathbf{r}'-E't)},$$

where $\mathbf{r} = \mathbf{r}'+\mathbf{V}t$; the particle momenta and energies in the two frames are related by

$$\mathbf{p} = \mathbf{p}'+m\mathbf{V},\ E = E'+\mathbf{V}.\mathbf{p}'+\tfrac{1}{2}mV^2$$

(see *Mechanics*, §8). Substitution of these expressions in Ψ gives

$$\Psi(\mathbf{r}, t) = \Psi'(\mathbf{r}', t)\exp\left[\frac{i}{\hbar}(m\mathbf{V}.\mathbf{r}'+\tfrac{1}{2}mV^2t)\right]$$

$$= \Psi'(\mathbf{r}-\mathbf{V}t, t)\exp\left[\frac{i}{\hbar}(m\mathbf{V}.\mathbf{r}-\tfrac{1}{2}mV^2t)\right]. \tag{1}$$

This formula does not contain the parameters of the free motion of the particle, and gives the required general transformation law for the wave function of any state of the particle. For a system of particles, the exponent in (1) contains a summation over the particles.

§18. The fundamental properties of Schrödinger's equation

The conditions which must be satisfied by solutions of Schrödinger's equation are very general in character. First of all, the wave function must be single-valued and continuous in all space. The requirement of continuity is maintained even in cases where the field $U(x, y, z)$ itself has a surface of discontinuity. At such a surface both the wave function and its derivatives must remain continuous. The continuity of the derivatives, however, does not hold if there is some surface beyond which the potential energy U becomes infinite. A particle cannot penetrate at all into a region of space where $U = \infty$, i.e. we must have $\psi = 0$ everywhere in this region. The continuity of ψ means that ψ vanishes at the boundary of this region; the derivatives of ψ, however, in general are discontinuous in this case.

If the field $U(x, y, z)$ nowhere becomes infinite, then the wave function also must be finite in all space. The same condition must hold in cases where U becomes infinite at some point but does so only as $1/r^s$ with $s < 2$ (see also §35).

Let $U_{\min}$ be the least value of the function $U(x, y, z)$. Since the Hamiltonian of a particle is the sum of two terms, the operators of the kinetic energy ($\hat{T}$) and of the potential energy, the mean value $\bar{E}$ of the energy in any state is equal to the sum $\bar{T}+\bar{U}$. But all the eigenvalues of the operator $\hat{T}$ (which is the Hamiltonian of a free particle) are positive; hence the mean value $\bar{T} \geqslant 0$. Recalling also the obvious inequality $\bar{U} > U_{\min}$, we find that $\bar{E} > U_{\min}$. Since this inequality holds for any state, it is clear that it is valid for all the eigenvalues of the energy:

$$E_n > U_{\min}. \qquad (18.1)$$

Let us consider a particle moving in an external field which vanishes at infinity; we define the function $U(x, y, z)$, in the usual way, so that it vanishes at infinity. It is easy to see that the spectrum of negative eigenvalues of the energy will then be discrete, i.e. all states with $E < 0$ in a field which vanishes at infinity are bound states. For, in the stationary states of a continuous spectrum, which correspond to infinite motion, the particle reaches infinity (see §10); however, at sufficiently large distances the field may be neglected, the motion of the particle may be regarded as free, and the energy of a freely moving particle can only be positive.

The positive eigenvalues, on the other hand, form a continuous spectrum and correspond to an infinite motion; for $E > 0$, Schrödinger's equation in general has no solutions (in the field concerned) for which the integral $\int |\psi|^2 \, dV$ converges.†

Attention must be drawn to the fact that, in quantum mechanics, a particle in a finite motion may be found in those regions of space where $E < U$; the probability $|\psi|^2$ of finding the particle tends rapidly to zero as the distance into such a region increases, yet it differs from zero at all finite distances.

† However, it must be mentioned that, for some particular mathematical forms of the function $U(x, y, z)$ (which have no physical significance), a discrete set of values may be absent from the otherwise continuous spectrum.

Here there is a fundamental difference from classical mechanics, in which a particle cannot penetrate into a region where $U > E$. In classical mechanics the impossibility of penetrating into this region is related to the fact that, for $E < U$, the kinetic energy would be negative, that is, the velocity would be imaginary. In quantum mechanics, the eigenvalues of the kinetic energy are likewise positive; nevertheless, we do not reach a contradiction here, since, if by a process of measurement a particle is localized at some definite point of space, the state of the particle is changed, as a result of this process, in such a way that it ceases in general to have any definite kinetic energy.

If $U(x, y, z) > 0$ in all space (and $U \to 0$ at infinity), then, by the inequality (18.1), we have $E_n > 0$. Since, on the other hand, for $E > 0$ the spectrum must be continuous, we conclude that, in this case, the discrete spectrum is absent altogether, i.e. only an infinite motion of the particle is possible.

Let us suppose that, at some point (which we take as origin), U tends to $-\infty$ in the manner

$$U \approx -\alpha r^{-s} \qquad (\alpha > 0). \tag{18.2}$$

We consider a wave function finite in some small region (of radius r_0) about the origin, and equal to zero outside this region. The uncertainty in the values of the coordinates of a particle in such a wave packet is of the order of r_0; hence the uncertainty in the value of the momentum is $\sim \hbar/r_0$. The mean value of the kinetic energy in this state is of the order of $\hbar^2/mr_0^2$, and the mean value of the potential energy is $\sim -\alpha/r_0^s$. Let us first suppose that $s > 2$. Then the sum

$$\hbar^2/mr_0^2 - \alpha/r_0^s$$

takes arbitrarily large negative values for sufficiently small r_0. If, however, the mean energy can take such values, this always means that the energy has negative eigenvalues which are arbitrarily large in absolute value. The motion of the particle in a very small region of space near the origin corresponds to the energy levels with large $|E|$. The "normal" state corresponds to a particle at the origin itself, i.e. the particle "falls" to the point $r = 0$.

If, however, $s < 2$, the energy cannot take arbitrarily large negative values. The discrete spectrum begins at some finite negative value. In this case the particle does not fall to the centre. It should be mentioned that, in classical mechanics, the fall of a particle to the centre would be possible in principle in any attractive field (i.e. for any positive s). The case $s = 2$ will be specially considered in §35.

Next, let us investigate how the nature of the energy spectrum depends on the behaviour of the field at large distances. We suppose that, as $r \to \infty$, the potential energy, which is negative, tends to zero according to the power law (18.2) (r is now large in this formula), and consider a wave packet "filling" a spherical shell of large radius r_0 and thickness $\Delta r \ll r_0$. Then the order of magnitude of the kinetic energy is again $\hbar^2/m(\Delta r)^2$, and of the potential energy, $-\alpha/r_0^s$. We increase r_0, at the same time increasing Δr, in such a way that Δr increases proportionally to r_0. If $s < 2$, then the sum $\hbar^2/m(\Delta r)^2 - \alpha/r_0^s$ becomes negative for sufficiently large r_0. Hence it follows that there are stationary states of negative energy, in which the particle may be found, with a fair probability, at large distances from the origin. This, however, means that there are levels of arbitrarily small negative energy (it must be recalled that the wave functions rapidly tend to zero in the region of space where $U > E$). Thus, in this case, the discrete spectrum contains an infinite number of levels, which become denser and denser towards the level $E = 0$.

If the field diminishes as $-1/r^s$ at infinity, with $s > 2$, then there are not levels of arbitrarily small negative energy. The discrete spectrum terminates at a level with a non-zero absolute value, so that the total number of levels is finite.

Schrödinger's equation for the wave functions ψ of stationary states is real, as are the conditions imposed on its solution. Hence its solutions can always be taken as real.† The eigenfunctions of non-degenerate values of the energy are automatically real, apart from the unimportant phase factor. For ψ^* satisfies the same equation as ψ, and therefore must also be an eigenfunction for the same value of the energy; hence, if this value is not degenerate, ψ and ψ^* must be essentially the same, i.e. they can differ only by a constant factor (of modulus unity). The wave functions corresponding to the same degenerate energy level need not be real, however, but by a suitable choice of linear combinations of them we can always obtain a set of real functions.

The complete (time-dependent) wave functions Ψ are determined by an equation in whose coefficients i appears. This equation, however, retains the same form if we replace t in it by $-t$ and at the same time take the complex conjugate.‡ Hence we can always choose the functions Ψ in such a way that Ψ and Ψ^* differ only by the sign of the time.

As is well known, the equations of classical mechanics are unchanged by *time reversal*, i.e. when the sign of the time is reversed. In quantum mechanics, the symmetry with respect to the two directions of time is expressed, as we see, in the invariance of the wave equation when the sign of t is changed and Ψ is simultaneously replaced by Ψ^*. However, it must be recalled that this symmetry here relates only to the equation, and not to the concept of measurement itself, which plays a fundamental part in quantum mechanics (as we have explained in detail in §7).

§19. The current density

In classical mechanics the velocity $\mathbf{v}$ of a particle is related to its momentum by $\mathbf{p} = m\mathbf{v}$. A similar relation holds between the corresponding operators in quantum mechanics, as we should expect. This is easily shown by calculating the operator $\hat{\mathbf{v}} = \hat{\dot{\mathbf{r}}}$ by the general rule (9.2) for the differentiation of operators with respect to time:

$$\hat{\mathbf{v}} = (i/\hbar)(\hat{H}\mathbf{r} - \mathbf{r}\hat{H}).$$

† These assertions are not valid for systems in a magnetic field.

‡ It is assumed that the potential energy U does not depend explicitly on the time: the system is either closed or in a constant (non-magnetic) field.

Using the expression (17.5) for $\hat{H}$ and formula (16.5), we obtain

$$\hat{\mathbf{v}} = \hat{\mathbf{p}}/m. \tag{19.1}$$

Similar relations will clearly hold between the eigenvalues of the velocity and momentum, and between their mean values in any state.

The velocity, like the momentum of a particle, cannot have a definite value simultaneously with the coordinates. But the velocity multiplied by an infinitely short time interval dt gives the displacement of the particle in the time dt. Hence the fact that the velocity cannot exist at the same time as the coordinates means that, if the particle is at a definite point in space at some instant, it has no definite position at an infinitely close subsequent instant.

We may notice a useful formula for the operator $\hat{\dot{f}}$ of the derivative, with respect to time, of some quantity $f(\mathbf{r})$ which is a function of the radius vector of the particle. Bearing in mind that f commutes with $U(\mathbf{r})$, we find

$$\hat{\dot{f}} = (i/\hbar)(\hat{H}f - f\hat{H}) = (i/2m\hbar)(\hat{\mathbf{p}}^2 f - f\hat{\mathbf{p}}^2).$$

Using (16.4), we can write

$$\hat{\mathbf{p}}^2 f = \hat{\mathbf{p}} \cdot (f\hat{\mathbf{p}} - i\hbar\nabla f),$$

$$f\hat{\mathbf{p}}^2 = (\hat{\mathbf{p}}f + i\hbar\nabla f).\hat{\mathbf{p}}.$$

Thus we obtain the required expression:

$$\hat{\dot{f}} = (1/2m)(\hat{\mathbf{p}} \cdot \nabla f + \nabla f \cdot \hat{\mathbf{p}}). \tag{19.2}$$

Next, let us find the acceleration operator. We have

$$\hat{\dot{\mathbf{v}}} = (i/\hbar)(\hat{H}\hat{\mathbf{v}} - \hat{\mathbf{v}}\hat{H}) = (i/m\hbar)(\hat{H}\hat{\mathbf{p}} - \hat{\mathbf{p}}\hat{H}) = (i/m\hbar)(U\hat{\mathbf{p}} - \hat{\mathbf{p}}U).$$

Using formula (16.4), we find

$$m\hat{\dot{\mathbf{v}}} = -\nabla U. \tag{19.3}$$

This operator equation is exactly the same in form as the equation of motion (Newton's equation) in classical mechanics.

The integral $\int |\Psi|^2 \,\mathrm{d}V$, taken over some finite volume V, is the probability of finding the particle in this volume. Let us calculate the derivative of this probability with respect to time. We have

$$\frac{\mathrm{d}}{\mathrm{d}t}\int |\Psi|^2 \,\mathrm{d}V = \int \left(\Psi\frac{\partial\Psi^*}{\partial t} + \Psi^*\frac{\partial\Psi}{\partial t}\right) \mathrm{d}V = \frac{i}{\hbar}\int (\Psi\hat{H}^*\Psi^* - \Psi^*\hat{H}\Psi)\,\mathrm{d}V.$$

Substituting here

$$\hat{H} = \hat{H}^* = -(\hbar^2/2m)\triangle + U(x, y, z)$$

and using the identity

$$\Psi\triangle\Psi^* - \Psi^*\triangle\Psi = \mathrm{div}\,(\Psi\nabla\Psi^* - \Psi^*\nabla\Psi),$$

we obtain

$$\frac{d}{dt}\int |\Psi|^2 \, dV = -\int \operatorname{div} \mathbf{j} \, dV,$$

where $\mathbf{j}$ denotes the vector‡

$$\mathbf{j} = (i\hbar/2m)(\Psi\nabla\Psi^* - \Psi^*\nabla\Psi). \tag{19.4}$$

$$= \frac{1}{2m}(\Psi\hat{\mathbf{p}}^*\Psi^* + \Psi^*\hat{\mathbf{p}}\Psi).$$

The integral of div $\mathbf{j}$ can be transformed by Gauss's theorem into an integral over the closed surface which bounds the volume V:

$$\frac{d}{dt}\int |\Psi|^2 \, dV = -\oint \mathbf{j} \, . \, d\mathbf{f}. \tag{19.5}$$

It is seen from this that the vector $\mathbf{j}$ may be called the *probability current density* vector, or simply the *current density*. The integral of this vector over a surface is the probability that the particle will cross the surface during unit time. The vector $\mathbf{j}$ and the probability density $|\Psi|^2$ satisfy the equation

$$\partial|\Psi|^2/\partial t + \operatorname{div} \mathbf{j} = 0, \tag{19.6}$$

which is analogous to the classical equation of continuity.

The wave function of free motion (the plane wave (17.9)) can be normalized so as to describe a flow of particles with unit current density (in which, on average, one particle crosses a unit cross-section of the flow per unit time). This function is then

$$\Psi = \frac{1}{\sqrt{v}} e^{-(i/\hbar)(Et - \mathbf{p}\cdot\mathbf{r})}, \tag{19.7}$$

where v is the velocity of the particle, since substitution of this in (19.4) gives $\mathbf{j} = \mathbf{p}/mv$, i.e. a unit vector in the direction of the motion.

It is useful to show how the orthogonality of the wave functions of states with different energies follows immediately from Schrödinger's equation. Let ψ_m and ψ_n be two such functions; they satisfy the equations

$$-(\hbar^2/2m)\triangle\psi_m + U\psi_m = E_m\psi_m,$$
$$-(\hbar^2/2m)\triangle\psi_n^* + U\psi_n^* = E_n\psi_n^*.$$

We multiply the first of these by ψ_n^* and the second by ψ_m and subtract corresponding terms; this gives

$$(E_m - E_n)\psi_m\psi_n^* = (\hbar^2/2m)(\psi_m\triangle\psi_n^* - \psi_n^*\triangle\psi_m)$$
$$= (\hbar^2/2m) \operatorname{div}(\psi_m\nabla\psi_n^* - \psi_n^*\nabla\psi_m).$$

‡ If ψ is written as $|\psi|e^{i\alpha}$, then

$$\mathbf{j} = (\hbar/m)|\psi|^2\nabla\alpha. \tag{19.4a}$$

If we now integrate both sides of this equation over all space, the right-hand side, on transformation by Gauss's theorem, reduces to zero, and we obtain

$$(E_m - E_n) \int \psi_m \psi_n^* \, dV = 0,$$

whence, by the hypothesis $E_m \neq E_n$, there follows the required orthogonality relation

$$\int \psi_m \psi_n^* \, dV = 0.$$

§20. The variational principle

Schrödinger's equation, in the general form $\hat{H}\psi = E\psi$, can be obtained from the variational principle

$$\delta \int \psi^* (\hat{H} - E) \psi \, dq = 0. \tag{20.1}$$

Since ψ is complex, we can vary ψ and ψ^* independently. Varying ψ^*, we have

$$\int \delta\psi^* (\hat{H} - E) \psi \, dq = 0,$$

whence, because $\delta\psi^*$ is arbitrary, we obtain the required equation $\hat{H}\psi = E\psi$. The variation of ψ gives nothing different. For, varying ψ and using the fact that the operator $\hat{H}$ is Hermitian, we have

$$\int \psi^* (\hat{H} - E) \delta\psi \, dq = \int \delta\psi (\hat{H}^* - E) \psi^* \, dq = 0,$$

from which we obtain the complex conjugate equation $\hat{H}^* \psi^* = E\psi^*$.

The variational principle (20.1) requires an unconditional extremum of the integral. It can be stated in a different form by regarding E as a Lagrangian multiplier in a problem with the conditional extremum requirement

$$\delta \int \psi^* \hat{H} \psi \, dq = 0, \tag{20.2}$$

the condition being

$$\int \psi \psi^* \, dq = 1. \tag{20.3}$$

The least value of the integral in (20.2) (with the condition (20.3)) is the first eigenvalue of the energy, i.e. the energy E_0 of the normal state. The function ψ which gives this minimum is accordingly the wave function ψ_0 of the normal state.† The wave functions ψ_n ($n > 0$) of the other stationary states correspond only to an extremum, and not to a true minimum of the integral.

In order to obtain, from the condition that the integral in (20.2) is a minimum, the wave function ψ_1 and the energy E_1 of the state next to the normal one, we must restrict our choice to those functions ψ which satisfy not only the

† In the rest of this section we shall suppose the wave functions ψ to be real; they can always be so chosen (if there is no magnetic field).

normalization condition (20.3) but also the condition of orthogonality with the wave function ψ_0 of the normal state: $\int \psi\psi_0\, dq = 0$. In general, if the wave functions $\psi_0, \psi_1, \ldots, \psi_{n-1}$ of the first n states (arranged in order of increasing energy) are known, the wave function of the next state gives a minimum of the integral in (20.2) with the additional conditions

$$\int \psi^2\, dq = 1, \quad \int \psi\psi_m\, dq = 0 \qquad (m = 0, 1, 2, \ldots, n-1). \tag{20.4}$$

We shall give here some general theorems which can be proved from the variational principle.†

The wave function ψ_0 of the normal state does not become zero (or, as we say, has no *nodes*) for any finite values of the coordinates.‡ In other words, it has the same sign in all space. Hence, it follows that the wave functions $\psi_n\,(n > 0)$ of the other stationary states, being orthogonal to ψ_0, must have nodes (if ψ_n is also of constant sign, the integral $\int \psi_0\psi_n\, dq$ cannot vanish).

Next, from the fact that ψ_0 has no nodes, it follows that the normal energy level cannot be degenerate. For, suppose the contrary to be true, and let ψ_0, ψ_0' be two different eigenfunctions corresponding to the level E_0. Any linear combination $c\psi_0 + c'\psi_0'$ will also be an eigenfunction; but by choosing the appropriate constants c, c', we can always make this function vanish at any given point in space, i.e. we can obtain an eigenfunction with nodes.

If the motion takes place in a bounded region of space, we must have $\psi = 0$ at the boundary of this region (see §18). To determine the energy levels, it is necessary to find, from the variational principle, the minimum value of the integral in (20.2) with this boundary condition. The theorem that the wave function of the normal state has no nodes means in this case that ψ_0 does not vanish anywhere inside this region.

We notice that, as the dimensions of the region containing the motion increase, all the energy levels E_n decrease; this follows immediately from the fact that an extension of the region increases the range of functions which can make the integral a minimum, and consequently the least value of the integral can only diminish.

The expression

$$\int \psi\hat{H}\psi\, dq = \int [-\sum_a (\hbar^2/2m_a)\psi\triangle_a\psi + U\psi^2]\, dq$$

for the states of the discrete spectrum of a particle system may be transformed into another expression which is more convenient in practice. In the first term of the integrand we write

$$\psi\triangle_a\psi = \mathrm{div}_a(\psi\nabla_a\psi) - (\nabla_a\psi)^2.$$

† The proof of theorems concerning the zeros of eigenfunctions (see also §21) is given by M. A. Lavrent'ev and L. A. Lyusternik, *The Calculus of Variations* (*Kurs variatsionnogo ischisleniya*), 2nd edition, chapter IX, Moscow, 1950; R. Courant and D. Hilbert, *Methods of Mathematical Physics*, volume I, chapter VI, Interscience, New York, 1953.

‡ This theorem and its consequences are not in general valid for the wave functions of systems consisting of several identical particles (see the end of §63).

The integral of $\mathrm{div}_a\,(\psi\nabla_a\psi)$ over all space is transformed into an integral over an infinitely distant closed surface, and since the wave functions of the states of a discrete spectrum tend to zero sufficiently rapidly at infinity, this integral vanishes. Thus

$$\int \psi\hat{H}\psi\,\mathrm{d}q = \int \Big[\sum_a (\hbar^2/2m_a)(\nabla_a\psi)^2 + U\psi^2\Big]\,\mathrm{d}q. \tag{20.5}$$

§21. General properties of motion in one dimension

If the potential energy of a particle depends on only one coordinate (x), then the wave function can be sought as the product of a function of y and z and a function of x only. The former of these is determined by Schrödinger's equation for free motion, and the second by the one-dimensional Schrödinger's equation

$$\frac{\mathrm{d}^2\psi}{\mathrm{d}x^2} + \frac{2m}{\hbar^2}[E - U(x)]\psi = 0. \tag{21.1}$$

Similar one-dimensional equations are evidently obtained for the problem of motion in a field whose potential energy is $U(x, y, z) = U_1(x) + U_2(y) + U_3(z)$, i.e. can be written as a sum of functions each of which depends on only one of the coordinates. In §§22–24 we shall discuss a number of actual examples of such "one-dimensional" motion. Here we shall obtain some general properties of the motion.

We shall show first of all that, in a one-dimensional problem, none of the energy levels of a discrete spectrum is degenerate. To prove this, suppose the contrary to be true, and let ψ_1 and ψ_2 be two different eigenfunctions corresponding to the same value of the energy. Since both of these satisfy the same equation (21.1), we have

$$\psi_1''/\psi_1 = (2m/\hbar^2)(U - E) = \psi_2''/\psi_2,$$

or $\psi_1''\psi_2 - \psi_1\psi_2'' = 0$ (the prime denotes differentiation with respect to x). Integrating this relation, we find

$$\psi_1'\psi_2 - \psi_1\psi_2' = \text{constant}. \tag{21.2}$$

Since $\psi_1 = \psi_2 = 0$ at infinity, the constant must be zero, and so

$$\psi_1'\psi_2 - \psi_1\psi_2' = 0,$$

or $\psi_1'/\psi_1 = \psi_2'/\psi_2$. Integrating again, we obtain $\psi_1 = \text{constant} \times \psi_2$, i.e. the two functions are essentially identical.

The following theorem (called the *oscillation theorem*) may be stated for the wave functions $\psi_n(x)$ of a discrete spectrum. The function $\psi_n(x)$ corresponding to the $(n+1)$th eigenvalue E_n (the eigenvalues being arranged in order of magnitude), vanishes n times (for finite† values of x).

---

† If the particle can be found only on a limited segment of the x-axis, we must consider the zeros of $\psi_n(x)$ within that segment.

We shall suppose that the function $U(x)$ tends to finite limiting values as $x \to \pm\infty$ (though it need not be a monotonic function). We take the limiting value $U(+\infty)$ as the zero of energy (i.e. we put $U(+\infty) = 0$), and we denote $U(-\infty)$ by U_0, supposing that $U_0 > 0$. The discrete spectrum lies in the range of energy values for which the particle cannot move off to infinity; for this to be so, the energy must be less than both limiting values $U(\pm\infty)$, i.e. it must be negative:

$$E < 0, \tag{21.3}$$

and we must, of course, have in any case $E > U_{\min}$, i.e. the function $U(x)$ must have at least one minimum with $U_{\min} < 0$.

Let us now consider the range of positive energy values less than U_0:

$$0 < E < U_0. \tag{21.4}$$

In this range the spectrum will be continuous, and the motion of the particle in the corresponding stationary states will be infinite, the particle moving off towards $x = +\infty$. It is easy to see that none of the eigenvalues of the energy in this part of the spectrum is degenerate either. To show this, it is sufficient to notice that the proof given above (for the discrete spectrum) still holds if the functions ψ_1, ψ_2 are zero at only one infinity (in the present case they tend to zero as $x \to -\infty$).

For sufficiently large positive values of x, we can neglect $U(x)$ in Schrödinger's equation (21.1):

$$\psi'' + (2m/\hbar^2)E\psi = 0.$$

This equation has real solutions in the form of a *stationary plane wave*

$$\psi = a\cos(kx+\delta), \tag{21.5}$$

where a and δ are constants, and the *wave number* $k = p/\hbar = \sqrt{(2mE)}/\hbar$. This formula determines the asymptotic form (for $x \to +\infty$) of the wave functions of the non-degenerate energy levels in the range (21.4) of the continuous spectrum. For large negative values of x, Schrödinger's equation is

$$\psi'' - (2m/\hbar^2)(U_0 - E)\psi = 0.$$

The solution which does not become infinite as $x \to -\infty$ is

$$\psi = be^{\kappa x}, \quad \text{where } \kappa = \sqrt{[2m(U_0 - E)]}/\hbar. \tag{21.6}$$

This is the asymptotic form of the wave function as $x \to -\infty$. Thus the wave function decreases exponentially in the region where $E < U$.

Finally, for

$$E > U_0 \tag{21.7}$$

the spectrum will be continuous, and the motion will be infinite in both directions. In this part of the spectrum all the levels are doubly degenerate. This follows from the fact that the corresponding wave functions are determined by the second-order equation (21.1), and both of the two independent solutions of this equation satisfy the necessary conditions at infinity (whereas, for instance, in the previous case one of the solutions became infinite as $x \to -\infty$, and therefore had to be rejected). The asymptotic form of the wave function as $x \to +\infty$ is

$$\psi = a_1 e^{ikx} + a_2 e^{-ikx}, \tag{21.8}$$

and similarly for $x \to -\infty$. The term e^{ikx} corresponds to a particle moving to the right, and e^{-ikx} corresponds to one moving to the left.

Let us suppose that the function $U(x)$ is even $[U(-x) = U(x)]$. Then Schrödinger's equation (21.1) is unchanged when the sign of the coordinate is reversed. It follows that, if $\psi(x)$ is some solution of this equation, then $\psi(-x)$ is also a solution, and coincides with $\psi(x)$ apart from a constant factor: $\psi(-x) = c\psi(x)$. Changing the sign of x again, we obtain $\psi(x) = c^2\psi(x)$, whence $c = \pm 1$. Thus, for a potential energy which is symmetrical (relative to $x = 0$), the wave functions of the stationary states must be either even $[\psi(-x) = \psi(x)]$ or odd $[\psi-(x) = -\psi(x)]$.† In particular, the wave function of the ground state is even, since it cannot have a node, while an odd function always vanishes for $x = 0$ $[\psi(0) = -\psi(0) = 0]$.

To normalize the wave functions of one-dimensional motion (in a continuous spectrum), there is a simple method of determining the normalization coefficient directly from the asymptotic expression for the wave function for large values of $|x|$.

Let us consider the wave function of a motion infinite in one direction, $x \to +\infty$. The normalization integral diverges as $x \to \infty$ (as $x \to -\infty$, the function decreases exponentially, so that the integral rapidly converges). Hence, to determine the normalization constant, we can replace ψ by its asymptotic value (for large positive x), and perform the integration, taking as the lower limit any finite value of x, say zero; this amounts to neglecting a finite quantity in comparison with an infinite one. We shall show that the wave function normalized by the condition

$$\int \psi_p{}^* \psi_{p'}\, dx = \delta\left(\frac{p-p'}{2\pi\hbar}\right) = 2\pi\hbar\, \delta(p-p'), \tag{21.9}$$

where p is the momentum of the particle at infinity, must have the asymptotic form (21.5) with $a = 2$:

$$\psi_p \approx 2\cos(kx+\delta) = e^{i(kx+\delta)} + e^{-i(kx+\delta)}. \tag{21.10}$$

Since we do not intend to verify the orthogonality of the functions corre-

---

† In this discussion it is assumed that the stationary state is not degenerate, i.e. the motion is not infinite in both directions. Otherwise, when the sign of x is changed, two wave functions belonging to the energy level concerned may be transformed into each other. In this case, however, although the wave functions of the stationary states need not be even or odd, they can always be made so (by choosing appropriate linear combinations of the original functions).

sponding to different p, on substituting the functions (21.10) in the normalization integral we shall suppose the momenta p and p' to be arbitrarily close; we can therefore put $\delta = \delta'$ (in general δ is a function of p). Next, we retain in the integrand only those terms which diverge for $p = p'$; in other words, we omit terms containing the factor $e^{\pm i(k+k')x}$. Thus we obtain

$$\int \psi_p{}^* \psi_{p'}\,\mathrm{d}x = \int_0^\infty e^{i(k'-k)x}\,\mathrm{d}x + \int_0^\infty e^{-i(k'-k)x}\,\mathrm{d}x = \int_{-\infty}^\infty e^{i(k'-k)x}\,\mathrm{d}x,$$

which, from (15.7), is the same as (21.9).

The change to normalization by the delta function of energy is effected, in accordance with (5.14), by multiplying ψ_p by

$$\left(\frac{\mathrm{d}(p/2\pi\hbar)}{\mathrm{d}E}\right)^{1/2} = \frac{1}{\sqrt{(2\pi\hbar v)}},$$

where v is the velocity of the particle at infinity. Thus

$$\psi_E = \frac{1}{\sqrt{(2\pi\hbar v)}}\psi_p$$

$$= \frac{1}{\sqrt{(2\pi\hbar v)}}(e^{i(kx+\delta)} + e^{-i(kx+\delta)}). \qquad (21.11)$$

The current density is $1/2\pi\hbar$ in each of the travelling waves that make up the stationary wave (21.11). Thus we can formulate the following rule for the normalization of the wave function for a motion infinite in one direction by the delta function of energy: having represented the asymptotic expression for the wave function in the form of a sum of two plane waves travelling in opposite directions, we must choose the normalization coefficient in such a way that the current density in the wave travelling towards (or away from) the origin is $1/2\pi\hbar$.

Similarly, we can obtain an analogous rule for normalizing the wave functions of a motion infinite in both directions. The wave function will be normalized by the delta function of energy if the sum of the probability currents in the waves travelling towards the origin from $x = +\infty$ and $x = -\infty$ is $1/2\pi\hbar$.

§22. The potential well

As a simple example of one-dimensional motion, let us consider motion in a square *potential well*, i.e. in a field where $U(x)$ has the form shown in Fig. 1 (p. 64) $U(x) = 0$ for $0 < x < a$, $U(x) = U_0$ for $x < 0$ and $x > a$. It is evident *a priori* that for $E < U_0$ the spectrum will be discrete, while for $E > U_0$ we have a continuous spectrum of doubly degenerate levels.

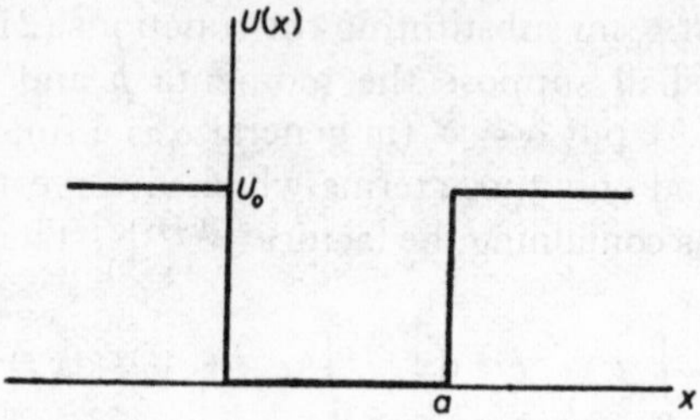

FIG. 1

In the region $0 < x < a$ we have Schrödinger's equation

$$\psi''+(2m/\hbar^2)E\psi = 0 \tag{22.1}$$

(the prime denotes differentiation with respect to x), while in the region outside the well

$$\psi''+(2m/\hbar^2)(E-U_0)\psi = 0. \tag{22.2}$$

For $x = 0$ and $x = a$ the solutions of these equations must be continuous together with their derivatives, while for $x = \pm\infty$ the solution of equation (22.2) must remain finite (for the discrete spectrum when $E < U_0$, it must vanish).

For $E < U_0$, the solution of equation (22.2) which vanishes at infinity is

$$\psi = \text{constant} \times e^{\mp\kappa x}, \quad \text{where } \kappa = \sqrt{[(2m/\hbar^2)(U_0-E)]}; \tag{22.3}$$

the signs $-$ and $+$ in the exponent refer to the regions $x > a$ and $x < 0$ respectively. The probability $|\psi|^2$ of finding the particle decreases exponentially in the region where $E < U(x)$. Instead of the continuity of ψ and ψ' at the edge of the potential well, it is convenient to require the continuity of ψ and of its logarithmic derivative ψ'/ψ. Taking account of (22.3), we obtain the boundary condition in the form

$$\psi'/\psi = \mp\kappa. \tag{22.4}$$

We shall not pause here to determine the energy levels in a well of arbitrary depth U_0 (see Problem 2), and shall analyse fully only the limiting case of infinitely high walls ($U_0 \to \infty$).

For $U_0 = \infty$, the motion takes place only between the points $x = 0$ and $x = a$ and, as was pointed out in §18, the boundary condition at these points is

$$\psi = 0. \tag{22.5}$$

(It is easy to see that this condition is also obtained from the general condition (22.4). For, when $U_0 \to \infty$, we have also $\kappa \to \infty$ and hence $\psi'/\psi \to \infty$; since ψ' cannot become infinite, it follows that $\psi = 0$.) We seek a solution of equation (22.1) inside the well in the form

$$\psi = c \sin(kx+\delta), \quad \text{where } k = \sqrt{(2mE/\hbar^2)}. \tag{22.6}$$

The condition $\psi = 0$ for $x = 0$ gives $\delta = 0$, and then the same condition for $x = a$ gives $\sin ka = 0$, whence $ka = n\pi$, n being a positive integer,† or

$$E_n = (\pi^2\hbar^2/2ma^2)n^2, \quad n = 1, 2, 3, \ldots. \tag{22.7}$$

This determines the energy levels of a particle in a potential well. The normalized wave functions of the stationary states are

$$\psi_n = \sqrt{(2/a)}\sin(\pi n x/a). \tag{22.8}$$

From these results we can immediately write down the energy levels for a particle in a rectangular "potential box", i.e. for three-dimensional motion in a field whose potential energy $U = 0$ for $0 < x < a$, $0 < y < b$, $0 < z < c$ and $U = \infty$ outside this region. In fact, these levels are given by the sums

$$E_{n_1 n_2 n_3} = \frac{\pi^2\hbar^2}{2m}\left(\frac{n_1^2}{a^2}+\frac{n_2^2}{b^2}+\frac{n_3^2}{c^2}\right) \quad (n_1, n_2, n_3 = 1, 2, 3, \ldots), \tag{22.9}$$

and the corresponding wave functions by the products

$$\psi_{n_1 n_2 n_3} = \sqrt{\frac{8}{abc}}\sin\frac{\pi n_1}{a}x \sin\frac{\pi n_2}{b}y \sin\frac{\pi n_3}{c}z. \tag{22.10}$$

It may be noted that the energy E_0 of the ground state is, by (22.7) or (22.9), of the order of $\hbar^2/ml^2$, where l is the linear dimension of the region in which the particle moves. This result is in accordance with the uncertainty relation; when the uncertainty in the coordinate is $\sim l$, the uncertainty in the momentum, and therefore the order of magnitude of the momentum itself, is $\sim \hbar/l$. The corresponding energy is $\sim (\hbar/l)^2/m$.

PROBLEMS

Problem 1. Determine the probability distribution for various values of the momentum for the normal state of a particle in an infinitely deep square potential well.

Solution. The coefficients $a(p)$ in the expansion of the function ψ_1 (22.8) in terms of the eigenfunctions of the momentum are

$$a(p) = \int \psi_p{}^*\psi_1 \, dx = \sqrt{\frac{2}{a}}\int_0^a \sin\left(\frac{\pi}{a}x\right)e^{-(i/\hbar)px} \, dx.$$

Calculating the integral and squaring its modulus, we obtain the required probability distribution:

$$|a(p)|^2 \frac{dp}{2\pi\hbar} = \frac{4\pi\hbar^3 a}{(p^2a^2-\pi^2\hbar^2)^2}\cos^2\frac{pa}{2\hbar}\, dp.$$

Problem 2. Determine the energy levels for the potential well shown in Fig. 2 (p. 66).

Solution. The spectrum of energy values $E < U_1$, which we shall consider, is discrete. In the region $x < 0$ the wave function is

$$\psi = c_1 e^{\kappa_1 x}, \text{ where } \kappa_1 = \sqrt{[(2m/\hbar^2)(U_1 - E)]},$$

† For $n = 0$ we should have $\psi = 0$ identically.

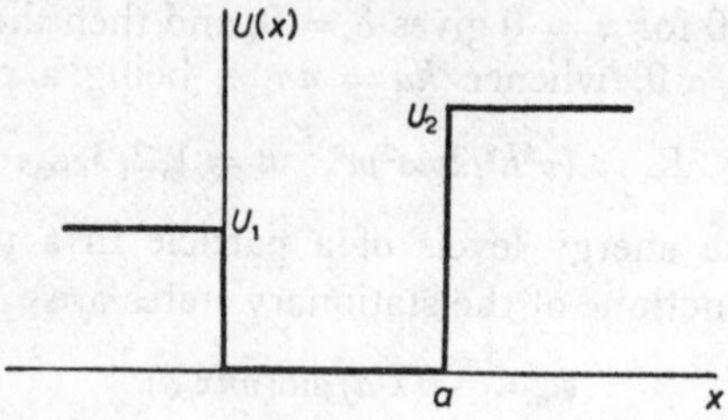

FIG. 2

while in the region $x > a$

$$\psi = c_2 e^{-\kappa_2 x}, \quad \text{where } \kappa_2 = \sqrt{[(2m/\hbar^2)(U_2-E)]}.$$

Inside the well ($0 < x < a$) we look for ψ in the form

$$\psi = c\sin(kx+\delta), \quad \text{where } k=\sqrt{(2mE/\hbar^2)}.$$

The condition of the continuity of ψ'/ψ at the edges of the well gives the equations

$$k\cot\delta = \kappa_1 = \sqrt{[(2m/\hbar^2)U_1-k^2]}, \; k\cot(ka+\delta) = -\kappa_2 = -\sqrt{[(2m/\hbar^2)U_2-k^2]},$$

or

$$\sin\delta = k\hbar/\sqrt{(2mU_1)}, \; \sin(ka+\delta) = -k\hbar/\sqrt{(2mU_2)}.$$

Eliminating δ, we obtain the transcendental equation

$$ka = n\pi - \sin^{-1}[k\hbar/\sqrt{(2mU_1)}] - \sin^{-1}[k\hbar/\sqrt{(2mU_2)}] \tag{1}$$

(where $n = 1, 2, 3, \ldots$, and the values of the inverse sine are taken between 0 and $\frac{1}{2}\pi$), whose roots determine the energy levels $E = k^2\hbar^2/2m$. For each n there is in general one root; the values of n number the levels in order of increasing energy.

Since the argument of the inverse sine cannot exceed unity, it is clear that the values of k can lie only in the range from 0 to $\sqrt{(2mU_1/\hbar^2)}$. The left-hand side of equation (1) increases monotonically with k, and the right-hand side decreases monotonically. Hence it is necessary, for a root of equation (1) to exist, that for $k = \sqrt{(2mU_1/\hbar^2)}$ the right-hand side should be less than the left-hand side. In particular, the inequality

$$a\sqrt{(2mU_1)}/\hbar \geqslant \tfrac{1}{2}\pi - \sin^{-1}\sqrt{(U_1/U_2)}, \tag{2}$$

which is obtained for $n = 1$, is the condition that at least one energy level exists in the well. We see that for given and unequal U_1, U_2 there are always widths a of the well which are so small that there is no discrete energy level. For $U_1 = U_2$, the condition (2) is evidently always satisfied.

For $U_1 = U_2 \equiv U_0$ (a symmetrical well), equation (1) reduces to

$$\sin^{-1}[\hbar k/\sqrt{(2mU_0)}] = \tfrac{1}{2}(n\pi - ka). \tag{3}$$

Introducing the variable $\xi = \frac{1}{2}ka$, we obtain for odd n the equation

$$\cos\xi = \pm\gamma\xi, \quad \text{where } \gamma = (\hbar/a)\sqrt{(2/mU_0)}, \tag{4}$$

and those roots of this equation must be taken for which $\tan\xi > 0$. For even n we obtain the equation

$$\sin\xi = \pm\gamma\xi, \tag{5}$$

and we must take those roots for which $\tan\xi < 0$. The roots of these two equations determine the energy levels $E = 2\xi^2\hbar^2/ma^2$. The number of levels is finite when $\gamma \neq 0$.

In particular, for a shallow well in which $U_0 \ll \hbar^2/ma^2$, we have $\gamma \gg 1$ and equation (5) has no root. Equation (4) has one root (with the upper sign on the right-hand side), $\xi \cong 1/\gamma - 1/2\gamma^3$. Thus the well contains only one energy level,

$$E_0 \cong U_0 - (ma^2/2\hbar^2)U_0^2,$$

which is near the top of the well.

PROBLEM 3. Determine the pressure exerted on the walls of a rectangular "potential box" by a particle inside it.

SOLUTION. The force on the the wall perpendicular to the x-axis is the mean value of the derivative $-\partial H/\partial a$ of the Hamilton's function of the particle with respect to the length of the box in the direction of the x-axis. The pressure is obtained by dividing this force by the area bc of the wall. According to the formula (11.16), the required mean value is found by differentiating the eigenvalue (22.9) of the energy. The result is

$$p^{(x)} = \pi^2\hbar^2 n_1^2/ma^3 bc.$$

§23. The linear oscillator

Let us consider a particle executing small oscillations in one dimension (what is called a *linear oscillator*). The potential energy of such a particle is $\frac{1}{2}m\omega^2x^2$, where ω is, in classical mechanics, the characteristic (angular) frequency of the oscillations. Accordingly, the Hamiltonian of the oscillator is

$$\hat{H} = \tfrac{1}{2}\hat{p}^2/m + \tfrac{1}{2}m\omega^2x^2. \tag{23.1}$$

Since the potential energy becomes infinite for $x = \pm\infty$, the particle can have only a finite motion, and the energy eigenvalue spectrum is entirely discrete.

Let us determine the energy levels of the oscillator, using the matrix method†. We shall start from the equations of motion in the form (19.3); in this case they give

$$\hat{\ddot{x}} + \omega^2 x = 0. \tag{23.2}$$

In matrix form, this equation reads

$$(\ddot{x})_{mn} + \omega^2 x_{mn} = 0.$$

For the matrix elements of the acceleration we have, according to (11.8), $(\ddot{x})_{mn} = i\omega_{mn}(\dot{x})_{mn} = -\omega_{mn}{}^2 x_{mn}$. Hence we obtain

$$(\omega_{mn}{}^2 - \omega^2)x_{mn} = 0.$$

Hence it is evident that all the matrix elements x_{mn} vanish except those for which $\omega_{mn} = \omega$ or $\omega_{mn} = -\omega$. We number all the stationary states so that the frequencies $\pm\omega$ correspond to transitions $n \to n \mp 1$, i.e. $\omega_{n,n\mp 1} = \pm\omega$. Then the only non-zero matrix elements are $x_{n,n\pm 1}$.

We shall suppose that the wave functions ψ_n are taken real. Since x is a real quantity, all the matrix elements x_{mn} are real. The Hermitian condition (11.10) now shows that the matrix x_{mn} is symmetrical:

$$x_{mn} = x_{nm}.$$

To calculate the matrix elements of the coordinate which are different

† This was done by Heisenberg in 1925, before Schrödinger's discovery of the wave equation.

from zero, we use the commutation rule

$$\hat{\dot{x}}\hat{x}-\hat{x}\hat{\dot{x}} = -i\hbar/m,$$

written in the matrix form

$$(\dot{x}x)_{mn}-(x\dot{x})_{mn} = -(i\hbar/m)\delta_{mn}.$$

By the matrix multiplication rule (11.12) we hence have for $m = n$

$$i \sum_l (\omega_{nl}x_{nl}x_{ln}-x_{nl}\omega_{ln}x_{ln}) = 2i \sum_l \omega_{nl}x_{nl}^2 = -i\hbar/m.$$

In this sum, only the terms with $l = n \pm 1$ are different from zero, so that we have

$$(x_{n+1,n})^2-(x_{n,n-1})^2 = \hbar/2m\omega. \tag{23.3}$$

From this equation we deduce that the quantities $(x_{n+1,n})^2$ form an arithmetic progression, which is unbounded above, but is certainly bounded below, since it can contain only positive terms. Since we have as yet fixed only the relative positions of the numbers n of the states, but not their absolute values, we can arbitrarily choose the value of n corresponding to the first (normal) state of the oscillator, and put this value equal to zero. Accordingly $x_{0,-1}$ must be regarded as being zero identically, and the application of equations (23.3) with $n = 0, 1, \ldots$ successively leads to the result

$$(x_{n,n-1})^2 = n\hbar/2m\omega.$$

Thus we finally obtain the following expression for the matrix elements of the coordinate which are different from zero:†

$$x_{n,n-1} = x_{n-1,n} = \sqrt{(n\hbar/2m\omega)}. \tag{23.4}$$

The matrix of the operator $\hat{H}$ is diagonal, and the matrix elements H_{nn} are the required eigenvalues E_n of the energy of the oscillator. To calculate them, we write

$$\begin{aligned} H_{nn} = E_n &= \tfrac{1}{2}m[(\dot{x}^2)_{nn}+\omega^2(x^2)_{nn}] \\ &= \tfrac{1}{2}m[\sum_l i\omega_{nl}x_{nl}i\omega_{ln}x_{ln}+\omega^2 \sum_l x_{nl}x_{ln}] \\ &= \tfrac{1}{2}m \sum_l (\omega^2+\omega_{nl}^2)x_{ln}^2. \end{aligned}$$

In the sum over l, only the terms with $l = n\pm1$ are different from zero; substituting (23.4), we obtain

$$E_n = (n+\tfrac{1}{2})\hbar\omega, \quad n = 0, 1, 2, \ldots. \tag{23.5}$$

Thus the energy levels of the oscillator lie at equal intervals of $\hbar\omega$ from one another. The energy of the normal state ($n = 0$) is $\frac{1}{2}\hbar\omega$; we call attention to the fact that it is not zero.

---

† We choose the indeterminate phases α_n (see the second footnote to §11) so as to obtain the plus sign in front of the radical in all the matrix elements (23.4). Such a choice is always possible for a matrix in which only those elements are different from zero which correspond to transitions between states with adjacent numbers.

The result (23.5) can also be obtained by solving Schrödinger's equation. For an oscillator, this has the form

$$\frac{d^2\psi}{dx^2}+\frac{2m}{\hbar^2}(E-\tfrac{1}{2}m\omega^2x^2)\psi = 0. \tag{23.6}$$

Here it is convenient to introduce, instead of the coordinate x, the dimensionless variable ξ by the relation

$$\xi = \sqrt{(m\omega/\hbar)}x. \tag{23.7}$$

Then we have the equation

$$\psi''+[(2E/\hbar\omega)-\xi^2]\psi = 0; \tag{23.8}$$

here the prime denotes differentiation with respect to ξ.

For large ξ, we can neglect $2E/\hbar\omega$ in comparison with ξ^2; the equation $\psi'' = \xi^2\psi$ has the asymptotic integrals $\psi = e^{\pm\frac{1}{2}\xi^2}$ (for differentiation of this function gives $\psi'' = \xi^2\psi$ on neglecting terms of order less than that of the term retained). Since the wave function ψ must remain finite as $\xi \to \pm\infty$, the index must be taken with the minus sign. It is therefore natural to make in equation (23.8) the substitution

$$\psi = e^{-\xi^2/2}\,\chi(\xi). \tag{23.9}$$

For the function $\chi(\xi)$ we obtain the equation (with the notation $(2E/\hbar\omega)-1 = 2n$; since we already know that $E > 0$, we have $n > -\frac{1}{2}$)

$$\chi''-2\xi\chi'+2n\chi = 0, \tag{23.10}$$

where the function χ must be finite for all finite ξ, and for $\xi \to \pm\infty$ must not tend to infinity more rapidly than every finite power of ξ (in order that the function ψ should tend to zero).

Such solutions of equation (23.10) exist only for positive integral (and zero) values of n (see §a of the Mathematical Appendices); this gives the eigenvalues (23.5) for the energy, which we know already. The solutions of equation (23.10) corresponding to various integral values of n are $\chi =$ constant$\times H_n(\xi)$, where $H_n(\xi)$ are what are called *Hermite polynomials*; these are polynomials of the nth degree in ξ, defined by the formula

$$H_n(\xi) = (-1)^n e^{\xi^2}\, d^n(e^{-\xi^2})/d\xi^n. \tag{23.11}$$

Determining the constants so that the functions ψ_n satisfy the normalization condition

$$\int_{-\infty}^{\infty} \psi_n^2(x)\, dx = 1,$$

we obtain (see (a.7))

$$\psi_n(x) = \left(\frac{m\omega}{\pi\hbar}\right)^{1/4} \frac{1}{2^{n/2}\sqrt{(n!)}} e^{-m\omega x^2/2\hbar} H_n(x\sqrt{[m\omega/\hbar]}). \tag{23.12}$$

Thus the wave function of the normal state is

$$\psi_0(x) = (m\omega/\pi\hbar)^{1/4}e^{-m\omega x^2/2\hbar}. \tag{23.13}$$

It has no zeros for finite x, which is as it should be.

By calculating the integrals $\int_{-\infty}^{\infty} \psi_n\psi_m\xi \,\mathrm{d}\xi$, we can determine the matrix elements of the coordinate; this calculation leads, of course, to the same values (23.4).

Finally, we shall show how the wave functions ψ_n may be calculated by the matrix method. We notice that, in the matrices of the operators $\hat{\dot{x}} \pm i\omega\hat{x}$, the only elements different from zero are

$$(\dot{x}-i\omega x)_{n-1,n} = -(\dot{x}+i\omega x)_{n,n-1} = -i\sqrt{(2\omega\hbar n/m)}. \tag{23.14}$$

Using the general formula (11.11), and taking into account the fact that $\psi_{-1} \equiv 0$, we conclude that

$$(\hat{\dot{x}}-i\omega x)\psi_0 = 0.$$

After substituting the expression $\hat{\dot{x}} = -i(\hbar/m)\mathrm{d}/\mathrm{d}x$, we obtain the equation

$$\mathrm{d}\psi_0/\mathrm{d}x = -(m\omega/\hbar)x\psi_0,$$

whose normalized solution is (23.13). And, since

$$(\hat{\dot{x}}+i\omega x)\psi_{n-1} = (\dot{x}+i\omega x)_{n,n-1}\psi_n = i\sqrt{(2\omega\hbar n/m)}\psi_n,$$

we obtain the recurrence formula

$$\begin{aligned}\psi_n &= \sqrt{(m/2\omega\hbar n)}[-(\hbar/m)\,\mathrm{d}/\mathrm{d}x+\omega x]\psi_{n-1}\\ &= \frac{1}{\sqrt{(2n)}}\left(-\frac{\mathrm{d}}{\mathrm{d}\xi}+\xi\right)\psi_{n-1} = -\frac{1}{\sqrt{(2n)}}e^{\xi^2/2}\frac{\mathrm{d}}{\mathrm{d}\xi}(e^{-\xi^2/2}\psi_{n-1});\end{aligned}$$

when this is applied n times to the function (23.13), we obtain the expression (23.12) for the normalized functions ψ_n.

PROBLEMS

PROBLEM 1. Determine the probability distribution of the various values of the momentum for an oscillator.

SOLUTION. Instead of expanding the wave function of the stationary state in terms of the eigenfunctions of momentum, it is simpler in the case of the oscillator to start directly from Schrödinger's equation in the momentum representation. Substituting in (23.1) the coordinate operator $\hat{x} = i\hbar\mathrm{d}/\mathrm{d}p$ (15.12), we obtain the Hamiltonian in the p representation,

$$\hat{H} = \tfrac{1}{2}p^2/m - \tfrac{1}{2}m\omega^2\hbar^2\,\mathrm{d}^2/\mathrm{d}p^2.$$

The corresponding Schrödinger's equation $\hat{H}a(p) = Ea(p)$ for the wave function $a(p)$ in the momentum representation is

$$\frac{\mathrm{d}^2a(p)}{\mathrm{d}p^2}+\frac{2}{m\omega^2\hbar^2}\left(E-\frac{p^2}{2m}\right)a(p) = 0.$$

This equation is of exactly the same form as (23.6); hence its solutions can be written down at once by analogy with (23.12). Thus we find the required probability distribution to be

$$|a_n(p)|^2 \frac{dp}{2\pi\hbar} = \frac{1}{2^n n!\sqrt{(\pi m\omega\hbar)}} e^{-p^2/m\omega\hbar} H_n^2(p/\sqrt{[m\omega\hbar]})\, dp.$$

PROBLEM 2. Determine the lower limit of the possible values of the energy of an oscillator, using the uncertainty relation (16.7).

SOLUTION. Since $\overline{x^2} = \bar{x}^2 + (\delta x)^2$, $\overline{p^2} = \bar{p}^2 + (\delta p)^2$, (16.7) gives for the mean value of the energy of the oscillator

$$\bar{E} = \tfrac{1}{2}m\omega^2\overline{x^2} + \tfrac{1}{2}\overline{p^2}/m \geqslant \tfrac{1}{2}m\omega^2(\delta x)^2 + \tfrac{1}{2}(\delta p)^2/m$$
$$\geqslant m\omega^2\hbar^2/8(\delta p)^2 + (\delta p)^2/2m.$$

On determining the minimum value of this expression (regarded as a function of δp), we find the lower limit of the mean values of the energy, and therefore that of all possible values:

$$E \geqslant \tfrac{1}{2}\hbar\omega.$$

PROBLEM 3. Find the wave functions of the states of a linear oscillator that minimize the uncertainty relation, i.e. in which the standard deviations of the coordinate and momentum in the wave packet are related by $\delta p\ \delta x = \frac{1}{2}\hbar$ (E. Schrödinger 1926).†

SOLUTION. The required wave functions must have the form

$$\Psi(x, t) = \frac{1}{(2\pi)^{1/4}(\delta x)^{1/2}} \exp\left\{\frac{i\bar{p}x}{\hbar} - \frac{(x-\bar{x})^2}{4(\delta x)^2} - i\phi(t)\right\}. \tag{1}$$

Their dependence on the coordinate at any instant is in accordance with (16.8), $\bar{x} = \bar{x}(t)$ and $\bar{p} = \bar{p}(t) = m\dot{\bar{x}}(t)$ being the mean values of the coordinate and the momentum; according to (19.3), we have, for a linear oscillator ($U = \frac{1}{2}m\omega^2x^2$), $\hat{\dot{p}} = -m\omega^2 x$, and therefore for the mean values $\dot{\bar{p}} = -m\omega^2\bar{x}$ or

$$\ddot{\bar{x}} + \omega^2\bar{x} = 0, \tag{2}$$

i.e. the function $\bar{x}(t)$ satisfies the classical equation of motion. The constant factor in (1) is determined by the normalization condition

$$\int_{-\infty}^{\infty} |\Psi|^2\, dx = 1;$$

in addition to this factor, Ψ may contain a phase factor with a time-dependent phase $\phi(t)$. The unknown constant δx and the unknown function $\phi(t)$ are found by substituting (1) in the wave equation

$$-\frac{\hbar^2}{2m}\frac{\partial^2\Psi}{\partial x^2} + \tfrac{1}{2}m\omega^2x^2\Psi = i\hbar\frac{\partial\Psi}{\partial t}.$$

With (2), the substitution gives

$$(\tfrac{1}{2}x^2 - x\bar{x})\left(\frac{m^2\omega^2}{\hbar^2} - \frac{1}{4(\delta x)^4}\right) + \left[\frac{m^2\dot{\bar{x}}^2}{2\hbar^2} - \frac{\bar{x}^2}{8(\delta x)^4} + \frac{1}{4(\delta x)^2} - \frac{m}{\hbar}\dot{\phi}(t)\right] = 0.$$

Hence $(\delta x)^2 = \hbar/2m\omega$ and

$$\dot{\phi} = \frac{m}{2\hbar}(\dot{\bar{x}}^2 - \omega^2\bar{x}^2) + \tfrac{1}{2}\omega,$$

$$\phi = \frac{1}{2\hbar}\bar{p}\bar{x} + \tfrac{1}{2}\omega t.$$

Thus we have finally

$$\Psi(x, t) = \left(\frac{m\omega}{\pi\hbar}\right)^{1/4} \exp\left\{\frac{i\bar{p}x}{\hbar} - \frac{m\omega(x-\bar{x})^2}{2\hbar}\right\} \exp\left\{-\tfrac{1}{2}i\omega t - \frac{i\bar{p}\bar{x}}{2\hbar}\right\}. \tag{3}$$

When $\bar{x} = 0$ and $\bar{p} = 0$, this becomes $\psi_0(x)e^{-i\omega t/2}$, the wave function of the oscillator ground state.

---

† These are called *coherent states*.

The mean energy of the oscillator in a coherent state is

$$\bar{E} = \frac{\overline{p^2}}{2m} + \tfrac{1}{2}m\omega^2\overline{x^2}$$

$$= \frac{\bar{p}^2}{2m} + \tfrac{1}{2}m\omega^2\bar{x}^2 + \tfrac{1}{2}\hbar\omega \equiv \hbar\omega(\bar{n}+\tfrac{1}{2}); \tag{4}$$

the quantity $\bar{n}$ is the mean "number" of quanta $\hbar\omega$ in the state. We see that the coherent state is completely specified by the function $\bar{x}(t)$ satisfying the classical equation (2). The general form of this function may be given as

$$\frac{m\omega\bar{x}+i\bar{p}}{\sqrt{(2m\hbar\omega)}} = ae^{-i\omega t}, \quad |a|^2 = \bar{n}. \tag{5}$$

The function (3) can be expanded in wave functions of the stationary states of the oscillator:

$$\Psi = \sum_{n=0}^{\infty} a_n\Psi_n,$$

$$\Psi_n(x, t) = \psi_n(x) \exp\{-i(n+\tfrac{1}{2})\omega t\}.$$

The coefficients in this expansion are (cf. §41, Problem 1)

$$a_n = \int_{-\infty}^{\infty} \Psi_n{}^*\Psi \, dx. \tag{6}$$

The probability for the oscillator to be in the nth state is therefore

$$w_n = |a_n|^2 = e^{-\bar{n}}\bar{n}^n/n!, \tag{7}$$

the Poisson distribution.

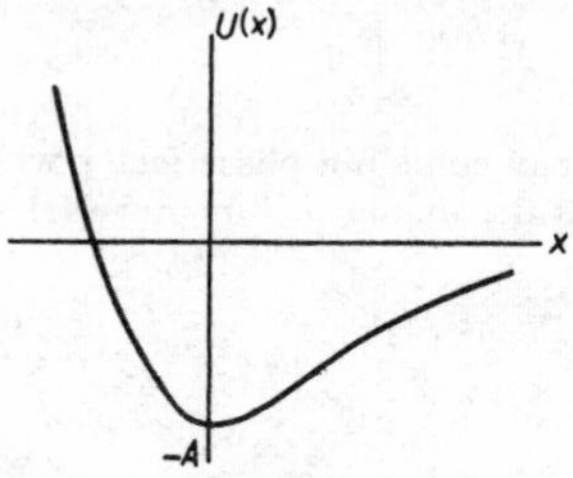

FIG. 3

PROBLEM 4. Determine the energy levels for a particle moving in a field of potential energy (Fig. 3)

$$U(x) = A(e^{-2\alpha x}-2e^{-\alpha x})$$

(P. M. Morse).

SOLUTION. The spectrum of positive eigenvalues of the energy is continuous (and the levels are not degenerate), while the spectrum of negative eigenvalues is discrete.

Schrödinger's equation reads

$$d^2\psi/dx^2+(2m/\hbar^2)(E-Ae^{-2\alpha x}+2Ae^{-\alpha x})\psi = 0.$$

We introduce a new variable

$$\xi = \frac{2\sqrt{(2mA)}}{\alpha\hbar}e^{-\alpha x}$$

(taking values from 0 to ∞) and the notation (we consider the discrete spectrum, so that $E < 0$)

$$s = \sqrt{(-2mE)}/\alpha\hbar, \quad n = \sqrt{(2mA)}/\alpha\hbar-(s+\tfrac{1}{2}). \tag{1}$$

Schrödinger's equation then takes the form

$$\psi''+\frac{1}{\xi}\psi'+\left(-\tfrac{1}{4}+\frac{n+s+\frac{1}{2}}{\xi}-\frac{s^2}{\xi^2}\right)\psi = 0.$$

As $\xi \to \infty$, the function ψ behaves asymptotically as $e^{\pm\frac{1}{2}\xi}$, while as $\xi \to 0$ it is proportional to $\xi^{\pm s}$. From considerations of finiteness we must choose the solution which behaves as $e^{-\frac{1}{2}\xi}$ as $\xi \to \infty$ and as ξ^s as $\xi \to 0$. We make the substitution

$$\psi = e^{-\xi/2}\xi^s w(\xi)$$

and obtain for w the equation

$$\xi w''+(2s+1-\xi)w'+nw = 0, \tag{2}$$

which has to be solved with the conditions that w is finite as $\xi \to 0$, while as $\xi \to \infty$, w tends to infinity not more rapidly than every finite power of ξ. Equation (2) is the equation for a confluent hypergeometric function (see §d of the Mathematical Appendices):

$$w = F(-n, 2s+1, \xi).$$

A solution satisfying the required conditions is obtained for non-negative integral n (when the function F reduces to a polynomial). According to the definitions (1), we thus obtain for the energy levels the values

$$-E_n = A\left[1-\frac{\alpha\hbar}{\sqrt{(2mA)}}(n+\tfrac{1}{2})\right]^2,$$

where n takes positive integral values from zero to the greatest value for which $\sqrt{(2mA)}/\alpha\hbar > n+\frac{1}{2}$ (so that the parameter s is positive in accordance with its definition). Thus the discrete spectrum contains only a limited number of levels. If $\sqrt{(2mA)}/\alpha\hbar < \frac{1}{2}$, there is no discrete spectrum at all.

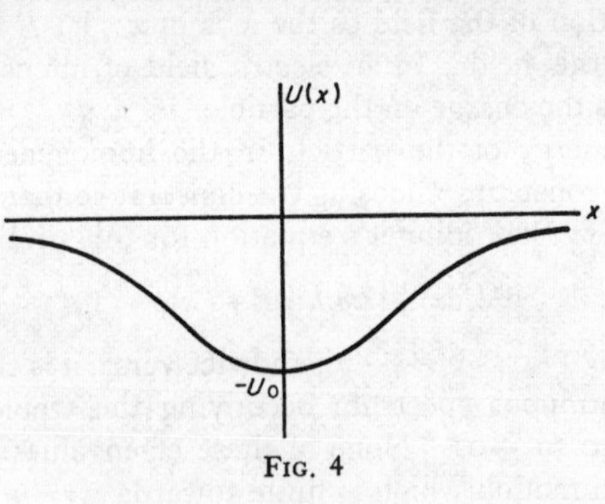

FIG. 4

PROBLEM 5. The same as Problem 4, but with $U = -U_0/\cosh^2 \alpha x$ (Fig. 4).

SOLUTION. The spectrum of positive eigenvalues of the energy is continuous, while that of negative values is discrete; we shall consider the latter. Schrödinger's equation is

$$\frac{d^2\psi}{dx^2}+\frac{2m}{\hbar^2}\left(E+\frac{U_0}{\cosh^2\alpha x}\right)\psi = 0.$$

We put $\xi = \tanh \alpha x$ and use the notation

$$\epsilon = \sqrt{(-2mE)}/\hbar\alpha, \qquad 2mU_0/\alpha^2\hbar^2 = s(s+1),$$

$$s = \tfrac{1}{2}\left(-1+\sqrt{\left[1+\frac{8mU_0}{\alpha^2\hbar^2}\right]}\right),$$

obtaining

$$\frac{d}{d\xi}\left[(1-\xi^2)\frac{d\psi}{d\xi}\right]+\left[s(s+1)-\frac{\epsilon^2}{1-\xi^2}\right]\psi = 0.$$

This is the equation of the associated Legendre polynomials; it can be brought to hypergeometric form by making the substitution $\psi = (1-\xi^2)^{\epsilon/2}\, w(\xi)$ and temporarily changing the variable to $u = \frac{1}{2}(1-\xi)$:

$$u(1-u)w''+(\epsilon+1)(1-2u)w'-(\epsilon-s)(\epsilon+s+1)w = 0.$$

The solution finite for $\xi = 1$ (i.e. for $x = \infty$) is

$$\psi = (1-\xi^2)^{\epsilon/2}F[\epsilon-s,\ \epsilon+s+1,\ \epsilon+1,\ \tfrac{1}{2}(1-\xi)].$$

If ψ remains finite for $\xi = -1$ (i.e. for $x = -\infty$), we must have $\epsilon-s = -n$, where $n = 0, 1, 2, \ldots$; then F is a polynomial of degree n, which is finite for $\xi = -1$.

Thus the energy levels are determined by $s-\epsilon = n$, or

$$E = -\frac{\hbar^2\alpha^2}{8m}\left[-(1+2n)+\sqrt{\left(1+\frac{8mU_0}{\alpha^2\hbar^2}\right)}\right]^2.$$

There is a finite number of levels, determined by the condition $\epsilon > 0$, i.e. $n < s$.

§24. Motion in a homogeneous field

Let us consider the motion of a particle in a homogeneous external field. We take the direction of the field as the axis of x; let F be the force acting on the particle in this field. In an electric field of intensity E, this force is $F = eE$, where e is the charge on the particle.

The potential energy of the particle in the homogeneous field is of the form $U = -Fx+\text{constant}$; choosing the constant so that $U = 0$ for $x = 0$, we have $U = -Fx$. Schrödinger's equation for this problem is

$$d^2\psi/dx^2+(2m/\hbar^2)(E+Fx)\psi = 0. \tag{24.1}$$

Since U tends to $+\infty$ as $x \to -\infty$, and vice versa, it is clear that the energy levels form a continuous spectrum occupying the whole range of energy values E from $-\infty$ to $+\infty$. None of these eigenvalues is degenerate, and they correspond to motion which is finite towards $x = -\infty$ and infinite towards $x = +\infty$.

Instead of the coordinate x, we introduce the dimensionless variable

$$\xi = (x+E/F)(2mF/\hbar^2)^{1/3}. \tag{24.2}$$

Equation (24.1) then takes the form

$$\psi''+\xi\psi = 0. \tag{24.3}$$

This equation does not contain the energy parameter. Hence, if we obtain a solution of it which satisfies the necessary conditions of finiteness, we at once have the eigenfunction for arbitrary values of the energy.

The solution of equation (24.3) which is finite for all x has the form (see §b of the Mathematical Appendices)

$$\psi(\xi) = A\Phi(-\xi), \tag{24.4}$$

where

$$\Phi(\xi) = \frac{1}{\sqrt{\pi}} \int_0^\infty \cos(\tfrac{1}{3}u^3 + u\xi)\, du$$

is called the *Airy function*, while A is a normalization factor which we shall determine below.

As $\xi \to -\infty$, the function $\psi(\xi)$ tends exponentially to zero. The asymptotic expression which determines $\psi(\xi)$ for large negative values of ξ is (see (b.4))

$$\psi(\xi) \approx \frac{A}{2|\xi|^{1/4}} \exp[-\tfrac{2}{3}|\xi|^{3/2}]. \tag{24.5}$$

For large positive values of ξ, the asymptotic expression for $\psi(\xi)$ is (see (b.5))†

$$\psi(\xi) = A\xi^{-1/4} \sin(\tfrac{2}{3}\xi^{3/2} + \tfrac{1}{4}\pi). \tag{24.6}$$

Using the general rule (5.4) for the normalization of eigenfunctions of a continuous spectrum, let us reduce the function (24.4) to the form normalized by the delta function of energy, for which

$$\int_{-\infty}^{\infty} \psi(\xi)\psi(\xi')\, dx = \delta(E'-E). \tag{24.7}$$

In §21 we gave a simple method of determining the normalization coefficient by means of the asymptotic expression for the wave functions. Following this method, we represent the function (24.6) as the sum of two travelling waves:

$$\psi(\xi) \approx \tfrac{1}{2}A\xi^{-1/4} \exp(i[\tfrac{2}{3}\xi^{3/2} - \tfrac{1}{4}\pi]) + \tfrac{1}{2}A\xi^{-1/4} \exp(-i[\tfrac{2}{3}\xi^{3/2} - \tfrac{1}{4}\pi]).$$

The current density, calculated from each of these two terms, is

$$v(A/2\xi^{1/4})^2 = \sqrt{[2(E+Fx)/m]}(A/2\xi^{1/4})^2 = A^2(2\hbar F)^{1/3}/4m^{2/3},$$

and equating this to $1/2\pi\hbar$ we find

$$A = \frac{(2m)^{1/3}}{\pi^{1/2}F^{1/6}\hbar^{2/3}}. \tag{24.8}$$

† It may be noted, by way of anticipation, that the asymptotic expressions (24.5) and (24.6) correspond to the quasi-classical expressions for the wave function in the classically inaccessible and accessible regions (§47).

PROBLEM

Determine the wave functions in the momentum representation for a particle in a homogeneous field.

SOLUTION. The Hamiltonian in the momentum representation is

$$\hat{H} = p^2/2m - i\hbar F\,\mathrm{d}/\mathrm{d}p,$$

so that Schrödinger's equation for the wave function $a(p)$ has the form

$$-i\hbar F\frac{\mathrm{d}a}{\mathrm{d}p}+\left(\frac{p^2}{2m}-E\right)a = 0.$$

Solving this equation, we find the required functions

$$a_E(p) = (2\pi\hbar F)^{-1/2}e^{(i/\hbar F)(Ep-p^3/6m)}.$$

These functions are normalized by the condition

$$\int_{-\infty}^{\infty} a_E^*(p)a_{E'}(p)\,\mathrm{d}p = \delta(E'-E).$$

§25. The transmission coefficient

Let us consider the motion of particles in a field of the type shown in Fig. 5: $U(x)$ increases monotonically from one constant limit ($U = 0$ as $x \to -\infty$) to another ($U = U_0$ as $x \to +\infty$). According to classical mechanics, a particle of energy $E < U_0$ moving in such a field from left to right, on reaching such a "potential wall", is reflected from it, and begins to move in the opposite direction; if, however, $E > U_0$, the particle continues to move in its original direction, though with diminished velocity. In quantum mechanics, a new phenomenon appears: even for $E > U_0$, the particle may be reflected from the potential wall. The probability of reflection must in principle be calculated as follows.

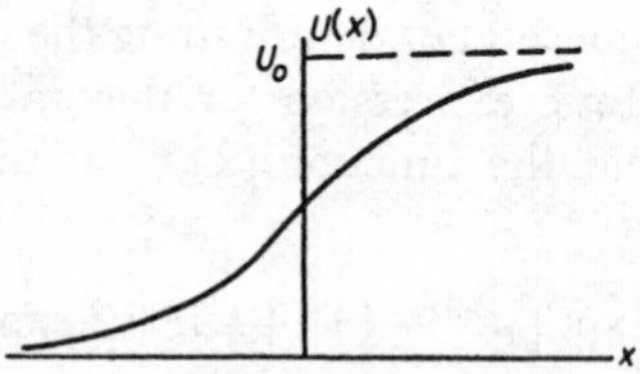

FIG. 5

Let the particle be moving from left to right. For large positive values of x, the wave function must describe a particle which has passed "above the wall" and is moving in the positive direction of x, i.e. it must have the asymptotic form

$$\text{for } x \to \infty,\ \psi \approx Ae^{ik_2x}, \quad \text{where } k_2 = (1/\hbar)\sqrt{[2m(E-U_0)]} \tag{25.1}$$

and A is a constant. To find the solution of Schrödinger's equation which satisfies this boundary condition, we calculate the asymptotic expression for

$x \to -\infty$; it is a linear combination of the two solutions of the equation of free motion, i.e. it has the form

$$\text{for } x \to -\infty, \psi \approx e^{ik_1x} + Be^{-ik_1x}, \quad \text{where } k_1 = \sqrt{(2mE)}/\hbar. \tag{25.2}$$

The first term corresponds to a particle incident on the wall (we suppose ψ normalized so that the coefficient of this term is unity); the second term represents a particle reflected from the wall. The probability current density in the incident wave is k_1, in the reflected wave $k_1|B|^2$, and in the transmitted wave $k_2|A|^2$. We define the *transmission coefficient* D of the particle as the ratio of the probability current density in the transmitted wave to that in the incident wave:

$$D = (k_2/k_1)|A|^2. \tag{25.3}$$

Similarly we can define the *reflection coefficient* R as the ratio of the density in the reflected wave to that in the incident wave. Evidently $R = 1-D$:

$$R = |B|^2 = 1-(k_2/k_1)|A|^2 \tag{25.4}$$

(this relation between A and B is automatically satisfied).

If the particle moves from left to right with energy $E < U_0$, then k_2 is purely imaginary, and the wave function decreases exponentially as $x \to +\infty$. The reflected current is equal to the incident one, i.e. we have "total reflection" of the particle from the potential wall. We emphasize, however, that in this case the probability of finding the particle in the region where $E < U$ is still different from zero, though it diminishes rapidly as x increases.

In the general case of an arbitrary stationary state (with energy $E > U_0$), the asymptotic form of the wave function is given, both for $x \to -\infty$ and for $x \to +\infty$, by a sum of waves propagated in each direction:

$$\left.\begin{aligned} \psi &= A_1e^{ik_1x} + B_1e^{-ik_1x} \quad \text{for} \quad x \to -\infty, \\ \psi &= A_2e^{ik_2x} + B_2e^{-ik_2x} \quad \text{for} \quad x \to +\infty. \end{aligned}\right\} \tag{25.5}$$

Since these expressions are asymptotic forms of the same solution of a linear differential equation, there must be a linear relation between the coefficients A_1, B_1 and A_2, B_2. Let $A_2 = \alpha A_1 + \beta B_1$, where α, β are constants (in general complex) which depend on the specific form of the field $U(x)$. The corresponding relation for B_2 can then be written down from the fact that Schrödinger's equation is real. This shows that, if ψ is a solution of a given Schrödinger's equation, the complex conjugate function ψ^* is also a solution. The asymptotic forms

$$\begin{aligned} \psi^* &= A_1^*e^{-ik_1x} + B_1^*e^{ik_1x} \quad \text{for} \quad x \to -\infty, \\ \psi^* &= A_2^*e^{-ik_2x} + B_2^*e^{ik_2x} \quad \text{for} \quad x \to +\infty \end{aligned}$$

differ from (25.5) only in the nomenclature of the constant coefficients; we therefore have $B_2^* = \alpha B_1^* + \beta A_1^*$ or $B_2 = \alpha^* B_1 + \beta^* A_1$. Thus the coefficients in (25.5) are related by equations of the form

$$A_2 = \alpha A_1+\beta B_1, \qquad B_2 = \beta^* A_1+\alpha^* B_1. \tag{25.6}$$

The condition of constant current along the x-axis leads to the relation

$$k_1(|A_1|^2-|B_1|^2) = k^2(|A_2|^2-|B_2|^2).$$

Expressing A_2, B_2 in terms of A_1, B_1 by (25.6), we find

$$|\alpha|^2-|\beta|^2 = k_1/k_2. \tag{25.7}$$

Using the relation (25.6), we can show, in particular, that the reflection coefficients are equal (for a given energy $E > U_0$) for particles moving in the positive and negative directions of the x-axis; the former case corresponds to putting $B_2 = 0$ in (25.5), and the latter case to $A_1 = 0$. In these two cases, $B_1/A_1 = -\beta^*/\alpha^*$ and $A_2/B_2 = \beta/\alpha^*$ respectively. The corresponding reflection coefficients are

$$R_1 = |B_1/A_1|^2 = |\beta^*/\alpha^*|^2,$$
$$R_2 = |A_2/B_2|^2 = |\beta/\alpha^*|^2,$$

whence it is clear that $R_1 = R_2$.

PROBLEMS

PROBLEM 1. Determine the reflection coefficient of a particle from a rectangular potential wall (Fig. 6); the energy of the particle $E > U_0$.

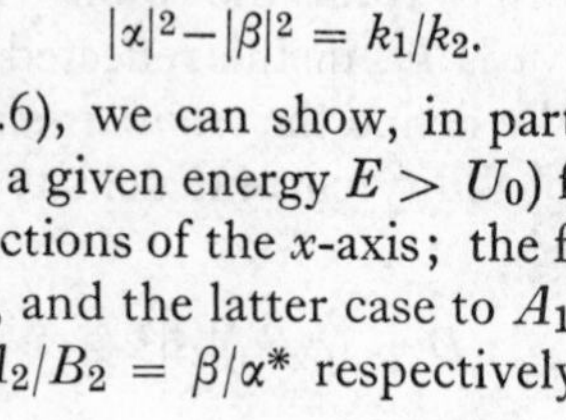

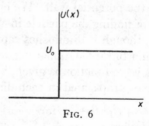

FIG. 6

SOLUTION. Throughout the region $x > 0$, the wave function has the form (25.1), while in the region $x < 0$ its form is (25.2). The constants A and B are determined from the condition that ψ and $d\psi/dx$ are continuous at $x = 0$:

$$1+B = A, \quad k_1(1-B) = k_2 A,$$

whence

$$A = 2k_1/(k_1+k_2), \quad B = (k_1-k_2)/(k_1+k_2).$$

The reflection coefficient† is (25.4)

$$R = \left(\frac{k_1-k_2}{k_1+k_2}\right)^2 = \left(\frac{p_1-p_2}{p_1+p_2}\right)^2.$$

For $E = U_0$ ($k_2 = 0$), R becomes unity, while for $E \to \infty$ it tends to zero as $(U_0/4E)^2$.

† In the limiting case of classical mechanics, the reflection coefficient must become zero. The expression obtained here, however, does not contain the quantum constant at all. This apparent contradiction is explained as follows. The classical limiting case is that in which the de Broglie wavelength of the particle $\lambda \sim \hbar/p$ is small in comparison with the characteristic dimensions of the problem, i.e. the distances over which the field $U(x)$ changes noticeably. In the schematic example considered, however, this distance is zero (at the point $x = 0$), so that the passage to the limit cannot be effected.

PROBLEM 2. Determine the transmission coefficient for a rectangular potential barrier (Fig. 7).

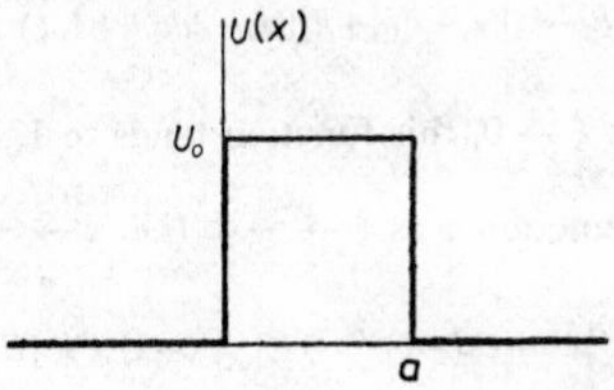

FIG. 7

SOLUTION. Let E be greater than U_0, and suppose that the incident particle is moving from left to right. Then we have for the wave function in the different regions expressions of the form

$$\text{for } x < 0, \quad \psi = e^{ik_1x}+Ae^{-ik_1x},$$
$$\text{for } 0 < x < a, \quad \psi = Be^{ik_2x}+B'e^{-ik_2x},$$
$$\text{for } x > a, \quad \psi = Ce^{ik_1x}$$

(on the side $x > a$ there can be only the transmitted wave, propagated in the positive direction of x). The constants A, B, B' and C are determined from the conditions of continuity of ψ and $d\psi/dx$ at the points $x = 0$ and a. The transmission coefficient is determined as $D = k_1|C|^2/k_1 = |C|^2$. On calculating this, we obtain

$$D = \frac{4k_1^2k_2^2}{(k_1^2-k_2^2)^2 \sin^2 ak_2+4k_1^2k_2^2}.$$

For $E < U_0$, k_2 is a purely imaginary quantity; the corresponding expression for D is obtained by replacing k_2 by $i\kappa_2$, where $\hbar\kappa_2 = \sqrt{[2m(U_0-E)]}$:

$$D = \frac{4k_1^2\kappa_2^2}{(k_1^2+\kappa_2^2)^2 \sinh^2 a\kappa_2+4k_1^2\kappa_2^2}.$$

PROBLEM 3. Determine the reflection coefficient for a potential wall defined by the formula $U(x) = U_0/(1+e^{-\alpha x})$ (Fig. 5); the energy of the particle is $E > U_0$.

SOLUTION. Schrödinger's equation is

$$\frac{d^2\psi}{dx^2}+\frac{2m}{\hbar^2}\left(E-\frac{U_0}{1+e^{-\alpha x}}\right)\psi = 0.$$

We have to find a solution which, as $x \to +\infty$, has the form

$$\psi = \text{constant} \times e^{ik_2x}.$$

We introduce a new variable

$$\xi = -e^{-\alpha x}$$

(which takes values from $-\infty$ to 0), and seek a solution of the form

$$\psi = \xi^{-ik_2/\alpha}w(\xi),$$

where $w(\xi)$ tends to a constant as $\xi \to 0$ (i.e. as $x \to \infty$). For $w(\xi)$ we find an equation of hypergeometric type:

$$\xi(1-\xi)w''+(1-2ik_2/\alpha)(1-\xi)w'+(k_2^2-k_1^2)w/\alpha^2 = 0,$$

which has as its solution the hypergeometric function

$$w = F(i[k_1-k_2]/\alpha, -i[k_1+k_2]/\alpha, -2ik_2/\alpha+1, \xi)$$

(we omit a constant factor). As $\xi \to 0$, this function tends to 1, i.e. it satisfies the condition imposed.

The asymptotic form of the function ψ as $\xi \to -\infty$ (i.e. $x \to -\infty$) is†

$$\psi \approx \xi^{-ik_2/\alpha}[C_1(-\xi)^{i(k_2-k_1)/\alpha}+C_2(-\xi)^{i(k_1+k_2)/\alpha}] = (-1)^{-ik_2/\alpha}[C_1e^{ik_1x}+C_2e^{-ik_1x}],$$

where

$$C_1 = \frac{\Gamma(-2ik_1/\alpha)\Gamma(-2ik_2/\alpha+1)}{\Gamma(-i(k_1+k_2)/\alpha)\Gamma(-i(k_1+k_2)/\alpha+1)},$$

$$C_2 = \frac{\Gamma(2ik_1/\alpha)\Gamma(-2ik_2/\alpha+1)}{\Gamma(i(k_1-k_2)/\alpha)\Gamma(i(k_1-k_2)/\alpha+1)}.$$

The required reflection coefficient is $R = |C_2/C_1|^2$; on calculating it by means of the well known formula

$$\Gamma(x)\Gamma(1-x) = \pi/\sin \pi x,$$

we have

$$R = \left(\frac{\sinh[\pi(k_1-k_2)/\alpha]}{\sinh[\pi(k_1+k_2)/\alpha]}\right)^2.$$

For $E = U_0$ ($k_2 = 0$), R becomes unity, while for $E \to \infty$ it tends to zero as

$$\left(\frac{\pi U_0}{\alpha\hbar}\right)^2 \frac{2m}{E} e^{-4\pi\sqrt{(2mE)}/\alpha\hbar}.$$

In the limiting case of classical mechanics, R becomes zero, as it should.

PROBLEM 4. Determine the transmission coefficient for a potential barrier defined by the formula

$$U(x) = U_0/\cosh^2\alpha x$$

(Fig. 8); the energy of the particle is $E < U_0$.

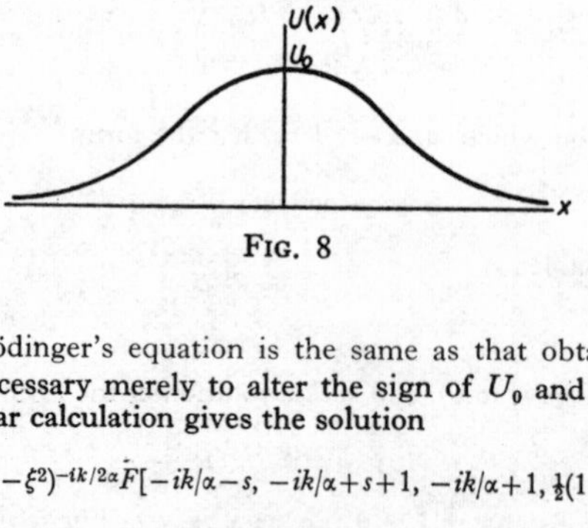

FIG. 8

SOLUTION. The Schrödinger's equation is the same as that obtained in the solution of Problem 5, §23; it is necessary merely to alter the sign of U_0 and to regard the energy E now as positive. A similar calculation gives the solution

$$\psi = (1-\xi^2)^{-ik/2\alpha}F[-ik/\alpha-s, -ik/\alpha+s+1, -ik/\alpha+1, \tfrac{1}{2}(1-\xi)],$$

† See formula (e.6), in each of whose two terms we must take only the first term of the expansion, i.e. replace the hypergeometric functions of $1/z$ by unity.

where

$$\xi = \tanh \alpha x, \qquad k = \sqrt{(2mE)}/\hbar,$$

$$s = \tfrac{1}{2}\left(-1+\sqrt{\left[1-\frac{8mU_0}{\alpha^2\hbar^2}\right]}\right).$$

This solution satisfies the condition that, as $x \to \infty$ (i.e. as $\xi \to 1$, $(1-\xi) \approx 2e^{-x}$), the wave function should include only the transmitted wave ($\sim e^{ikx}$). The asymptotic form of the wave function as $x \to -\infty$ ($\xi \to -1$) is found by transforming the hypergeometric function with the aid of formula (e.7):

$$\psi \sim e^{-ikx}\frac{\Gamma(ik/\alpha)\Gamma(1-ik/\alpha)}{\Gamma(-s)\Gamma(1+s)} + e^{ikx}\frac{\Gamma(-ik/\alpha)\Gamma(1-ik/\alpha)}{\Gamma(-ik/\alpha-s)\Gamma(-ik/\alpha+s+1)}. \tag{2}$$

Taking the squared modulus of the ratio of coefficients in this function, we obtain the following expression for the transmission coefficient $D = 1 - R$:

$$D = \frac{\sinh^2(\pi k/\alpha)}{\sinh^2(\pi k/\alpha)+\cos^2[\frac{1}{2}\pi\sqrt{(1-8mU_0/\hbar^2\alpha^2)}]}$$

(if $8mU_0/\hbar^2\alpha^2 < 1$), or

$$D = \frac{\sinh^2(\pi k/\alpha)}{\sinh^2(\pi k/\alpha)+\cosh^2[\frac{1}{2}\pi\sqrt{(8mU_0/\hbar^2\alpha^2-1)}]}$$

(if $8mU_0/\hbar^2\alpha^2 > 1$). The first of these formulae holds also for the case $U_0 < 0$, i.e. when the particle is passing over a potential well instead of a potential barrier. It is interesting to note that in that case $D = 1$ if $1+8m|U_0|/\hbar^2\alpha^2 = (2n+1)^2$; thus, for certain values of the depth $|U_0|$ of the well, particles passing over it are not reflected. This is evident from equation (2), where the term in e^{-ikx} vanishes for positive integral s.

CHAPTER IV

ANGULAR MOMENTUM

§26. Angular momentum

IN §15, to derive the law of conservation of momentum, we have made use of the homogeneity of space relative to a closed system of particles. Besides its homogeneity, space has also the property of isotropy: all directions in it are equivalent. Hence the Hamiltonian of a closed system cannot change when the system rotates as a whole through an arbitrary angle about an arbitrary axis. It is sufficient to require the fulfilment of this condition for an infinitely small rotation.

Let $\delta\boldsymbol{\varphi}$ be the vector of an infinitely small rotation, equal in magnitude to the angle $\delta\phi$ of the rotation and directed along the axis about which the rotation takes place. The changes $\delta\mathbf{r}_a$ (in the radius vectors $\mathbf{r}_a$ of the particles) in such a rotation are

$$\delta\mathbf{r}_a = \delta\boldsymbol{\varphi}\times\mathbf{r}_a.$$

An arbitrary function $\psi(\mathbf{r}_1, \mathbf{r}_2, \ldots)$ is thereby transformed into the function

$$\begin{aligned}\psi(\mathbf{r}_1+\delta\mathbf{r}_1, \mathbf{r}_2+\delta\mathbf{r}_2, \ldots) &= \psi(\mathbf{r}_1, \mathbf{r}_2, \ldots)+\sum_a \delta\mathbf{r}_a\cdot\nabla_a\psi \\ &= \psi(\mathbf{r}_1, \mathbf{r}_2, \ldots)+\sum_a \delta\boldsymbol{\varphi}\times\mathbf{r}_a\cdot\nabla_a\psi \\ &= (1+\delta\boldsymbol{\varphi}\cdot\sum_a\mathbf{r}_a\times\nabla_a)\psi(\mathbf{r}_1, \mathbf{r}_2, \ldots).\end{aligned}$$

The expression

$$1+\delta\boldsymbol{\varphi}\cdot\sum_a \mathbf{r}_a\times\nabla_a$$

is the operator of an infinitely small rotation. The fact that an infinitely small rotation does not alter the Hamiltonian of the system is expressed (cf. §15) by the commutability of the "rotation operator" with the operator $\hat{H}$. Since $\delta\boldsymbol{\varphi}$ is a constant vector, this condition reduces to the relation

$$(\sum_a \mathbf{r}_a\times\nabla_a)\hat{H}-\hat{H}(\sum_a \mathbf{r}_a\times\nabla_a) = 0, \tag{26.1}$$

which expresses a certain law of conservation.

The quantity whose conservation for a closed system follows from the property of isotropy of space is the *angular momentum* of the system (cf. *Mechanics*, §9). Thus the operator $\Sigma\, \mathbf{r}_a\times\nabla_a$ must correspond exactly, apart from a constant factor, to the total angular momentum of the system, and each of the terms $\mathbf{r}_a\times\nabla_a$ of this sum corresponds to the angular momentum of an individual particle.

The coefficient of proportionality must be put equal to $-i\hbar$; then the expression for the angular momentum operator of a particle is $-i\hbar\mathbf{r}\times\nabla = \mathbf{r}\times\hat{\mathbf{p}}$ and corresponds exactly to the classical expression $\mathbf{r}\times\mathbf{p}$. Henceforward we shall always use the angular momentum measured in units of $\hbar$. The angular momentum operator of a particle, so defined, will be denoted by $\hat{\mathbf{l}}$, and that of the whole system by $\hat{\mathbf{L}}$. Thus the angular momentum operator of a particle is

$$\hbar\hat{\mathbf{l}} = \mathbf{r}\times\hat{\mathbf{p}} = -i\hbar\mathbf{r}\times\nabla, \tag{26.2}$$

or, in components,

$$\hbar\hat{l}_x = y\hat{p}_z - z\hat{p}_y, \quad \hbar\hat{l}_y = z\hat{p}_x - x\hat{p}_z, \quad \hbar\hat{l}_z = x\hat{p}_y - y\hat{p}_x.$$

For a system which is in an external field, the angular momentum is in general not conserved. However, it may still be conserved if the field has a certain symmetry. Thus, if the system is in a centrally symmetric field, all directions in space at the centre are equivalent, and hence the angular momentum about this centre will be conserved. Similarly, in an axially symmetric field, the component of angular momentum along the axis of symmetry is conserved. All these conservation laws holding in classical mechanics are valid in quantum mechanics also.

In a system where angular momentum is not conserved, it does not have definite values in the stationary states. In such cases the mean value of the angular momentum in a given stationary state is sometimes of interest. It is easily seen that, in any non-degenerate stationary state, the mean value of the angular momentum is zero. For, when the sign of the time is changed, the energy does not alter, and, since only one stationary state corresponds to a given energy level, it follows that when t is changed into $-t$ the state of the system must remain the same. This means that the mean values of all quantities, and in particular that of the angular momentum, must remain unchanged. But when the sign of the time is changed, so is that of the angular momentum, and we have $\bar{\mathbf{L}} = -\bar{\mathbf{L}}$, whence it follows that $\bar{\mathbf{L}} = 0$. The same result can be obtained by starting from the mathematical definition of the mean value $\bar{\mathbf{L}}$ as being the integral of $\psi^*\hat{\mathbf{L}}\psi$. The wave functions of non-degenerate states are real (see the end of §18). Hence the expression

$$\bar{\mathbf{L}} = -i\hbar\int\psi(\sum_a \mathbf{r}_a\times\nabla_a)\psi\,dq$$

is purely imaginary, and since $\bar{\mathbf{L}}$ must, of course, be real, it is evident that $\bar{\mathbf{L}} = 0$.

Let us derive the rules for commutation of the angular momentum operators with those of coordinates and linear momenta. By means of the relations (16.2) we easily find

$$\left.\begin{aligned}\{\hat{l}_x, x\} &= 0, \quad \{\hat{l}_x, y\} = iz, \quad \{\hat{l}_x, z\} = -iy,\\ \{\hat{l}_y, y\} &= 0, \quad \{\hat{l}_y, z\} = ix, \quad \{\hat{l}_y, x\} = -iz,\\ \{\hat{l}_z, z\} &= 0, \quad \{\hat{l}_z, x\} = iy, \quad \{\hat{l}_z, y\} = -ix.\end{aligned}\right\} \tag{26.3}$$

For instance,

$$\hat{l}_x y - y\hat{l}_x = (1/\hbar)(y\hat{p}_z - z\hat{p}_y)y - y(y\hat{p}_z - z\hat{p}_y)(1/\hbar)$$
$$= -(z/\hbar)\{\hat{p}_y, y\} = iz.$$

All the relations (26.3) can be written in tensor form as follows:

$$\{\hat{l}_i, x_k\} = ie_{ikl}x_l, \tag{26.4}$$

where e_{ikl} is the antisymmetric unit tensor of rank three,† and summation is implied over those suffixes which appear twice (called *dummy suffixes*).

It is easily seen that a similar commutation rule holds for the angular momentum and linear momentum operators:

$$\{\hat{l}_i, \hat{p}_k\} = ie_{ikl}\hat{p}_l. \tag{26.5}$$

By means of these formulae, it is easy to find the rules for commutation of the operators $\hat{l}_x$, $\hat{l}_y$, $\hat{l}_z$ with one another. We have

$$\hbar(\hat{l}_x\hat{l}_y - \hat{l}_y\hat{l}_x) = \hat{l}_x(z\hat{p}_x - x\hat{p}_z) - (z\hat{p}_x - x\hat{p}_z)\hat{l}_x$$
$$= (\hat{l}_x z - z\hat{l}_x)\hat{p}_x - x(\hat{l}_x\hat{p}_z - \hat{p}_z\hat{l}_x)$$
$$= -iy\hat{p}_x + ix\hat{p}_y = i\hbar\hat{l}_z.$$

Thus

$$\{\hat{l}_y, \hat{l}_z\} = i\hat{l}_x, \quad \{\hat{l}_z, \hat{l}_x\} = i\hat{l}_y, \quad \{\hat{l}_x, \hat{l}_y\} = i\hat{l}_z, \tag{26.6}$$

or

$$\{\hat{l}_i, \hat{l}_k\} = ie_{ikl}\hat{l}_l. \tag{26.7}$$

Exactly the same relations hold for the operators $\hat{L}_x$, $\hat{L}_y$, $\hat{L}_z$ of the total angular momentum of the system. For, since the angular momentum operators of different individual particles commute, we have, for instance,

$$\sum_a \hat{l}_{ay} \sum_a \hat{l}_{az} - \sum_a \hat{l}_{az} \sum_a \hat{l}_{ay} = \sum_a (\hat{l}_{ay}\hat{l}_{az} - \hat{l}_{az}\hat{l}_{ay}) = i\sum_a \hat{l}_{ax}.$$

Thus

$$\{\hat{L}_y, \hat{L}_z\} = i\hat{L}_x, \quad \{\hat{L}_z, \hat{L}_x\} = i\hat{L}_y, \quad \{\hat{L}_x, \hat{L}_y\} = i\hat{L}_z. \tag{26.8}$$

The relations (26.8) show that the three components of the angular momentum cannot simultaneously have definite values (except in the case where all three components simultaneously vanish: see below). In this respect the angular momentum is fundamentally different from the linear momentum, whose three components are simultaneously measurable.

---

† The *antisymmetric unit tensor* of rank three, e_{ikl} (also called the *unit axial tensor*), is defined as a tensor antisymmetric in all three suffixes, with $e_{123} = 1$. It is evident that, of its 27 components, only 6 are not zero, namely those in which the suffixes i, k, l form some permutation of 1, 2, 3. Such a component is $+1$ if the permutation i, k, l is obtained from 1, 2, 3 by an even number of transpositions of pairs of figures, and is -1 if the number of transpositions is odd. Clearly $e_{ikl}e_{ikm} = 2\delta_{lm}$, $e_{ikl}e_{ikl} = 6$. The components of the vector $\mathbf{C} = \mathbf{A}\times\mathbf{B}$ which is the vector product of the two vectors $\mathbf{A}$ and $\mathbf{B}$ can be written by means of the tensor e_{ikl} in the form

$$C_i = e_{ikl}A_kB_l.$$

From the operators $\hat{L}_x$, $\hat{L}_y$, $\hat{L}_z$ we can form the operator of the square of the modulus of the angular momentum vector, and which we denote by $\hat{\mathbf{L}}^2$:

$$\hat{\mathbf{L}}^2 = \hat{L}_x^2+\hat{L}_y^2+\hat{L}_z^2. \tag{26.9}$$

This operator commutes with each of the operators $\hat{L}_x$, $\hat{L}_y$, $\hat{L}_z$:

$$\{\hat{\mathbf{L}}^2, \hat{L}_x\} = 0, \quad \{\hat{\mathbf{L}}^2, \hat{L}_y\} = 0, \quad \{\hat{\mathbf{L}}^2, \hat{L}_z\} = 0. \tag{26.10}$$

Using (26.8), we have

$$\begin{aligned}
\{\hat{L}_x^2, \hat{L}_z\} &= \hat{L}_x\{\hat{L}_x, \hat{L}_z\}+\{\hat{L}_x, \hat{L}_z\}\hat{L}_x \\
&= -i(\hat{L}_x\hat{L}_y+\hat{L}_y\hat{L}_x), \\
\{\hat{L}_y^2, \hat{L}_z\} &= i(\hat{L}_x\hat{L}_y+\hat{L}_y\hat{L}_x), \\
\{\hat{L}_z^2, \hat{L}_z\} &= 0.
\end{aligned}$$

Adding these equations, we have $\{\hat{\mathbf{L}}^2, \hat{L}_z\} = 0$. Physically, the relations (26.10) mean that the square of the angular momentum, i.e. its modulus, can have a definite value at the same time as one of its components.

Instead of the operators $\hat{L}_x$, $\hat{L}_y$ it is often more convenient to use the complex combinations

$$\hat{L}_+ = \hat{L}_x+i\hat{L}_y, \qquad \hat{L}_- = \hat{L}_x-i\hat{L}_y. \tag{26.11}$$

It is easily verified by direct calculation using (26.8) that the following commutation rules hold:

$$\left.\begin{aligned}
\{\hat{L}_+, \hat{L}_-\} = 2\hat{L}_z, \qquad \{\hat{L}_z, \hat{L}_+\} = \hat{L}_+, \\
\{\hat{L}_z, \hat{L}_-\} = -\hat{L}_-,
\end{aligned}\right\} \tag{26.12}$$

and it is also not difficult to see that

$$\begin{aligned}
\mathbf{L}^2 &= \hat{L}_+\hat{L}_-+\hat{L}_z^2-\hat{L}_z \\
&= \hat{L}_-\hat{L}_++\hat{L}_z^2+\hat{L}_z.
\end{aligned} \tag{26.13}$$

Finally, we shall give some frequently used expressions for the angular momentum operator of a single particle in spherical polar coordinates. Defining the latter by means of the usual relations

$$x = r\sin\theta\cos\phi, \qquad y = r\sin\theta\sin\phi, \qquad z = r\cos\theta,$$

we have after a simple calculation

$$\hat{l}_z = -i\frac{\partial}{\partial\phi}, \tag{26.14}$$

$$\hat{l}_\pm = e^{\pm i\phi}\left(\pm\frac{\partial}{\partial\theta}+i\cot\theta\frac{\partial}{\partial\phi}\right). \tag{26.15}$$

Substitution in (26.13) gives the squared angular momentum operator of the particle:

$$\hat{\mathbf{l}}^2 = -\left[\frac{1}{\sin^2\theta}\frac{\partial^2}{\partial\phi^2}+\frac{1}{\sin\theta}\frac{\partial}{\partial\theta}\left(\sin\theta\frac{\partial}{\partial\theta}\right)\right]. \tag{26.16}$$

It should be noticed that this is, apart from a factor, the angular part of the Laplacian operator.

§27. Eigenvalues of the angular momentum

In order to determine the eigenvalues of the component, in some direction, of the angular momentum of a particle, it is convenient to use the expression for its operator in spherical polar coordinates, taking the direction in question as the polar axis. According to formula (26.14), the equation $\hat{l}_z\psi = l_z\psi$ can be written in the form

$$-i\,\partial\psi/\partial\phi = l_z\psi. \tag{27.1}$$

Its solution is

$$\psi = f(r,\theta)e^{il_z\phi},$$

where $f(r,\theta)$ is an arbitrary function of r and θ. If the function ψ is to be single-valued, it must be periodic in ϕ, with period 2π. Hence we find†

$$l_z = m, \quad \text{where } m = 0, \pm 1, \pm 2, \ldots. \tag{27.2}$$

Thus the eigenvalues l_z are the positive and negative integers, including zero. The factor depending on ϕ, which characterizes the eigenfunctions of the operator $\hat{l}_z$, is denoted by

$$\Phi_m(\phi) = (2\pi)^{-1/2}e^{im\phi}. \tag{27.3}$$

These functions are normalized so that

$$\int_0^{2\pi}\Phi_m^*(\phi)\Phi_{m'}(\phi)\,d\phi = \delta_{mm'}. \tag{27.4}$$

The eigenvalues of the z-component of the total angular momentum of the system are evidently also equal to the positive and negative integers:

$$L_z = M, \quad \text{where } M = 0, \pm 1, \pm 2, \ldots \tag{27.5}$$

(this follows at once from the fact that the operator $\hat{L}_z$ is equal to the sum of the commuting operators $\hat{l}_z$ for the individual particles).

Since the direction of the z-axis is in no way distinctive, it is clear that the same result is obtained for $\hat{L}_x$, $\hat{L}_y$ and in general for the component of the angular momentum in any direction: they can all take integral values only. At first sight this result may appear paradoxical, particularly if we apply it to two directions infinitely close to each other. In fact, however, it must

† The customary notation for the eigenvalues of the angular momentum component is m, which also denotes the mass of a particle, but this should not lead to any confusion.

be remembered that the only common eigenfunction of the operators $\hat{L}_x$, $\hat{L}_y$, $\hat{L}_z$ corresponds to the simultaneous values

$$L_x = L_y = L_z = 0;$$

in this case the angular momentum vector is zero, and consequently so is its projection upon any direction. If even one of the eigenvalues L_x, L_y, L_z is not zero, the operators $\hat{L}_x$, $\hat{L}_y$, $\hat{L}_z$ have no common eigenfunctions. In other words, there is no state in which two or three of the angular momentum components in different directions simultaneously have definite values different from zero, so that we can say only that one of them is integral.

The stationary states of a system which differ only in the value of M have the same energy; this follows from general considerations, based on the fact that the direction of the z-axis is in no way distinctive. Thus the energy levels of a system whose angular momentum is conserved (and is not zero) are always degenerate.†

Let us now look for the eigenvalues of the square of the angular momentum. We shall show how these values may be found, starting from the commutation rules (26.8) only. We denote by ψ_M the wave functions of the stationary states with the same value of $\mathbf{L}^2$, belonging to one degenerate energy level, and distinguished by the value of M.‡

First of all we note that, since the two directions of the z-axis are physically equivalent, for every possible positive value $M = |M|$ there is a corresponding negative value $M = -|M|$. Let L (a positive integer or zero) denote the greatest possible value of $|M|$ for a given $\mathbf{L}^2$. The existence of this upper limit follows because $\hat{\mathbf{L}}^2 - \hat{L}_z^2 = \hat{L}_x^2 + \hat{L}_y^2$ is the operator of the essentially positive physical quantity $L_x^2 + L_y^2$, and its eigenvalues therefore cannot be negative.

Applying the operator $\hat{L}_z \hat{L}_\pm$ to the eigenfunction ψ_M of the operator $\hat{L}_z$ and using the commutation rule $\{\hat{L}_z, \hat{L}_\pm\} = \pm \hat{L}_\pm$ (26.12), we obtain

$$\hat{L}_z \hat{L}_\pm \psi_M = (M \pm 1) \hat{L}_\pm \psi_M. \tag{27.6}$$

Hence we see that the function $\hat{L}_\pm \psi_M$ is (apart from a normalization constant) the eigenfunction corresponding to the value $M \pm 1$ of the quantity L_z:

$$\psi_{M+1} = \text{constant} \times \hat{L}_+ \psi_M, \qquad \psi_{M-1} = \text{constant} \times \hat{L}_- \psi_M. \tag{27.7}$$

---

† This is a particular case of the general theorem, mentioned in §10, which states that the levels are degenerate when two or more conserved quantities exist whose operators do not commute. Here the components of the angular momentum are such quantities.

‡ Here it is supposed that there is no additional degeneracy leading to the same value of the energy for different values of the squared angular momentum. This is true for a discrete spectrum (except for the case of what is called *accidental degeneracy* in a Coulomb field; see §36) and in general untrue for the energy levels of a continuous spectrum. However, even when such additional degeneracy is present, we can always choose the eigenfunctions so that they correspond to states with definite values of $\mathbf{L}^2$, and then we can choose from these the states with the same values of E and $\mathbf{L}^2$. This is mathematically expressed by the fact that the matrices of commuting operators can always be simultaneously brought into diagonal form. In what follows we shall, in such cases, speak, for the sake of brevity, as if there were no additional degeneracy, bearing in mind that the results obtained do not in fact depend on this assumption, by what we have just said.

If we put $M = L$ in the first of these equations, we must have identically

$$\hat{L}_+\psi_L = 0, \tag{27.8}$$

since there is by definition no state with $M > L$. Applying the operator $\hat{L}_-$ to this equation and using the relation (26.13), we obtain

$$\hat{L}_-\hat{L}_+\psi_L = (\hat{\mathbf{L}}^2 - \hat{L}_z{}^2 - \hat{L}_z)\psi_L = 0.$$

Since, however, the ψ_M are common eigenfunctions of the operators $\hat{\mathbf{L}}^2$ and $\hat{L}_z$, we have

$$\hat{\mathbf{L}}^2\psi_L = \mathbf{L}^2\psi_L, \quad \hat{L}_z{}^2\psi_L = L^2\psi_L, \quad \hat{L}_z\psi_L = L\psi_L,$$

so that the equation found above gives

$$\mathbf{L}^2 = L(L+1). \tag{27.9}$$

Formula (27.9) determines the required eigenvalues of the square of the angular momentum; the number L takes all positive integral values, including zero. For a given value of L, the component $L_z = M$ of the angular momentum can take the values

$$M = L, L-1, \ldots, -L, \tag{27.10}$$

i.e. $2L+1$ different values in all. The energy level corresponding to the angular momentum L thus has $(2L+1)$-fold degeneracy; this is usually called degeneracy with respect to the direction of the angular momentum. A state with angular momentum $L = 0$ (when all three components are also zero) is not degenerate. The wave function of such a state is spherically symmetric, as is evident from the fact that the change in it under any infinitesimal rotation, given by $\hat{\mathbf{L}}\psi$, is in this case zero.

We shall often, for the sake of brevity, and in accordance with custom, speak of the "angular momentum" L of a system, understanding by this an angular momentum whose square is $L(L+1)$; the z-component of the angular momentum is usually called just the "angular momentum component".

The angular momentum of a single particle is denoted by the small letter l, for which formula (27.9) becomes

$$\mathbf{l}^2 = l(l+1). \tag{27.11}$$

Let us calculate the matrix elements of the quantities L_x and L_y in a representation in which L_z and $\mathbf{L}^2$, as well as the energy, are diagonal (M. Born, W. Heisenberg and P. Jordan 1926). First of all, we note that, since the operators $\hat{L}_x$ and $\hat{L}_y$ commute with the Hamiltonian, their matrices are diagonal with respect to the energy, i.e. all matrix elements for transitions between states of different energy (and different angular momentum L) are zero. Thus it is sufficient to consider the matrix elements for transitions within a group of states with different values of M, corresponding to a single degenerate energy level.

It is seen from formulae (27.7) that, in the matrices of the operators $\hat{L}_+$ and $\hat{L}_-$, only those elements are different from zero which correspond to transitions $M-1 \to M$ and $M \to M-1$ respectively. Taking this into account, we find the diagonal matrix elements on both sides of the equation (26.13), obtaining†

$$L(L+1) = \langle M|L_+|M-1\rangle\langle M-1|L_-|M\rangle + M^2 - M.$$

Noticing that, since the operators $\hat{L}_x$ and $\hat{L}_y$ are Hermitian,

$$\langle M-1|L_-|M\rangle = \langle M|L_+|M-1\rangle^*,$$

we can rewrite this equation in the form

$$\begin{aligned} |\langle M|L_+|M-1\rangle|^2 &= L(L+1) - M(M-1) \\ &= (L-M+1)(L+M), \end{aligned}$$

whence‡

$$\begin{aligned} \langle M|L_+|M-1\rangle &= \langle M-1|L_-|M\rangle \\ &= \sqrt{[(L+M)(L-M+1)]}. \end{aligned} \tag{27.12}$$

Hence we have for the non-zero matrix elements of the quantities L_x and L_y themselves

$$\left.\begin{aligned} \langle M|L_x|M-1\rangle &= \langle M-1|L_x|M\rangle = \tfrac{1}{2}\sqrt{[(L+M)(L-M+1)]}, \\ \langle M|L_y|M-1\rangle &= -\langle M-1|L_y|M\rangle = -\tfrac{1}{2}i\sqrt{[(L+M)(L-M+1)]}. \end{aligned}\right\} \tag{27.13}$$

The diagonal elements in the matrices of the quantities L_x and L_y are zero. Since a diagonal matrix element gives the mean value of the quantity in the corresponding state, it follows that the mean values $\bar{L}_x$ and $\bar{L}_y$ are zero in states having definite values of L_z. Thus, if the angular-momentum component in a given direction in space has a definite value, the vector $\bar{\mathbf{L}}$ itself is in that direction.

§28. Eigenfunctions of the angular momentum

The wave function of a particle is not completely determined when the values of l and m are prescribed. This is seen from the fact that the expressions for the operators of these quantities in spherical polar coordinates contain only the angles θ and ϕ, so that their eigenfunctions can contain an arbitrary factor depending on r. We shall here consider only the angular part of the wave function which characterizes the eigenfunctions of the angular momentum, and denote this by $Y_{lm}(\theta, \phi)$, with the normalization condition

$$\int |Y_{lm}|^2 \, do = 1,$$

where $do = \sin\theta \, d\theta d\phi$ is an element of solid angle.

---

† In the symbols for the matrix elements, we omit for brevity all suffixes with respect to which they are diagonal (including L).

‡ The choice of sign in this formula corresponds to the choice of the phase factors in the eigenfunctions of the angular momentum.

We shall see that the problem of determining the common eigenfunctions of the operators $\hat{\mathbf{l}}^2$ and $\hat{l}_z$ admits of separation of the variables θ and ϕ, and these functions can be sought in the form

$$Y_{lm} = \Phi_m(\phi)\Theta_{lm}(\theta), \tag{28.1}$$

where $\Phi_m(\phi)$ are the eigenfunctions of the operator $\hat{l}_z$, which are given by formula (27.3). Since the functions Φ_m are already normalized by the condition (27.4), the Θ_{lm} must be normalized by the condition

$$\int_0^\pi |\Theta_{lm}|^2 \sin\theta \, d\theta = 1. \tag{28.2}$$

The functions Y_{lm} with different l or m are automatically orthogonal:

$$\int_0^{2\pi}\int_0^\pi Y_{l'm'}{}^* Y_{lm} \sin\theta \, d\theta d\phi = \delta_{ll'}\delta_{mm'}, \tag{28.3}$$

as being the eigenfunctions of angular momentum operators corresponding to different eigenvalues. The functions $\Phi_m(\phi)$ separately are themselves orthogonal (see (27.4)), as being the eigenfunctions of the operator $\hat{l}_z$ corresponding to different eigenvalues m of this operator. The functions $\Theta_{lm}(\theta)$ are not themselves eigenfunctions of any of the angular momentum operators; they are mutually orthogonal for different l, but not for different m.

The most direct method of calculating the required functions is by directly solving the problem of finding the eigenfunctions of the operator $\hat{\mathbf{l}}^2$ written in spherical polar coordinates (formula (26.16)). The equation $\hat{\mathbf{l}}^2\psi = \mathbf{l}^2\psi$ is

$$\frac{1}{\sin\theta}\frac{\partial}{\partial\theta}\left(\sin\theta\frac{\partial\psi}{\partial\theta}\right) + \frac{1}{\sin^2\theta}\frac{\partial^2\psi}{\partial\phi^2} + l(l+1)\psi = 0.$$

Substituting in this equation the form (28.1) for ψ, we obtain for the function Θ_{lm} the equation

$$\frac{1}{\sin\theta}\frac{d}{d\theta}\left(\sin\theta\frac{d\Theta_{lm}}{d\theta}\right) - \frac{m^2}{\sin^2\theta}\Theta_{lm} + l(l+1)\Theta_{lm} = 0. \tag{28.4}$$

This equation is well known in the theory of spherical harmonics. It has solutions satisfying the conditions of finiteness and single-valuedness for positive integral values of $l \geqslant |m|$, in agreement with the eigenvalues of the angular momentum obtained above by the matrix method. The corresponding solutions are what are called *associated Legendre polynomials* $P_l^m(\cos\theta)$ (see §c of the Mathematical Appendices). Using the normalization condition (28.2), we find†

$$\Theta_{lm}(\theta) = (-1)^m i^l \sqrt{[\tfrac{1}{2}(2l+1)(l-m)!/(l+m)!]} P_l^m(\cos\theta). \tag{28.5}$$

† The choice of the phase factor is not, of course, determined by the normalization condition. The definition (28.5) used in this book is the most natural from the viewpoint of the theory of addition of angular momenta. It differs by a factor i^l from the one usually adopted. The advantages of this choice will be clear from the footnotes in §§60, 106 and 107.

Here it is supposed that $m \geqslant 0$. For negative m, we use the definition

$$\Theta_{l,-|m|} = (-1)^m \Theta_{l|m|}. \tag{28.6}$$

In other words, Θ_{lm} for $m < 0$ is given by (28.5) with $|m|$ instead of m and the factor $(-1)^m$ omitted.

Thus the angular momentum eigenfunctions are mathematically just spherical harmonic functions normalized in a particular way. For reference, the complete expression embodying the above definitions is

$$Y_{lm}(\theta, \phi) = (-1)^{(m+|m|)/2}\, i^l \left[\frac{2l+1}{4\pi}\frac{(l-|m|)!}{(l+|m|)!}\right]^{1/2} P_l^{|m|}(\cos\theta) e^{im\phi}. \tag{28.7}$$

In particular,

$$Y_{l0} = i^l \sqrt{\frac{2l+1}{4\pi}}\, P_l(\cos\theta). \tag{28.8}$$

It is evident that the functions differing in the sign of m are related by

$$(-1)^{l-m} Y_{l,-m} = Y_{lm}^*. \tag{28.9}$$

For $l = 0$ (so that $m = 0$ also) the spherical harmonic function reduces to a constant. In other words, the wave functions of the states of a particle with zero angular momentum depend only on r, i.e. they have complete spherical symmetry, in agreement with the general statement in §27.

For a given m, the values of l starting from $|m|$ denumerate the successive eigenvalues of the quantity $\mathbf{l}^2$ in order of increasing magnitude. Hence, from the general theory of the zeros of eigenfunctions (§21), we can deduce that the function Θ_{lm} becomes zero for $l-|m|$ different values of the angle θ; in other words, it has as nodal lines $l-|m|$ "lines of latitude" on the sphere. If the complete angular functions are taken with the real factors $\cos m\phi$ or $\sin m\phi$ instead of† $e^{\pm i|m|\phi}$, they have as further nodal lines $|m|$ "lines of longitude"; the total number of nodal lines is thus l.

Finally, we shall show how the functions Θ_{lm} may be calculated by the matrix method. This is done similarly to the calculation of the wave functions of an oscillator in §23. We start from the equation (27.8):

$$\hat{l}_+ Y_{ll} = 0.$$

Using the expression (26.15) for the operator $\hat{l}_+$ and substituting $Y_{ll} = (2\pi)^{-\frac{1}{2}} e^{il\phi}\Theta_{ll}(\theta)$, we obtain for Θ_{ll} the equation

$$d\Theta_{ll}/d\theta - l\cot\theta\, \Theta_{ll} = 0,$$

whence $\Theta_{ll} = \text{constant} \times \sin^l\theta$. Determining the constant from the normalization condition, we find

$$\Theta_{ll} = (-i)^l \sqrt{[\tfrac{1}{2}(2l+1)!]}\, 2^{-l}(1/l!) \sin^l\theta. \tag{28.10}$$

---

† Each such function corresponds to a state in which l_z does not have a definite value, but can have the values $\pm m$ with equal probability.

Next, using (27.12), we write

$$\hat{l}_- Y_{l,m+1} = (l_-)_{m,m+1} Y_{lm}$$
$$= \sqrt{[(l-m)(l+m+1)]}\, Y_{lm}.$$

A repeated application of this formula gives

$$\sqrt{[(l-m)!/(l+m)!]}\, Y_{lm} = [(2l)!]^{-1/2} (\hat{l}_-)^{l-m} Y_{ll}.$$

The right-hand side of this equation is easily calculated by means of the expression (26.15) for the operator $\hat{l}_-$. We have

$$\hat{l}_-[f(\theta)e^{im\phi}] = e^{i(m-1)\phi} \sin^{1-m}\theta \, \mathrm{d}(f \sin^m\theta)/\mathrm{d}(\cos\theta).$$

A repeated application of this formula gives

$$(\hat{l}_-)^{l-m} e^{il\phi}\Theta_{ll} = e^{im\phi} \sin^{-m}\theta \, \mathrm{d}^{l-m}(\sin^l\theta \, . \, \Theta_{ll})/\mathrm{d}(\cos\theta)^{l-m}.$$

Finally, using these relations and the expression (28.10) for Θ_{ll}, we obtain the formula

$$\Theta_{lm}(\theta) = (-i)^l \sqrt{\left[\frac{(2l+1)(l+m)!}{2(l-m)!}\right]} \frac{1}{2^l l! \sin^m\theta} \frac{\mathrm{d}^{l-m}}{\mathrm{d}(\cos\theta)^{l-m}} \sin^{2l}\theta, \quad (28.11)$$

which is the same as (28.5).

§29. Matrix elements of vectors

Let us again consider a closed system of particles;† let f be any scalar physical quantity characterizing the system, and $\hat{f}$ the operator corresponding to this quantity. Every scalar is invariant with respect to rotation of the coordinate system. Hence the scalar operator $\hat{f}$ does not vary when acted on by a rotation operator, i.e. it commutes with a rotation operator. We know, however, that the operator of an infinitely small rotation is the same, apart from a constant factor, as the angular momentum operator, so that

$$\{\hat{f}, \hat{\mathbf{L}}\} = 0. \quad (29.1)$$

From the commutability of $\hat{f}$ with the angular momentum operator it follows that the matrix of f with respect to transitions between states with definite values of L and M is diagonal with respect to these suffixes. Moreover, since the specification of M defines only the orientation of the system relative to the axes of coordinates, and the value of a scalar is independent of this orientation, we can say that the matrix elements $\langle n'LM|\, f\, |nLM\rangle$ are independent of the value of M; n conventionally denotes all the quantum numbers other than L and M which define the state of the system. A formal

† All the results in this section are valid also for a particle in a centrally symmetric field (and in general whenever the total angular momentum of the system is conserved).

proof of this assertion can be obtained from the commutativity of the operators f and $\hat{L}_+$:

$$f\hat{L}_+ - \hat{L}_+ f = 0. \tag{29.2}$$

Let us write down the matrix element of this equation corresponding to the transition $n, L, M \to n', L, M+1$. Taking into account the fact that the matrix of the quantity L_+ has only elements with $n, L, M \to n, L, M+1$, we obtain

$$\langle n', L, M+1|\, f\, |n, L, M+1\rangle \langle n, L, M+1|L_+|n, L, M\rangle = \langle n', L, M+1|L_+|n', L, M\rangle \langle n', L, M|\, f\, |n, L, M\rangle,$$

and since the matrix elements of the quantity L_+ are independent of the suffix n, we find

$$\langle n', L, M+1|\, f\, |n, L, M+1\rangle = \langle n', L, M|\, f\, |n, L, M\rangle, \tag{29.3}$$

whence it follows that all the quantities $\langle n', L, M|\, f\, |n, L, M\rangle$ for different M (the other suffixes being the same) are equal.

If we apply this result to the Hamiltonian itself, we obtain our previous result that the energy of the stationary states is independent of M, i.e. that the energy levels have $(2L+1)$-fold degeneracy.

Next, let **A** be some vector physical quantity characterizing a closed system. When the system of coordinates is rotated (and, in particular, in an infinitely small rotation, i.e. when the angular momentum operator is applied), the components of a vector are transformed into linear functions of one another. Hence, as a result of the commutation of the operators $\hat{L}_i$ with the operators $\hat{A}_i$, we must again obtain components of the same vector, $\hat{A}_i$. The exact form can be found by noticing that, in the particular case where **A** is the radius vector of the particle, the formulae (26.4) must be obtained. Thus we find the commutation rules

$$\{\hat{L}_i, \hat{A}_k\} = ie_{ikl}\hat{A}_l. \tag{29.4}$$

These relations enable us to obtain several results concerning the form of the matrices of the components of the vector **A** (M. Born, W. Heisenberg and P. Jordan 1926). First of all, it is possible to derive *selection rules* which determine the transitions for which the matrix elements can be different from zero. We shall not go through the fairly lengthy calculations here, however, since it will appear later (§107) that these rules are actually a direct consequence of the general transformation properties of vector quantities and can be derived from the latter with hardly any calculation at all. Here we shall merely give the rules, without proof.

The matrix elements of all the components of a vector can be different from zero only for transitions in which the angular momentum L changes by not more than one unit:

$$L \to L \quad \text{or} \quad L \pm 1. \tag{29.5}$$

There is a further selection rule which forbids transitions between any two states with $L = 0$. This rule is an obvious consequence of the complete spherical symmetry of states with angular momentum zero.

The selection rules for the angular momentum component M are different for the different components of a vector: the matrix elements can be different from zero for transitions where M changes as follows:

$$\left.\begin{array}{lll} \text{for} & A_+ = A_x + iA_y, & M \to M+1, \\ \text{for} & A_- = A_x - iA_y, & M \to M-1, \\ \text{for} & A_z, & M \to M. \end{array}\right\} \tag{29.6}$$

Moreover, it is possible to determine a general form for the matrix elements of a vector as functions of the number M. These important and frequently used formulae are given here, also without proof, since they are actually a particular case of more general relations derived in §107 for any tensor quantities.

The non-zero matrix elements of the quantity A_z are given by the formulae

$$\left.\begin{aligned} \langle n'LM|A_z|nLM\rangle &= \frac{M}{\sqrt{[L(L+1)(2L+1)]}}\langle n'L||A||nL\rangle, \\ \langle n'LM|A_z|n, L-1, M\rangle &= \sqrt{\frac{L^2-M^2}{L(2L-1)(2L+1)}}\langle n'L||A||n, L-1\rangle, \\ \langle n', L-1, M|A_z|nLM\rangle &= \sqrt{\frac{L^2-M^2}{L(2L-1)(2L+1)}}\langle n', L-1||A||nL\rangle. \end{aligned}\right\} \tag{29.7}$$

Here the symbol $\langle n'L'||A||nL\rangle$ denotes a *reduced matrix element*, a quantity independent of the quantum number M.† These matrix elements are related by

$$\langle n'L'||A||nL\rangle = \langle nL||A||n'L'\rangle^*, \tag{29.8}$$

which follows directly from the fact that the operator $\hat{A}_z$ is Hermitian.

The matrix elements of the quantities A_- and A_+ are also determined by

† The appearance in formulae (29.7) and (29.9) of denominators which depend on L is in accordance with the general notation used in §107. The convenience of these denominators is shown, in particular, by the simple form of equation (29.12) for the matrix elements of the scalar product of two vectors.

The symbol for the reduced matrix element is to be taken as a whole, in contrast to the matrix element symbol (see the comments following (11.17)).

the reduced matrix elements. The non-zero matrix elements of A_- are

$$\left.\begin{aligned}
&\langle n',|L, M-1|A_-|nLM\rangle \\
&\qquad = \sqrt{\frac{(L-M+1)(L+M)}{L(L+1)(2L+1)}}\langle n'L||A||nL\rangle, \\
&\langle n', L, M-1|A_-|n, L-1M\rangle \\
&\qquad = \sqrt{\frac{(L-M+1)(L-M)}{L(2L-1)(2L+1)}}\langle n'L||A||n, L-1\rangle, \\
&\langle n', L-1, M-1|A_-|nLM\rangle \\
&\qquad = -\sqrt{\frac{(L+M-1)(L+M)}{L(2L-1)(2L+1)}}\langle n', L-1||A||nL\rangle.
\end{aligned}\right\} \quad (29.9)$$

The matrix elements of A_+ need not be written out separately: since A_x and A_y are real we have

$$\langle n'L'M'|A_+|nLM\rangle = \langle nLM|A_-|n'L'M'\rangle^*. \quad (29.10)$$

There is a formula which expresses the matrix elements of the scalar **A . B** in terms of the reduced matrix elements of the two vector quantities **A** and **B**. The calculation is conveniently carried out by writing the operator $\hat{\mathbf{A}} \cdot \hat{\mathbf{B}}$ in the form

$$\hat{\mathbf{A}} \cdot \hat{\mathbf{B}} = \tfrac{1}{2}(\hat{A}_+\hat{B}_- + \hat{A}_-\hat{B}_+) + \hat{A}_z\hat{B}_z. \quad (29.11)$$

The matrix of **A . B** (like that of any scalar) is diagonal with respect to L and M. A calculation by means of formulae (29.7)–(29.9) gives the result

$$\langle n'LM|\mathbf{A}.\mathbf{B}|nLM\rangle = \frac{1}{2L+1}\sum_{n'',L''}\langle n'L||A||n''L''\rangle\langle n''L''||B||nL\rangle, \quad (29.12)$$

where L'' takes the values L, $L \pm 1$.

For reference, we shall give the reduced matrix elements for the vector **L** itself. A comparison of (29.9) and (27.12) shows that

$$\left.\begin{aligned}
\langle L||L||L\rangle &= \sqrt{[L(L+1)(2L+1)]}, \\
\langle L-1||L||L\rangle &= \langle L||L||L-1\rangle = 0.
\end{aligned}\right\} \quad (29.13)$$

A quantity that often occurs in applications is the unit vector **n** along the radius vector of the particle. Its reduced matrix elements can be calculated by finding, for example, the matrix elements of $n_z = \cos\theta$ for a zero angular-momentum component, $m = 0$;

$$\langle l-1, 0|n_z|l0\rangle = \int_0^\pi \Theta_{l-1,0}^* \cos\theta . \Theta_{l0} \sin\theta d\theta,$$

with the functions Θ_{l0} given by (28.11). The evaluation of the integral† gives

$$\langle l-1, 0|n_z|l0\rangle = il/\sqrt{[(2l-1)(2l+1)]}.$$

The matrix elements for transitions $l \to l$ are zero (as for any polar vector of an individual particle; see (30.8) below). Comparison with (29.7) then gives

$$\left.\begin{aligned}\langle l-1||n||l\rangle &= -\langle l||n||l-1\rangle = i\sqrt{l},\\ \langle l||n||l\rangle &= 0.\end{aligned}\right\} \quad (29.14)$$

PROBLEM

Average the tensor $n_i n_k - \frac{1}{3}\delta_{ik}$ (where $\mathbf{n}$ is a unit vector along the radius vector of a particle) over a state where the magnitude but not the direction of the vector $\mathbf{l}$ is given (i.e. l_z is indeterminate).

SOLUTION. The required mean value is an operator which can be expressed in terms of the operator $\hat{\mathbf{l}}$ alone. We seek it in the form

$$\overline{n_i n_k} - \tfrac{1}{3}\delta_{ik} = a[\hat{l}_i\hat{l}_k + \hat{l}_k\hat{l}_i - \tfrac{2}{3}\delta_{ik} l(l+1)];$$

this is the most general symmetrical tensor of rank two with zero trace that can be formed from the components of $\hat{\mathbf{l}}$. To determine the constant a we multiply this equation on the left by $\hat{l}_i$ and on the right by $\hat{l}_k$ (summing over i and k). Since the vector $\mathbf{n}$ is perpendicular to the vector $\hbar\hat{\mathbf{l}} = \hat{\mathbf{r}}\times\hat{\mathbf{p}}$, we have $n_i\hat{l}_i = 0$. The product $\hat{l}_i\hat{l}_i\hat{l}_k\hat{l}_k = (\hat{\mathbf{l}}^2)^2$ is replaced by its eigenvalue $l^2(l+1)^2$, and the product $\hat{l}_i\hat{l}_k\hat{l}_i\hat{l}_k$ is transformed by means of the commutation relations (26.7) as follows:

$$\begin{aligned}\hat{l}_i\hat{l}_k\hat{l}_i\hat{l}_k &= \hat{l}_i\hat{l}_i\hat{l}_k\hat{l}_k - ie_{ikl}\hat{l}_i\hat{l}_l\hat{l}_k\\ &= (\hat{\mathbf{l}}^2)^2 - \tfrac{1}{2}ie_{ikl}\hat{l}_i(\hat{l}_l\hat{l}_k - \hat{l}_k\hat{l}_l)\\ &= (\hat{\mathbf{l}}^2)^2 + \tfrac{1}{2}e_{ikl}e_{lkm}\hat{l}_i\hat{l}_m\\ &= (\hat{\mathbf{l}}^2)^2 - \hat{\mathbf{l}}^2\\ &= l^2(l+1)^2 - l(l+1)\end{aligned}$$

(using the fact that $e_{ikl}e_{mkl} = 2\delta_{im}$). After a simple reduction we obtain the result

$$a = -1/(2l-1)(2l+3).$$

§30. Parity of a state

Besides the parallel displacements and rotations of the coordinate system, the invariance under which represents the homogeneity and isotropy of space

† By $l-1$ times integrating by parts with $d\cos\theta$; the general formula for integrals of this type is (107.14).

respectively, there is another transformation which leaves unaltered the Hamiltonian of a closed system. This is what is called the *inversion transformation*, which consists in simultaneously changing the sign of all the coordinates, i.e. a reversal of the direction of each coordinate axis; a right-handed coordinate system then becomes left-handed, and vice versa. The invariance of the Hamiltonian under this transformation expresses the symmetry of space under mirror reflections.† In classical mechanics, the invariance of Hamilton's function with respect to inversion does not lead to a conservation law, but the situation is different in quantum mechanics.

Let us denote by $\hat{P}$ (for "parity") an inversion operator whose effect on a wave function $\psi(\mathbf{r})$ is to change the sign of the coordinates:

$$\hat{P}\psi(\mathbf{r}) = \psi(-\mathbf{r}). \tag{30.1}$$

It is easy to find the eigenvalues P of this operator, which are determined by the equation

$$\hat{P}\psi(\mathbf{r}) = P\psi(\mathbf{r}). \tag{30.2}$$

To do so, we notice that a double application of the inversion operator amounts to identity: the argument of the function is unchanged. In other words, we have $\hat{P}^2\psi = P^2\psi = \psi$, i.e. $P^2 = 1$, whence

$$P = \pm 1. \tag{30.3}$$

Thus the eigenfunctions of the inversion operator are either unchanged or change in sign when acted upon by this operator. In the first case, the wave function (and the corresponding state) is said to be *even*, and in the second it is said to be *odd*.

The invariance of the Hamiltonian under inversion (i.e. the fact that the operators $\hat{H}$ and $\hat{P}$ commute) thus expresses the *law of conservation of parity*: if the state of a closed system has a definite parity (i.e. if it is even, or odd) then this parity is conserved in the course of time.‡

The angular momentum operator also is invariant under inversion, which changes the sign of the coordinates and of the operators of differentiation with respect to them; the operator (26.2) thus remains unaltered. In other words, the inversion operator commutes with the angular momentum operator, and this means that the system can have a definite parity simultaneously with definite values of the angular momentum L and its component M. All states that differ only in the value of M have the same parity; this is evident because the properties of a closed system are independent of its orientation in space,

† Invariance under inversion exists also for the Hamiltonian of a system of particles in a centrally symmetric field with the centre at the origin.

‡ To avoid misunderstanding, it should be mentioned that this refers to the non-relativistic theory. There exist interactions in Nature, falling in the realm of relativistic theory, which violate the conservation of parity.

and it can be formally demonstrated from the commutation rule $\hat{L}_+\hat{P}-\hat{P}\hat{L}_+ = 0$ by the same method as in deriving (29.3) from (29.2).

There are specific *parity selection rules* for the matrix elements of various physical quantities. Let us first consider scalars. Here we must distinguish *true scalars*, which are unchanged by inversion, from *pseudoscalars*, which change sign, for instance the scalar product of an axial and a polar vector. The operator of a true scalar f commutes with $\hat{P}$; hence it follows that, if the matrix of P is diagonal, then the matrix of f is diagonal also as regards the parity suffix, i.e. the matrix elements are zero except for transitions $g \to g$ and $u \to u$ (where g and u denote even and odd states respectively). For the operator of a pseudoscalar quantity, we have $\hat{P}\hat{f} = -\hat{f}\hat{P}$; the operators $\hat{P}$ and $\hat{f}$ *anticommute*. The matrix element of this equation for a transition $g \to g$ is $P_{gg}f_{gg} = -f_{gg}P_{gg}$, and so $f_{gg} = 0$ since $P_{gg} = 1$. Similarly we find that $f_{uu} = 0$. Thus, in the matrix of a pseudoscalar quantity, only those elements can be different from zero which correspond to transitions with change of parity. The selection rules for the matrix elements of scalars are therefore:

$$\left.\begin{array}{ll} \text{true scalars} & g \to g,\ u \to u; \\ \text{pseudoscalars} & g \to u,\ u \to g. \end{array}\right\} \tag{30.4}$$

These rules can also be obtained directly from the definition of the matrix elements. Let us consider, for example, the integral $f_{ug} = \int \psi_u^* \hat{f} \psi_g \, dq$, where the function ψ_g is even and ψ_u odd. When all the coordinates change sign, the integrand does so if f is a true scalar; on the other hand, the integral taken over all space cannot change when the variables of integration are renamed. Hence it follows that $f_{ug} = -f_{ug}$, i.e. $f_{ug} \equiv 0$.

We can similarly derive selection rules for vector quantities. Here it must be recalled that ordinary (polar) vectors change sign on inversion, while axial vectors (such as the angular momentum vector, which is the vector product of the two polar vectors $\mathbf{p}$ and $\mathbf{r}$) are unchanged by inversion. The selection rules are found to be:

$$\left.\begin{array}{ll} \text{polar vectors} & g \to u,\ u \to g; \\ \text{axial vectors} & g \to g,\ u \to u. \end{array}\right\} \tag{30.5}$$

Let us determine the parity of the state of a single particle with angular momentum l. The inversion transformation $(x \to -x, y \to -y, z \to -z)$ is, in spherical polar coordinates, the transformation

$$r \to r, \quad \theta \to \pi - \theta, \quad \phi \to \pi + \phi. \tag{30.6}$$

The dependence of the wave function of the particle on the angle is given by the spherical harmonic Y_{lm}, which, apart from a constant that is here unimportant, has the form $P_l^m(\cos\theta)e^{im\phi}$. When ϕ is replaced by $\pi+\phi$, the factor $e^{im\phi}$ is multiplied by $(-1)^m$, and when θ is replaced by $\pi-\theta$, $P_l^m(\cos\theta)$ becomes $P_l^m(-\cos\theta) = (-1)^{l-m}P_l^m(\cos\theta)$. Thus the whole function is multiplied by $(-1)^l$ (independent of m, in agreement with what was said above), i.e. the parity of a state with a given value of l is

$$P = (-1)^l. \tag{30.7}$$

We see that all states with even l are even, and all those with odd l are odd.

A vector physical quantity relating to an individual particle can have non-zero matrix elements only for transitions with $l \to l$ or $l \pm 1$ (§29). Remembering this, and comparing formula (30.7) with what was said above regarding the change of parity in the matrix elements of vectors, we reach the result that the matrix elements of vectors relating to an individual particle are zero except for the transitions:

$$\left.\begin{array}{ll} \text{polar vectors} & l \to l \pm 1; \\ \text{axial vectors} & l \to l. \end{array}\right\} \tag{30.8}$$

§31. Addition of angular momenta

Let us consider a system composed of two parts whose interaction is weak. If the interaction is entirely neglected, then for each part the law of conservation of angular momentum holds. The angular momentum $\mathbf{L}$ of the whole system can be regarded as the sum of the angular momenta $\mathbf{L}_1$ and $\mathbf{L}_2$ of its parts. In the next approximation, when the weak interaction is taken into account, $\mathbf{L}_1$ and $\mathbf{L}_2$ are not exactly conserved, but the numbers L_1 and L_2 which determine their squares remain "good" quantum numbers suitable for an approximate description of the state of the system. Regarding the angular momenta in a classical manner, we can say that in this approximation $\mathbf{L}_1$ and $\mathbf{L}_2$ rotate round the direction of $\mathbf{L}$ while remaining unchanged in magnitude.

For such systems the question arises regarding the "law of addition" of angular momenta: what are the possible values of L for given values of L_1 and L_2? The law of addition for the components of angular momentum is evident: since $\hat{L}_z = \hat{L}_{1z} + \hat{L}_{2z}$, it follows that

$$M = M_1 + M_2. \tag{31.1}$$

There is no such simple relation for the operators of the squared angular momenta, however, and to derive their "law of addition" we reason as follows.

If we take the quantities $\mathbf{L}_1^2$, $\mathbf{L}_2^2$, L_{1z}, L_{2z} as a complete set of physical quantities,† every state will be determined by the values of the numbers L_1, L_2, M_1, M_2. For given L_1 and L_2, the numbers M_1 and M_2 take $(2L_1+1)$ and $(2L_2+1)$ different values respectively, so that there are altogether $(2L_1+1)(2L_2+1)$ different states with the same L_1 and L_2. We denote the wave functions of the states for this representation by $\phi_{L_1 L_2 M_1 M_2}$.

Instead of the above four quantities, we can take the four quantities $\mathbf{L}_1^2$, $\mathbf{L}_2^2$, $\mathbf{L}^2$, L_z as a complete set. Then every state is characterized by

† Together with such other quantities as form a complete set when combined with these four. These other quantities play no part in the subsequent discussion, and for brevity we shall ignore them entirely, and conventionally call the above four quantities a complete set.

the values of the numbers L_1, L_2, L, M (we denote the corresponding wave functions by $\psi_{L_1L_2LM}$). For given L_1 and L_2, there must of course be $(2L_1+1)(2L_2+1)$ different states as before, i.e. for given L_1 and L_2 the pair of numbers L and M must take $(2L_1+1)(2L_2+1)$ pairs of values. These values can be determined as follows.

By adding the various possible values of M_1 and M_2, we get the corresponding values of M, as shown below:

M_1	M_2	M
L_1	L_2	L_1+L_2
L_1	L_2-1	L_1+L_2-1
L_1-1	L_2	
L_1	L_2-2	
L_1-1	L_2-1	L_1+L_2-2
L_1-2	L_2	
...	...	...

We see that the greatest possible value of M is $M = L_1+L_2$, corresponding to one state ϕ (one pair of values of M_1 and M_2). The greatest possible value of M in the states ψ, and hence the greatest possible value of L also, is therefore L_1+L_2. Next, there are two states ϕ with $M = L_1+L_2-1$. Consequently, there must also be two states ψ with this value of M; one of them is the state with $L = L_1+L_2$ (and $M = L-1$), and the other is that with $L = L_1+L_2-1$ (and $M = L$). For the value $M = L_1+L_2-2$ there are three different states ϕ. This means that, besides the values $L = L_1+L_2$, $L = L_1+L_2-1$, the value $L = L_1+L_2-2$ can occur.

The argument can be continued in this way so long as a decrease of M by 1 increases by 1 the number of states with a given M. It is easily seen that this is so until M reaches the value $|L_1-L_2|$. When M decreases further, the number of states no longer increases, remaining equal to $2L_2+1$ (if $L_2 \leqslant L_1$). Thus $|L_1-L_2|$ is the least possible value of L, and we arrive at the result that, for given L_1 and L_2, the number L can take the values

$$L = L_1+L_2, L_1+L_2-1, \ldots, |L_1-L_2|, \tag{31.2}$$

that is $2L_2+1$ different values altogether (supposing that $L_2 \leqslant L_1$). It is easy to verify that we do in fact obtain $(2L_1+1)(2L_2+1)$ different values of the pair of numbers M, L. Here it is important to note that, if we ignore the $2L+1$ values of M for a given L, then only one state will correspond to each of the possible values (31.2) of L.

This result can be illustrated by means of what is called the *vector model.* If we take two vectors $\mathbf{L}_1$, $\mathbf{L}_2$ of lengths L_1 and L_2, then the values of L are represented by the integral lengths of the vectors $\mathbf{L}$ which are obtained by vector addition of $\mathbf{L}_1$ and $\mathbf{L}_2$; the greatest value of L is L_1+L_2, which is obtained when $\mathbf{L}_1$ and $\mathbf{L}_2$ are parallel, and the least value is $|L_1-L_2|$, when $\mathbf{L}_1$ and $\mathbf{L}_2$ are antiparallel.

In states with definite values of the angular momenta L_1, L_2 and of the total angular momentum L, the scalar products $\mathbf{L}_1 \cdot \mathbf{L}_2$, $\mathbf{L} \cdot \mathbf{L}_1$ and $\mathbf{L} \cdot \mathbf{L}_2$ also have definite values. These values are easily found. To calculate $\mathbf{L}_1 \cdot \mathbf{L}_2$, we write $\hat{\mathbf{L}} = \hat{\mathbf{L}}_1 + \hat{\mathbf{L}}_2$ or, squaring and transposing,

$$2\hat{\mathbf{L}}_1 \cdot \hat{\mathbf{L}}_2 = \hat{\mathbf{L}}^2 - \hat{\mathbf{L}}_1{}^2 - \hat{\mathbf{L}}_2{}^2.$$

Replacing the operators on the right-hand side of this equation by their eigenvalues, we obtain the eigenvalue of the operator on the left-hand side:

$$\mathbf{L}_1 \cdot \mathbf{L}_2 = \tfrac{1}{2}\{L(L+1) - L_1(L_1+1) - L_2(L_2+1)\}. \tag{31.3}$$

Similarly we find

$$\mathbf{L} \cdot \mathbf{L}_1 = \tfrac{1}{2}\{L(L+1) + L_1(L_1+1) - L_2(L_2+1)\}. \tag{31.4}$$

Let us now determine the "addition rule for parities". The wave function Ψ of a system consisting of two independent parts is the product of the wave functions Ψ_1 and Ψ_2 of these parts. Hence it is clear that if the latter are of the same parity (i.e. both change sign, or both do not change sign, when the sign of all the coordinates is reversed), then the wave function of the whole system is even. On the other hand, if Ψ_1 and Ψ_2 are of opposite parity, then the function Ψ is odd. These statements may be written

$$P = P_1 P_2, \tag{31.5}$$

where P is the parity of the whole system and P_1, P_2 those of its parts. This rule can, of course, be generalized at once to the case of a system composed of any number of non-interacting parts.

In particular, if we are concerned with a system of particles in a centrally symmetric field (the mutual interaction of the particles being supposed weak), then the parity of the state of the whole system is given by

$$P = (-1)^{l_1 + l_2 + \dots}; \tag{31.6}$$

see (30.7). We emphasize that the exponent here contains the algebraic sum of the angular momenta l_i, and this is not in general the same as their "vector sum", i.e. the angular momentum L of the system.

If a closed system disintegrates (under the action of internal forces), the total angular momentum and parity must be conserved. This circumstance may render it impossible for a system to disintegrate, even if this is energetically possible.

For instance, let us consider an atom in an even state with angular momentum $L = 0$, which is able, so far as energy considerations go, to disintegrate into a free electron and an ion in an odd state with the same angular momentum $L = 0$. It is easy to see that in fact no such disintegration can occur (it is, as we say, *forbidden*). For, by virtue of the law of conservation of angular momentum, the free electron would also have to have zero angular momentum, and therefore be in an even state ($P = (-1)^0 = +1$); the state of the system ion+electron would then be odd, however, whereas the original state of the atom was even.

CHAPTER V

MOTION IN A CENTRALLY SYMMETRIC FIELD

§32. Motion in a centrally symmetric field

THE problem of the motion of two interacting particles can be reduced in quantum mechanics to that of one particle, as can be done in classical mechanics. The Hamiltonian of the two particles (of masses m_1, m_2) interacting in accordance with the law $U(r)$ (where r is the distance between the particles) is of the form

$$\hat{H} = -\frac{\hbar^2}{2m_1}\triangle_1 - \frac{\hbar^2}{2m_2}\triangle_2 + U(r), \tag{32.1}$$

where $\triangle_1$ and $\triangle_2$ are the Laplacian operators with respect to the coordinates of the particles. Instead of the radius vectors $\mathbf{r}_1$ and $\mathbf{r}_2$ of the particles, we introduce new variables $\mathbf{R}$ and $\mathbf{r}$:

$$\mathbf{r} = \mathbf{r}_2 - \mathbf{r}_1, \quad \mathbf{R} = (m_1\mathbf{r}_1 + m_2\mathbf{r}_2)/(m_1+m_2); \tag{32.2}$$

$\mathbf{r}$ is the vector of the distance between the particles, and $\mathbf{R}$ the radius vector of their centre of mass. A simple calculation gives

$$\hat{H} = -\frac{\hbar^2}{2(m_1+m_2)}\triangle_R - \frac{\hbar^2}{2m}\triangle + U(r), \tag{32.3}$$

where $\triangle_R$ and $\triangle$ are the Laplacian operators with respect to the components of the vectors $\mathbf{R}$ and $\mathbf{r}$ respectively, m_1+m_2 is the total mass of the system, and $m = m_1m_2/(m_1+m_2)$ is the *reduced mass*. Thus the Hamiltonian falls into the sum of two independent parts. Hence we can look for $\psi(\mathbf{r}_1, \mathbf{r}_2)$ in the form of a product $\phi(\mathbf{R})\psi(\mathbf{r})$, where the function $\phi(\mathbf{R})$ describes the motion of the centre of mass (as a free particle of mass m_1+m_2), and $\psi(\mathbf{r})$ describes the relative motion of the particles (as a particle of mass m moving in the centrally symmetric field $U(r)$).

Schrödinger's equation for the motion of a particle in a centrally symmetric field is

$$\triangle\psi + (2m/\hbar^2)[E - U(r)]\psi = 0. \tag{32.4}$$

Using the familiar expression for the Laplacian operator in spherical polar coordinates, we can write this equation in the form

$$\frac{1}{r^2}\frac{\partial}{\partial r}\left(r^2\frac{\partial\psi}{\partial r}\right)+\frac{1}{r^2}\left[\frac{1}{\sin\theta}\frac{\partial}{\partial\theta}\left(\sin\theta\frac{\partial\psi}{\partial\theta}\right)+\frac{1}{\sin^2\theta}\frac{\partial^2\psi}{\partial\phi^2}\right]+\frac{2m}{\hbar^2}[E-U(r)]\psi = 0. \quad (32.5)$$

If we introduce here the operator (26.16) of the squared angular momentum we obtain†

$$\frac{\hbar^2}{2m}\left[-\frac{1}{r^2}\frac{\partial}{\partial r}\left(r^2\frac{\partial\psi}{\partial r}\right)+\frac{\hat{\mathbf{l}}^2}{r^2}\psi\right]+U(r)\psi = E\psi. \quad (32.6)$$

The angular momentum is conserved during motion in a centrally symmetric field. We shall consider stationary states in which the angular momentum l and the component m have definite values. These values determine the angular dependence of the wave functions. We thus seek solutions of equation (32.6) in the form

$$\psi = R(r)Y_{lm}(\theta,\phi), \quad (32.7)$$

where the $Y_{lm}(\theta, \phi)$ are spherical harmonic functions.

Since $\hat{\mathbf{l}}^2 Y_{lm} = l(l+1)Y_{lm}$, we obtain for the *radial function* $R(r)$ the equation

$$\frac{1}{r^2}\frac{\mathrm{d}}{\mathrm{d}r}\left(r^2\frac{\mathrm{d}R}{\mathrm{d}r}\right)-\frac{l(l+1)}{r^2}R+\frac{2m}{\hbar^2}[E-U(r)]R = 0. \quad (32.8)$$

This equation does not contain the value of $l_z = m$ at all, in accordance with the $(2l+1)$-fold degeneracy of the levels with respect to the directions of the angular momentum, with which we are already familiar.

Let us investigate the radial part of the wave functions. By the substitution

$$R(r) = \chi(r)/r \quad (32.9)$$

equation (32.8) is brought to the form

$$\frac{\mathrm{d}^2\chi}{\mathrm{d}r^2}+\left[\frac{2m}{\hbar^2}(E-U)-\frac{l(l+1)}{r^2}\right]\chi = 0. \quad (32.10)$$

If the potential energy $U(r)$ is everywhere finite, the wave function ψ must also be finite in all space, including the origin, and consequently so must its

---

† If we introduce the operator of the radial component p_r of the linear momentum, in the form

$$\hat{p}_r\psi = -i\hbar\frac{1}{r}\frac{\partial}{\partial r}(r\psi) = -i\hbar\left(\frac{\partial}{\partial r}+\frac{1}{r}\right)\psi,$$

the Hamiltonian can be written in the form

$$\hat{H} = (1/2m)(\hat{p}_r^2+\hbar^2\hat{\mathbf{l}}^2/r^2)+U(r),$$

which is the same in form as the classical Hamilton's function in spherical polar coordinates.

radial part $R(r)$. Hence it follows that $\chi(r)$ must vanish for $r = 0$:

$$\chi(0) = 0. \tag{32.11}$$

This condition actually holds also (see §35) for a field which becomes infinite as $r \to 0$.

Equation (32.10) is formally identical with Schrödinger's equation for one-dimensional motion in a field of potential energy

$$U_l(r) = U(r)+\frac{\hbar^2}{2m}\frac{l(l+1)}{r^2}, \tag{32.12}$$

which is the sum of the energy $U(r)$ and a term

$$\hbar^2 l(l+1)/2mr^2 = \hbar^2 \mathbf{l}^2/2mr^2,$$

which may be called the *centrifugal energy*. Thus the problem of motion in a centrally symmetric field reduces to that of one-dimensional motion in a region bounded on one side (the boundary condition for $r = 0$). The normalization condition for the function χ is also "one-dimensional", being determined by the integral

$$\int_0^\infty |R|^2 r^2 \, dr = \int_0^\infty |\chi|^2 \, dr.$$

In one-dimensional motion in a region bounded on one side, the energy levels are not degenerate (§21). Hence we can say that, if the energy is given, the solution of equation (32.10), i.e. the radial part of the wave function, is completely determined. Bearing in mind also that the angular part of the wave function is completely determined by the values of l and m, we reach the conclusion that, for motion in a centrally symmetric field, the wave function is completely determined by the values of E, l and m. In other words, the energy, the squared angular momentum and the z-component of the angular momentum together form a complete set of physical quantities for such a motion.

The reduction of the problem of motion in a centrally symmetric field to a one-dimensional problem enables us to apply the oscillation theorem (see §21). We arrange the eigenvalues of the energy (discrete spectrum) for a given l in order of increasing magnitude, and give them numbers n_r, the lowest level being given the number $n_r = 0$. Then n_r determines the number of nodes of the radial part of the wave function for finite values of r (excluding the point $r = 0$). The number n_r is called the ***radial quantum number***. The number l for motion in a centrally symmetric field is sometimes called the ***azimuthal quantum number***, and m the ***magnetic quantum number***.

There is an accepted notation for states with various values of the angular momentum l of the particle: they are denoted by Latin letters, as follows:

$$\begin{array}{ccccccccccc} l = & 0 & 1 & 2 & 3 & 4 & 5 & 6 & 7 & \ldots \\ & s & p & d & f & g & h & i & k & \ldots \end{array} \tag{32.13}$$

The normal state of a particle moving in a centrally symmetric field is always the s state; for, if $l \neq 0$, the angular part of the wave function invariably has nodes, whereas the wave function of the normal state can have no nodes. We can also say that the least possible eigenvalue of the energy, for a given l, increases with l. This follows from the fact that the presence of an angular momentum involves the addition of the essentially positive term $\hbar^2 l(l+1)/2mr^2$, which increases with l, to the Hamiltonian.

Let us determine the form of the radial function near the origin. Here we shall suppose that

$$\lim_{r\to 0} U(r)r^2 = 0. \tag{32.14}$$

We seek $R(r)$ in the form of a power series in r, retaining only the first term of the series for small r; in other words, we seek $R(r)$ in the form $R = \text{constant} \times r^s$. Substituting this in the equation

$$\mathrm{d}(r^2\, \mathrm{d}R/\mathrm{d}r)/\mathrm{d}r - l(l+1)R = 0,$$

which is obtained from (32.8) by multiplying by r^2 and taking the limit as $r \to 0$, we find

$$s(s+1) = l(l+1).$$

Hence

$$s = l \quad \text{or} \quad s = -(l+1).$$

The solution with $s = -(l+1)$ does not satisfy the necessary conditions; it becomes infinite for $r = 0$ (we recall that $l \geqslant 0$). Thus the solution with $s = l$ remains, i.e. near the origin the wave functions of states with a given l are proportional to r^l:

$$R_l \approx \text{constant} \times r^l. \tag{32.15}$$

The probability of a particle's being at a distance between r and $r+\mathrm{d}r$ from the centre is determined by the value of $r^2|R|^2$ and is thus proportional to $r^{2(l+1)}$. We see that it becomes zero at the origin the more rapidly, the greater the value of l.

§33. Spherical waves

The plane wave

$$\psi_{\mathbf{p}} = \text{constant} \times e^{(i/\hbar)\mathbf{p}\cdot\mathbf{r}}$$

describes a stationary state in which a free particle has a definite momentum $\mathbf{p}$ (and energy $E = p^2/2m$). Let us now consider stationary states of a free particle in which it has a definite value, not only of the energy, but also of the absolute value and component of the angular momentum. Instead of the energy, it is convenient to introduce the *wave number*

$$k = p/\hbar = \sqrt{(2mE)}/\hbar. \tag{33.1}$$

The wave function of a state with angular momentum l and projection thereof m has the form

$$\psi_{klm} = R_{kl}(r)Y_{lm}(\theta, \phi), \tag{33.2}$$

where the radial function is determined by the equation

$$R_{kl}'' + \frac{2}{r}R_{kl}' + \left[k^2 - \frac{l(l+1)}{r^2}\right]R_{kl} = 0 \tag{33.3}$$

(equation (32.8) with $U(r) \equiv 0$). The wave functions ψ_{klm} for the continuous (with respect to k) spectrum satisfy the conditions of normalization and orthogonality:

$$\int \psi_{k'l'm'}{}^*\psi_{klm}\, dV = \delta_{ll'}\delta_{mm'}\delta\left(\frac{k'-k}{2\pi}\right).$$

The orthogonality for different l, l' and m, m' is ensured by the angular functions. The radial functions must be normalized by the condition

$$\int_0^\infty r^2 R_{k'l}R_{kl}\, dr = \delta\left(\frac{k'-k}{2\pi}\right) = 2\pi\delta(k'-k). \tag{33.4}$$

If we normalize the wave functions, not on the "$k/2\pi$ scale", but on the "energy scale", i.e. by the condition

$$\int_0^\infty r^2 R_{E'l}R_{El}\, dr = \delta(E'-E),$$

then, by the general formula (5.14), we have

$$R_{El} = R_{kl}\sqrt{(1/2\pi)}\sqrt{(dk/dE)} = (1/\hbar)\sqrt{(m/2\pi k)}R_{kl}. \tag{33.5}$$

For $l = 0$, equation (33.3) can be written

$$\frac{d^2(rR_{k0})}{dr^2} + k^2 rR_{k0} = 0;$$

its solution finite for $r = 0$ and normalized by the condition (33.4) is (cf. (21.9))

$$R_{k0} = \frac{2\sin kr}{r}. \tag{33.6}$$

To solve equation (33.3) with $l \neq 0$, we make the substitution

$$R_{kl} = r^l\chi_{kl}. \tag{33.7}$$

For χ_{kl} we have the equation

$$\chi_{kl}'' + 2(l+1)\chi_{kl}'/r + k^2\chi_{kl} = 0.$$

If we differentiate this equation with respect to r, we obtain

$$\chi_{kl}'''+\frac{2(l+1)}{r}\chi_{kl}''+\left[k^2-\frac{2(l+1)}{r^2}\right]\chi_{kl}' = 0.$$

By the substitution $\chi_{kl}' = r\chi_{k,l+1}$ it becomes

$$\chi_{k,l+1}''+\frac{2(l+2)}{r}\chi_{k,l+1}'+k^2\chi_{k,l+1} = 0,$$

which is in fact the equation satisfied by $\chi_{k,l+1}$. Thus the successive functions χ_{kl} are related by

$$\chi_{k,l+1} = \chi_{kl}'/r, \tag{33.8}$$

and hence

$$\chi_{kl} = \left(\frac{1}{r}\frac{\mathrm{d}}{\mathrm{d}r}\right)^l \chi_{k0},$$

where $\chi_{k0} = R_{k0}$ is determined by formula (33.6) (this expression can, of course, be multiplied by an arbitrary constant).

Thus we finally have the following expression for the radial functions in the free motion of a particle:

$$R_{kl} = (-1)^l \frac{2r^l}{k^l}\left(\frac{1}{r}\frac{\mathrm{d}}{\mathrm{d}r}\right)^l \frac{\sin kr}{r} \tag{33.9}$$

(the factor k^{-l} is introduced for normalization purposes—see below— and the factor $(-1)^l$ for convenience). The functions (33.9) can be expressed in terms of Bessel functions of half-integral order, in the form

$$R_{kl} = \sqrt{(2\pi k/r)}\, J_{l+1/2}(kr) = 2kj_l(kr); \tag{33.10}$$

the functions

$$j_l(x) = \sqrt{(\pi/2x)}J_{l+1/2}(x) \tag{33.11}$$

are called *spherical Bessel functions*.†

To obtain an asymptotic expression for the radial function (33.9) at large distances, we notice that the term which decreases least rapidly as $r \to \infty$ is obtained by differentiating the sine l times. Since each differentiation $-\mathrm{d}/\mathrm{d}r$ of the sine adds $-\frac{1}{2}\pi$ to its argument, we have the following asymptotic expression:

$$R_{kl} \approx \frac{2\sin\left(kr-\frac{1}{2}l\pi\right)}{r}. \tag{33.12}$$

---

† The first few of these are

$$j_0 = \frac{\sin x}{x}, \quad j_1 = \frac{\sin x}{x^2} - \frac{\cos x}{x}, \quad j_2 = \left(\frac{3}{x^3}-\frac{1}{x}\right)\sin x - \frac{3\cos x}{x^2}.$$

Functions defined as x times these are also sometimes used.

The normalization of the functions R_{kl} can be effected by means of their asymptotic expressions, as was explained in §21. Comparing the asymptotic formula (33.12) with the normalized function R_{k0} (33.6), we see that the functions R_{kl}, with the coefficient used in (33.9), are in fact normalized as they should be.

Near the origin (r small) we have, expanding $\sin kr$ in series and retaining only the term containing the lowest power of r after the differentiation,†

$$\left(\frac{1}{r}\frac{\mathrm{d}}{\mathrm{d}r}\right)^l \frac{\sin kr}{r}$$

$$\approx (-1)^l\left(\frac{1}{r}\frac{\mathrm{d}}{\mathrm{d}r}\right)^l \frac{k^{2l+1}r^{2l}}{(2l+1)!}$$

$$= (-1)^l k^{2l+1}/(2l+1)!!.$$

Thus the functions R_{kl} near the origin have the form

$$R_{kl} = \frac{2k^{l+1}}{(2l+1)!!}r^l, \tag{33.13}$$

in agreement with the general result (32.15).

In some problems (of scattering theory) it is necessary to consider wave functions which do not satisfy the usual conditions of finiteness, but correspond to a flux of particles from or to the centre. The wave function which describes such a flux of particles with angular momentum $l = 0$ is obtained by taking, instead of the "stationary spherical wave" (33.6), a solution in the form of an outgoing spherical wave $R_{k0}{}^{+}$ or an ingoing spherical wave $R_{k0}{}^{-}$, with

$$R_{k0}{}^{\pm} = (A/r)e^{\pm ikr}. \tag{33.14}$$

In the general case of an angular momentum l which is not zero, we obtain a solution of equation (33.3) in the form

$$R_{kl}{}^{\pm} = (-1)^l A\frac{r^l}{k^l}\left(\frac{1}{r}\frac{\mathrm{d}}{\mathrm{d}r}\right)^l \frac{e^{\pm ikr}}{r}. \tag{33.15}$$

These functions can be expressed in terms of Hankel functions:

$$R_{kl}{}^{\pm} = \pm iA\sqrt{(k\pi/2r)}H^{(1,2)}_{l+1/2}(kr), \tag{33.16}$$

of the first and second kinds for the signs + and − respectively.

† The symbol !! denotes the product of all integers of the same parity up to and including the number in question.

The asymptotic expression for these functions is

$$R_{kl}^{\pm} \approx A e^{\pm i(kr - l\pi/2)}/r. \tag{33.17}$$

Near the origin, it has the form

$$R_{kl}^{\pm} \approx A \frac{(2l-1)!!}{k^l} r^{-l-1}. \tag{33.18}$$

We normalize these functions so that they correspond to the emission (or absorption) of one particle per unit time. To do so, we notice that, at large distances, the spherical wave can be regarded as plane in any small interval, and the current density in it is $j = v\psi\psi^*$, where $v = k\hbar/m$ is the velocity of a particle. The normalization is determined by the condition $\oint j\,\mathrm{d}f = 1$, where the integration is carried out over a spherical surface of large radius r, i.e. $\int jr^2\,\mathrm{d}o = 1$, where d$o$ is an element of solid angle. If the angular functions are normalized as before, the coefficient A in the radial function must be put equal to

$$A = 1/\sqrt{v} = \sqrt{(m/k\hbar)}. \tag{33.19}$$

An asymptotic expression similar to (33.12) holds, not only for the radial part of the wave function of free motion, but also for motion (with positive energy) in any field which falls off sufficiently rapidly with distance.† At large distances we can neglect both the field and the centrifugal energy in Schrödinger's equation, and there remains the approximate equation

$$\frac{1}{r}\frac{\mathrm{d}^2(rR_{kl})}{\mathrm{d}r^2} + k^2 R_{kl} = 0.$$

The general solution of this equation is

$$R_{kl} \approx \frac{2\sin(kr - \frac{1}{2}l\pi + \delta_l)}{r}, \tag{33.20}$$

where δ_l is a constant, called the *phase shift*, and the common factor is chosen in accordance with the normalization of the wave function on the "$k/2\pi$ scale".‡ The constant phase shift δ_l is determined by the boundary condition (R_{kl} is finite as $r \to 0$); to do this, the exact Schrödinger's equation must be solved, and δ_l cannot be calculated in a general form. The phase shifts δ_l are, of course, functions of both l and k, and are an important property of the eigenfunctions of the continuous spectrum.

---

† As we shall show in §124, the field must decrease more rapidly than $1/r$.

‡ The term $-\frac{1}{2}l\pi$ in the argument of the sine is added so that $\delta_l = 0$ when the field is absent. Since the sign of the wave function as a whole is not significant, the phase shifts δ_l are determined to within $n\pi$ (not $2n\pi$). Their values may therefore always be chosen in the range between 0 and π.

PROBLEMS

PROBLEM 1. Determine the energy levels for the motion of a particle with angular momentum $l = 0$ in a spherical square potential well:

$$U(r) = -U_0 \text{ for } r < a, \; U(r) = 0 \text{ for } r > a.$$

SOLUTION. For $l = 0$ the wave functions depend only on r. Inside the well, Schrödinger's equation has the form

$$\frac{1}{r}\frac{d^2}{dr^2}(r\psi)+k^2\psi = 0, \quad k = \frac{1}{\hbar}\sqrt{[2m(U_0-|E|)]}.$$

The solution finite for $r = 0$ is

$$\psi = A\frac{\sin kr}{r}.$$

For $r > a$, we have the equation

$$\frac{1}{r}\frac{d^2}{dr^2}(r\psi)-\kappa^2\psi = 0, \quad \kappa = \frac{1}{\hbar}\sqrt{(2m|E|)}.$$

The solution vanishing at infinity is

$$\psi = A'e^{-\kappa r}/r.$$

The condition of the continuity of the logarithmic derivative of $r\psi$ at $r = a$ gives

$$k \cot ka = -\kappa = -\sqrt{[(2mU_0/\hbar^2)-k^2]}, \tag{1}$$

or

$$\sin ka = \pm\sqrt{(\hbar^2/2ma^2U_0)}ka. \tag{2}$$

This equation determines in implicit form the required energy levels (we must take those roots of the equation for which $\cot ka < 0$, as follows from (1)). The first of these levels (with $l = 0$) is at the same time the deepest of all energy levels whatsoever, i.e. it corresponds to the normal state of the particle.

If the depth U_0 of the potential well is small enough, there are no levels of negative energy, and the particle cannot "stay" in the well. This is easily seen from equation (2), by means of the following graphical construction. The roots of an equation of the form $\pm\sin x = \alpha x$ are given by the points of intersection of the line $y = \alpha x$ with the curves $y = \pm\sin x$, and we must take only those points of intersection for which $\cot x < 0$; the corresponding parts of the curve $y = \sin x$ are shown in Fig. 9 by a continuous line. We see that, if α is sufficiently large (U_0 small), there are no such points of intersection. The first such point appears when the line $y = \alpha x$ occupies the position shown, i.e. for $\alpha = 2/\pi$, and is at $x = \frac{1}{2}\pi$.

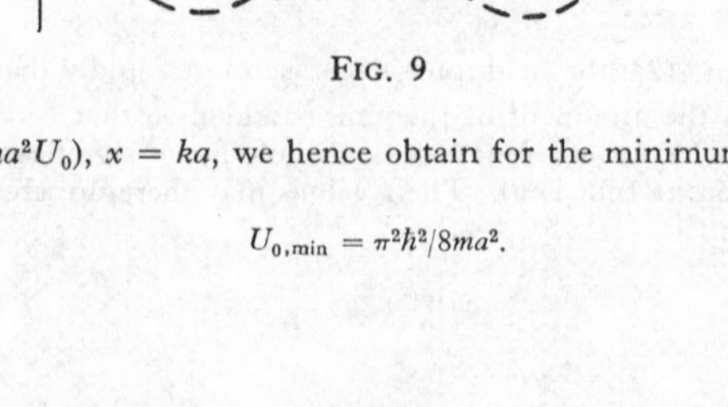

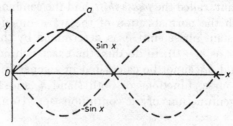

FIG. 9

Putting $\alpha = \hbar/\sqrt{(2ma^2U_0)}$, $x = ka$, we hence obtain for the minimum well depth to give a single negative level

$$U_{0,\min} = \pi^2\hbar^2/8ma^2. \tag{3}$$

This quantity increases with decreasing well radius a. The position of the first level E_1 at the point where it first appears is determined from $ka = \frac{1}{2}\pi$ and is $E_1 = 0$, as we should expect. As the well depth increases further, the normal level E_1 descends. When the difference $\Delta = (U_0/U_{0,\min}) - 1$ is small,

$$-E_1 = (\pi^2/16)U_{0,\min}\,\Delta^2. \tag{4}$$

PROBLEM 2. Determine the order of the energy levels with various values of the angular momentum l in a very deep spherical potential well ($U_0 \gg \hbar^2/ma^2$) (W. Elsasser 1933).

SOLUTION. The condition at the boundary of the well requires that $\psi \to 0$ as $U_0 \to \infty$ (see §22). Writing the radial wave function within the well in the form (33.10), we thus have the equation

$$J_{l+1/2}(ka) = 0,$$

whose roots give the position of the levels above the bottom of the well ($U_0 - |E| = \hbar^2k^2/2m$) for various values of l. The order of the levels from the ground state is found to be

1s, 1p, 1d, 2s, 1f, 2p, 1g, 2d, 1h, 3s, 2f, ...

The numbers preceding the letters give the sequence of levels for each l.†

PROBLEM 3. Determine the order of appearance of levels with various l as the depth U_0 of the well increases.

SOLUTION. When it first appears, each new level has energy $E = 0$. The corresponding wave function in the region outside the well, which vanishes as $r \to \infty$, is

$$R_l = \text{constant} \times r^{-(l+1)}$$

(the solution of equation (33.3) with $k = 0$). From the continuity of R_l and R_l' at the boundary of the well it follows, in particular, that the derivative $(r^{l+1}R_l)'$ is continuous, and so we have the following condition for the wave function within the well:

$$(r^{l+1}R_l)' = 0 \qquad \text{for} \qquad r = a.$$

This is equivalent‡ to the condition for the function R_{l-1} to vanish and, from (33.10), we obtain the equation

$$J_{l-1/2}(a\sqrt{(2mU_0)}/\hbar) = 0;$$

for $l = 0$ the function $J_{l-1/2}$ must be replaced by the cosine. This gives the following order of appearance of new levels as U_0 increases:

1s, 1p, 1d, 2s, 1f, 2p, 1g, 2d, 3s, 1h, 2f, ...

It may be noted that differences from the order of levels in a deep well occur only for comparatively high levels.

PROBLEM 4. Determine the energy levels of a three-dimensional oscillator (a particle in a field $U = \frac{1}{2}\mu\omega^2r^2$), their degrees of degeneracy, and the possible values of the orbital angular momentum in the corresponding stationary states.

SOLUTION. Schrödinger's equation for a particle in a field $U = \frac{1}{2}\mu\omega^2(x^2+y^2+z^2)$ allows separation of the variables, leading to three equations like that of a linear oscillator. The energy levels are therefore

$$E_n = \hbar\omega(n_1+n_2+n_3+\tfrac{3}{2}) \equiv \hbar\omega(n+\tfrac{3}{2}).$$

The degree of degeneracy of the nth level is equal to the number of ways in which n can be divided into the sum of three positive integral (or zero) numbers;‖ this is

$$\tfrac{1}{2}(n+1)(n+2).$$

---

† This notation is customary for particle levels in the nucleus (see §118).

‡ According to (33.7) and (33.8) we have $(r^{-l}R_l)' = r^{-l}R_{l+1}$. Since the equation (33.3) is unaltered when l is replaced by $-l-1$, we also have $(r^{l+1}R_{-l-1})' = r^{l+1}R_{-l}$. Finally, since the functions R_{-l} and R_{l-1} satisfy the same equation, we obtain $(r^{l+1}R_l)' = r^{l+1}R_{l-1}$, the formula used in the text.

‖ In other words, this is the number of ways in which n similar balls can be distributed among three urns.

The wave functions of the stationary states are

$$\psi_{n_1 n_2 n_3} = \text{constant} \times e^{-\alpha^2 r^2/2} H_{n_1}(\alpha x) H_{n_2}(\alpha y) H_{n_3}(\alpha z), \tag{1}$$

where $\alpha = \sqrt{(m\omega/\hbar)}$ and m is the mass of the particle. When the sign of the coordinate is changed, the polynomial H_n is multiplied by $(-1)^n$. The parity of the function (1) is therefore $(-1)^{n_1+n_2+n_3} = (-1)^n$. Taking linear combinations of these functions with a given sum $n_1+n_2+n_3 = n$, we can form the functions

$$\psi_{nlm} = \text{constant} \times e^{-\alpha^2 r^2/2} r^l Y_{lm}(\theta, \phi) F(-\tfrac{1}{2}n - \tfrac{1}{2}l, l+\tfrac{3}{2}, \alpha^2 r^2), \tag{2}$$

where $|m| = 0, 1, \ldots, l$ and l takes the values $0, 2, \ldots, n$ for even n and $1, 3, \ldots, n$ for odd n; F is the confluent hypergeometric function. This is evident from a comparison of the parities $(-1)^n$ of the functions (1) and $(-1)^l$ of the functions (2), which must be the same. This determines the possible values of the orbital angular momentum corresponding to the energy levels considered.

The order of levels of the three-dimensional oscillator is, therefore, with the same notation as in Problems 2 and 3,

$$(1s), (1p), (1d, 2s), (1f, 2p), (1g, 2d, 3s), \ldots,$$

where the parentheses enclose sets of degenerate states.†

§34. Resolution of a plane wave

Let us consider a free particle moving with a given momentum $p = k\hbar$ in the positive direction of the z-axis. The wave function of such a particle is of the form

$$\psi = \text{constant} \times e^{ikz}.$$

Let us expand this function in terms of the wave functions ψ_{klm} of free motion with various angular momenta. Since, in the state considered, the energy has the definite value $k^2\hbar^2/2m$, it is clear that only functions with this k will appear in the required expansion. Moreover, since the function e^{ikz} has axial symmetry about the z-axis, its expansion can contain only functions independent of the angle ϕ, i.e. functions with $m = 0$. Thus we must have

$$e^{ikz} = \sum_{l=0}^{\infty} a_l \psi_{kl0} = \sum_{l=0}^{\infty} a_l R_{kl} Y_{l0},$$

where the a_l are constants. Substituting the expressions (28.8) and (33.9) for the functions Y_{l0} and R_{kl}, we obtain

$$e^{ikz} = \sum_{l=0}^{\infty} C_l P_l(\cos\theta) \left(\frac{r}{k}\right)^l \left(\frac{1}{r}\frac{d}{dr}\right)^l \frac{\sin kr}{kr} \qquad (z = r\cos\theta),$$

where the C_l are other constants. These constants are conveniently determined by comparing the coefficients of $(r\cos\theta)^n$ in the expansions of the two sides of the equation in powers of r. On the right-hand side of the equation this term occurs only in the nth summand; for $l > n$, the expansion of the radial function begins at a higher power of r, while for $l < n$ the polynomial

† Note that levels with different angular momenta l are mutually degenerate; see the footnote at the end of §36.

$P_l(\cos\theta)$ contains only lower powers of $\cos\theta$. The term in $\cos^l\theta$ in $P_l(\cos\theta)$ has the coefficient $(2l)!/2^l(l!)^2$ (see formula (c.1)). Using also formula (33.13), we find the desired term of the expansion of the right-hand side of the equation to be

$$(-1)^l C_l \frac{(2l)!\,(kr\cos\theta)^l}{2^l(l!)^2(2l+1)!!}.$$

On the left-hand side of the equation the corresponding term (in the expansion of $e^{ikr\cos\theta}$) is

$$(ikr\cos\theta)^l/l!.$$

Equating these two quantities, we find $C_l = (-i)^l(2l+1)$. Thus we finally obtain the required expansion:

$$e^{ikz} = \sum_{l=0}^{\infty}(-i)^l(2l+1)P_l(\cos\theta)\left(\frac{r}{k}\right)^l\left(-\frac{1}{r}\frac{\mathrm{d}}{\mathrm{d}r}\right)^l\frac{\sin kr}{kr}. \tag{34.1}$$

At large distances this relation takes the asymptotic form

$$e^{ikz} \approx \frac{1}{kr}\sum_{l=0}^{\infty} i^l(2l+1)P_l(\cos\theta)\sin(kr-\tfrac{1}{2}l\pi). \tag{34.2}$$

In (34.1) the z-axis is in the direction of the wave vector $\mathbf{k}$ of the plane wave. This expansion can also be written in a more general form which does not presuppose a particular choice of the coordinate axes. For this purpose we must use the addition theorem for spherical harmonics (see (c.11)) to express the polynomials $P_l(\cos\theta)$ in terms of spherical harmonic functions of the directions of $\mathbf{k}$ and $\mathbf{r}$ (the angle between which is θ). The result is

$$e^{i\mathbf{k}.\mathbf{r}} = 4\pi\sum_{l=0}^{\infty}\sum_{m=-l}^{l} i^l j_l(kr)Y_{lm}^*(\mathbf{k}/k)Y_{lm}(\mathbf{r}/r). \tag{34.3}$$

The functions $j_l(kr)$ (defined by (33.11)) depend only on the product kr, and this makes evident the symmetry of the formula with respect to the vectors $\mathbf{k}$ and $\mathbf{r}$; it does not matter which of the two spherical harmonics is labelled as the complex conjugate.

We normalize the wave function e^{ikz} to give a probability current density of unity, i.e. so that it corresponds to a flux of particles (parallel to the z-axis) with one particle passing through unit area in unit time. This function is

$$\psi = v^{-1/2}e^{ikz} = \sqrt{(m/k\hbar)}e^{ikz}, \tag{34.4}$$

where v is the velocity of the particles; see (19.7). Multiplying both sides of equation (34.1) by $\sqrt{(m/k\hbar)}$ and introducing on the right-hand side the normalized functions $\psi_{klm}^{\pm} = R_{kl}^{\pm}(r)Y_{lm}(\theta,\phi)$, we obtain

$$\psi = \sum_{l=0}^{\infty}\sqrt{[\pi(2l+1)]}\frac{1}{ik}(\psi_{kl0}^{+}-\psi_{kl0}^{-}).$$

The squared modulus of the coefficient of ψ_{kl0}^{-} (or ψ_{kl0}^{+}) in this expansion determines, according to the usual rules, the probability that a particle in a current converging to (or diverging from) the centre has an angular momentum l (about the origin). Since the wave function $v^{-\frac{1}{2}}e^{ikz}$ corresponds to a current of particles of unit density, this "probability" has the dimensions of length squared; it can be conveniently interpreted as the magnitude of the "cross-section" (in the xy-plane) on which the particle must fall if its angular momentum is l. Denoting this quantity by σ_l, we have

$$\sigma_l = \pi(2l+1)/k^2. \tag{34.5}$$

For large values of l, the sum of the cross-sections over a range Δl of l (such that $1 \ll \Delta l \ll l$) is

$$\sum_{\Delta l} \sigma_l \approx \frac{\pi}{k^2} 2l\Delta l = 2\pi \frac{l\hbar^2}{p^2} \Delta l.$$

On substituting the classical expression for the angular momentum, $\hbar l = \rho p$ (where ρ is what is called the *impact parameter*), this expression becomes

$$2\pi\rho\Delta\rho,$$

in agreement with the classical result. This is no accident; we shall see below that, for large values of l, the motion is quasi-classical (see §49).

§35. Fall of a particle to the centre

To reveal certain properties of quantum-mechanical motion it is useful to examine a case which, it is true, has no direct physical meaning: the motion of a particle in a field where the potential energy becomes infinite at some point (the origin) according to the law $U(r) \approx -\beta/r^2$, $\beta > 0$; the form of the field at large distances from the origin is here immaterial. We have seen in §18 that this is a case intermediate between those where there are ordinary stationary states and those where a "fall" of the particle to the origin takes place.

Near the origin, Schrödinger's equation in the present case is

$$R'' + 2R'/r + \gamma R/r^2 = 0, \tag{35.1}$$

where $R(r)$ is the radial part of the wave function, and we have introduced the constant

$$\gamma = 2m\beta/\hbar^2 - l(l+1) \tag{35.2}$$

and have omitted all terms of lower orders in $1/r$; the value of the energy E is supposed finite, and so the corresponding term in the equation is omitted also.

Let us seek R in the form $R \sim r^s$; we then obtain for s the quadratic equation

$$s(s+1)+\gamma = 0,$$

which has the two roots

$$s_1 = -\tfrac{1}{2}+\sqrt{(\tfrac{1}{4}-\gamma)}, \quad s_2 = -\tfrac{1}{2}-\sqrt{(\tfrac{1}{4}-\gamma)}. \tag{35.3}$$

For further investigations it is convenient to proceed as follows. We draw a small region of radius r_0 round the origin, and replace the function $-\gamma/r^2$ in this region by the constant $-\gamma/r_0^2$. After determining the wave functions in this "cut off" field, we then examine the result of passing to the limit $r_0 \to 0$.

Let us first suppose that $\gamma < \frac{1}{4}$. Then s_1 and s_2 are real negative quantities, and $s_1 > s_2$. For $r > r_0$, the general solution of Schrödinger's equation has the form (always restricting ourselves to small r),

$$R = Ar^{s_1}+Br^{s_2}, \tag{35.4}$$

A and B being constants. For $r < r_0$, the solution of the equation

$$R''+2R'/r+\gamma R/r_0^2 = 0$$

which is finite at the origin has the form

$$R = C\frac{\sin kr}{r}, \quad k = \sqrt{\gamma}/r_0. \tag{35.5}$$

For $r = r_0$, the function R and its derivative R' must be continuous. It is convenient to write one of the conditions as a condition of continuity of the logarithmic derivative of rR. This gives the equation

$$\frac{A(s_1+1)r_0^{s_1}+B(s_2+1)r_0^{s_2}}{Ar_0^{s_1+1}+Br_0^{s_2+1}} = k\cot kr_0,$$

or

$$\frac{A(s_1+1)r_0^{s_1}+B(s_2+1)r_0^{s_2}}{Ar_0^{s_1}+Br_0^{s_2}} = \sqrt{\gamma}\cot\sqrt{\gamma}.$$

On solving for the ratio B/A, this equation gives an expression of the form

$$B/A = \text{constant}\times r_0^{s_1-s_2}. \tag{35.6}$$

Passing now to the limit $r_0 \to 0$, we find that $B/A \to 0$ (recalling that $s_1 > s_2$). Thus, of the two solutions of Schrödinger's equation (35.1) which diverge at the origin, we must choose that which becomes infinite less rapidly:

$$R = A/r^{|s_1|}. \tag{35.7}$$

Next, let $\gamma > \frac{1}{4}$. Then s_1 and s_2 are complex:

$$s_1 = -\tfrac{1}{2}+i\sqrt{(\gamma-\tfrac{1}{4})}, \quad s_2 = s_1^*.$$

Repeating the above analysis, we again arrive at equation (35.6), which, on substituting the values of s_1 and s_2, gives

$$B/A = \text{constant} \times r_0^{\,i\sqrt{(4\gamma-1)}}. \tag{35.8}$$

On passing to the limit $r_0 \to 0$, this expression does not tend to any definite limit, so that a direct passage to the limit is not possible. Using (35.8), the general form of the real solution can be written

$$R = \text{constant} \times r^{-1/2} \cos(\sqrt{(\gamma-\tfrac{1}{4})} \log(r/r_0) + \text{constant}). \tag{35.9}$$

This function has a number of zeros which increases without limit as r_0 decreases. Since, on the one hand, the expression (35.9) is valid for the wave function (when r is sufficiently small) with any finite value of the energy E of the particle, and, on the other hand, the wave function of the normal state can have no zeros, we can infer that the "normal state" of a particle in the field considered corresponds to the energy $E = -\infty$. In every state of a discrete spectrum, however, the particle is mainly in a region of space where $E > U$. Hence, for $E \to -\infty$, the particle is in an infinitely small region round the origin, i.e. the particle falls to the centre.

The "critical" field U_{cr} for which the fall of a particle to the centre becomes possible corresponds to the value $\gamma = \frac{1}{4}$. The smallest value of the coefficient of $-1/r^2$ is obtained for $l = 0$, i.e.

$$U_{cr} = -\hbar^2/8mr^2. \tag{35.10}$$

It is seen from formula (35.3) (for s_1) that the permissible solution of Schrödinger's equation (near the point where $U \sim 1/r^2$) diverges, as $r \to 0$, not more rapidly than $1/\sqrt{r}$. If the field becomes infinite, as $r \to 0$, more slowly than $1/r^2$, we can neglect $U(r)$, in Schrödinger's equation near the origin, in comparison with the other terms, and we obtain the same solutions as for free motion, i.e. $\psi \sim r^l$ (see §33). Finally, if the field becomes infinite more rapidly than $1/r^2$ (as $-1/r^s$ with $s > 2$), the wave function near the origin is proportional to $r^{\frac{1}{4}s-1}$ (see §49, Problem). In all these cases the product $r\psi$ tends to zero at $r = 0$.

Next, let us investigate the properties of the solutions of Schrödinger's equation in a field which diminishes at large distances according to the law $U \approx -\beta/r^2$, and has any form at small distances. We first suppose that $\gamma < \frac{1}{4}$. It is easy to see that in this case only a finite number of negative energy levels can exist.† For with energy $E = 0$ Schrödinger's equation at large distances has the form (35.1), with the general solution (35.4). The function (35.4), however, has no zeros (for $r \neq 0$); hence all zeros of the required radial wave function lie at finite distances from the origin, and their

† It is assumed that for small r the field is such that the particle does not fall.

number is always finite. In other words, the ordinal number of the level $E = 0$ which terminates the discrete spectrum is finite.

If $\gamma > \frac{1}{4}$, on the other hand, the discrete spectrum contains an infinite number of negative energy levels. For the wave function of the state with $E = 0$ has, at large distances, the form (35.9), with an infinite number of zeros, so that its ordinal number is always infinite.

Finally, let the field be $U = -\beta/r^2$ in all space. Then, for $\gamma > \frac{1}{4}$, the particle falls, but if $\gamma < \frac{1}{4}$ there are no negative energy levels. For the wave function of the state with $E = 0$ is of the form (35.7) in all space; it has no zeros at finite distances, i.e. it corresponds to the lowest energy level (for the given l).

§36. Motion in a Coulomb field (spherical polar coordinates)

A very important case of motion in a centrally symmetric field is that of motion in a *Coulomb field*

$$U = \pm\alpha/r$$

(where α is a positive constant). We shall first consider a Coulomb attraction, and shall therefore write $U = -\alpha/r$. It is evident from general considerations that the spectrum of negative eigenvalues of the energy will be discrete (with an infinite number of levels), while that of the positive eigenvalues will be continuous.

Equation (32.8) for the radial functions has the form

$$\frac{\mathrm{d}^2R}{\mathrm{d}r^2}+\frac{2}{r}\frac{\mathrm{d}R}{\mathrm{d}r}-\frac{l(l+1)}{r^2}R+\frac{2m}{\hbar^2}\left(E+\frac{\alpha}{r}\right)R = 0. \tag{36.1}$$

If we are concerned with the relative motion of two attracting particles, m must be taken as the reduced mass.

In calculations connected with the Coulomb field it is convenient to use, instead of the ordinary units, special units for the measurement of all quantities, which we shall call *Coulomb units*. As the units of measurement of mass, length and time, we take respectively

$$m,\quad \hbar^2/m\alpha,\quad \hbar^3/m\alpha^2.$$

All the remaining units are derived from these; thus the unit of energy is

$$m\alpha^2/\hbar^2.$$

From now on, in this section and the following one, we shall always (unless explicitly stated otherwise) use these units.†

† If $m = 9{\cdot}11\times10^{-28}$ g is the mass of the electron, and $\alpha = e^2$ (where e is the charge on the electron), the Coulomb units are the same as what are called *atomic units*. The atomic unit of length is

$$\hbar^2/me^2 = 0{\cdot}529\times10^{-8}\text{ cm}$$

(what is called the *Bohr radius*). The atomic unit of energy is

$$me^4/\hbar^2 = 4{\cdot}36\times10^{-11}\text{ erg} = 27{\cdot}21\text{ electron-volts}.$$

A quantity equal to one half of this unit is called a *rydberg*. The atomic unit of charge is $e = 4{\cdot}80\times10^{-10}$ esu. We formally obtain the formulae in atomic units by putting $e = m = \hbar = 1$. For $\alpha = Ze^2$ the Coulomb and atomic units are not the same.

Equation (36.1) in the new units is

$$\frac{d^2R}{dr^2}+\frac{2}{r}\frac{dR}{dr}-\frac{l(l+1)}{r^2}R+2\left(E+\frac{1}{r}\right)R = 0. \tag{36.2}$$

Discrete spectrum

Instead of the parameter E and the variable r, we introduce the new quantities

$$n = 1/\sqrt{(-2E)}, \quad \rho = 2r/n. \tag{36.3}$$

For negative energies, n is a real positive number. The equation (36.2), on making the substitutions (36.3), becomes

$$R''+\frac{2}{\rho}R'+\left[-\tfrac{1}{4}+\frac{n}{\rho}-\frac{l(l+1)}{\rho^2}\right]R = 0 \tag{36.4}$$

(the primes denote differentiation with respect to ρ).

For small ρ, the solution which satisfies the necessary conditions of finiteness is proportional to ρ^l (see (32.15)). To calculate the asymptotic behaviour of R for large ρ, we omit from (36.4) the terms in $1/\rho$ and $1/\rho^2$ and obtain the equation

$$R'' = \tfrac{1}{4}R,$$

whence $R = e^{\pm\frac{1}{2}\rho}$. The solution in which we are interested, which vanishes at infinity, consequently behaves as $e^{-\frac{1}{2}\rho}$ for large ρ.

It is therefore natural to make the substitution

$$R = \rho^l e^{-\rho/2} w(\rho), \tag{36.5}$$

when equation (36.4) becomes

$$\rho w''+(2l+2-\rho)w'+(n-l-1)w = 0. \tag{36.6}$$

The solution of this equation must diverge at infinity not more rapidly than every finite power of ρ, while for $\rho = 0$ it must be finite. The solution which satisfies the latter condition is the confluent hypergeometric function

$$w = F(-n+l+1, 2l+2, \rho) \tag{36.7}$$

(see §d of the Mathematical Appendices).† A solution which satisfies the condition at infinity is obtained only for negative integral (or zero) values of $-n+l+1$, when the function (36.7) reduces to a polynomial of degree $n-l-1$. Otherwise it diverges at infinity as e^ρ (see (d.14)).

Thus we reach the conclusion that the number n must be a positive integer, and for a given l we must have

$$n \geqslant l+1. \tag{36.8}$$

† The second solution of equation (36.6) diverges as ρ^{-2l-1} as $\rho \to 0$.

Recalling the definition (36.3) of the parameter n, we find

$$E = -1/2n^2, \quad n = 1, 2, \ldots. \tag{36.9}$$

This solves the problem of determining the energy levels of the discrete spectrum in a Coulomb field. We see that there are an infinite number of levels between the normal level $E_1 = -\frac{1}{2}$ and zero. The distances between successive levels diminish as n increases; the levels become more crowded as we approach the value $E = 0$, where the discrete spectrum closes up into the continuous spectrum. In ordinary units, formula (36.9) is†

$$E = -m\alpha^2/2\hbar^2 n^2. \tag{36.10}$$

The integer n is called the *principal quantum number*. The radial quantum number defined in §32 is

$$n_r = n-l-1.$$

For a given value of the principal quantum number, l can take the values

$$l = 0, 1, \ldots, n-1, \tag{36.11}$$

i.e. n different values in all. Only n appears in the expression (36.9) for the energy. Hence all states with different l but the same n have the same energy. Thus each eigenvalue is degenerate, not only with respect to the magnetic quantum number m (as in any motion in a centrally symmetric field) but also with respect to the number l. This latter degeneracy (called *accidental* or *Coulomb* degeneracy) is a specific property of the Coulomb field. To each value of l there correspond $2l+1$ different values of m. Hence the degree of degeneracy of the nth energy level is

$$\sum_{l=0}^{n-1} (2l+1) = n^2. \tag{36.12}$$

The wave functions of the stationary states are determined by formulae (36.5) and (36.7). The confluent hypergeometric functions with both parameters integral are the same, apart from a factor, as what are called the *generalized Laguerre polynomials* (see §d of the Mathematical Appendices). Hence

$$R_{nl} = \text{constant} \times \rho^l e^{-\rho/2} L_{n+l}^{2l+1}(\rho).$$

The radial functions must be normalized by the condition

$$\int_0^\infty R_{nl}^2 r^2 \, dr = 1.$$

† Formula (36.10) was first derived by N. Bohr in 1913, before the discovery of quantum mechanics. In quantum mechanics it was derived by W. Pauli in 1926 using the matrix method, and a few months later by E. Schrödinger using the wave equation.

Their final form is†

$$R_{nl} = -\frac{2}{n^2}\sqrt{\frac{(n-l-1)!}{[(n+l)!]^3}}e^{-r/n}\left(\frac{2r}{n}\right)^l L_{n+l}^{2l+1}\left(\frac{2r}{n}\right)$$

$$= \frac{2}{n^{l+2}(2l+1)!}\sqrt{\frac{(n+l)!}{(n-l-1)!}}(2r)^l e^{-r/n}F(-n+l+1, 2l+2, 2r/n); \quad (36.13)$$

the normalization integral is calculated by (f.6).‡

Near the origin, R_{nl} has the form

$$R_{nl} \cong r^l \frac{2^{l+1}}{n^{2+l}(2l+1)!}\sqrt{\frac{(n+l)!}{(n-l-1)!}}. \quad (36.14)$$

At large distances,

$$R_{nl} \approx (-1)^{n-l-1}\frac{2^n}{n^{n+1}\sqrt{[(n+l)!(n-l-1)!]}}r^{n-1}e^{-r/n}. \quad (36.15)$$

The wave function R_{10} of the normal state decreases exponentially at distances of the order $r \sim 1$, i.e. $r \sim \hbar^2/m\alpha$ in ordinary units.

The mean values of the various powers of r are calculated from the formula

$$\overline{r^k} = \int_0^\infty r^{k+2}R_{nl}{}^2\, dr.$$

The general formula for $\overline{r^k}$ can be obtained by means of formula (f.7). Here we shall give the first few values of $\overline{r^k}$ (for positive and negative k):

$$\left.\begin{aligned} \bar{r} &= \tfrac{1}{2}[3n^2-l(l+1)], \quad \overline{r^2} = \tfrac{1}{2}n^2[5n^2+1-3l(l+1)], \\ \overline{r^{-1}} &= 1/n^2, \quad \overline{r^{-2}} = 1/n^3(l+\tfrac{1}{2}). \end{aligned}\right\} \quad (36.16)$$

† We give the first few functions R_{nl} explicitly:

$$R_{10} = 2e^{-r},$$

$$R_{20} = (1/\sqrt{2})e^{-r/2}(1-\tfrac{1}{2}r),$$

$$R_{21} = (1/2\sqrt{6})e^{-r/2}r,$$

$$R_{30} = (2/3\sqrt{3})e^{-r/3}\left(1-\frac{2}{3}r+\frac{2}{27}r^2\right),$$

$$R_{31} = (8/27\sqrt{6})e^{-r/3}r\left(1-\frac{1}{6}r\right),$$

$$R_{32} = (4/81\sqrt{30})e^{-r/3}r^2.$$

‡ The normalization integral can also be calculated by substituting the expression (d.13) for the Laguerre polynomials and integrating by parts (similarly to the calculation of the integral (c.8) for the Legendre polynomials).

CONTINUOUS SPECTRUM

The spectrum of positive eigenvalues of the energy is continuous and extends from zero to infinity. Each of these eigenvalues is infinitely degenerate; to each value of E there corresponds an infinite number of states, with l taking all integral values from 0 to ∞ (and with all possible values of m for the given l).

The number n and the variable ρ, defined by the formulae (36.3), are now purely imaginary:

$$n = -i/\sqrt{(2E)} = -i/k, \quad \rho = 2ikr, \tag{36.17}$$

where $k = \sqrt{(2E)}$.† The radial eigenfunctions of the continuous spectrum are of the form

$$R_{kl} = \frac{C_{kl}}{(2l+1)!}(2kr)^l e^{-ikr} F(i/k+l+1, 2l+2, 2ikr), \tag{36.18}$$

where the C_{kl} are normalization factors. They can be represented as a complex integral (see §d):

$$R_{kl} = C_{kl}(2kr)^l e^{-ikr} \frac{1}{2\pi i} \oint e^{\xi}\left(1-\frac{2ikr}{\xi}\right)^{-i/k-l-1} \xi^{-2l-2}\, d\xi, \tag{36.19}$$

which is taken along the contour‡ shown in Fig. 10. The substitution

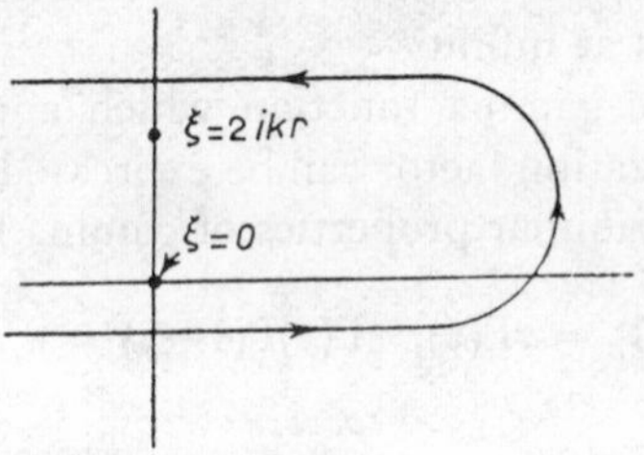

FIG. 10

$\xi = 2ikr(t+\frac{1}{2})$ converts this integral to the more symmetrical form

$$R_{kl} = C_{kl} \frac{(-2kr)^{-l-1}}{2\pi} \oint e^{2ikrt}(t+\tfrac{1}{2})^{i/k-l-1}(t-\tfrac{1}{2})^{-i/k-l-1}\, dt; \tag{36.20}$$

the path of integration passes in the positive direction round the points $t = \pm\frac{1}{2}$. It is seen at once from this representation that the functions R_{kl} are real.

The asymptotic expansion (d.14) of the confluent hypergeometric function

† It would be possible to define n and ρ by the complex conjugate expressions $n = i/k$, $\rho = -2ikr$; the real functions R_{kl} do not, of course, depend on which definition is used.

‡ Instead of this contour we could use any closed loop passing round the singular points $\xi = 0$ and $\xi = 2ikr$ in the positive direction. For integral l, the function $V(\xi) = \xi^{-n-l}(\xi-2ikr)^{n-l}$ (see §d) returns to its initial value on passing round such a contour.

enables us to obtain immediately a similar expansion for the wave functions R_{kl}. The two terms in (d.14) give two complex conjugate expressions in the function R_{kl}, and as a result we obtain

$$R_{kl} = C_{kl}\frac{e^{-\pi/2k}}{kr}\mathrm{re}\left\{\frac{e^{-i[kr-\pi(l+1)/2+(1/k)\log 2kr]}}{\Gamma(l+1-i/k)}G(l+1+i/k, i/k-l, -2ikr)\right\}. \tag{36.21}$$

If we normalize the wave functions on the "$k/2\pi$ scale" (i.e. by the condition (33.4)), the normalization coefficient is

$$C_{kl} = 2ke^{\pi/2k}|\Gamma(l+1-i/k)|. \tag{36.22}$$

For the asymptotic expression for R_{kl} when r is large (the first term of the expansion (36.21)) is then of the form

$$\left.\begin{aligned} R_{kl} &\approx \frac{2}{r}\sin\left(kr + \frac{1}{k}\log 2kr - \tfrac{1}{2}l\pi + \delta_l\right), \\ \delta_l &= \arg\Gamma(l+1-i/k), \end{aligned}\right\} \tag{36.23}$$

in agreement with the general form (33.20) of the normalized wave functions of the continuous spectrum in a centrally symmetric field. The expression (36.23) differs from (33.20) by the presence of a logarithmic term in the argument of the sine; however, since $\log r$ increases only slowly compared with r itself, the presence of this term is immaterial in calculating a normalization integral which diverges at infinity.

The modulus of the gamma function which appears in the expression (36.22) for the normalization factor can be expressed in terms of elementary functions. Using the familiar properties of gamma functions:

$$\Gamma(z+1) = z\Gamma(z), \quad \Gamma(z)\Gamma(1-z) = \pi/\sin\pi z,$$

we have

$$\begin{aligned} \Gamma(l+1+i/k) &= (l+i/k)\ldots(1+i/k)(i/k)\Gamma(i/k), \\ \Gamma(l+1-i/k) &= (l-i/k)\ldots(1-i/k)\Gamma(1-i/k), \end{aligned}$$

and also

$$\begin{aligned} |\Gamma(l+1-i/k)| &= [\Gamma(l+1-i/k)\Gamma(l+1+i/k)]^{1/2} \\ &= \sqrt{\frac{\pi}{k}}\prod_{s=1}^{l}\sqrt{\left(s^2+\frac{1}{k^2}\right)}\sinh^{-\frac{1}{2}}\frac{\pi}{k}. \end{aligned}$$

Thus

$$C_{kl} = \frac{\sqrt{(8\pi k)}}{\sqrt{(1-e^{-2\pi/k})}}\prod_{s=1}^{l}\sqrt{\left(s^2+\frac{1}{k^2}\right)}; \tag{36.24}$$

for $l = 0$ the product is replaced by unity.

The radial function for the special case of zero energy can be obtained by taking the limit $k \to 0$, for which

$$F\left(\frac{i}{k}+l+1,\ 2l+2,\ 2ikr\right) \to F\left(\frac{i}{k},\ 2l+2,\ 2ikr\right)$$

$$= 1 - \frac{2r}{(2l+2)\,.\,1!} + \frac{(2r)^2}{(2l+2)(2l+3)\,.\,2!} - \dots$$

$$= (2l+1)!\,(2r)^{-l-1/2} J_{2l+1}(\sqrt{(8r)}),$$

where J_{2l+1} is a Bessel function. The coefficients C_{kl} (36.24) for $k \to 0$ become

$$C_{kl} \approx \sqrt{(8\pi)}k^{-l+1/2}.$$

Hence

$$[R_{kl}/\sqrt{k}]_{k\to 0} = \sqrt{(4\pi/r)}J_{2l+1}(\sqrt{(8r)}). \qquad (36.25)$$

The asymptotic form of this function for large r is†

$$[R_{kl}/\sqrt{k}]_{k\to 0} = (8/r^3)^{1/4} \sin\left(\sqrt{(8r)} - l\pi - \tfrac{1}{4}\pi\right). \qquad (36.26)$$

The factor $\sqrt{k}$ disappears if we change to normalization on the energy scale, i.e. from the functions R_{kl} to R_{El} given by (33.5); the latter remains finite as $E \to 0$.

In a repulsive Coulomb field ($U = \alpha/r$) there is only a continuous spectrum of positive eigenvalues of the energy. Schrödinger's equation in this field can be formally obtained from the equation for an attractive field by changing the sign of r. Hence the wave functions of the stationary states are found immediately from (36.18) by the same alteration. The normalization coefficient is again determined from the asymptotic expression, and as a result we obtain

$$R_{kl} = \frac{C_{kl}}{(2l+1)!}(2kr)^l e^{ikr} F(i/k+l+1,\ 2l+2,\ -2ikr),$$

$$C_{kl} = 2\,ke^{-\pi/2k}|\Gamma(l+1+i/k)|$$

$$= \frac{\sqrt{(8\pi k)}}{\sqrt{(e^{2\pi/k}-1)}} \prod_{s=1}^{l} \sqrt{\left(s^2+\frac{1}{k^2}\right)}. \qquad (36.27)$$

† It may be noted that this function corresponds to the quasi-classical approximation (§49) applied to motion in the region $(l+\frac{1}{2})^2 \ll r \ll k^{-2}$.

The asymptotic expression for this function for large r is

$$\left.\begin{aligned} R_{kl} &\approx \frac{2}{r}\sin\left(kr-\frac{1}{k}\log 2kr-\tfrac{1}{2}l\pi+\delta_l\right),\\ \delta_l &= \arg\Gamma(l+1+i/k). \end{aligned}\right\} \tag{36.28}$$

THE NATURE OF THE COULOMB DEGENERACY

In classical motion of a particle in a Coulomb field, there is a conservation law peculiar to this type of field; if the field is an attractive one,

$$\mathbf{A} = \mathbf{r}/r-\mathbf{p}\times\mathbf{l} = \text{constant} \tag{36.29}$$

(see *Mechanics*, §15). In quantum mechanics, the corresponding operator is

$$\hat{\mathbf{A}} = \mathbf{r}/r-\tfrac{1}{2}(\hat{\mathbf{p}}\times\hat{\mathbf{l}}-\hat{\mathbf{l}}\times\hat{\mathbf{p}}), \tag{36.30}$$

and is easily seen to commute with the Hamiltonian $\hat{H} = \frac{1}{2}\hat{\mathbf{p}}^2-1/r$.

Direct calculation gives the following commutation rules for the operators $\hat{A}_i$ with one another and with the angular momentum operators:

$$\{\hat{l}_i, \hat{A}_k\} = ie_{ikl}\hat{A}_l, \quad \{\hat{A}_i, \hat{A}_k\} = -2i\hat{H}e_{ikl}\hat{l}_l. \tag{36.31}$$

The non-commutativity of the $\hat{A}_i$ means that the quantities A_x, A_y, A_z cannot simultaneously have definite values in quantum mechanics. Any one of the operators, say $\hat{A}_z$, commutes with the corresponding angular momentum component $\hat{l}_z$, but not with the squared angular momentum operator $\hat{\mathbf{l}}^2$. The existence of a further conserved quantity, which cannot be measured simultaneously with the others, leads (see §10) to an additional degeneracy of the levels, and this is the "accidental" degeneracy of the discrete energy levels, peculiar to the Coulomb field.

The origin of this degeneracy may also be formulated in terms of the increased symmetry of the Coulomb problem in quantum mechanics, in comparison with the symmetry relative to spatial rotations (V. A. Fok 1935). For this purpose we note that, in discrete-spectrum states with a fixed negative energy, we can replace $\hat{H}$ on the right of the second equation (36.31) by E, and use instead of the $\hat{A}_i$ the operators $\hat{u}_i = \hat{A}_i/\sqrt{(-2E)}$. The commutation rules for these operators are

$$\{\hat{l}_i, \hat{u}_k\} = ie_{ikl}\hat{u}_l, \quad \{\hat{u}_i, \hat{u}_k\} = ie_{ikl}\hat{l}_l. \tag{36.32}$$

These, together with the rule $\{\hat{\mathbf{l}}_i, \hat{\mathbf{l}}_k\} = ie_{ikl}\hat{\mathbf{l}}_l$, are formally identical with the commutation rules for the operators of infinitesimal rotations in four-dimensional Euclidean space.† This is the symmetry of the Coulomb problem in quantum mechanics.‡

From the commutation rules (36.32) we can again derive an expression for the energy levels in a Coulomb field.|| They can be rewritten by using instead of $\hat{\mathbf{l}}$ and $\hat{\mathbf{u}}$ the operators

$$\hat{\mathbf{j}}_1 = \tfrac{1}{2}(\hat{\mathbf{l}}+\hat{\mathbf{u}}), \quad \hat{\mathbf{j}}_2 = \tfrac{1}{2}(\hat{\mathbf{l}}-\hat{\mathbf{u}}). \tag{36.33}$$

For these,

$$\{\hat{j}_{1i}, \hat{j}_{1k}\} = ie_{ikl}\hat{j}_{1l}, \quad \{\hat{j}_{2i}, \hat{j}_{2k}\} = ie_{ikl}\hat{j}_{2l}, \quad \{\hat{j}_{1i}, \hat{j}_{2k}\} = 0. \tag{36.34}$$

These are formally identical with the commutation rules for two independent three-dimensional angular momentum vectors. The eigenvalues of $\mathbf{j}_1^2$ and $\mathbf{j}_2^2$ are therefore $j_1(j_1+1)$ and $j_2(j_2+1)$, where $j_1, j_2 = 0, \tfrac{1}{2}, 1, \tfrac{3}{2}, \ldots$.†† On the other hand, the definition of the operators $\hat{\mathbf{u}}$ and $\hat{\mathbf{l}} = \mathbf{r}\times\hat{\mathbf{p}}$ shows by a simple calculation that

$$\hat{\mathbf{l}}\,.\,\hat{\mathbf{u}} = \hat{\mathbf{u}}\,.\,\hat{\mathbf{l}} = 0,$$

$$\hat{\mathbf{l}}^2 + \hat{\mathbf{u}}^2 = -1 - \frac{1}{2E}$$

with $\hat{H}$ again being replaced by E in calculating $\hat{\mathbf{l}}^2+\hat{\mathbf{u}}^2$. Hence

$$\mathbf{j}_1^2 = \mathbf{j}_2^2 = -\tfrac{1}{4}\left(1 + \frac{1}{2E}\right) = j(j+1)$$

(where $j \equiv j_1 \equiv j_2$), and then $E = -1/2(2j+1)^2$. With the notation

$$2j+1 = n, \quad n = 1, 2, 3, \ldots, \tag{36.35}$$

we get the required result $E = 1/2n^2$. The degree of degeneracy of the levels is $(2j_1+1)(2j_2+1) = (2j+1)^2 = n^2$, as it should be. Lastly, since $\hat{\mathbf{l}} = \hat{\mathbf{j}}_1+\hat{\mathbf{j}}_2$,

---

† Here $\hat{l}_x, \hat{l}_y, \hat{l}_z$ represent the operators of infinitesimal rotations in the yz, zx and xy planes in four-dimensional Cartesian coordinates x, y, z, u; $\hat{u}_x, \hat{u}_y, \hat{u}_z$ are the operators of infinitesimal rotations in the xu, yu and zu planes.

‡ The symmetry appears explicitly in the wave functions in the momentum representation: see V. A. Fok, *Zeitschrift für Physik* **98,** 145, 1935.

|| This derivation is essentially as given by W. Pauli (1926).

†† Here we anticipate the properties of the angular momentum that are to be described in §54 (the possibility of integral and half-integral j).

for a given value of $j_1 = j_2 = \frac{1}{2}(n-1)$ the orbital angular momentum l takes values from 0 to $2j = n-1$.†

PROBLEMS

PROBLEM 1. Determine the probability distribution of various values of the momentum in the ground state of the hydrogen atom.

SOLUTION.‡ The wave function of the ground state is $\psi = R_{10}Y_{00} = (1/\sqrt{\pi})e^{-r}$. The wave function of this state in the **p** representation is then given by the integral

$$a(\mathbf{p}) = \int \psi(\mathbf{r})e^{-i\mathbf{p}.\mathbf{r}}\, dV$$

(see (15.10)). The integral is calculated by changing to spherical polar coordinates with the polar axis along **p**; the result is

$$a(\mathbf{p}) = \frac{8\sqrt{\pi}}{(1+p^2)^2},$$

and the probability density in **p**-space is $|a(\mathbf{p})|^2/(2\pi)^3$.

PROBLEM 2. Determine the mean potential of the field created by the nucleus and the electron in the ground state of the hydrogen atom.

SOLUTION. The mean potential ϕ_e created by an "electron cloud" at an arbitrary point **r** is most simply found as the spherically symmetric solution of Poisson's equation with charge density $\rho = -|\psi|^2$:

$$\frac{1}{r}\frac{d^2}{dr^2}(r\phi_e) = 4e^{-2r}.$$

Integrating this equation, and choosing the constants so that $\phi_e(0)$ is finite and $\phi_e(\infty) = 0$, and adding the potential of the field of the nucleus, we obtain

$$\phi = \frac{1}{r} + \phi_e(r) = \left(\frac{1}{r}+1\right)e^{-2r}.$$

For $r \ll 1$ we have $\phi \approx 1/r$ (the field of the nucleus), and for $r \gg 1$ the potential $\phi \approx e^{-2r}$ (the nucleus is screened by the electron).

---

† The "accidental" degeneracy of levels with different values of l occurs also for motion in a centrally symmetric field $U = \frac{1}{2}m\omega^2 r^2$ (a three-dimensional oscillator; see §33, Problem 4). This degeneracy is likewise due to the extra symmetry of the Hamiltonian. In this case, the symmetry arises because in $\hat{H} = \hat{\mathbf{p}}^2/2m + \frac{1}{2}m\omega^2\mathbf{r}^2$ both the operators $\hat{p}_i$ and the coordinates x_i occur as sums of squares. If they are replaced by the operators

$$\hat{a}_i = \frac{m\omega x_i + i\hat{p}_i}{\sqrt{(2m\hbar\omega)}},$$

$$\hat{a}_i{}^+ = \frac{m\omega x_i - i\hat{p}_i}{\sqrt{(2m\hbar\omega)}},$$

we obtain

$$\hat{H} = \hbar\omega[\hat{\mathbf{a}}^+ . \hat{\mathbf{a}} + \tfrac{3}{2}].$$

This is invariant under any unitary transformations of the operators $\hat{a}_i{}^+$ and $\hat{a}_i$ forming a group that is wider than that of the three-dimensional rotations (under which the particle Hamiltonian is invariant in any centrally symmetric field).

The specific property of the Coulomb and oscillator fields in quantum mechanics (presence of accidental degeneracy) is in correspondence with the fact that in classical mechanics closed particle trajectories exist in these (and only these) fields.

‡ In Problems 1 and 2, atomic units are used.

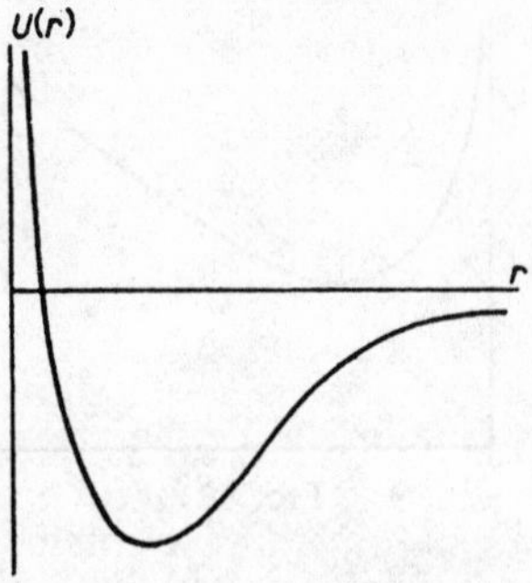

FIG. 11

PROBLEM 3. Determine the energy levels of a particle moving in a centrally symmetric field with potential energy $U = A/r^2 - B/r$ (Fig. 11).

SOLUTION. The spectrum of positive energy levels is continuous, while that of negative levels is discrete; we shall consider the latter. Schrödinger's equation for the radial function is

$$\frac{d^2R}{dr^2}+\frac{2}{r}\frac{dR}{dr}+\frac{2m}{\hbar^2}\left(E-\frac{\hbar^2}{2m}l(l+1)\frac{1}{r^2}-\frac{A}{r^2}+\frac{B}{r}\right)R = 0. \tag{1}$$

We introduce the new variable

$$\rho = 2\sqrt{(-2mE)}\,r/\hbar,$$

and the notation

$$2mA/\hbar^2+l(l+1) = s(s+1), \tag{2}$$

$$B\sqrt{(m/-2E)}/\hbar = n. \tag{3}$$

Then equation (1) takes the form

$$R''+\frac{2}{\rho}R'+\left(-\frac{1}{4}+\frac{n}{\rho}-\frac{s(s+1)}{\rho^2}\right)R = 0,$$

which is formally identical with (36.4). Hence we can at once conclude that the solution satisfying the necessary conditions is

$$R = \rho^s e^{-\rho/2}F(-n+s+1,\ 2s+2,\ \rho),$$

where $n-s-1 = p$ must be a positive integer (or zero), and s must be taken as the positive root of equation (2). From the definition (3) we consequently obtain the energy levels

$$-E_p = \frac{2B^2m}{\hbar^2}[2p+1+\sqrt{\{(2l+1)^2+8mA/\hbar^2\}}]^{-2}.$$

PROBLEM 4. The same as Problem 3, but with $U = A/r^2+Br^2$ (Fig. 12).

SOLUTION. There is only a discrete spectrum. Schrödinger's equation is

$$\frac{d^2R}{dr^2}+\frac{2}{r}\frac{dR}{dr}+\frac{2m}{\hbar^2}\left[E-\frac{\hbar^2 l(l+1)}{2mr^2}-\frac{A}{r^2}-Br^2\right]R = 0.$$

Introducing the variable

$$\xi = \sqrt{(2mB)}\,r^2/\hbar$$

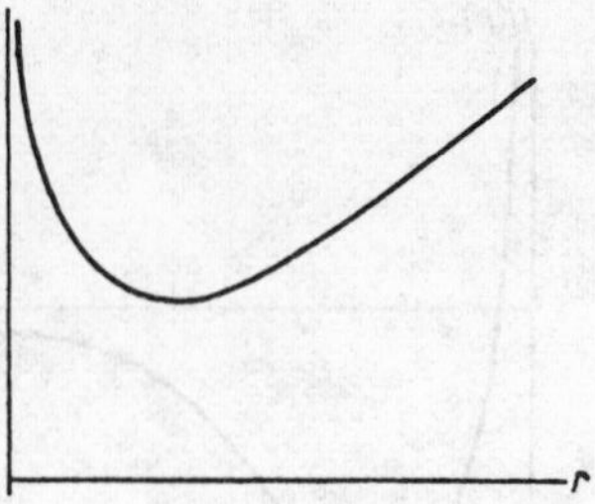

FIG. 12

and the notation

$$l(l+1)+2mA/\hbar^2 = 2s(2s+1),$$
$$\sqrt{(2m/B)}E/\hbar = 4(n+s)+3,$$

we obtain the equation

$$\xi R''+\frac{3}{2}R'+[n+s+\tfrac{3}{4}-\tfrac{1}{4}\xi-s(s+\tfrac{1}{2})/\xi]R = 0.$$

The solution required behaves asymptotically as $e^{-\frac{1}{2}\xi}$ when $\xi \to \infty$, while for small ξ it is proportional to ξ^s, where s must be taken as the positive quantity

$$s = \tfrac{1}{4}[-1+\sqrt{\{(2l+1)^2+8mA/\hbar^2\}}].$$

Hence we seek a solution in the form

$$R = e^{-\xi/2}\xi^s w,$$

obtaining for w the equation

$$\xi w''+\left(2s+\frac{3}{2}-\xi\right)w'+nw = 0,$$

whence

$$w = F\left(-n,\ 2s+\frac{3}{2},\ \xi\right),$$

where n must be a non-negative integer. We consequently find as the energy levels the infinite set of equidistant values

$$E_n = \hbar\sqrt{(B/2m)}[4n+2+\sqrt{\{(2l+1)^2+8mA/\hbar^2\}}],\ n = 0, 1, 2, \ldots.$$

§37. Motion in a Coulomb field (parabolic coordinates)

The separation of the variables in Schrödinger's equation written in spherical polar coordinates is always possible for motion in any centrally symmetric field. In the case of a Coulomb field, the separation of the variables is also possible in what are called *parabolic coordinates*. The solution of the problem of motion in a Coulomb field in terms of parabolic coordinates is useful in investigating a number of problems where a certain direction in space is distinctive; for example, for an atom in an external electric field (see §77).

The parabolic coordinates ξ, η, ϕ are defined by the formulae

$$\left.\begin{aligned} x &= \sqrt{(\xi\eta)}\cos\phi, \quad y = \sqrt{(\xi\eta)}\sin\phi, \quad z = \tfrac{1}{2}(\xi-\eta), \\ r &= \sqrt{(x^2+y^2+z^2)} = \tfrac{1}{2}(\xi+\eta), \end{aligned}\right\} \tag{37.1}$$

or conversely

$$\xi = r+z, \quad \eta = r-z, \quad \phi = \tan^{-1}(y/x); \tag{37.2}$$

ξ and η take values from 0 to ∞, and ϕ from 0 to 2π. The surfaces ξ = constant and η = constant are paraboloids of revolution about the z-axis, with focus at the origin. This system of coordinates is orthogonal. The element of length is given by the expression

$$(\mathrm{d}l)^2 = \frac{\xi+\eta}{4\xi}(\mathrm{d}\xi)^2 + \frac{\xi+\eta}{4\eta}(\mathrm{d}\eta)^2 + \xi\eta(\mathrm{d}\phi)^2, \tag{37.3}$$

and the element of volume is

$$\mathrm{d}V = \tfrac{1}{4}(\xi+\eta)\mathrm{d}\xi\mathrm{d}\eta\mathrm{d}\phi. \tag{37.4}$$

From (37.3) we have the Laplacian operator

$$\triangle = \frac{4}{\xi+\eta}\left[\frac{\partial}{\partial\xi}\left(\xi\frac{\partial}{\partial\xi}\right) + \frac{\partial}{\partial\eta}\left(\eta\frac{\partial}{\partial\eta}\right)\right] + \frac{1}{\xi\eta}\frac{\partial^2}{\partial\phi^2}. \tag{37.5}$$

Schrödinger's equation for a particle in an attractive Coulomb field with $U = -1/r = -2/(\xi+\eta)$ is

$$\frac{4}{\xi+\eta}\left[\frac{\partial}{\partial\xi}\left(\xi\frac{\partial\psi}{\partial\xi}\right) + \frac{\partial}{\partial\eta}\left(\eta\frac{\partial\psi}{\partial\eta}\right)\right] + \frac{1}{\xi\eta}\frac{\partial^2\psi}{\partial\phi^2} + 2\left(E + \frac{2}{\xi+\eta}\right)\psi = 0. \tag{37.6}$$

Let us seek the eigenfunctions ψ in the form

$$\psi = f_1(\xi)f_2(\eta)e^{im\phi}, \tag{37.7}$$

where m is the magnetic quantum number. Substituting this expression in equation (37.6) multiplied by $\frac{1}{4}(\xi+\eta)$, and separating the variables ξ and η, we obtain for f_1 and f_2 the equations

$$\left.\begin{aligned} \frac{\mathrm{d}}{\mathrm{d}\xi}\left(\xi\frac{\mathrm{d}f_1}{\mathrm{d}\xi}\right) + [\tfrac{1}{2}E\xi - \tfrac{1}{4}m^2/\xi + \beta_1]f_1 &= 0, \\ \frac{\mathrm{d}}{\mathrm{d}\eta}\left(\eta\frac{\mathrm{d}f_2}{\mathrm{d}\eta}\right) + [\tfrac{1}{2}E\eta - \tfrac{1}{4}m^2/\eta + \beta_2]f_2 &= 0, \end{aligned}\right\} \tag{37.8}$$

where the *separation parameters* β_1, β_2 are related by

$$\beta_1 + \beta_2 = 1. \tag{37.9}$$

Let us consider the discrete energy spectrum ($E < 0$). We introduce in place of E, ξ, η the quantities

$$n = 1/\sqrt{(-2E)}, \quad \rho_1 = \xi\sqrt{(-2E)} = \xi/n, \quad \rho_2 = \eta/n, \tag{37.10}$$

whereupon we obtain the equation for f_1

$$\frac{d^2 f_1}{d\rho_1^2}+\frac{1}{\rho_1}\frac{df_1}{d\rho_1}+\left[-\tfrac{1}{4}+\frac{1}{\rho_1}\left(\frac{|m|+1}{2}+n_1\right)-\frac{m^2}{4\rho_1^2}\right]f_1 = 0, \tag{37.11}$$

and a similar equation for f_2, with the notation

$$n_1 = -\tfrac{1}{2}(|m|+1)+n\beta_1, \quad n_2 = -\tfrac{1}{2}(|m|+1)+n\beta_2. \tag{37.12}$$

Similarly to the calculation for equation (36.4), we find that f_1 behaves as $e^{-\frac{1}{2}\rho_1}$ for large ρ_1 and as $\rho_1^{\frac{1}{2}|m|}$ for small ρ_1. Accordingly, we seek a solution of equation (37.11) in the form

$$f_1(\rho_1) = e^{-\rho_1/2}\rho_1^{|m|/2}w_1(\rho_1),$$

and similarly for f_2, obtaining for w_1 the equation

$$\rho_1 w_1''+(|m|+1-\rho_1)w_1'+n_1 w_1 = 0.$$

This is again the equation for a confluent hypergeometric function. The solution satisfying the conditions of finiteness is

$$w_1 = F(-n_1, |m|+1, \rho_1),$$

where n_1 must be a non-negative integer.

Thus each stationary state of the discrete spectrum is determined in parabolic coordinates by three integers: the *parabolic quantum numbers* n_1 and n_2, and the magnetic quantum number m. For n, the principal quantum number, we have from (37.9) and (37.12)

$$n = n_1+n_2+|m|+1. \tag{37.13}$$

For the energy levels, of course, we obtain our previous result (36.9).

For given n, the number $|m|$ can take n different values from 0 to $n-1$. For fixed n and $|m|$ the number n_1 takes $n-|m|$ values, from 0 to $n-|m|-1$. Taking into account also that for given $|m|$ we can choose the functions with $m = \pm|m|$, we find that for a given n there are altogether

$$2\sum_{m=1}^{n-1}(n-m)+(n-0) = n^2$$

different states, in agreement with the result obtained in §36.

The wave functions $\psi_{n_1 n_2 m}$ of the discrete spectrum must be normalized by the condition

$$\int |\psi_{n_1 n_2 m}|^2\, dV = \tfrac{1}{4}\int_0^\infty\int_0^\infty\int_0^{2\pi} |\psi_{n_1 n_2 m}|^2(\xi+\eta)\, d\phi\, d\xi\, d\eta = 1. \tag{37.14}$$

The normalized functions are

$$\psi_{n_1 n_2 m} = \frac{\sqrt{2}}{n^2} f_{n_1 m}\left(\frac{\xi}{n}\right) f_{n_2 m}\left(\frac{\eta}{n}\right) \frac{e^{im\phi}}{\sqrt{(2\pi)}}, \tag{37.15}$$

where

$$f_{pm}(\rho) = \frac{1}{|m|!}\sqrt{\frac{(p+|m|)!}{p!}} F(-p, |m|+1, \rho) e^{-\rho/2}\rho^{|m|/2}. \tag{37.16}$$

The wave functions in parabolic coordinates, unlike those in spherical polar coordinates, are not symmetrical about the plane $z = 0$. For $n_1 > n_2$ the probability of finding the particle in the direction $z > 0$ is greater than that for $z < 0$, and *vice versa* for $n_1 < n_2$.

To the continuous spectrum ($E > 0$) there corresponds a continuous spectrum of real values of the parameters β_1, β_2 in equations (37.8) (connected as before, of course, by the relation (37.9)). We shall not pause to write out here the corresponding wave functions, since it is not usually necessary to employ them. Equations (37.8), regarded as equations for the "eigenvalues" of the quantities β_1, β_2, have also (for $E > 0$) a spectrum of complex values. The corresponding wave functions are written out in §135, where we shall use them to solve a problem of scattering in a Coulomb field.

The existence of stationary states $|n_1 n_2 m\rangle$ leads to an additional conservation law (36.29). In these states, the quantities $l_z = m$ and A_z, as well as the energy, have definite values. Calculating the diagonal matrix elements of the operator $\hat{A}_z$, we find that

$$A_z = (n_1 - n_2)/n. \tag{37.17}$$

Here $u_z = n_1 - n_2$, and the components of the "angular momenta" $\mathbf{j}_1$ and $\mathbf{j}_2$ are

$$\left.\begin{aligned} j_{1z} &= \tfrac{1}{2}(m + n_1 - n_2) \equiv \mu_1, \\ j_{2z} &= \tfrac{1}{2}(m - n_1 + n_2) \equiv \mu_2. \end{aligned}\right\} \tag{37.18}$$

These properties of the states $|n_1 n_2 m\rangle$ (or, equivalently, $|n\mu_1\mu_2\rangle$) make it easy to establish the relation between their wave functions and those of the states $|nlm\rangle$. Since $\mathbf{l} = \mathbf{j}_1 + \mathbf{j}_2$, the change from one of these descriptions to the other is essentially the construction of wave functions with addition of two angular momenta, discussed in §106. In terms of the "angular momenta" $\mathbf{j}_1$ and $\mathbf{j}_2$, the states $|nlm\rangle$ and $|n_1 n_2 m\rangle$ are described as $|j_1 j_2 l m\rangle$ and $|j_1 j_2 \mu_1 \mu_2\rangle$, where, from (36.35) and (37.13),

$$j_1 = j_2 = \tfrac{1}{2}(n-1) = \tfrac{1}{2}(n_1 + n_2 + |m|). \tag{37.19}$$

According to the general formulae (106.9)–(106.11),

$$\left.\begin{aligned} \psi_{nlm} &= \sum_{\mu_1+\mu_2=m} \langle lm|\mu_1\mu_2\rangle \psi_{n\mu_1\mu_2}, \\ \psi_{n\mu_1\mu_2} &= \sum_{l=0}^{n-1} \langle l, \ \mu_1+\mu_2|\mu_1\mu_2\rangle \psi_{nlm} \end{aligned}\right\} \quad (37.20)$$

(D. Park 1960).

CHAPTER VI

PERTURBATION THEORY

§38. Perturbations independent of time

THE exact solution of Schrödinger's equation can be found only in a comparatively small number of the simplest cases. The majority of problems in quantum mechanics lead to equations which are too complex to be solved exactly. Often, however, quantities of different orders of magnitude appear in the conditions of the problem; among them there may be small quantities such that, when they are neglected, the problem is so much simplified that its exact solution becomes possible. In such cases, the first step in solving the physical problem concerned is to solve exactly the simplified problem, and the second step is to calculate approximately the errors due to the small terms that have been neglected in the simplified problem. There is a general method of calculating these errors; it is called *perturbation theory*.

Let us suppose that the Hamiltonian of a given physical system is of the form

$$\hat{H} = \hat{H}_0 + \hat{V},$$

where $\hat{V}$ is a small correction (or *perturbation*) to the *unperturbed* operator $\hat{H}_0$. In §§38, 39 we shall consider perturbations $\hat{V}$ which do not depend explicitly on time (the same is assumed regarding $\hat{H}_0$ also). The conditions which are necessary for it to be permissible to regard the operator $\hat{V}$ as "small" compared with the operator $\hat{H}$ will be derived below.

The problem of perturbation theory for a discrete spectrum can be formulated as follows. It is assumed that the eigenfunctions $\psi_n^{(0)}$ and eigenvalues $E_n^{(0)}$ of the discrete spectrum of the unperturbed operator $\hat{H}_0$ are known, i.e. the exact solutions of the equation

$$\hat{H}_0\psi^{(0)} = E^{(0)}\psi^{(0)} \tag{38.1}$$

are known. It is desired to find approximate solutions of the equation

$$\hat{H}\psi = (\hat{H}_0 + \hat{V})\psi = E\psi, \tag{38.2}$$

i.e. approximate expressions for the eigenfunctions ψ_n and eigenvalues E_n of the perturbed operator $\hat{H}$.

In this section we shall assume that no eigenvalue of the operator $\hat{H}_0$ is degenerate. Moreover, to simplify our results, we shall at first suppose that there is only a discrete spectrum of energy levels.

The calculations are conveniently performed in matrix form throughout.

To do this, we expand the required function ψ in terms of the functions $\psi_n^{(0)}$:

$$\psi = \sum_m c_m \psi_m^{(0)}. \tag{38.3}$$

Substituting this expansion in (38.2) we obtain

$$\sum_m c_m(E_m^{(0)}+\hat{V})\psi_m^{(0)} = \sum_m c_m E\psi_m^{(0)};$$

multiplying both sides of this equation by $\psi_k^{(0)*}$ and integrating, we find

$$(E-E_k^{(0)})c_k = \sum_m V_{km}c_m. \tag{38.4}$$

Here we have introduced the matrix V_{km} of the perturbation operator $\hat{V}$, defined with respect to the unperturbed functions $\psi_m^{(0)}$:

$$V_{km} = \int \psi_k^{(0)*}\hat{V}\psi_m^{(0)}\,dq. \tag{38.5}$$

We shall seek the values of the coefficients c_m and the energy E in the form of series

$$E = E^{(0)}+E^{(1)}+E^{(2)}+\ldots,\quad c_m = c_m^{(0)}+c_m^{(1)}+c_m^{(2)}+\ldots,$$

where the quantities $E^{(1)}$ and $c_m^{(1)}$ are of the same order of smallness as the perturbation $\hat{V}$, the quantities $E^{(2)}$ and $c_m^{(2)}$ are of the second order of smallness, and so on.

Let us determine the corrections to the *n*th eigenvalue and eigenfunction, putting accordingly $c_n^{(0)} = 1$, $c_m^{(0)} = 0$ for $m \neq n$. To find the first approximation, we substitute in equation (38.4) $E = E_n^{(0)}+E_n^{(1)}$, $c_k = c_k^{(0)}+c_k^{(1)}$, and retain only terms of the first order. The equation with $k = n$ gives

$$E_n^{(1)} = V_{nn} = \int \psi_n^{(0)*}\hat{V}\psi_n^{(0)}\,dq. \tag{38.6}$$

Thus the first-order correction to the eigenvalue $E_n^{(0)}$ is equal to the mean value of the perturbation in the state $\psi_n^{(0)}$.

The equation (38.4) with $k \neq n$ gives

$$c_k^{(1)} = V_{kn}/(E_n^{(0)}-E_k^{(0)}) \text{ for } k \neq n, \tag{38.7}$$

while $c_n^{(1)}$ remains arbitrary; it must be chosen so that the function $\psi_n = \psi_n^{(0)} + \psi_n^{(1)}$ is normalized up to and including terms of the first order. For this we must put $c_n^{(1)} = 0$. For the functions

$$\psi_n^{(1)} = \sum_m{}' \frac{V_{mn}}{E_n^{(0)}-E_m^{(0)}}\psi_m^{(0)} \tag{38.8}$$

(the prime means that the term with $m = n$ is omitted from the sum) are orthogonal to $\psi_n^{(0)}$, and hence the integral of $|\psi_n^{(0)}+\psi_n^{(1)}|^2$ differs from unity only by a quantity of the second order of smallness.

Formula (38.8) determines the correction to the wave functions in the first approximation. Incidentally, we see from this formula the condition for the applicability of the above method. This condition is that the inequality

$$|V_{mn}| \ll |E_n^{(0)} - E_m^{(0)}| \tag{38.9}$$

must hold, i.e. the matrix elements of the perturbation must be small compared with the corresponding differences between the unperturbed energy levels.

Next, let us determine the correction to the eigenvalue $E_n^{(0)}$ in the second approximation. To do this, we substitute in (38.4) $E = E_n^{(0)} + E_n^{(1)} + E_n^{(2)}$, $c_k = c_k^{(0)} + c_k^{(1)} + c_k^{(2)}$, and examine the terms of the second order of smallness. The equation with $k = n$ gives

$$E_n^{(2)} c_n^{(0)} = \sum_m{}' V_{nm} c_m^{(1)},$$

whence

$$E_n^{(2)} = \sum_m{}' \frac{|V_{mn}|^2}{E_n^{(0)} - E_m^{(0)}} \tag{38.10}$$

(we have substituted $c_m^{(1)}$ from (38.7), and used the fact that, since the operator $\hat{V}$ is Hermitian, $V_{mn} = V_{nm}^*$).

We notice that the correction in the second approximation to the energy of the normal state is always negative; for, since $E_n^{(0)}$ then corresponds to the lowest value of the energy, all the terms in the sum (38.10) are negative.

The further approximations can be calculated in a similar manner.

The results obtained can be generalized at once to the case where the operator $\hat{H}_0$ has also a continuous spectrum (but the perturbation is applied, as before, to a state of the discrete spectrum). To do so, we need only add to the sums over the discrete spectrum the corresponding integrals over the continuous spectrum. We shall distinguish the various states of the continuous spectrum by the suffix ν, which takes a continuous range of values; by ν we conventionally understand an assembly of values of quantities sufficient for a complete description of the state (if the states of the continuous spectrum are degenerate, which is almost always the case, the value of the energy alone does not suffice to determine the state).† Then, for instance, we must write instead of (38.8)

$$\psi_n^{(1)} = \sum_m{}' \frac{V_{mn}}{E_n^{(0)} - E_m^{(0)}} \psi_m^{(0)} + \int \frac{V_{\nu n}}{E_n^{(0)} - E_\nu} \psi_\nu^{(0)}\, d\nu, \tag{38.11}$$

and similarly for the other formulae.

It is useful to note also the formula for the perturbed value of the matrix element of a physical quantity f, calculated as far as terms of the first order by using the functions $\psi_n = \psi_n^{(0)} + \psi_n^{(1)}$, with $\psi_n^{(1)}$ given by (38.8). The

† Here the wave functions $\psi_\nu^{(0)}$ must be normalized by delta functions of the quantities ν.

following expression is easily obtained:

$$f_{nm} = f_{nm}^{(0)} + \sum_k{}' \frac{V_{nk} f_{km}^{(0)}}{E_n^{(0)} - E_k^{(0)}} + \sum_k{}' \frac{V_{km} f_{nk}^{(0)}}{E_m^{(0)} - E_k^{(0)}}. \tag{38.12}$$

In the first sum $k \neq n$, while in the second $k \neq m$.

PROBLEMS

PROBLEM 1. Determine the correction $\psi_n^{(2)}$ in the second approximation to the eigenfunctions.

SOLUTION. The coefficients $c_k^{(2)}$ ($k \neq n$) are calculated from equations (38.4) with $k \neq n$, written out up to terms of the second order, and the coefficient $c_n^{(2)}$ is chosen so that the function $\psi_n = \psi_n^{(0)} + \psi_n^{(1)} + \psi_n^{(2)}$ is normalized up to terms of the second order. As a result we find

$$\psi_n^{(2)} = \sum_m{}' \sum_k{}' \frac{V_{mk} V_{kn}}{\hbar^2 \omega_{nk} \omega_{nm}} \psi_m^{(0)} - \sum_m{}' \frac{V_{nn} V_{mn}}{\hbar^2 \omega_{nm}^2} \psi_m^{(0)} - \tfrac{1}{2} \psi_n^{(0)} \sum_m{}' \frac{|V_{mn}|^2}{\hbar^2 \omega_{nm}^2},$$

where we have introduced the frequencies

$$\omega_{nm} = (E_n^{(0)} - E_m^{(0)})/\hbar.$$

PROBLEM 2. Determine the correction in the third approximation to the eigenvalues of the energy.

SOLUTION. Writing out the terms of the third order of smallness in equation (38.4) with $k = n$, we obtain

$$E_n^{(3)} = \sum_k{}' \sum_m{}' \frac{V_{nm} V_{mk} V_{kn}}{\hbar^2 \omega_{mn} \omega_{kn}} - V_{nn} \sum_m{}' \frac{|V_{nm}|^2}{\hbar^2 \omega_{mn}^2}.$$

PROBLEM 3. Determine the energy levels of an anharmonic linear oscillator whose Hamiltonian is

$$\hat{H} = \tfrac{1}{2} \hat{p}^2/m + \tfrac{1}{2} m x^2 \omega^2 + \alpha x^3 + \beta x^4.$$

SOLUTION. The matrix elements of x^3 and x^4 can be obtained directly according to the rule of matrix multiplication, using the expression (23.4) for the matrix elements of x. We find for the matrix elements of x^3 that are not zero

$$(x^3)_{n-3,n} = (x^3)_{n,n-3} = (\hbar/m\omega)^{3/2} \sqrt{[\tfrac{1}{8} n(n-1)(n-2)]},$$
$$(x^3)_{n-1,n} = (x^3)_{n,n-1} = (\hbar/m\omega)^{3/2} \sqrt{(9n^3/8)}.$$

The diagonal elements in this matrix vanish, so that the correction in the first approximation due to the term αx^3 in the Hamiltonian (regarded as a perturbation of the harmonic oscillator) is zero. The correction in the second approximation due to this term is of the same order as that in the first approximation due to the term βx^4. The diagonal matrix elements of x^4 are

$$(x^4)_{n,n} = (\hbar/m\omega)^2 \cdot \tfrac{3}{4}(2n^2+2n+1).$$

Using the general formulae (38.6) and (38.10), we find the following approximate expression for the energy levels of the anharmonic oscillator:

$$E_n = \hbar\omega(n+\tfrac{1}{2}) - \frac{15}{4} \frac{\alpha^2}{\hbar\omega} \left(\frac{\hbar}{m\omega}\right)^3 \left(n^2+n+\frac{11}{30}\right) + \frac{3}{2} \beta \left(\frac{\hbar}{m\omega}\right)^2 (n^2+n+\tfrac{1}{2}).$$

PROBLEM 4. A spherical potential well with infinitely high walls is subjected to a small deformation (without change of volume) which gives it the form of a slightly prolate or oblate spheroid with semi-axes $a = b$ and c. Find the splitting of the energy levels of a particle in the deformed well (A. B. Migdal 1959).

SOLUTION. The equation of the well boundary is

$$\frac{x^2+y^2}{a^2} + \frac{z^2}{c^2} = 1,$$

and by the change of variables $x \to ax/R$, $y \to ay/R$, $z \to cz/R$ it is converted into $x^2+y^2+z^2 = R^2$, the equation of a sphere with radius R. The same change of variables converts the Hamiltonian of the particle, $\hat{H} = \hat{\mathbf{p}}^2/2M = -\hbar^2\triangle/2M$ (where M is the mass of the particle and the energy is measured from the bottom of the well) into $\hat{H} = \hat{H}_0 + \hat{V}$, where

$$\hat{H}_0 = -\hbar^2\triangle/2M,$$

$$\hat{V} = -\frac{\hbar^2}{2M}\left[\left(\frac{R^2}{a^2}-1\right)\left(\frac{\partial^2}{\partial x^2}+\frac{\partial^2}{\partial y^2}\right)+\left(\frac{R^2}{c^2}-1\right)\frac{\partial^2}{\partial z^2}\right].$$

Thus the problem of motion in an ellipsoidal well reduces to that of motion in a spherical well. If the ellipsoid is almost a sphere of radius $R = (a^2c)^{1/3}$, $\hat{V}$ may be regarded as a small perturbation. If the ellipsoidality β ($|\beta| \ll 1$) is defined by

$$a \approx R(1-\tfrac{1}{3}\beta), \qquad c \approx R(1+\tfrac{2}{3}\beta),$$

the perturbation operator may be written

$$\hat{V} = (\beta/3M)(\hat{\mathbf{p}}^2 - 3\hat{p}_z^2).$$

In the first order of perturbation theory, the change in the energy levels of the particle from their values in the spherical well is

$$\Delta E_{nlm} = E_{nlm} - E_{nl}^{(0)}$$
$$= \langle nlm|V|nlm\rangle,$$

where l and m are the angular momentum of the particle and its component along the axis of the spheroid; n numbers the levels in the spherical well for a given l, which are independent of m. Since $\mathbf{p}^2 - 3p_z^2$ is the zz-component of an irreducible tensor, $\delta_{ik}\mathbf{p}^2 - 3p_ip_k$, with zero trace, we find from (107.2) and (107.6) that the matrix element $\langle nlm|V|nlm\rangle$ is proportional to

$$(-1)^m\begin{pmatrix} l & 2 & l \\ -m & 0 & m \end{pmatrix},$$

and therefore

$$\langle nlm|V|nlm\rangle = \left(1 - \frac{3m^2}{l(l+1)}\right)\langle nl0|V|nl0\rangle.$$

A table of $3j$-symbols is given in §106.

Next,

$$\langle nl0|V|nl0\rangle = \tfrac{2}{3}\beta E_{nl}^{(0)} + \beta\frac{\hbar^2}{M}\langle nl0\left|\frac{\partial^2}{\partial z^2}\right|nl0\rangle$$
$$= \tfrac{2}{3}\beta E_{nl}^{(0)} - \beta\frac{\hbar^2}{M}\int\left|\frac{\partial\psi_{n0}}{\partial z}\right|^2 r^2\,dr\,do;$$

in the first term we have used Schrödinger's equation $\hat{H}_0\psi_{nlm} = E_{nl}^{(0)}\psi_{nlm}$ for a spherical well, and in the second term integrated by parts. With Y_{l0} in the form (28.11), we find the derivative of $\psi_{nl0} = R_{nl}(r) Y_{l0}(\theta, \phi)$ to be

$$\frac{\partial}{\partial z}\psi_{nl0} = \left(\cos\theta\frac{\partial}{\partial r} - \frac{\sin\theta}{r}\frac{\partial}{\partial\theta}\right)\psi_{nl0}$$

$$= -\frac{i(l+1)}{[4(l+1)^2-1]^{1/2}}\left(R_{nl}' - \frac{l}{r}R_{nl}\right)Y_{l+1,\,0} +$$

$$+\frac{il}{[4l^2-1]^{1/2}}\left(R_{nl}' + \frac{l+1}{r}R_{nl}\right)Y_{l-1,\,0}.$$

The radial integrals are calculated by means of the formulae

$$\int_0^\infty R_{nl}R_{nl}'\, r\, dr = -\tfrac{1}{2}\int_0^\infty R_{nl}^2\, dr,$$

$$\int_0^\infty R_{nl}'^2 r^2\, dr = \frac{2M}{\hbar^2}E_{nl}^{(0)} - l(l+1)\int_0^\infty R_{nl}^2\, dr,$$

which are derived by integrating by parts and using the radial Schrödinger's equation (33.3)

$$R_{nl}'' + \frac{2}{r}R_{nl}' - \frac{l(l+1)}{r^2}R_{nl} = -\frac{2M}{\hbar^2}E_{nl}^{(0)}.$$

The terms containing integrals of R_{nl}^2 cancel, and the final result is

$$\Delta E_{nlm} = 4\beta\frac{l(l+1)}{(2l-1)(2l+3)}\left[\frac{m^2}{l(l+1)} - \frac{1}{3}\right]E_{nl}^{(0)}.$$

Note that

$$\frac{1}{2l+1}\sum_{m=-l}^{l} E_{nlm} = E_{nl}^{(0)},$$

i.e. the "centre of gravity" of the multiplet is not shifted.

§39. The secular equation

Let us now turn to the case where the unperturbed operator $\hat{H}_0$ has degenerate eigenvalues. We denote by $\psi_n^{(0)}$, $\psi_{n'}^{(0)}$, ... the eigenfunctions belonging to the same eigenvalue $E_n^{(0)}$ of the energy. The choice of these functions is, as we know, not unique; instead of them we can choose any s (where s is the degree of degeneracy of the level $E_n^{(0)}$) independent linear combinations of these functions. The choice ceases to be arbitrary, however, if we subject the wave functions to the requirement that the change in them under the action of the small applied perturbation should be small.

At present we shall understand by $\psi_n^{(0)}$, $\psi_{n'}^{(0)}$, ... some arbitrarily selected unperturbed eigenfunctions. The correct functions in the zeroth approximation are linear combinations of the form

$$c_n^{(0)}\psi_n^{(0)} + c_{n'}^{(0)}\psi_{n'}^{(0)} + \dots\,.$$

The coefficients in these combinations are determined, together with the corrections in the first approximation to the eigenvalues, as follows.

We write out equations (38.4) with $k = n, n', \ldots$, and substitute in them, in the first approximation, $E = E_n^{(0)} + E^{(1)}$; for the quantities c_k it suffices to take the zero-order values $c_n = c_n^{(0)}$, $c_{n'} = c_{n'}^{(0)}, \ldots$; $c_m = 0$ for $m \neq n$, $n', \ldots$. We then obtain

$$E^{(1)} c_n^{(0)} = \sum_{n'} V_{nn'} c_{n'}^{(0)}$$

or

$$\sum_{n'} (V_{nn'} - E^{(1)} \delta_{nn'}) c_{n'}^{(0)} = 0, \tag{39.1}$$

where n, n' take all values denumerating states belonging to the given unperturbed eigenvalue $E_n^{(0)}$. This system of homogeneous linear equations for the quantities $c_n^{(0)}$ has solutions which are not all zero if the determinant of the coefficients of the unknowns vanishes. Thus we obtain the equation

$$|V_{nn'} - E^{(1)} \delta_{nn'}| = 0. \tag{39.2}$$

This equation is of the sth degree in $E^{(1)}$ and has, in general, s different real roots. These roots are the required corrections to the eigenvalues in the first approximation. Equation (39.2) is called the *secular equation*.† We notice that the sum of its roots is equal to the sum of the diagonal matrix elements $V_{nn}, V_{n'n'}, \ldots$ (this being the coefficient of $[E^{(1)}]^{s-1}$ in the equation).

Substituting in turn the roots of equation (39.2) in the system (39.1) and solving, we find the coefficients $c_n^{(0)}$ and so determine the eigenfunctions in the zeroth approximation.

As a result of the perturbation, an originally degenerate energy level ceases in general to be degenerate (the roots of equation (39.2) are in general distinct); the perturbation *removes* the degeneracy, as we say. The removal of the degeneracy may be either total or partial (in the latter case, after the perturbation has been applied, there remains a degeneracy of degree less than the original one).

It may happen that for some reason all the matrix elements are particularly small (or even zero) for transitions within a group of mutually degenerate states $n, n', \ldots$. It may then be useful to take into account not only in the first order the matrix elements $V_{nn'}$ but also in the higher orders the matrix elements V_{nm} ($m \neq n, n', \ldots$) for transitions to states with a different energy. Let us do this for the matrix elements V_{mn} in the second order.

In equation (38.4) with $k = n$ we put on the left $E = E_n^{(0)} + E_n^{(1)}$ (retaining the notation $E^{(1)}$ for the correction to the energy in the approximation considered), and replace c_n by $c_n^{(0)}$. Since $c_m^{(0)} = 0$ for all $m \neq n, n', \ldots$, we have

$$E^{(1)} c_n^{(0)} = \sum_m V_{nm} c_m^{(1)} + \sum_{n'} V_{nn'} c_{n'}^{(0)}. \tag{39.3}$$

† The name is taken from celestial mechanics.

The equations (38.4) with $k = m \neq n, n', \ldots$ give as far as the first-order terms

$$(E_n^{(0)} - E_m^{(0)})c_m^{(1)} = \sum_{n'} V_{mn'}c_{n'}^{(0)},$$

whence

$$c_m^{(1)} = \sum_{n'} \frac{V_{mn'}}{E_n^{(0)} - E_m^{(0)}} c_{n'}^{(0)}.$$

Substitution in (39.3) gives

$$E^{(1)}c_n^{(0)} = \sum_{n'} c_{n'}^{(0)} \left(V_{nn'} + \sum_m \frac{V_{nm}V_{mn'}}{E_n^{(0)} - E_m^{(0)}} \right).$$

These equations replace (39.1); the condition for them to be compatible again leads to the secular equation, which differs from (39.2) by the change

$$V_{nn'} \to V_{nn'} + \sum_m \frac{V_{nm}V_{mn'}}{E_n^{(0)} - E_m^{(0)}}. \tag{39.4}$$

PROBLEMS

PROBLEM 1. Determine the corrections to the eigenvalue in the first approximation and the correct functions in the zeroth approximation, for a doubly degenerate level.

SOLUTION. Equation (39.2) here has the form

$$\begin{vmatrix} V_{11}-E^{(1)} & V_{12} \\ V_{21} & V_{22}-E^{(1)} \end{vmatrix} = 0$$

(the suffixes 1 and 2 correspond to two arbitrarily chosen unperturbed eigenfunctions $\psi_1^{(0)}$ and $\psi_2^{(0)}$ of the degenerate level in question). Solving, we find

$$E^{(1)} = \tfrac{1}{2}[(V_{11}+V_{22}) \pm \hbar\omega^{(1)}], \tag{1}$$

with the notation

$$\hbar\omega^{(1)} = \sqrt{\{(V_{11}-V_{22})^2 + 4|V_{12}|^2\}}$$

for the difference between the two values of the correction $E^{(1)}$. Solving also equations (39.1) with these values of $E^{(1)}$, we obtain for the coefficients in the correct normalized function in the zeroth approximation, $\psi^{(0)} = c_1^{(0)}\psi_1^{(0)} + c_2^{(0)}\psi_2^{(0)}$, the values

$$\left.\begin{aligned} c_1^{(0)} &= \left\{ \frac{V_{12}}{2|V_{12}|} \left[1 \pm \frac{V_{11}-V_{22}}{\hbar\omega^{(1)}} \right] \right\}^{1/2}, \\ c_2^{(0)} &= \pm \left\{ \frac{V_{21}}{2|V_{12}|} \left[1 \mp \frac{V_{11}-V_{22}}{\hbar\omega^{(1)}} \right] \right\}^{1/2}. \end{aligned}\right\} \tag{2}$$

PROBLEM 2. Derive the formulae for the correction to the eigenfunctions in the first approximation and to the eigenvalues in the second approximation.

SOLUTION. We shall suppose that the correct functions in the zeroth approximation are chosen as the functions $\psi_n^{(0)}$. The matrix $V_{nn'}$ defined with respect to these is clearly diagonal with respect to the suffixes n, n' (belonging to the same group of functions of a degenerate level), and the diagonal elements $V_{nn}, V_{n'n'}$ are equal to the corresponding corrections $E_n^{(1)}, E_{n'}^{(1)}, \ldots$ in the first approximation.

Let us consider a perturbation of the eigenfunction $\psi_n^{(0)}$, so that in the zeroth approximation $E = E_n^{(0)}$, $c_n^{(0)} = 1$, $c_m^{(0)} = 0$ for $m \neq n$. In the first approximation $E = E_n^{(0)} + V_{nn}$, $c_n = 1 + c_n^{(1)}$, $c_m = c_m^{(1)}$. We write out from the system (38.4) the equation with $k \neq n, n', \ldots$, retaining in it terms of the first order:

$$(E_n^0 - E_k^{(0)})c_k^{(1)} = V_{kn}c_n^{(0)} = V_{kn},$$

whence

$$c_k^{(1)} = V_{kn}/(E_n^{(0)} - E_k^{(0)}) \text{ for } k \neq n, n', \ldots . \tag{1}$$

Next we write out the equation with $k = n'$, retaining in it terms of the second order:

$$E_n^{(1)}c_{n'}^{(1)} = V_{n'n'}c_{n'}^{(1)} + \sum_m{}' V_{n'm}c_m^{(1)}$$

(the terms with $m = n, n', \ldots$ are omitted in the sum over m). Substituting $E_n^{(1)} = V_{nn}$ and the expression (1) for $c_m^{(1)}$, we obtain for $n' \neq n$

$$c_{n'}^{(1)} = \frac{1}{V_{nn} - V_{n'n'}} \sum_m{}' \frac{V_{n'm}V_{mn}}{E_n^{(0)} - E_m^{(0)}}. \tag{2}$$

(In this approximation the coefficient $c_n^{(1)}$ is zero.) Formulae (1) and (2) determine the correction $\psi_n^{(1)} = \Sigma c_m^{(1)}\psi_m^{(0)}$ to the eigenfunctions in the first approximation.†

Finally, writing out the second-order terms in equation (38.4) with $k = n$, we obtain for the second-order corrections to the energy the formula

$$E_n^{(2)} = \sum_m{}' \frac{V_{nm}V_{mn}}{E_n^{(0)} - E_m^{(0)}}, \tag{3}$$

which is formally identical with (38.10).

Problem 3. At the initial instant $t = 0$, a system is in a state $\psi_1^{(0)}$ which belongs to a doubly degenerate level. Determine the probability that, at a subsequent instant t, the system will be in the state $\psi_2^{(0)}$ with the same energy; the transition occurs under the action of a constant perturbation.

Solution. We form the correct functions in the zeroth approximation,

$$\psi = c_1\psi_1 + c_2\psi_2, \qquad \psi' = c_1'\psi_1 + c_2'\psi_2,$$

where $c_1, c_2; c_1', c_2'$ are two pairs of coefficients determined by formulae (2) of Problem 1 (for brevity, we omit the index $^{(0)}$ on all quantities).

Conversely,

$$\psi_1 = \frac{c_2'\psi - c_2\psi'}{c_1c_2' - c_1'c_2}.$$

The functions ψ and ψ' belong to states with perturbed energies $E + E^{(1)}$ and $E + E^{(1)\prime}$, where $E^{(1)}$ and $E^{(1)\prime}$ are the two values of the correction (1) in Problem 1. On introducing the time factors we pass to the time-dependent wave functions:

$$\Psi_1 = \frac{e^{-(i/\hbar)Et}}{c_1c_2' - c_1'c_2}[c_2'\psi e^{-(i/\hbar)E^{(1)}t} - c_2\psi' e^{-(i/\hbar)E^{(1)\prime}t}]$$

† Note that the condition for the quantities (1) and (2) to be small (and therefore the condition for this method of perturbation theory to be applicable) again requires the conditions (38.9) to be satisfied only for transitions between states belonging to different energy levels. Transitions between states belonging to the same degenerate level are taken into account exactly (in a certain sense) by the secular equation.

(at time $t = 0$, $\Psi_1 = \psi_1$). Finally, again expressing ψ, ψ' in terms of ψ_1, ψ_2, we obtain Ψ_1 as a linear combination of ψ_1 and ψ_2, with coefficients depending on time. The squared modulus of the coefficient of ψ_2 determines the required transition probability w_{21}. Calculation with (1) and (2) from Problem 1 gives

$$w_{21} = \frac{2|V_{12}|^2}{(\hbar\omega^{(1)})^2}\{1-\cos\omega^{(1)}t\}.$$

We see that the probability varies periodically with time, with frequency $\omega^{(1)}$.

For times t which are small compared with the period in question, the expression in the braces, and therefore w_{21}, is proportional to t^2: $w_{21} = |V_{12}|^2t^2/\hbar^2$. This formula can be very simply obtained by the method given in the next section (using equation (40.4)).

§40. Perturbations depending on time

Let us now go on to study perturbations depending explicitly on time. We cannot speak in this case of corrections to the eigenvalues, since, when the Hamiltonian is time-dependent (as will be the perturbed operator $\hat{H} = \hat{H}_0 + \hat{V}(t)$), the energy is not conserved, so that there are no stationary states. The problem here consists in approximately calculating the wave functions from those of the stationary states of the unperturbed system.

To do this, we shall apply a method analogous to the well-known method of varying the constants to solve linear differential equations (P. A. M. Dirac 1926). Let $\Psi_k^{(0)}$ be the wave functions (including the time factor) of the stationary states of the unperturbed system. Then an arbitrary solution of the unperturbed wave equation can be written in the form of a sum $\Psi = \Sigma a_k \Psi_k^{(0)}$. We shall now seek the solution of the perturbed equation

$$i\hbar\,\partial\Psi/\partial t = (\hat{H}_0 + \hat{V})\Psi \tag{40.1}$$

in the form of a sum

$$\Psi = \sum_k a_k(t)\Psi_k^{(0)}, \tag{40.2}$$

where the expansion coefficients are functions of time. Substituting (40.2) in (40.1), and recalling that the functions $\Psi_k^{(0)}$ satisfy the equation

$$i\hbar\,\partial\Psi_k^{(0)}/\partial t = \hat{H}_0\Psi_k^{(0)},$$

we obtain

$$i\hbar\sum_k \Psi_k^{(0)}\frac{da_k}{dt} = \sum_k a_k\hat{V}\Psi_k^{(0)}.$$

Multiplying both sides of this equation on the left by $\Psi_m^{(0)*}$ and integrating, we have

$$i\hbar\frac{da_m}{dt} = \sum_k V_{mk}(t)a_k, \tag{40.3}$$

where

$$V_{mk}(t) = \int \Psi_m^{(0)*}\hat{V}\Psi_k^{(0)}\,dq$$

$$= V_{mk}e^{i\omega_{mk}t}, \qquad \omega_{mk} = \frac{E_m^{(0)}-E_k^{(0)}}{\hbar},$$

are the matrix elements of the perturbation, including the time factor (and it must be borne in mind that, when V depends explicitly on time, the quantities V_{mk} also are functions of time).

As the unperturbed wave function we take the wave function of the nth stationary state, for which the corresponding values of the coefficients in (40.2) are $a_n^{(0)} = 1$, $a_k^{(0)} = 0$ for $k \neq n$. To find the first approximation, we seek a_k in the form $a_k = a_k^{(0)}+a_k^{(1)}$, substituting $a_k = a_k^{(0)}$ on the right-hand side of equation (40.3), which already contains the small quantities V_{mk}. This gives

$$i\hbar\, da_k^{(1)}/dt = V_{kn}(t). \tag{40.4}$$

In order to show the unperturbed function to which the correction is being calculated, we introduce a second suffix in the coefficients a_k, writing

$$\Psi_n = \sum_k a_{kn}(t)\Psi_k^{(0)}.$$

Accordingly, we write the result of integrating equation (40.4) in the form

$$a_{kn}^{(1)} = -(i/\hbar)\int V_{kn}(t)\,dt = -(i/\hbar)\int V_{kn}e^{i\omega_{kn}t}\,dt. \tag{40.5}$$

This determines the wave functions in the first approximation.

Let us now consider in more detail the important case of a perturbation which is periodic with respect to time, of the form

$$\hat{V} = \hat{F}e^{-i\omega t}+\hat{G}e^{i\omega t}, \tag{40.6}$$

where $\hat{F}$ and $\hat{G}$ are operators independent of time. Since $\hat{V}$ is Hermitian, we must have

$$\hat{F}e^{-i\omega t}+\hat{G}e^{i\omega t} = \hat{F}^+e^{i\omega t}+\hat{G}^+e^{-i\omega t},$$

whence $\hat{G} = F^+$, i.e.

$$G_{nm} = F_{mn}^*. \tag{40.7}$$

This relation shows that

$$V_{kn}(t) = V_{kn}e^{i\omega_{kn}t} = F_{kn}e^{i(\omega_{kn}-\omega)t}+F_{nk}^*e^{i(\omega_{kn}+\omega)t}. \tag{40.8}$$

Substituting in (40.5) and integrating, we obtain the following expression for the expansion coefficients of the wave functions:

$$a_{kn}^{(1)} = -\frac{F_{kn}e^{i(\omega_{kn}-\omega)t}}{\hbar(\omega_{kn}-\omega)} - \frac{F_{nk}^{*}e^{i(\omega_{kn}+\omega)t}}{\hbar(\omega_{kn}+\omega)}. \tag{40.9}$$

These expressions are applicable if none of the denominators vanishes,† i.e. if for all k (and the given n)

$$E_k^{(0)} - E_n^{(0)} \neq \pm\hbar\omega. \tag{40.10}$$

In a number of applications it is useful to have expressions for the matrix elements of an arbitrary quantity f, defined with respect to the perturbed wave functions. In the first approximation we have

$$f_{nm}(t) = f_{nm}^{(0)}(t) + f_{nm}^{(1)}(t),$$

where

$$f_{nm}^{(0)}(t) = \int \Psi_n^{(0)*} f \Psi_m^{(0)}\, dq = f_{nm}^{(0)} e^{i\omega_{nm}t},$$

$$f_{nm}^{(1)}(t) = \int [\Psi_n^{(0)*} f \Psi_m^{(1)} + \Psi_n^{(1)*} f \Psi_m^{(0)}]\, dq.$$

Substituting here $\Psi_n^{(1)} = \Sigma\, a_{kn}^{(1)}\Psi_k^{(0)}$, with $a_{kn}^{(1)}$ determined by formula (40.9), it is easy to obtain the required expression

$$f_{nm}^{(1)}(t) = -e^{i\omega_{nm}t} \sum_k \left\{ \left[\frac{f_{nk}^{(0)}F_{km}}{\hbar(\omega_{km}-\omega)} + \frac{f_{km}^{(0)}F_{nk}}{\hbar(\omega_{kn}+\omega)} \right] e^{-i\omega t} + \left[\frac{f_{nk}^{(0)}F_{mk}^{*}}{\hbar(\omega_{km}+\omega)} + \frac{f_{km}^{(0)}F_{kn}^{*}}{\hbar(\omega_{kn}-\omega)} \right] e^{i\omega t} \right\}. \tag{40.11}$$

This formula is applicable if none of its terms becomes large, i.e. if none of the frequencies ω_{kn}, ω_{km} is too close to ω. For $\omega = 0$ we return to formula (38.12).

In all the formulae given here, it is understood that there is only a discrete spectrum of unperturbed energy levels. However, these formulae can be immediately generalized to the case where there is also a continuous spectrum (as before, we are concerned with the perturbation of states of the discrete spectrum); this is done by simply adding to the sums over the levels of the discrete spectrum the corresponding integrals over the continuous spectrum. Here it is necessary for the denominators $\omega_{kn} \pm \omega$ in formulae (40.9), (40.11) to be non-zero when the energy $E_k^{(0)}$ takes all values, not only of the discrete but also of the continuous spectrum. If, as usually happens, the continuous

† More precisely, if none is so small that the quantities $a_{kn}^{(1)}$ are no longer small compared with unity.

spectrum lies above all the levels of the discrete spectrum, then, for instance, the condition (40.10) must be supplemented by the condition

$$E_{\min}^{(0)} - E_n^{(0)} > \hbar\omega, \tag{40.12}$$

where $E_{\min}^{(0)}$ is the energy of the lowest level of the continuous spectrum.

PROBLEM

Determine the change in the nth and mth solutions of Schrödinger's equation in the presence of a periodic perturbation (of the form (40.6)), of frequency ω such that $E_m^{(0)} - E_n^{(0)} = \hbar(\omega+\epsilon)$, where ϵ is a small quantity.

SOLUTION. The method developed in the text is here inapplicable, since the coefficient $a_{mn}^{(1)}$ in (40.9) becomes large. We start afresh from the exact equations (40.3), with $V_{mk}(t)$ given by (40.8). It is evident that the most important effect is due to those terms, in the sums on the right-hand side of equations (40.3), in which the time dependence is determined by the small frequency $\omega_{mn} - \omega$. Omitting all other terms, we obtain a system of two equations:

$$i\hbar \mathrm{d}a_m/\mathrm{d}t = F_{mn}e^{i(\omega_{mn}-\omega)t}a_n = F_{mn}e^{i\epsilon t}a_n,$$

$$i\hbar \mathrm{d}a_n/\mathrm{d}t = F_{mn}{}^*e^{-i\epsilon t}a_m.$$

We make the substitution

$$a_n e^{i\epsilon t} = b_n$$

and obtain the equations

$$i\hbar\dot{a}_m = F_{mn}b_n, \quad i\hbar(\dot{b}_n - i\epsilon b_n) = F_{mn}{}^*a_m.$$

Eliminating a_m, we have

$$\ddot{b}_n - i\epsilon\dot{b}_n + |F_{mn}|^2 b_n/\hbar^2 = 0.$$

We can take as two independent solutions of these equations

$$a_n = Ae^{i\alpha_1 t}, \quad a_m = -A\hbar\alpha_1 e^{i\alpha_2 t}/F_{mn}{}^* \tag{1}$$

and

$$a_n = Be^{-i\alpha_2 t}, \quad a_m = B\hbar\alpha_2 e^{-i\alpha_1 t}/F_{mn}{}^*, \tag{2}$$

where A and B are constants (which have to be determined from the normalization condition), and we have used the notation

$$\alpha_1 = -\tfrac{1}{2}\epsilon + \Omega, \qquad \alpha_2 = \tfrac{1}{2}\epsilon + \Omega,$$

$$\Omega = \sqrt{(\tfrac{1}{4}\epsilon^2 + |\eta|^2)}, \qquad \eta = F_{mn}/\hbar.$$

Thus, under the action of the perturbation, the functions $\Psi_n^{(0)}$, $\Psi_m^{(0)}$ become $a_n\Psi_n^{(0)} + a_m\Psi_m^{(0)}$, with a_n and a_m given by (1) and (2).

Let the system be in the state $\Psi_m^{(0)}$ at the initial instant ($t = 0$). The state of the system at subsequent instants is given by a linear combination of the two functions which we have obtained, which becomes $\Psi_m^{(0)}$ for $t = 0$:

$$\Psi = e^{i\epsilon t/2}\left(\cos\Omega t - \frac{i\epsilon}{2\pi}\sin\Omega t\right)\Psi_m^{(0)} - (i\eta^*/\Omega)e^{-i\epsilon t/2}\sin\Omega t \,.\, \Psi_n^{(0)}. \tag{3}$$

The squared modulus of the coefficient of $\Psi_n^{(0)}$ is

$$\frac{|\eta|^2}{2\Omega^2}(1 - \cos 2\Omega t). \tag{4}$$

This gives the probability of finding the system in the state $\Psi_n^{(0)}$ at time t. We see that it is a periodic tunction with frequency 2Ω, and varies from 0 to $|\eta|^2/\Omega^2$.

For $\epsilon = 0$ (exact resonance) the probability (4) becomes

$$\tfrac{1}{2}(1 - \cos 2|\eta|t).$$

It varies periodically between 0 and 1; in other words, the system makes periodic transitions from the state $\Psi_m^{(0)}$ to the state $\Psi_n^{(0)}$.

§41. Transitions under a perturbation acting for a finite time

Let us suppose that the perturbation $V(t)$ acts only during some finite interval of time (or that $V(t)$ diminishes sufficiently rapidly as $t \to \pm\infty$). Let the system be in the nth stationary state (of a discrete spectrum) before the perturbation begins to act (or in the limit as $t \to -\infty$). At any subsequent instant the state of the system will be determined by the function

$$\Psi = \sum_k a_{kn}\Psi_k^{(0)},$$

where, in the first approximation,

$$\left.\begin{aligned} a_{kn} &= a_{kn}^{(1)} = -\frac{i}{\hbar}\int_{-\infty}^{t} V_{kn}e^{i\omega_{kn}t}\,dt \quad \text{for} \quad k \neq n, \\ a_{nn} &= 1 + a_{nn}^{(1)} = 1 - \frac{i}{\hbar}\int_{-\infty}^{t} V_{nn}\,dt; \end{aligned}\right\} \tag{41.1}$$

the limits of integration in (40.5) are taken so that, as $t \to -\infty$, all the $a_{kn}^{(1)}$ tend to zero. After the perturbation has ceased to act (or in the limit $t \to +\infty$), the coefficients a_{kn} take constant values $a_{kn}(\infty)$, and the system is in the state with wave function

$$\Psi = \sum_k a_{kn}(\infty)\Psi_k^{(0)},$$

which again satisfies the unperturbed wave equation, but is different from the original function $\Psi_n^{(0)}$. According to the general rule, the squared modulus of the coefficient $a_{kn}(\infty)$ determines the probability for the system to have an energy $E_k^{(0)}$, i.e. to be in the kth stationary state.

Thus, under the action of the perturbation, the system may pass from its initial stationary state to any other. The probability of a transition from the initial (ith) to the final (fth) stationary state is†

$$w_{fi} = \frac{1}{\hbar^2}\left|\int_{-\infty}^{\infty} V_{fi} e^{i\omega_{fi}t}\,dt\right|^2. \tag{41.2}$$

Let us now consider a perturbation which, once having begun, continues to act for an indefinite time (always, of course, remaining small). In other words, $V(t)$ tends to zero as $t \to -\infty$ and to a finite non-zero limit as $t \to +\infty$. Formula (41.2) cannot be applied directly here, since the integral in it diverges. This divergence, however, is physically unimportant and can easily be removed. To do this, we integrate by parts:

$$a_{fi} = -\frac{i}{\hbar}\int_{-\infty}^{t} V_{fi} e^{i\omega_{fi}t}\,dt = -\left[\frac{V_{fi} e^{i\omega_{fi}t}}{\hbar\omega_{fi}}\right]_{-\infty}^{t} + \int_{-\infty}^{t} \frac{\partial V_{fi}}{\partial t}\frac{e^{i\omega_{fi}t}}{\hbar\omega_{fi}}\,dt.$$

The value of the first term vanishes at the lower limit, while at the upper limit it is formally identical with the expansion coefficients in formula (38.8); the presence of an additional periodic factor $e^{i\omega_{fi}t}$ is merely due to the fact that the a_{fi} are the expansion coefficients of the complete wave function Ψ, while the c_{fi} in §38 are the expansion coefficients of the time-independent function ψ. Hence it is clear that its limit as $t \to \infty$ gives simply the change in the original wave function $\Psi_i^{(0)}$ under the action of the "constant" part $V(+\infty)$ of the perturbation, and consequently has no relation to transitions into other states. The probability of a transition is given by the squared modulus of the second term and is

$$w_{fi} = \frac{1}{\hbar^2\omega_{fi}^2}\left|\int_{-\infty}^{\infty} \frac{\partial V_{fi}}{\partial t} e^{i\omega_{fi}t}\,dt\right|^2. \tag{41.3}$$

The derivation is also valid when the transition is from a state of the discrete spectrum to a state of the continuous spectrum. The only difference is that here we have the probability of the transition from a given (ith) state to states in a range of values of ν_f (see the end of §38) from ν_f to $\nu_f + d\nu_f$, so that, for example, formula (41.2) must be written

$$dw_{fi} = \frac{1}{\hbar^2}\left|\int_{-\infty}^{\infty} V_{fi} e^{i\omega_{fi}t}\,dt\right|^2 d\nu_f. \tag{41.4}$$

If the perturbation $V(t)$ varies little during time intervals of the order of the period $1/\omega_{fi}$ the value of the integral in (41.2) or (41.3) will be very

† For uniformity, the initial and final states will henceforward be denoted by i and f when transition probabilities are discussed. The suffixes of these probabilities will be written in the order fi, the same as for matrix elements.

small. In the limit when the applied perturbation varies arbitrarily slowly, the probability of any transition with change of energy (i.e. with a non-zero frequency ω_{fi}) tends to zero. Thus, when the applied perturbation changes sufficiently slowly (*adiabatically*), a system in any non-degenerate stationary state will remain in that state (see also §53).

In the opposite limiting case of a very rapid, "instantaneous" application of the perturbation, the derivatives $\partial V_{fi}/\partial t$ become infinite at the "instant of application". In the integral of $(\partial V_{fi}/\partial t)e^{i\omega_{fi}t}$, we can take outside the integral the comparatively slowly varying factor $e^{i\omega_{fi}t}$ and use its value at this instant. The integral is then found at once, and we obtain

$$w_{fi} = |V_{fi}|^2/\hbar^2\omega_{fi}^2. \tag{41.5}$$

The transition probabilities in instantaneous perturbations can also be found in cases where the perturbation is not small. Let the system be in a state described by one of the eigenfunctions $\psi_i^{(0)}$ of the original Hamiltonian $\hat{H}_0$. If the change in the Hamiltonian occurs instantaneously (i.e. in a time short compared with the periods $1/\omega_{fi}$ of transitions from the given state i to other states), then the wave function of the system is "unable" to vary and remains the same as before the perturbation. It will no longer, however, be an eigenfunction of the new Hamiltonian $\hat{H}$ of the system, i.e. the state $\psi_i^{(0)}$ will not be a stationary state. The probabilities w_{fi} for transitions of the system into the new stationary states are determined, according to the general rules of quantum mechanics, by the coefficients in the expansion of the function $\psi_i^{(0)}$ in terms of the eigenfunctions ψ_f of the Hamiltonian $\hat{H}$:

$$w_{fi} = \left|\int \psi_i^{(0)}\psi_f^* \, dq\right|^2. \tag{41.6}$$

We shall show how this general formula becomes (41.5) if the change $\hat{V} = \hat{H} - \hat{H}_0$ in the Hamiltonian is small. We multiply the equations

$$\hat{H}_0\psi_i^{(0)} = E_i^{(0)}\psi_i^{(0)}, \qquad \hat{H}^*\psi_f^* = E_f\psi_f^*$$

by ψ_f^* and $\psi_i^{(0)}$ respectively, integrate with respect to q and subtract. Using also the self-conjugacy of the operator $\hat{H}$, we obtain

$$(E_f - E_i^{(0)}) \int \psi_f^*\psi_i^{(0)} \, dq = \int \psi_f^*\hat{V}\psi_i^{(0)} \, dq.$$

If the perturbation $\hat{V}$ is small, in the first approximation we can replace E_f by the adjoining unperturbed level $E_f^{(0)}$, and the wave function ψ_f (on the right-hand side of the equation) by the corresponding function $\psi_f^{(0)}$. This gives

$$\int \psi_f^*\psi_i^{(0)} \, dq = \frac{1}{\hbar\omega_{fi}} \int \psi_f^{(0)*}\hat{V}\psi_i^{(0)} \, dq,$$

and formula (41.6) becomes (41.5).

PROBLEMS

PROBLEM 1. A uniform electric field is suddenly applied to a charged oscillator in the ground state. Determine the probabilities of transitions of the oscillator to excited states under the action of this perturbation.

SOLUTION. The potential energy of the oscillator in the uniform field (which exerts a force F on it) is

$$U(x) = \tfrac{1}{2}m\omega^2x^2 - Fx$$
$$= \tfrac{1}{2}m\omega^2(x-x_0)^2 + \text{constant}$$

(where $x_0 = F/m\omega^2$), i.e. has still the pure oscillator form but with the equilibrium position shifted. Hence the wave functions of the stationary states of the perturbed oscillator are $\psi_k(x-x_0)$, where $\psi_k(x)$ are the oscillator functions (23.12); the initial wave function is $\psi_0(x)$ (23.13). Using these functions and the expression (23.11) for the Hermite polynomials, we find

$$\int_{-\infty}^{\infty} \psi_0{}^{(0)}\psi_k \,dx = \frac{(-1)^k}{\sqrt{(2^k\pi k!)}}e^{-\xi_0^2/2}\int_{-\infty}^{\infty} e^{-\xi\xi_0}\frac{d^k}{d\xi^k}e^{-\xi^2+2\xi\xi_0}\,d\xi,$$

with the notation $\xi_0 = x_0\sqrt{(m\omega/\hbar)}$. On integrating k times by parts, the integral on the right becomes

$$\xi_0{}^k\int_{-\infty}^{\infty} e^{-\xi^2+\xi\xi_0}\,d\xi = \xi_0{}^k\sqrt{\pi}e^{\xi_0^2/4}.$$

Thus the transition probability (41.6) is

$$w_{0k} = \frac{\bar{k}^k}{k!}e^{-\bar{k}}, \quad \bar{k} = \tfrac{1}{2}\xi_0{}^2 = F^2/2m\hbar\omega^3.$$

As a function of the number k it represents a Poisson distribution for which the mean value of k is $\bar{k}$.

Perturbation theory is applicable when F is small, so that $\bar{k} \ll 1$. Then the excitation probabilities are small, and decrease rapidly with increasing k. The largest is $w_{10} \approx \bar{k}$.

In the opposite case of large F ($\bar{k} \gg 1$), excitation of the oscillator occurs with very high probability: the probability that the oscillator will remain in the normal state is $w_{00} = e^{-\bar{k}}$.

PROBLEM 2. The nucleus of an atom in the normal state receives an impulse which gives it a velocity v; the duration τ of the impulse is assumed short in comparison both with the electron periods and with a/v, where a is the dimension of the atom. Determine the probability of excitation of the atom under the influence of such a "jolt" (A. B. Migdal 1939).

SOLUTION. We use a frame of reference K' moving with the nucleus after the impact. By virtue of the condition $\tau \ll a/v$, the nucleus may be regarded as practically stationary during the impact, so that the coordinates of the electrons in K' and in the original frame K immediately after the perturbation are the same. The initial wave function in K' is

$$\psi_0' = \psi_0 \exp(-i\mathbf{q}\cdot\sum_a \mathbf{r}_a), \qquad \mathbf{q} = m\mathbf{v}/\hbar,$$

where ψ_0 is the wave function of the normal state with the nucleus at rest, and the summation in the exponent is over all Z electrons in the atom. The required probability of transition to the kth excited state is now given, according to (41.6), by

$$w_{k0} = |\langle k|\exp(-i\mathbf{q}\cdot\sum_a \mathbf{r}_a)|0\rangle|^2.$$

In particular, if $qa \ll 1$, then by expanding the exponential factor in the integrand and noting that the integral of $\psi_k{}^*\psi_0$ is zero because the functions ψ_0 and ψ_k are orthogonal, we obtain

$$w_{k0} = |\langle k|(\mathbf{q}\cdot\sum_a \mathbf{r}_a)|0\rangle|^2.$$

PROBLEM 3. Determine the total probability of excitation and ionization of an atom of hydrogen which receives a sudden "jolt" (see Problem 2).

SOLUTION. The required probability can be calculated as the difference

$$1-w_{00} = 1-\left|\int \psi_0{}^2 e^{-i\mathbf{q}\cdot\mathbf{r}}\, dV\right|^2,$$

where w_{00} is the probability that the atom will remain in the ground state ($\psi_0 = (\pi a^3)^{-1/2}e^{-r/a}$ being the wave function of the ground state of the hydrogen atom, with a the Bohr radius). Calculation of the integral gives

$$1-w_{00} = 1-1/(1+\tfrac{1}{4}q^2a^2)^4.$$

In the limiting case $qa \ll 1$ this probability tends to zero as q^2a^2, while for $qa \gg 1$ it tends to unity as $1-(2/qa)^8$.

PROBLEM 4. Determine the probability that an electron will leave the K-shell of an atom with large atomic number Z when the nucleus undergoes β-decay. The velocity of the β-particle is assumed large in comparison with that of the K-electron (A. B. Migdal and E. L. Feinberg 1941).

SOLUTION.† In the conditions stated the time taken by the β-particle to pass through the K-shell is small compared with the period of revolution of the electron, so that the change in the nuclear charge can be regarded as instantaneous. The perturbation is here represented by the change $V = 1/r$ in the field of the nucleus when the change in its charge is small (1 compared with Z). According to (41.5) the transition probability for one of the two K-shell electrons with energy $E_0 = -\tfrac{1}{2}Z^2$ (here and below we use the fact that the state of the K-electrons is hydrogen-like; see §74) to a state of the continuous spectrum with energy $E = \tfrac{1}{2}k^2$ in the range $dE = k\,dk$ is

$$dw = 2\frac{4|V_{0k}|^2}{(k^2+Z^2)^2}\,dk.$$

In the range which determines the matrix element V_{0k}, the important part is that of short distances ($\sim 1/Z$) from the nucleus, in which the hydrogen-like expression can again be used for the wave function of a state of the continuous spectrum. The final state of the electron must have angular momentum $l = 0$ (the same as that of the initial state). By means of the functions R_{10}, and R_{k0} (normalized on the $k/2\pi$ scale), derived in §36 and formula (f.3) in the Mathematical Appendices we find‡

$$\left(\frac{1}{r}\right)_{0k} = \frac{4\sqrt{(2\pi k)}}{\sqrt{(1-e^{-2\pi Z/k})}}\,\frac{(1+ik/Z)^{iZ/k}(1-ik/Z)^{-iZ/k}}{1+k^2/Z^2}$$

and, since

$$|(1+i\alpha)^{i/\alpha}|^2 = \exp[-(2/\alpha)\tan^{-1}\alpha],$$

we obtain finally

$$dw = \frac{2^7}{Z^4(1+k^2/Z^2)^4}f(k/Z)k\,dk,$$

with

$$f(\alpha) = \frac{1}{1-e^{-2\pi/\alpha}}\exp[-(4/\alpha)\tan^{-1}\alpha].$$

The limiting values of the function $f(\alpha)$ are e^{-4} for $\alpha \ll 1$ and $\alpha/2\pi$ for $\alpha \gg 1$.

The total probability of ionization of the K-shell is obtained by integration of dw over all energies of the emergent electron. A numerical evaluation gives $w = 0{\cdot}65Z^2$.

PROBLEM 5. Determine the probability of emergence of an electron from the K-shell of an atom with large Z in α-decay of the nucleus. The velocity of the α-particle is small

† In Problems 4 and 5, atomic units are used.

‡ In the calculation it is convenient to use Coulomb units and then return to atomic units in the final result.

compared with that of the K-electron, but the time which it takes to leave the nucleus is small in comparison with the time of revolution of the electron (A. B. Migdal 1941, J. S. Levinger 1953).

SOLUTION. After the emergence of the α-particle, the perturbation acting on the electron is adiabatic. The required effect is therefore determined essentially by the interval of time close to the "instant of application" of the perturbation which destroys the adiabaticity, when the α-particle, leaving the nucleus and moving freely, is still at a distance small compared with the radius of the K-orbit. The perturbation V which causes the ionization of the atom is here represented by the deviation of the combined field of the nucleus and the α-particle from the purely Coulomb field Z/r. The dipole moment of two particles with atomic weights 4 and $A-4$, and charges 2 and $Z-2$, at a distance $\mathbf{v}t$ apart (where v is the relative velocity of the nucleus and the α-particle), is

$$\frac{2(A-4)-(Z-2)4}{A}vt = \frac{2(A-2Z)}{A}vt.$$

Hence the dipole term in the field of the nucleus and the α-particle is†

$$V = \frac{2(A-2Z)}{A}vt\frac{z}{r^3},$$

where the z-axis is in the direction of the velocity $\mathbf{v}$. The matrix element of this perturbation reduces to that of z: taking the matrix element of the equation of motion of the electron $\ddot{z} = -Zz/r^3$, we obtain

$$(z/r^3)_{0k} = (E-E_0)^2 z_{0k}/Z.$$

The required transition probability for one of the two electrons in the K-shell is, by (41.2),

$$\mathrm{d}w = 2\left|\int_0^\infty V_{0k}e^{i(E_0-E)t}\,\mathrm{d}t\right|^2 \mathrm{d}k$$

$$= \frac{8(A-2Z)^2v^2}{A^2Z^2}|z_{0k}|^2\frac{\mathrm{d}k}{2\pi};$$

to calculate the integral, we include in the integrand an additional damping factor $e^{-\lambda t}$ with $\lambda > 0$, and then make $\lambda \to 0$ in the result. To calculate the matrix element of $z = r\cos\theta$, we note that, since the orbital angular momentum in the initial state is $l = 0$, $\cos\theta$ has a non-zero matrix element only for the transition to a state with $l = 1$, and

$$|(\cos\theta)_{01}|^2 = (\cos\theta)_{00} = \tfrac{1}{3}$$

and

$$|z_{0k}|^2 = \tfrac{1}{3}|r_{0k}|^2.$$

Calculating r_{0k} by means of the radial functions R_{00} and R_{k1}, we find

$$\mathrm{d}w = \frac{2^{11}(A-2Z)^2v^2}{3A^2Z^6(1+k^2/Z^2)^5}f(k/Z)k\,\mathrm{d}k,$$

the function f being as in Problem 4.

§42. Transitions under the action of a periodic perturbation

The results are different for the probability of transitions to the states of the continuous spectrum under the action of a periodic perturbation. Let

† If the difference $A-2Z$ is small, it may be necessary to take account of the next (quadrupole) term also.

us suppose that, at some initial instant $t = 0$, the system is in the ith stationary state of the discrete spectrum. We shall assume that the frequency ω of the periodic perturbation is such that

$$\hbar\omega > E_{\min} - E_i^{(0)}, \tag{42.1}$$

where $E_{\min}$ is the value of the energy where the continuous spectrum begins.

It is evident from the results of §40 that the chief part will be played by states of the continuous spectrum with energies E_f very close to the *resonance* energy $E_i^{(0)} + \hbar\omega$, i.e. those for which the difference $\omega_{fi} - \omega$ is small. For this reason it is sufficient to consider, in the matrix elements (40.8) of the perturbation, only the first term (with the frequency $\omega_{fi} - \omega$ close to zero). Substituting this term in (40.5) and integrating, we obtain

$$a_{fi} = -\frac{i}{\hbar}\int_0^t V_{fi}(t)\,dt = -F_{fi}\frac{e^{i(\omega_{fi}-\omega)t} - 1}{\hbar(\omega_{fi}-\omega)}. \tag{42.2}$$

The lower limit of integration is chosen so that $a_{fi} = 0$ for $t = 0$, in accordance with the initial condition imposed.

Hence we find for the squared modulus of a_{fi}

$$|a_{fi}|^2 = |F_{fi}|^2 \,.\, 4\sin^2[\tfrac{1}{2}(\omega_{fi}-\omega)t]/\hbar^2(\omega_{fi}-\omega)^2. \tag{42.3}$$

It is easy to see that, for large t, this function can be regarded as proportional to t. To show this, we notice that

$$\lim_{t\to\infty}\frac{\sin^2\alpha t}{\pi t\alpha^2} = \delta(\alpha). \tag{42.4}$$

For when $\alpha \neq 0$ this limit is zero, while for $\alpha = 0$ we have $(\sin^2\alpha t)/t\alpha^2 = t$, so that the limit is infinite; finally, integrating over α from $-\infty$ to $+\infty$, we have (with the substitution $\alpha t = \xi$)

$$\frac{1}{\pi}\int_{-\infty}^{\infty}\frac{\sin^2\alpha t}{t\alpha^2}\,d\alpha = \frac{1}{\pi}\int_{-\infty}^{\infty}\frac{\sin^2\xi}{\xi^2}\,d\xi = 1.$$

Thus the function on the left-hand side of equation (42.4) in fact satisfies all the conditions which define the delta function. Accordingly, we can write for large t

$$|a_{fi}|^2 = (1/\hbar^2)|F_{fi}|^2\pi t\delta(\tfrac{1}{2}\omega_{fi} - \tfrac{1}{2}\omega),$$

or, substituting $\hbar\omega_{fi} = E_f - E_i^{(0)}$ and using the fact that $\delta(ax) = (1/a)\delta(x)$,

$$|a_{fi}|^2 = (2\pi/\hbar)|F_{fi}|^2\delta(E_f - E_i^{(0)} - \hbar\omega)t.$$

The expression $|a_{fi}|^2\,d\nu_f$ is the probability of a transition from the original state to one in the interval $d\nu_f$. We see that, for large t, it is proportional to the time interval elapsed since $t = 0$. The probability dw_{fi} of the transition per unit time is†

$$dw_{fi} = (2\pi/\hbar)|F_{fi}|^2\delta(E_f - E_i^{(0)} - \hbar\omega)\, d\nu_f. \tag{42.5}$$

As we should expect, it is zero except for transitions to states with energy $E_f = E_i^{(0)} + \hbar\omega$. If the energy levels of the continuous spectrum are not degenerate, so that ν_f can be taken as the value of the energy alone, then the whole "interval" of states $d\nu_f$ reduces to a single state with energy $E = E_i^{(0)} + \hbar\omega$, and the probability of a transition to this state is

$$w_{Ei} = (2\pi/\hbar)|F_{Ei}|^2. \tag{42.6}$$

There is another method of deriving formula (42.5) that is methodologically instructive, in which the periodic perturbation is assumed not to be applied at a time $t = 0$ but to increase slowly from $t = -\infty$ by an exponential law $e^{\lambda t}$ with a positive constant λ which is then made to tend to zero (adiabatic switch-on). The initial condition $a_{fi} = 0$ is accordingly applied at $t = -\infty$. The matrix element of the perturbation now has the form

$$V_{fi}(f) = F_{fi}e^{i(\omega_{fi}-\omega)t+\lambda t},$$

and (42.2) becomes

$$a_{fi} = -\frac{i}{\hbar}\int_{-\infty}^{t} V_{fi}(t)\, dt$$

$$= -F_{fi}\frac{e^{i(\omega_{fi}-\omega)t+\lambda t}}{\hbar(\omega_{fi}-\omega-i\lambda)}. \tag{42.7}$$

Hence

$$|a_{fi}|^2 = \frac{1}{\hbar^2}|F_{fi}|^2\frac{e^{2\lambda t}}{(\omega_{fi}-\omega)^2+\lambda^2}.$$

The transition probability per unit time is given by the derivative

$$d|a_{fi}|^2/dt = 2\lambda|a_{fi}|^2.$$

There is a formula

$$\lim_{\lambda\to 0}\frac{\lambda}{\pi(\alpha^2+\lambda^2)} = \delta(\alpha), \tag{42.8}$$

valid in the same sense as (42.4); with this we find, taking the limit $\lambda \to 0$,

$$\frac{d}{dt}|a_{fi}|^2 \to \frac{2\pi}{\hbar^2}|F_{fi}|^2\delta(\omega_{fi}-\omega),$$

and thus return to (42.5).

† It is easy to verify that, on taking account of the second term in (40.8), which we have omitted, additional expressions are obtained which, on being divided by t, tend to zero as $t \to +\infty$.

§43. Transitions in the continuous spectrum

One of the most important applications of perturbation theory is to calculate the probability of a transition in the continuous spectrum under the action of a constant (time-independent) perturbation. We have already mentioned that the states of the continuous spectrum are almost always degenerate. Having chosen in some manner the set of unperturbed wave functions corresponding to some given energy level, we can put the problem as follows. It is known that, at the initial instant, the system is in one of these states; it is required to determine the probability of the transition to another state with the same energy. For transitions from the initial state i to states between ν_f and $\nu_f + d\nu_f$ we have at once from (42.5) (putting $\omega = 0$ and changing the notation)

$$dw_{fi} = (2\pi/\hbar)|V_{fi}|^2 \delta(E_f - E_i)\, d\nu_f. \tag{43.1}$$

This expression is, as we should expect, zero except for $E_f = E_i$: under the action of a constant perturbation, transitions occur only between states with the same energy. It must be noticed that, for transitions from states of the continuous spectrum, the quantity dw_{fi} cannot be regarded directly as the transition probability; it is not even of the right dimensions (1/time). Formula (43.1) represents the number of transitions per unit time, and its dimensions depend on the chosen method of normalization of the wave functions of the continuous spectrum.†

Let us calculate the perturbed wave function, which before the action of the perturbation is the same as the original unperturbed function $\psi_i^{(0)}$. Using the method given at the end of §42, we can regard the perturbation as being adiabatically applied according to $e^{\lambda t}$ with $\lambda \to 0$. From (42.7), putting $\omega = 0$ and changing the notation, we have

$$a_{fi}^{(1)} = V_{fi} \frac{\exp\{(i/\hbar)(E_f - E_i)t + \lambda t\}}{E_i - E_f + i\lambda}. \tag{43.2}$$

The perturbed wave function is

$$\Psi_i = \Psi_i^{(0)} + \int a_{fi}^{(1)} \Psi_f^{(0)}\, d\nu_f,$$

where the integration is extended over the whole of the continuous spectrum.‡ Substitution of (43.2) gives

$$\Psi_i = \left[\psi_i^{(0)} + \int V_{fi}\psi_f^{(0)} \frac{d\nu_f}{E_i - E_f + i0}\right] \exp\left(-\frac{i}{\hbar}E_i t\right). \tag{43.3}$$

---

† The phenomena comprised within the theory here discussed include, for example, various types of collision; the system in its initial and final states is a set of free particles and the perturbation is the interaction between them. With appropriate normalization of the wave functions, (43.1) may then be the collision cross-section (see §126).

‡ If there is also a discrete spectrum, then we must add to the integral in this formula (and subsequent ones) the appropriate sum over the states of the discrete spectrum.

In the limit as $\lambda \to 0$, the factor $e^{\lambda t}$ becomes unity. The term $+i0$, denoting the limit of $i\lambda$ as λ tends to zero from positive values, determines the manner of integration with respect to the variable E_f (dE_f occurs as a factor in $d\nu_f$ together with the differentials of other quantities which describe the states of the continuous spectrum). Without the term $i\lambda$, the integrand in (43.3) would have a pole at $E_f = E_i$, near which the integral would diverge. The term $i\lambda$ moves this pole into the upper half-plane of the complex variable E_f. After the limit $\lambda \to 0$ is taken, the pole returns to the real axis, but we know that the path of integration must pass beneath it:

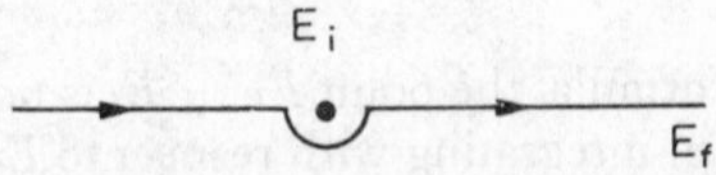

(43.4)

The time factor in (43.3) shows that this function belongs, as it should, to the same energy E_i as the original unperturbed function. In other words, the function

$$\psi_i = \psi_i^{(0)} + \int \frac{V_{fi}}{E_i - E_f + i0} \psi_f^{(0)} \, d\nu_f$$

satisfies Schrödinger's equation

$$(\hat{H}_0 + \hat{V})\psi_i = E_i \psi_i.$$

It is therefore natural that the expression obtained should correspond exactly to (38.8).†

The calculations given above correspond to the first approximation of perturbation theory. It is not difficult to calculate the second approximation as well. To do this, we must derive the formula for the next approximation to Ψ_i; this is easily effected by using the method of §38 (now that we know the method of dealing with the "divergent" integrals). A simple calculation gives the formula

$$\Psi_i = \left\{ \psi_i^{(0)} + \int \left[V_{fi} + \int \frac{V_{f\nu} V_{\nu i}}{E_i - E_\nu + i0} d\nu \right] \times \right.$$

$$\left. \times \frac{\psi_f^{(0)} \, d\nu_f}{E_i - E_f + i0} \right\} e^{-(i/\hbar)E_i t}. \qquad (43.5)$$

Comparing this expression with formula (43.3), we can write down the corresponding formula for the probability (or, more precisely, the number)

† With this formula, the way in which the integral is to be taken can be found from the condition that the asymptotic expression for ψ_i at large distances should contain only an outgoing (and not an ingoing) wave (see §136).

of transitions, by direct analogy with (43.1):

$$dw_{fi} = \frac{2\pi}{\hbar}\left| V_{fi} + \int \frac{V_{f\nu}V_{\nu i}}{E_i - E_\nu + i0}\, d\nu \right|^2 \delta(E_i - E_f)\, d\nu_f. \qquad (43.6)$$

It may happen that the matrix element V_{fi} for the transition considered vanishes. The effect is then zero in the first approximation, and (43.6) becomes

$$dw_{fi} = \frac{2\pi}{\hbar}\left| \int \frac{V_{f\nu}V_{\nu i}}{E_i - E_\nu}\, d\nu \right|^2 \delta(E_f - E_i)\, d\nu_f. \qquad (43.7)$$

In applications of this formula, the point $E_\nu = E_i$ is not usually a pole of the integrand; the manner of integrating with respect to E_ν is then unimportant, and the integral can be taken along the real axis.

The states ν for which $V_{f\nu}$ and $V_{\nu i}$ are not zero are usually called *intermediate* states for the transition $i \to f$. Intuitively, we may say that this transition takes place as if in two steps $i \to \nu$ and $\nu \to f$ (but such a description must not be taken literally, of course). It may happen that the transition $i \to f$ can take place not through one but only through several successive intermediate states. Formula (43.7) can be at once generalized to such cases. For example, if two intermediate states are needed, we have

$$dw_{fi} = \frac{2\pi}{\hbar}\left| \int \frac{V_{f\nu'}V_{\nu'\nu}V_{\nu i}}{(E_i - E_{\nu'})(E_i - E_\nu)}\, d\nu\, d\nu' \right|^2 \delta(E_f - E_i)\, d\nu_f. \qquad (43.8)$$

Lastly, to clarify the mathematical significance of the integrals taken along a path of the form (43.4), we shall prove the formula

$$\int \frac{f(x)\, dx}{x - a - i0} = P \int \frac{f(x)\, dx}{x - a} + i\pi f(a), \qquad (43.9)$$

where the integration is along a segment of the real axis including the point $x = a$. If we pass round the pole $x = a$ along a semicircle of radius ρ, we find that the whole integral is equal to the sum of the integrals along the real axis from the lower limit to $a - \rho$ and from $a + \rho$ to the upper limit, together with $i\pi$ times the residue of the integrand at the pole. In the limit $\rho \to 0$, the integrals along the real axis make the integral along the complete segment, taken as a principal value (denoted by P), and the result is (43.9), which may also be symbolically written

$$\frac{1}{x - a - i0} = P\frac{1}{x - a} + i\pi\delta(x - a); \qquad (43.10)$$

P here denotes the taking of the principal value when integrating the function $f(x)/(x - a)$.

§44. The uncertainty relation for energy

Let us consider a system composed of two weakly interacting parts. We suppose that it is known that at some instant these parts have definite values of the energy, which we denote by E and ϵ respectively. Let the energy be measured again after some time interval Δt; the values E', ϵ' obtained are in general different from E, ϵ. It is easy to determine the order of magnitude of the most probable value of the difference $E'+\epsilon'-E-\epsilon$ which is found as a result of the measurement.

According to formula (42.3) with $\omega = 0$, the probability of a transition of the system (after time t), under the action of a time-independent perturbation, from a state with energy E to one with energy E' is proportional to

$$\sin^2[(E'-E)t/2\hbar]/(E'-E)^2.$$

Hence we see that the most probable value of the difference $E'-E$ is of the order of $\hbar/t$.

Applying this result to the case we are considering (the perturbation being the interaction between the parts of the system), we obtain the relation

$$|E+\epsilon-E'-\epsilon'|\Delta t \sim \hbar. \tag{44.1}$$

Thus the smaller the time interval Δt, the greater the energy change that is observed. It is important to notice that its order of magnitude $\hbar/\Delta t$ is independent of the amount of the perturbation. The energy change determined by the relation (44.1) will be observed, however weak the interaction between the two parts of the system. This result is peculiar to quantum theory and has a deep physical significance. It shows that, in quantum mechanics, the law of conservation of energy can be verified by means of two measurements only to an accuracy of the order of $\hbar/\Delta t$, where Δt is the time interval between the measurements.

The relation (44.1) is often called the *uncertainty relation for energy*. However, it must be emphasized that its significance is entirely different from that of the uncertainty relation $\Delta p \Delta x \sim \hbar$ for the coordinate and momentum. In the latter, Δp and Δx are the uncertainties in the values of the momentum and coordinate at the same instant; they show that these two quantities can never have entirely definite values simultaneously. The energies E, ϵ, on the other hand, can be measured to any degree of accuracy at any instant. The quantity $(E+\epsilon)-(E'+\epsilon')$ in (44.1) is the difference between two exactly measured values of the energy $E+\epsilon$ at two different instants, and not the uncertainty in the value of the energy at a given instant.

If we regard E as the energy of some system and ϵ as that of a "measuring apparatus", we can say that the energy of interaction between them can be taken into account only to within $\hbar/\Delta t$. Let us denote by ΔE, $\Delta\epsilon$, ... the errors in the measurements of the corresponding quantities. In the favourable case when ϵ, ϵ' are known exactly ($\Delta\epsilon = \Delta\epsilon' = 0$), we have

$$\Delta(E-E') \sim \hbar/\Delta t. \tag{44.2}$$

From this relation we can derive important consequences concerning the measurement of momentum. The process of measuring the momentum of a particle (for definiteness, we shall speak of an electron) consists in a collision of the electron with some other ("measuring") particle, whose momenta before and after the collision can be regarded as known exactly.† If we apply to this collision the law of conservation of momentum, we obtain three equations (the three components of a single vector equation) in six unknowns (the components of the momentum of the electron before and after the collision). The number of equations can be increased by bringing about a series of further collisions between the electron and "measuring" particles, and applying to each collision the law of conservation of momentum. This, however, increases the number of unknowns also (the momenta of the electron between collisions), and it is easy to see that, whatever the number of collisions, the number of unknowns will always be three more than the number of equations. Hence, in order to measure the momentum of the electron, it is necessary to bring in the law of conservation of energy at each collision, as well as that of momentum. The former, however, can be applied, as we have seen, only to an accuracy of the order of $\hbar/\Delta t$, where Δt is the time between the beginning and end of the process in question.

To simplify the subsequent discussion, it is convenient to consider an imaginary idealized experiment in which the "measuring particle" is a perfectly reflecting plane mirror; only one momentum component is then of importance, namely that perpendicular to the plane of the mirror. To determine the momentum P of the particle, the laws of conservation of momentum and energy give the equations

$$p'+P'-p-P = 0, \tag{44.3}$$

$$|\epsilon'+E'-\epsilon-E| \sim \hbar/\Delta t, \tag{44.4}$$

where P, E are the momentum and energy of the particle, and p, ϵ those of the mirror; the unprimed and primed quantities refer to the instants before and after the collision respectively. The quantities p, p', ϵ, ϵ' relating to the "measuring particle" can be regarded as known exactly, i.e. the errors in them are zero. Then we have for the errors in the remaining quantities, from the above equations:

$$\Delta P = \Delta P', \quad \Delta E'-\Delta E \sim \hbar/\Delta t.$$

But $\Delta E = (\partial E/\partial P)\Delta P = v\Delta P$, where v is the velocity of the electron (before the collision), and similarly $\Delta E' = v'\Delta P' = v'\Delta P$. Hence we obtain

$$(v'_x-v_x)\Delta P_x \sim \hbar/\Delta t. \tag{44.5}$$

We have here added the suffix x to the velocity and momentum, in order to emphasize that this relation holds for each of their components separately.

This is the required relation. It shows that the measurement of the

† In the present analysis it is of no importance how the energy of the "measuring" particle is ascertained.

momentum of the electron (with a given degree of accuracy ΔP) necessarily involves a change in its velocity (i.e. in the momentum itself). This change becomes greater as the duration of the measuring process becomes shorter. The change in velocity can be made arbitrarily small only as $\Delta t \to \infty$, but measurements of momentum occupying a long time can be significant only for a free particle. The non-repeatability of a measurement of momentum after short intervals of time, and the "two-faced" nature of measurement in quantum mechanics—the necessity of a distinction between the measured value of a quantity and the value resulting from the process of measurement—are here exhibited with particular clarity.†

The conclusion reached at the beginning of this section, which was based on perturbation theory, can also be derived from another standpoint by considering the decay of a system under the action of some perturbation. Let E_0 be some energy level of the system, calculated without any allowance for the possibility of its decay. We denote by τ the *lifetime* of this state of the system, i.e. the reciprocal of the probability of decay per unit time. Then we find by the same method that

$$|E_0-E-\epsilon| \sim \hbar/\tau, \tag{44.6}$$

where E, ϵ are the energies of the two parts into which the system decays. The sum $E+\epsilon$, however, gives us an estimate of the energy of the system before it decays. Hence the above relation shows that the energy of a system, in some "quasi-stationary" state, which is free to decay can be determined only to within a quantity of the order of $\hbar/\tau$. This quantity is usually called the *width* Γ of the level. Thus

$$\Gamma \sim \hbar/\tau. \tag{44.7}$$

§45. Potential energy as a perturbation

The case where the total potential energy of the particle in an external field can be regarded as a perturbation merits special consideration. The unperturbed Schrödinger's equation is then the equation of free motion of the particle:

$$\triangle\psi^{(0)}+k^2\psi^{(0)} = 0, \qquad k = \sqrt{(2mE/\hbar^2)} = p/\hbar, \tag{45.1}$$

and has solutions which represent plane waves. The energy spectrum of free motion is continuous, so that we are concerned with an unusual case of perturbation theory in a continuous spectrum. The solution of the problem is here more conveniently obtained directly, without having recourse to general formulae.

The equation for the correction $\psi^{(1)}$ to the wave function in the first approximation is

$$\triangle\psi^{(1)}+k^2\psi^{(1)} = (2mU/\hbar^2)\psi^{(0)}, \tag{45.2}$$

† The relation (44.5) and the elucidation of the physical significance of the uncertainty relation for energy are due to N. Bohr (1928).

where U is the potential energy. The solution of this equation, as we know from electrodynamics, can be written in the form of *retarded potentials*, i.e. in the form†

$$\psi^{(1)}(x, y, z) = -(m/2\pi\hbar^2) \int \psi^{(0)}U(x', y', z')e^{ikr}\,\mathrm{d}V'/r, \tag{45.3}$$

where

$$\mathrm{d}V' = \mathrm{d}x'\mathrm{d}y'\mathrm{d}z', \quad r^2 = (x-x')^2+(y-y')^2+(z-z')^2.$$

Let us find what conditions must be satisfied by the field U in order that it may be regarded as a perturbation. The condition of applicability of perturbation theory is contained in the requirement that $\psi^{(1)} \ll \psi^{(0)}$. Let a be the order of magnitude of the dimensions of the region of space in which the field is noticeably different from zero. We shall first suppose that the energy of the particle is so small that ka is at most of the order of unity. Then the factor e^{ikr} in the integrand of (45.3) is unimportant in an order-of-magnitude estimate, and the integral is of the order of $\psi^{(0)}|U|a^2$, so that

$$\psi^{(1)} \sim m|U|a^2\psi^{(0)}/\hbar^2,$$

and we have the condition

$$|U| \ll \hbar^2/ma^2 \qquad \text{(for } ka \gtrsim 1\text{)}. \tag{45.4}$$

We notice that the expression on the right has a simple physical meaning; it is the order of magnitude of the kinetic energy which the particle would have if enclosed in a volume of linear dimensions a (since, by the uncertainty relation, its momentum would be of the order of $\hbar/a$).

Let us consider, in particular, a potential well so shallow that the condition (45.4) holds for it. It is easy to see that in such a well there are no negative energy levels (R. Peierls 1929); this has been shown, for the particular case of a spherically symmetric well, in §33, Problem. For, when $E = 0$, the unperturbed wave function reduces to a constant, which can be arbitrarily taken as unity: $\psi^{(0)} = 1$. Since $\psi^{(1)} \ll \psi^{(0)}$, it is clear that the wave function $\psi = 1+\psi^{(1)}$ for motion in the well nowhere vanishes; the eigenfunction, being without nodes, belongs to the normal state, so that $E = 0$ remains the least possible value of the energy of the particle. Thus, if the well is sufficiently shallow, only an infinite motion of the particle is possible: the particle cannot be "captured" by the well. Note that this result is peculiar to quantum theory; in classical mechanics a particle can execute a finite motion in any potential well.

It must be emphasized that all that has been said refers only to a three-dimensional well. In a one- or two-dimensional well (i.e. one in which the field is a function of only one or two coordinates), there are always negative

† This is a particular integral of equation (45.2), to which we may add any solution of the same equation with zero on the right-hand side, i.e. the unperturbed equation (45.1).

energy levels (see the Problems at the end of this section). This is related to the fact that, in the one- and two-dimensional cases, the perturbation theory under consideration is inapplicable for an energy E which is zero (or very small).†

For large energies, when $ka \gg 1$, the factor e^{ikr} in the integrand plays an important part, and markedly reduces the value of the integral. The solution (45.3) in this case can be transformed; the alternative form, however, is more conveniently derived by returning to equation (45.2). We take as x-axis the direction of the unperturbed motion; the unperturbed wave function then has the form $\psi^{(0)} = e^{ikx}$ (the constant factor is arbitrarily taken as unity). Let us seek a solution of the equation

$$\triangle\psi^{(1)} + k^2\psi^{(1)} = (2m/\hbar^2)Ue^{ikx}$$

in the form $\psi^{(1)} = e^{ikx}f$; in view of the assumed large value of k, it is sufficient to retain in $\triangle\psi^{(1)}$ only those terms in which the factor e^{ikx} is differentiated one or more times. We then obtain for f the equation

$$2ik\,\partial f/\partial x = 2mU/\hbar^2,$$

whence

$$\psi^{(1)} = e^{ikx}f = -(im/\hbar^2 k)e^{ikx}\int U\,dx. \tag{45.5}$$

An estimation of this integral gives $|\psi^{(1)}| \sim m|U|a/\hbar^2 k$, so that the condition of applicability of perturbation theory in this case is

$$|U| \ll (\hbar^2/ma^2)ka = \hbar v/a \qquad (ka \gg 1), \tag{45.6}$$

where $v = k\hbar/m$ is the velocity of the particle. It is to be observed that this condition is weaker than (45.4). Hence, if the field can be regarded as a perturbation at small energies of the particle, it can always be so regarded at large energies, whereas the converse is not necessarily true.‡

The applicability of the perturbation theory developed here to a Coulomb field requires special consideration. In a field where $U = \alpha/r$, it is impossible to separate a finite region of space outside which U is considerably less than inside it. The required condition can be obtained by writing in (45.6) a variable distance r instead of the parameter a; this leads to the inequality

$$\alpha/\hbar v \ll 1. \tag{45.7}$$

† In the two-dimensional case $\psi^{(1)}$ is expressed (as is known from the theory of the two-dimensional wave equation) as an integral similar to (45.3), in which, instead of $e^{ikr}\,dx'dy'dz'/r$ we have $i\pi H_0^{(1)}(kr)\,dx'dy'$, where $H_0^{(1)}$ is the Hankel function and $r^2 = (x-x')^2+(y-y')^2$. As $k \to 0$, the Hankel function, and therefore the whole integral, tend logarithmically to infinity.

Similarly, in the one-dimensional case, we have, in the integrand, $2\pi ie^{ikr}\,dx'/k$, where $r = |x-x'|$, and as $k \to 0$ $\psi^{(1)}$ tends to infinity as $1/k$.

‡ In the one-dimensional case the condition for perturbation theory to be applicable is given by the inequality (45.6) for all ka. The derivation of the condition (45.4) given above for the three dimensional case is not valid in the one dimensional case, owing to the divergence of the resulting function $\psi^{(1)}$ (see the preceding footnote).

Thus, for large energies of the particle, a Coulomb field can be regarded as a perturbation.†

Finally, we shall derive a formula which approximately determines the wave function of a particle whose energy E everywhere considerably exceeds the potential energy U (no other conditions being imposed). In the first approximation, the wave function depends on the coordinates in the same way as for free motion (whose direction is taken as the x-axis). Accordingly, let us look for ψ in the form $\psi = e^{ikx}F$, where F is a function of the coordinates which varies slowly in comparison with the factor e^{ikx} (but we cannot in general say that it is close to unity). Substituting in Schrödinger's equation, we obtain for F the equation

$$2ik\,\partial F/\partial x = (2m/\hbar^2)UF, \tag{45.8}$$

whence

$$\psi = e^{ikx}F = \text{constant}\times e^{ikx}e^{-(i/\hbar v)\int U\,dx}. \tag{45.9}$$

This is the required expression. It should, however, be borne in mind that this formula is not valid at large distances. In equation (45.8) a term $\triangle F$ has been omitted which contains second derivatives of F. The derivative $\partial^2 F/\partial x^2$, together with the first derivative $\partial F/\partial x$, tends to zero at large distances, but the derivatives with respect to the transverse coordinates y and z do not tend to zero, and can be neglected only if $x \ll ka^2$.

PROBLEMS

PROBLEM 1. Determine the energy level in a one-dimensional potential well whose depth is small. It is assumed that the condition (45.4) is satisfied.

SOLUTION. We make the hypothesis, which will be confirmed by the result, that the energy level $|E| \ll |U|$. Then, on the right-hand side of Schrödinger's equation

$$d^2\psi/dx^2 = (2m/\hbar^2)[U(x)-E]\psi,$$

we can neglect E in the region of the well, and regard ψ as a constant, which without loss of generality can be taken as unity:

$$d^2\psi/dx^2 = 2mU/\hbar^2.$$

We integrate this equation with respect to x between two points $\pm x_1$ such that $a \ll x_1 \ll 1/\kappa$, where a is the width of the well and $\kappa = \sqrt{(2m|E|/\hbar^2)}$. Since the integral of $U(x)$ converges, the integration on the right can be extended to the whole range from $-\infty$ to $+\infty$:

$$\left[\frac{d\psi}{dx}\right]_{-x_1}^{x_1} = \frac{2m}{\hbar^2}\int_{-\infty}^{\infty} U\,dx. \tag{1}$$

At large distances from the well, the wave function is of the form $\psi = e^{\pm\kappa x}$. Substituting this in (1), we find

$$-2\kappa = (2m/\hbar^2)\int_{-\infty}^{\infty} U\,dx$$

† It must be borne in mind that the integral (45.5) with a field $U = \alpha/r$ diverges (logarithmically) when $x/\sqrt{(y^2+z^2)}$ is large. Hence the wave function in a Coulomb field, obtained by means of perturbation theory, is inapplicable within a narrow cone about the x-axis.

or

$$|E| = (m/2\hbar^2)\left[\int_{-\infty}^{\infty} U\,dx\right]^2.$$

We see that, in accordance with the hypothesis, the energy of the level is a small quantity of a higher order (the second) than the depth of the well.

PROBLEM 2. Determine the energy level in a two-dimensional potential well $U(r)$ (where r is the polar coordinate in the plane) of small depth; it is assumed that the integral $\int_0^\infty rU\,dr$ converges.

SOLUTION. Proceeding as in the previous problem, we have in the region of the well the equation

$$\frac{1}{r}\frac{d}{dr}\left(r\frac{d\psi}{dr}\right) = \frac{2m}{\hbar^2}U.$$

Integrating this with respect to r from 0 to r_1 (where $a \ll r_1 \ll 1/\kappa$), we find

$$\left[\frac{d\psi}{dr}\right]_{r=r_1} = \frac{2m}{\hbar^2 r_1}\int_0^\infty rU(r)\,dr. \tag{1}$$

At large distances from the well, the equation of free motion in two dimensions is

$$\frac{1}{r}\frac{d}{dr}\left(r\frac{d\psi}{dr}\right) + \frac{2m}{\hbar^2}E\psi = 0,$$

and has a solution (vanishing at infinity) $\psi = \text{constant} \times H_0^{(1)}(i\kappa r)$; for small values of the argument, the leading term in this function is proportional to $\log \kappa r$. Bearing this in mind, we equate the logarithmic derivatives of ψ for $r \sim a$ inside the well (the right-hand side of (1)) and outside it, obtaining

$$\frac{1}{a \log \kappa a} \approx \frac{2m}{\hbar^2 a}\int_0^\infty U(r)r\,dr,$$

whence

$$|E| \sim \frac{\hbar^2}{ma^2}\exp\left\{-\frac{\hbar^2}{m}\left|\int_0^\infty Ur\,dr\right|^{-1}\right\}.$$

We see that the energy of the level is exponentially small compared with the depth of the well.

CHAPTER VII

THE QUASI-CLASSICAL CASE

§46. The wave function in the quasi-classical case

If the de Broglie wavelengths of particles are small in comparison with the characteristic dimensions L which determine the conditions of a given problem, then the properties of the system are close to being classical, just as wave optics passes into geometrical optics as the wavelength tends to zero.

Let us now investigate more closely the properties of *quasi-classical* systems. To do this, we make in Schrödinger's equation

$$\sum_a \frac{\hbar^2}{2m_a}\triangle_a\psi+(E-U)\psi = 0$$

the substitution

$$\psi = e^{(i/\hbar)\sigma}. \tag{46.1}$$

For the function σ we obtain the equation

$$\sum_a \frac{1}{2m_a}(\nabla_a\sigma)^2-\sum_a \frac{i\hbar}{2m_a}\triangle_a\sigma = E-U. \tag{46.2}$$

Since the system is supposed almost classical in its properties, we seek σ in the form of a series:

$$\sigma = \sigma_0+(\hbar/i)\sigma_1+(\hbar/i)^2\sigma_2+ \dots , \tag{46.3}$$

expanded in powers of $\hbar$.

We begin by considering the simplest case, that of one-dimensional motion of a single particle. Equation (46.2) then reduces to

$$\sigma'^2/2m-i\hbar\sigma''/2m = E-U(x), \tag{46.4}$$

where the prime denotes differentiation with respect to the coordinate x.

In the first approximation we write $\sigma = \sigma_0$ and omit from the equation the term containing $\hbar$:

$$\sigma_0'^2/2m = E-U(x).$$

Hence we find

$$\sigma_0 = \pm \int \sqrt{\{2m[E-U(x)]\}}\, dx.$$

The integrand is simply the classical momentum $p(x)$ of the particle, expressed as a function of the coordinate. Defining the function $p(x)$ with the + sign in front of the radical, we have

$$\sigma_0 = \pm \int p \, \mathrm{d}x, \quad p = \sqrt{[2m(E-U)]}, \tag{46.5}$$

as we should expect from the limiting expression (6.1) for the wave function.† The approximation made in equation (46.4) is legitimate only if the second term on the left-hand side is small compared with the first, i.e. we must have $\hbar|\sigma''/\sigma'^2| \ll 1$ or

$$|\mathrm{d}(\hbar/\sigma')/\mathrm{d}x| \ll 1.$$

In the first approximation we have, according to (46.5), $\sigma' = p$, so that the condition obtained can be written

$$|\mathrm{d}(\lambda/2\pi)/\mathrm{d}x| \ll 1, \tag{46.6}$$

where $\lambda(x) = 2\pi\hbar/p(x)$ is the de Broglie wavelength of the particle, expressed as a function of x by means of the classical function $p(x)$. Thus we have obtained a quantitative *quasi-classicality condition*: the wavelength of the particle must vary only slightly over distances of the order of itself. The formulae here derived are not applicable in regions of space where this condition is not satisfied.

The condition (46.6) can be written in another form by noticing that

$$\frac{\mathrm{d}p}{\mathrm{d}x} = \frac{\mathrm{d}}{\mathrm{d}x}\sqrt{[2m(E-U)]} = -\frac{m}{p}\frac{\mathrm{d}U}{\mathrm{d}x} = \frac{mF}{p},$$

where $F = -\mathrm{d}U/\mathrm{d}x$ is the classical force acting on the particle in the external field. In terms of this force we find

$$m\hbar|F|/p^3 \ll 1. \tag{46.7}$$

It is seen from this that the quasi-classical approximation becomes inapplicable if the momentum of the particle is too small. In particular, it is clearly inapplicable near *turning points*, i.e. near points where the particle, according to classical mechanics, would stop and begin to move in the opposite direction. These points are given by the equation $p(x) = 0$, i.e. $E = U(x)$. As $p \to 0$, the de Broglie wavelength tends to infinity, and hence cannot possibly be supposed small.

It must be emphasized, however, that the condition (46.6) or (46.7) alone may be insufficient for the quasi-classical approximation to be valid. The reason is that this condition has been derived from estimates of the various terms in the differential equation (46.4), the term omitted containing a higher derivative. It would be necessary, in fact, to stipulate the smallness of the

---

† As is well known, $\int p \, \mathrm{d}x$ is the time-independent part of the action. The total mechanical action S of a particle is $S = -Et \pm \int p \, \mathrm{d}x$. The term $-Et$ is absent from σ_0, since we are considering a time-independent wave function ψ.

subsequent expansion terms in the solution of this equation, and this need not be ensured by the smallness of the term omitted. For example, if the solution for $\sigma(x)$ contains a term which increases almost linearly with the coordinate x, the smallness of the second derivative in the equation will not prevent this term from becoming large at sufficiently great distances. Such a situation occurs, in general, when the field extends to distances large in comparison with the characteristic length L over which it varies by an appreciable amount; see the discussion of (46.11) below. The quasi-classical approximation is then invalid for investigating the behaviour of the wave function at large distances.

Let us now calculate the next term in the expansion (46.3). The first-order terms in $\hbar$ in equation (46.4) give

$$\sigma_0'\sigma_1' + \tfrac{1}{2}\sigma_0'' = 0,$$

whence

$$\sigma_1' = -\sigma_0''/2\sigma_0' = -p'/2p.$$

Integrating, we find

$$\sigma_1 = -\tfrac{1}{2}\log p, \tag{46.8}$$

omitting the constant of integration.

Substituting this expression in (46.1) and (46.3), we find the wave function in the form

$$\psi = C_1 p^{-1/2} e^{(i/\hbar)\int p\,dx} + C_2 p^{-1/2} e^{-(i/\hbar)\int p\,dx}. \tag{46.9}$$

The factor $1/\sqrt{p}$ in this function has a simple interpretation. The probability of finding the particle at a point with coordinate between x and $x+dx$ is given by the square $|\psi|^2$, i.e. is essentially proportional to $1/p$. This is exactly what we should expect for a "quasi-classical" particle, since, in classical motion, the time spent by a particle in the segment dx is inversely proportional to the velocity (or momentum) of the particle.

In the "classically inaccessible" parts of space, where $E < U(x)$, the function $p(x)$ is purely imaginary, so that the exponents are real. The general form of the solution of the wave equation in these regions is

$$\psi = \frac{C_1}{\sqrt{|p|}} e^{-(1/\hbar)\int |p|\,dx} + \frac{C_2}{\sqrt{|p|}} e^{(1/\hbar)\int |p|\,dx}. \tag{46.10}$$

It must, however, be borne in mind that the accuracy of the quasi-classical approximation is not such as to allow the retention in the wave function of exponentially small terms superimposed on exponentially large ones, and in this sense it is usually not permissible to retain both terms in (46.10).

Although there is, as a rule, no need to use the higher-order terms in the wave function, we shall derive the next term in the expansion (46.3), with a view to noting some aspects of the accuracy of the quasi-classical approximation.

The terms of order $\hbar^2$ in equation (46.4) give

$$\sigma_0'\sigma_2'+\tfrac{1}{2}\sigma_1'^2+\tfrac{1}{2}\sigma_1'' = 0,$$

whence (substituting (46.5) and (46.8) for σ_0 and σ_1)

$$\sigma_2' = p''/4p^2-3p'^2/8p^3.$$

Integrating (by parts in the first term) and introducing the force $F = pp'/m$, we obtain

$$\sigma_2 = \tfrac{1}{4}mF/p^3+\tfrac{1}{8}m^2\int (F^2/p^5)\,\mathrm{d}x.$$

The wave function in this approximation is of the form

$$\psi = e^{(i/\hbar)\sigma} = e^{(i/\hbar)\sigma_0+\sigma_1}(1-i\hbar\sigma_2)$$

or

$$\psi = \frac{\text{constant}}{\sqrt{p}}\left[1-\tfrac{1}{4}im\hbar F/p^3-\tfrac{1}{8}i\hbar m^2\int (F^2/p^5)\,\mathrm{d}x\right]e^{(i/\hbar)\int p\,\mathrm{d}x}. \tag{46.11}$$

The occurrence of imaginary correction terms in the coefficient of the exponential is equivalent to the presence of a similar correction in the phase of the wave function, i.e. of an addition to the integral $(1/\hbar)\int p\,\mathrm{d}x$ in its exponent. This correction is proportional to $\hbar$, i.e. is of order λ/L.

The second and third terms in the brackets in (46.11) must be small in comparison with unity. For the second term, this condition is the same as (46.7); for the third term, an estimate of the integral gives (46.7) only if F^2 tends to zero sufficiently rapidly at distances $\sim L$.

§47. Boundary conditions in the quasi-classical case

Let $x = a$ be a turning point, so that $U(a) = E$, and let $U > E$ for all $x > a$, so that the region to the right of the turning point is classically inaccessible. The wave function must be damped in this region. Sufficiently far from the turning point, it has the form

$$\psi = \frac{C}{2\sqrt{|p|}}\exp\left(-\frac{1}{\hbar}\left|\int_a^x p\,\mathrm{d}x\right|\right) \quad \text{for} \quad x > a, \tag{47.1}$$

corresponding to the first term in (46.10). To the left of the turning point, the wave function must be represented by a real combination (46.9) of two quasi-classical solutions of Schrödinger's equation:

$$\psi = \frac{C_1}{\sqrt{p}}\exp\left(\frac{i}{\hbar}\int_a^x p\,\mathrm{d}x\right)+\frac{C_2}{\sqrt{p}}\exp\left(-\frac{i}{\hbar}\int_a^x p\,\mathrm{d}x\right) \quad \text{for} \quad x < a. \tag{47.2}$$

To determine the coefficients in this combination we must follow the variation in the wave function from positive $x-a$ (where (47.1) holds) to negative $x-a$. In doing so, however, it is necessary to pass through a region near the turning point where the quasi-classical approximation is invalid, and the exact solution of Schrödinger's equation must be considered. For small $|x-a|$ we have

$$E-U(x) \approx F_0(x-a),\ F_0 = -[\mathrm{d}U/\mathrm{d}x]_{x=a} < 0; \tag{47.3}$$

that is, the problem in this region is one of movement in a homogeneous field. The exact solution of Schrödinger's equation for this problem has been found in §24, and the relation between the coefficients in (47.1) and (47.2) can be derived by comparison with the asymptotic forms (24.5) and (24.6) of this exact solution on either side of the turning point. Here it must be noted that (47.3) gives $p(x) = \sqrt{[2mF_0(x-a)]}$, so that the integral

$$\frac{1}{\hbar}\int_a^x p\,\mathrm{d}x = \frac{2}{3\hbar}\sqrt{(2mF_0)}(x-a)^{3/2}$$

is equal to the argument of the exponential in (24.5) or the sine in (24.6). In this discussion it is important that the region where the expansion (47.3) is valid and the quasi-classical region partly overlap: if the motion is quasi-classical in almost the whole of the field region (as we assume), then there exist values of $|x-a|$ small enough for the expansion (47.3) to be valid but also large enough for the quasi-classicality condition to be satisfied and for the asymptotic forms (24.5) and (24.6) to be applicable.†

There is, however, another approach that is methodologically more instructive and does not make use of the exact solution. For this, $\psi(x)$ must be formally regarded as a function of a complex variable x, and the passage from positive to negative $x-a$ must be along a path which is always sufficiently far from the point $x = a$, so that the quasi-classicality condition is formally satisfied along the whole path (A. Zwaan 1929). We then again consider values of $|x-a|$ such that the expansion (47.3) is also valid, so that the wave function (47.1) has the form

$$\psi(x) = \frac{C}{2\sqrt{(2m|F_0|)}}\frac{1}{(x-a)^{1/4}}\exp\left\{-\frac{1}{\hbar}\int_a^x \sqrt{[2m|F_0|(x-a)]}\,\mathrm{d}x\right\}. \tag{47.4}$$

Let us first examine the variation of this function on passing round the point $x = a$ from right to left along a semicircle of radius ρ in the upper half-

† The expansion (47.3) is valid for $|x-a| \ll L$, where L is the characteristic distance for variation of the field $U(x)$. The quasi-classicality condition (46.7) requires that $|x-a|^{3/2} \gg \hbar/\sqrt{(m|F_0|)}$. These two conditions are compatible, since the quasi-classicality of the motion far from the turning-point (i.e. for $|x-a| \sim L$) implies that $L^{3/2} \gg \hbar/\sqrt{(m|F_0|)}$.

plane of the complex variable x. On this semicircle,

$$x-a = \rho e^{i\phi}, \int_a^x \sqrt{(x-a)}\,dx = \tfrac{2}{3}\rho^{3/2}(\cos\tfrac{3}{2}\phi + i\sin\tfrac{3}{2}\phi),$$

the phase ϕ varying from 0 to π. The exponential factor in (47.4) at first (for $0 < \phi < \frac{2}{3}\pi$) increases in modulus, and then decreases to modulus 1. At the end of the semicircle the exponent becomes purely imaginary, equal to

$$-\frac{i}{\hbar}\int_a^x \sqrt{[2m|F_0|(a-x)]}\,dx = -\frac{i}{\hbar}\int_a^x p(x)\,dx.$$

In the coefficient of the exponential in (47.4), the change along the semicircle is

$$(x-a)^{-1/4} \to (a-x)^{-1/4}e^{-i\pi/4}.$$

Thus the whole function (47.4) becomes the second term in (47.2) with coefficient $C_2 = \frac{1}{2}Ce^{-i\pi/4}$.

The fact that by passing through the upper half-plane it is possible to determine only the coefficient C_2 in (47.2) has a simple explanation. If we follow the variation of the function (47.2) along the same semicircle in the opposite direction (from left to right), we see that at the beginning the first term rapidly becomes exponentially small in comparison with the second term. But the quasi-classical approximation does not allow us to include exponentially small terms in ψ superimposed on the large principal term, and this is why the first term in (47.2) is "lost" in the passage along the semicircle.

To determine the coefficient C_1, we must pass from right to left along a semicircle in the lower half-plane of the complex variable x. In a similar manner, we find that formula (47.4) then becomes the first term in (47.2) with coefficient $C_1 = \frac{1}{2}Ce^{i\pi/4}$.

Thus the wave function (47.1) for $x > a$ corresponds to the function

$$\psi = \frac{C}{\sqrt{p}}\cos\left(\frac{1}{\hbar}\int_a^x p\,dx + \tfrac{1}{4}\pi\right)$$

for $x < a$. This rule of correspondence may be written in a form independent of the side of the turning-point on which the classically inaccessible region lies:

$$\frac{C}{2\sqrt{|p|}}\exp\left\{-\left|\frac{1}{\hbar}\int_a^x p\,dx\right|\right\} \to \frac{C}{\sqrt{p}}\cos\left\{\left|\frac{1}{\hbar}\int_a^x p\,dx\right| - \tfrac{1}{4}\pi\right\} \qquad (47.5)$$

$$\text{for } U(x) > E \qquad\qquad \text{for } U(x) < E$$

(H. A. Kramers 1926).

Let us once again emphasize what is obvious from the proof, namely that this rule is associated with a particular boundary condition imposed on one side of the turning-point, and in this sense it can be applied only in a particular direction. The rule (47.5) is derived with the boundary condition that $\psi \to 0$ into the classically inaccessible region, and must be applied to a passage from the latter to the classically allowed region, as is shown by the arrow.†

If the classically accessible region is bounded (at $x = a$) by an infinitely high "potential wall", the boundary condition for the wave function at $x = a$ is $\psi = 0$ (see §18). The quasi-classical approximation is then valid up to the wall itself, and the wave function is

$$\left.\begin{aligned} \psi &= \frac{C}{\sqrt{p}} \sin \frac{1}{\hbar} \int\limits_a^x p\,dx \qquad \text{for} \quad x < a, \\ \psi &= 0 \qquad\qquad\qquad\qquad \text{for} \quad x > a. \end{aligned}\right\} \tag{47.6}$$

§48. Bohr and Sommerfeld's quantization rule

States that belong to the discrete energy spectrum are quasi-classical for high values of the quantum number n, the ordinal number of the state, since this gives the number of nodes of the eigenfunction (see §21), and the distance between adjacent nodes is equal in order of magnitude to the de Broglie wavelength. For large n this distance is small, and the wavelength is therefore small in comparison with the dimensions of the region of the motion.

Let us derive the condition which determines the quantum energy levels in the quasi-classical case. To do this we consider a finite one-dimensional motion of a particle in a potential well; the classically accessible region $b \leqslant x \leqslant a$ is bounded by two turning points.‡

According to the rule (47.5), the boundary condition at $x = b$ gives (in the region right of this point) the wave function

$$\psi = \frac{C}{\sqrt{p}} \cos \left[\frac{1}{\hbar} \int\limits_b^x p\,dx - \tfrac{1}{4}\pi\right]. \tag{48.1}$$

† A passage in the opposite direction is meaningless in that even a small change of the wave function on the right in (47.5) may give rise to an exponentially increasing term in the function on the left.

‡ In classical mechanics, a particle in such a field would execute a periodic motion with period (time taken in moving from $x = b$ to $x = a$ and back)

$$T = 2 \int\limits_b^a dx/v = 2m \int\limits_b^a \frac{dx}{p},$$

where v is the velocity of the particle.

Applying the same rule to the region left of the point $x = a$, we obtain the same function in the form

$$\psi = \frac{C'}{\sqrt{p}} \cos\left[\frac{1}{\hbar}\int_x^a p\,dx - \tfrac{1}{4}\pi\right].$$

If these two expressions are the same throughout the region, the sum of their phases (which is a constant) must be an integral multiple of π:

$$\frac{1}{\hbar}\int_b^a p\,dx - \tfrac{1}{2}\pi = n\pi,$$

with $C = (-1)^n C'$. Hence

$$\frac{1}{2\pi\hbar}\oint p\,dx = n + \tfrac{1}{2} \tag{48.2}$$

where $\oint p\,dx = 2\int_b^a p\,dx$ is the integral taken over the whole period of the classical motion of the particle. This is the condition which determines the stationary states of the particle in the quasi-classical case. It corresponds to Bohr and Sommerfeld's quantization rule in the old quantum theory.

It is easy to see that the integer n is equal to the number of zeros of the wave function, and hence it is the ordinal number of the stationary state. For the phase of the wave function (48.1) increases from $-\frac{1}{4}\pi$ at $x = b$ to $(n+\frac{1}{4})\pi$ at $x = a$, so that the cosine vanishes n times in this range (outside the range $b \leqslant x \leqslant a$, the wave function decreases monotonically and has no zeros at a finite distance).†

As has been shown previously, the number n is large in the quasi-classical case. It must be emphasized, however, that the retention of the term $\frac{1}{2}$ added to n in (48.2) is nevertheless legitimate: to take account of the subsequent correction terms in the phase of the wave functions would give only terms $\sim\lambda/L$ on the right of (48.2), which are small in comparison with unity; see the remark at the end of §46.‡

In normalizing these wave functions, the integration of $|\psi|^2$ can be restricted to the range $b \leqslant x \leqslant a$, since outside this range ψ decreases exponentially. Since the argument of the cosine in (48.1) is a rapidly varying function, we can with sufficient accuracy replace the squared cosine by its mean value $\frac{1}{2}$.

† Strictly speaking, the zeros should be counted by means of the exact form of the wave function near the turning points. If this is done, the result given in the text is confirmed.

‡ In some cases the exact expression for the energy levels $E(n)$ (as a function of the quantum number n), obtained from the exact Schrödinger's equation, is such that it retains its form as $n \to \infty$; examples are the energy levels in a Coulomb field, and those of a harmonic oscillator. In these cases, of course, the quantization rule (48.2), although really applicable only for large n, gives for the function $E(n)$ an expression which is the exact one.

This gives

$$\int |\psi|^2 \, \mathrm{d}x \approx \tfrac{1}{2}C^2 \int_b^a \frac{\mathrm{d}x}{p(x)}$$

$$= \pi C^2/2m\omega = 1,$$

where $\omega = 2\pi/T$ is the frequency of the classical periodic motion. Thus the normalized quasi-classical function is

$$\psi = \sqrt{\frac{2\omega}{\pi v}} \cos\left[\frac{1}{\hbar}\int_b^x p \, \mathrm{d}x - \tfrac{1}{4}\pi\right]. \tag{48.3}$$

It must be recalled that the frequency ω is in general different for different levels, being a function of energy.

The relation (48.2) can also be interpreted in another manner. The integral $\oint p \, \mathrm{d}x$ is the area enclosed by the closed classical phase trajectory of the particle (i.e. the curve in the px-plane, which is the phase space of the particle). Dividing this area into cells, each of area $2\pi\hbar$, we have n cells altogether; n, however, is the number of states with energies not exceeding the given value (corresponding to the phase trajectory considered). Thus we can say that, in the quasi-classical case, there corresponds to each quantum state a *cell* in phase space of area $2\pi\hbar$. In other words, the number of states belonging to the volume element $\Delta p \Delta x$ of phase space is

$$\Delta p \Delta x / 2\pi\hbar. \tag{48.4}$$

If we introduce, instead of the momentum, the *wave number* $k = p/\hbar$, this number can be written

$$\Delta k \Delta x / 2\pi.$$

It is, as we should expect, the same as the familiar expression for the number of characteristic vibrations of a wave field (see *Fields*, §52).

Starting from the quantization rule (48.2), we can ascertain the general nature of the distribution of levels in the energy spectrum. Let ΔE be the distance between two neighbouring levels, i.e. levels whose quantum numbers n differ by unity. Since ΔE is small (for large n) compared with the energy itself of the levels, we can write, from (48.2),

$$\Delta E \oint (\partial p/\partial E) \, \mathrm{d}x = 2\pi\hbar.$$

But $\partial E/\partial p = v$, so that

$$\oint (\partial p/\partial E) \, \mathrm{d}x = \oint \mathrm{d}x/v = T.$$

Hence we have

$$\Delta E = 2\pi\hbar/T = \hbar\omega. \tag{48.5}$$

Thus the distance between two neighbouring levels is $\hbar\omega$. The frequencies ω may be regarded as approximately the same for several adjacent levels (the difference in whose numbers n is small compared with n itself). Hence we reach the conclusion that, in any small range of a quasi-classical part of the spectrum, the levels are equidistant, at intervals of $\hbar\omega$. This result could have been foreseen, since, in the quasi-classical case, the frequencies corresponding to transitions between different energy levels must be integral multiples of the classical frequency ω.

It is of interest to investigate what the matrix elements of any physical quantity f become in the limit of classical mechanics. To do this, we start from the fact that the mean value $\bar{f}$ in any quantum state must become, in the limit, simply the classical value of the quantity, provided that the state itself gives, in the limit, a motion of the particle in a definite path. A wave packet (see §6) corresponds to such a state; it is obtained by superposition of a number of stationary states with nearly the same energy. The wave function of such a state is of the form

$$\Psi = \sum_n a_n \Psi_n,$$

where the coefficients a_n are noticeably different from zero only in some range Δn of values of the quantum number n such that $1 \ll \Delta n \ll n$; the numbers n are supposed large, because the stationary states are quasi-classical. The mean value of f is, by definition,

$$\bar{f} = \int \Psi^* f \Psi \, dx = \sum_n \sum_m a_m^* a_n f_{mn} e^{i\omega_{mn} t},$$

or, replacing the summation over n and m by a summation over n and the difference $m - n = s$,

$$\bar{f} = \sum_n \sum_s a_{n+s}^* a_n f_{n+s,\,n} e^{i\omega s t},$$

where we have put $\omega_{mn} = s\omega$ in accordance with (48.5).

The matrix elements f_{nm} calculated by means of the quasi-classical wave functions decrease rapidly in magnitude as the difference $m-n$ increases, though at the same time they vary only slowly with n itself ($m-n$ being fixed). Hence we can write approximately

$$\bar{f} = \sum_n \sum_s a_n^* a_n f_s e^{i\omega s t} = \sum_n |a_n|^2 \sum_s f_s e^{i\omega s t},$$

where we have introduced the notation $f_s = f_{\bar{n}+s,\bar{n}}$, $\bar{n}$ being some mean value of the quantum number in the range Δn. But $\sum |a_n|^2 = 1$; hence

$$\bar{f} = \sum_s f_s e^{i\omega s t}.$$

The sum obtained is in the form of an ordinary Fourier series. Since $\bar{f}$ must, in the limit, coincide with the classical quantity $f(t)$, we arrive at the result that the matrix elements f_{mn} in the limit become the components f_{m-n} in the expansion of the classical function $f(t)$ as a Fourier series.

Similarly, the matrix elements for transitions between states of the continuous spectrum become the components in the expansion of $f(t)$ as a Fourier integral. Here the wave functions of the stationary states must be normalized by $(1/\hbar)$ times the delta function of energy.

All the above results can be generalized immediately to systems with several degrees of freedom, executing a finite motion for which the problem in classical mechanics allows a complete separation of the variables in the Hamilton–Jacobi method (called a *conditionally periodic* motion; see *Mechanics*, §52). After separation of the variables for each degree of freedom, the problem reduces to a one-dimensional problem, and the corresponding quantization conditions are

$$\oint p_i \, dq_i = 2\pi\hbar(n_i+\gamma_i), \tag{48.6}$$

where the integral is taken over the period of variation of the generalized coordinate q_i, and γ_i is a number of the order of unity which depends on the nature of the boundary conditions for the degree of freedom considered.†

In the general case of an arbitrary (not conditionally periodic) motion in several dimensions the formulation of the quasi-classical conditions of quantization calls for more far-reaching considerations.‡ The concept of "cells" in phase space is, however, applicable (in the quasi-classical approximation) in the same form always. This is clear from the above-mentioned relationship between it and the number of characteristic vibrations of the wave field in a given volume of space. In the general case of a system with s degrees of freedom, there are

$$\Delta N = \Delta q_1 \ldots \Delta q_s \Delta p_1 \ldots \Delta p_s/(2\pi\hbar)^s \tag{48.7}$$

quantum states in a volume element in phase space.||

PROBLEMS

PROBLEM 1. Determine (approximately) the number of discrete energy levels of a particle moving in an arbitrary (not central) field $U(\mathbf{r})$ which satisfies the quasi-classical condition.

SOLUTION. The number of states belonging to a volume of phase space which corresponds to momenta in the range $0 \leqslant p \leqslant p_{max}$ and particle coordinates in the volume element dV

† For example, in motion in a centrally symmetric field we have

$$\oint p_r \, dr = 2\pi\hbar(n_r+\tfrac{1}{2}), \quad \oint p_\theta \, d\theta = 2\pi\hbar(l-m+\tfrac{1}{2}), \quad \oint p_\phi \, d\phi = 2\pi\hbar m,$$

where $n_r = n-l-1$ is the radial quantum number. The last of the three equations simply expresses the fact that p_ϕ is the z-component of the angular momentum, equal to $\hbar m$.

‡ See J. B. Keller, *Annals of Physics* **4**, 180, 1958.

|| In particular, for one particle, $d^3p/(2\pi\hbar)^3$ is the number of states for a range d^3p of values of the momentum in unit volume of coordinate space. This explains the agreement of the two methods of normalizing the plane wave (15.8), mentioned in the footnote to that formula.

is $\frac{4}{3}\pi p_{max}^3\, dV/(2\pi\hbar)^3$. For given $\mathbf{r}$ the particle can have (in its classical motion) a momentum satisfying the condition $E = p^2/2m + U(\mathbf{r}) \leqslant 0$. Substituting $p_{max} = \sqrt{[-2mU(\mathbf{r})]}$, we obtain the total number of states of the discrete spectrum:

$$\frac{\sqrt{2}\, m^{3/2}}{3\pi^2\, \hbar^3} \int (-U)^{3/2}\, dV,$$

where the integration is over the region of space in which $U < 0$. This integral diverges (i.e. the number of states is infinite) if U decreases at infinity as r^{-s} with $s < 2$, in accordance with the results of §18.

Problem 2. The same as Problem 1, but for a quasi-classical centrally symmetric field $U(r)$ (V. L. Pokrovskiĭ).

Solution. In a centrally symmetric field the number of states is not the same as the number of energy levels, on account of the degeneracy of the latter with respect to the direction of the angular momentum. The required number can be found by noting that the number of levels with a given value of the angular momentum M is the same as the number of (non-degenerate) levels for a one-dimensional motion in a field with potential energy $U_{eff} = U(r) + M^2/2mr^2$. The maximum possible value of the momentum p_r for given r and energies $E \leqslant 0$ is $p_{r,max} = \sqrt{(-2mU_{eff})}$. The number of states (i.e. the required number of levels) is therefore

$$\int \frac{dr\, dp_r}{2\pi\hbar} = \frac{\sqrt{(2m)}}{2\pi\hbar} \int \sqrt{\left(-U - \frac{M^2}{2mr^2}\right)}\, dr.$$

The required total number of discrete levels is obtained from this by integration with respect to $M/\hbar$ (which replaces in the quasi-classical case the summation with respect to l), and is

$$(m/4\hbar^2) \int (-U) r\, dr.$$

§49. Quasi-classical motion in a centrally symmetric field

In motion in a centrally symmetric field the wave function of a particle falls, as we know, into an angular and a radial part. Let us first consider the former.

The dependence of the angular wave function on the angle ϕ (determined by the quantum number m) is so simple that the question of finding approximate formulae for it does not arise. The dependence on the polar angle θ is, according to the general rule, quasi-classical if the corresponding quantum number l is large (this condition will be more precisely formulated below).

We shall here confine ourselves to deriving the quasi-classical expression for the angular function for the case (the most important one in applications) of states whose magnetic quantum number is zero ($m = 0$).† This function is, apart from a constant factor, the Legendre polynomial $P_l(\cos\theta)$ (see (28.8)), and satisfies the differential equation

$$d^2P_l/d\theta^2 + \cos\theta\, dP_l/d\theta + l(l+1)P_l = 0. \qquad (49.1)$$

The substitution

$$P_l(\cos\theta) = \chi(\theta)/\sqrt{\sin\theta} \qquad (49.2)$$

† The opposite case, $m = l$, must correspond in the limit to motion in a classical orbit lying in the equatorial plane $\theta = \frac{1}{2}\pi$, since $P_l^l(\cos\theta) = \text{constant} \times \sin^l\theta$, and as $l \to \infty$ this function (and therefore $|\psi|^2$) tends to zero for all $\theta \neq \frac{1}{2}\pi$.

reduces this to

$$\chi''+[(l+\tfrac{1}{2})^2+\tfrac{1}{4}\operatorname{cosec}^2\theta]\chi = 0, \tag{49.3}$$

which does not contain the first derivative and is similar in appearance to the one-dimensional Schrödinger's equation.

In equation (49.3), the part of the de Broglie wavelength is played by

$$\lambda = 2\pi\,[(l+\tfrac{1}{2})^2+\tfrac{1}{4}\operatorname{cosec}^2\theta]^{-1/2}.$$

The requirement that the derivative $d(\lambda/2\pi)/dx$ is small (the condition (46.6)) gives the inequalities

$$\theta l \gg 1, \qquad (\pi-\theta)l \gg 1, \tag{49.4}$$

which are the conditions that the angular part of the wave function is quasi-classical. For large l these conditions hold for almost all values of θ, excluding only a range of angles very close to 0 or π.

When the conditions (49.4) are satisfied, we can neglect the second term in the brackets in (49.3) compared with the first:

$$\chi''+(l+\tfrac{1}{2})^2\chi = 0.$$

The solution of this equation is

$$\chi = \sqrt{\sin\theta}\,P_l(\cos\theta) = A\sin[(l+\tfrac{1}{2})\theta+\alpha], \tag{49.5}$$

where A and α are constants.

For angles $\theta \ll 1$, we can put in equation (49.1) $\cos\theta \approx 1/\theta$; replacing also $l(l+1)$ by the approximation $(l+\frac{1}{2})^2$, we obtain the equation

$$\frac{d^2P_l}{d\theta^2}+\frac{1}{\theta}\frac{dP_l}{d\theta}+(l+\tfrac{1}{2})^2P_l = 0,$$

which has as solution the Bessel function of zero order:

$$P_l(\cos\theta) = J_0[(l+\tfrac{1}{2})\theta], \quad \theta \ll 1. \tag{49.6}$$

The constant factor is put equal to unity, since we must have $P_l = 1$ for $\theta = 0$. The approximate expression (49.6) for P_l is valid for all angles $\theta \ll 1$. In particular, it can be applied for angles in the range $1/l \ll \theta \ll 1$, where it must agree with the expression (49.5), which holds for all $\theta \gg 1/l$. For $\theta l \gg 1$ the Bessel function can be replaced by its asymptotic expression for large values of the argument, and we obtain

$$P_l \approx \sqrt{\frac{2}{\pi l}}\,\frac{\sin[(l+\frac{1}{2})\theta+\frac{1}{4}\pi]}{\sqrt{\theta}}$$

(we can neglect $\frac{1}{2}$ in the coefficient compared with l). On comparison with (49.5), we find that $A = \sqrt{(2/\pi l)}$, $\alpha = \frac{1}{4}\pi$. Thus we obtain finally the following expression for $P_l(\cos\theta)$, applicable in the quasi-classical case:†

$$P_l(\cos\theta) \approx \sqrt{\frac{2}{\pi l}}\frac{\sin[(l+\frac{1}{2})\theta+\frac{1}{4}\pi]}{\sqrt{\sin\theta}}. \tag{49.7}$$

The normalized spherical harmonic function Y_{l0} is obtained from this as (cf. (28.8))

$$Y_{l0} \approx \frac{i^l}{\pi}\frac{\sin[(l+\frac{1}{2})\theta+\frac{1}{4}\pi]}{\sqrt{\sin\theta}}. \tag{49.8}$$

Let us now turn to the radial part of the wave function. It has been shown in §32 that the function $\chi(r) = rR(r)$ satisfies an equation identical with the one-dimensional Schrödinger's equation, with the potential energy

$$U_l(r) = U(r)+\frac{\hbar^2}{2m}\frac{l(l+1)}{r^2}.$$

Hence we can apply the results obtained in the previous sections, if the potential energy is understood to be the function $U_l(r)$.

The case $l = 0$ is the simplest. The centrifugal energy vanishes and, if the field $U(r)$ satisfies the necessary condition (46.6), the radial wave function will be quasi-classical in all space. For $r = 0$ we must have $\chi = 0$, and hence the quasi-classical function $\chi(r)$ is determined by formulae (47.6).

If $l \neq 0$, the centrifugal energy also must satisfy the condition (46.6). In the region of small r, where the centrifugal energy is of the same order as the total energy, the wavelength $\lambda = 2\pi\hbar/p \sim r/l$, and the condition (46.6) gives $l \gg 1$. Thus, if l is small, the quasi-classical condition is violated by the centrifugal energy in the region of small r. It is easily seen that we obtain the correct value of the phase of the quasi-classical wave function $\chi(r)$ by calculating it from the formulae for one-dimensional motion, replacing the coefficient $l(l+1)$ in the potential energy $U_l(r)$ by $(l+\frac{1}{2})^2$:‡

$$U_l(r) = U(r)+\frac{\hbar^2}{2m}\frac{(l+\frac{1}{2})^2}{r^2}. \tag{49.9}$$

The question of the applicability of the quasi-classical approximation to a Coulomb field $U = \pm\alpha/r$ requires special consideration. The most important part of the whole region of the motion is that corresponding to distances r for which $|U| \sim |E|$, i.e. $r \sim \alpha/|E|$. The condition for quasi-classical motion in this region amounts to the requirement that the wavelength $\lambda \sim \hbar/\sqrt{(2m|E|)}$ is small compared with the dimensions $\alpha/|E|$ of the

† Note that, as a result of replacing $l(l+1)$ by $(l+\frac{1}{2})^2$, we have obtained an expression which is multiplied by $(-1)^l$ when θ is replaced by $\pi-\theta$; this is as it should be for the function $P_l(\cos\theta)$.

‡ For example, in the simple case of free motion ($U = 0$) the phase of the function calculated from formula (48.1) with U_l from (49.9) will be the same as the phase of (33.12) for large r, as it should be.

region; this gives

$$|E| \ll m\alpha^2/\hbar^2, \tag{49.10}$$

i.e. the absolute value of the energy must be small compared with the energy of the particle in the first Bohr orbit. This condition can also be written in the form

$$\alpha/\hbar v \gg 1, \tag{49.11}$$

where $v \sim \sqrt{(|E|/m)}$ is the velocity of the particle. It should be noticed that this condition is the opposite of the condition (45.7) for the applicability of perturbation theory to a Coulomb field.

The region of small distances ($|U(r)| \gg E$) is without interest in a repulsive Coulomb field, since for $U > E$ the quasi-classical wave functions diminish exponentially. In an attractive field, however, when l is small it is possible for the particle to penetrate into the region where $|U| \gg E$, so that we have to consider the limits of applicability of the quasi-classical approximation in this case. We use the general condition (46.7), putting there

$$F = -\mathrm{d}U/\mathrm{d}r = -\alpha/r^2, \quad p \approx \sqrt{(2m|U|)} \sim \sqrt{(m\alpha/r)}.$$

As a result, we find that the region of applicability of the quasi-classical approximation is restricted to distances such that

$$r \gg \hbar^2/m\alpha, \tag{49.12}$$

i.e. distances large in comparison with the "radius" of the first Bohr orbit.

PROBLEM

Determine the behaviour of the wave function near the origin, if the field becomes infinite as $\pm\alpha/r^s$, with $s > 2$, when $r \to 0$.

SOLUTION. For sufficiently small r, the wavelength $\lambda \sim \hbar/\sqrt{(m|U|)} \sim \hbar r^{s/2}/\sqrt{(m\alpha)}$, so that $\mathrm{d}\lambda/\mathrm{d}r \sim \hbar r^{s/2-1}/\sqrt{(m\alpha)} \ll 1$; thus the quasi-classical condition is satisfied. In an attractive field $U_l \to -\infty$ when $r \to 0$. The region near the origin is in this case classically accessible, and the radial wave function $\chi \sim 1/\sqrt{p}$, whence

$$\psi \sim r^{s/4-1}.$$

In a repulsive field, the region of small r is classically inaccessible. In this case the wave function tends exponentially to zero as $r \to 0$. Omitting the coefficient of the exponential function, we have

$$\psi \sim \exp\left[-\frac{1}{\hbar}\int_{r_0}^{r} p\,\mathrm{d}r\right], \text{ or } \psi \sim \exp\left\{-\frac{2\sqrt{(2m\alpha)}}{(s-2)\hbar} r^{-(s/2-1)}\right\}.$$

§50. Penetration through a potential barrier

Let us consider the motion of a particle in a field of the type shown in Fig. 13, characterized by the presence of a *potential barrier*, i.e. a region in which the potential energy $U(x)$ exceeds the total energy E of the particle. In classical mechanics, a potential barrier is "impenetrable" to a particle; in quantum mechanics, however, a particle can pass "through the barrier":

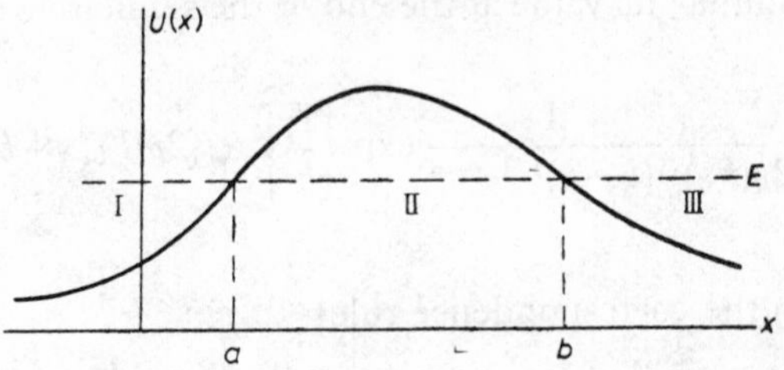

FIG. 13

the probability of this is not zero. The phenomenon is also called the *tunnel effect*.† If the field $U(x)$ satisfies the quasi-classical conditions, the transmission coefficient for the barrier can be calculated in a general form. We may remark that, in particular, these conditions give the result that the barrier must be "wide", and hence the transmission coefficient is small in the quasi-classical case.

In order not to interrupt the subsequent calculations, we shall first solve the following problem. Let the quasi-classical wave function in the region to the right of the turning point $x = b$ (where $U(x) < E$) have the form of a travelling wave:

$$\psi = \frac{C}{\sqrt{p}} \exp\left[\frac{i}{\hbar}\int_b^x p\,dx + \tfrac{1}{4}i\pi\right]. \tag{50.1}$$

We require to find the wave function of this state in the region $x < b$. This can be done by the same procedure as in §47, using the plane of the complex variable x. Putting

$$E - U(x) \approx F_0(x-b),\ F_0 > 0,$$

we can write the function (50.1) as

$$\psi(x) = \frac{C}{\sqrt{(2mF_0)}}\frac{1}{(x-b)^{1/4}} \exp\left\{\frac{i}{\hbar}\sqrt{(2mF_0)}\int_b^x \sqrt{(x-b)}\,dx + \tfrac{1}{4}i\pi\right\},$$

and pass from right to left along a semicircle in the upper half-plane:

$$x - b = \rho e^{i\phi},\ i\int_b^x \sqrt{(x-b)}\,dx = \tfrac{2}{3}\rho^{3/2}(-\sin\tfrac{3}{2}\phi + i\cos\tfrac{3}{2}\phi),$$

the phase ϕ varying from 0 to π. The function $\psi(x)$ at first decreases and then

† An example of this type has already occurred in §25, Problem 2.

increases in modulus, its value at the end of the semicircle being

$$\psi(x) = \frac{C}{\sqrt{(2mF_0)}} \frac{1}{(b-x)^{1/4} e^{i\pi/4}} \exp\left\{\frac{1}{\hbar}\int_x^b \sqrt{[2mF_0(x-b)]}\, dx + \tfrac{1}{4}i\pi\right\}.$$

Thus we obtain the correspondence rule†

$$\frac{C}{\sqrt{p}} \exp\left\{\frac{i}{\hbar}\int_b^x p\, dx + \tfrac{1}{4}i\pi\right\} \rightarrow \frac{C}{\sqrt{|p|}} \exp\left\{\frac{1}{\hbar}\left|\int_b^x p\, dx\right|\right\}. \tag{50.2}$$

$$\text{for } x > b \qquad\qquad \text{for } x < b$$

It must be emphasized that this rule presupposes a particular form of the wave function (a wave travelling to the right) in the classically allowed region, and must be applied to go from the latter to the classically inaccessible region.

Let us now go on to calculate the coefficient for the penetration of the potential barrier. Let the particle be incident on the barrier from left to right, coming from region I. Then, in region III beyond the barrier, there will be only the wave that has passed through the barrier and is propagated to the right; the wave function in this region may be written

$$\psi = \sqrt{\frac{D}{v}} \exp\left(\frac{i}{\hbar}\int_b^x p\, dx + \tfrac{1}{4}i\pi\right), \tag{50.3}$$

where $v = p/m$ is the particle velocity and D the current density in the wave. Using the rule (50.2), we can now find the wave function in region II, within the barrier:

$$\psi = \sqrt{\frac{D}{|v|}} \exp\left(\frac{1}{\hbar}\left|\int_x^b p\, dx\right|\right)$$

$$= \sqrt{\frac{D}{|v|}} \exp\left(\frac{1}{\hbar}\left|\int_a^b p\, dx\right| - \frac{1}{\hbar}\left|\int_a^x p\, dx\right|\right). \tag{50.4}$$

† In a passage from right to left through the lower half-plane, the function $\psi(x)$ at first increases and then decreases in modulus, becoming an exponentially small quantity on the left-hand axis ($\phi \rightarrow -\pi$), which it would not be legitimate to keep superimposed on the exponentially large function (50.2). In the region where $\psi(x)$ is exponentially large, the inexactness of the quasi-classical approximation loses the exponentially small correction which for $\phi \rightarrow -\pi$ could become an exponentially large term, and the latter is therefore lost also.

Finally, applying the rule (47.5), we have in region I in front of the barrier

$$\psi = 2\sqrt{\frac{D}{v}} \exp\left(\frac{1}{\hbar}\int_a^b |p|\,dx\right) \cos\left(\frac{1}{\hbar}\int_x^a p\,dx - \tfrac{1}{4}\pi\right).$$

If we put here

$$D = \exp\left(-\frac{2}{\hbar}\int_a^b |p|\,dx\right), \tag{50.5}$$

this becomes

$$\psi = \frac{2}{\sqrt{v}} \cos\left(\frac{1}{\hbar}\int_a^x p\,dx + \tfrac{1}{4}\pi\right)$$

$$= \frac{1}{\sqrt{v}} \exp\left(\frac{i}{\hbar}\int_a^x p\,dx + \tfrac{1}{4}i\pi\right) + \frac{1}{\sqrt{v}} \exp\left(-\frac{i}{\hbar}\int_a^x p\,dx - \tfrac{1}{4}i\pi\right).$$

This first term (which becomes a plane wave $\psi = e^{(i/\hbar)px}$ as $x \to -\infty$) represents a wave incident on the barrier, and the second a reflected wave. The normalization chosen corresponds to a unit current density in the incident wave, and therefore D, the current density in the transmitted wave, is equal to the required transmission coefficient for the barrier. Note that this formula is applicable only if the exponent is large, so that D itself is small.†

It has been assumed in the foregoing that the field $U(x)$ satisfies the quasi-classical condition over the whole extent of the barrier (excluding only the immediate neighbourhood of the turning points). In practice, however, we often have to deal with barriers where the potential energy curve on one side drops so steeply that the quasi-classical approximation is inapplicable. The exponential factor in D remains the same in this case as in formula (50.5), but the coefficient of the exponential (equal to unity in (50.5)) is different. To calculate it we must, essentially, calculate the exact wave function in the non-quasi-classical region and determine the quasi-classical wave function inside the barrier in accordance with this.

PROBLEMS

PROBLEM 1. Determine the transmission coefficient for the potential barrier shown in Fig. 14 (p. 182): $U(x) = 0$ for $x < 0$, $U(x) = U_0 - Fx$ for $x > 0$; only the exponential factor need be calculated.

---

† The exponential smallness of D is related to the fact that the amplitudes of the incident and reflected waves in region I are found to be the same; the exponentially small difference between them is lost in the quasi-classical approximation.

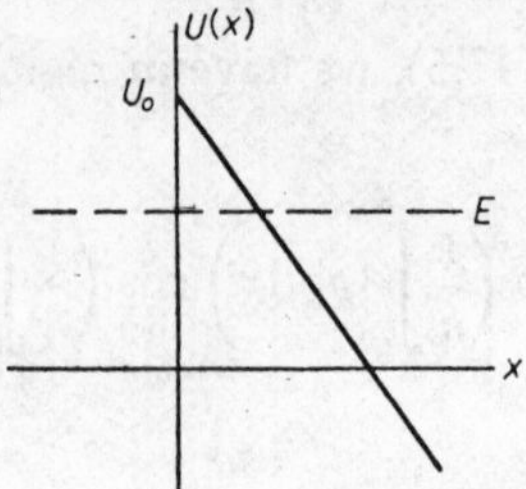

FIG. 14

SOLUTION. A simple calculation gives the result

$$D \sim \exp\left\{-\frac{4\sqrt{(2m)}}{3\hbar F}(U_0-E)^{3/2}\right\}.$$

PROBLEM 2. Determine the probability that a particle (with zero angular momentum) will emerge from a centrally symmetric potential well with $U(r) = -U_0$ for $r < r_0$, $U(r) = \alpha/r$ for $r > r_0$ (Fig. 15).†

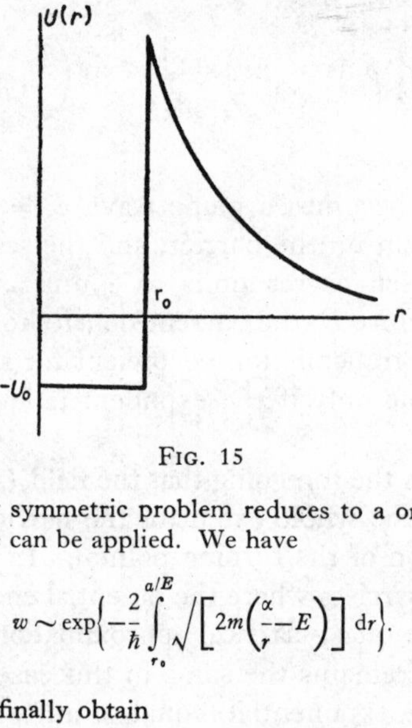

FIG. 15

SOLUTION. The centrally symmetric problem reduces to a one-dimensional one, so that the formulae obtained above can be applied. We have

$$w \sim \exp\left\{-\frac{2}{\hbar}\int_{r_0}^{\alpha/E}\sqrt{\left[2m\left(\frac{\alpha}{r}-E\right)\right]}\,dr\right\}.$$

Evaluating the integral, we finally obtain

$$w \sim \exp\left\{-\frac{2\alpha}{\hbar}\sqrt{\frac{2m}{E}}\left[\cos^{-1}\sqrt{\frac{Er_0}{\alpha}}-\sqrt{\left\{\frac{Er_0}{\alpha}\left(1-\frac{Er_0}{\alpha}\right)\right\}}\right]\right\}.$$

In the limiting case $r_0 \to 0$, this formula becomes

$$w \sim e^{-(\pi\alpha/\hbar)\sqrt{(2m/E)}} = e^{-2\pi\alpha/\hbar v}.$$

These formulae are applicable when the exponent is large, i.e. when $\alpha/\hbar v \gg 1$. This condition agrees, as it should, with the condition (49.11) for quasi-classical motion in a Coulomb field.

† This problem was first discussed by G. Gamow (1928) and by R. W. Gurney and E. U. Condon (1929) in connection with the theory of radioactive α-decay.

PROBLEM 3. The field $U(x)$ consists of two symmetrical potential wells (I and II in Fig. 16), separated by a barrier. If the barrier were impenetrable to a particle, there would be energy levels corresponding to the motion of the particle in one or other well, the same for both wells. The fact that a passage through the barrier is possible results in a splitting of each of these levels into two neighbouring ones, corresponding to states in which the particle moves simultaneously in both wells. Determine the magnitude of the splitting (the field $U(x)$ is supposed quasi-classical).

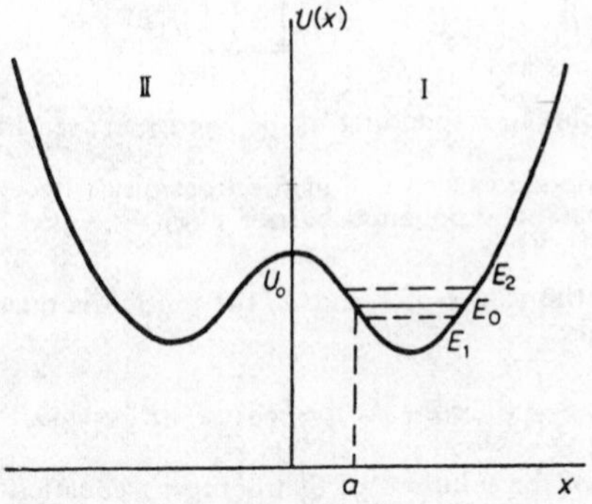

FIG. 16

SOLUTION. An approximate solution of Schrödinger's equation in the field $U(x)$, neglecting the probability of passage through the barrier, can be constructed with the quasi-classical wave function $\psi_0(x)$ which describes the motion with a certain energy E_0 in one well, say I, i.e. which is exponentially damped on both sides of this well; the function $\psi_0(x)$ is assumed to be normalized so that the integral of $\psi_0{}^2$ over well I is unity. When the small probability of tunnelling is taken into account, the level E_0 splits into levels E_1 and E_2. The correct zero-approximation wave functions corresponding to these levels are the symmetric and anti-symmetric combinations of $\psi_0(x)$ and $\psi_0(-x)$:

$$\left.\begin{aligned}\psi_1(x) &= \frac{1}{\sqrt{2}}[\psi_0(x)+\psi_0(-x)],\\ \psi_2(x) &= \frac{1}{\sqrt{2}}[\psi_0(x)-\psi_0(-x)].\end{aligned}\right\} \quad (1)$$

In well I, the function $\psi_0(-x)$ is vanishingly small in comparison with $\psi_0(x)$; in well II the opposite is true. The product $\psi_0(x)\psi_0(-x)$ is therefore vanishingly small everywhere, and the functions (1) are normalized so that the integrals of their squares over wells I and II are unity.

Schrödinger's equations are

$$\psi_0''+(2m/\hbar^2)(E_0-U)\psi_0 = 0, \quad \psi_1''+(2m/\hbar^2)(E_1-U)\psi_1 = 0;$$

we multiply the former by ψ_1 and the latter by ψ_0, subtract corresponding terms, and integrate over x from 0 to ∞. Bearing in mind that, for $x = 0$, $\psi_1 = \sqrt{2}\psi_0$ and $\psi_1' = 0$, and that

$$\int_0^\infty \psi_0\psi_1\,dx \approx \frac{1}{\sqrt{2}}\int_0^\infty \psi_0{}^2\,dx = 1/\sqrt{2},$$

we find

$$E_1-E_0 = -(\hbar^2/m)\psi_0(0)\psi_0'(0).$$

Similarly, we find for E_2-E_0 the same expression with the sign changed. Thus

$$E_2-E_1 = (2\hbar^2/m)\psi_0(0)\psi_0'(0).$$

By means of formula (47.1), with the coefficient C from (48.3), we find that

$$\psi_0(0) = \sqrt{\frac{\omega}{2\pi v_0}} \exp\left[-\frac{1}{\hbar}\int_0^a |p|\,dx\right], \quad \psi_0'(0) = \frac{mv_0}{\hbar}\psi_0(0),$$

where $v_0 = \sqrt{[2(U_0-E_0)/m]}$. Thus

$$E_2-E_1 = \frac{\omega\hbar}{\pi}\exp\left[-\frac{1}{\hbar}\int_{-a}^{a}|p|\,dx\right].$$

where a is the turning point corresponding to the energy E_0; see Fig. 16.

PROBLEM 4. Determine the exact value of the transmission coefficient D for the passage of a particle through a parabolic potential barrier $U(x) = -\frac{1}{2}kx^2$ (supposing that D is *not* small) (E. C. Kemble 1935).†

SOLUTION. Whatever the values of k and E, the motion is quasi-classical at sufficiently large distances $|x|$, with

$$p = \sqrt{[2m(E+\tfrac{1}{2}kx^2)]} \approx x\sqrt{(mk)}+E\sqrt{(m/k)}/x,$$

and the asymptotic form of the solutions of Schrödinger's equation is

$$\psi = \text{constant}\times e^{\pm i\xi^2/2}\xi^{\pm i\epsilon-1/2},$$

where we have introduced the notation

$$\xi = x(mk/\hbar^2)^{1/4}, \qquad \epsilon = (E/\hbar)\sqrt{(m/k)}.$$

We are interested in the solution which, as $x \to +\infty$, contains only a wave which has passed the barrier, i.e. is propagated from left to right. We put

$$\text{as } x \to \infty, \quad \psi = Be^{i\xi^2/2}\xi^{i\epsilon-1/2}, \tag{1}$$

$$\text{as } x \to -\infty, \quad \psi = e^{-i\xi^2/2}(-\xi)^{-i\epsilon-1/2} + Ae^{i\xi^2/2}(-\xi)^{i\epsilon-1/2}. \tag{2}$$

In the expression (2), the first term represents the incident wave, and the second the reflected wave (the direction of propagation of a wave is that in which its phase increases). The relation between A and B can be found by using the fact that in this case the asymptotic expression for ψ is valid in the whole of a sufficiently distant region of the plane of the complex variable ξ. Let us follow the variation of the function (1) as we go round a semicircle of large radius ρ in the upper half-plane of ξ:

$$\xi = \rho e^{i\phi}, i\xi^2 = \rho^2(-\sin 2\phi + i\cos 2\phi),$$

with ϕ varying from 0 to π. As a result of traversing this semicircle, the function (1) becomes the second term in (2), with coefficient

$$A = B(e^{i\pi})^{i\epsilon-1/2} = -iB^{-\pi\epsilon}; \tag{3}$$

in the part of the path ($\frac{1}{2}\pi < \phi < \pi$) where the modulus $|e^{i\xi^2/2}|$ is exponentially large, the exponentially small quantity which should give the first term in (2) is lost.‡

† The solution of this problem can also be applied to penetration sufficiently near the top of any barrier $U(x)$ whose dependence on x near the maximum is quadratic.

‡ The passage through the lower half-plane to determine A would be unsuitable, since on the part of the path ($-\pi < \phi < -\frac{1}{2}\pi$) that adjoins its left-hand end (where ψ is given by (2)), the term in $e^{i\xi^2/2}$ is exponentially small in comparison with $e^{-i\xi^2/2}$.

With the normalization of the incident wave chosen in (2), the condition of conservation of number of particles is

$$|A|^2+|B|^2 = 1. \tag{4}$$

From (3) and (4) we find the required transmission coefficient:

$$D = |B|^2 = 1/(1+e^{-2\pi\epsilon}).$$

This formula holds for any E. If the energy is large and negative, it gives $D \approx e^{-2\pi|\epsilon|}$ in accordance with formula (50.5). For $E > 0$, the quantity

$$R = 1-D = 1/(1+e^{2\pi\epsilon})$$

is the coefficient of reflection above the barrier.

§51. Calculation of the quasi-classical matrix elements

A direct calculation of the matrix elements of any physical quantity f with respect to the quasi-classical wave functions presents great difficulty. We may suppose that the energies of the states between which the matrix element is calculated are not close to each other, so that the element does not reduce to the Fourier component of the quantity f (§48). The difficulties arise because, owing to the fact that the wave functions are exponential (with a large imaginary exponent), the integrand oscillates rapidly.

We shall consider a one-dimensional case (motion in a field $U(x)$), and suppose for simplicity that the operator of the physical quantity is merely a function $f(x)$ of the coordinate. Let ψ_1 and ψ_2 be the wave functions corresponding to some values E_1 and E_2 of the energy of the particle (with $E_2 > E_1$, Fig. 17); we shall suppose that ψ_1 and ψ_2 are taken real. We have to calculate the integral

$$f_{12} = \int_{-\infty}^{\infty} \psi_1 f \psi_2 \, dx. \tag{51.1}$$

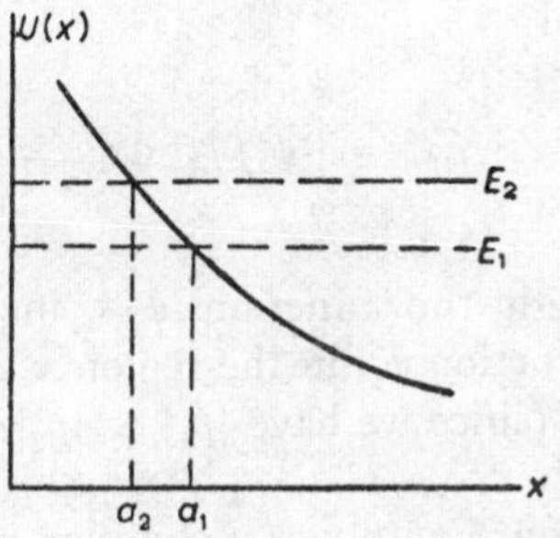

FIG. 17

According to (47.5), the wave function ψ_1 in the regions on both sides of the turning-point $x = a_1$, but not in its immediate neighbourhood, is of the form

$$\left.\begin{aligned} &\text{for } x < a_1, \quad \psi_1 = \frac{C_1}{2\sqrt{|p_1|}} \exp\left[-\frac{1}{\hbar}\int_{a_1}^{x} p_1\, dx\right], \\ &\text{for } x > a_1, \quad \psi_1 = \frac{C_1}{\sqrt{p_1}} \cos\left(\frac{1}{\hbar}\int_{a_1}^{x} p_1\, dx - \tfrac{1}{4}\pi\right), \end{aligned}\right\} \tag{51.2}$$

and similarly for ψ_2 (replacing the suffix 1 by 2).

However, the calculation of the integral (51.1) by substituting in it these asymptotic expressions for the wave functions would not give the correct result. The reason is, as we shall see below, that this integral is an exponentially small quantity, whereas the integrand is not itself small. Hence even a relatively small change in the integrand will in general change the order of magnitude of the integral. This difficulty can be circumvented as follows.

We represent the function ψ_2 as a sum $\psi_2 = \psi_2^+ + \psi_2^-$, expressing the cosine (in the region $x > a_2$) as the sum of two exponentials. According to (50.2), we have

$$\left.\begin{aligned} &\text{for } x < a_2, \quad \psi_2^+ = \frac{C_2}{2\sqrt{|p_2|}} \exp\left[\frac{1}{\hbar}\left|\int_{a_2}^{x} p_2\, dx\right|\right], \\ &\text{for } x > a_2, \quad \psi_2^+ = \frac{C_2}{2\sqrt{p_2}} \exp\left[\frac{i}{\hbar}\int_{a_2}^{x} p_2\, dx + \tfrac{1}{4}i\pi\right]; \end{aligned}\right\} \tag{51.3}$$

the function ψ_2^- is the complex conjugate of $\psi_2^+ : \psi_2^- = (\psi_2^+)^*$.

The integral (51.1) is also divided into the sum of two complex conjugate integrals $f_{12} = f_{12}^+ + f_{12}^-$, which we shall proceed to calculate. First of all, we note that the integral

$$f_{12}^+ = \int_{-\infty}^{\infty} \psi_1 f \psi_2^+\, dx$$

converges. For, although the function ψ_2^+ increases exponentially in the region $x < a_2$, the function ψ_1, in the region $x < a_1$, tends exponentially to zero still more rapidly (since we have $|p_1| > |p_2|$ everywhere in the region $x < a_2$).

We shall regard the coordinate x as a complex variable, and displace the path of integration off the real axis into the upper half-plane. When x receives a positive imaginary increment, an increasing term appears in the function ψ_1 (in the region $x > a_1$), but the function ψ_2^+ decreases still more rapidly, since we have $p_2 > p_1$ everywhere in the region $x > a_1$. Hence the integrand decreases.

The displaced path of integration does not pass through the points $x = a_1$, a_2 on the real axis, near which the quasi-classical approximation is inapplicable. Hence we can use for ψ_1 and ψ_2^+, over the whole path, the functions which are their asymptotic expressions in the upper half-plane. These are

$$\left.\begin{aligned} \psi_1 &= \frac{C_1}{2[2m(U-E_1)]^{1/4}} \exp\left[\frac{1}{\hbar}\int_{a_1}^{x} \sqrt{\{2m(U-E_1)\}}\, dx\right], \\ \psi_2^+ &= \frac{C_2}{2[2m(U-E_2)]^{1/4}} \exp\left[-\frac{1}{\hbar}\int_{a_2}^{x} \sqrt{\{2m(U-E_2)\}}\, dx\right], \end{aligned}\right\} \qquad (51.4)$$

where the roots are taken so as to be positive on the real axis for $x < a_1, a_2$. In the integral

$$f_{12}^+ = \frac{C_1 C_2}{4\sqrt{(2m)}} \int \exp\left[\frac{1}{\hbar}\int_{a_1}^{x} \sqrt{\{2m(U-E_1)\}}\, dx - \frac{1}{\hbar}\int_{a_2}^{x} \sqrt{\{2m(U-E_2)\}}\, dx\right] \times$$
$$\times \frac{f(x)\, dx}{[(U-E_1)(U-E_2)]^{1/4}} \qquad (51.5)$$

we desire to displace the path of integration in such a way that the exponential factor is diminished as much as possible. The exponent has an extreme value only where $U(x) = \infty$ (for $E_1 \neq E_2$, its derivative with respect to x vanishes at no other point). Hence the displacement of the contour of integration into the upper half-plane is restricted only by the necessity of passing round the singular points of the function $U(x)$; according to the general theory of linear differential equations, these coincide with the singular points of the wave function $\psi(x)$. The actual choice of the contour depends on the actual form of the field $U(x)$. Thus, if the function $U(x)$ has only one singular point $x = x_0$ in the upper half-plane, the integration can be effected along the type of path shown in Fig. 18. The immediate neighbourhood of the singular point plays the important part in the integral, so that the matrix element $f_{12} = 2\,\mathrm{re}\, f_{12}^+$ required is practically proportional to an exponentially small expression of the form

$$f_{12} \sim \exp\left\{-\frac{1}{\hbar}\mathrm{im}\left[\int^{x_0} \sqrt{[2m(E_2-U)]}\, dx - \int^{x_0} \sqrt{[2m(E_1-U)]}\, dx\right]\right\} \qquad (51.6)$$

(L. D. Landau 1932).†

† In deriving formulae (51.5) and (51.6), we have replaced the wave functions by their asymptotic expressions, since, in the integral taken along the contour shown in Fig. 18 (p. 188), the order of magnitude of the integral is determined by that of the integrand; hence a relatively small change in the latter does not have any great effect on the value of the integral.

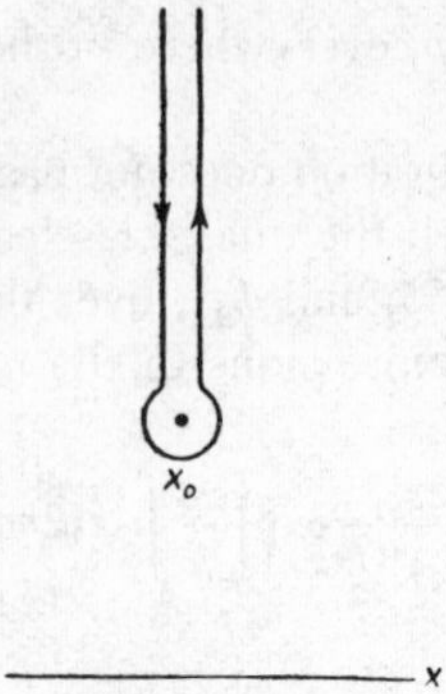

FIG. 18

The lower limits of the integrals may be any points in the classically accessible regions; their particular values evidently do not affect the imaginary parts of the integrals. If the function $U(x)$ has several singular points in the upper half-plane, x_0 in (51.6) must be taken as that for which the exponent is smallest in absolute value.†

The quasi-classical matrix elements for motion in a centrally symmetric field must be calculated by the same method. However, we must now replace $U(r)$ by the effective potential energy (the sum of the potential energy and the centrifugal energy), which will be different for states with different l. In view of further applications of the method in question, we shall write the effective potential energies in the two states in a general form, as $U_1(r)$ and $U_2(r)$. Then the exponent in the exponential factor in the integrand in (51.5) has an extreme value not only at the points where $U_1(r)$ or $U_2(r)$ becomes infinite, but also at those where

$$U_2(r) - U_1(r) = E_2 - E_1. \tag{51.7}$$

Hence, in the formula

$$f_{12} \sim \exp\left\{-\frac{1}{\hbar}\,\mathrm{im}\left[\int^{r_0} \sqrt{[2m(E_2-U_2)]}\,dr - \int^{r_0} \sqrt{[2m(E_1-U_1)]}\,dr\right]\right\} \tag{51.8}$$

the possible values of r_0 include not only the singular points of $U_1(r)$ and $U_2(r)$, but also the roots of equation (51.7).

The centrally symmetric case differs also in that the integration over r in (51.1) is taken from 0 (and not from $-\infty$) to ∞:

$$f_{12} = \int_0^\infty \chi_1 f \chi_2\,dr.$$

Here two cases must be distinguished. If the integrand is an even function of r, the integration can be formally extended to the whole range from $-\infty$

† We assume that the quantity $f(x)$ itself has no singular points.

to ∞, so that there is no difference from the previous case. This may occur if $U_1(r)$ and $U_2(r)$ are even functions of r $[U(-r) = U(r)]$. Then the wave functions $\chi_1(r)$ and $\chi_2(r)$ are either even or odd functions† (see §21), and, if the function $f(r)$ is also even or odd, the product $\chi_1 f \chi_2$ may be even.

If, on the other hand, the integrand is not even (as always happens if $U(r)$ is not even), the start of the path of integration cannot be moved away from the point $r = 0$, and this point must be included among the possible values of r_0 in (51.8).

PROBLEMS

PROBLEM 1. Calculate the quasi-classical matrix elements (exponential factor only) in a field $U = U_0 e^{-\alpha x}$.

SOLUTION. $U(x)$ becomes infinite only for $x \to -\infty$. Accordingly, we put $x_0 = -\infty$ in (51.6). We can extend the integration to $+\infty$.

Each of the integrals diverges at the lower limit. Hence we first calculate them from $-x$ to ∞, and then pass to the limit $x \to \infty$. We find

$$f_{12} \sim e^{-(\pi m/\alpha\hbar)(v_2 - v_1)},$$

where $v_1 = \sqrt{(2E_1/m)}$, $v_2 = \sqrt{(2E_2/m)}$ are the velocities of the particle at infinity ($x \to \infty$), where the motion is free.

PROBLEM 2. The same as Problem 1, but in a Coulomb field $U = \alpha/r$, for transitions between states with $l = 0$.

SOLUTION. The only singular point of the function $U(r)$ is $r = 0$. The corresponding integral has been calculated in §50, Problem 2. As a result we have by formula (51.8)

$$f_{12} \sim \exp\left[\frac{\pi\alpha}{\hbar}\left(\frac{1}{v_2} - \frac{1}{v_1}\right)\right].$$

§52. The transition probability in the quasi-classical case

Penetration through a potential barrier is an example of a process which is entirely impossible in classical mechanics. In the quasi-classical case the probability of such processes is exponentially small. The relevant exponent can be determined as follows.

Considering a transition of any system from one state to another, we solve the corresponding classical equations of motion and find the "path" of the transition; this, however, is complex, in accordance with the fact that the process cannot occur in classical mechanics. In particular, it is found that, in general, the "transition point" q_0 at which the formal transition of the system from one state to the other occurs is complex; the position of this point is determined by the classical conservation laws. We next calculate the action $S_1(q_1, q_0) + S_2(q_0, q_2)$ for the motion of the system in the first state from some initial position q_1 to the "transition point" q_0, and then in the second state from q_0 to the final position q_2. The required probability of the process is then given by the formula

$$w \sim \exp\left\{-\frac{2}{\hbar}\,\mathrm{im}\,[S_1(q_1,q_0) + S_2(q_0,q_2)]\right\}. \tag{52.1}$$

† For even $U(r)$, the radial wave function $R(r)$ is even (or odd) when l is even (or odd), as is seen from its behaviour for small r (where $R \sim r^l$).

If the position of the "transition point" is not unique, it must be chosen so that the exponent in (52.1) has the smallest absolute value (which must yet, of course, be sufficiently large for formula (52.1) to be valid).†

Formula (52.1) is in accordance with the rule derived in §51 for calculating the quasi-classical matrix elements. It should be emphasized, however, that it would not be correct to use the square of the matrix element in calculating the coefficient before the exponential in the probability of such transitions.

The method of *complex classical paths* based on (52.1) is a general one, applicable to transitions in systems with any number of degrees of freedom (L. D. Landau 1932). If the transition point is real, but lies in the classically inaccessible region, then (in the simple case of one-dimensional motion) formula (52.1) is the same as (50.5) for the probability of penetration through the potential barrier.

Reflection above the barrier

Let us apply (52.1) to the one-dimensional problem of reflection above the barrier, i.e. reflection of a particle whose energy exceeds the height of the barrier. In this case, q_0 is to be taken as the complex coordinate x_0 of the "turning point" at which the particle reverses its direction of motion, i.e. the complex root of the equation $U(x) = E$. We shall show how the reflection coefficient may then be calculated more precisely, including the coefficient of the exponential.

We must again (as in §50) establish the relation between the wave functions far to the right of the barrier (the transmitted wave) and far to the left (the incident and reflected waves). This is easily done by a method similar to that used in §§47 and 50, regarding ψ as a function of the complex variable x.

We write the transmitted wave in the form

$$\psi_+ = \frac{1}{\sqrt{p}} \exp\left(\frac{i}{\hbar}\int_{x_1}^{x} p \, dx\right),$$

where x_1 is any point on the real axis, and follow its variation on passing along a path C in the upper half-plane which encloses (at a sufficient distance) the turning point x_0 (Fig. 19); the whole of the latter part of this path must lie so far to the left that the error in the approximate (quasi-classical) wave function of the incident wave is less than the required small quantity ψ_-. Passage round the point x_0 causes a change in the sign of the root $\sqrt{[E-U(x)]}$, and after the return to the real axis the function ψ_+ therefore becomes ψ_-, a wave propagated to the left (i.e. the reflected wave).‡ Since the amplitudes of the incident and transmitted waves may be regarded as equal, the

† If the potential energy of the system has itself singular points, these also must be considered as possible values of q_0.

‡ A passage along a path below the point x_0 (simply going along the real axis, for example) converts the function ψ_+ into the incident wave.

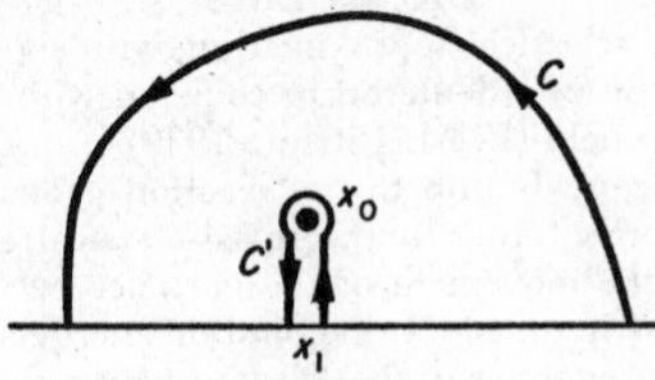

FIG. 19

required reflection coefficient R is simply the ratio of the squared moduli of ψ_- and ψ_+:

$$R = \left|\frac{\psi_-}{\psi_+}\right|^2 = \exp\left(-\frac{2}{\hbar}\,\mathrm{im}\int\limits_C p\,\mathrm{d}x\right). \tag{52.2}$$

Having derived this formula, we can deform the path of integration in the exponent in any manner; if we convert it into the path C' shown in Fig. 19, the integral reduces to twice the integral from x_1 to x_0, giving

$$R = \exp\left(-4\sigma(x_1, x_0)/\hbar\right), \quad \sigma(x_1, x_0) = \mathrm{im}\int\limits_{x_1}^{x_0} p(x)\,\mathrm{d}x; \tag{52.3}$$

since $p(x)$ is real everywhere on the real axis, the choice of x_1 is immaterial. Note that the coefficient of the exponential in (52.3) is unity (V. L. Pokrovskii, S. K. Savvinykh and F. R. Ulinich 1958).†

As already mentioned, among the possible values of x_0 we must select the one for which the exponent in (52.3) is smallest in absolute magnitude (and this value must be large compared with unity).‡ It is also implied that, if the potential energy $U(x)$ itself has singularities in the upper half-plane, the integral $\sigma(x_1, x_0)$ has larger values for such points; otherwise the exponent would be determined by one of these points, but the coefficient of the exponential would not be unity as in (52.3). This condition is certainly not satisfied with increasing energy E if $U(x)$ becomes infinite anywhere in the upper half-plane: ultimately the point x_0 at which $U = E$ becomes so close to the point x_∞ where $U = \infty$ that the two points give comparable contributions to the reflection coefficient (the integral $\sigma(x_\infty, x_0) \sim 1$), and formula (52.3) becomes invalid. In the limit where E is so large that this integral is small compared with unity, perturbation theory becomes applicable (see Problem 2).‖

† The proof given here is due to L. D. Landau (1961).

‡ Of course, only points x_0 are considered for which $\sigma > 0$, i.e. points lying in the upper half-plane.

‖ An intermediate case is discussed by V. L. Pokrovskii and I. M. Khalatnikov, *Soviet Physics JETP* **13**, 1207, 1961.

PROBLEMS

PROBLEM 1. Using the quasi-classical approximation, with exponential accuracy, determine the probability of disintegration of a deuteron in collision with a heavy nucleus regarded as the fixed centre of a Coulomb field (E. M. Lifshitz 1939).

SOLUTION. The principal contribution to the reaction probability comes from collisions with zero orbital angular momentum. In the quasi-classical approximation these are the head-on collisions, in which the movement of the particles becomes one-dimensional.

Let E be the deuteron energy in units of ϵ, the binding energy of the proton and the neutron in the deuteron; E_n and E_p the energies of the released neutron and proton in the same units. We shall also use the dimensionless coordinate $q = \epsilon r/Ze^2$ (where Ze is the charge on the nucleus), and denote by q_0 its value (which is in general complex) at the "transition point", i.e. at the "moment of disintegration" of the deuteron. We can write

$$E_n = \tfrac{1}{2}v_n^2, \quad E_p = \tfrac{1}{2}v_p^2 + \frac{1}{q_0}, \quad E = v_d^2 + \frac{1}{q_0}; \tag{1}$$

here v_n, v_p and v_d are the velocities of the particles at the moment of disintegration, in units of $\sqrt{(\epsilon/m)}$, where m is the nucleon mass; v_n is real and is the same as the velocity of the released neutron, but v_p and v_d are complex. The conditions for the conservation of energy and momentum at the transition point give

$$E_p + E_n = E - 1, \quad v_p + v_n = 2v_d, \tag{2}$$

whence

$$v_p = 2i + v_n, \quad v_d = i + v_n, \quad \frac{1}{q_0} = E + 1 - v_n^2 + 2iv_n.$$

The action of the system before the transition corresponds to the motion of the deuteron in the field of the nucleus up to the point of disintegration; its imaginary part is

$$\begin{aligned} \operatorname{im} S_1 &= Ze^2 \sqrt{\frac{m}{\epsilon}} \operatorname{im} \int_{\infty}^{q_0} \sqrt{\left[4\left(E - \frac{1}{q}\right)\right]} dq \\ &= Ze^2 \sqrt{\frac{m}{\epsilon}} \operatorname{im} \left\{ 2q_0 v_d - \frac{2}{\sqrt{E}} \cosh^{-1}\sqrt{(q_0 E)} \right\}. \end{aligned} \tag{3}$$

After the transition, the action corresponds to the motion of the neutron and the proton away from the point of disintegration:

$$\begin{aligned} \operatorname{im} S_2 &= Ze^2 \sqrt{\frac{m}{\epsilon}} \operatorname{im} \left\{ \int_{q_0}^{\infty} v_n \, dq + \int_{q_0}^{\infty} \sqrt{\left[2\left(E_p - \frac{1}{q}\right)\right]} dq \right\} \\ &= Ze^2 \sqrt{\frac{m}{\epsilon}} \operatorname{im} \left\{ -v_n q_0 - v_p q_0 + \sqrt{\frac{2}{E_p}} \cosh^{-1}\sqrt{(q_0 E_p)} \right\}. \end{aligned} \tag{4}$$

According to (52.1), the probability of the process is

$$w \sim \exp\left\{ -\frac{2Ze^2}{\hbar} \sqrt{\frac{m}{\epsilon}} \operatorname{im} \left[\sqrt{\frac{2}{E_p}} \cosh^{-1}\sqrt{(q_0 E_p)} - \frac{2}{\sqrt{E}} \cosh^{-1}\sqrt{(q_0 E)} \right] \right\}. \tag{5}$$

In accordance with the fact that the two inverse hyperbolic cosines here come from (4) and (3), the signs of their imaginary parts must be the same as those of im v_p and im v_d respectively, and the signs of the latter in the solution of equations (2) are chosen so as to make $\operatorname{im}(S_1 + S_2) > 0$.

Because w depends exponentially on E_n, the total probability of disintegration (with any values of E_n and $E_p = E - 1 - E_n$) is given by the minimum absolute value of the exponent as a function of E_n. Analysis shows that this occurs when $E_n \to 0$. Then $q_0 = 1/(E+1)$, and from (5) we find

$$w \sim \exp\left\{-\frac{2Ze^2}{\hbar}\sqrt{\frac{m}{\epsilon}}\left[\sqrt{\frac{2}{E-1}}\cos^{-1}\sqrt{\frac{E-1}{E+1}}-\frac{2}{\sqrt{E}}\cos^{-1}\sqrt{\frac{E}{E+1}}\right]\right\}.$$

The condition for this formula to be valid is that the exponent should be large compared with unity.

Having calculated the imaginary part of the action $S = S_1+S_2$ for non-zero values of E_n, we can find the energy distribution of the particles released. Near $E_n = 0$, we have†

$$\operatorname{im} S(E_n) - \operatorname{im} S(0) \approx E_n\left[\frac{d \operatorname{im} S}{dE_n}\right]_{E_n = 0}.$$

A calculation of the derivative gives

$$\frac{dw}{dE_n} \sim \exp\left\{-\frac{2Ze^2}{\hbar}\sqrt{\frac{m}{\epsilon}}E_n\left[\frac{3-E}{(E-1)(E+1)^2}+\frac{1}{\sqrt{[2(E-1)^3]}}\cos^{-1}\sqrt{\frac{E-1}{E+1}}\right]\right\}.$$

PROBLEM 2. Determine the coefficient of reflection above the barrier for particle energies such that perturbation theory is applicable.

SOLUTION. Formula (43.1) is used, the initial and final wave functions being plane waves propagated in opposite directions and normalized respectively by unit current density and the delta function of momentum divided by $2\pi\hbar$, with $dv = dp'/2\pi\hbar$ and p' the momentum after reflection. Carrying out the integration with respect to p' (taking account of the delta function), we obtain

$$R = \frac{m^2}{\hbar^2 p^2}\left|\int_{-\infty}^{\infty} U(x)e^{2ipx/\hbar}\,dx\right|^2. \tag{1}$$

This formula is valid if the conditions for perturbation theory to be applicable are satisfied: $Ua/\hbar v \ll 1$, where a is the width of the barrier (see the third footnote to §45), and also $pa/\hbar \lesssim 1$. The latter condition ensures that the function $R(p)$ is not exponential; otherwise the question of the validity of formula (1) would require further investigation.

PROBLEM 3. Determine the coefficient of reflection above the barrier for a quasi-classical barrier when the function $U(x)$ has a discontinuity of slope at $x = x_0$.

SOLUTION. If the function $U(x)$ has a singularity for real x, the reflection coefficient is determined mainly by the field near that point, and perturbation theory can be formally applied to calculate it, without having to be valid for all x; the fulfilment of the quasi-classical condition is sufficient. We then have formula (1) of Problem 2, the only difference being that the momentum of the incident particle must be replaced by the value of $p(x)$ at the singular point.

In this case we take the point of discontinuous slope as $x = 0$, and thus have near this point

$$U = -F_1 x \quad \text{for} \quad x > 0, \qquad U = -F_2 x \quad \text{for} \quad x < 0,$$

with different F_1 and F_2. The integration with respect to x is effected by including in the integrand a damping factor $e^{\pm\lambda x}$ and then letting $\lambda \to 0$. The result is

$$R = \frac{m^2\hbar^2}{16p_0^6}(F_2-F_1)^2,$$

where $p_0 = p(0)$.

† When $E_n = 0$, the function $\operatorname{im} S(E_n)$ has a cusp from which it increases for both positive and negative E_n (the negative values corresponding to the capture of the neutron by the nucleus).

§53. Transitions under the action of adiabatic perturbations

It has already been mentioned in §41 that, in the limit of a perturbation which varies arbitrarily slowly with time, the probability of a transition of a system from one state to another tends to zero. Let us now consider this problem quantitatively, by calculating the transition probability under the action of a slowly varying (adiabatic) perturbation (L. D. Landau 1961).

Let the Hamiltonian of the system be a slowly varying function of time, tending to definite limits as $t \to \pm\infty$, and let $\psi_n(q, t)$ and $E_n(t)$ be the eigenfunctions and the eigenvalues of the energy (depending on time as a parameter) obtained by solving Schrödinger's equation $\hat{H}(t)\psi_n = E_n\psi_n$; on account of the adiabatic variation of $\hat{H}$ with time, the time variation of E_n and ψ_n with time will also be slow. The problem is to determine the probability w_{21} of finding the system in a certain state ψ_2 as $t \to +\infty$, if it was in the state ψ_1 as $t \to -\infty$.

The slow variation of the perturbation means that the duration of the "transition process" is very long, and therefore the change in the action during this time (given by the integral $-\int E(t)\,\mathrm{d}t$) is large. In this sense the problem is quasi-classical, and the required probability is mainly determined by the values t_0 of t for which

$$E_1(t_0) = E_2(t_0) \tag{53.1}$$

and which correspond, as it were, to the "instant of transition" in classical mechanics (cf. §52); in reality, of course, such a transition is classically impossible, as is shown by the fact that the roots of equation (53.1) are complex. It is therefore necessary to examine the properties of the solutions of Schrödinger's equation for complex values of the parameter t in the neighbourhood of the point $t = t_0$ at which the two eigenvalues of the energy become equal.

As we shall see, the eigenfunctions ψ_1, ψ_2 vary rapidly with t near this point. To determine this dependence, we first define linear combinations ϕ_1, ϕ_2 of ψ_1, ψ_2 which satisfy the conditions

$$\int \phi_1^2\,\mathrm{d}q = \int \phi_2^2\,\mathrm{d}q = 0, \qquad \int \phi_1\phi_2\,\mathrm{d}q = 1. \tag{53.2}$$

This can always be achieved by suitable choice of the complex coefficients (which are functions of t). The functions ϕ_1, ϕ_2 have no singularity at $t = t_0$.

We now seek the eigenfunctions as linear combinations

$$\psi = a_1\phi_1 + a_2\phi_2. \tag{53.3}$$

Here it must be borne in mind that, when the "time" t is complex, the operator $\hat{H}(t)$ (of the form (17.4)) is still equal to its transpose ($\hat{H} = \tilde{\hat{H}}$), but is no longer Hermitian ($\hat{H} \neq \tilde{\hat{H}}^*$), since the potential energy $U(t) \neq U^*(t)$.

We substitute (53.3) in Schrödinger's equation, multiply on the left by

ϕ_1 or ϕ_2, and integrate with respect to q. With the notation

$$H_{ik}(t) = \int \phi_i \hat{H} \phi_k \, dq, \tag{53.4}$$

and using the fact that $H_{12} = H_{21}$ owing to the above-mentioned property of the Hamiltonian, we obtain the equations

$$\left.\begin{aligned} H_{11}a_1 + H_{12}a_2 &= Ea_2, \\ H_{12}a_1 + H_{22}a_2 &= Ea_1. \end{aligned}\right\} \tag{53.5}$$

The condition for these equations to have non-zero solutions is $(H_{12}-E)^2 = H_{11}H_{22}$, and the roots of this give the energy eigenvalues

$$E = H_{12} \pm \sqrt{(H_{11}H_{22})}. \tag{53.6}$$

Then (53.5) gives

$$a_2/a_1 = \pm\sqrt{(H_{11}/H_{22})}. \tag{53.7}$$

It is seen from (53.6) that, for a coincidence at the point $t = t_0$ of the two eigenvalues, either H_{11} or H_{22} must vanish at that point; let H_{11} vanish there. At a regular point, a function in general vanishes as $t - t_0$, and therefore

$$E(t) - E(t_0) = \pm \text{constant} \times \sqrt{(t-t_0)}, \tag{53.8}$$

i.e. $E(t)$ has a branch point at $t = t_0$. We also have $a_2 \sim \sqrt{(t-t_0)}$, and so there is at the point $t = t_0$ only one eigenfunction, ϕ_1.

We now see that the problem is formally completely analogous to the problem of reflection above the barrier discussed in §52. We have a wave function $\Psi(t)$ which is "quasi-classical with respect to time", instead of the function quasi-classical with respect to the coordinate in §52, and wish to find the term of the form $c_2\psi_2 e^{-iE_2t/\hbar}$ in the wave function for $t \to +\infty$, if the wave function $\Psi(t) = \psi_1 e^{-iE_1t/\hbar}$ as $t \to -\infty$. This is analogous to the problem of determining the reflected wave for $x \to -\infty$ from the transmitted wave for $x \to +\infty$. The required transition probability $w_{21} = |c_2|^2$. The action $S = -\int E(t)\,dt$ is given by the time integral of a function having complex branch points (just as the function $p(x)$ in the integral $\int p\,dx$ had complex branch points). The problem under consideration is therefore dealt with by means of a contour in the plane of the complex variable t from large negative to large positive values, just as in §52 for the plane of the variable x, and we shall not repeat the derivation here.

We shall suppose that $E_2 > E_1$ on the real axis. Then the contour must lie in the upper half-plane of the complex variable t (where the ratio $e^{-iE_2t/\hbar}/e^{-iE_1t/\hbar}$ increases). The resulting formula (analogous to (52.2)) is

$$w_{21} = \exp\left(\frac{2}{\hbar}\,\text{im}\int_{C'} E(t)\,dt\right), \tag{53.9}$$

where the integration is along the contour shown in Fig. 19 (from left to right).

On the left-hand branch of this contour $E = E_1$, and on the right-hand branch $E = E_2$. We can therefore write (53.9) in the form

$$w_{21} = \exp\left(-2\,\mathrm{im}\int_{t_1}^{t_0} \omega_{21}(t)\,dt\right), \tag{53.10}$$

where $\omega_{21} = (E_2 - E_1)/\hbar$, and t_1 is any point on the real axis of t; t_0 must be taken as that root of equation (53.1) lying in the upper half-plane for which the exponent in (53.10) is smallest in absolute value.† In addition, besides the direct transition from state 1 to state 2, there may be possible paths through various intermediate states; the probabilities of these are given by analogous formulae. For example, for a transition $1 \to 3 \to 2$ the integral in (53.10) is replaced by a sum of integrals:

$$\int^{t_0^{(31)}} \omega_{31}(t)\,dt + \int^{t_0^{(23)}} \omega_{23}(t)\,dt,$$

where the upper limits are the "points of intersection" of the terms $E_1(t)$, $E_3(t)$ and $E_3(t)$, $E_2(t)$ respectively. This result is obtained by means of a contour which encloses both these complex points.‡

† The possible values of t_0 must include any points at which $E(t)$ becomes infinite; for such points the coefficient of the exponential in (53.9) will not be unity.

‡ The intermediate states of a continuous spectrum require a special discussion.

CHAPTER VIII

SPIN

§54. Spin

IN BOTH classical and quantum mechanics, the law of conservation of angular momentum is a consequence of the isotropy of space with respect to a closed system. This already demonstrates the relation between the angular momentum and the symmetry properties under rotation. In quantum mechanics, however, the relation in question is a particularly far-reaching one, and essentially constitutes the basic content of the concept of angular momentum, especially as the classical definition of the angular momentum of a particle as the product $\mathbf{r} \times \mathbf{p}$ has no direct significance in quantum mechanics, owing to the fact that position and momentum cannot be simultaneously measured.

We have seen in §28 that, if the values of l and m are specified, the angular dependence of the wave function of the particle is determined, and therefore so are all its symmetry properties under rotation. The most general formulation of these properties involves specifying the transformation of the wave functions when the coordinate system is rotated.

The wave function ψ_{LM} of a system of particles (with specified values of the angular momentum L and its component M) remains unchanged† only in a rotation of the coordinate system about the z-axis. Any rotation that alters the direction of this axis has the result that the z-component of the angular momentum does not have a definite value. This means that, in the new coordinates, the wave function in general becomes a superposition (a linear combination) of $2L+1$ functions corresponding to the different possible values of M for the given L. We can say that the $2L+1$ functions ψ_{LM} are transformed into linear combinations of one another when the coordinate system is rotated.‡ The law governing this transformation (i.e. the coefficients in the superposition as functions of the angles of rotation of the coordinate axes) is entirely determined by specifying the value of L. Thus the angular momentum acquires the significance of a quantum number which classifies the states of the system according to their transformation properties under rotation of the coordinate system. This aspect of the concept of angular momentum in quantum mechanics is particularly important because it is not directly related to the explicit angular dependence of the wave functions; the law of mutual transformation of these functions can be stated without reference to that dependence.

---

† Apart from an unimportant phase factor.

‡ In mathematical terms, these functions are the *irreducible representations* of the rotation group. The number of functions which are transformed into linear combinations of one another is called the *dimension* of the representation; it is assumed that this number cannot be made smaller by taking any other linear combinations of these functions.

Let us consider a composite particle, such as an atomic nucleus, which is at rest as a whole and is in a definite internal state. In addition to an internal energy, it has also an angular momentum of definite magnitude L, due to the motion of the particles within the nucleus. This angular momentum can have $2L+1$ different orientations in space. Thus, in considering the movement of a complex particle as a whole, we must assign to it, as well as its coordinates, another discrete variable: the projection of its internal angular momentum on some chosen direction in space.

However, with the preceding understanding of the concept of angular momentum, the origin of it becomes unimportant, and we naturally arrive at the concept of an "intrinsic" angular momentum which must be ascribed to the particle regardless of whether it is "composite" or "elementary".

Thus, in quantum mechanics an elementary particle must be assigned a certain "intrinsic" angular momentum unconnected with its motion in space. This property of elementary particles is peculiar to quantum theory (it disappears in the limit $\hbar \to 0$), and therefore has in principle no classical interpretation.†

The intrinsic angular momentum of a particle is called its *spin*, as distinct from the angular momentum due to the motion of the particle in space, called the *orbital angular momentum*.‡ The particle concerned may be either elementary, or composite but behaving in some respect as an elementary particle (e.g. an atomic nucleus). The spin of a particle (measured, like the orbital angular momentum, in units of $\hbar$) will be denoted by s.

For particles having spin, the description of the state by means of the wave function must determine the probability not only of its different positions in space but also of the possible orientations of the spin. Thus the wave function must depend not only on three continuous variables, the coordinates of the particle, but also on a discrete *spin variable*, which gives the value of the projection of the spin on a selected direction in space (the z-axis) and takes a limited number of discrete values, which we shall denote by σ.

Let $\psi(x, y, z; \sigma)$ be such a wave function. It is essentially a set of several different functions of the coordinates, corresponding to different values of σ; these functions will be called the *spin components* of the wave function. The integral

$$\int |\psi(x, y, z; \sigma)|^2 \, dV$$

determines the probability that the particle has a certain value of σ. The probability that the particle is in the volume element dV with any value of σ is

$$dV \sum_\sigma |\psi(x, y, z; \sigma)|^2.$$

† In particular, it would be wholly meaningless to imagine the "intrinsic" angular momentum of an elementary particle as being the result of its rotation "about its own axis".

‡ The physical idea that an electron has an intrinsic angular momentum was put forward by G. Uhlenbeck and S. Goudsmit in 1925. Spin was introduced into quantum mechanics in 1927 by W. Pauli.

The quantum-mechanical spin operator, on being applied to the wave function, acts on the spin variable σ. In other words, it in some way linearly transforms the components of the wave function into one another. The form of this operator will be established later. However, it is easy to see from very general considerations that the operators $\hat{s}_x$, $\hat{s}_y$, $\hat{s}_z$ satisfy the same commutation conditions as the operators of the orbital angular momentum.

The angular momentum operator is essentially the same as that of an infinitely small rotation. In deriving, in §26, the expression for the orbital angular momentum operator, we considered the result of applying the rotation operator to a function of the coordinates. In the case of the spin, this derivation becomes invalid, since the spin operator acts on the spin variable, and not on the coordinates. Hence, to obtain the required commutation relations, we must consider the operation of an infinitely small rotation in a general form, as a rotation of the system of coordinates. If we successively perform infinitely small rotations about the x-axis and the y-axis, and then about the same axes in the reverse order, it is easy to see by direct calculation that the difference between the results of these two operations is equivalent to an infinitely small rotation about the z-axis (through an angle equal to the product of the angles of rotation about the x and y-axes). We shall not pause here to carry out these simple calculations, as a result of which we again obtain the usual commutation relations between the operators of the components of angular momentum; these must therefore hold for the spin operators also:

$$\{\hat{s}_y, \hat{s}_z\} = i\hat{s}_x, \quad \{\hat{s}_z, \hat{s}_x\} = i\hat{s}_y, \quad \{\hat{s}_x, \hat{s}_y\} = i\hat{s}_z, \tag{54.1}$$

together with all the physical consequences resulting from them.

The commutation relations (54.1) enable us to determine the possible values of the absolute magnitude and components of the spin. All the results derived in §27 (formulae (27.7)–(27.9)) were based only on the commutation relations, and hence are fully applicable here also; we need only replace $\mathbf{L}$ in these formulae by $\mathbf{s}$. It follows from formula (27.7) that the eigenvalues of the component of the spin form a sequence of numbers differing by unity. However, we cannot now assert that these values must be integral, as we could for the component L_z of the orbital angular momentum (the derivation given at the beginning of §27 is invalid here, since it was based on the expression (26.14) for the operator $\hat{l}_z$, which holds only for the orbital angular momentum).

Moreover, we find that the sequence of eigenvalues s_z is limited above and below by values equal in absolute magnitude and opposite in sign, which we denote by $\pm s$. The difference $2s$ between the greatest and least values of s_z must be an integer or zero. Consequently s can take the values $0, \frac{1}{2}, 1, \frac{3}{2}, \ldots$.

Thus the eigenvalues of the square of the spin are

$$\mathbf{s}^2 = s(s+1), \tag{54.2}$$

where s can be either an integer (including zero) or half an integer. For given s, the component s_z of the spin can take the values $s, s-1, \ldots, -s$, i.e. $2s+1$

values in all. Accordingly, the wave function of a particle with spin s has $2s+1$ components.†

Experiment shows that the majority of the elementary particles (electrons, positrons, protons, neutrons, μ-mesons and all hyperons (Λ, Σ, Ξ)) have a spin of $\frac{1}{2}$. There are also elementary particles, the π-mesons and the K-mesons, whose spin is zero.

The total angular momentum of a particle is composed of its orbital angular momentum $\mathbf{l}$ and its spin $\mathbf{s}$. Their operators act on functions of different variables, and therefore, of course, commute. The eigenvalues of the total angular momentum

$$\mathbf{j} = \mathbf{l}+\mathbf{s} \tag{54.3}$$

are determined by the same "vector model" rule as the sum of the orbital angular momenta of two different particles (§31). That is, for given values of l and s, the total angular momentum can take the values $l+s$, $l+s-1, \ldots, |l-s|$. Thus, for an electron (spin $\frac{1}{2}$) with non-zero orbital angular momentum l, the total angular momentum can be $j = l\pm\frac{1}{2}$; for $l = 0$ the angular momentum j has, of course, only the one value $j = \frac{1}{2}$.

The operator of the total angular momentum $\mathbf{J}$ of a system of particles is equal to the sum of the operators of the angular momentum $\mathbf{j}$ of each particle, so that its values are again determined by the vector model rules. The angular momentum $\mathbf{J}$ can be put in the form

$$\mathbf{J} = \mathbf{L}+\mathbf{S}, \qquad \mathbf{L} = \sum_a \mathbf{l}_a, \qquad \mathbf{S} = \sum_a \mathbf{s}_a, \tag{54.4}$$

where $\mathbf{S}$ may be called the *total spin* and $\mathbf{L}$ the *total orbital angular momentum* of the system. We notice that, if the total spin of the system is half-integral (or integral), the same is true of the total angular momentum, since the orbital angular momentum is always integral. In particular, if the system consists of an even number of similar particles, its total spin is always integral, and therefore so is the total angular momentum.

The operators of the total angular momentum $\mathbf{j}$ of a particle (or $\mathbf{J}$, of a system of particles) satisfy the same commutation rules as the operators of the orbital angular momentum or the spin, since these rules are general commutation rules holding for any angular momentum. The formulae (27.13) for the matrix elements of angular momentum, which follow from the commutation rules, are also valid for any angular momentum, provided that the matrix elements are defined with respect to the eigenstates of this angular momentum. Formulae (29.7)—(29.10) for the matrix elements of arbitrary vector quantities also remain valid (with appropriate change of notation).

---

† Since s is fixed for each kind of particle, the spin angular momentum $\hbar s$ becomes zero in the limit of classical mechanics ($\hbar \to 0$). This consideration does not apply to the orbital angular momentum, since l can take any value. The transition to classical mechanics is represented by $\hbar$ tending to zero and l simultaneously tending to infinity, in such a way that the product $\hbar l$ remains finite.

PROBLEM

A particle with spin $\frac{1}{2}$ is in a state with a definite value $s_z = \frac{1}{2}$. Determine the probabilities of the possible values of the component of the spin along an axis z' at an angle θ to the z-axis.

SOLUTION. The mean spin vector $\bar{\mathbf{s}}$ is evidently along the z-axis and has magnitude $\frac{1}{2}$. Taking the component along the z'-axis, we find that the mean value of the spin in that direction is $\overline{s_{z'}} = \frac{1}{2}\cos\theta$. We also have $\overline{s_{z'}} = \frac{1}{2}(w_+ - w_-)$, where $w_\pm$ are the probabilities of the values $s_{z'} = \pm\frac{1}{2}$. Since $w_+ + w_- = 1$, we find $w_+ = \cos^2\frac{1}{2}\theta$, $w_- = \sin^2\frac{1}{2}\theta$.

§55. The spin operator

In the rest of this chapter we shall not be interested in the dependence of the wave functions on the coordinates. For example, in speaking of the behaviour of the functions $\psi(x, y, z; \sigma)$ when the system of coordinates is rotated, we can suppose that the particle is at the origin, so that its coordinates remain unchanged by such a rotation, and the results obtained will characterize the behaviour of the function ψ with regard to the spin variable σ.

The variable σ differs from the ordinary variables (the coordinates) by being discrete. The most general form of a linear operator acting on functions of a discrete variable σ is

$$(\hat{f}\psi)(\sigma) = \sum_{\sigma'} f_{\sigma\sigma'}\psi(\sigma'), \tag{55.1}$$

where the $f_{\sigma\sigma'}$ are constants. We put $\hat{f}\psi$ in parentheses in order to emphasize that the spin argument following it is not that of the original function ψ but that of the function resulting from it under the action of the operator $\hat{f}$. It is easy to see that the quantities $f_{\sigma\sigma'}$ are the same as the matrix elements of the operator, defined by the usual rule (11.5).†

The integration over the coordinates in (11.5) is here replaced by summation over the discrete variable, so that the definition of the matrix element is

$$f_{\sigma_2\sigma_1} = \sum_{\sigma} \psi_{\sigma_2}{}^*(\sigma)[\hat{f}\psi_{\sigma_1}(\sigma)]. \tag{55.2}$$

Here $\psi_{\sigma_1}(\sigma)$ and $\psi_{\sigma_2}(\sigma)$ are the eigenfunctions of the operator $\hat{s}_z$ corresponding to the eigenvalues $s_z = \sigma_1$ and σ_2; each such function corresponds to a state in which the particle has a definite value of s_z, i.e. in which only one component of the wave function is non-zero:‡

$$\psi_{\sigma_1}(\sigma) = \delta_{\sigma\sigma_1}, \quad \psi_{\sigma_2}(\sigma) = \delta_{\sigma\sigma_2}. \tag{55.3}$$

---

† Note that the suffixes in the matrix elements on the right of (55.1) are written in an order which is, in a sense, the reverse of the usual order in (11.11).

‡ More precisely, we should write

$$\psi_{\sigma_1}(\sigma) = \psi(x, y, z)\delta_{\sigma_1\sigma};$$

in (55.3) the coordinate factors are omitted, being unimportant in this connection.

We must once again emphasize the distinction between the specified eigenvalue σ_1 or σ_2 of s_z and the independent variable σ.

According to (55.1),

$$(\hat{f}\psi_{\sigma_1})(\sigma) = \sum_{\sigma'} f_{\sigma\sigma'}\psi_{\sigma_1}(\sigma')$$
$$= \sum_{\sigma'} f_{\sigma\sigma'}\delta_{\sigma'\sigma_1}$$
$$= f_{\sigma\sigma_1},$$

and on substitution of this and $\psi_{\sigma_2}(\sigma)$ the equation (55.2) is satisfied identically; this completes the proof.

Thus the operators acting on functions of σ can be represented in the form of $(2s+1)$-rowed matrices. In particular, we have for the operator of the spin itself, acting on the wave function, by (55.1),

$$(\hat{\mathbf{s}}\psi)(\sigma) = \sum_{\sigma'} \mathbf{s}_{\sigma\sigma'}\psi(\sigma'). \tag{55.4}$$

According to what has been said at the end of §54, the matrices $\hat{s}_x$, $\hat{s}_y$, $\hat{s}_z$ are identical with the matrices $\hat{L}_x$, $\hat{L}_y$, $\hat{L}_z$ obtained in §27, where the letters L and M need only be replaced by s and σ:

$$\left.\begin{aligned} (s_x)_{\sigma,\sigma-1} &= (s_x)_{\sigma-1,\sigma} = \tfrac{1}{2}\sqrt{[(s+\sigma)(s-\sigma+1)]},\\ (s_y)_{\sigma,\sigma-1} &= -(s_y)_{\sigma-1,\sigma} = -\tfrac{1}{2}i\sqrt{[(s+\sigma)(s-\sigma+1)]},\\ (s_z)_{\sigma\sigma} &= \sigma. \end{aligned}\right\} \tag{55.5}$$

This determines the spin operator.

In the important case of a spin of $\frac{1}{2}$ $(s = \frac{1}{2}, \sigma = \pm\frac{1}{2})$, these matrices have two rows, and are of the form

$$\hat{\mathbf{s}} = \tfrac{1}{2}\hat{\boldsymbol{\sigma}}, \tag{55.6}$$

where†

$$\hat{\sigma}_x = \begin{pmatrix} 0 & 1 \\ 1 & 0 \end{pmatrix}, \quad \hat{\sigma}_y = \begin{pmatrix} 0 & -i \\ i & 0 \end{pmatrix}, \quad \hat{\sigma}_z = \begin{pmatrix} 1 & 0 \\ 0 & -1 \end{pmatrix}. \tag{55.7}$$

These are called *Pauli matrices*. The matrix $\hat{s}_z = \frac{1}{2}\hat{\sigma}_z$ is diagonal, as it should be, since it is defined in terms of the eigenfunctions of the quantity s_z itself.‡

The following are some specific properties of the Pauli matrices. Direct multiplication of the matrices (55.7) gives the equations

$$\left.\begin{aligned} \hat{\sigma}_x^2 = \hat{\sigma}_y^2 = \hat{\sigma}_z^2 = 1,\\ \hat{\sigma}_y\hat{\sigma}_z = i\hat{\sigma}_x, \quad \hat{\sigma}_z\hat{\sigma}_x = i\hat{\sigma}_y, \quad \hat{\sigma}_x\hat{\sigma}_y = i\hat{\sigma}_z. \end{aligned}\right\} \tag{55.8}$$

† In the tabular matrices (55.7) the rows and columns are numbered by the values of σ, the row number corresponding to the first and the column number to the second suffix of the matrix element. In the present case, these numbers are $+\frac{1}{2}$ and $-\frac{1}{2}$. The action of the operator shown by (55.4) multiplies row σ of the matrix by a column matrix containing the components of the wave function:

$$\psi = \begin{pmatrix} \psi(\frac{1}{2}) \\ \psi(-\frac{1}{2}) \end{pmatrix}.$$

‡ There should be no misunderstanding because of the use of the same letter to denote the spin component and the Pauli matrices, since the latter always have the circumflex.

Combining these with the general commutation rules (54.1), we find that

$$\hat{\sigma}_i\hat{\sigma}_k + \hat{\sigma}_k\hat{\sigma}_i = 2\delta_{ik}, \tag{55.9}$$

i.e. the Pauli matrices anticommute with one another. By means of these equations, we can easily verify the following useful formulae:

$$\hat{\boldsymbol{\sigma}}^2 = 3, \quad (\hat{\boldsymbol{\sigma}}\,.\,\mathbf{a})(\hat{\boldsymbol{\sigma}}\,.\,\mathbf{b}) = \mathbf{a}\,.\,\mathbf{b} + i\hat{\boldsymbol{\sigma}}\,.\,\mathbf{a}\times\mathbf{b}. \tag{55.10}$$

where **a** and **b** are any vectors.† According to these relations, any scalar polynomial formed from the matrices $\hat{\sigma}_i$ can be reduced to terms independent of $\hat{\boldsymbol{\sigma}}$ and terms linear in $\hat{\boldsymbol{\sigma}}$; hence it follows that any scalar function of the operator $\hat{\boldsymbol{\sigma}}$ reduces to a linear function (see Problem 1). Lastly, the values of the traces (sums of diagonal elements) of the Pauli matrices and their products are

$$\mathrm{tr}\ \sigma_i = 0, \qquad \mathrm{tr}\ \sigma_i\sigma_k = 2\delta_{ik}. \tag{55.11}$$

Subsequent sections of this chapter give a more detailed account of the spin properties of wave functions, including their behaviour under any rotation of the coordinate system, but we may note immediately an important property of these functions, namely their behaviour in respect of rotations about the z-axis.

Let there be an infinitesimal rotation through an angle $\delta\phi$ about the z-axis. The operator of such a rotation is expressed in terms of the angular momentum operator (in this case, the spin operator) as $1 + i\delta\phi\,.\,\hat{s}_z$. As a result of the rotation, the functions $\psi(\sigma)$ therefore become $\psi(\sigma) + \delta\psi(\sigma)$, where

$$\delta\psi(\sigma) = i\delta\phi\,.\,\hat{s}_z\psi(\sigma) = i\sigma\psi(\sigma)\delta\phi.$$

Writing this relation in the form $\mathrm{d}\psi/\mathrm{d}\phi = i\sigma\psi(\sigma)$ and integrating, we find that a rotation through a finite angle ϕ changes the functions $\psi(\sigma)$ into

$$\psi(\sigma)' = \psi(\sigma)e^{i\sigma\phi}. \tag{55.12}$$

In particular, a rotation through 2π multiplies them by a factor $e^{2\pi i\sigma}$, which is the same for all σ and is equal to $(-1)^{2s}$ (since 2σ always has the same parity as $2s$). Thus, in a complete rotation of the coordinate system about the z-axis, the wave functions of particles with integral spin return to their original values, and those of particles with half-integral spin change sign.

PROBLEMS

PROBLEM 1. Reduce an arbitrary function of the scalar $a + \mathbf{b}\,.\,\hat{\boldsymbol{\sigma}}$ linear in the Pauli matrices to another linear function.

SOLUTION. To determine the coefficients in the required formula $f(a + \mathbf{b}\,.\,\hat{\boldsymbol{\sigma}}) = A + \mathbf{B}\,.\,\hat{\boldsymbol{\sigma}}$, we note that, when the z-axis is taken in the direction of **b**, the eigenvalues of the operator

† The terms on the right of (55.8)–(55.10) which are independent of $\hat{\boldsymbol{\sigma}}$ must, of course, be understood as constants multiplying the unit two-by-two matrix.

$a+\mathbf{b}\cdot\hat{\boldsymbol{\sigma}}$ are $a\pm b$, and the corresponding eigenvalues of the operator $f(a+\mathbf{b}\cdot\hat{\boldsymbol{\sigma}})$ are $f(a\pm b)$. Hence we find $A = \frac{1}{2}[f(a+b)+f(a-b)]$, $\mathbf{B} = (\mathbf{b}/2b)[f(a+b)-f(a-b)]$.

PROBLEM 2. Determine the values of the scalar product $\mathbf{s}_1\cdot\mathbf{s}_2$ of spins ($\frac{1}{2}$) of two particles in states in which the total spin of the system, $\mathbf{S} = \mathbf{s}_1+\mathbf{s}_2$, has definite values (0 or 1).

SOLUTION. From the general formula (31.3), which is valid for the addition of any two angular momenta, we find $\mathbf{s}_1\cdot\mathbf{s}_2 = \frac{1}{4}$ for $S = 1$, $\mathbf{s}_1\cdot\mathbf{s}_2 = -\frac{3}{4}$ for $S = 0$.

PROBLEM 3. Which powers of the operator $\hat{\mathbf{s}}$ of an arbitrary spin s are independent?

SOLUTION. The operator

$$(\hat{s}_z-s)(\hat{s}_z-s+1)\ldots(\hat{s}_z+s),$$

formed from the differences between $\hat{s}_z$ and all possible eigenvalues s_z, gives zero when it is applied to any wave function, and is therefore itself zero. Hence it follows that $(\hat{s}_z)^{2s+1}$ is expressed in terms of lower powers of the operator $\hat{s}_z$, so that only its powers from 1 to $2s$ are independent.

§56. Spinors

When the spin is zero, the wave function has only one component, $\psi(0)$. The effect of the spin operator is to reduce it to zero: $\hat{\mathbf{s}}\psi = 0$. The relation between $\hat{\mathbf{s}}$ and the operator of an infinitesimal rotation implies that the wave function of a particle with zero spin is invariant under rotation of the co-ordinate system, i.e. it is a scalar.

The wave function of a particle with spin $\frac{1}{2}$ has two components, $\psi(\frac{1}{2})$ and $\psi(-\frac{1}{2})$. For convenience in later generalizations, we shall distinguish these components by the superscripts 1 and 2 respectively. The two-component quantity

$$\psi = \begin{pmatrix}\psi^1\\ \psi^2\end{pmatrix} \equiv \begin{pmatrix}\psi(\frac{1}{2})\\ \psi(-\frac{1}{2})\end{pmatrix} \tag{56.1}$$

is called a *spinor*.

In any rotation of the coordinate system, the components of the spinor undergo a linear transformation:

$$\psi^{1\prime} = a\psi^1+b\psi^2, \quad \psi^{2\prime} = c\psi^1+d\psi^2. \tag{56.2}$$

This may be written

$$\psi^{\lambda\prime} = (\hat{U}\psi)^\lambda, \quad \hat{U} = \begin{pmatrix}a & b\\ c & d\end{pmatrix}, \tag{56.3}$$

where $\hat{U}$ is the transformation matrix.† Its elements are in general complex functions of the angles of rotation of the coordinate axes. They are connected by relations which follow directly from the physical conditions imposed on the spinor as the wave function of a particle.

Let us consider the bilinear form

$$\psi^1\phi^2-\psi^2\phi^1, \tag{56.4}$$

† The notation $\hat{U}\psi$ implies that the rows of the matrix $\hat{U}$ are multiplied by the column ψ.

where ψ and ϕ are two spinors. A simple calculation gives

$$\psi^{1'}\phi^{2'} - \psi^{2'}\phi^{1'} = (ad-bc)(\psi^1\phi^2 - \psi^2\phi^1),$$

i.e. (56.4) is transformed into itself when the coordinate system is rotated. If, however, there is only one function which is transformed into itself, it can be regarded as corresponding to zero spin, and therefore must be a scalar, i.e. must remain unchanged when the coordinate system is rotated in any manner. Hence we have

$$ad-bc = 1: \tag{56.5}$$

the determinant of the transformation matrix is unity.†

Further relations follow from the requirement that the expression

$$\psi^1\psi^{1*} + \psi^2\psi^{2*}, \tag{56.6}$$

which determines the probability of finding the particle at a given point in space, should be a scalar. A transformation which leaves unchanged the sum of the squared moduli of the quantities is a unitary transformation, i.e. we must have $\hat{U}^+ = \hat{U}^{-1}$ (see §12). With the condition (56.5) the inverse matrix is

$$\hat{U}^{-1} = \begin{pmatrix} d & -b \\ -c & a \end{pmatrix}.$$

Equating this to the Hermitian conjugate matrix

$$\hat{U}^+ = \begin{pmatrix} a^* & c^* \\ b^* & d^* \end{pmatrix},$$

we find

$$a = d^*, \quad b = -c^*. \tag{56.7}$$

By virtue of the relations (56.5) and (56.7), the four complex quantities a, b, c, d actually contain only three independent real parameters, corresponding to the three angles which define a rotation of a three-dimensional system of coordinates.

Comparison of the expressions for the scalars (56.4) and (56.6) shows that ψ^{1*} and ψ^{2*} must be transformed as ψ^2 and $-\psi^1$ respectively. It is easy to verify that this is in fact so, using (56.5) and (56.7).‡

It is possible to put the algebra of spinors in a form analogous to that of tensor algebra. This is done by introducing, in addition to *contravariant*

† Such a transformation of two quantities is called a *binary transformation*.

‡ This property is closely associated with symmetry under time reversal. The latter (see §18) involves the replacement of the wave function by its complex conjugate. Under time reversal, the angular momentum components also change sign. Hence the functions that are the complex conjugates of the components $\psi^1 \equiv \psi(\frac{1}{2})$ and $\psi^2 \equiv \psi(-\frac{1}{2})$ must have properties equivalent to those of the components corresponding to spin projections $-\frac{1}{2}$ and $\frac{1}{2}$ respectively.

spinor components ψ^1, ψ^2 (with superscripts), the *covariant* components (with subscripts), defined by

$$\psi_1 = \psi^2, \quad \psi_2 = -\psi^1. \tag{56.8}$$

The invariant combination (56.4) of the two spinors may also be written as a scalar product

$$\psi^\lambda \phi_\lambda = \psi^1\phi_1 + \psi^2\phi_2 = \psi^1\phi^2 - \psi^2\phi^1; \tag{56.9}$$

here and below, summation over repeated (*dummy*) indices is implied, as in tensor algebra. We may note the following rule which has to be borne in mind in spinor algebra. We have

$$\psi^\lambda \phi_\lambda = \psi^1\phi_1 + \psi^2\phi_2 = -\psi_2\phi^2 - \psi_1\phi^1.$$

Thus

$$\psi^\lambda \phi_\lambda = -\psi_\lambda \phi^\lambda. \tag{56.10}$$

Hence it is evident that the scalar product of any spinor with itself is zero:

$$\psi^\lambda \psi_\lambda = 0. \tag{56.11}$$

According to the foregoing discussion, the quantities ψ_1 and ψ_2 are transformed as ψ^{1*} and ψ^{2*}, i.e.

$$\psi'_\lambda = (\hat{U}^*\psi)_\lambda. \tag{56.12}$$

The product $\hat{U}^*\psi$ may also be written as $\psi\tilde{\hat{U}}^*$, with the transposed matrix $\tilde{\hat{U}}^*$. Since $\hat{U}$ is unitary, we have $\tilde{\hat{U}}^* = \hat{U}^{-1}$, so that $\psi'_\lambda = (\psi\hat{U}^{-1})_\lambda$ or†

$$\psi_\lambda = (\psi'\hat{U})_\lambda. \tag{56.13}$$

Analogously to the transition from vectors to tensors in ordinary tensor algebra, we can introduce the idea of spinors of higher rank. Thus, a quantity $\psi^{\lambda\mu}$, having four components which are transformed as the products $\psi^\lambda\phi^\mu$ of the components of two spinors of rank one, is called a spinor of rank two. Besides the contravariant components $\psi^{\lambda\mu}$ we can consider the covariant components $\psi_{\lambda\mu}$ and the mixed components $\psi_\lambda{}^\mu$ which are transformed as the products $\psi_\lambda\phi_\mu$ and $\psi_\lambda\phi^\mu$ respectively. Spinors of any rank are similarly defined.

The transition from contravariant to covariant spinor components and *vice versa* may be written

$$\psi_\lambda = g_{\lambda\mu}\psi^\mu, \quad \psi^\lambda = g^{\mu\lambda}\psi_\mu, \tag{56.14}$$

where

$$(g_{\lambda\mu}) = (g^{\lambda\mu}) = \begin{pmatrix} 0 & 1 \\ -1 & 0 \end{pmatrix} \tag{56.15}$$

† The notation $\psi\hat{U}$ (with ψ to the left of $\hat{U}$) denotes that the components $(\psi_1 \psi_2)$ as a row are multiplied by the columns of the matrix $\hat{U}$.

is the *metric spinor* in a vector space of two dimensions. Thus we have, for example,

$$\psi_{\lambda}{}^{\mu} = g_{\lambda\nu}\psi^{\nu\mu}, \qquad \psi_{\lambda\mu} = g_{\lambda\nu}g_{\mu\rho}\psi^{\nu\rho},$$

so that $\psi_{12} = -\psi_1{}^1 = -\psi^{21}$, $\psi_{11} = \psi_1{}^2 = \psi^{22}$, and so on.

The quantities $g_{\lambda\mu}$ themselves form an antisymmetric unit spinor of rank two. It is easy to see that the values of its components remain unchanged under transformations of the coordinates, and that

$$g_{\lambda\nu}g^{\mu\nu} = \delta_{\lambda}{}^{\mu}, \tag{56.16}$$

where $\delta_1{}^1 = \delta_2{}^2 = 1$, $\delta_2{}^1 = \delta_1{}^2 = 0$.

As in ordinary tensor algebra, there are two fundamental operations in spinor algebra: multiplication, and contraction with respect to a pair of indices. The *multiplication* of two spinors gives a spinor of higher rank; thus, from two spinors of ranks two and three, $\psi_{\lambda\mu}$ and $\psi^{\nu\rho\sigma}$, we can form a spinor of rank five, $\psi_{\lambda\mu}\psi^{\nu\rho\sigma}$. *Contraction* with respect to a pair of indices (i.e. summation of the components over corresponding values of one covariant and one contravariant index) decreases the rank of a spinor by two. Thus, a contraction of the spinor $\psi_{\lambda\mu}{}^{\nu\rho\sigma}$ with respect to the indices μ and ν gives the spinor $\psi_{\lambda\mu}{}^{\mu\rho\sigma}$ of rank three; the contraction of the spinor $\psi_{\lambda}{}^{\mu}$ gives the scalar $\psi_{\lambda}{}^{\lambda}$. Here there is a rule similar to that expressed by formula (56.10): if we interchange the upper and lower indices with respect to which the contraction is effected, the sign is changed (i.e. $\psi_{\lambda}{}^{\lambda} = -\psi^{\lambda}{}_{\lambda}$). Hence, in particular, it follows that, if a spinor is symmetrical with respect to any two of its indices, the result of a contraction with respect to these indices is zero. Thus, for a symmetrical spinor $\psi_{\lambda\mu}$ of rank two, we have $\psi_{\lambda}{}^{\lambda} = 0$.

A spinor of rank n symmetrical with respect to all its indices is called a *symmetrical spinor* of rank n. From an asymmetrical spinor we can construct a symmetrical one by the process of *symmetrization*, i.e. summation of the components obtained by all possible interchanges of the indices. From what has been said above, it is impossible to construct (by contraction) a spinor of lower rank from the components of a symmetrical spinor.

Only a spinor of rank two can be antisymmetrical with respect to all its indices. For, since each index can take only two values, at least two out of three or more indices must have the same value, and therefore the components of the spinor are zero identically. Any antisymmetrical spinor of rank two is a scalar multiple of the unit spinor $g_{\lambda\mu}$. We may notice here the following relation:

$$g_{\lambda\mu}\psi_{\nu}+g_{\mu\nu}\psi_{\lambda}+g_{\nu\lambda}\psi_{\mu} = 0 \tag{56.17}$$

(where ψ_{λ} is any spinor), which follows from the above; this rule is simply a consequence of the fact that the expression on the left is (as we may easily verify) an antisymmetrical spinor of rank three.

The spinor which is the product of a spinor $\psi_{\lambda\mu}$ with itself, on contraction with respect to one pair of indices, becomes antisymmetrical with respect to the other pair:

$$\psi_{\lambda\nu}\psi_{\mu}{}^{\nu} = -\psi_{\lambda}{}^{\nu}\psi_{\mu\nu}.$$

Hence, from what was said above, this spinor must be a scalar multiple of the spinor $g_{\lambda\mu}$. Defining the scalar factor so that contraction with respect to the second pair of indices gives the correct result, we find

$$\psi_{\lambda\nu}\psi_{\mu}{}^{\nu} = -\tfrac{1}{2}\psi_{\rho\sigma}\psi^{\rho\sigma}g_{\lambda\mu}. \tag{56.18}$$

The components of the spinor $\psi_{\lambda\mu\ldots}{}^{*}$ which is the complex conjugate of $\psi_{\lambda\mu\ldots}$ are transformed as the components of a contravariant spinor $\phi^{\lambda\mu\ldots}$, and conversely. The sum of the squared moduli of the components of any spinor is consequently invariant.

§57. The wave functions of particles with arbitrary spin

Having developed a formal algebra for spinors of any rank, we can now turn to our immediate problem, to study the properties of wave functions of particles with arbitrary spin.

This subject is conveniently approached by considering an assembly of particles with spin $\frac{1}{2}$. The greatest possible value of the z-component of the total spin is $\frac{1}{2}n$, which is obtained when $s_z = \frac{1}{2}$ for every particle (i.e. all the spins are directed the same way, along the z-axis). In this case we can evidently say that the total spin S of the system is also $\frac{1}{2}n$.

All the components of the wave function $\psi(\sigma_1, \sigma_2, \ldots, \sigma_n)$ of the system of particles are then zero, except for $\psi(\frac{1}{2}, \frac{1}{2}, \ldots, \frac{1}{2})$. If we write the wave function as a product of n spinors $\psi^{\lambda}\phi^{\mu}\ldots$, each of which refers to one of the particles, only the component with $\lambda, \mu, \ldots = 1$ in each spinor is not zero. Thus only the product $\psi^1\phi^1\ldots$ is not zero. The set of all these products, however, is a spinor of rank n which is symmetrical with respect to all its indices. If we transform the coordinate system (so that the spins are not directed along the z-axis), we obtain a spinor of rank n, general in form except that it is symmetrical as before.

The spin properties of wave functions, being essentially their properties with respect to rotations of the coordinate system, are identical for a particle with spin s and for a system of $n = 2s$ particles each with spin $\frac{1}{2}$ directed so that the total spin of the system is s. Hence we conclude that the wave function of a particle with spin s is a symmetrical spinor of rank $n = 2s$.

It is easy to see that the number of independent components of a symmetrical spinor of rank $2s$ is equal to $2s+1$, as it should be. For all those components are the same whose indices include $2s$ ones and 0 twos; so are all those with $2s-1$ ones and 1 two, and so on up to 0 ones and $2s$ twos.

Mathematically, the symmetrical spinors provide a classification of the possible types of transformation of quantities when the coordinate system is rotated. If there are $2s+1$ different quantities which are transformed linearly into one another (and which cannot be reduced in number by any choice of linear combinations of them), then we can assert that their law of transformation is equivalent to that of the components of a symmetrical spinor of rank $2s$. Any set of any number of functions which are transformed linearly into one another when the coordinate system is rotated can be reduced (by an appropriate linear transformation) to one or more symmetrical spinors.†

Thus an arbitrary spinor $\psi_{\lambda\mu\nu\ldots}$ of rank n can be reduced to symmetrical spinors of ranks $n, n-2, n-4, \ldots$. In practice, such a reduction can be made as follows. By symmetrizing the spinor $\psi_{\lambda\mu\nu\ldots}$ with respect to all its indices, we form a symmetrical spinor of the same rank n. Next, by contracting the original spinor $\psi_{\lambda\mu\nu\ldots}$ with respect to various pairs of indices, we obtain spinors of rank $n-2$, of the form $\psi^{\lambda}{}_{\lambda\nu\ldots}$, which, in turn, we symmetrize, so that symmetrical spinors of rank $n-2$ are obtained. By symmetrizing the spinors obtained by contracting $\psi_{\lambda\mu\ldots}$ with respect to two pairs of indices, we obtain symmetrical spinors of rank $n-4$, and so on.

We have still to establish the relation between the components of a symmetrical spinor of rank $2s$ and the $2s+1$ functions $\psi(\sigma)$, where $\sigma = s, s-1, \ldots, -s$. The component

$$\psi^{\overbrace{11\ldots1}^{s+\sigma}\overbrace{22\ldots2}^{s-\sigma}},$$

in whose indices 1 occurs $s+\sigma$ times and 2 $s-\sigma$ times, corresponds to a value σ of the projection of the spin on the z-axis. For, if we again consider a system of $n = 2s$ particles with spin $\frac{1}{2}$, instead of one particle with spin s, the product $\underbrace{\psi^1\phi^1\ldots}_{s+\sigma}\underbrace{\chi^2\rho^2\ldots}_{s-\sigma}$ corresponds to the above component; this product belongs to a state in which $s+\sigma$ particles have a projection of the spin equal to $\frac{1}{2}$, and $s-\sigma$ a projection of $-\frac{1}{2}$, so that the total projection is $\frac{1}{2}(s+\sigma)-\frac{1}{2}(s-\sigma) = \sigma$. Finally, the proportionality coefficient between the above component of the spinor and $\psi(\sigma)$ is chosen so that the equation

$$\sum_{\sigma=-s}^{s} |\psi(\sigma)|^2 = \sum_{\lambda,\mu,\ldots=1}^{2} |\psi^{\lambda\mu\ldots}|^2 \tag{57.1}$$

holds; this sum is a scalar, as it should be, since it determines the probability of finding the particle at a given point in space. In the sum on the right-hand side, the components with $(s+\sigma)$ indices 1 occur

$$\frac{(2s)!}{(s+\sigma)!\,(s-\sigma)!}$$

times. Hence it is clear that the relation between the functions $\psi(\sigma)$ and the components of the spinor is given by the formula

$$\psi(\sigma) = \sqrt{\left[\frac{(2s)!}{(s+\sigma)!\,(s-\sigma)!}\right]}\,\psi^{\overbrace{11\ldots1}^{s+\sigma}\overbrace{22\ldots2}^{s-\sigma}}. \tag{57.2}$$

† In other words, the symmetrical spinors form what are called irreducible representations of the rotation group (see §98).

The relation (57.2) ensures the fulfilment not only of the condition (57.1), but also, as we easily see, of the more general condition

$$\psi^{\lambda\mu\ldots}\phi_{\lambda\mu\ldots} = \sum_{\sigma} (-1)^{s-\sigma}\psi(\sigma)\phi(-\sigma), \tag{57.3}$$

where $\psi^{\lambda\mu\ldots}$ and $\phi_{\lambda\mu\ldots}$ are two different spinors of the same rank, while $\psi(\sigma), \phi(\sigma)$ are functions derived from these spinors by formula (57.2); the factor $(-1)^{s-\sigma}$ is due to the fact that, when all the indices of the spinor components are raised, the sign changes as many times as there are twos among the indices.

Formulae (55.5) determine the result of the action of the spin operator on the wave functions $\psi(\sigma)$. It is not difficult to find how these operators act on a wave function written in the form of a spinor of rank $2s$. For a spin $\frac{1}{2}$, the functions $\psi(\frac{1}{2}), \psi(-\frac{1}{2})$ are the same as the components ψ^1, ψ^2 of the spinor. According to (55.6) and (55.7), the result of the spin operators' acting on them will be

$$\left.\begin{aligned}
(\hat{s}_x\psi)^1 &= \tfrac{1}{2}\psi^2, \quad (\hat{s}_y\psi)^1 = -\tfrac{1}{2}i\psi^2, \quad (\hat{s}_z\psi)^1 = \tfrac{1}{2}\psi^1,\\
(\hat{s}_x\psi)^2 &= \tfrac{1}{2}\psi^1, \quad (\hat{s}_y\psi)^2 = \tfrac{1}{2}i\psi^1, \quad (\hat{s}_z\psi)^2 = -\tfrac{1}{2}\psi^2.
\end{aligned}\right\} \tag{57.4}$$

To pass to the general case of arbitrary spin, we again consider a system of $2s$ particles with spin $\frac{1}{2}$, and write its wave function as a product of $2s$ spinors. The spin operator of the system is the sum of the spin operators of each particle, acting only on the corresponding spinor, the result of this action being given by formulae (57.4). Next, returning to arbitrary symmetrical spinors, i.e. to the wave functions of a particle with spin s, we obtain

$$\left.\begin{aligned}
(\hat{s}_x\psi)^{\overbrace{11\ldots}^{s+\sigma}\overbrace{22\ldots}^{s-\sigma}} &= \tfrac{1}{2}(s+\sigma)\,\psi^{\overbrace{11\ldots}^{s+\sigma-1}\overbrace{22\ldots}^{s-\sigma+1}} + \tfrac{1}{2}(s-\sigma)\psi^{\overbrace{11\ldots}^{s+\sigma+1}\overbrace{22\ldots}^{s-\sigma-1}},\\
(\hat{s}_y\psi)^{\overbrace{11\ldots}^{s+\sigma}\overbrace{22\ldots}^{s-\sigma}} &= -\tfrac{1}{2}i(s+\sigma)\psi^{\overbrace{11\ldots}^{s+\sigma-1}\overbrace{22\ldots}^{s-\sigma+1}} + \tfrac{1}{2}i(s-\sigma)\psi^{\overbrace{11\ldots}^{s+\sigma+1}\overbrace{22\ldots}^{s-\sigma-1}},\\
(\hat{s}_z\psi)^{\overbrace{11\ldots}^{s+\sigma}\overbrace{22\ldots}^{s-\sigma}} &= \sigma\psi^{\overbrace{11\ldots}^{s+\sigma}\overbrace{22\ldots}^{s-\sigma}}.
\end{aligned}\right\} \tag{57.5}$$

Hitherto we have spoken of spinors as wave functions of the intrinsic angular momentum of elementary particles. Formally, however, there is no difference between the spin of a single particle and the total angular momentum of any system regarded as a whole, neglecting its internal structure. It is therefore evident that the transformation properties of spinors apply equally to the behaviour, with respect to rotations in space, of the wave functions ψ_{jm} of any particle or system of particles with total angular momentum j, independent of whether orbital or spin angular momentum is concerned. There must therefore be some definite relation between the laws of transformation for the eigenfunctions ψ_{jm} under rotations of the coordinate system and those for the components of a symmetrical spinor of rank $2j$.

In establishing this relation we must, however, make a clear distinction between two aspects of the dependence of the wave functions on the component m (for a given value of j). The wave function may be regarded as the probability amplitude for various values of m, or may be considered for a given value of m.

These two aspects have already been discussed at the beginning of §55, where we dealt with the eigenfunction $\delta_{\sigma\sigma_0}$ of the operator $\hat{s}_z$ which corresponds to $s_z = \sigma_0$. The mathematical difference between them is especially clear for a particle of spin $s = \frac{1}{2}$. In this case the spin function is, with respect to the variable σ, a contravariant spinor of rank 1, i.e. must be written in spinor notation as $\delta^{\sigma}{}_{\sigma_0}$. With respect to σ_0 it is therefore a covariant spinor.

This is evidently a general result: the eigenfunctions ψ_{jm} can be put in correspondence with the components of a covariant symmetrical spinor of rank $2j$ by means of formulae analogous to (57.2):†

$$\psi_{jm} = \sqrt{\frac{(2j)!}{(j+m)!(j-m)!}}\,\psi_{\underbrace{11\ldots}_{j+m}\,\underbrace{22\ldots}_{j-m}}. \tag{57.6}$$

The eigenfunctions of integral angular momentum j are spherical harmonics Y_{jm}. The case $j = 1$ is of particular importance. The three spherical harmonics Y_{1m} are

$$Y_{10} = i\sqrt{\frac{3}{4\pi}}\cos\theta = i\sqrt{\frac{3}{4\pi}}\,n_z,$$

$$Y_{1,\pm1} = \mp i\sqrt{\frac{3}{8\pi}}\sin\theta\, e^{\pm i\phi} = \mp i\sqrt{\frac{3}{8\pi}}(n_x \pm i n_y),$$

where $\mathbf{n}$ is a unit vector along the radius vector. It is seen that these three functions are equivalent, as regards their transformation properties, to the components of a vector $\mathbf{a}$, with the relations

$$\psi_{10} = i a_z, \qquad \psi_{11} = -\frac{i}{\sqrt{2}}(a_x + i a_y), \qquad \psi_{1,-1} = \frac{i}{\sqrt{2}}(a_x - i a_y). \tag{57.7}$$

Comparing with (57.6), we see that the components of a symmetrical spinor of rank two can be brought into correspondence with the components of the

---

† This result can also be regarded somewhat differently. If the wave function ψ of a particle in a state with angular momentum j is expanded in terms of the eignfunctions ψ_{jm}:

$$\psi = \sum_m a_m \psi_{jm},$$

then the coefficients a_m are the probability amplitudes for various values of m. In this sense they correspond to the "components" $\psi(m)$ of a spin wave function, and this gives their law of transformation. On the other hand, the value of ψ at a given point in space cannot depend on the choice of the coordinate system, i.e. the sum $\Sigma\, a_m \psi_{jm}$ must be a scalar. Comparing with the scalar (57.3), we see that a_m must transform as $(-1)^{j-m}\psi_{j,-m}$.

vector by the formulae

$$\psi_{12} = \frac{i}{\sqrt{2}}a_z, \quad \psi_{11} = -\frac{i}{\sqrt{2}}(a_x+ia_y), \quad \psi_{22} = \frac{i}{\sqrt{2}}(a_x-ia_y), \tag{57.8}$$

$$\psi^{12} = -\frac{i}{\sqrt{2}}a_z, \quad \psi^{11} = \frac{i}{\sqrt{2}}(a_x-ia_y), \quad \psi^{22} = -\frac{i}{\sqrt{2}}(a_x+ia_y). \tag{57.9}$$

Conversely

$$a_z = i\sqrt{2}\psi^{12}, \quad a_x = \frac{i}{\sqrt{2}}(\psi^{22}-\psi^{11}), \quad a_y = \frac{1}{\sqrt{2}}(\psi^{11}+\psi^{22}). \tag{57.10}$$

It is easily verified that with these definitions we have

$$\psi_{\lambda\mu}\phi^{\lambda\mu} = \mathbf{a}.\mathbf{b}, \tag{57.11}$$

where **a** and **b** are vectors corresponding to the symmetrical spinors $\psi^{\lambda\mu}$ and $\phi^{\lambda\mu}$. It is also not difficult to see that there is a correspondence between the spinor and the vector†

$$\psi^{\lambda}_{\nu}\phi^{\mu\nu} + \psi^{\mu}_{\nu}\phi^{\lambda\nu} \text{ and } \sqrt{2}\mathbf{a}\times\mathbf{b}. \tag{57.12}$$

Formulae (57.10) may be compactly written by means of the Pauli matrices:

$$\mathbf{a} = \frac{i}{\sqrt{2}}\boldsymbol{\sigma}^{\lambda}{}_{\mu}\psi^{\mu}_{\lambda}, \quad \psi^{\mu}_{\lambda} = -\frac{i}{\sqrt{2}}\mathbf{a}\,.\,\boldsymbol{\sigma}^{\mu}{}_{\lambda}; \tag{57.13}$$

the matrix indices of $\hat{\boldsymbol{\sigma}}$ are written as superscript and subscript in correspondence with the position of the spinor indices in ψ^{μ}_{λ}. The origin of this formula is easily understood by considering the particular case where the spinor of rank two ψ^{μ}_{λ} reduces to a product of a spinor of rank one ψ^{μ} and its complex conjugate $\psi^{\lambda *}$. Then the quantity $\frac{1}{2}\psi^{\lambda *}\boldsymbol{\sigma}^{\lambda}{}_{\mu}\psi^{\mu}$ is the mean value of the spin (for a particle with wave function ψ^{μ}) and it is therefore evidently a vector.

The relations (57.8) or (57.9) are a particular case of a general rule: any symmetrical spinor of even rank $2j$, where j is integral, can be correlated with a symmetrical tensor of half the rank (j) which gives zero on contraction with respect to any pair of indices; we call this an *irreducible* tensor.

This follows from the fact that the numbers of independent components of the spinor and of the tensor are the same $(2j+1)$, as may easily be seen.‡ The relation between the components of the spinor and of the tensor can be found by means of formulae (57.8)–(57.10), if we consider a spinor of the rank concerned as the product of several spinors of rank two, and the tensor as a product of vectors.

---

† The mixed components of a symmetrical spinor may be written in the form ψ^{λ}_{μ}, without distinction between $\psi^{\lambda}{}_{\mu}$ and $\psi_{\mu}{}^{\lambda}$.

‡ We can say that the $2j+1$ components of an irreducible tensor of rank j (an integer), the $2j+1$ spherical harmonics Y_{jm}, and the $2j+1$ components of a symmetrical spinor of rank $2j$ give the same irreducible representation of the rotation group.

PROBLEMS

PROBLEM 1. Rewrite the definition (57.4) of the operator of spin $\frac{1}{2}$ in terms of the spinor components of the vector $\hat{\mathbf{s}}$.

SOLUTION. By means of formulae (57.9), which give the relation between the vector $\hat{\mathbf{s}}$ and the spinor $\hat{s}^{\lambda\mu}$, the definition (57.4) can be written as

$$\hat{s}^{\lambda\mu}\psi^{\nu} = \frac{i}{2\sqrt{2}}(\psi^{\lambda}g^{\mu\nu}+\psi^{\mu}g^{\lambda\nu}).$$

PROBLEM 2. Derive formulae which determine the effect of the spin operator on a vector wave function of a particle with spin 1.

SOLUTION. The relation between the components of the vector function $\boldsymbol{\psi}$ and the components of the spinor $\psi^{\lambda\mu}$ is given by formulae (57.9), and from (57.5) we have

$$\hat{s}_z\psi_+ = -\psi_+, \qquad \hat{s}_z\psi_- = \psi_-, \qquad \hat{s}_z\psi_z = 0$$

(where $\psi_{\pm} = \psi_x \pm i\psi_y$) or

$$\hat{s}_z\psi_x = -i\psi_y, \qquad \hat{s}_z\psi_y = i\psi_x, \qquad \hat{s}_z\psi_z = 0.$$

The remaining formulae are derived from these by cyclic permutation of the suffixes x, y, z. They can be written together as

$$\hat{s}_i\psi_k = -ie_{ikl}\psi_l.$$

The complex vector $\boldsymbol{\psi}$ can be put in the form $\boldsymbol{\psi} = e^{i\alpha}(\mathbf{u}+i\mathbf{v})$, where $\mathbf{u}$ and $\mathbf{v}$ are real vectors, which can be taken to be mutually perpendicular if the common phase α is suitably chosen. The two vectors $\mathbf{u}$ and $\mathbf{v}$ determine a plane which has the property that the spin component perpendicular to it can take only the values ± 1.

§58. The operator of finite rotations

Let us now return to the transformation of spinors, and show how the coefficients of this transformation can in fact be expressed in terms of the angles of rotation of the coordinate axes.

By the definition of the angular momentum operator (in this case, the spin operator), $1+i\delta\phi\,.\,\mathbf{n}\,.\,\hat{\mathbf{s}}$ is the operator of a rotation through an angle $\delta\phi$ about a direction specified by the unit vector $\mathbf{n}$; for application to the wave function of a particle with spin $\frac{1}{2}$, i.e. a spinor of rank one, we must take $\hat{\mathbf{s}} = \frac{1}{2}\hat{\boldsymbol{\sigma}}$ in this operator. The operator of a rotation through a finite angle ϕ about the same direction will be correspondingly given by

$$\hat{U}_{\mathbf{n}} = \exp(\tfrac{1}{2}i\phi\mathbf{n}\,.\,\hat{\boldsymbol{\sigma}}); \tag{58.1}$$

cf. (15.13). Like any function of the Pauli matrices (see §55, Problem 1), this expression reduces to one that is linear in these matrices:

$$\hat{U}_{\mathbf{n}} = \cos\tfrac{1}{2}\phi + i\mathbf{n}\,.\,\hat{\boldsymbol{\sigma}}\sin\tfrac{1}{2}\phi. \tag{58.2}$$

For example, with a rotation about the z-axis,

$$\hat{U}_z(\phi) = \cos\tfrac{1}{2}\phi + i\hat{\sigma}_z \sin\tfrac{1}{2}\phi = \begin{pmatrix} e^{i\phi/2} & 0 \\ 0 & e^{-i\phi/2} \end{pmatrix}. \tag{58.3}$$

This means that the components of the spinor are transformed in such a rotation according to

$$\psi^{1'} = \psi^1 e^{i\phi/2}, \quad \psi^{2'} = \psi^2 e^{-i\phi/2}.$$

In particular, in a rotation through an angle 2π the spinor components change sign; spinors of any odd rank must therefore have the same property (cf. the end of §55).

Similarly, we can find the matrices of transformations consisting of a rotation through an angle ϕ about the x-axis or the y-axis:

$$\left.\begin{aligned} \hat{U}_x(\phi) &= \begin{pmatrix} \cos\frac{1}{2}\phi & i\sin\frac{1}{2}\phi \\ i\sin\frac{1}{2}\phi & \cos\frac{1}{2}\phi \end{pmatrix}, \\ \hat{U}_y(\phi) &= \begin{pmatrix} \cos\frac{1}{2}\phi & \sin\frac{1}{2}\phi \\ -\sin\frac{1}{2}\phi & \cos\frac{1}{2}\phi \end{pmatrix}. \end{aligned}\right\} \tag{58.4}$$

We may note the particular case of a rotation through an angle π about the y-axis, for which

$$\psi^{1'} = \psi^2, \quad \psi^{2'} = -\psi^1,$$

i.e.

$$\psi^{1'} = \psi_1, \quad \psi^{2'} = \psi_2. \tag{58.5}$$

It is now easy to write down the transformation matrix for any rotation of the coordinate axes, as a function of the Eulerian angles which specify the rotation.

A rotation of the axes, defined by the Eulerian angles α, β, γ, is carried out in three stages: (1) a rotation through the angle α ($0 \leqslant \alpha \leqslant 2\pi$) about the z-axis, (2) a rotation through the angle β ($0 \leqslant \beta \leqslant \pi$) about the new position of the y-axis (ON in Fig. 20, called the *line of nodes*), (3) a rotation through the angle γ ($0 \leqslant \gamma \leqslant 2\pi$) about the resulting final position z' of the z-axis.†

It is evident that the angles α and β are the spherical polar angles ϕ and θ of the new z'-axis with respect to the xyz axes: $\alpha = \phi$, $\beta = \theta$.

In accordance with this manner of rotating the axes, the matrix of the

---

† The systems xyz and $x'y'z'$ are, as always, right-handed, and a positive angle corresponds to the movement of a corkscrew advanced in the positive direction of the axis of rotation. The definition of the Eulerian angles given here (and usual in quantum mechanics) differs from that in *Mechanics*, §35, in that the second rotation is about the y-axis and not the x-axis. The angles α, β, γ are related to the angles ϕ, θ, ψ used in *Mechanics* (*not* the spherical polar angles ϕ and θ) by $\phi = \alpha + \frac{1}{2}\pi$, $\theta = \beta$, $\psi = \gamma - \frac{1}{2}\pi$.

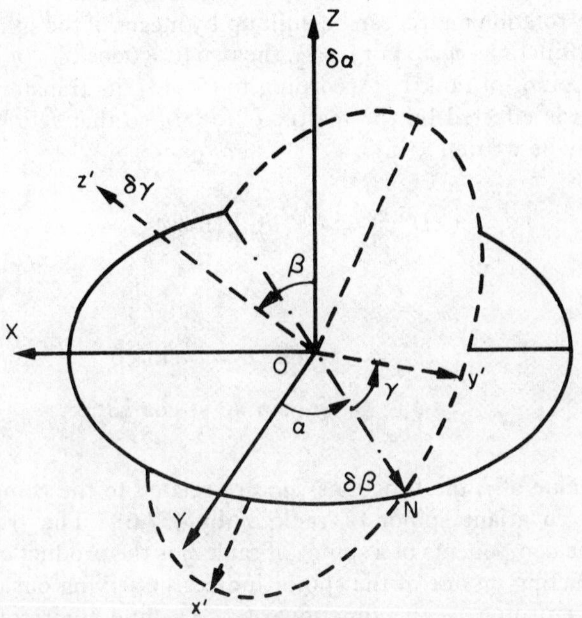

FIG. 20

complete transformation is equal to the product of three matrices (58.3) and (58.4):

$$\hat{U}(\alpha, \beta, \gamma) = \hat{U}_z(\gamma)\hat{U}_y(\beta)\hat{U}_z(\alpha).$$

By direct multiplication of the matrices we finally obtain

$$\hat{U}(\alpha, \beta, \gamma) = \begin{pmatrix} \cos \tfrac{1}{2}\beta \,.\, e^{i(\alpha+\gamma)/2} & \sin \tfrac{1}{2}\beta \,.\, e^{-i(\alpha-\gamma)/2} \\ -\sin \tfrac{1}{2}\beta \,.\, e^{i(\alpha-\gamma)/2} & \cos \tfrac{1}{2}\beta \,.\, e^{-i(\alpha+\gamma)/2} \end{pmatrix}. \tag{58.6}$$

Spinors of higher ranks are, by definition, transformed as products of components of a spinor of rank one. In physical applications, however, we are interested in the wave functions ψ_{jm} rather than the transformation laws of the spinors themselves.

Let the functions ψ_{jm} ($m = j, j-1, \ldots, -j$) describe, in a coordinate system xyz, a state having a definite value of the angular momentum j, and $\psi_{jm'}$ the same state for the axes $x'y'z'$; in the first case m is the value of j_z, and in the second case m' is the value of $j_{z'}$. The two sets of functions are connected by linear relations, which we write in the form

$$\psi_{jm} = \sum_{m'} D^{(j)}_{m'm}(\alpha, \beta, \gamma)\, \psi_{jm'}. \tag{58.7}$$

The coefficients $D^{(j)}{}_{m'm}$ form a matrix of order $2j+1$ with respect to m' and m, called the *finite-rotation matrix* $\hat{D}^{(j)}$; its elements are functions of the angles α, β, γ of rotation of the system $x'y'z'$ relative to xyz.

The finite-rotation matrix can be built up by means of the spinor representation of the functions ψ_{jm}. For $j = \frac{1}{2}$, the two functions $\psi_{1/2\,m}(m = \pm\frac{1}{2})$ form a covariant spinor of rank 1. According to (56.13), its transformation from $x'y'z'$ to xyz is effected by the matrix $\hat{U}$ (58.6), so that $\hat{D}^{(1/2)} = \hat{U}$.† Its elements may be written

$$D^{(1/2)}_{m'm} = e^{im'\gamma} d^{(1/2)}_{m'm}(\beta) e^{im\alpha},$$

where

$$d^{(1/2)}_{m'm} = \begin{array}{c|cc} m' = \; m = & \frac{1}{2} & -\frac{1}{2} \\ \hline \frac{1}{2} & \cos\frac{1}{2}\beta & \sin\frac{1}{2}\beta \\ -\frac{1}{2} & -\sin\frac{1}{2}\beta & \cos\frac{1}{2}\beta \end{array} \tag{58.8}$$

For any value of j, the functions ψ_{jm} are related to the components of a symmetrical covariant spinor of rank $2j$ by (57.6). The transformation matrix for the components of a spinor of rank $2j$ is the product of $2j$ matrices $\hat{D}^{(1/2)}$, each acting on one of the spinor indices. Carrying out the multiplication and returning to the functions ψ_{jm}, we find their transformation matrix:

$$D^{(j)}_{m'm}(\alpha, \beta, \gamma) = e^{im'\gamma} d^{(j)}_{m'm}(\beta) e^{im\alpha}, \tag{58.9}$$

the functions $d^{(j)}_{m'm}(\beta)$ being given by‡

$$d^{(j)}_{m'm}(\beta) = \left[\frac{(j+m')!(j-m')!}{(j+m)!(j-m)!}\right]^{1/2} (\cos\tfrac{1}{2}\beta)^{m'+m} \times$$

$$\times (\sin\tfrac{1}{2}\beta)^{m'-m} P^{(m'-m,\,m'+m)}_{j-m'}(\cos\beta), \tag{58.10}$$

where

$$P_n^{(a,\,b)}(\cos\beta) = \frac{(-1)^n}{2^n n!}(1-\cos\beta)^{-a}(1+\cos\beta)^{-b} \times$$

$$\times \left(\frac{d}{d\cos\beta}\right)^n [(1-\cos\beta)^{a+n}(1+\cos\beta)^{b+n}] \tag{58.11}$$

---

† Note that the matrix indices in (58.7) are placed in the order that corresponds to multiplying the columns of the matrix $\hat{D}^{(j)}$ by the functions ψ_{jm} arranged in a row. In the symbolic notation, (58.7) would have to be written $\psi_{jm} = (\psi_j' \hat{D}^{(j)})_m$ in accordance with (56.13).

‡ The calculations are described by A. R. Edmonds, *Angular Momentum in Quantum Mechanics*, Princeton, 1957. The definition of the functions $D_{m'm}{}^{(j)}$ by (58.9) differs from that used in Edmonds's book by the interchange of α and γ, this being the more natural treatment in the approach given here.

are called *Jacobi polynomials*.† We may note that

$$P_n^{(a,\,b)}(-\cos\beta) = (-1)^n P_n^{(b,\,a)}(\cos\beta). \qquad (58.12)$$

The functions $d^{(j)}_{m'm}$ possess a number of symmetry properties which might be derived from the expressions (58.11) and (58.12), but it is simpler to obtain them directly from the definition as coefficients in the rotational transformation.

The matrix $\hat{D}^{(j)}$ is unitary, being the matrix of a rotational transformation. Since the transformation inverse to the rotation (α, β, γ) is the rotation $(-\gamma, -\beta, -\alpha)$, we have for the real matrix $d^{(j)}$ the relations

$$d^{(j)}_{m'm}(-\beta) = d^{(j)}_{mm'}(\beta). \qquad (58.13)$$

The following equations are also valid:

$$d^{(j)}_{m'm}(\beta) = d^{(j)}_{-m,\,-m'}(\beta), \qquad (58.14)$$

$$\left.\begin{aligned} d^{(j)}_{m'm}(\pi) &= (-1)^{j+m}\delta_{m',\,-m},\\ d^{(j)}_{m'm}(-\pi) &= (-1)^{j-m}\delta_{m',\,-m},\\ d^{(j)}_{m'm}(0) &= \delta_{m'm}. \end{aligned}\right\} \qquad (58.15)$$

When $j = \frac{1}{2}$ these are evident from (58.8); the generalization to arbitrary j is evident from the manner of construction of the transformation matrix, described above.

A rotation through an angle $\pi-\beta$ can be carried out as two successive rotations through π and $-\beta$:

$$\begin{aligned} d^{(j)}_{m'm}(\pi-\beta) &= \sum_{m''} d^{(j)}_{m'm''}(\pi) d^{(j)}_{m''m}(-\beta)\\ &= (-1)^{j-m'} d^{(j)}_{-m'm}(-\beta), \end{aligned}$$

or, using (58.13),

$$d^{(j)}_{m'm}(\pi-\beta) = (-1)^{j-m'} d^{(j)}_{m,-m'}(\beta). \qquad (58.16)$$

The result of two rotations about the same axis is independent of the sequence in which they occur. We must therefore arrive at the same result by carrying out the rotations through $-\beta$ and π in the opposite order. Comparison of the result with (58.16) gives the relation

$$d^{(j)}_{m'm}(\beta) = (-1)^{m'-m} d^{(j)}_{-m',\,-m}(\beta). \qquad (58.17)$$

From (58.17), (58.14) and (58.13), it follows that

$$d^{(j)}_{m'm}(\beta) = (-1)^{m'-m} d^{(j)}_{mm'}(\beta) = (-1)^{m'-m} d^{(j)}_{m'm}(-\beta). \qquad (58.18)$$

† See §e of the Mathematical Appendices, formula (e.11), for the relation between these polynomials and the hypergeometric series.

Using (58.13)–(58.18), we can deduce various symmetry properties of the complete matrix elements $D^{(j)}_{m'm}$. In particular, the complex conjugate function is given by

$$D^{(j)*}_{m'm}(\alpha, \beta, \gamma) = D^{(j)}_{m'm}(-\alpha, \beta, -\gamma)$$
$$= (-1)^{m'-m} D^{(j)}_{-m', -m}(\alpha, \beta, \gamma). \quad (58.19)$$

Mathematically, the matrices $\hat{D}^{(j)}$ give the unitary irreducible representations of the rotation group having dimension $2j+1$ (see §98 below). Hence we have immediately the orthonormality relation

$$\int D^{(j_1)*}_{m'_1 m_1}(\alpha, \beta, \gamma)\, D^{(j_2)}_{m'_2 m_2}(\alpha, \beta, \gamma) \frac{d\omega}{8\pi^2} = \frac{1}{2j_1+1} \delta_{j_1 j_2} \delta_{m_1 m_2} \delta_{m'_1 m'_2}, \quad (58.20)$$

where $d\omega = \sin\beta \, d\alpha \, d\beta \, d\gamma$.

The orthogonality of the functions with respect to the suffixes m and m' is ensured by the factor $e^{i(m\alpha + m'\gamma)}$; that with respect to the index j arises from the functions $d^{(j)}_{m'm}$, for which we have

$$\int_0^\pi d^{(j_1)}_{m'm}(\beta) d^{(j_2)}_{m'm}(\beta) \,.\, \tfrac{1}{2} \sin\beta \, d\beta = \frac{1}{2j_1+1} \delta_{j_1 j_2}. \quad (58.21)$$

Lastly, we shall give for reference the expressions for the functions $d^{(j)}_{m'm}$ for various particular values of the parameters. For $j = 1$, we have

$$d^{(1)}_{m'm}(\beta) = \begin{array}{c|ccc} & m = 1 & 0 & -1 \\ m' = & & & \\ 1 & \tfrac{1}{2}(1+\cos\beta) & \frac{1}{\sqrt{2}}\sin\beta & \tfrac{1}{2}(1-\cos\beta) \\ 0 & -\frac{1}{\sqrt{2}}\sin\beta & \cos\beta & \frac{1}{\sqrt{2}}\sin\beta \\ -1 & \tfrac{1}{2}(1-\cos\beta) & -\frac{1}{\sqrt{2}}\sin\beta & \tfrac{1}{2}(1+\cos\beta) \end{array} \quad (58.22)$$

For integral $j = l$ and $m' = 0$, formulae (58.10) and (58.11) give

$$d^{(l)}_{0m}(\beta) = (-1)^m d^{(l)}_{m0}(\beta) = (-1)^m \sqrt{\frac{(l-m)!}{(l+m)!}} P_l^m(\cos\beta). \quad (58.23)$$

The derivation of this formula is easily seen from the original definition (58.7).

We shall assign the values of the functions $\psi_{jm'}$ on the right of (58.7) to the z-axis, on which (for $j = l$)

$$Y_{lm'}(\mathbf{n}_{z'}) = i^l \sqrt{\frac{2l+1}{4\pi}}\delta_{m'0}. \tag{58.24}$$

The function ψ_{jm} on the left is then the spherical harmonic function $Y_{lm}(\beta, \alpha)$ of the spherical polar angles $\phi \equiv \alpha$, $\theta \equiv \beta$ giving the direction of the z'-axis. Substitution of (58.24) in (58.7) leads to

$$Y_{lm}(\beta, \alpha) = i^l \sqrt{\frac{2l+1}{4\pi}} D^{(l)}_{0m}(\alpha, \beta, \gamma), \tag{58.25}$$

which is equivalent to (58.23).

Lastly, there is the following expression for the function with the maximum possible value of m or m':

$$\begin{aligned} d^{(j)}_{jm}(\beta) &= (-1)^{j-m} d^{(j)}_{mj} \\ &= \left[\frac{(2j)!}{(j+m)!(j-m)!}\right]^{1/2} \cos^{j+m}\tfrac{1}{2}\beta \sin^{j-m}\tfrac{1}{2}\beta. \end{aligned} \tag{58.26}$$

§59. Partial polarization of particles

By a suitable choice of the direction of the z-axis, we can always cause one component (e.g. ψ^2) of a given spinor ψ^λ, the wave function of a particle with spin $\frac{1}{2}$, to vanish. This is evident from the fact that a direction in space is determined by two quantities (angles), i.e. the number of disposable parameters is just equal to the number of quantities (the real and imaginary parts of the complex ψ^2) which it is desired to make zero.

Physically this means that, if a particle with spin $\frac{1}{2}$ (for definiteness, we shall speak of an electron) is in a state described by a spin wave function, then there is a direction in space in which the component of the particle spin has the definite value $\sigma = \frac{1}{2}$. We can say that in such a state the electron is *completely polarized.*

There are also, however, states of an electron which may be said to be *partially polarized.* Such states are not described by wave functions but only by density matrices, i.e. they are mixed states (with respect to spin) (see §14).

The spin or *polarization* density matrix of an electron is a spinor $\rho^{\lambda\mu}$ of rank two normalized by the condition

$$\rho^\lambda{}_\lambda = \rho^1{}_1 + \rho^2{}_2 = 1, \tag{59.1}$$

and satisfying the "Hermitian" condition

$$(\rho^\lambda{}_\mu)^* = \rho^\mu{}_\lambda. \tag{59.2}$$

For a pure (i.e. completely polarized) spin state of the electron the spinor $\rho^{\lambda}{}_{\mu}$ reduces to a product of components of the wave function ψ^{λ}:

$$\rho^{\lambda}{}_{\mu} = \psi^{\lambda}(\psi^{\mu})^{*}. \tag{59.3}$$

The diagonal components of the density matrix determine the probabilities of the values $+\frac{1}{2}$ and $-\frac{1}{2}$ of the z-component of the electron spin. The mean value of this component is therefore

$$\overline{s_z} = \tfrac{1}{2}(\rho^{1}{}_{1} - \rho^{2}{}_{2}),$$

or, using (59.1),

$$\rho^{1}{}_{1} = \tfrac{1}{2} + \overline{s_z}, \qquad \rho^{2}{}_{2} = \tfrac{1}{2} - \overline{s_z}. \tag{59.4}$$

In a pure state the mean value of the quantities $s_{\pm} = s_x \pm i s_y$ is calculated as

$$\overline{s_+} = \psi^{\lambda *}\hat{s}_+\psi^{\lambda},$$

$$\overline{s_-} = \psi^{\lambda *}\hat{s}_-\psi^{\lambda}.$$

Since, according to (55.6) and (55.7), the operators $\hat{s}_{\pm}$ are given by the matrices

$$\hat{s}_+ = \begin{pmatrix} 0 & 1 \\ 0 & 0 \end{pmatrix}, \qquad \hat{s}_- = \begin{pmatrix} 0 & 0 \\ 1 & 0 \end{pmatrix},$$

we find that

$$\overline{s_+} = \psi^{1*}\psi^{2}, \qquad \overline{s_-} = \psi^{2*}\psi^{1}.$$

Accordingly we have in a mixed state

$$\rho^{1}{}_{2} = \overline{s_-}, \qquad \rho^{2}{}_{1} = \overline{s_+}. \tag{59.5}$$

Using the Pauli matrices, formulae (59.4) and (59.5) can be combined as

$$\rho^{\lambda}{}_{\mu} = \tfrac{1}{2}(\delta^{\lambda}{}_{\mu} + 2\hat{\boldsymbol{\sigma}}^{\lambda}{}_{\mu} \cdot \overline{\mathbf{s}}). \tag{59.6}$$

Thus all the components of the polarization density matrix of the electron are expressed in terms of the mean values of components of its spin vector. In other words, the real vector $\overline{\mathbf{s}}$ entirely determines the polarization properties of a particle with spin $\frac{1}{2}$. In the limit of complete polarization one of the components of this vector (with an appropriate choice of the directions of the axes) is $\frac{1}{2}$ and the other two are zero. In the opposite case of an unpolarized state all three components are zero. In the general case of an arbitrary partial polarization and any choice of the coordinate system we have $0 \leqslant \rho \leqslant 1$, where

$$\rho = 2(\overline{s_x}^2 + \overline{s_y}^2 + \overline{s_z}^2)^{1/2}$$

is a quantity which may be called the *degree of polarization* of the electron.

For a particle of arbitrary spin s, the density matrix is a spinor $\rho^{\lambda\mu\dots}{}_{\rho\sigma\dots}$ of rank $4s$, symmetrical in the first $2s$ and the last $2s$ indices and satisfying the conditions

$$\rho^{\lambda\mu\dots}{}_{\lambda\mu\dots} = 1, \tag{59.7}$$

$$(\rho^{\lambda\mu\dots}{}_{\rho\sigma\dots})^* = \rho^{\rho\sigma\dots}{}_{\lambda\mu\dots}. \tag{59.8}$$

To calculate the number of independent components of the density matrix, we note that, among the possible sets of values of the indices λ, μ, ... (or ρ, σ, ...) there are only $2s+1$ which are essentially different. Using also the fact that the components of the spinor $\rho^{\lambda\mu\dots}{}_{\rho\sigma\dots}$ are related by (59.7), we find that the number of different components is $(2s+1)^2-1 = 4s(s+1)$. Although these components are complex, the relation (59.8) shows that this does not increase the total number of independent quantities describing the state of partial polarization of the particle, which is therefore $4s(s+1)$.† For comparison, it may be remarked that the state of complete polarization of the particle is described by only $4s$ quantities (the $2s+1$ complex components of the wave function $\psi^{\lambda\mu\dots}$, related by one normalization condition and containing one common phase which is unimportant in the description of the state).

Like any spinor of rank $4s$, the spinor $\rho^{\lambda\mu\dots}{}_{\rho\sigma\dots}$ is equivalent to a set of irreducible tensors of ranks $4s$, $4s-2$, ... , 0. In the present case there is only one tensor of each rank, since, on account of the symmetry properties of the spinor $\rho^{\lambda\mu\dots}{}_{\rho\sigma\dots}$, each contraction of it can be carried out in only one way: with respect to any one of the indices λ, μ, ... , and one of ρ, σ, ... In addition, the scalar (tensor of rank 0) does not appear, reducing to unity by virtue of the condition (59.7).

§60. Time reversal and Kramers' theorem

The symmetry of motion with respect to a change in the sign of the time is expressed in quantum mechanics by the fact that, if ψ is the wave function of a stationary state of the system, the "time-reversed" wave function (which we denote by ψ^{rev}) describes a possible state with the same energy. At the end of §18 it has been pointed out that ψ^{rev} is the same as the complex conjugate function ψ^*. In this simple form the statement applies to wave functions where the spin of particles is neglected. When spin is present, a refinement is necessary.

Let us take the wave function of a particle of spin s in the form of the contravariant spinor $\psi^{\lambda\mu\dots}$ (of rank $2s$). On taking the complex conjugate function $\psi^{\lambda\mu\dots *}$ we obtain a set of quantities which are transformed as components of a covariant spinor. Hence the operation of time reversal corresponds to a change from the wave function $\psi^{\lambda\mu\dots}$ to a new wave function whose covariant components are given by

$$\psi^{\text{rev}}_{\lambda\mu\dots} = \psi^{\lambda\mu\dots *}. \tag{60.1}$$

† When these quantities are given, so are the mean values of the components of the vector **s** and all their powers and products 2, 3, ..., $2s$ at a time, which do not reduce to lower powers (see §55, Problem 3).

For a given set of values of the indices $\lambda, \mu, \ldots$, the components of covariant and contravariant spinors correspond to values of the angular-momentum component which differ in sign. In terms of the functions $\psi_{s\sigma}$, therefore, time reversal corresponds to a change from $\psi_{s\sigma}$ to $\psi_{s,-\sigma}$, as it should, since a change in the sign of the time changes the direction of the angular momentum. The exact relation is given by (60.1):

$$\psi_{s,-\sigma}^{\text{rev}} = \psi_{s\sigma}^*(-1)^{s-\sigma}.$$

Thus the change $\psi_{s\sigma} \to \psi_{s\sigma}^*$ required by the operation of time reversal signifies the change†

$$\psi_{s,-\sigma}^{\text{rev}} = \psi_{s\sigma}^*(-1)^{s-\sigma}. \tag{60.2}$$

When this operation is repeated, we have

$$\psi_{s\sigma} \to \psi_{s,-\sigma}(-1)^{s-\sigma} \to \psi_{s\sigma}(-1)^{s-\sigma}(-1)^{s+\sigma} = \psi_{s\sigma}(-1)^{2s}.$$

Thus a twofold time reversal restores the wave function to its original value only if the spin is integral; if the spin is half-integral, the sign of the wave function is changed.

Let us consider an arbitrary system of interacting particles. The orbital and spin angular momenta of such a system are not in general separately conserved when relativistic interactions are taken into account. Only the total angular momentum **J** is conserved. If there is no external field, each energy level of the system has $(2J+1)$-fold degeneracy. When an external field is applied, the degeneracy is removed. The question arises whether the degeneracy can be removed completely, i.e. so that the system has only simple levels. This is closely related to the symmetry with respect to time reversal.

In classical electrodynamics the equations are invariant with respect to a change in the sign of the time, if the electric field is left unchanged and the sign of the magnetic field is reversed.‡ This fundamental property of motion must be preserved in quantum mechanics. Hence, not only in a closed system but in any external electric field (there being no magnetic field), there is symmetry with respect to time reversal.

The wave functions of the system are spinors $\psi^{\lambda\mu\cdots}$, whose rank n is twice the sum of the spins s_a of all the particles ($n = 2\sum s_a$); this sum may not be equal to the total spin S of the system.

According to what was said above, we can assert that, in any electric field, the wave function and its time reversal must correspond to states with the same energy. If a level is non-degenerate, it is necessary that these states should be identical, i.e. the corresponding wave functions must be the same apart from a

† Note that the rule for the complex conjugate of a spherical harmonic function, according to (28.9), coincides with the general rule (60.3).

‡ See, for example, *Fields*, §17, and the end of §111 below.

constant factor (both, of course, being expressed as similar (covariant or contravariant) spinors).

We write $\psi^{\text{rev}}_{\lambda\mu\ldots} = C\psi_{\lambda\mu\ldots}$ or, by (60.1),

$$\psi^{\lambda\mu\ldots *} = C\psi_{\lambda\mu\ldots}, \tag{60.4}$$

where C is a constant.

Taking the complex conjugate of both sides of this equation, we obtain

$$\psi^{\lambda\mu\ldots} = C^*\psi_{\lambda\mu\ldots}{}^*.$$

We lower the indices on the left-hand side of the equation and correspondingly raise them on the right. This means that we multiply both sides of the equation by $g_{\alpha\lambda}g_{\beta\mu}\ldots$ and sum over the indices $\lambda, \mu, \ldots$; on the right-hand side we must use the fact that

$$g_{\alpha\lambda}g_{\beta\mu}\ldots = (-1)^n g^{\lambda\alpha}g^{\mu\beta}\ldots.$$

As a result we have

$$\psi_{\lambda\mu\ldots} = C^*(-1)^n\psi^{\lambda\mu\ldots *}.$$

Substituting $\psi^{\lambda\mu\ldots *}$ from (60.4), we find

$$\psi_{\lambda\mu\ldots} = (-1)^n CC^*\psi_{\lambda\mu\ldots}.$$

This equation must be satisfied identically, i.e. we must have $(-1)^n CC^* = 1$. Since, however, $|C|^2$ is always positive, it is clear that this is possible only for even n (i.e. for integral values of the sum $\Sigma\, s_a$). For odd n (half-integral values of $\Sigma\, s_a$) the condition (60.4) cannot be fulfilled.†

Thus we reach the result that an electric field can completely remove the degeneracy only for a system with an integral value of the sum of the spins of the particles. For a system with a half-integral value of this sum, in an arbitrary electric field, all the levels must be doubly degenerate, and complex conjugate spinors correspond to two different states with the same energy‡ (H. A. Kramers 1930).

One further, mathematical, comment may be made. A relation of the form (60.4) with a real constant C is mathematically the condition that the components of the spinor may be put in correspondence with a set of real quantities, and may be called the condition for the spinor to be "real".|| The impossibility of fulfilling the condition (60.4) for odd n signifies that no real quantity can correspond to a spinor of odd rank. For even n, on the other hand, the condition (60.4) can be satisfied, and C can be real. In particular, a

---

† When the sum $\Sigma\, s_a$ is integral (or half-integral), all possible values of the total spin S of the system are also integral (or half-integral).

‡ If the electric field possesses a high (cubic) symmetry, fourfold degeneracy may occur (see §99, including the Problem).

|| It is meaningless to call the spinor real in the literal sense, since complex conjugate spinors have different laws of transformation.

real vector can correspond to a symmetrical spinor of rank two if the condition (60.4) is satisfied with $C = 1$:

$$\psi^{\lambda\mu *} = \psi_{\lambda\mu}$$

(as is easily seen by means of (57.8) and (57.9)). The condition (60.4) with $C = 1$ is in fact the condition for a symmetrical spinor of any even rank to be "real".

CHAPTER IX

IDENTITY OF PARTICLES

§61. The principle of indistinguishability of similar particles

In classical mechanics, identical particles (electrons, say) do not lose their "individuality", despite the identity of their physical properties. For we can imagine the particles at some instant to be "numbered", and follow the subsequent motion of each of these in its path; then at any instant the particles can be identified.

In quantum mechanics the situation is entirely different. We have already mentioned several times that, by virtue of the uncertainty principle, the concept of the path of an electron ceases to have any meaning. If the position of an electron is exactly known at a given instant, its coordinates have no definite values even at the next instant. Hence, by localizing and numbering the electrons at some instant, we make no progress towards identifying them at subsequent instants; if we localize one of the electrons, at some other instant, at some point in space, we cannot say which of the electrons has arrived at this point.

Thus, in quantum mechanics, there is in principle no possibility of separately following each of a number of similar particles and thereby distinguishing them. We may say that, in quantum mechanics, identical particles entirely lose their "individuality". The identity of the particles with respect to their physical properties is here very far-reaching: it results in the complete indistinguishability of the particles.

This principle of the *indistinguishability of similar particles*, as it is called, plays a fundamental part in the quantum theory of systems composed of identical particles. Let us start by considering a system of only two particles. Because of the identity of the particles, the states of the system obtained from each other by merely interchanging the two particles must be completely equivalent physically. This means that, as a result of this interchange, the wave function of the system can change only by an unimportant phase factor. Let $\psi(\xi_1, \xi_2)$ be the wave function of the system, ξ_1 and ξ_2 conventionally denoting the three coordinates and the spin projection for each particle. Then we must have

$$\psi(\xi_1, \xi_2) = e^{i\alpha}\psi(\xi_2, \xi_1),$$

where α is some real constant. By repeating the interchange, we return to the original state, while the function ψ is multiplied by $e^{2i\alpha}$. Hence it follows that $e^{2i\alpha} = 1$, or $e^{i\alpha} = \pm 1$. Thus

$$\psi(\xi_1, \xi_2) = \pm\psi(\xi_2, \xi_1).$$

We thus reach the result that there are only two possibilities: the wave function is either *symmetrical* (i.e. it is unchanged when the particles are interchanged) or *antisymmetrical* (i.e. it changes sign when this interchange is made). It is obvious that the wave functions of all the states of a given system must have the same symmetry; otherwise, the wave function of a state which was a superposition of states of different symmetry would be neither symmetrical nor antisymmetrical.

This result can be immediately generalized to systems consisting of any number of identical particles. For it is clear from the identity of the particles that, if any pair of them has the property of being described by, say, symmetrical wave functions, any other pair of such particles has the same property. Hence the wave function of identical particles must either be unchanged when any pair of particles are interchanged (and hence when the particles are permuted in any manner), or change sign when any pair are interchanged. In the first case we speak of a *symmetrical* wave function, and in the second case of an *antisymmetrical* one.

The property of being described by symmetrical or antisymmetrical wave functions depends on the nature of the particles. Particles described by antisymmetrical functions are said to obey *Fermi–Dirac statistics* (or to be *fermions*), while those which are described by symmetrical functions are said to obey *Bose–Einstein statistics* (or to be *bosons*).†

From the laws of relativistic quantum mechanics it can be shown (see *RQT*, §25) that the statistics obeyed by particles is uniquely related to their spin: particles with half-integral spin are fermions, and those with integral spin are bosons.

The statistics of complex particles is determined by the parity of the number of elementary fermions entering into their composition. For an interchange of two identical complex particles is equivalent to the simultaneous interchange of several pairs of identical elementary particles. The interchange of bosons does not change the wave function, while the interchange of fermions changes its sign. Hence complex particles containing an odd number of elementary fermions obey Fermi statistics, while those containing an even number obey Bose statistics. This result is, of course, in agreement with the above rule, since a complex particle has an integral or a half-integral spin according as the number of particles with half-integral spin entering into its composition is even or odd.

Thus atomic nuclei of odd atomic weight (i.e. containing an odd number of neutrons and protons) obey Fermi statistics, and those of even atomic weight obey Bose statistics. For atoms, which contain both nuclei and electrons, the statistics is evidently determined by the parity of the sum of the atomic weight and the atomic number.

---

† This terminology refers to the statistics which describes a perfect gas composed of particles with antisymmetrical and symmetrical wave functions respectively. In actual fact we are concerned here not only with a different statistics, but essentially with a different mechanics. Fermi statistics was proposed by E. Fermi for electrons in 1926, and its relation to quantum mechanics was elucidated by P. A. M. Dirac (1926). Bose statistics was proposed by S. N. Bose for light quanta, and generalized by A. Einstein (1924).

Let us consider a system composed of N identical particles, whose mutual interaction can be neglected. Let $\psi_1, \psi_2, \ldots$ be the wave functions of the various stationary states which each of the particles separately may occupy. The state of the system as a whole can be defined by giving the numbers of the states which the individual particles occupy. The question arises how the wave function ψ of the whole system should be constructed from the functions $\psi_1, \psi_2, \ldots$.

Let $p_1, p_2, \ldots, p_N$ be the numbers of the states occupied by the individual particles (some of these numbers may be the same). For a system of bosons, the wave function $\psi(\xi_1, \xi_2, \ldots, \xi_N)$ is given by a sum of products of the form

$$\psi_{p_1}(\xi_1)\psi_{p_2}(\xi_2)\ldots\psi_{p_N}(\xi_N), \tag{61.1}$$

with all possible permutations of the different suffixes $p_1, p_2, \ldots$; this sum clearly possesses the required symmetry property. For example, for a system of two particles in different states ($p_1 \neq p_2$),

$$\psi(\xi_1, \xi_2) = [\psi_{p_1}(\xi_1)\psi_{p_2}(\xi_2)+\psi_{p_1}(\xi_2)\psi_{p_2}(\xi_1)]/\sqrt{2}. \tag{61.2}$$

The factor $1/\sqrt{2}$ is introduced for normalization purposes; all the functions $\psi_1, \psi_2, \ldots$ are orthogonal and are supposed normalized.

In the general case of a system containing an arbitrary number N of particles, the normalized wave function is

$$\psi = \left(\frac{N_1!N_2!\ldots}{N!}\right)^{1/2} \Sigma\, \psi_{p_1}(\xi_1)\psi_{p_2}(\xi_2)\ldots\psi_{p_N}(\xi_N), \tag{61.3}$$

where the sum is taken over all permutations of the different suffixes $p_1, p_2, \ldots, p_N$ and the numbers N_i show how many of these suffixes have the same value i (with $\Sigma N_i = N$). In the integration of $|\psi|^2$ over $\xi_1, \xi_2, \ldots, \xi_N$, all terms vanish except the squared modulus of each term of the sum;† since the total number of terms in the sum (61.3) is evidently $N_1!N_2!N_3!\ldots$, the normalization factor in (61.3) is obtained.

For a system of fermions, the wave function ψ is an antisymmetrical combination of the products (61.1). For a system of two particles we have

$$\psi(\xi_1, \xi_2) = [\psi_{p_1}(\xi_1)\psi_{p_2}(\xi_2)-\psi_{p_1}(\xi_2)\psi_{p_2}(\xi_1)]/\sqrt{2}. \tag{61.4}$$

For the general case of N particles, the wave function can be written in the

† The integration over ξ is conventionally understood in §§63–65 as including integration over the coordinates and summation over σ.

form of a determinant

$$\psi = \frac{1}{\sqrt{N!}} \begin{vmatrix} \psi_{p_1}(\xi_1) & \psi_{p_1}(\xi_2) & \dots & \psi_{p_1}(\xi_N) \\ \psi_{p_2}(\xi_1) & \psi_{p_2}(\xi_2) & \dots & \psi_{p_2}(\xi_N) \\ \dots & \dots & \dots & \dots \\ \psi_{p_N}(\xi_1) & \psi_{p_N}(\xi_2) & \dots & \psi_{p_N}(\xi_N) \end{vmatrix}. \quad (61.5)$$

Here an interchange of two particles corresponds to an interchange of two columns of the determinant, as a result of which the latter changes sign.

The following important result is a consequence of the expression (61.5). If among the numbers $p_1, p_2, \dots$ two are the same, two rows of the determinant are the same, and it therefore vanishes identically. It will be different from zero only when all the numbers $p_1, p_2, \dots$ are different. Thus, in a system consisting of identical fermions, no two (or more) particles can be in the same state at the same time. This is called *Pauli's principle* (1925).

§62. Exchange interaction

The fact that Schrödinger's equation does not take account of the spin of particles does not invalidate this equation or the results obtained by means of it. This is because the electrical interaction of the particles does not depend on their spins.† Mathematically, this means that the Hamiltonian of a system of electrically interacting particles (in the absence of a magnetic field) does not contain the spin operators, and hence, when it is applied to the wave function, it has no effect on the spin variables. Hence Schrödinger's equation is actually satisfied by each component of the wave function; in other words, the wave function of the system of particles can be written in the form of a product

$$\psi(\xi_1, \xi_2) = \chi(\sigma_1, \sigma_2, \dots)\phi(\mathbf{r}_1, \mathbf{r}_2, \dots),$$

where the function ϕ depends only on the coordinates of the particles and the function χ only on their spins. We call the former a *coordinate* or *orbital* wave function, and the latter a *spin* wave function. Schrödinger's equation essentially determines only the coordinate function ϕ, the function χ remaining arbitrary. In any instance where we are not interested in the actual spin of the particles, we can therefore use Schrödinger's equation and regard as the wave function the coordinate function alone, as we have done hitherto.

However, despite the fact that the electrical interaction of the particles is independent of their spin, there is a peculiar dependence of the energy

† This is true only so long as we consider the non-relativistic approximation. When relativistic effects are taken into account, the interaction of charged particles does depend on their spin.

of the system on its total spin, arising ultimately from the principle of indistinguishability of similar particles.

Let us consider a system consisting of only two identical particles. By solving Schrödinger's equation we find a series of energy levels, to each of which there corresponds a definite symmetrical or antisymmetrical co-ordinate wave function $\phi(\mathbf{r}_1, \mathbf{r}_2)$. For, by virtue of the identity of the particles, the Hamiltonian (and therefore the Schrödinger's equation) of the system is invariant with respect to interchange of the particles. If the energy levels are not degenerate, the function $\phi(\mathbf{r}_1, \mathbf{r}_2)$ can change only by a constant factor when the coordinates $\mathbf{r}_1$ and $\mathbf{r}_2$ are interchanged; repeating this interchange, we see that this factor can only be† ± 1.

Let us first suppose that the particles have zero spin. The spin factor for such particles is absent altogether, and the wave function reduces to the coordinate function $\phi(\mathbf{r}_1, \mathbf{r}_2)$, which must be symmetrical (since particles with zero spin obey Bose statistics). Thus not all the energy levels obtained by a formal solution of Schrödinger's equation can actually exist; those to which antisymmetrical functions ϕ correspond are not possible for the system under consideration.

The interchange of two similar particles is equivalent to the operation of inversion of the coordinate system (the origin being taken to bisect the line joining the two particles). On the other hand, the result of inversion is to multiply the wave function ϕ by $(-1)^l$, where l is the orbital angular momentum of the relative motion of the two particles (see §30). By comparing these considerations with those given above, we conclude that a system of two identical particles with zero spin can have only an even orbital angular momentum.

Next, let the system consist of two particles with spin $\frac{1}{2}$ (say, electrons). Then the complete wave function of the system (i.e. the product of the function $\phi(\mathbf{r}_1, \mathbf{r}_2)$ and the spin function $\chi(\sigma_1, \sigma_2)$) must certainly be antisymmetrical with respect to an interchange of the two particles. Hence, if the coordinate function is symmetrical, the spin function must be antisymmetrical, and *vice versa*. We shall write the spin function in spinor form, i.e. as a spinor $\chi^{\lambda\mu}$ of rank two, each of whose indices corresponds to the spin of one of the electrons. A symmetrical spinor ($\chi^{\lambda\mu} = \chi^{\mu\lambda}$) corresponds to a function symmetrical with respect to the spins of the two particles, and an antisymmetrical spinor ($\chi^{\lambda\mu} = -\chi^{\mu\lambda}$) to an antisymmetrical function. We know, however, that a symmetrical spinor of rank two describes a system with total spin unity, while an antisymmetrical spinor reduces to a scalar, corresponding to zero spin.

Thus we reach the following conclusion. The energy levels to which there correspond symmetrical solutions $\phi(\mathbf{r}_1, \mathbf{r}_2)$ of Schrödinger's equation can actually occur when the total spin of the system is zero, i.e. when the spins of the two electrons are "antiparallel", giving a sum of zero. The values of the energy belonging to antisymmetrical functions $\phi(\mathbf{r}_1, \mathbf{r}_2)$, on the other hand,

† When there is degeneracy we can always choose linear combinations of the functions belonging to a given level, such that this condition is again satisfied.

require a value of unity for the total spin, i.e. the spins of the two electrons must be "parallel".

In other words, the possible values of the energy of a system of electrons depend on their total spin. For this reason we can speak of a peculiar interaction of the particles which results in this dependence. This is called *exchange interaction*. It is a purely quantum effect, which entirely vanishes (like the spin itself) in the passage to the limit of classical mechanics.

The following situation is characteristic of the case of a system of two electrons which we have discussed. To each energy level there corresponds one definite value of the total spin, 0 or 1. This one-to-one correspondence between the spin values and the energy levels is preserved, as we shall see below (§63), in systems containing any number of electrons. It does not hold, however, for systems composed of particles whose spin exceeds $\frac{1}{2}$.

Let us consider a system of two particles, each with arbitrary spin s. Its spin wave function is a spinor of rank $4s$:

$$\chi^{\overbrace{\lambda\mu\ldots}^{2s}\overbrace{\rho\sigma\ldots}^{2s}},$$

half ($2s$) of whose indices correspond to the spin of one particle, and the other half to that of the other particle. The spinor is symmetrical with respect to the indices in each group. An interchange of the two particles corresponds to an interchange of all the indices λ, μ, ... of the first group with the indices ρ, σ, ... of the second group. In order to obtain the spin function of a state of the system with total spin S, we must contract this spinor with respect to $2s-S$ pairs of indices (each pair containing one index from λ, μ, ... and one from ρ, σ, ...), and symmetrize it with respect to the remainder; as a result we obtain a symmetrical spinor of rank $2S$. However, the contraction of a spinor with respect to a pair of indices means, as we know, the construction of a combination antisymmetrical with respect to these indices. Hence, when the particles are interchanged, the spin wave function is multiplied by $(-1)^{2s-S}$.

On the other hand, the complete wave function of a system of two particles must be multiplied by $(-1)^{2s}$ when they are interchanged (i.e. by $+1$ for integral s and by -1 for half-integral s). Hence it follows that the symmetry of the coordinate wave function with respect to an interchange of the particles is given by the factor $(-1)^S$, which depends only on S. Thus we reach the result that the coordinate wave function of a system of two identical particles is symmetrical when the total spin is even, and antisymmetrical when it is odd.

Recalling what was said above concerning the relation between interchange of the particles and inversion of the coordinate system, we conclude also that, when the spin S is even (odd), the system can have only an even (odd) orbital angular momentum.

We see that here also a certain dependence is revealed between the possible values of the energy of the system and the total spin, but this dependence is not necessarily one-to-one. The energy levels to which there correspond symmetrical (antisymmetrical) coordinate wave functions can occur for any even (odd) value of S.

Let us calculate how many different states of the system there are with even and odd S. The quantity S takes $2s+1$ values: $2s$, $2s-1$, ..., 0. For any given S there are $2S+1$ states differing in the value of the z-component of the spin ($(2s+1)^2$ different states altogether). Let s be integral. Then, among the $2s+1$ values of S, $s+1$ are even and s odd. The total number of states with even S is equal to the sum

$$\sum_{S=0,2,\ldots,2s}(2S+1) = (2s+1)(s+1);$$

the remaining $s(2s+1)$ states have odd S. Similarly, we find that, when s is half-integral, there are $s(2s+1)$ states with even values of S and $(s+1)(2s+1)$ with odd values.

PROBLEMS

PROBLEM 1. Determine the exchange splitting of the energy levels of a system of two electrons, regarding the interaction of the electrons as a perturbation.

SOLUTION. Let the particles be (when their interaction is neglected) in states with orbital wave functions $\phi_1(\mathbf{r})$ and $\phi_2(\mathbf{r})$. The states of the system with total spin $S = 0$ and $S = 1$ correspond to symmetrized and antisymmetrized products respectively:

$$\phi = \frac{1}{\sqrt{2}}[\phi_1(\mathbf{r}_1)\phi_2(\mathbf{r}_2)\pm\phi_1(\mathbf{r}_2)\phi_2(\mathbf{r}_1)].$$

The mean value of the operator of the interaction $U(\mathbf{r}_2-\mathbf{r}_1)$ of the particles in these states is $A\pm J$, where

$$A = \int\int U|\phi_1(\mathbf{r}_1)|^2|\phi_2(\mathbf{r}_2)|^2\, dV_1\, dV_2,$$

$$J = \int\int U\phi_1(\mathbf{r}_1)\phi_1^*(\mathbf{r}_2)\phi_2(\mathbf{r}_2)\phi_2^*(\mathbf{r}_1)\, dV_1\, dV_2,$$

the latter being called the *exchange integral*. Omitting the additive constant A, which is not an exchange term, we therefore find the level shifts $\Delta E_0 = J$, $\Delta E_1 = -J$ (where the suffix indicates the value of S). These quantities can be represented as the eigenvalues of the spin *exchange operator*†

$$\hat{V}_{\text{exch}} = -\tfrac{1}{2}J(1+4\hat{\mathbf{s}}_1\cdot\hat{\mathbf{s}}_2); \tag{1}$$

the eigenvalues of the product $\mathbf{s}_1\cdot\mathbf{s}_2$ are derived in §55, Problem 2.

If the electrons belong to different atoms, for example, the exchange integral decreases exponentially with increasing distance R between the atoms. It is clear from the form of the integrand that this integral is determined by the "overlap" of the wave functions of the states $\phi_1(\mathbf{r}_1)$ and $\phi_2(\mathbf{r}_2)$; using the asymptotic law of decrease of the wave functions of states of a discrete spectrum (cf. (21.6)), we find that

$$J \sim e^{-(\kappa_1+\kappa_2)R}, \qquad \kappa_1 = \sqrt{(2m|E_1|)}/\hbar, \qquad \kappa_2 = \sqrt{(2m|E_2|)}/\hbar,$$

where E_1 and E_2 are the energy levels of the electron in the two atoms.

PROBLEM 2. The same as Problem 1, but for a system of three electrons.

SOLUTION. Using formula (1), Problem 1, we can write the operator of pairwise exchange interaction in a system of three electrons as

$$\hat{V}_{\text{exch}} = -\sum J_{ab}(\tfrac{1}{2}+2\hat{\mathbf{s}}_a\cdot\hat{\mathbf{s}}_b), \tag{1}$$

† First used by Dirac.

where the summation is over pairs of particles 12, 13 and 23. The matrix elements of the operators $\hat{\mathbf{s}}_a \cdot \hat{\mathbf{s}}_b$ between states with different values of the pair of numbers σ_a, σ_b are given by formulae (55.6) as

$$\langle \tfrac{1}{2}, \tfrac{1}{2}|\mathbf{s}_a \cdot \mathbf{s}_b|\tfrac{1}{2}, \tfrac{1}{2}\rangle = \tfrac{1}{4}, \quad \langle \tfrac{1}{2}, -\tfrac{1}{2}|\mathbf{s}_a \cdot \mathbf{s}_b|\tfrac{1}{2}, -\tfrac{1}{2}\rangle = -\tfrac{1}{4}, \quad \langle \tfrac{1}{2}, -\tfrac{1}{2}|\mathbf{s}_a . \mathbf{s}_b| -\tfrac{1}{2}, \tfrac{1}{2}\rangle = \tfrac{1}{2}.$$

We first determine the energy corresponding to the greatest possible value of the total-spin component $M_S = \sigma_1 + \sigma_2 + \sigma_3$, viz. $M_S = 3/2$. This gives the energy of the state with total spin $S = 3/2$. On calculating the corresponding diagonal matrix element of the operator (1), we find

$$\Delta E_{3/2} = -(J_{12} + J_{13} + J_{23}).$$

Next we take states with $M_S = \tfrac{1}{2}$. This value can occur in three ways, depending on which of the numbers σ_1, σ_2, σ_3 is $-\tfrac{1}{2}$ (the other two being $\tfrac{1}{2}$). Thus for these states we should have a secular equation of the third degree. The calculation can, however, be simplified immediately by noting that one of the roots of this equation must correspond to the energy already found for the state with $S = 3/2$, and the secular equation must therefore have the factor $\Delta E - \Delta E_{3/2}$. In this way the calculation of the free term in the cubic equation can be avoided.†

The leading terms of the equation are found to be

$$(\Delta E)^3 + (J_{12} + J_{13} + J_{23})(\Delta E^2) + [J_{12}J_{13} + J_{12}J_{23} + J_{13}J_{23} -$$
$$-(J_{12}^2 + J_{13}^2 + J_{23}^2)]\Delta E + \ldots = 0.$$

Dividing by $\Delta E + J_{12} + J_{13} + J_{23}$, we find the two energy levels corresponding to states with spins $S = \tfrac{1}{2}$:

$$\Delta E_{1/2} = \pm[(J_{12}^2 + J_{13}^2 + J_{23}^2) - J_{12}J_{13} - J_{12}J_{23} - J_{13}J_{23}]^{1/2}.$$

Thus there are three energy levels, in accordance with the calculation in §63, Problem 1.

PROBLEM 3. In which states can the ^{8}Be nucleus decay into two α-particles?

SOLUTION. Since the α-particle has no spin, a system of two α-particles can only have an even orbital angular momentum (equal to the total angular momentum), and its states are even. The decay in question is therefore possible only from even states of the ^{8}Be nucleus with even total angular momentum.

§63. Symmetry with respect to interchange

By considering a system composed of only two particles, we have been able to show that its coordinate wave functions $\phi(\mathbf{r}_1, \mathbf{r}_2)$ for the stationary states must be either symmetrical or antisymmetrical. In the general case of a system of an arbitrary number of particles, the solutions of Schrödinger's equation (the coordinate wave functions) need not necessarily be either symmetrical or antisymmetrical with respect to the interchange of any pair of particles, as the complete wave functions (which include the spin factor) must be. This is because an interchange of only the coordinates of two particles does not correspond to a physical interchange of them. The physical identity of the particles here leads only to the fact that the Hamiltonian of the system is invariant with respect to the interchange of the particles, and hence, if some function is a solution of Schrödinger's equation, the functions obtained from it by various interchanges of the variables will also be solutions.

Let us first of all make some remarks regarding interchanges in general.

† This device is particularly useful in similar calculations for systems with a larger number of particles.

In a system of N particles, $N!$ different permutations in all are possible. If we imagine all the particles to be numbered, each permutation can be represented by a definite sequence of the numbers 1, 2, 3, ... Every such sequence can be obtained from the natural sequence 1, 2, 3, ... by successive interchanges of pairs of particles. The permutation is called *even* or *odd*, according as it is brought about by an even or odd number of such interchanges. We denote by $\hat{P}$ the operators of permutations of N particles, and introduce a quantity δ_P which is $+1$ if $\hat{P}$ is an even permutation and -1 if it is odd. If ϕ is a function symmetrical with respect to all the particles, we have

$$\hat{P}\phi = \phi,$$

while, if ϕ is antisymmetrical with respect to all the particles, then

$$\hat{P}\phi = \delta_P\phi.$$

From an arbitrary function $\phi(\mathbf{r}_1, \mathbf{r}_2, \dots, \mathbf{r}_N)$, we can form a symmetrical function by the operation of *symmetrization*, which can be written

$$\phi_{\text{sym}} = \text{constant} \times \sum_P \hat{P}\phi, \tag{63.1}$$

where the summation extends over all possible permutations. The formation of an antisymmetrical function (an operation sometimes called *alternation*) can be written as

$$\phi_{\text{ant}} = \text{constant} \times \sum_P \delta_P\hat{P}\phi. \tag{63.2}$$

Let us return to considering the behaviour, with respect to permutations, of the wave functions ϕ of a system of identical particles.† The fact that the Hamiltonian $\hat{H}$ of the system is symmetrical with respect to all the particles means, mathematically, that $\hat{H}$ commutes with all the permutation operators $\hat{P}$. These operators, however, do not commute with one another, and so they cannot be simultaneously brought into diagonal form. This means that the wave functions ϕ cannot be so chosen that each of them is either symmetrical or antisymmetrical with respect to all interchanges separately.‡

Let us try to determine the possible types of symmetry of the functions $\phi(\mathbf{r}_1, \mathbf{r}_2, \dots, \mathbf{r}_N)$ of N variables (or of sets of several such functions) with respect to permutations of the variables. The symmetry must be such that it cannot be increased, i.e. such that any additional operation of symmetrization or alternation, on being applied to these functions, would reduce them either to linear combinations of themselves or to zero identically.

We already know two operations which give functions with the greatest

† From the mathematical point of view, the problem is to find irreducible representations of the permutation group. A detailed account of the mathematical theory of permutation (or symmetry) groups is given by H. Weyl, *The Theory of Groups and Quantum Mechanics*, Methuen, London 1931; M. Hamermesh, *Group Theory and its Application to Physical Problems*, Pergamon, London, 1962; I. G. Kaplan, *Symmetry of Many-Electron Systems*, Academic Press, New York, 1974.

‡ Except for a system of only two particles, where there is a single interchange operator, which can be brought into diagonal form simultaneously with the Hamiltonian.

possible symmetry: symmetrization with respect to all the variables, and alternation with respect to all the variables. These operations can be generalized as follows.

We divide the set of all the N variables $\mathbf{r}_1, \mathbf{r}_2, \ldots, \mathbf{r}_N$ (or, what is the same thing, the suffixes 1, 2, 3, ..., N) into several sets, containing $N_1, N_2, \ldots$ elements (variables); $N_1+N_2+\ldots = N$. This division can be conveniently shown by a diagram (known as a *Young diagram*) in which each of the numbers $N_1, N_2, \ldots$ is represented by a line of several cells (thus, Fig. 21 gives a diagram of the divisions $6+4+4+3+3+1+1$ and $7+5+5+3+1+1$ for $N = 22$); one of the numbers 1, 2, 3, ... is to be placed in each square. If we place the lines in order of decreasing length (as in Fig. 21), the diagram contains not only successive horizontal rows, but also vertical columns.

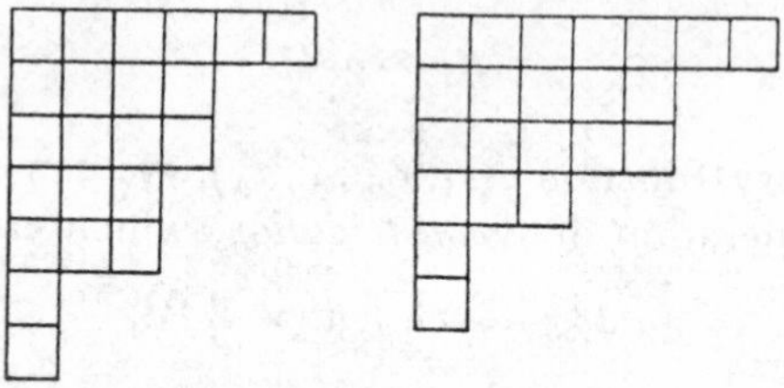

Fig. 21

Let us symmetrize an arbitrary function $\phi(\mathbf{r}_1, \mathbf{r}_2, \ldots, \mathbf{r}_N)$ with respect to the variables in each row. The alternation operation can then be performed only with respect to the variables in different rows; alternation with respect to a pair of variables in the same row clearly gives zero identically.

Having chosen one variable from each row, we can, without loss of generality, regard them as being in the first cells in each row (after symmetrization, the order of the variables among the cells in each row is immaterial); let us alternate with respect to these variables. Having then deleted the first column, we alternate with respect to variables chosen one from each row in the thus "curtailed" diagram; these variables can again be regarded as being in the first cells of the "curtailed" rows. Continuing this process, we finally have the function first symmetrized with respect to the variables in each row and then alternated with respect to the variables in each column. After alternation, of course, the function in general ceases to be symmetrical with respect to the variables in each row. The symmetry is preserved only with respect to the variables in the cells of the first row which project beyond the other rows.

Having distributed the N variables in various ways among the rows of a Young diagram (the distribution among the cells in each row is immaterial), we thus obtain a series of functions, which are transformed linearly into one another when the variables are permuted in any manner.† However, it must

† It would be possible to perform the symmetrization and alternation in the reverse order: to alternate with respect to the variables in each column, and then to symmetrize with respect to those in the rows. This, however, would give effectively the same thing, since the functions obtained by the two methods are linear combinations of one another.

be emphasized that not all these functions are linearly independent; the number of independent functions is in general less than the number of possible distributions of the variables among the rows of the diagram. We shall not pause here, however, to discuss this more closely.†

Thus any Young diagram determines some type of symmetry of functions with respect to permutations. By constructing all the possible Young diagrams (for a given N), we find all possible types of symmetry. This amounts to dividing the number N in all possible ways into a sum of smaller terms, including the number N itself; thus for $N = 4$ the possible partitions are $4, 3+1, 2+2, 2+1+1, 1+1+1+1$.

To each energy level of the system we can make correspond a Young diagram which determines the permutational symmetry of the appropriate solutions of Schrödinger's equation; in general, several different functions correspond to each value of the energy, and these are transformed linearly into each other by permutations. The existence of this "permutational degeneracy" is related to the fact that the operators $\hat{P}$ each commute with the Hamiltonian but not with one another (see the middle of §10). However, it must be emphasized that this does not signify any additional physical degeneracy of the energy levels. All these different coordinate wave functions, multiplied by the spin functions, enter into a single definite combination—the complete wave function—which satisfies (according to the spin of the particles) the condition of symmetry or antisymmetry.

Among the various types of symmetry there are always (for any given N) two to each of which only one function corresponds. One of these corresponds to a function symmetrical with respect to all the variables, and the other to one which is similarly antisymmetrical; in the first case, the Young diagram consists of a single row of N cells, and in the second case of a single column.

Let us now consider the spin wave functions $\chi(\sigma_1, \sigma_2, \dots, \sigma_N)$. Their kinds of symmetry with respect to permutations of the particles are given by the same Young diagrams, with the components of the spins of the particles taking the part of variables. There arises the question of what diagram must correspond to the spin function for a given diagram of the coordinate function. Let us first suppose that the spin of the particles is integral. Then the complete wave function ψ must be symmetrical with respect to all the particles. For this to be so, the symmetry of the spin and coordinate functions must be given by the same Young diagram, and the complete wave function ψ is expressed as definite bilinear combinations of the two; we shall not here pause to examine more closely the problem of constructing these combinations.

Next, suppose the spin of the particles to be half-integral. Then the complete wave function must be antisymmetrical with respect to all the particles. It can be shown that, for this to be so, the Young diagrams for the coordinate

† The independent functions that are transformed into linear combinations of one another form the basis of an irreducible representation of the permutation group. Their number is the dimension of the representation. For particles with spin $\frac{1}{2}$ the number is derived in Problem 1 below.

and spin functions must be in dual relation, i.e. obtained from each other by interchanging rows and columns (as in the two diagrams shown in Fig. 21).

Let us consider in more detail the important case of particles with spin $\frac{1}{2}$ (electrons, for instance). Each of the spin variables $\sigma_1, \sigma_2, \ldots$ here takes only the two values $\pm\frac{1}{2}$. Since a function antisymmetrical with respect to any two variables vanishes when these variables take the same value, it is clear that the function χ can be alternated only with respect to pairs of variables; if we alternate with respect to even three variables, two of them must always take the same value, so that we have zero identically.

Thus, for a system of electrons, the Young diagrams for the spin functions can contain columns of only one or two cells (i.e. only one or two rows); in the Young diagrams for the coordinate functions, the same is true of the number of columns. The number of possible types of permutational symmetry for a system of N electrons is therefore equal to the number of possible partitions of the number N into a sum of ones and twos. When N is even, this number is $\frac{1}{2}N+1$ (partitions with 0, 1, ... , $\frac{1}{2}N$ twos), while if N is odd it is $\frac{1}{2}(N+1)$ (partitions with 0, 1, ... , $\frac{1}{2}(N-1)$ twos). Thus, for instance, Fig. 22 shows the possible Young diagrams (coordinate and spin) for $N = 4$.

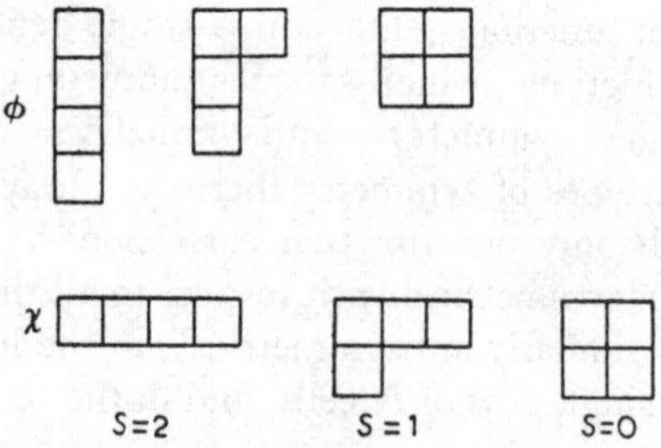

FIG. 22

It is easy to see that each of these types of symmetry (i.e. each of the Young diagrams) corresponds to a definite total spin S of the system of electrons. We shall consider the spin functions in spinor form, i.e. as spinors $\chi^{\lambda\mu\ldots}$ of rank N, whose indices (each of which corresponds to the spin of an individual particle) will be the variables that are arranged in the cells of the Young diagrams. Let us examine the Young diagram consisting of two rows with N_1 and N_2 cells ($N_1+N_2 = N$, and $N_1 \geqslant N_2$). In each of the first N_2 columns there are two cells, and the spinor must be antisymmetrical with respect to the corresponding pairs of indices. With respect to the indices in the last $n = N_1 - N_2$ cells in the first row, however, it must be symmetrical. As we know, such a spinor of rank N reduces to a symmetrical spinor of rank n, to which there corresponds a total spin $S = \frac{1}{2}n$. Returning to the Young diagrams for the coordinate functions, we can say that the diagram with n rows each of one cell corresponds to a total spin $S = \frac{1}{2}n$. For even N, the total spin can take integral values from 0 to $\frac{1}{2}N$, while for odd N it can take half-integral values from $\frac{1}{2}$ to $\frac{1}{2}N$, as it should.

We emphasize that this one-to-one correspondence between the Young diagrams and the total spin holds only for systems of particles with spin $\frac{1}{2}$; we have seen this, for a system of two particles, in the previous section. For a system of N particles with spin s, the spin wave function is made up of a product of N symmetrical spinors of rank $2s$, i.e. is a spinor of rank $2Ns$. If this spinor is symmetrized according to a particular Young diagram of N cells, we can usually construct from the independent components of the symmetrized spinor several sets of linear combinations, each set corresponding to a different total spin S of the system.

In the same way as the Young diagram for the spin functions of particles with spin $\frac{1}{2}$ cannot contain columns of more than two cells, so for particles with any spin s the columns cannot contain more than $2s+1$ cells.

If the number N of particles in the system is an integral multiple of $2s+1$, the possible Young diagrams include a rectangle with $2s+1$ cells in each column. This corresponds to one definite value of the total spin, $S = 0$. Hence we can conclude that the same value of S corresponds to any two (spin) Young diagrams which can be fitted together to form a rectangle of height $2s+1$.† This is a simple consequence of the fact that the addition of two angular momenta can give zero only if they have the same absolute magnitude.

To conclude this section, let us return to the fact already mentioned in the footnote at the end of §20 that, for a system of several identical particles, we cannot assert that the wave function of the stationary state of lowest energy is without nodes. We can now amplify this statement and elucidate its origin.

The wave function (that is, the coordinate function), if it has no nodes, must certainly be symmetrical with respect to all the particles; for, if it were antisymmetrical with respect to the interchange of any pair of particles 1, 2, it would vanish for $\mathbf{r}_1 = \mathbf{r}_2$. If, however, the system consists of three or more electrons, no completely symmetrical coordinate wave function is possible (the Young diagram of the coordinate function cannot have rows with more than two cells). Thus, although the solution of Schrödinger's equation which corresponds to the lowest eigenvalue is without nodes (by the theorem of the variational calculus), this solution may be physically inadmissible; the smallest eigenvalue of Schrödinger's equation will not then correspond to the normal state of the system, and the wave function of this state will in general have nodes. For particles with a half-integral spin s, this situation occurs in systems with more than $2s+1$ particles. For systems of bosons, a completely symmetrical coordinate wave function is always possible.

PROBLEMS

PROBLEM 1. Determine the number of energy levels with different values of the total spin S, for a system of N particles with spin $\frac{1}{2}$.

---

† For example, the two diagrams (for $s = 1$)

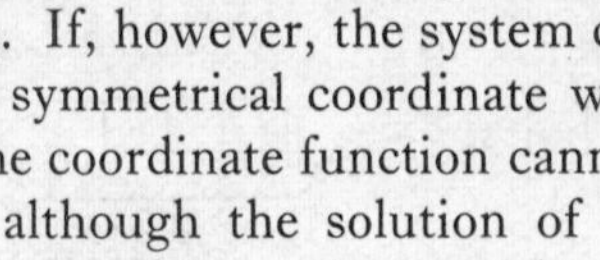

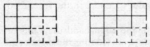

The continuous and broken lines show the complementary diagrams.

SOLUTION. A given value of the projection of the total spin of the system, $M_S = \Sigma\, \sigma$, can be obtained in

$$f(M_S) = N!/(\tfrac{1}{2}N+M_S)!(\tfrac{1}{2}N-M_S)!$$

ways, with $\frac{1}{2}N+M_S$ particles taken to have $\sigma = \frac{1}{2}$ and the remainder $\sigma = -\frac{1}{2}$. To each energy level with a given S, there correspond $2S+1$ states with $M_S = S, S-1, \ldots, -S$. Hence it is easy to see that the number of different energy levels with a given value of S is

$$n(S) = f(S)-f(S+1) = N!(2S+1)/(\tfrac{1}{2}N+S+1)!(\tfrac{1}{2}N-S)!.$$

The total number of different energy levels is

$$n = \sum_S n(S) = f(0) = N!/[(\tfrac{1}{2}N)!]^2$$

for even N, and

$$n = f(\tfrac{1}{2}) = N!/(\tfrac{1}{2}N+\tfrac{1}{2})!(\tfrac{1}{2}N-\tfrac{1}{2})!$$

for odd N.

PROBLEM 2. Find the values of the total spin S that occur for various types of symmetry of the spin functions of a system of two, three or four particles with spin 1.

SOLUTION. For two particles, the correspondence is established by the fact that the factor by which the spin function is multiplied when the particles are interchanged must be $(-1)^{2s-S}$ (see the end of §62). For particles with spin $s = 1$ this gives

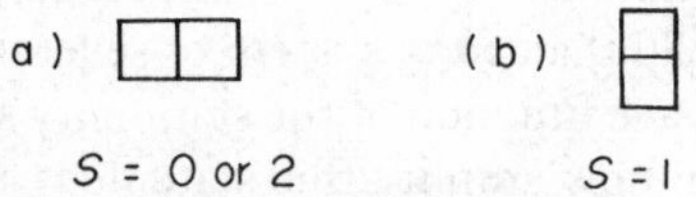

(1)

The Young diagrams for a system of three particles are obtained by adding to the diagrams (1) one cell in every possible way. The result may be written as the symbolic equations

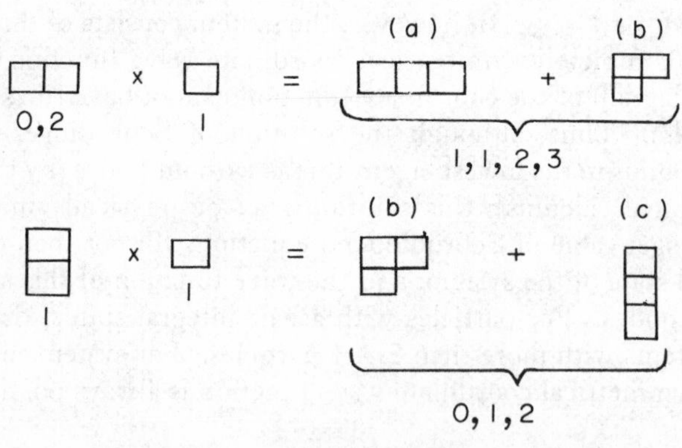

The values of S are shown beneath each diagram, and the values of the total spin of the system of three particles (the diagrams on the right) are found from the spins of the two-particle and one-particle systems (the diagrams on the left) by the rule of addition of angular momenta.† The distribution of the resulting values of S among the diagrams on the right is established by noting that diagram (c) (a column of three cells) corresponds to $S = 0$, and (b) therefore to the remaining values 1 and 2 in the second equation, while (a) belongs to the

† The repetition of 1 beneath the right-hand diagrams occurs because this value of the angular momentum comes firstly from adding the angular momenta 0 and 1, and secondly from adding 2 and 1.

values 1 and 3 that are left after (b) has been labelled in the first equation:

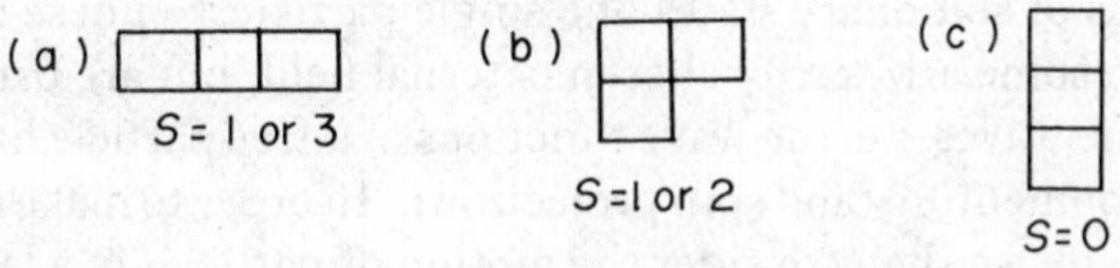

(2)

The Young diagrams for a system of four particles are obtained by adding one cell to the diagrams (2), with the condition that no column should contain more than three cells:

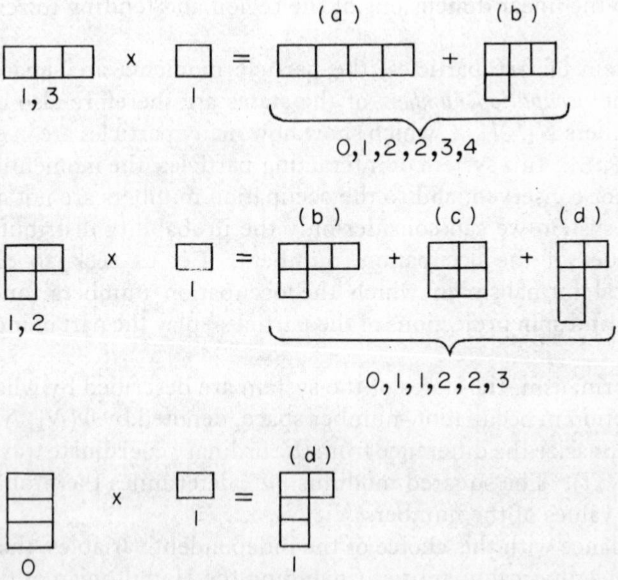

Diagram (c) can be added to (a) in (1) to form a rectangle with three-cell columns, and therefore corresponds to the same values $S = 0, 2$. The values of S for diagram (b) are found from the remainder of the second equation, and then those for (a) from the remainder of the first equation:

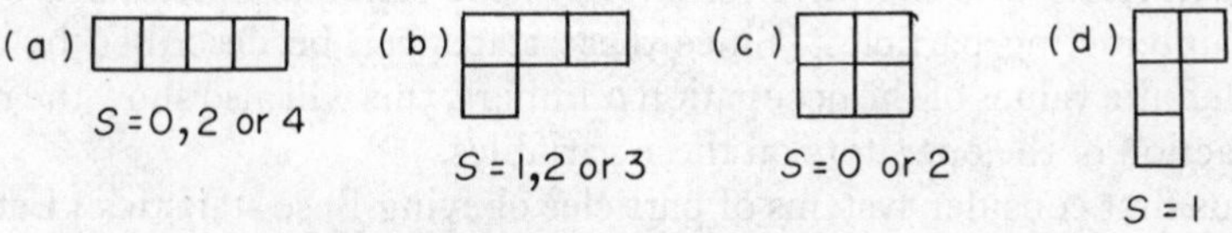

§64. Second quantization. The case of Bose statistics

In the theory of systems consisting of a large number of identical particles, there is a widely used method of considering the problem, known as *second quantization*. This method is especially necessary in relativistic theory, where we have to deal with systems in which the number of particles is itself variable.†

---

† The method of second quantization was developed by P. A. M. Dirac (1927) for photons in radiation theory, and later extended to fermions by E. Wigner and P. Jordan (1928).

Let $\psi_1(\xi)$, $\psi_2(\xi)$, ... be some complete set of orthogonal and normalized wave functions of stationary states of a single particle.† These may be states of a particle in some arbitrarily chosen external field, but are usually taken to be simply plane waves, i.e. the wave functions of a free particle having definite values of the momentum (and spin projection). In order to make the spectrum of states discrete, we shall consider the motion of particles in a large but finite region, for which the eigenvalues of the momentum components form a discrete series, the intervals between adjacent values being inversely proportional to the linear dimensions of the region and tending to zero as these increase.

In a system of free particles, the particle momenta are separately conserved. The *occupation numbers* of the states are therefore also conserved, i.e. the numbers N_1, N_2, ... which show how many particles are in each of the states ψ_1, ψ_2, ... In a system of interacting particles, the momentum of each particle is not conserved, and so the occupation numbers are not conserved. For such a system we can consider only the probability distribution of the various values of the occupation numbers. Let us seek to construct a mathematical formalism in which the occupation numbers (and not the coordinates and spin projections of the particles) play the part of independent variables.

In this formalism, the states of the system are described by what is called a wave function in occupation-number space, denoted by $\Phi(N_1, N_2, ...; t)$ in order to emphasize the difference from the ordinary coordinate wave function $\Psi(\xi_1, \xi_2, ...; t)$. The squared modulus $|\Phi|^2$ determines the probabilities of the various values of the numbers N_1, N_2, ...

In accordance with this choice of the independent variables, the operators of the various physical quantities (including the Hamiltonian of the system) must be formulated in terms of their action on functions of the occupation numbers. Such a formulation can be obtained on the basis of the usual matrix representation of operators. The operator matrix elements must be considered in relation to the wave functions of the stationary states of a system of non-interacting particles. Since these states can be described by specifying definite values of the occupation numbers, this will also show the nature of the action of the operators on these variables.

Let us first consider systems of particles obeying Bose statistics. Let $\hat{f}^{(1)}{}_a$ be the operator of some quantity pertaining to the ath particle, i.e. acting only on functions of the variables ξ_a. We introduce the operator

$$\hat{F}^{(1)} = \sum_a \hat{f}^{(1)}{}_a, \tag{64.1}$$

which is symmetrical with respect to all the particles (the summation being over all particles), and determine its matrix elements with respect to the wave functions (61.3). First of all, it is easy to see that the matrix elements will

† As in §61, ξ denotes the set of the coordinates and the spin projection σ of the particle, and integration with respect to ξ is taken to mean integration over the coordinates and summation over σ.

be different from zero only for transitions which leave the numbers $N_1, N_2, \ldots$ unchanged (diagonal elements) and for transitions where one of these numbers is increased, and another decreased, by unity. For, since each of the operators $\hat{f}^{(1)}_a$ acts only on one function in the product $\psi_{p_1}(\xi_1)\psi_{p_2}(\xi_2) \ldots \psi_{p_N}(\xi_N)$, its matrix elements can be different from zero only for transitions whereby the state of a single particle is changed; this, however, means that the number of particles in one state is diminished by unity, while the number in another state is correspondingly increased. The calculation of these matrix elements is in principle very simple; it is easier to do it oneself than to follow an account of it. Hence we shall give only the result of this calculation. The non-diagonal elements are

$$\langle N_i, N_k - 1|F^{(1)}|N_i - 1, N_k\rangle = f^{(1)}_{ik}\sqrt{(N_i N_k)}. \tag{64.2}$$

We shall indicate only those suffixes with respect to which the matrix element is non-diagonal, omitting the remainder for brevity. Here $f^{(1)}_{ik}$ is the matrix element

$$f^{(1)}_{ik} = \int \psi_i^*(\xi)\hat{f}^{(1)}\psi_k(\xi)\, d\xi; \tag{64.3}$$

since the operators $\hat{f}^{(1)}_a$ differ only in the naming of the variables on which they act, the integrals (64.3) are independent of a, which is therefore omitted. The diagonal matrix elements of $F^{(1)}$ are the mean values of the quantity $F^{(1)}$ in the states $\Psi_{N_1 N_2 \ldots}$. Calculation gives

$$\overline{F^{(1)}} = \sum_i f^{(1)}_{ii} N_i. \tag{64.4}$$

We now introduce the operators $\hat{a}_i$, which play a leading part in the method of second quantization; they act, not on functions of the coordinates, but on functions of the occupation numbers. By definition, the operator $\hat{a}_i$ acting on the function $\Phi(N_1, N_2, \ldots)$ decreases the value of the variable N_i by unity, and at the same time it multiplies the function by $\sqrt{N_i}$:

$$\hat{a}_i\Phi(N_1, N_2, \ldots, N_i, \ldots) = \sqrt{N_i}\Phi(N_1, N_2, \ldots, N_i - 1, \ldots). \tag{64.5}$$

We can say that the operator $\hat{a}_i$ diminishes by one the number of particles in the ith state; it is therefore called a particle *annihilation operator*. It can be represented in the form of a matrix whose only non-zero element is

$$\langle N_i - 1|a_i|N_i\rangle = \sqrt{N_i}. \tag{64.6}$$

The operator $\hat{a}_i^+$ which is the Hermitian conjugate of $\hat{a}_i$ is, by definition (see (11.9)), represented by a matrix whose only non-zero element is

$$\langle N_i|a_i^+|N_i - 1\rangle = \langle N_i - 1|a_i|N_i\rangle^* = \sqrt{N_i}. \tag{64.7}$$

This means that, when acting on the function $\Phi(N_1, N_2, \ldots)$, it increases the number N_i by unity:

$$\hat{a}_i\Phi(N_1, N_2, \ldots, N_i, \ldots) = \sqrt{(N_i+1)}\Phi(N_1, N_2, \ldots, N_i+1, \ldots). \quad (64.8)$$

In other words, the operator $\hat{a}_i^+$ increases by one the number of particles in the ith state, and is therefore called a particle *creation operator*.

The product of the operators $\hat{a}_i^+\hat{a}_i$, acting on the wave function, must multiply it by a constant simply, leaving unchanged all the variables N_1, N_2, ...: the operator $\hat{a}_i$ diminishes N_i by unity, and $\hat{a}_i^+$ then restores it to its original value. Direct multiplication of the matrices (64.6) and (64.7) shows that $\hat{a}_i^+\hat{a}_i$ is represented, as we should expect, by a diagonal matrix whose diagonal elements are N_i. We can write

$$\hat{a}_i^+\hat{a}_i = N_i. \quad (64.9)$$

Similarly, we find that

$$\hat{a}_i\hat{a}_i^+ = N_i+1. \quad (64.10)$$

The difference of these equations gives the commutation rule for the operators $\hat{a}_i$ and $\hat{a}_i^+$:

$$\hat{a}_i\hat{a}_i^+ - \hat{a}_i^+\hat{a}_i = 1. \quad (64.11)$$

The operators with i and k different act on different variables (N_i and N_k), and commute:

$$\hat{a}_i\hat{a}_k - \hat{a}_k\hat{a}_i = 0, \quad \hat{a}_i\hat{a}_k^+ - \hat{a}_k^+\hat{a}_i = 0 \quad (i \neq k). \quad (64.12)$$

From the above properties of the operators $\hat{a}_i$, $\hat{a}_i^+$ it is easy to see that the operator

$$\hat{F}^{(1)} = \sum_{i,k} f^{(1)}_{ik}\hat{a}_i^+\hat{a}_k \quad (64.13)$$

is the same as the operator (64.1). For all the matrix elements calculated from (64.6), (64.7) are the same as the elements (64.2), (64.4). This is a very important result. In formula (64.13), the quantities $f^{(1)}_{ik}$ are simply numbers. Thus we have been able to express an ordinary operator, acting on functions of the coordinates, in the form of an operator acting on functions of new variables, the occupation numbers N_i.

The result which we have obtained is easily generalized to operators of other forms. Let

$$\hat{F}^{(2)} = \sum_{a>b} \hat{f}^{(2)}_{ab}, \quad (64.14)$$

where $f^{(2)}_{ab}$ is the operator of a physical quantity pertaining to two particles at once, and hence acts on functions of ξ_a and ξ_b. Similar calculations show that this operator can be expressed in terms of the operators $\hat{a}_i$, $\hat{a}_i^+$ by

$$\hat{F}^{(2)} = \tfrac{1}{2} \sum_{i,k,l,m} \langle ik|f^{(2)}|lm\rangle \hat{a}_i^+ \hat{a}_k^+ \hat{a}_m \hat{a}_l, \tag{64.15}$$

where

$$\langle ik|f^{(2)}|lm\rangle = \iint \psi_i^*(\xi_1)\psi_k^*(\xi_2)\hat{f}^{(2)}\psi_l(\xi_1)\psi_m(\xi_2)\, d\xi_1 d\xi_2.$$

The generalization of these formulae to operators of any other form symmetrical with respect to all the particles (of the form $\hat{F}^{(3)} = \sum \hat{f}^{(3)}_{abc}$, etc.) is obvious.

These formulae can be used to express, in terms of the operators $\hat{a}_i$ and $\hat{a}_i^+$, the Hamiltonian of the physical system of N identical interacting particles that is being considered. The Hamiltonian of such a system is, of course, symmetrical with respect to all the particles. In the non-relativistic approximation,† it is independent of the spins of the particles, and can be represented in a general form as follows:

$$\hat{H} = \sum_a \hat{H}^{(1)}_a + \sum_{a>b} U^{(2)}(\mathbf{r}_a, \mathbf{r}_b) + \sum_{a>b>c} U^{(3)}(\mathbf{r}_a, \mathbf{r}_b, \mathbf{r}_c) + \dots. \tag{64.16}$$

Here $\hat{H}^{(1)}_a$ is the part of the Hamiltonian which depends on the coordinates of the ath particle only:

$$\hat{H}^{(1)}_a = -(\hbar^2/2m)\triangle_a + U^{(1)}(\mathbf{r}_a), \tag{64.17}$$

where $U^{(1)}(\mathbf{r}_a)$ is the potential energy of a single particle in the external field. The remaining terms in (64.16) correspond to the mutual interaction energy of the particles; the terms depending on the coordinates of two, three, etc. particles have been separated.

This representation of the Hamiltonian enables us to apply formulae (64.13), (64.15) and their analogues directly. Thus

$$\hat{H} = \sum_{i,k} H^{(1)}_{ik} \hat{a}_i^+ \hat{a}_k + \tfrac{1}{2} \sum_{i,k,l,m} \langle ik|U^{(2)}|lm\rangle \hat{a}_i^+ \hat{a}_k^+ \hat{a}_m \hat{a}_l + \dots. \tag{64.18}$$

This gives the required expression for the Hamiltonian in the form of an operator acting on functions of the occupation numbers.

For a system of non-interacting particles, only the first term in the expression (64.18) remains:

$$\hat{H} = \sum_{i,k} H^{(1)}_{ik} \hat{a}_i^+ \hat{a}_k. \tag{64.19}$$

If the functions ψ_i are taken to be the eigenfunctions of the Hamiltonian $\hat{H}^{(1)}$ of an individual particle, the matrix $H^{(1)}_{ik}$ is diagonal, and its diagonal elements are the eigenvalues ϵ_i of the energy of the particle. Thus

$$\hat{H} = \sum_i \epsilon_i \hat{a}_i^+ \hat{a}_i;$$

† In the absence of a magnetic field.

replacing the operator $\hat{a}_i^+\hat{a}_i$ by its eigenvalues (64.9), we have for the energy levels of the system the expression

$$E = \sum_i \epsilon_i N_i,$$

a trivial result which could have been foreseen.

The formalism which we have developed can be put in a more compact form by introducing the *ψ-operators*†

$$\hat{\psi}(\xi) = \sum_i \psi_i(\xi)\hat{a}_i, \quad \hat{\psi}^+(\xi) = \sum_i \psi_i^*(\xi)\hat{a}_i^+, \tag{64.20}$$

where the variables ξ are regarded as parameters. By what has been said above concerning the operators $\hat{a}_i$, $\hat{a}_i^+$, it is clear that the operator $\hat{\psi}$ decreases the total number of particles in the system by one, while $\hat{\psi}^+$ increases it by one.

It is easy to see that the operator $\hat{\psi}^+(\xi_0)$ creates a particle at the point ξ_0. For the result of the action of the operator $\hat{a}_i^+$ is to create a particle in a state with wave function $\psi_i(\xi)$. Hence it follows that the result of the action of the operator $\hat{\psi}^+(\xi_0)$ is to create a particle in a state with wave function $\sum \psi_i^*(\xi)\psi_i(\xi_0) = \delta(\xi-\xi_0)$, which corresponds to a particle with definite values of the coordinates (and spin). Here we have used formula (5.12).‡

The commutation rules for the ψ operators are obtained at once from those for the operators $\hat{a}_i$, $\hat{a}_i^+$:

$$\hat{\psi}(\xi)\hat{\psi}(\xi') - \hat{\psi}(\xi')\hat{\psi}(\xi) = 0, \tag{64.21}$$

$$\hat{\psi}(\xi)\hat{\psi}^+(\xi') - \hat{\psi}^+(\xi')\hat{\psi}(\xi) = \sum_i \psi_i(\xi)\psi_i^*(\xi') = \delta(\xi-\xi'). \tag{64.22}$$

The second-quantized operator $\hat{F}^{(1)}$ can be written by means of the ψ operators in the form

$$\hat{F}^{(1)} = \int \hat{\psi}^+(\xi) f^{(1)} \hat{\psi}(\xi)\, d\xi \tag{64.23}$$

where it is understood that the operator $f^{(1)}$ acts on functions of the parameters ξ in $\hat{\psi}(\xi)$. For, substituting $\hat{\psi}$ and $\hat{\psi}^+$ in the form (64.20) and using the definition (64.3), we return to (64.13). Similarly, (64.15) becomes

$$\hat{F}^{(2)} = \tfrac{1}{2}\iint \hat{\psi}^+(\xi)\hat{\psi}^+(\xi') f^{(2)} \hat{\psi}(\xi')\hat{\psi}(\xi)\, d\xi\, d\xi'. \tag{64.24}$$

† Note the analogy between (64.20) and the expansion

$$\psi = \Sigma a_i \psi_i$$

of any wave function in terms of a complete set of functions. Here it is "re-quantized", and this is the reason for the term *second quantization method*.

‡ $\delta(\xi-\xi_0)$ conventionally denotes the product

$$\delta(x-x_0)\delta(y-y_0)\delta(z-z_0)\delta_{\sigma\sigma_0}.$$

In particular, the Hamiltonian of the system, expressed in terms of the ψ operators, is

$$\hat{H} = \int \left\{ -\frac{\hbar^2}{2m} \hat{\psi}^+(\xi) \triangle \hat{\psi}(\xi) + \hat{\psi}^+(\xi) U^{(1)}(\xi) \hat{\psi}(\xi) \right\} d\xi$$
$$+ \tfrac{1}{2} \iint \hat{\psi}^+(\xi') \psi^+(\xi) U^{(2)}(\xi, \xi') \hat{\psi}(\xi') \hat{\psi}(\xi) \, d\xi \, d\xi' + \ldots . \quad (64.25)$$

The operator $\hat{\psi}^+(\xi)\hat{\psi}(\xi)$, constructed from the ψ operators by analogy with the product $\psi^*\psi$ which determines the probability density for a particle in a state with wave function ψ, is called the *particle density operator*. The integral

$$\hat{N} = \int \hat{\psi}^+ \hat{\psi} \, d\xi \quad (64.26)$$

represents in the second-quantization formalism the operator of the total number of particles in the system. For, substituting the ψ operators in the form (64.20) and using the normalization and the orthogonality of the wave functions, we have

$$\hat{N} = \Sigma \, \hat{a}_i^+ \hat{a}_i.$$

Each term in this sum is the operator of the number of particles in the ith state; according to (64.9), its eigenvalues are equal to the occupation numbers N_i, and the sum of all these numbers is the total number of particles in the system.†

Lastly, if the system consists of bosons of various kinds, operators $\hat{a}$ and $\hat{a}^+$ for each kind of particle must be defined in the second quantization method. It is evident that operators pertaining to particles of different kinds commute.

§65. Second quantization. The case of Fermi statistics

The basic theory of the method of second quantization remains wholly unchanged for systems of identical fermions, but the actual formulae for the matrix elements of quantities and for the operators $\hat{a}_i$ are naturally different.

The wave function $\Psi_{N_1 N_2 \ldots}$ now has the form (61.5). Because of the antisymmetry of this function, the question of its sign arises first of all. This question did not arise in the case of Bose statistics, since, because of the symmetry of the wave function, its sign, once chosen, was preserved under all permutations of the particles. In order to make definite the sign of the function (61.5), we shall agree to choose it as follows. We number successively, once and for all, all the states ψ_i. We then complete the rows

† For systems containing a specified number of particles these statements are trivial, as are the properties of the Hamiltonian (64.19) of a system of free particles. Their generalization in the relativistic theory, however, yields new results that are by no means trivial (cf. *RQT*, §11).

of the determinant (61.5) so that always

$$p_1 < p_2 < p_3 < \dots < p_N, \tag{65.1}$$

whilst in the successive columns we have functions of the different variables in the order $\xi_1, \xi_2, \dots, \xi_N$. No two of the numbers $p_1, p_2, \dots$ can be equal, since otherwise the determinant would vanish. In other words, the occupation numbers N_i can take only the values 0 and 1.

Let us again consider an operator of the form (64.1), $\hat{F}^{(1)} = \Sigma \hat{f}^{(1)}{}_a$. As in §64, its matrix elements will be non-zero only for transitions where all the occupation numbers remain unchanged and for those where one occupation number (N_i) is diminished by unity (becoming zero instead of one) and another (N_k) is increased by unity (becoming one instead of zero). We easily find that, for $i < k$,

$$\langle 1_i, 0_k | F^{(1)} | 0_i, 1_k \rangle = f^{(1)}{}_{ik}(-1)^{\Sigma(i+1,\, k-1)}, \tag{65.2}$$

where by 0_i, 1_i we signify $N_i = 0$, $N_i = 1$ and the symbol $\Sigma(k, l)$ denotes the sum of the occupation numbers of all states from the kth to the lth:†

$$\Sigma(k, l) = \sum_{n=k}^{l} N_n.$$

For the diagonal elements we obtain our previous formula (64.4):

$$\overline{F^{(1)}} = \sum_i f^{(1)}{}_{ii} N_i. \tag{65.3}$$

In order to represent the operator $\hat{F}^{(1)}$ in the form (64.13), the operators $\hat{a}_i$ must be defined as matrices whose elements are

$$\langle 0_i | a_i | 1_i \rangle = \langle 1_i | a_i{}^+ | 0_i \rangle = (-1)^{\Sigma(1,\, i-1)}. \tag{65.4}$$

On multiplying these matrices, we find, for $k > i$,

$$\langle 1_i, 0_k | a_i{}^+ a_k | 0_i, 1_k \rangle = \langle 1_i, 0_k | a_i{}^+ | 0_i, 0_k \rangle \langle 0_i, 0_k | a_k | 0_i, 1_k \rangle$$

$$= (-1)^{\Sigma(1,\, i-1)} (-1)^{\Sigma(1,\, i-1) + \Sigma(i+1,\, k-1)},$$

or

$$\langle 1_i, 0_k | a_i{}^+ a_k | 0_i, 1_k \rangle = (-1)^{\Sigma(i+1,\, k-1)}. \tag{65.5}$$

If $i = k$, the matrix of $\hat{a}_i{}^+ \hat{a}_i$ is diagonal, and its elements are unity for $N_i = 1$, and zero for $N_i = 0$; this can be written

$$\hat{a}_i{}^+ \hat{a}_i = N_i. \tag{65.6}$$

† For $i > k$ the exponent in (65.2) becomes $\Sigma(k+1, i-1)$. The sum must be taken as zero when $i = k \pm 1$.

On substituting these expressions in (64.13), we in fact obtain (65.2), (65.3).

Multiplying $\hat{a}_i^+$, $\hat{a}_k$ in the opposite order, we have

$$\langle 1_i, 0_k|a_k a_i^+|0_i, 1_k\rangle = \langle 1_i, 0_k|a_k|1_i, 1_k\rangle\langle 1_i, 1_k|a_i^+|0_i, 1_k\rangle$$
$$= (-1)^{\Sigma(1,\, i-1)+\Sigma(i+1,\, k-1)+\Sigma(1,\, i-1)+1},$$

or

$$\langle 1_i, 0_k|a_k a_i^+|0_i, 1_k\rangle = -(-1)^{\Sigma(i+1,\, k-1)}. \tag{65.7}$$

Comparing (65.7) with (65.5), we see that these quantities have opposite signs, i.e.

$$\hat{a}_i^+\hat{a}_k+\hat{a}_k\hat{a}_i^+ = 0 \qquad (i \neq k).$$

For the diagonal matrix $\hat{a}_i\hat{a}_i^+$, we find

$$\hat{a}_i\hat{a}_i^+ = 1-N_i. \tag{65.8}$$

Adding this to (65.6), we obtain

$$\hat{a}_i\hat{a}_i^+ + \hat{a}_i^+\hat{a}_i = 1.$$

Both the above equations can be written in the form

$$\hat{a}_i\hat{a}_k^+ + \hat{a}_k^+\hat{a}_i = \delta_{ik}. \tag{65.9}$$

On carrying out similar calculations, we find for the products $\hat{a}_i\hat{a}_k$ the relations

$$\hat{a}_i\hat{a}_k+\hat{a}_k\hat{a}_i = 0, \tag{65.10}$$

and in particular $\hat{a}_i\hat{a}_i = 0$.

Thus we see that the operators $\hat{a}_i$ and $\hat{a}_k$ (or $\hat{a}_k^+$) for $i \neq k$ anticommute, whereas in the case of Bose statistics they commuted with one another. This difference is perfectly natural. In the case of Bose statistics, the operators $\hat{a}_i$ and $\hat{a}_k$ were completely independent; each of the operators $\hat{a}_i$ acted only on a single variable N_i, and the result of this action did not depend on the values of the other occupation numbers. In the case of Fermi statistics, however, the result of the action of the operator $\hat{a}_i$ depends not only on the number N_i itself, but also on the occupation numbers of all the preceding states, as we see from the definition (65.4). Hence the action of the various operators $\hat{a}_i$, $\hat{a}_k$ cannot be considered independent.

The properties of the operators $\hat{a}_i$, $\hat{a}_i^+$ having been thus defined, all the remaining formulae (64.13)–(64.18) remain valid. The formulae (64.23)–(64.25), which express the operators of physical quantities in terms of the ψ-operators defined by (64.20), also hold good. The commutation rules (64.21), (64.22), however, are now replaced by

$$\hat{\psi}^+(\xi')\hat{\psi}(\xi)+\hat{\psi}(\xi)\hat{\psi}^+(\xi') = \delta(\xi-\xi'), \tag{65.11}$$

$$\hat{\psi}(\xi')\hat{\psi}(\xi)+\hat{\psi}(\xi)\hat{\psi}(\xi') = 0. \tag{65.12}$$

If the system consists of particles of different kinds, second quantization operators must be defined for each kind of particle (as already mentioned at the end of §64). Operators belonging to bosons and fermions commute; those belonging to different fermions may formally be regarded as either commutative or anticommutative within the limits of non-relativistic theory. On either assumption the results obtained by means of the second quantization method are the same.

However, with a view to later applications in the relativistic theory, which allows different particles to be transformed into one another, we should assume that the creation and annihilation operators for different fermions anticommute. This becomes evident if we regard as "different" particles two different internal states of a single complex particle.

CHAPTER X

THE ATOM

§66. Atomic energy levels

In the non-relativistic approximation, the stationary states of the atom are determined by Schrödinger's equation for the system of electrons, which move in the Coulomb field of the nucleus and interact electrically with one another; the spin operators of the electrons do not appear in this equation. As we know, for a system of particles in a centrally symmetric external field the total orbital angular momentum L and the parity of the state are conserved. Hence each stationary state of the atom will be characterized by a definite value of the orbital angular momentum L and by its parity. Moreover, the coordinate wave functions of the stationary states of a system of identical particles have a certain permutational symmetry. We have seen in §63 that, for a system of electrons, a definite value of the total spin of the system corresponds to each type of permutational symmetry (i.e. to each Young diagram). Hence every stationary state of the atom is characterized also by the total spin S of the electrons.

The energy level having given values of S and L is degenerate to a degree equal to the number of different possible directions in space of the vectors **S** and **L**. The degree of the degeneracy from the directions of **L** and **S** is respectively $2L+1$ and $2S+1$. Consequently, the total degree of the degeneracy of a level with given L and S is equal to the product $(2L+1)(2S+1)$.

In fact, however, the electromagnetic interaction of the electrons contains relativistic effects, which depend on their spins. These effects have the result that the energy of the atom depends not only on the absolute magnitudes of the vectors **L** and **S** but also on their relative positions. Strictly speaking, when the relativistic interactions are taken into account the orbital angular momentum **L** and the spin **S** of the atom are not separately conserved. Only the total angular momentum $\mathbf{J} = \mathbf{L} + \mathbf{S}$ is conserved; this is a universal and exact law which follows from the isotropy of space relative to a closed system. For this reason, the exact energy levels must be characterized by the values J of the total angular momentum.

However, if the relativistic effects are comparatively small (as often happens), they can be allowed for as a perturbation. Under the action of this perturbation, a degenerate level with given L and S is "split" into a number of distinct (though close) levels, which differ in the value of the total angular momentum J. These levels are determined (in the first approximation) by the appropriate secular equation (§39), while their wave functions (in the zeroth approximation) are definite linear combinations of the wave functions of the initial degenerate level with the given L and S. In this approximation we can therefore, as before, regard the absolute values of the orbital angular momentum and spin (but not their directions) as being conserved, and characterize the levels by the values of L and S also.

Thus, as a result of the relativistic effects, a level with given values of L and S is split into a number of levels with different values of J. This splitting is called the *fine structure* (or the *multiplet splitting*) of the level. As we know, J takes values from $L+S$ to $|L-S|$; hence a level with given L and S is split into $2S+1$ (if $L > S$) or $2L+1$ (if $L < S$) distinct levels. Each of these is still degenerate with respect to the directions of the vector $\mathbf{J}$; the degree of this degeneracy is $2J+1$. It is easily verified that the sum of the numbers $2J+1$ for all possible values of J is equal to $(2L+1)(2S+1)$, as it should be.

There is a generally accepted notation to denote the atomic energy levels (or, as they are called, the *spectral terms* of the atoms), similar to that used for the states of individual particles with definite values of the angular momentum (§32): states with different values of the total orbital angular momentum L are denoted by capital Latin letters, as follows:

$$\begin{array}{lcccccccccccc} L = & 0 & 1 & 2 & 3 & 4 & 5 & 6 & 7 & 8 & 9 & 10 & \dots \\ & S & P & D & F & G & H & I & K & L & M & N & \dots \end{array}$$

Above and to the left of this letter is placed the number $2S+1$, called the *multiplicity* of the term (though it must be borne in mind that this number is the number of fine-structure components of the level only when $L \geqslant S$).† Below and to the right of the letter is placed the value of the total angular momentum J. Thus the symbols ${}^2P_{1/2}$, ${}^2P_{3/2}$ denote levels with $L = 1$, $S = \frac{1}{2}$, $J = \frac{1}{2}$ and $\frac{3}{2}$.

§67. Electron states in the atom

An atom with more than one electron is a complex system of mutually interacting electrons moving in the field of the nucleus. For such a system we can, strictly speaking, consider only states of the system as a whole. Nevertheless, it is found that we can, with fair accuracy, introduce the idea of the states of each individual electron in the atom, as being the stationary states of the motion of each electron in some effective centrally symmetric field due to the nucleus and to all the other electrons. These fields are in general different for different electrons in the atom, and they must all be defined simultaneously, since each of them depends on the states of all the other electrons. Such a field is said to be *self-consistent*.

Since the self-consistent field is centrally symmetric, each state of the electron is characterized by a definite value of its orbital angular momentum l. The states of an individual electron with a given l are numbered (in order of increasing energy) by the ***principal quantum number n***, which takes the values $n = l+1, l+2, \dots$; this choice of the order of numbering is made in

† The levels with $2S+1 = 1, 2, 3, \dots$ are called *singlet, doublet, triplet*, etc., levels.

accordance with what is usual for the hydrogen atom. However, the sequence of levels of increasing energy for various l in complex atoms is in general different from that found in the hydrogen atom. In the latter, the energy is independent of l, so that the states with larger values of n always have higher energies. In complex atoms, on the other hand, the level with $n = 5$, $l = 0$, for example, is found to lie below that with $n = 4$, $l = 2$ (this is discussed in more detail in §73).

The states of individual electrons with different values of n and l are customarily denoted by a figure which gives the value of the principal quantum number, followed by a letter which gives the value of l:† thus $4d$ denotes the state with $n = 4$, $l = 2$. A complete description of the atom demands that, besides the values of the total L, S and J, the states of all the electrons should also be enumerated. Thus the symbol $1s\ 2p\ {}^3P_0$ denotes a state of the helium atom in which $L = 1$, $S = 1$, $J = 0$ and the two electrons are in the $1s$ and $2p$ states. If several electrons are in states with the same l and n, this is usually shown for brevity by means of an index: thus $3p^2$ denotes two electrons in the $3p$ state. The distribution of the electrons in the atom among states with different l and n is called the *electron configuration.*

For given values of n and l, the electron can have different values of the projections of the orbital angular momentum (m) and of the spin (σ) on the z-axis. For a given l, the number m takes $2l+1$ values; the number σ is restricted to only two values, $\pm\frac{1}{2}$. Hence there are altogether $2(2l+1)$ different states with the same n and l; these states are said to be *equivalent.* According to Pauli's principle there can be only one electron in each such state. Thus at most $2(2l+1)$ electrons in an atom can simultaneously have the same n and l. An assembly of electrons occupying all the states with the given n and l is called a *closed shell* of the type concerned.

The difference in energy between atomic levels having different L and S but the same electron configuration‡ is due to the electrostatic interaction of the electrons. These energy differences are usually small, and several times less than the distances between the levels of different configurations. The following empirical principle (*Hund's rule*; F. Hund 1925) is known concerning the relative position of levels with the same configuration but different L and S:

The term with the greatest possible value of S (for the given electron configuration) and the greatest possible value of L (for this S) has the lowest energy.‖

We shall show how the possible atomic terms can be found for a given electron configuration. If the electrons are not equivalent, the possible values

---

† Another terminology often used is that in which electrons with principal quantum numbers $n = 1, 2, 3, \ldots$ are said to belong to the $K, L, M, \ldots$ shells (see §74).

‡ We here ignore the fine structure of each multiplet level.

‖ The requirement that S should be as large as possible can be explained as follows. Let us consider, for example, a system of two electrons. Here we can have $S = 0$ or $S = 1$; the spin 1 corresponds to an antisymmetrical coordinate wave function $\phi(\mathbf{r}_1, \mathbf{r}_2)$. For $\mathbf{r}_1 = \mathbf{r}_2$, this function vanishes; in other words, in the state with $S = 1$ the probability of finding the two electrons close together is small. This means that their electrostatic repulsion is comparatively small, and hence the energy is less. Similarly, for a system of several electrons, the "most antisymmetrical" coordinate wave function corresponds to the greatest spin.

of L and S are determined immediately from the rule for the addition of angular momenta. Thus, for instance, with the configurations np, $n'p$ (n, n' being different) the total angular momentum L can take the values 2, 1, 0, and the total spin $S = 0, 1$; combining these, we obtain the terms $^{1,3}S$, $^{1,3}P$, $^{1,3}D$.

If we are concerned with equivalent electrons, however, restrictions imposed by Pauli's principle make their appearance. Let us consider, for example, a configuration of three equivalent p electrons. For $l = 1$ (the p state), the projection m of the orbital angular momentum can take the values $m = 1, 0, -1$, so that there are six possible states, with the following values of m and σ:

$$(a)\ 1, \tfrac{1}{2} \quad (b)\ 0, \tfrac{1}{2} \quad (c)\ -1, \tfrac{1}{2}$$

$$(a')\ 1, -\tfrac{1}{2} \quad (b')\ 0, -\tfrac{1}{2} \quad (c')\ -1, -\tfrac{1}{2}.$$

The three electrons can be one in each of any three of these states. As a result we obtain states of the atom with the following values of the projections $M_L = \Sigma m$, $M_S = \Sigma \sigma$ of the total orbital angular momentum and spin:

$$(a+a'+b)\ 2, \tfrac{1}{2} \quad (a+a'+c)\ 1, \tfrac{1}{2} \quad (a+b+c)\ 0, \tfrac{3}{2}$$

$$(a+b+b')\ 1, \tfrac{1}{2} \quad (a+b+c')\ 0, \tfrac{1}{2}$$

$$(a+b'+c)\ 0, \tfrac{1}{2}$$

$$(a'+b+c)\ 0, \tfrac{1}{2}.$$

The states with M_L or M_S negative need not be written out, since they give nothing different. The presence of a state with $M_L = 2$, $M_S = \frac{1}{2}$ shows that there must be a ^2D term, and to this term there must correspond one state $(1, \frac{1}{2})$ and one $(0, \frac{1}{2})$. Next, there remains one state with $(1, \frac{1}{2})$, so that there must be a ^2P term; one of the states $(0, \frac{1}{2})$ corresponds to this. Finally, there remain the states $(0, \frac{3}{2})$ and $(0, \frac{1}{2})$, corresponding to a ^4S term. Thus, for a configuration of three equivalent p electrons, the only possibilities are one term of each of the types ^2D, ^2P, ^4S.

Table 1 gives the possible terms for various configurations of equivalent p and d electrons. The figures below the letters of the terms show the number of terms of the type concerned that exist for the given configuration, if this number is more than one. For the configuration with the greatest possible number of equivalent electrons (s^2, p^6, d^{10}, ...), the term is always ^1S. Like terms always correspond to configurations which differ in that one of them has as many electrons as the other lacks to form a closed shell. This is an evident result of the fact that the absence of an electron from the shell can be regarded as a "hole", whose state is defined by the same quantum numbers as the state of the missing electron.

When Hund's rule is applied to determine the ground term of an atom from a known electron configuration, only the unfilled shell need be considered, since the moments of electrons in closed shells cancel out. For

TABLE 1

Possible terms for configurations of equivalent electrons

p, p^5	${}^2P$		
p^2, p^4	${}^1SD$	${}^3P$	
p^3	${}^2PD$	${}^4S$	
d, d^9	${}^2D$		
d^2, d^8	${}^1SDG$	${}^3PF$	
d^3, d^7	${}^2P\underset{2}{D}FGH$	${}^4PF$	
d^4, d^6	${}^1\underset{2}{S}\underset{2}{D}F\underset{2}{G}I$	${}^3\underset{2}{P}D\underset{2}{F}GH$	${}^5D$
d^5	${}^2SP\underset{3}{D}\underset{2}{F}\underset{2}{G}HI$	${}^4PDFG$	${}^6S$

example, let there be four d electrons outside the closed shells in an atom. The magnetic quantum number of the d electron can take five values: $0, \pm 1, \pm 2$. Hence all four electrons can have the same spin component $\sigma = \frac{1}{2}$, and the maximum possible total spin is $S = 2$. We must then assign to the electrons different values of m so as to give the maximum value of $M_L = \Sigma m$, namely 2, 1, 0, -1, $M_L = 2$. This means that the maximum value of L for $S = 2$ is also 2, and the term is ${}^5D$.

PROBLEM

Find the orbital wave functions of the possible states of a system of three equivalent p electrons.

SOLUTION. In the states ${}^4S$ the spin projections σ of all the electrons are the same, and the values of m are therefore different. The wave function is given by a determinant of the form (61.5) composed of the functions ψ_0, ψ_1, ψ_{-1} (where the suffix shows the value of m).

For the ${}^2D$ term we consider the state with the maximum possible value $M_L = 2$. Two of the components m will be 1 and the other 0. Let electrons 2 and 3 have $\sigma = +\frac{1}{2}$ and electron 1 have $\sigma = -\frac{1}{2}$ (corresponding to total spin $S = \frac{1}{2}$). The orbital wave function having the required symmetry is

$$\psi = \frac{1}{\sqrt{2}}\psi_1(1)[\psi_0(2)\psi_1(3) - \psi_0(3)\psi_1(2)],$$

the argument of each function ψ being the number of the electron to which it refers.

For the ${}^2P$ term we consider the state with $M_L = 1$ and the same values of the electron spin components as previously. This state can be obtained with two different sets of values of m, so that the orbital wave function is given by the linear combination

$$\psi = a\psi_{-111} + b\psi_{100},$$

$$\psi_{-111} = \psi_1(1)[\psi_{-1}(2)\psi_1(3) - \psi_{-1}(3)\psi_1(2)],$$

$$\psi_{100} = \psi_0(1)[\psi_1(2)\psi_0(3) - \psi_1(3)\psi_0(2)].$$

To determine the coefficients, we use the relation

$$\hat{L}_+\psi = (\hat{l}_+^{(1)} + \hat{l}_+^{(2)} + \hat{l}_+^{(3)})\psi = 0,$$

which must be satisfied by the wave function with $M_L = L$ (see (27.8)). Using the matrix elements (27.12), we find that

$$\hat{l}_+\psi_1 = 0, \quad \hat{l}_+\psi_{-1} = \sqrt{2}\psi_0, \quad \hat{l}_+\psi_0 = \sqrt{2}\psi_1,$$

and so

$$\hat{L}_+\psi = \sqrt{2}(a-b)\psi_{011} = 0.$$

Hence $a-b = 0$, and using also the normalization condition, we have $a = b = \frac{1}{2}$.

The wave functions of states with $M_L < L$ are obtained from those found above by applying to them the operator $\hat{L}_-$.

§68. Hydrogen-like energy levels

The only atom for which Schrödinger's equation can be exactly solved is the simplest of all atoms, that of hydrogen. The energy levels of the hydrogen atom, and of the ions He^+, Li^{++}, ... which each have only one electron, are given by Bohr's formula (36.10)

$$E = -\frac{mZ^2e^4}{2\hbar^2(1+m/M)} \cdot \frac{1}{n^2}. \tag{68.1}$$

Here Ze is the charge on the nucleus, M its mass, and m the mass of the electron. We notice that the dependence on the mass of the nucleus is only very slight.

The formula (68.1) does not take account of any relativistic effects. In this approximation there is an additional (*accidental*) degeneracy, peculiar to the hydrogen atom, of which we have already spoken in §36; for a given principal quantum number n, the energy is independent of the orbital angular momentum l.

Other atoms have states whose properties recall those of hydrogen. We refer to highly excited states, in which one of the electrons has a large principal quantum number, and so is mostly at large distances from the nucleus. The motion of such an electron can be regarded, to a certain approximation, as motion in the Coulomb field of the rest of the atom, whose effective charge is unity. The values of the energy levels thus obtained are, however, too inexact; it is necessary to apply to them a correction to take account of the deviation of the field from the pure Coulomb field at small distances. The nature of this correction is easily ascertained from the following considerations.

Since the states with large quantum numbers are quasi-classical, the energy levels can be determined from Bohr and Sommerfeld's quantization rule (48.6). The deviation from the Coulomb field at distances from the nucleus small compared with the "orbit radius" can be formally allowed for by an alteration in the boundary condition imposed on the wave function at $r = 0$. This brings about a change in the constant γ in the quantization condition for radial motion. Since this condition is otherwise unchanged, we can conclude that we obtain for the energy levels an expression which differs from that for hydrogen in that the radial, that is, the principal, quantum number n is replaced by $n + \Delta_l$, where Δ_l is some constant (known as *Rydberg's correction*):

$$E = -\frac{me^4}{2\hbar^2}\frac{1}{(n+\Delta_l)^2}. \tag{68.2}$$

Rydberg's correction is (by definition) independent of n, but it is of course a function of the azimuthal quantum number l of the excited electron (which we add as a suffix to Δ), and of the angular momenta L and S of the whole atom. For given L and S, Δ_l decreases rapidly as l increases. The greater l, the less time the electron spends near the nucleus, and hence the energy levels must approach more and more closely those of hydrogen as l increases.†

PROBLEM

Find the asymptotic form of the wave function for the hydrogen-like s state of an electron at large distances from the rest of the atom.

SOLUTION. At large distances, where the field $U = -1/r$ (in atomic units), the required function $\psi(r)$ satisfies Schrödinger's equation

$$\psi'' + \frac{2}{r}\psi' - \kappa^2\psi + \frac{2}{r}\psi = 0,$$

where $\kappa = \sqrt{(2|E|)}$. Seeking the solution in the form $\psi = \text{constant} \times r^{\gamma}e^{-\kappa r}$ and neglecting terms in the equation that decrease more rapidly than ψ/r, we find

$$\psi = \text{constant} \times r^{1/\kappa - 1}e^{-\kappa r}.$$

§69. The self-consistent field

Schrödinger's equation for atoms containing more than one electron cannot be solved in an analytical form. Approximate methods of calculating the energies and wave functions of the stationary states of the atoms are therefore important. The most important of these methods is what is called the *self-consistent field method*. The idea of this method consists in regarding each electron in the atom as being in motion in the "self-consistent field" due to the nucleus together with all the other electrons.

As an example, let us consider the helium atom, restricting ourselves to those terms in which both the electrons are in s states (with or without the same n); the states of the whole atom will then be S states also. Let $\psi_1(r_1)$ and $\psi_2(r_2)$ be the wave functions of the electrons; in the s states they are functions only of the distances r_1, r_2 of the electrons from the nuclei. The wave function $\psi(r_1, r_2)$ of the atom as a whole is a symmetrized

$$\psi = \psi_1(r_1)\psi_2(r_2) + \psi_1(r_2)\psi_2(r_1) \tag{69.1}$$

or antisymmetrized

$$\psi = \psi_1(r_1)\psi_2(r_2) - \psi_1(r_2)\psi_2(r_1) \tag{69.2}$$

† As an illustration, we may give the experimental values of Rydberg's correction for the highly excited states of the helium atom. The total spin of this atom can have the values $S = 0$ and 1, while the total orbital angular momentum L is, in the states considered, the same as the angular momentum l of the excited electron (the other electron being in the state $1s$). Rydberg's corrections are

$$\text{for } S = 0: \Delta_0 = -0{\cdot}140,\ \Delta_1 = +0{\cdot}012,\ \Delta_2 = -0{\cdot}0022;$$

$$\text{for } S = 1: \Delta_0 = -0{\cdot}296,\ \Delta_1 = -0{\cdot}068,\ \Delta_2 = -0{\cdot}0029.$$

product of the two functions, according as we are concerned with states of total spin† $S = 0$ or $S = 1$. We shall consider the second of these. The functions ψ_1 and ψ_2 can then be regarded as orthogonal.‡

Let us try to determine the function of the form (69.2) which is the best approximation to the true wave function of the atom. To do so, it is natural to start from the variational principle, allowing only functions of the form (69.2) to be considered; this method was proposed by V. A. Fok (1930).

As we know, Schrödinger's equation can be obtained from the variational principle

$$\iint \psi^* \hat{H} \psi \, dV_1 dV_2 = \text{minimum},$$

with the additional condition

$$\iint |\psi|^2 \, dV_1 dV_2 = 1$$

(the integration is extended over the coordinates of both electrons in the helium atom). The variation gives the equation

$$\iint \delta\psi^* (\hat{H} - E)\psi \, dV_1 dV_2 = 0, \tag{69.3}$$

and hence, with an arbitrary variation of the wave function ψ, we obtain the usual Schrödinger's equation. In the self-consistent field method, the expression (69.2) for ψ is substituted in (69.3), and the variation is effected with respect to the functions ψ_1 and ψ_2 separately. In other words, we seek an extremum of the integral with respect to functions ψ of the form (69.2); as a result we obtain, of course, an inexact eigenvalue of the energy and an inexact wave function, but the best of the functions that can be represented in this form.

The Hamiltonian for the helium atom is of the form||

$$\hat{H} = \hat{H}_1 + \hat{H}_2 + 1/r_{12}, \quad \hat{H}_1 = -\tfrac{1}{2}\triangle_1 - 2/r_1, \tag{69.4}$$

where r_{12} is the distance between the electrons. Substituting (69.2) in (69.3), carrying out the variation, and equating to zero the coefficients of $\delta\psi_1$ and $\delta\psi_2$ in the integrand, we easily obtain the following equations:

$$\left.\begin{aligned} [\tfrac{1}{2}\triangle + 2/r + E - H_{22} - G_{22}(r)]\psi_1(r) + [H_{12} + G_{12}(r)]\psi_2(r) &= 0, \\ [\tfrac{1}{2}\triangle + 2/r + E - H_{11} - G_{11}(r)]\psi_2(r) + [H_{12} + G_{12}(r)]\psi_1(r) &= 0, \end{aligned}\right\} \tag{69.5}$$

where

$$G_{ab}(r_1) = \int \psi_a(r_2)\psi_b(r_2) \, dV_2/r_{12},$$

$$H_{ab} = \int \psi_a[-\tfrac{1}{2}\triangle - 2/r]\psi_b \, dV \quad (a, b = 1, 2). \tag{69.6}$$

---

† The states of the helium atom with $S = 0$ are usually called ***parahelium*** states, and those with $S = 1$ *orthohelium* states.

‡ The wave functions $\psi_1, \psi_2, \ldots$ of the various states of the electron which are obtained by the self-consistent field method are not in general orthogonal, since they are solutions of different equations, not of the same equation. In (69.2), however, without altering the function ψ of the whole atom, we can replace ψ_2 by $\psi_2' = \psi_2 + \text{constant} \times \psi_1$; by an appropriate choice of the constant, we can always ensure that ψ_1 and ψ_2' are orthogonal.

|| In this section (including the Problems) we use atomic units.

These are the final equations resulting from the self-consistent field method; they can, of course, be solved only numerically.†

The equations are similarly derived in more complex cases. The wave function of the atom to be substituted in the integral in the variational principle is in the form of a linear combination of products of the wave functions of the individual electrons. This combination must be so chosen that, firstly, its permutational symmetry corresponds to the total spin S of the state of the atom considered and, secondly, it corresponds to the given value of the total orbital angular momentum L of the atom.‡

By using, in the variational principle, the wave function having the necessary permutational symmetry, we automatically take account of the exchange interaction of the electrons in the atom. Simpler equations (though leading to less accurate results) are obtained if we neglect the exchange interaction and also the dependence on L of the energy of the atom for a given electron configuration (D. R. Hartree 1928). As an example, let us again consider the helium atom; we can then write the equations for the wave functions of the electrons immediately in the form of ordinary Schrödinger's equations:

$$[\tfrac{1}{2}\triangle_a + E_a - V_a(r_a)]\psi_a(r_a) = 0 \quad (a = 1, 2), \tag{69.7}$$

where V_a is the potential energy of one electron moving in the field of the nucleus and in that of the distributed charge of the other electron:

$$V_1(r_1) = -2/r_1 + \int (1/r_{12})\psi_2^2(r_2)\, dV_2, \tag{69.8}$$

and similarly for V_2. In order to find the energy E of the whole atom, we must notice that, in the sum E_1+E_2, the electrostatic interaction between the two electrons is counted twice, since it appears in the potential energy $V_1(r_1)$ of the first electron and in that—$V_2(r_2)$—of the second. Hence E is obtained from the sum E_1+E_2 by subtracting once the mean energy of this interaction; that is.

$$E = E_1+E_2 - \iint (1/r_{12})\psi_1^2(r_1)\psi_2^2(r_2)\, dV_1 dV_2. \tag{69.9}$$

To refine the results obtained by this simplified method, the exchange interaction and the dependence of the energy on L can afterwards be taken into account as perturbations.

PROBLEMS

PROBLEM 1. Determine approximately the energy of the ground level of the helium atom and helium-like ions (a nucleus of charge Z and two electrons), regarding the interaction between the electrons as a perturbation.

---

† A comparison of the energy levels of light atoms, calculated by the self-consistent field method, with spectroscopic data enables us to estimate the accuracy of the method at about 5 per cent, and in some cases even better. For complex atoms, however, the error may become comparable with the intervals between adjacent levels, and hence give an incorrect sequence of levels.

‡ An account of general methods of constructing wave functions for a system of electrons in a central field is given in Kaplan's book quoted in §63.

SOLUTION. In the ground state of the ion, both electrons are in s states. The unperturbed value of the energy is twice the ground level of a hydrogen-like ion (because of the two electrons):

$$E^{(0)} = 2(-\tfrac{1}{2}Z^2) = -Z^2.$$

The correction in the first approximation is given by the mean value of the electron interaction energy in a state with wave function

$$\psi = \psi_1(r_1)\psi_2(r_2) = \frac{Z^3}{\pi}e^{-Z(r_1+r_2)} \tag{1}$$

(the product of two hydrogen functions with $l = 0$). The integral

$$E^{(1)} = \iint \psi^2 \frac{1}{r_{12}} \mathrm{d}V_1 \mathrm{d}V_2$$

is most simply calculated as

$$E^{(1)} = 2\int_0^\infty \mathrm{d}V_2 \,.\, \rho_2 \frac{1}{r_2}\int_0^{r_2} \rho_1 \mathrm{d}V_1, \quad \mathrm{d}V_1 = 4\pi r_1^2 \mathrm{d}r_1,$$
$$\mathrm{d}V_2 = 4\pi r_2^2 \mathrm{d}r_2,$$

the energy of the charge distribution $\rho_2 = |\psi_2|^2$ in the field of the spherically symmetric distribution $\rho_1 = |\psi_1|^2$; the integrand with $\mathrm{d}V_2$ is the energy of the charge $\rho_2(r_2)$ in the field of the sphere $r_1 < r_2$, and the factor 2 takes account of the contribution from configurations in which $r_1 > r_2$. Thus we find $E^{(1)} = 5Z/8$, and finally

$$E = E^{(0)} + E^{(1)} = -Z^2 + \tfrac{5}{8}Z.$$

For the helium atom ($Z = 2$) this gives $-E = 11/4 = 2{\cdot}75$; the actual value of the ground-state energy of this atom is $-E = 2{\cdot}90$ atomic units $= 78{\cdot}9$ eV.

PROBLEM 2. The same as Problem 1, but using the variational principle, approximating the wave function by a product of two hydrogen functions with some effective nuclear charge.

SOLUTION. We calculate the integral

$$\iint \psi \hat{H} \psi \mathrm{d}V_1 \mathrm{d}V_2, \quad \hat{H} = -\tfrac{1}{2}(\triangle_1 + \triangle_2) - \frac{Z}{r_1} - \frac{Z}{r_2} + \frac{1}{r_{12}}$$

with the function ψ given by (1), Problem 1, but with Z_{eff} instead of Z. The integral of ψ^2/r_{12} is calculated as in Problem 1; the integral of $\psi\triangle_1\psi$ can be reduced to that of ψ^2/r_1, since, by Schrödinger's equation,

$$(-\tfrac{1}{2}\triangle_1 - \frac{Z_{\text{eff}}}{r_1})\psi_1 = -\tfrac{1}{2}Z_{\text{eff}}^2\psi_1.$$

The result is

$$\iint \psi \hat{H} \psi \mathrm{d}V_1 \mathrm{d}V_2 = Z_{\text{eff}}^2 - 2ZZ_{\text{eff}} + \tfrac{5}{8}Z_{\text{eff}}.$$

This expression as a function of Z_{eff} has a minimum at $Z_{\text{eff}} = Z - \frac{5}{16}$. The corresponding value of the energy is

$$E = -(Z - \tfrac{5}{16})^2.$$

For the helium atom this gives $-E = 2{\cdot}85$.

It may be noted that the wave function (1) with the above value of Z_{eff} is in fact the best not only of all functions of the form (1) but of all functions which depend only on the sum $r_1 + r_2$.

§70. The Thomas–Fermi equation

Numerical calculations of the charge distribution and field in the atom by the self-consistent field method are extremely cumbersome, especially for complex atoms. For these, however, there is another approximate method, whose value lies in its simplicity; its results are admittedly much less accurate than those of the self-consistent field method.

The basis of this method (E. Fermi, and L. Thomas, 1927) is the fact that, in complex atoms with a large number of electrons, the majority of the electrons have comparatively large principal quantum numbers. In these conditions the quasi-classical approximation is applicable. Hence we can apply the concept of "cells in phase space" (§48) to the states of the individual electrons.

The volume of phase space corresponding to electrons which have momenta less than p and are in the volume element $\mathrm{d}V$ of physical space is $\frac{4}{3}\pi p^3\,\mathrm{d}V$. The number of cells, i.e. possible states, corresponding to this volume is† $4\pi p^3\,\mathrm{d}V/3(2\pi)^3$, and in these states there cannot at any one time be more than

$$2\frac{4\pi p^3}{3(2\pi)^3}\mathrm{d}V = \frac{p^3}{3\pi^2}\mathrm{d}V$$

electrons (two electrons, with opposite spins, in each cell). In the normal state of the atom, the electrons in each volume element $\mathrm{d}V$ must occupy (in phase space) the cells corresponding to momenta from zero up to some maximum value p_0. Then the kinetic energy of the electrons will have its smallest possible value at every point. If we write the number of electrons in the volume $\mathrm{d}V$ as $n\mathrm{d}V$ (where n is the number density of electrons), we can say that the maximum value p_0 of the momenta of the electrons at every point is related to n by

$$p_0^3/3\pi^2 = n.$$

The greatest value of the kinetic energy of an electron at a point where the electron density is n is therefore

$$\tfrac{1}{2}p_0^2 = \tfrac{1}{2}(3\pi^2 n)^{2/3}. \tag{70.1}$$

Next, let $\phi(r)$ be the electrostatic potential, which we suppose zero at infinity. The total energy of the electron is $\frac{1}{2}p^2-\phi$. It is evident that the total energy of each electron must be negative, since otherwise the electron moves off to infinity. We denote the maximum value of the total energy of the electron at each point by $-\phi_0$, where ϕ_0 is a positive constant; if this quantity were not constant, the electrons would move from points with smaller ϕ_0 to those with greater ϕ_0. Thus we can write

$$\tfrac{1}{2}p_0^2 = \phi-\phi_0. \tag{70.2}$$

† In this section we use atomic units.

Equating the expressions (70.1) and (70.2), we obtain

$$n = [2(\phi-\phi_0)]^{3/2}/3\pi^2, \tag{70.3}$$

a relation between the electron density and the potential at every point in the atom.

For $\phi = \phi_0$ the density n vanishes; n must clearly be put equal to zero also in the whole of the region where $\phi < \phi_0$, and where the relation (70.2) would give a negative maximum kinetic energy. Thus the equation $\phi = \phi_0$ determines the boundary of the atom. There is, however, no field outside a centrally symmetric system of charges whose total charge is zero. Hence we must have $\phi = 0$ at the boundary of a neutral atom. It follows from this that, for a neutral atom, the constant ϕ_0 must be put equal to zero. On the other hand, ϕ_0 is not zero for an ion.

Below we shall consider a neutral atom, putting accordingly $\phi_0 = 0$. According to Poisson's electrostatic equation, we have $\triangle\phi = 4\pi n$; substituting (70.3) in this, we obtain the fundamental equation of the Thomas–Fermi method:

$$\triangle\phi = (8\sqrt{2}/3\pi)\phi^{3/2}. \tag{70.4}$$

The field distribution in the normal state of the atom is determined by the centrally symmetric solution of this equation that satisfies the following boundary conditions: for $r \to 0$ the field must become the Coulomb field of the nucleus, i.e. $\phi r \to Z$, while for $r \to \infty$ we must have $\phi r \to 0$. Introducing here, in place of the variable r, a new variable x according to the definitions

$$r = xbZ^{-1/3}, \quad b = \tfrac{1}{2}(\tfrac{3}{4}\pi)^{2/3} = 0{\cdot}885, \tag{70.5}$$

and, in place of ϕ, a new unknown function χ by †

$$\phi(r) = \frac{Z}{r}\chi\left(\frac{rZ^{1/3}}{b}\right) = \frac{Z^{4/3}}{b}\frac{\chi(x)}{x}, \tag{70.6}$$

we obtain the equation

$$x^{1/2}\,d^2\chi/dx^2 = \chi^{3/2}, \tag{70.7}$$

with the boundary conditions $\chi = 1$ for $x = 0$ and $\chi = 0$ for $x = \infty$. This equation contains no parameters, and thus defines a universal function $\chi(x)$. Table 2 gives values of this function obtained by numerical integration of equation (70.7). The function $\chi(x)$ decreases monotonically, and vanishes only at infinity.‡ In other words, the atom has no boundaries in the Thomas-Fermi model, and formally extends to infinity.

† In ordinary units,

$$\phi(r) = (Ze/r)\chi(rZ^{1/3}me^2/0{\cdot}885\hbar^2).$$

‡ The equation (70.7) has the exact solution $\chi(x) = 144x^{-3}$, which vanishes at infinity but does not satisfy the boundary condition at $x = 0$. It could be used as an asymptotic expression for the function $\chi(x)$ for large x. However, this expression gives fairly exact values only for very large x, whilst the Thomas–Fermi equation becomes inapplicable at large distances (see below).

TABLE 2

Values of the function $\chi(x)$

x	$\chi(x)$	x	$\chi(x)$	x	$\chi(x)$
0·00	1·000	1·4	0·333	6	0·0594
0·02	0·972	1·6	0·298	7	0·0461
0·04	0·947	1·8	0·268	8	0·0366
0·06	0·924	2·0	0·243	9	0·0296
0·08	0·902	2·2	0·221	10	0·0243
0·10	0·882	2·4	0·202	11	0·0202
0·2	0·793	2·6	0·185	12	0·0171
0·3	0·721	2·8	0·170	13	0·0145
0·4	0·660	3·0	0·157	14	0·0125
0·5	0·607	3·2	0·145	15	0·0108
0·6	0·561	3·4	0·134	20	0·0058
0·7	0·521	3·6	0·125	25	0·0035
0·8	0·485	3·8	0·116	30	0·0023
0·9	0·453	4·0	0·108	40	0·0011
1·0	0·424	4·5	0·0919	50	0·00063
1·2	0·374	5·0	0·0788	60	0·00039

The value of the derivative $\chi'(x)$ for $x = 0$ is $\chi'(0) = -1{\cdot}59$. Hence, as $x \to 0$, the function $\chi(x)$ is of the form $\chi \approx 1-1{\cdot}59x$, and accordingly the potential $\phi(r)$ is

$$\phi(r) \approx Z/r - 1{\cdot}80Z^{4/3}. \tag{70.8}$$

The first term is the potential of the field of the nucleus, while the second is the potential at the origin due to the electrons.

Substituting (70.6) in (70.3), we find for the electron density an expression of the form

$$n = Z^2 f(rZ^{1/3}/b),\; f(x) = (32/9\pi^3)(\chi/x)^{3/2}. \tag{70.9}$$

We see that, in the Thomas–Fermi model, the charge density distribution in different atoms is similar, with $Z^{-1/3}$ as the characteristic length (in ordinary units $\hbar^2/me^2Z^{1/3}$, i.e. the Bohr radius divided by $Z^{1/3}$). If we measure distances in atomic units, the distances at which the electron density has its maximum value are the same for all Z. Hence we can say that the majority of the electrons in an atom of atomic number Z are at distances from the nucleus of the order of $Z^{-1/3}$. A numerical calculation shows that half the total electron charge in an atom lies inside a sphere of radius $1{\cdot}33Z^{-1/3}$.

Similar considerations show that the mean velocity of the electrons in the atom (taken, as an order of magnitude, as the square root of the energy) is of the order of $Z^{2/3}$.

The Thomas–Fermi equation becomes inapplicable both at very small and at very large distances from the nucleus. Its range of applicability for small r is restricted by the inequality (49.12); at smaller distances the quasi-classical approximation becomes invalid in the Coulomb field of the nucleus. Putting in (49.12) $\alpha = Z$, we find $1/Z$ as the lower limit of distance. The quasi-classical approximation becomes invalid for large r also in a complex atom. In fact, it is easy to see that, for $r \sim 1$, the de Broglie wavelength of the electron becomes of the same order of magnitude as the distance itself, so that the quasi-classical condition is undoubtedly violated. This can be seen by estimating the values of the terms in equations (70.2) and (70.4); indeed, the result is obvious without calculation, since equation (70.4) does not involve Z. Thus the applicability of the Thomas–Fermi equation is limited to distances large compared with $1/Z$ and small compared with unity. In complex atoms, however, the majority of the electrons in fact lie in this region.

This means that the "outer boundary" of the atom in the Thomas–Fermi model is at $r \sim 1$, i.e. the dimensions of the atom do not depend on Z. The energy of the outer electrons, i.e. the ionization potential of the atom, is likewise independent of Z.†

By means of the Thomas–Fermi method we can calculate the total ionization energy E, i.e. the energy needed to remove all the electrons from the neutral atom. To do this, we must calculate the electrostatic energy of the Thomas–Fermi distribution for the charges in the atom; the required total energy is half this electrostatic energy, since the mean kinetic energy in a system of particles interacting in accordance with Coulomb's law is (by the virial theorem; see *Mechanics*, §10) minus half the mean potential energy. The dependence of E on Z can be determined *a priori* from simple considerations: the electrostatic energy of Z electrons at a mean distance $Z^{-1/3}$ from a nucleus of charge Z, and moving in its field, is proportional to $Z \cdot Z/Z^{-1/3} = Z^{7/3}$. A numerical calculation gives the result $E = 20{\cdot}8Z^{7/3}$ eV. The dependence on Z is in good agreement with the experimental data, though the empirical value of the coefficient is close to 16.

We have already mentioned that positive (non-zero) values of the constant ϕ_0 correspond to ionized atoms. If we define the function χ by $\phi - \phi_0 = Z\chi/r$, we obtain the same equation (70.7) for χ as previously. We must now, however, consider only solutions which vanish not at infinity as for the neutral atom, but for finite values x_0 of x. Such solutions exist for any x_0. At the point $x = x_0$, the charge density vanishes together with χ, but the potential remains finite. The value of x_0 is related to the degree of ionization in the following manner. The total charge inside a sphere of radius r is, by Gauss's theorem, $-r^2\partial\phi/\partial r = Z[\chi(x) - x\chi'(x)]$. The total charge z on the ion is obtained by putting $x = x_0$ in this; since $\chi(x_0) = 0$, we have

$$z = -Zx_0\chi'(x_0). \tag{70.10}$$

† This model does not, of course, show the periodic dependence of the dimensions and ionization potential of the atom on Z, which appears in the periodic system of the elements. Moreover, experimental data indicate the existence of a slight but steady increase in dimensions and decrease in the ionization potential as Z increases.

The thick line in Fig. 23 shows the curve of $\chi(x)$ for a neutral atom; below it are two curves for ions of different degrees of ionization. The quantity z/Z is shown graphically by the length of the segment intercepted on the axis of ordinates by the tangent to the curve at $x = x_0$.

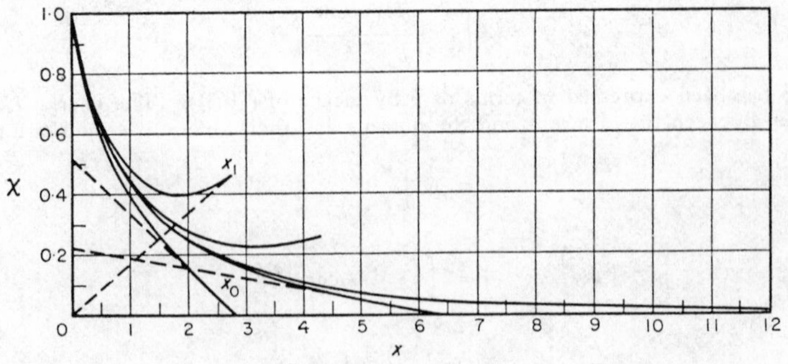

FIG. 23

Equation (70.7) also has solutions which are nowhere zero; these diverge at infinity. They can be regarded as corresponding to negative values of the constant ϕ_0. Figure 23 also shows two such curves of $\chi(x)$; they lie above the curve for the neutral atom. At the point $x = x_1$, where

$$\chi(x_1) - x_1\chi'(x_1) = 0, \tag{70.11}$$

the total charge inside the sphere $x < x_1$ is zero (graphically, this point is evidently the one where the tangent to the curve passes through the origin). If we cut off the curve at this point, we can say that it defines $\chi(x)$ for a neutral atom at whose boundary the charge density remains non-zero. Physically, this corresponds to a "compressed" atom confined to some given finite volume.†

The Thomas–Fermi equation does not take account of the exchange interaction between electrons. The effects which this involves are of the next order of magnitude with respect to $Z^{-2/3}$. Hence an allowance for the exchange interaction in the Thomas–Fermi method requires a simultaneous consideration of both these effects and others of the same order of magnitude.‡

PROBLEM

Find the relation between the energy of the electrostatic interaction between electrons and that of their interaction with the nucleus in a neutral atom, using the Thomas–Fermi model.

SOLUTION. The potential ϕ_e of the field due to the electrons is found by subtracting the potential Z/r of the nucleus from the total potential ϕ. The energy of the interaction between

† This approach may be useful in studying the equation of state of highly compressed matter.

‡ This has been done by A. S. Kompaneets and E. S. Pavlovskii (*Soviet Physics JETP* **4**, 328, 1957) and by D. Kirzhnits (*ibid.* 5, 64, 1338, 1957).

the electrons is therefore

$$U_{ee} = -\tfrac{1}{2}\int \phi_e n \mathrm{d}V$$
$$= \tfrac{1}{2}Z\int \frac{n}{r}\mathrm{d}V - \tfrac{1}{2}\int \phi n \mathrm{d}V$$
$$= \tfrac{1}{2}Z\int \frac{n}{r}\mathrm{d}V - \frac{(3\pi^2)^{2/3}}{4}\int n^{5/3}\mathrm{d}V$$

(where ϕ has been expressed in terms of n by means of (70.3)). The energy U_{en} of the interaction between the electrons and the nucleus and their kinetic energy T are therefore

$$U_{en} = -Z\int \frac{n}{r}\mathrm{d}V,$$

$$T = 2\int\int_0^{p_0} \tfrac{1}{2}p^2 . 4\pi p^2 \mathrm{d}p \mathrm{d}V$$
$$= 3\frac{(3\pi^2)^{2/3}}{10}\int n^{5/3}\mathrm{d}V.$$

Comparing these expressions with the previous equation, we find

$$U_{ee} = -\tfrac{1}{2}U_{en} - \tfrac{5}{6}T.$$

According to the virial theorem (see *Mechanics*, §10), for a system of particles interacting according to Coulomb's law we have $2T = -U = -U_{en} - U_{ee}$. Thus finally

$$U_{ee} = -\tfrac{1}{7}U_{en}.$$

§71. Wave functions of the outer electrons near the nucleus

We have seen, on the basis of the Thomas–Fermi model, that the outer electrons in complex atoms (Z large) are mainly at distances $r \sim 1$ from the nucleus.† A number of properties of atoms, however, depend significantly on the electron density near the nucleus; such properties will be considered in §§72 and 120. To determine the order of magnitude of this density we may examine the variation of the wave function $\psi(r)$ of the electron in the atom when r varies from large ($r \sim 1$) to small distances.

In the region $r \sim 1$, the field of the nucleus is screened by the remaining electrons, so that the potential energy $U(r) \sim 1/r \sim 1$. The energy of the electron level in this field $E \sim 1$. At distances of the order of the Bohr radius in the field of a charge Z, $r \sim 1/Z$, the field of the nucleus may be regarded as unscreened, and $U = -Z/r$. In the transitional region, $1/Z \ll r \ll 1$, the potential energy $|U|$ is large compared with the electron energy E, and the condition

$$\frac{\mathrm{d}}{\mathrm{d}r}\left(\frac{1}{p}\right) \sim \frac{\mathrm{d}}{\mathrm{d}r}\frac{1}{\sqrt{|U|}} \ll 1$$

† In this section we use atomic units.

holds (where p is the momentum), so that the motion of the electron is quasi-classical. The spherically symmetrical quasi-classical wave function is

$$|\psi(r)| \sim \frac{1}{r\sqrt{p}} \sim \frac{1}{r|U|^{1/4}} \quad \text{for} \quad \frac{1}{Z} \ll r \ll 1, \tag{71.1}$$

the order of magnitude of the coefficient (~ 1) being determined by the condition $\psi \sim 1$ for "joining" to the wave function for $r \sim 1$.

Applying the expression (71.1) in order of magnitude for $r \sim 1/Z$ (substituting $U = -Z/r$), we obtain the required value of the wave function near the nucleus:†

$$\psi(1/Z) \sim \sqrt{Z}. \tag{71.2}$$

In accordance with the general properties of wave functions in a central field (§32), when the distance decreases further $\psi(r)$ either remains constant in order of magnitude (for an s electron) or begins to decrease (for $l \neq 0$).

The probability of finding the electron in the region $r \lesssim 1/Z$ is

$$w \sim |\psi|^2 r^3 \sim 1/Z^2. \tag{71.3}$$

The formulae (71.2) and (71.3) of course determine only the systematic variation with increasing Z, and do not take into account non-systematic variations from one element to the next.

§72. Fine structure of atomic levels

A consistent derivation of the formulae for relativistic effects in the interaction of electrons belongs to the next volume (see *RQT*, §§33 and 83). Here we shall give only a general account of these effects as they relate to atomic terms. It is found that the relativistic terms in the Hamiltonian of an atom fall into two classes. One of these contains terms linear with respect to the spin operators of the electrons, while the other includes quadratic terms. The former correspond to the interaction between the orbital motion of the electrons and their spin (this interaction is called *spin–orbit* interaction), while the latter correspond to the interaction between the spins of the electrons (*spin–spin* interaction). Both interactions are of the same order (the second) with respect to v/c, the ratio of the velocity of the electrons to that of light; in practice, the spin–orbit interaction considerably exceeds the spin–spin interaction in heavy atoms. This is because the spin–orbit interaction increases rapidly with the atomic number, whereas the spin–spin interaction is essentially independent of Z (see below).

The spin–orbit interaction operator is of the form

$$\hat{V}_{sl} = \sum_a \hat{\mathbf{A}}_a \cdot \hat{\mathbf{s}}_a \tag{72.1}$$

(the summation being over all the electrons in the atom), where $\hat{\mathbf{s}}_a$ are the

† To determine the coefficient in this formula (when the wave function is known in the region $r \sim 1$), we should have to use the expression (36.25) in the range $r \lesssim 1/Z$.

spin operators of the electrons, and $\hat{\mathbf{A}}_a$ are some "orbital" operators, i.e. operators acting on functions of the coordinates. In the self-consistent field approximation the operators $\hat{\mathbf{A}}_a$ are proportional to the operators $\hat{\mathbf{l}}_a$ of the orbital angular momentum of the electrons, and $\hat{V}_{sl}$ can then be written in the form

$$\hat{V}_{sl} = \Sigma \alpha_a \hat{\mathbf{l}}_a \cdot \hat{\mathbf{s}}_a. \tag{72.2}$$

The coefficients in the sum are given in terms of the potential energy $U(r)$ of the electron in the self-consistent field by

$$\alpha_a = \frac{\hbar^2}{2m^2c^2r_a}\frac{\mathrm{d}U(r_a)}{\mathrm{d}r_a}. \tag{72.3}$$

Since $|U(r)|$ decreases away from the nucleus, all the $\alpha_a > 0$.

Regarding the interaction (72.2) as a perturbation, we should, in order to calculate the energy, average it with respect to the unperturbed state. The main contribution to the energy is given by distances close to the nucleus, of the order of the Bohr radius ($\sim \hbar^2/Zme^2$) for a nucleus with charge Ze. In this region the field of the nucleus is almost unscreened and the potential energy is

$$|U(r)| \sim Ze^2/r \sim Z^2me^4/\hbar^2,$$

so that

$$\begin{aligned}\alpha &\sim \hbar^2 U/m^2c^2r^2 \\ &\sim Z^4(e^2/\hbar c)^2 me^4/\hbar^2.\end{aligned}$$

The mean value of α is obtained by multiplying by the probability w of finding the electron near the nucleus. According to (71.3), $w \sim Z^{-2}$, so that we have finally that the energy of the spin–orbit interaction of the electron is given by

$$\bar{\alpha} \sim \left(\frac{Ze^2}{\hbar c}\right)^2 \frac{me^4}{\hbar^2},$$

i.e. differs from the fundamental energy of the outer electrons in the atom ($\sim me^4/\hbar^2$) only by the factor $(Ze^2/\hbar c)^2$. This factor increases rapidly with the atomic number, and reaches values of the order of unity in heavy atoms.

The actual averaging of the operator (72.2) over the unperturbed states of the electron envelope is done in two steps. First of all, we average over electron states of the atom with given values L and S of the total orbital angular momentum and spin, but not with given directions of these. After this averaging $\hat{V}_{sl}$ is still an operator, which, however, we must now express only in terms of operators of quantities characterizing the atom as a whole, not its individual electrons. These are the operators $\hat{\mathbf{L}}$ and $\hat{\mathbf{S}}$. We denote by $\hat{V}_{SL}$ the operator of the spin–orbit interaction thus averaged. Being linear in $\hat{\mathbf{S}}$, it has the form

$$\hat{V}_{SL} = A\,\hat{\mathbf{S}}.\hat{\mathbf{L}}, \tag{72.4}$$

where A is a constant characterizing a given (unsplit) term, i.e. depending on S and L but not on the total angular momentum J of the atom.†

To calculate the energy of the splitting of a degenerate level, we must now solve the secular equation formed from the matrix elements of the operator (72.4). In this case, however, we already know the correct functions in the zeroth approximation, in which the matrix of V_{SL} is diagonal. These are the wave functions of states with definite values of the total angular momentum J. The averaging with respect to such a state involves replacing the operator $\hat{\mathbf{S}}.\hat{\mathbf{L}}$ by its eigenvalues, which, according to (31.3), are

$$\mathbf{L}.\mathbf{S} = \tfrac{1}{2}[J(J+1)-L(L+1)-S(S+1)].$$

Since the values of L and S are the same for all the components of a multiplet, and we are interested only in their relative position, we can write the energy of the multiplet splitting in the form

$$\tfrac{1}{2}AJ(J+1). \tag{72.5}$$

The intervals between adjacent components (with numbers J and $J-1$) are consequently

$$\Delta E_{J,J-1} = AJ. \tag{72.6}$$

This formula gives what is called *Landé's interval rule* (1923).

The constant A can be either positive or negative. For $A > 0$ the lowest component of the multiplet level is the one with the smallest possible J, i.e. $J = |L-S|$; such multiplets are said to be *normal.* If $A < 0$, on the other hand, the lowest level of the multiplet is that with $J = L+S$; these multiplets are said to be *inverted.*

It is easy to determine the sign of A for the normal states of atoms if the electron configuration is such that there is only one shell not completely filled. If this shell is not more than half filled, then according to Hund's rule (§67) all n electrons in it have parallel spins, so that the total spin has the greatest possible value, $S = \tfrac{1}{2}n$. Substituting in (72.2) $\mathbf{s}_a = \mathbf{S}/n$ and taking α_a (which is the same for all electrons in a given shell) outside the sum we obtain

$$\hat{V}_{SL} = (\alpha/2S)\hat{\mathbf{S}}.\hat{\mathbf{L}},$$

i.e. $A = \alpha/2S > 0$. If the shell is more than half full, we first add and sub-

† In order to clarify the meaning of this operation, it may be noted that averaging in quantum mechanics has the general significance of taking the appropriate diagonal matrix element. A partial averaging consists in taking a set of matrix elements that are diagonal with respect to only some of the quantum numbers describing the state of the system. For example, in this case the averaging of the operator (72.2) denotes the construction of a matrix with elements $\langle nM'_LM'_S|V_{sl}|nM_LM_S\rangle$ with all possible M_L, M'_L and M_S, M'_S and diagonal with respect to all the other quantum numbers (the assembly of which we denote by n). Correspondingly, the operators $\hat{\mathbf{S}}$ and $\hat{\mathbf{L}}$ are to be regarded as matrices $\langle M'_S|\mathbf{S}|M_S\rangle$ and $\langle M'_L|\mathbf{L}|M_L\rangle$, whose elements are given by (27.13). A similar device of stepwise averaging will be needed in several subsequent treatments.

tract in (72.2) the same sum taken over the unoccupied places or holes in the incomplete shell. Since, for a completely filled shell, we should have $V_{sl} = 0$, the operator $\hat{V}_{sl}$ is thereby represented as a sum

$$\hat{V}_{sl} = -\Sigma \alpha_a \hat{\mathbf{l}}_a.\hat{\mathbf{s}}_a,$$

taken only over the holes, the total spin and orbital angular momentum of the atom being $\mathbf{S} = -\Sigma \mathbf{s}_a$, $\mathbf{L} = -\Sigma \mathbf{l}_a$. By the same method as previously we therefore find $A = -\alpha/2S$, i.e. $A < 0$.

From the above we have a simple rule which gives the value of J in the normal state of an atom with one incompletely filled shell. If this shell contains not more than half the greatest possible number of electrons for that shell, then $J = |L-S|$; if the shell is more than half full, $J = L+S$.

As already mentioned, the spin–spin interaction, unlike the spin–orbit interaction, is essentially independent of Z. This is evident from the fact that it is a direct interaction between electrons and does not involve the field of the nucleus.

For the averaged spin–spin interaction operator we should obtain, analogously to formula (72.4), an expression quadratic in $\hat{\mathbf{S}}$. The expressions $\hat{\mathbf{S}}^2$ and $(\hat{\mathbf{S}}.\hat{\mathbf{L}})^2$ are quadratic in $\hat{\mathbf{S}}$. The former has eigenvalues independent of J, and therefore does not give any splitting of the term. Hence it can be omitted, and we can write

$$\hat{V}_{SS} = B(\hat{\mathbf{S}}.\hat{\mathbf{L}})^2, \tag{72.7}$$

where B is a constant. The eigenvalues of this operator contain terms independent of J, terms proportional to $J(J+1)$, and finally a term proportional to $J^2(J+1)^2$. The first of these do not give any splitting and hence are without interest; the second can be included in the expression (72.5), which simply means a change in the constant A. Finally, the last term gives an energy

$$\tfrac{1}{4}BJ^2(J+1)^2. \tag{72.8}$$

The scheme for the construction of the atomic levels discussed in §§66–67 is based on the supposition that the orbital angular momenta of the electrons combine to give the total orbital angular momentum L of the atom, and their spins to give the total spin S. As has already been mentioned, this supposition is legitimate only when the relativistic effects are small; more exactly, the intervals in the fine structure must be small compared with the differences between levels with different L and S. This approximation is called the *Russell–Saunders case* (H. N. Russell and F. A. Saunders 1925), and we speak also of *LS coupling*.

In practice, however, this approximation has a limited range of applicability. The levels of the light atoms are arranged in accordance with the *LS* model, but as the atomic number increases the relativistic interactions in the atom become stronger, and the Russell–Saunders approximation becomes inapplicable.† It must also be noticed that this approximation is, in particular, inapplicable to highly excited levels, in which the atom contains an electron which is in a state with large n, and which is therefore mainly at large distances from the nucleus (see §68). The electrostatic interaction of this electron with the motion of the other electrons is comparatively weak, but the relativistic interaction in the rest of the atom is not diminished.

In the opposite limiting case the relativistic interaction is large compared with the electrostatic (or, more precisely, compared with that part of it which governs the dependence of the energy on L and S). In this case we cannot speak of the orbital angular momentum and spin separately, since they are not conserved. The individual electrons are characterized by their total angular momenta j, which combine to give the total angular momentum J of the atom. This scheme of arrangement of the atomic levels is called *jj coupling*. In practice, this coupling is not found in the pure state, but various types of coupling intermediate between LS and jj are observed among the levels of very heavy atoms.‡

A peculiar type of coupling is observed in certain highly excited states. Here the rest of the atom may be in a Russell–Saunders state, i.e. may be characterized by the values of L and S, while its coupling with the highly excited electron is of the jj type; this is again due to the weakness of the electrostatic interaction for this electron.

The fine structure of the energy levels of the hydrogen atom has certain characteristic properties. It will be calculated exactly in *RQT* (§34), but here we shall only mention that, for a given principal quantum number n, the energy depends only on the total angular momentum j of the electron. Thus the degeneracy of the levels is not completely removed; to a level with given n and j there correspond two states with orbital angular momenta $l = j \pm \frac{1}{2}$ (unless j has the value $n-\frac{1}{2}$, which is the greatest possible for a given n). Thus the level with $n = 3$ is split into three levels, of which the states $s_{1/2}$, $p_{1/2}$ correspond to one, $p_{3/2}$ and $d_{3/2}$ to another, and $d_{5/2}$ to the third.

§73. The Mendeleev periodic system

The elucidation of the nature of the periodic variation of properties, observed in the series of elements when they are placed in order of increasing atomic number (D. I. Mendeleev 1869), requires an examination of the peculiarities in the successive completion of the electron shells of atoms. The theory of the periodic system is due to N. Bohr (1922).

When we pass from one atom to the next, the charge is increased by unity

---

† Nevertheless, although the quantitative formulae which describe this type of coupling become inapplicable, the method of classifying levels according to this scheme may itself remain meaningful for heavier atoms, especially for the lowest states (including the normal state).

‡ For further details regarding types of coupling and the quantitative aspect of the problem, see, for instance, E. U. Condon and G. H. Shortley, *The Theory of Atomic Spectra*, Cambridge University Press, 1935.

and one electron is added to the envelope. At first sight we might expect the binding energy of each of the successively added electrons to vary monotonically as the atomic number increases. The actual variation, however, is entirely different.

In the normal state of the hydrogen atom there is only one electron, in the 1*s* state. In the atom of the next element, helium, another 1*s* electron is added; the binding energy of the 1*s* electrons in the helium atom is, however, considerably greater than in the hydrogen atom. This is a natural consequence of the difference between the field in which the electron moves in the hydrogen atom and the field encountered by an electron added to the He^+ ion. At large distances these fields are approximately the same, but near the nucleus with charge $Z = 2$ the field of the He^+ ion is stronger than that of the hydrogen nucleus with $Z = 1$. In the lithium atom ($Z = 3$), the third electron enters the 2*s* state, since no more than two electrons can be in 1*s* states at the same time. For a given Z the 2*s* level lies above the 1*s* level; as the nuclear charge increases, both levels become lower. In the transition from $Z = 2$ to $Z = 3$, however, the former effect is predominant, and so the binding energy of the third electron in the lithium atom is considerably less than those of the electrons in the helium atom. Next, in the atoms from Be ($Z = 4$) to Ne ($Z = 10$), first one more 2*s* electron and then six 2*p* electrons are successively added. The binding energies of these electrons increase on the average, owing to the increasing nuclear charge. The next electron added, on going to the sodium atom ($Z = 11$), enters the 3*s* state, and the binding energy again diminishes markedly, since the effect of going to a higher shell predominates over that of the increase of the nuclear charge.

This picture of the filling up of the electron envelope is characteristic of the whole sequence of elements. All the electron states can be divided into successively occupied groups such that, as the states of each group are occupied in a series of elements, the binding energy increases on the average, but when the states of the next group begin to be occupied the binding energy decreases noticeably. Figure 24 shows those ionization potentials of elements that are known from spectroscopic data; they give the binding energies of the electrons added as we pass from each element to the next.

The different states are distributed as follows into successively occupied groups:

$$
\left.\begin{array}{ll}
1s & 2 \text{ electrons} \\
2s,\ 2p & 8 \quad ,, \\
3s,\ 3p & 8 \quad ,, \\
4s,\ 3d,\ 4p & 18 \quad ,, \\
5s,\ 4d,\ 5p & 18 \quad ,, \\
6s,\ 4f,\ 5d,\ 6p & 32 \quad ,, \\
7s,\ 6d,\ 5f, \ldots &
\end{array}\right\} \qquad (73.1)
$$

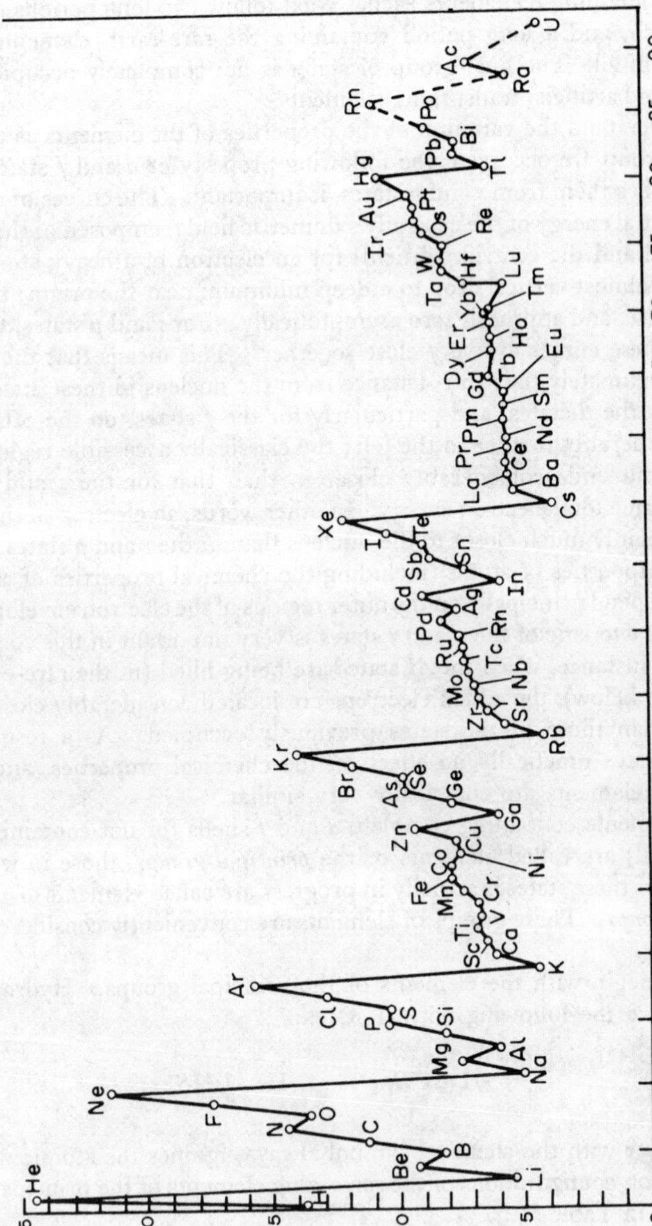
eV
25
20
15
10
5
0
1
5
10
15
20
25
30
35
40
45
50
55
60
65
70
75
80
85
90
He
H
Li
Be
B
C
N
O
F
Ne
Na
Mg
Al
Si
P
S
Cl
Ar
K
Ca
Sc
Ti
V
Cr
Mn
Fe
Co
Ni
Cu
Zn
Ga
Ge
As
Se
Br
Kr
Rb
Sr
Y
Zr
Nb
Mo
Tc
Ru
Rh
Pd
Ag
Cd
In
Sn
Sb
Te
I
Xe
Cs
Ba
La
Ce
Pr
Nd
Pm
Sm
Eu
Gd
Tb
Dy
Ho
Er
Tm
Yb
Lu
Hf
Ta
W
Re
Os
Ir
Pt
Au
Hg
Tl
Pb
Bi
Po
Rn
Ra
Ac
U

FIG. 24

The first group is occupied in H and He; the occupation of the second and third groups corresponds to the first two (short) periods of the periodic system, containing 8 elements each. Next follow two long periods of 18 elements each, and a long period containing the rare-earth elements and 32 elements in all. The final group of states is not completely occupied in the natural (and artificial transuranic) elements.

To understand the variation of the properties of the elements as the states of each group are occupied, the following property of *d* and *f* states, which distinguishes them from *s* and *p* states, is important. The curves of the effective potential energy of the centrally symmetric field (composed of the electrostatic field and the centrifugal field) for an electron in a heavy atom have a rapid and almost vertical drop to a deep minimum near the origin; they then begin to rise, and approach zero asymptotically. For *s* and *p* states, the rising parts of these curves are very close together. This means that the electron is at approximately the same distance from the nucleus in these states. The curves for the *d* states, and particularly for the *f* states, on the other hand, pass considerably further to the left; the classically accessible region which they delimit ends considerably closer in than that for the *s* and *p* states with the same total electron energy. In other words, an electron in the *d* and *f* states is mainly much closer to the nucleus than in the *s* and *p* states.

Many properties of atoms (including the chemical properties of elements; see §81) depend principally on the outer regions of the electron envelopes. The above characteristic of the *d* and *f* states is very important in this connection. Thus, for instance, when the 4*f* states are being filled (in the rare-earth elements; see below), the added electrons are located considerably closer to the nucleus than those in the states previously occupied. As a result, these electrons have practically no effect on the chemical properties, and all the rare-earth elements are chemically very similar.

The elements containing complete *d* and *f* shells (or not containing these shells at all) are called elements of the ***principal groups***; those in which the filling up of these states is actually in progress are called elements of the ***intermediate groups***. These groups of elements are conveniently considered separately.

Let us begin with the elements of the principal groups. Hydrogen and helium have the following normal states:

$$ {}_{1}\mathrm{H} : 1s\ {}^{2}S_{1/2} \qquad {}_{2}\mathrm{He} : 1s^{2}\ {}^{1}S_{0} $$

(the number with the chemical symbol always signifies the atomic number). The electron configurations of the remaining elements of the principal groups are shown in Table 3.

In each atom, the shells shown on the right of the table in the same line and above are completely filled. The electron configuration in the shells that are being filled is shown at the top, while the principal quantum number of the electrons in these states is shown by the figure on the left of the table in the same line. The normal states of the whole atom are shown at the bottom. Thus, the aluminium atom has the electron configuration $1s^2\,2s^2\,2p^6\,3s^2\,3p\ {}^2P_{1/2}$.

The values of L and S in the normal state of the atom can be determined (the electron configuration being known) by means of Hund's rule (§67), and the value of J is determined by the rule given in §72.

TABLE 3

Electron configurations of the atoms of elements in the principal groups

	s	s^2	s^2p	s^2p^2	s^2p^3	s^2p^4	s^2p^5	s^2p^6	
$n = 2$	$_3$Li	$_4$Be	$_5$B	$_6$C	$_7$N	$_8$O	$_9$F	$_{10}$Ne	$1s^2$
3	$_{11}$Na	$_{12}$Mg	$_{13}$Al	$_{14}$Si	$_{15}$P	$_{16}$S	$_{17}$Cl	$_{18}$Ar	$2s^2\,2p^6$
4	$_{19}$K	$_{20}$Ca							$3s^2\,3p^6$
4	$_{29}$Cu	$_{30}$Zn	$_{31}$Ga	$_{32}$Ge	$_{33}$As	$_{34}$Se	$_{35}$Br	$_{36}$Kr	$3d^{10}$
5	$_{37}$Rb	$_{38}$Sr							$4s^2\,4p^6$
5	$_{47}$Ag	$_{48}$Cd	$_{49}$In	$_{50}$Sn	$_{51}$Sb	$_{52}$Te	$_{53}$I	$_{54}$Xe	$4d^{10}$
6	$_{55}$Cs	$_{56}$Ba							$5s^2\,5p^6$
6	$_{79}$Au	$_{80}$Hg	$_{81}$Tl	$_{82}$Pb	$_{83}$Bi	$_{84}$Po	$_{85}$At	$_{86}$Rn	$4f^{14}\,5d^{10}$
7	$_{87}$Fr	$_{88}$Ra							$6s^2\,6p^6$
	${}^2S_{1/2}$	${}^1S_0$	${}^2P_{1/2}$	${}^3P_0$	${}^4S_{3/2}$	${}^3P_2$	${}^2P_{3/2}$	${}^1S_0$	

The atoms of the inert gases (He, Ne, Ar, Kr, Xe, Rn) occupy a special position in the table: the filling up of one of the groups of states listed in (73.1) is completed in each of them. Their electron configurations have unusual stability (their ionization potentials are the greatest in their respective series). This causes the chemical inertness of these elements.

We see that the occupation of different states occurs very regularly in the series of elements of the principal groups: first the s states and then the p states are occupied for each principal quantum number n. The electron configurations of the ions of these elements are also regular (until electrons from the d and f shells are removed in the ionization): each ion has the configuration corresponding to the preceding atom. Thus, the Mg^+ ion has the configuration of the sodium atom, and the Mg^{++} ion that of neon.

Let us now turn to the elements of the intermediate groups. The filling up of the $3d$, $4d$, and $5d$ shells takes place in groups of elements called respectively the *iron group*, the *palladium group* and the *platinum group*. Table 4 gives those electron configurations and terms of the atoms in these groups that are known from experimental spectroscopic data. As is seen from this table, the d shells are filled up with considerably less regularity than the s and p shells in the atoms of elements of the principal groups. Here a characteristic feature is the "competition" between the s and d states. It is seen in the fact that, instead of a regular sequence of configurations of the type $d^p\,s^2$ with increasing p, configurations of the type $d^{p+1}s$ or d^{p+2} are often found. Thus, in the iron group, the chromium atom has the configuration $3d^5\,4s$, and not $3d^4\,4s^2$; after nickel with 8 d electrons, there follows at once the copper atom with a completely filled d shell (and hence we place this

TABLE 4

Electron configurations of the atoms of elements in the iron, palladium and platinum groups

	Iron group							
	$_{21}$Sc	$_{22}$Ti	$_{23}$V	$_{24}$Cr	$_{25}$Mn	$_{26}$Fe	$_{27}$Co	$_{28}$Ni
Ar envelope +	$3d\,4s^2$	$3d^2\,4s^2$	$3d^3\,4s^2$	$3d^5\,4s$	$3d^5\,4s^2$	$3d^6\,4s^2$	$3d^7\,4s^2$	$3d^8\,4s^2$
	$^2D_{3/2}$	$^3F_2$	$^4F_{3/2}$	$^7S_3$	$^6S_{5/2}$	$^5D_4$	$^4F_{9/2}$	$^3F_4$
	Palladium group							
	$_{39}$Y	$_{40}$Zr	$_{41}$Nb	$_{42}$Mo	$_{43}$Tc	$_{44}$Ru	$_{45}$Rh	$_{46}$Pd
Kr envelope +	$4d\,5s^2$	$4d^2\,5s^2$	$4d^4\,5s$	$4d^5\,5s$	$4d^5\,5s^2$	$4d^7\,5s$	$4d^8\,5s$	$4d^{10}$
	$^2D_{3/2}$	$^3F_2$	$^6D_{1/2}$	$^7S_3$	$^6S_{5/2}$	$^5F_5$	$^4F_{9/2}$	$^1S_0$
	Platinum group							
	$_{57}$La							
Xe envelope +	$5d\,6s^2$							
	$^2D_{3/2}$							
	$_{71}$Lu	$_{72}$Hf	$_{73}$Ta	$_{74}$W	$_{75}$Re	$_{76}$Os	$_{77}$Ir	$_{78}$Pt
Xe envelope $+4f^{14}+$	$5d\,6s^2$	$5d^2\,6s^2$	$5d^3\,6s^2$	$5d^4\,6s^2$	$5d^5\,6s^2$	$5d^6\,6s^2$	$5d^7\,6s^2$	$5d^9\,6s$
	$^2D_{3/2}$	$^3F_2$	$^4F_{3/2}$	$^5D_0$	$^6S_{5/2}$	$^5D_4$	$^4F_{9/2}$	$^3D_3$

element in the principal groups). This lack of regularity is observed in the terms of ions also: the electron configurations of the ions do not usually agree with those of the preceding atoms. For instance, the V^+ ion has the configuration $3d^4$ (and not $3d^2\,4s^2$ like titanium); the Fe^+ ion has $3d^6\,4s$ (instead of $3d^5\,4s^2$ as in manganese). We may remark that all ions found naturally in crystals and solutions contain only *d* (not *s* or *p*) electrons in their incomplete shells. Thus iron is found in crystals or solutions only as the ions Fe^{++} and Fe^{+++}, whose configurations are $3d^6$ and $3d^5$ respectively.

A similar situation occurs in the filling up of the $4f$ shell; this takes place in the series of elements known as the *rare earths* (Table 5).† The filling up of the $4f$ shell also occurs in a slightly irregular manner characterized by the competition between $4f$, $5d$ and $6s$ states.

† In books on chemistry, lutetium is also usually placed with the rare-earth elements. This, however, is incorrect, since the $4f$ shell is complete in lutetium; it must therefore be placed in the platinum group, as in Table 4.

The last group of intermediate elements begins with actinium. In this group the $6d$ and $5f$ shells are filled, similarly to what happens in the group of rare-earth elements (Table 6).

To conclude this section, let us examine an interesting application of the Thomas–Fermi method. We have seen that the electrons in the p shell first appear in the fifth element (boron), the d electrons for $Z = 21$ (scandium), and the f electrons for $Z = 58$ (cerium). These values of Z can be predicted by the Thomas–Fermi method, as follows.

An electron with orbital angular momentum l in a complex atom moves with an "effective potential energy"† of

$$U_l(r) = -\phi(r)+\tfrac{1}{2}(l+\tfrac{1}{2})^2/r^2.$$

The first term is the potential energy in an electric field described by the Thomas–Fermi potential $\phi(r)$. The second term is the centrifugal energy, in which we put $(l+\frac{1}{2})^2$ instead of $l(l+1)$, since the motion is quasi-classical. Since the total energy of the electron in the atom is negative, it is clear that, if (for given values of Z and l) $U_l(r) > 0$ for all r, there can be no electrons in the atom concerned with the given value of the angular momentum l. If we consider any definite value of l and vary Z, it is found that in fact $U_l(r) > 0$ everywhere when Z is sufficiently small. As Z is increased, a value is reached for which the curve of $U_l(r)$ touches the axis of abscissae, while for larger Z there is a region where $U_l(r) < 0$. Thus the value of Z at which electrons with the given l appear in the atom is determined by the condition that the curve of $U_l(r)$ touches the axis of abscissae, i.e. by the equations

$$U_l(r) = -\phi+\tfrac{1}{2}(l+\tfrac{1}{2})^2/r^2 = 0, \quad U_l'(r) = -\phi'(r)-(l+\tfrac{1}{2})^2/r^3 = 0.$$

Substituting here the expression (70.6) for the potential, we obtain the equations

$$\left.\begin{aligned} Z^{2/3}\chi(x)/x &= (4/3\pi)^{2/3}(l+\tfrac{1}{2})^2/x^2, \\ Z^{2/3}[x\chi'(x)-\chi(x)]/x &= -2(4/3\pi)^{2/3}(l+\tfrac{1}{2})^2/x^2. \end{aligned}\right\} \quad (73.2)$$

Dividing each side of the second equation by the corresponding side of the first, we find for x the equation

$$\chi'(x)/\chi(x) = -1/x,$$

and we then calculate Z from the first of equations (73.2). A numerical calculation gives

$$Z = 0{\cdot}155(2l+1)^3.$$

This formula determines the value of Z for which electrons with a given l first appear in the atom; the error is about 10 per cent.

Very accurate values are obtained by taking the coefficient as 0·17 instead of 0·155:

$$Z = 0{\cdot}17(2l+1)^3. \quad (73.3)$$

† As in §70, we use atomic units.

TABLE 5

Electron configurations of the atoms of the rare-earth elements

	$_{58}$Ce	$_{59}$Pr	$_{60}$Nd	$_{61}$Pm	$_{62}$Sm	$_{63}$Eu	$_{64}$Gd	$_{65}$Tb	$_{66}$Dy	$_{67}$Ho	$_{68}$Er	$_{69}$Tm	$_{70}$Yb
Xe envelope +	$4f5d\,6s^2$	$4f^3\,6s^2$	$4f^4\,6s^2$	$4f^5\,6s^2$	$4f^6\,6s^2$	$4f^7\,6s^2$	$4f^7\,5d\,6s^2$	$4f^9\,6s^2$	$4f^{10}\,6s^2$	$4f^{11}\,6s^2$	$4f^{12}\,6s^2$	$4f^{13}\,6s^2$	$4f^{14}\,6s^2$
	$^1G_4$	$^4I_{9/2}$	$^5I_4$	$^6H_{5/2}$	$^7F_0$	$^8S_{7/2}$	$^9D_2$	$^6H_{15/2}$	$^5I_8$	$^4I_{15/2}$	$^3H_6$	$^2F_{7/2}$	$^1S_0$

TABLE 6

Electron configurations of the atoms of elements in the actinide group

	$_{89}$Ac	$_{90}$Th	$_{91}$Pa	$_{92}$U	$_{93}$Np	$_{94}$Pu	$_{95}$Am	$_{96}$Cm
Rn envelope +	$6d\,7s^2$	$6d^2\,7s^2$	$5f^2\,6d\,7s^2$	$5f^3\,6d\,7s^2$	$5f^4\,6d\,7s^2$	$5f^6\,7s^2$	$5f^7\,7s^2$	$5f^7\,6d\,7s^2$
	$^2D_{3/2}$	$^3F_2$	$^4K_{11/2}$	$^5L_6$	$^6L_{11/2}$	$^7F_0$	$^8S_{7/2}$	$^9D_2$

For $l = 1, 2, 3$ this formula gives respectively, after rounding to the nearest integer, the correct values 5, 21, 58. For $l = 4$, formula (73.3) gives $Z = 124$; this means that g electrons should first appear only in the 124th element.

§74. X-ray terms

The binding energy of the inner electrons in the atom is so large that, if such an electron makes a transition into an outer unfilled shell (or is removed from the atom), the excited atom (or ion) is mechanically unstable with respect to ionization, which is accompanied by the reconstruction of the electron envelope and the formation of a stable ion. However, because of the comparatively weak interaction between the electrons in the atom, the probability of such a transition is comparatively small, so that the lifetime τ of the excited state is long. Hence the "width" $\hbar/\tau$ of the level (see §44) is so small that it is reasonable to regard the energies of an atom with an excited inner electron as discrete energy levels of "quasi-stationary" states of the atom. These levels are called *X-ray terms*.†

The X-ray terms are primarily classified according to the shell from which the electron is removed, or in which, as we say, a hole is formed. Where the electron goes has almost no effect on the energy of the atom, and hence is unimportant.

The total angular momentum of the set of electrons occupying any shell is zero. When one electron has been removed, the shell acquires some angular momentum j. For the (n, l) shell, the angular momentum j can take the values $l \pm \frac{1}{2}$. Thus we obtain levels which might be denoted by $1s_{1/2}$, $2s_{1/2}$, $2p_{1/2}$, $2p_{3/2}$, ..., where the value of j is added as a suffix to the letter giving the position of the hole. It is usual, however, to employ special symbols as follows:

$1s_{1/2}$	$2s_{1/2}$	$2p_{1/2}$	$2p_{3/2}$	$3s_{1/2}$	$3p_{1/2}$	$3p_{3/2}$	$3d_{3/2}$	$3d_{5/2}$	...
K	L_{I}	L_{II}	L_{III}	M_{I}	M_{II}	M_{III}	M_{IV}	M_{V}	...

The levels with $n = 4, 5, 6$ are similarly denoted by the letters N, O, P.

Levels with the same n (denoted by the same capital letter) lie close together and at a distance from levels with a different n. The reason for this is that, owing to the relative nearness of the inner electrons to the nucleus, they are in the almost unscreened field of the nucleus, and hence their states are hydrogen-like; the energy is, to a first approximation, $-Z^2/2n^2$ (in atomic units), i.e. depends only on n. If relativistic effects are taken into account, terms with different j are separated (cf. the discussion in §72 of the fine structure of the hydrogen levels), such as, for example, L_{I} and L_{II} from L_{III}, and M_{I} and M_{II} from M_{III} and M_{IV}. These pairs of levels are said to be *relativistic* doublets. The separation of terms with different l and the same j (for instance L_{I} and L_{II}, M_{I} and M_{II}) is due to the deviation of the field in

† The name is due to the fact that transitions between these levels cause the emission of X-rays by the atom.

which the inner electrons move from the Coulomb field of the nucleus, i.e. to the taking into account of the interaction of the electron with other electrons. These are said to be *screening* doublets. The main correction term to the "hydrogen-like" energy of the electron results from the potential due to the remaining electrons in the region near the nucleus, and is proportional to $Z^{4/3}$ (see (70.8)). However, since this correction does not depend on either n or l, it does not affect the level spacings. The principal correction terms in the level differences are therefore due to the interaction of one electron with those adjoining it. Since the distances between the inner electrons are $r \sim 1/Z$ (the Bohr radius in the field of a charge Z), the energy of this interaction is $\sim 1/r \sim Z$. Taking this correction into account, we can write the energy of an X-ray term, to the same accuracy, as $-(Z-\delta)^2/2n^2$, where $\delta = \delta(n,l)$ is a quantity small compared with Z, and may be regarded as a measure of the screening of the nuclear charge.

Terms with two and three holes may exist in the electron shells together with the X-ray terms with one hole. Since the spin–orbit interaction is strong for the inner electrons, the holes are subject to jj coupling.

The width of an X-ray term is determined by the total probability of all possible processes by rearrangement of the electron envelope of the atom so as to fill the hole in question. In the heavy atoms, transitions of the hole from a given shell to a higher one (i.e. electron transitions in the opposite direction) are the most important, and are accompanied by the emission of X-ray quanta. The probability of these "radiative" transitions, and therefore the corresponding part of the level width, increase very rapidly with the atomic number (as Z^4) but decrease towards higher levels for a given Z.

For lighter atoms (and higher levels) an important or even predominant part is played by radiationless transitions, in which the energy liberated when a hole is filled by an electron from above goes to remove another inner electron from the atom (called the *Auger effect*). As a result of this process the atom is in a state with two holes. The probabilities of these processes and the corresponding contribution to the level width are independent of the atomic number to a first approximation with respect to $1/Z$ (see the Problem)†.

PROBLEM

Find the limiting law of dependence of the Auger width of X-ray terms on atomic number when the latter is sufficiently large.

SOLUTION. The Auger transition probability is proportional to the square of a matrix element of the form

$$M = \int\int \psi_1'^* \psi_2'^* V \psi_1 \psi_2 \, dV_1 \, dV_2,$$

where ψ_1, ψ_2 and ψ_1', ψ_2' are the initial and final wave functions of the two electrons involved in the transition, and $V = e^2/r_{12}$ is their interaction energy. When Z is sufficiently large, the wave functions of the inner electrons may be regarded as hydrogen-like and the screening of the field of the nucleus by other electrons may be neglected (the wave function of the ioniza-

† As an example it may be mentioned that the Auger width of the K level is about 1 eV, and reaches values of the order of 10 eV for higher levels.

tion electron is also hydrogen-like in the region within the atom which is of importance in the integral M). If we carry out the calculations and all quantities are expressed in Coulomb units (with the constant $\alpha = Ze^2$; see §36), then the only quantity in the integral M which depends on Z is $V = 1/Zr_{12}$, so that $M \sim 1/Z$. The transition probability, and therefore the Auger width ΔE of the level, are proportional to $1/Z^2$. On returning to ordinary units (the Coulomb unit of energy being $Z^2me^4/\hbar^2$), we find that ΔE is independent of Z.

§75. Multipole moments

In classical theory, the electrical properties of a system of particles are described by its multipole moments of various orders, expressed in terms of the charges and coordinates of the particles. In the quantum theory, the definitions of these quantities are the same in form, but they must now be regarded as operators.

The first multipole moment is the *dipole moment*, defined as the vector

$$\mathbf{d} = \Sigma e\mathbf{r},$$

where the summation is over all the particles, and the suffix which numbers the particles is omitted for brevity. The matrix of this operator, like that of any polar vector (see §30), has non-zero elements only for transitions between states of different parity. The diagonal elements are therefore always zero. In other words, the mean values of the dipole moment of any system of particles (e.g. an atom) in stationary states are zero.†

The same is evidently true of all 2^l-pole moments with odd l. The components of such a moment are polynomials of odd degree l in the coordinates, which, like the components of a polar vector, change sign on inversion of the coordinates. The same parity selection rule therefore applies.

The *quadrupole moment* of a system is defined as the symmetrical tensor

$$Q_{ik} = \Sigma e(3x_ix_k - \delta_{ik}\mathbf{r}^2), \tag{75.1}$$

the sum of whose diagonal terms is zero. The determination of the values of these quantities in a particular state of a system (an atom, say) requires an averaging of the operator (75.1) over the corresponding wave function. This averaging is conveniently carried out in two stages (cf. §72).

Let $\hat{Q}_{ik}$ denote the quadrupole moment operator averaged over the electron states with a given value of the total angular momentum J (but not of its component M_J).

The operator thus averaged must be expressible in terms of operators of quantities describing the state of the atom as a whole. The only such vector

† To avoid misunderstanding it should be emphasized that this refers to a closed system of particles or to a system of particles in a centrally symmetric external electric field. For example, if the nuclei are regarded as "fixed", the above statement is valid for the electrons in an atom, but not for those in a molecule.

It is also assumed that there is no additional ("accidental") degeneracy of the energy level other than that with respect to directions of the total angular momentum. If this is not so, wave functions of stationary states can be constructed which do not have any definite parity, and the corresponding diagonal elements of the dipole moment need not vanish.

is the "vector" $\hat{\mathbf{J}}$. Thus the operator $\hat{\mathbf{Q}}_{ik}$ must have the form

$$\hat{Q}_{ik} = \frac{3Q}{2J(2J-1)}(\hat{J}_i\hat{J}_k+\hat{J}_k\hat{J}_i-\tfrac{2}{3}\hat{\mathbf{J}}^2\delta_{ik}), \tag{75.2}$$

where the expression in parentheses is constructed so as to be symmetrical in the suffixes i and k and to vanish on contraction with respect to them; the significance of the coefficient Q will be explained later. The operators $\hat{J}_i$ must here be understood as the familiar (§§27 and 54) matrices with respect to states having different values of M_J. The operator $\hat{\mathbf{J}}^2$ can, of course, be simply replaced by its eigenvalue $J(J+1)$.

Since the three components of the angular momentum $\mathbf{J}$ cannot simultaneously have definite values, the same is true of the components of the tensor Q_{ik}. For the component Q_{zz}, we have

$$\hat{Q}_{zz} = \frac{3Q}{J(2J-1)}(\hat{J}_z{}^2-\tfrac{1}{3}\hat{\mathbf{J}}^2).$$

In a state with given values of $\mathbf{J}^2 = J(J+1)$ and $J_z = M_J$, Q_{zz} also has a definite value:

$$Q_{zz} = \frac{3Q}{J(2J-1)}[M_J{}^2-\tfrac{1}{3}J(J+1)]. \tag{75.3}$$

For $M_J = J$ (when the angular momentum is "entirely" in the z-direction), we have $Q_{zz} = Q$; this quantity is usually called simply the quadrupole moment.

For $J = 0$, all the elements of the angular-momentum matrix are zero, and the operators (75.2) therefore also vanish. They likewise vanish identically when $J = \frac{1}{2}$. This is easily seen by direct multiplication of the Pauli matrices (55.7), which are the matrices of the components of any angular momentum equal to $\frac{1}{2}$.

This is no accident, but is a particular case of the general rule that the tensor of a 2^l-pole moment (with even l) is non-zero only for states of the system with total angular momentum

$$J \geqslant \tfrac{1}{2}l. \tag{75.4}$$

The tensor of a 2^l-pole moment is an irreducible tensor of rank l (see *Fields*, §41), and the condition (75.4) follows from the general angular-momentum selection rules for the matrix elements of such tensors – the condition for the diagonal matrix elements to be non-zero (§107). As already mentioned, the parity selection rules then require that l should be even.

It should also be noted that the electric multipole moments are purely "orbital" quantities; their operators do not involve the spin operators.

Hence, if the spin–orbit interaction is negligible, so that L and S are separately conserved, the matrix elements of the multipole moments are subject to selection rules with respect to the quantum number L as well as J.

PROBLEMS

PROBLEM 1. Find the relation between the operators of the quadrupole moment of an atom in states corresponding to various components of the fine structure of a level (i.e. states with different values of J but given values of L and S).

SOLUTION. In states with given values of L and S, the operator of the quadrupole moment, a purely orbital quantity, depends only on the operator $\hat{\mathbf{L}}$, and so is given by the same formula (75.2) with $\hat{\mathbf{J}}$ replaced by $\hat{\mathbf{L}}$ and with a different constant Q. The operator (75.2) is obtained by a further averaging over the state with a given value of J:

$$\hat{Q}_{ik} = \frac{3Q_J}{2J(2J-1)}[\hat{J}_i\hat{J}_k+\hat{J}_k\hat{J}_i-\tfrac{2}{3}J(J+1)\delta_{ik}]$$
$$= \frac{3Q_L}{2L(L-1)}[\overline{\hat{L}_i\hat{L}_k}+\overline{\hat{L}_k\hat{L}_i}-\tfrac{2}{3}L(L+1)\delta_{ik}]. \qquad (1)$$

It is required to find the relation between the coefficients Q_J and Q_L. To do so, we multiply equation (1) on the left by $\hat{J}_i$ and on the right by $\hat{J}_k$, sum over i and k, and take the eigenvalues of the diagonal operators. We have

$$\hat{J}_i\hat{L}_i\hat{L}_k\hat{J}_k = (\mathbf{J}.\mathbf{L})^2,$$

where, by (31.4),

$$2\mathbf{J}.\mathbf{L} = J(J+1)+L(L+1)-S(S+1).$$

The product $\hat{J}_i\,\hat{L}_k\,\hat{L}_i\,\hat{J}_k$ can be transformed by means of the formulae

$$\{\hat{L}_i\hat{L}_k\} = ie_{ikl}\hat{L}_l, \quad \{\hat{J}_i\hat{L}_l\} = ie_{ilm}\hat{L}_m,$$

as in §29, Problem; the result is

$$\hat{J}_i\hat{L}_k\hat{L}_i\hat{J}_k = (\mathbf{J}.\mathbf{L})^2-\mathbf{J}.\mathbf{L}.$$

Similary

$$\hat{J}_i\hat{J}_i\hat{J}_k\hat{J}_k = (\mathbf{J}^2)^2,$$
$$\hat{J}_i\hat{J}_k\hat{J}_i\hat{J}_k = \mathbf{J}^2(\mathbf{J}^2-1).$$

Thus we obtain from (1) the relation

$$Q_J = Q_L\frac{3\mathbf{J}.\mathbf{L}(2\mathbf{J}.\mathbf{L}-1)-2J(J+1)L(L+1)}{(J+1)(2J+3)L(2L-1)}. \qquad (2)$$

In particular, for $S = \frac{1}{2}$ this formula gives

$$\left.\begin{aligned} Q_J &= Q_L \qquad \text{for } J = L+\tfrac{1}{2},\\ Q_J &= Q_L\frac{(L-1)(2L+3)}{L(2L+1)} \qquad \text{for } J = L-\tfrac{1}{2}. \end{aligned}\right\} \qquad (3)$$

PROBLEM 2. Express the quadrupole moment of an electron (charge $-|e|$) with orbital angular momentum l in terms of the mean square of its distance from the centre.

SOLUTION. We have to average the expression

$$Q_{zz} = -|e|r^2(3\cos^2\theta-1) = -|e|r^2(3n_z^2-1)$$

over a state with given angular momentum l and component $m = l$. The mean value of the angle factor is found immediately from the formula derived in §29, Problem (where $\hat{l}_z$ must be replaced by l); the result is

$$Q_l = |e|\overline{r^2}\frac{2l}{2l+3}. \qquad (4)$$

The sign of this quantity is opposite to that of the electron charge, as it should be: a particle moving with an angular momentum in the z-direction is mainly near the plane $z = 0$, and hence $\overline{\cos^2\theta} < \frac{1}{3}$.

For an electron with a given value of $j = l \pm \frac{1}{2}$, formulae (3) give

$$Q_j = |e|\overline{r^2}(2j-1)/(2j+2). \tag{5}$$

PROBLEM 3. Determine the quadrupole moment of an atom (in the ground state) in which all ν electrons in excess of closed shells are in equivalent states with orbital angular momentum l.

SOLUTION. Since the total quadrupole moment of completed shells is zero, the quadrupole moment operator of the atom is given by the sum

$$\hat{Q}_{ik} = \frac{3|e|\overline{r^2}}{(2l-1)(2l+3)}\sum\left[\hat{l}_i\hat{l}_k+\hat{l}_k\hat{l}_i-\tfrac{2}{3}l(l+1)\delta_{ik}\right],$$

taken over the ν outer electrons (here we have used formula (4)).

Let us first suppose that $\nu \leqslant 2l+1$, i.e. at most half the places in the shell are occupied. Then, by Hund's rule (§67), the spins of all the ν electrons are parallel (so that $S = \frac{1}{2}\nu$). This means that the spin wave function of the atom is symmetrical, and the coordinate wave function therefore antisymmetrical, with respect to these electrons. Thus the electrons must all have different values of m, so that the greatest possible value of M_L (and the value of L, which is the same) is

$$L = (M_L)_{\max} = \sum_{m=l-\nu+1}^{l} m = \tfrac{1}{2}\nu(2l-\nu+1).$$

The required Q_L is the eigenvalue Q_{zz} for $M_L = L$. We therefore have

$$Q_L = \frac{6|e|\overline{r^2}}{(2l-1)(2l+3)}\sum_{m=l-\nu+1}^{l}[m^2-\tfrac{1}{3}l(l+1)],$$

whence, on calculating the sum,

$$Q_L = \frac{2l(2l-2\nu+1)}{(2l-1)(2l+3)}|e|\overline{r^2}. \tag{6}$$

The final change from Q_L to Q_J is effected by means of formula (2).

The case of an atom whose outer shell is more than half filled is reduced to the previous one by considering holes instead of electrons: the result is therefore given by the same formula (6) with the opposite sign (the "hole" charge being $+|e|$), ν being now taken not as the number of electrons but as the number of unoccupied places in the shell.

§76. An atom in an electric field

If an atom is placed in an external electric field, its energy levels are altered; this phenomenon is known as the *Stark effect*.

In an atom placed in a uniform external electric field, we have a system of electrons in an axially symmetric field (the field of the nucleus together with the external field). The total angular momentum of the atom is therefore, strictly speaking, no longer conserved; only the projection M_J of the total angular momentum $\mathbf{J}$ on the direction of the field is conserved. The states with different values of M_J have different energies, i.e. the electric field removes the degeneracy with respect to directions of the angular momentum. The removal is, however, incomplete: the states differing only in the sign of M_J are degenerate as before. For an atom in a homogeneous external electric field is symmetrical with respect to reflection in any plane passing through the axis of symmetry (i.e. the axis passing through the nucleus in the direction of the field; we shall take this as the z-axis). Hence the states obtained from one another by such a reflection must have the same energy. On reflection in a plane passing through some axis, however, the angular momentum about this axis changes sign (the direction of a positive revolution about the axis becomes that of a negative one).

We shall suppose that the electric field is so weak that the additional energy due to it is small compared with the distances between neighbouring energy levels of the atom, including the fine-structure intervals. Then, in order to calculate the displacement of the levels in the electric field, we can use the perturbation theory developed in §§38 and 39. Here the perturbation operator is the energy of the system of electrons in the homogeneous field $\mathscr{E}$, and this is

$$V = -\mathbf{d} \cdot \mathscr{E} = -\mathscr{E} d_z, \tag{76.1}$$

where $\mathbf{d}$ is the dipole moment of the system. In the zeroth approximation, the energy levels are degenerate (with respect to directions of the total angular momentum); in the present case, however, this degeneracy is unimportant, and in applying perturbation theory we can proceed as if we were dealing with non-degenerate levels. This follows from the fact that, in the matrix of the quantity d_z (as in that of the z-component of any vector), only the elements for transitions without change of M_J are not zero (see §29), and hence states with different values of M_J behave independently when perturbation theory is applied.

The displacement of the energy levels is determined, in the first approximation, by the diagonal matrix elements of the perturbation. But all the diagonal matrix elements of the dipole moment vanish (§75). Thus the splitting of the levels in an electric field is a second-order effect with respect to the field.†

Being quadratic in the field, the displacement ΔE_n of the level E_n must be of the form

$$\Delta E_n = -\tfrac{1}{2}\alpha_{ik}^{(n)}\mathscr{E}_i\mathscr{E}_k, \tag{76.2}$$

where $\alpha_{ik}^{(n)}$ is a symmetrical tensor of rank two; taking the z-axis in the direction of the field, we obtain

$$\Delta E_n = -\tfrac{1}{2}\alpha_{zz}^{(n)}\mathscr{E}^2. \tag{76.3}$$

The tensor $\alpha_{ik}^{(n)}$ is also the *polarizability* of the atom in the external field: taking the parameters λ in the general formula (11.16) to be the components of the vector $\mathscr{E}_i$, and putting $\hat{H} = \hat{H}_0 - \mathscr{E}_i d_i$, we find that the mean value

† The hydrogen atom forms an exception; here the Stark effect is linear in the field (see the next section). The atoms of other elements, when in highly excited states (and therefore hydrogen-like; see §68), behave like hydrogen in sufficiently strong fields.

of the dipole moment of the atom, induced by the field, is

$$\overline{d_i}^{(n)} = \partial \Delta E_n / \partial \mathscr{E}_i.$$

Substitution of (76.2) gives

$$\overline{d_i}^{(n)} = \alpha_{ik}^{(n)} \mathscr{E}_k. \tag{76.4}$$

The polarizability must be calculated by the usual rules of perturbation theory. According to the second-approximation formula (38.10), we have

$$\alpha_{ik}^{(n)} = -2\sum_m{}' \frac{(d_i)_{nm}(d_k)_{mn}}{E_n - E_m}. \tag{76.5}$$

The polarizability of the atom depends on its (unperturbed) state, and in particular on the quantum number M_J. The latter dependence can be written in a general form. The values of $\alpha_{ik}^{(n)}$ for various values of M_J may be regarded as the eigenvalues of the operator

$$\hat{\alpha}_{ik}^{(n)} = \alpha_n \delta_{ik} + \beta_n(\hat{J}_i\hat{J}_k + \hat{J}_k\hat{J}_i - \tfrac{2}{3}\delta_{ik}\hat{\mathbf{J}}^2). \tag{76.6}$$

This is the most general symmetrical tensor of rank 2 depending on the vector $\hat{\mathbf{J}}$ (cf. §75). From (76.3) and (76.6) we have

$$\Delta E_n = -\tfrac{1}{2}\mathscr{E}^2\{\alpha_n + 2\beta_n[M_J^2 - \tfrac{1}{3}J(J+1)]\}. \tag{76.7}$$

On summation over all values of M_J, the second term in the braces vanishes, so that the first term is the displacement of the "centre of gravity" of the split level. Moreover, according to (76.7) a level with $J = \frac{1}{2}$ remains unsplit, in accordance with Kramers' theorem (§60).

If the atom is in a non-uniform external field (which varies only slightly over the dimensions of the atom), there can also exist a splitting effect linear in the field, due to the quadrupole moment of the atom. The operator of the quadrupole interaction between the system and the field has the form which corresponds to the classical expression (*Fields*, §42) for the quadrupole energy:

$$\hat{V} = \tfrac{1}{6}\frac{\partial^2 \phi}{\partial x_i \partial x_k}\hat{Q}_{ik}, \tag{76.8}$$

where ϕ is the potential of the electric field (the derivatives being understood to be taken at the position of the atom).

PROBLEMS

PROBLEM 1. Determine the Stark splitting of the different components of a multiplet level as a function of J.

SOLUTION. The problem is conveniently solved by changing the order in which the perturbations are applied; we first consider the Stark splitting of the level in the absence of fine structure, and then bring in the spin–orbit interaction. Since the spin of the atom does not interact with the external electric field, the Stark splitting of a level with orbital angular momentum L is given by a formula of the same form (76.2), with a tensor $\hat{\alpha}_{ik}$ which is expressed in terms of the operator $\hat{\mathbf{L}}$ in the same way as $\hat{\alpha}_{ik}$ in (76.6) is expressed in terms of $\hat{\mathbf{J}}$:

$$\hat{\alpha}_{ik} = a\delta_{ik} + b(\hat{L}_i\hat{L}_k + \hat{L}_k\hat{L}_i - \tfrac{2}{3}\delta_{ik}\hat{\mathbf{L}}^2),$$

the suffixes n being everywhere omitted. When the spin–orbit interaction is included, the states of the atom must be described by the total angular momentum J. The averaging of the operator $\hat{\alpha}_{ik}$ over states with a given value of the angular momentum J (but not of its component M_J) is formally identical with the averaging carried out in §75, Problem 1. We thus return to formulae (76.6), (76.7), with constants α, β which are given in terms of the constants a, b by

$$\alpha = a,\ \beta = b\frac{3\mathbf{J}.\mathbf{L}[2\mathbf{J}.\mathbf{L}-1] - 2J(J+1)L(L+1)}{J(J+1)(2J-1)(2J+3)}.$$

This determines the splitting as a function of J (but not of L and S, of course; these are characteristics of the unsplit term on which the constants a and b also depend).

PROBLEM 2. Determine the splitting of a doublet level (spin $S = \frac{1}{2}$) in an arbitrary (not weak) electric field.

SOLUTION. If the splitting is not small in comparison with the interval between the components of the doublet, the perturbation from the electric field and the spin–orbit interaction must be taken into account simultaneously, i.e. the perturbation operator is the sum

$$\hat{V} = A\hat{\mathbf{S}}.\hat{\mathbf{L}} - \tfrac{1}{2}\mathscr{E}^2\{a + 2b[\hat{L}_z^2 - \tfrac{1}{3}L(L+1)]\}$$

(cf. (72.4) and Problem 1). Omitting the constant terms which do not affect the splitting, we can write this operator in the form

$$\hat{V} = \tfrac{1}{2}A[\hat{S}_+\hat{L}_- + \hat{S}_-\hat{L}_+ + 2\hat{S}_z\hat{L}_z] - b\mathscr{E}^2\hat{L}_z^2$$

(see (29.11)). For each given value of $M \equiv M_J$ the eigenvalues of this operator are determined by the roots of the secular equation formed from its matrix elements with respect to the states $|M_LM_S\rangle = |M \mp \frac{1}{2}, \pm\frac{1}{2}\rangle$. From formulae (27.12) we find

$$\langle M-\tfrac{1}{2}, \tfrac{1}{2}|V|M-\tfrac{1}{2}, \tfrac{1}{2}\rangle = \tfrac{1}{2}A(M-\tfrac{1}{2}) - b\mathscr{E}^2(M-\tfrac{1}{2})^2,$$
$$\langle M+\tfrac{1}{2}, -\tfrac{1}{2}|V|M+\tfrac{1}{2}, -\tfrac{1}{2}\rangle = -\tfrac{1}{2}A(M+\tfrac{1}{2}) - b\mathscr{E}^2(M+\tfrac{1}{2})^2,$$
$$\langle M-\tfrac{1}{2}, \tfrac{1}{2}|V|M+\tfrac{1}{2}, -\tfrac{1}{2}\rangle = \tfrac{1}{2}A\sqrt{[(L+M+\tfrac{1}{2})(L-M+\tfrac{1}{2})]}.$$

Thus (see §39, Problem 1) the level displacement is

$$\Delta E = -b\mathscr{E}^2M^2 \pm \sqrt{[\tfrac{1}{4}A^2(L+\tfrac{1}{2})^2 + b\mathscr{E}^2(b\mathscr{E}^2 + A)M^2]}, \tag{1}$$

where all terms which are the same for all components of the split doublet are omitted. This formula (with both signs of the root) applies to all levels with $|M| \leqslant L-\frac{1}{2}$. For $|M| = L-\frac{1}{2}$ there is only one state $|M_LM_S\rangle$, and the displacement of the level is given simply by the corresponding diagonal matrix element, i.e., with the same choice of the additive constant as in (1),

$$\Delta E = (\tfrac{1}{2}A + b\mathscr{E}^2)(L+\tfrac{1}{2}) - b\mathscr{E}^2(L+\tfrac{1}{2})^2. \tag{2}$$

This is the same as the result obtained from formula (1) with only one sign of the root.

PROBLEM 3. Determine the quadrupole splitting of levels in an axially symmetric electric field.†

† A similar problem for any field is in §103, Problem 6.

SOLUTION. In a field symmetrical about the z-axis we have $\partial^2\phi/\partial x^2 = \partial^2\phi/\partial y^2 \equiv a$, $\partial^2\phi/\partial z^2 = -2a$, the remaining second derivatives being zero. The quadrupole energy operator (76.8) is

$$\frac{a}{6}(\hat{Q}_{xx}+\hat{Q}_{yy}-2\hat{Q}_{zz}) = \frac{Qa}{2J(2J-1)}(\hat{\mathbf{J}}^2-3\hat{J}_z^2).$$

Replacing the operators by their eigenvalues, we obtain for the displacement of the levels

$$\Delta E = a\frac{Q}{2J(2J-1)}[J(J+1)-3M_J^2].$$

PROBLEM 4. Calculate the polarizability of the hydrogen atom in the ground state.

SOLUTION. Owing to the spherical symmetry of the s state, the polarizability tensor is a scalar ($\alpha_{ik} = \alpha\delta_{ik}$), for which we have, according to (76.5),

$$\alpha = -2e^2 \sum_k{}' \frac{|z_{0k}|^2}{E_0-E_k};$$

the dipole moment of the electron is $d_z = ez$, and E_0 is the energy of the ground state. We define an auxiliary operator $\hat{b}$ by

$$z = \frac{m}{\hbar}\frac{d\hat{b}}{dt},$$

where m is the electron mass. Then $z_{0k} = (im/\hbar^2)(E_0-E_k)b_{0k}$, and

$$\alpha = \frac{2ime^2}{\hbar^2}\sum_k z_{0k}b_{k0} = \frac{2ime^2}{\hbar^2}(z\hat{b})_{00}. \tag{1}$$

To calculate this quantity, we need only know the result of applying $\hat{b}$ to the wave function $\psi_0(r)$.

According to (9.2),

$$z\psi_0 = \frac{m}{\hbar}\frac{d\hat{b}}{dt}\psi_0 = \frac{im}{\hbar}(\hat{H}\hat{b}-\hat{b}\hat{H})\psi_0.$$

Denoting the function $\hat{b}\psi_0$ by $b(\mathbf{r})\psi_0$ and noticing that ψ_0 satisfies the equation $\hat{H}\psi_0 = E_0\psi_0$, where $\hat{H} = -\hbar^2\triangle/2m+U(\mathbf{r})$, we obtain for $b(\mathbf{r})$ the differential equation

$$\tfrac{1}{2}\psi_0\triangle b+\nabla b.\nabla\psi_0 = iz\psi_0.$$

By substituting $b = f(r)\cos\theta$ (where θ is the polar angle in spherical polar coordinates and $z = r\cos\theta$), this becomes

$$\tfrac{1}{2}f'' + \frac{f'}{r} - \frac{f}{r^2} + \frac{\psi_0'}{\psi_0}f' = ir. \tag{2}$$

Its solution must satisfy the condition that $f\psi_0$ is finite as $r \to 0$ and as $r \to \infty$.

For the ground state of the hydrogen atom, $\psi_0 = (1/\sqrt{\pi})\exp(-r/a_B)$, where $a_B = \hbar^2/me^2$ is the Bohr radius. The solution of equation (2) which satisfies the condition stated is $f = -ira_B(a_B+\frac{1}{2}r)$. From formula (1) we now have†

$$\alpha = \frac{2i}{a_B}(rf\cos^2\theta)_{00} = \frac{2i}{3a_B}(rf)_{00} = \frac{9}{2}a_B^3.$$

PROBLEM 5. Calculate the polarizability of an electron in a bound s state in a potential well with force range a such that $a\kappa \ll 1$, with $\kappa = \sqrt{(2m|E_0|)}/\hbar$ and E_0 the electron binding energy.

SOLUTION. From the condition $a\kappa \ll 1$, in calculating the matrix element $(z\hat{b})_{00}$ we can neglect the region within the well, and use in all space the wave function

$$\psi_0 = \sqrt{\frac{\kappa}{2\pi}}\frac{e^{-\kappa r}}{r},$$

† This result will be derived by a different method in §77.

which relates to the region outside the well. (The normalization of this function also uses the condition $a\kappa \ll 1$; see §133.) Equation (2) in Problem 4 becomes

$$\tfrac{1}{2}f'' - \kappa f' - \frac{f}{r^2} = ir,$$

and its solution satisfying the boundary conditions is $f = -ir^2/2\kappa$. A calculation using formula (1) gives

$$\alpha = me^2/4\hbar^2\kappa^4.$$

§77. A hydrogen atom in an electric field

The levels of the hydrogen atom, unlike those of other atoms, undergo a splitting proportional to the field (the *linear Stark effect*) in a uniform electric field. This is due to the occurrence of an accidental degeneracy in the hydrogen terms, whereby states with different l (for a given principal quantum number n) have the same energy. The matrix elements of the dipole moment for transitions between these states are not zero, and hence the secular equation gives a non-zero displacement of the levels, even in the first approximation.†

For purposes of calculation it is convenient to choose the unperturbed wave functions so that the perturbation matrix is diagonal with respect to each group of mutually degenerate states. It is found that this is achieved by quantizing the hydrogen atom in parabolic coordinates. The wave functions $\psi_{n_1 n_2 m}$ of the stationary states of the hydrogen atom in parabolic coordinates are given by formulae (37.15) and (37.16).

The perturbation operator (the energy of an electron in the field $\mathscr{E}$) is $\mathscr{E}z = \frac{1}{2}\mathscr{E}(\xi-\eta)$; the field is directed along the positive z-axis, and the force on the electron along the negative z-axis.‡ We are interested in the matrix elements for transitions $n_1 n_2 m \rightarrow n_1' n_2' m'$, for which the energy (i.e. the principal quantum number n) is unaltered. It is easy to see that, of these, only the diagonal matrix elements

$$\int |\psi_{n_1 n_2 m}|^2 \mathscr{E}z \,\mathrm{d}V = \tfrac{1}{8}\mathscr{E}\int_0^\infty\int_0^\infty\int_0^{2\pi} (\xi^2-\eta^2)|\psi_{n_1 n_2 m}|^2 \,\mathrm{d}\phi\,\mathrm{d}\xi\,\mathrm{d}\eta$$

$$= \tfrac{1}{4}\mathscr{E}\int_0^\infty\int_0^\infty f_{n_1 m}^2(\rho_1) f_{n_2 m}^2(\rho_2)(\rho_1^2-\rho_2^2)\,\mathrm{d}\rho_1\mathrm{d}\rho_2 \qquad (77.1)$$

are non-zero (we have made the substitution $\xi = n\rho_1$, $\eta = n\rho_2$). The matrix concerned is evidently diagonal with respect to the number m, while its diagonality with respect to the numbers n_1, n_2 follows from the orthogonality of the functions $f_{n_1 m}$ for different n_1 and the same n (see below). The integrations over ρ_1 and ρ_2 in (77.1) are separable; the integrals obtained are calcu-

† In the following calculations we do not take account of the fine structure of the hydrogen levels. Hence the field must be, though not strong (for perturbation theory to be applicable), yet such that the Stark splitting is large in comparison with the fine structure. For the opposite case see *RQT*, §52, Problem.

‡ In this section we use atomic units.

lated in §f of the Mathematical Appendices (integral (f.6)). After a simple calculation, we find for the corrections to the energy levels in the first approximation†

$$E^{(1)} = \tfrac{3}{2}\mathscr{E}n(n_1-n_2), \tag{77.2}$$

or, in ordinary units,

$$E^{(1)} = \tfrac{3}{2}n(n_1-n_2)|e|\mathscr{E}\hbar^2/me^2.$$

The two extreme components of the split level correspond to $n_1 = n-1$, $n_2 = 0$ and $n_1 = 0$, $n_2 = n-1$. The distance between these two extreme levels is, by (77.2),

$$3\mathscr{E}n(n-1),$$

i.e. the total splitting of the level by the Stark effect is approximately proportional to n^2. It is natural that the splitting should increase with the principal quantum number: the further the electrons are from the nucleus, the greater the dipole moment of the atom.

The presence of the linear effect means that, in the unperturbed state, the atom has a dipole moment whose mean value is

$$\overline{d_z} = -\tfrac{3}{2}n(n_1-n_2). \tag{77.3}$$

This is in accordance with the fact that, in a state determined by parabolic quantum numbers, the distribution of the charges in the atom is not symmetrical about the plane $z = 0$ (see §37). Thus, for $n_1 > n_2$, the electron is predominantly on the side of positive z, and hence the atom has a dipole moment opposite to the external field (the charge on the electron being negative).

In the previous section we have shown that a uniform electric field cannot entirely remove degeneracy: there always remains a twofold degeneracy of states differing in the sign of the projection of the angular momentum on the direction of the field (in this case, states whose projected angular momenta are $\pm m$). However, we see from formula (77.2) that even this removal of the degeneracy does not occur in the linear Stark effect in hydrogen: the displacement of the levels (for given n and n_1-n_2) is independent of m and n_2. A further removal of the degeneracy occurs in the second approximation; the calculation of this effect is the more interesting in that the linear Stark effect is altogether absent in states with $n_1 = n_2$.

To calculate the quadratic effect, it is not convenient to use ordinary perturbation theory, since it would be necessary to deal with infinite sums of complicated form. Instead we use the following slightly modified method.

Schrödinger's equation for the hydrogen atom in a uniform electric field is of the form

$$(\tfrac{1}{2}\triangle+E+1/r-\mathscr{E}z)\psi = 0.$$

† This result was derived by K. Schwarzschild and P. Epstein (1916), using the old quantum theory, and by W. Pauli and E. Schrödinger (1926) using quantum mechanics.

Like the equation with $\mathscr{E} = 0$, it allows separation of the variables in parabolic coordinates. The same substitution (37.7) as was used in §37 gives the two equations

$$\left.\begin{aligned} \frac{\mathrm{d}}{\mathrm{d}\xi}\left(\xi\frac{\mathrm{d}f_1}{\mathrm{d}\xi}\right)+\left(\tfrac{1}{2}E\xi-\tfrac{1}{4}\frac{m^2}{\xi}-\tfrac{1}{4}\mathscr{E}\xi^2\right)f_1 &= -\beta_1 f_1,\\ \frac{\mathrm{d}}{\mathrm{d}\eta}\left(\eta\frac{\mathrm{d}f_2}{\mathrm{d}\eta}\right)+\left(\tfrac{1}{2}E\eta-\tfrac{1}{4}\frac{m^2}{\eta}+\tfrac{1}{4}\mathscr{E}\eta^2\right)f_2 &= -\beta_2 f_2,\\ \beta_1+\beta_2 &= 1, \end{aligned}\right\} \tag{77.4}$$

which differ from (37.8) by the presence of the terms in $\mathscr{E}$. We shall regard the energy E in these equations as a parameter which has a definite value, and the quantities β_1, β_2 as eigenvalues of corresponding operators; it is easy to see that these operators are self-conjugate. These quantities are determined, by solving the equations, as functions of E and $\mathscr{E}$, and then the condition $\beta_1+\beta_2 = 1$ gives the energy as a function of the external field.

For an approximate solution of equations (77.4), we regard the terms containing the field $\mathscr{E}$ as a small perturbation. In the zeroth approximation ($\mathscr{E} = 0$), the equations have the familiar solutions

$$\left.\begin{aligned} f_1 &= \sqrt{\epsilon}\, f_{n_1 m}(\xi\epsilon),\\ f_2 &= \sqrt{\epsilon}\, f_{n_2 m}(\eta\epsilon), \end{aligned}\right\} \tag{77.5}$$

where the functions $f_{n_1 m}$ are the same as in (37.16), and instead of the energy we have introduced the parameter

$$\epsilon = \sqrt{(-2E)}. \tag{77.6}$$

The corresponding values of β_1, β_2 (from the equations (37.12), in which n must be replaced by $1/\epsilon$) are

$$\beta_1^{(0)} = (n_1+\tfrac{1}{2}|m|+\tfrac{1}{2})\epsilon, \quad \beta_2^{(0)} = (n_2+\tfrac{1}{2}|m|+\tfrac{1}{2})\epsilon. \tag{77.7}$$

The functions f_1 with different n_1 for a given ϵ are orthogonal, as are the eigenfunctions of any self-conjugate operator; we have already used this fact above in discussing the linear effect. In (77.5) these functions are normalized by the conditions

$$\int_0^\infty f_1^2\,\mathrm{d}\xi = 1, \qquad \int_0^\infty f_2^2\,\mathrm{d}\eta = 1.$$

The corrections to β_1 and β_2 in the first approximation are determined by the diagonal matrix elements of the perturbation:

$$\beta_1^{(1)} = \tfrac{1}{4}\mathscr{E}\int_0^\infty \xi^2 f_1^2\,\mathrm{d}\xi, \qquad \beta_2^{(1)} = -\tfrac{1}{4}\mathscr{E}\int_0^\infty \eta^2 f_2^2\,\mathrm{d}\eta.$$

Calculation gives

$$\beta_1^{(1)} = \tfrac{1}{4}\mathscr{E}(6n_1^2+6n_1|m|+m^2+6n_1+3|m|+2)/\epsilon^2.$$

The expression for $\beta_2^{(1)}$ is obtained by replacing n_1 by n_2 and changing the sign.

In the second approximation we have, by the general formulae of perturbation theory,

$$\beta_1^{(2)} = \frac{\mathscr{E}^2}{16}\sum_{n_1' \neq n_1} \frac{|(\xi^2)_{n_1 n_1'}|^2}{\beta_1^{(0)}(n_1)-\beta_1^{(0)}(n_1')}.$$

The integrals appearing in the matrix elements $(\xi^2)_{n_1 n_1'}$ are calculated in §f of the Mathematical Appendices. The only non-zero elements are

$$(\xi^2)_{n_1,n_1-1} = (\xi^2)_{n_1-1,n_1} = -2(2n_1+|m|)\sqrt{[n_1(n_1+|m|)]}/\epsilon^2,$$

$$(\xi^2)_{n_1,n_1-2} = (\xi^2)_{n_1-2,n_1} = \sqrt{[n_1(n_1-1)(n_1+|m|)(n_1+|m|-1)]}/\epsilon^2.$$

The differences occurring in the denominators are

$$\beta_1^{(0)}(n_1)-\beta_1^{(0)}(n_1') = \epsilon(n_1-n_1').$$

As a result of the calculations we have

$$\beta_1^{(2)} = -\mathscr{E}^2(|m|+2n_1+1)[4m^2+17(2|m|n_1+2n_1^2+|m|+2n_1)+18]/16\epsilon^5;$$

the expression for $\beta_2^{(2)}$ is obtained by replacing n_1 by n_2. Combining the expressions obtained and substituting in the relation $\beta_1+\beta_2 = 1$, we have the equation

$$\epsilon n-\mathscr{E}^2 n[17n^2+51(n_1-n_2)^2-9m^2+19]/16\epsilon^5+\tfrac{3}{2}\mathscr{E}n(n_1-n_2)/\epsilon^2 = 1.$$

Solving by successive approximations, we have in the second approximation for the energy $E = -\frac{1}{2}\epsilon^2$ the expression

$$E = -\frac{1}{2n^2}+\tfrac{3}{2}\mathscr{E}n(n_1-n_2)-\frac{\mathscr{E}^2}{16}n^4[17n^2-3(n_1-n_2)^2-9m^2+19]. \qquad (77.8)$$

The second term is the already familiar linear Stark effect, and the third is the required quadratic effect (G. Wentzel, I. Waller and P. Epstein 1926). We notice that this quantity is always negative, i.e. the terms are always displaced downwards by the quadratic effect. The mean value of the dipole moment is obtained by differentiating (77.8) with respect to the field; in the states with $n_1 = n_2$ it is

$$\overline{d_z} = \tfrac{1}{8}n^4(17n^2-9m^2+19)\mathscr{E}. \qquad (77.9)$$

Thus the polarizability of the hydrogen atom in the normal state ($n = 1$, $m = 0$) is 9/2 (see also §76, Problem 4).

The absolute value of the energy of the hydrogen terms falls rapidly as the principal quantum number n increases, while the Stark splitting is increased. Hence it is of interest to examine the Stark effect for highly excited levels in fields so strong that the splitting they cause is comparable with the energy of the level itself, and perturbation theory is inapplicable.† This can be done by using the fact that states with large values of n are quasi-classical.

By the substitution

$$f_1 = \chi_1/\sqrt{\xi}, \qquad f_2 = \chi_2/\sqrt{\eta} \tag{77.10}$$

the equations (77.4) are brought into the form

$$\left.\begin{aligned} \frac{d^2\chi_1}{d\xi^2}+\left(\tfrac{1}{2}E+\frac{\beta_1}{\xi}-\frac{m^2-1}{4\xi^2}-\tfrac{1}{4}\mathscr{E}\xi\right)\chi_1 &= 0,\\ \frac{d^2\chi_2}{d\eta^2}+\left(\tfrac{1}{2}E+\frac{\beta_2}{\eta}-\frac{m^2-1}{4\eta^2}+\tfrac{1}{4}\mathscr{E}\eta\right)\chi_2 &= 0. \end{aligned}\right\} \tag{77.11}$$

Each of these equations, however, is the same in form as the one-dimensional Schrödinger's equation, the part of the total energy of the particle being taken by $\frac{1}{4}E$, and that of the potential energy by the functions

$$\left.\begin{aligned} U_1(\xi) &= -\frac{\beta_1}{2\xi}+\frac{m^2-1}{8\xi^2}+\tfrac{1}{8}\mathscr{E}\xi,\\ U_2(\eta) &= -\frac{\beta_2}{2\eta}+\frac{m^2-1}{8\eta^2}-\tfrac{1}{8}\mathscr{E}\eta \end{aligned}\right\} \tag{77.12}$$

respectively.

Figures 25 and 26 respectively show the approximate form of these functions (for $m > 1$). By Bohr and Sommerfeld's quantization rule (48.2) we write

$$\left.\begin{aligned} \int_{\xi_1}^{\xi_2} \sqrt{\{2[\tfrac{1}{4}E-U_1(\xi)]\}}\, d\xi &= (n_1+\tfrac{1}{2})\pi,\\ \int_{\eta_1}^{\eta_2} \sqrt{\{2[\tfrac{1}{4}E-U_2(\eta)]\}}\, d\eta &= (n_2+\tfrac{1}{2})\pi, \end{aligned}\right\} \tag{77.13}$$

† The applicability of perturbation theory to high levels requires the perturbation to be small only in comparison with the energy of the level itself (the binding energy of the electron), and not with the intervals between the levels. For in the quasi-classical case (which corresponds to highly excited states) the perturbation can be regarded as small if the force due to it is small in comparison with those acting on the particle in the unperturbed system; and this condition is equivalent to the one given above.

where n_1, n_2 are integers.† These equations determine implicitly the dependence of the parameters β_1 and β_2 on E. Together with the equation $\beta_1+\beta_2=1$, they therefore give the energies of the levels when displaced by the electric field. The integrals in equations (77.13) can be reduced to elliptic integrals; these equations can be solved only numerically.

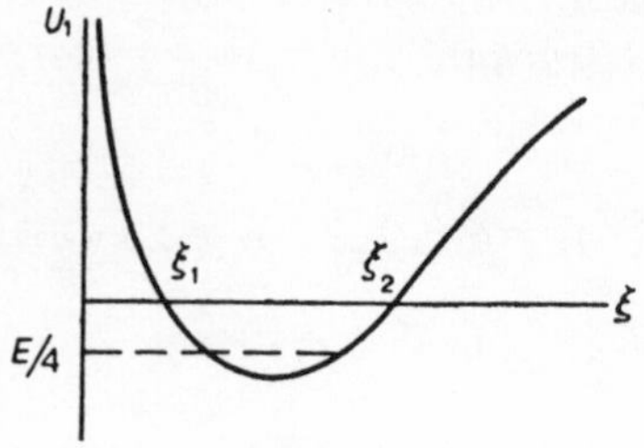

FIG. 25

The Stark effect in strong fields is complicated by another phenomenon, the ionization of the atom by the electric field (C. Lanczos 1931). The potential energy $\mathscr{E}z$ of an electron in the external field takes arbitrarily large negative values as $z \to -\infty$. Added to the potential energy of the electron within the atom, it has the effect that the region of possible motion for the electron (whose total energy E is negative) includes, besides the region inside the atom, the region of large distances from the nucleus in the direction of the anode. These two regions are separated by a potential barrier, whose width diminishes as the field increases. However, in quantum mechanics there is always a certain non-zero probability that a particle will penetrate a potential barrier. In the case we are considering, the emergence of the electron from the region within the atom, through the barrier, is simply the ionization of the atom. In weak fields the probability of this ionization is

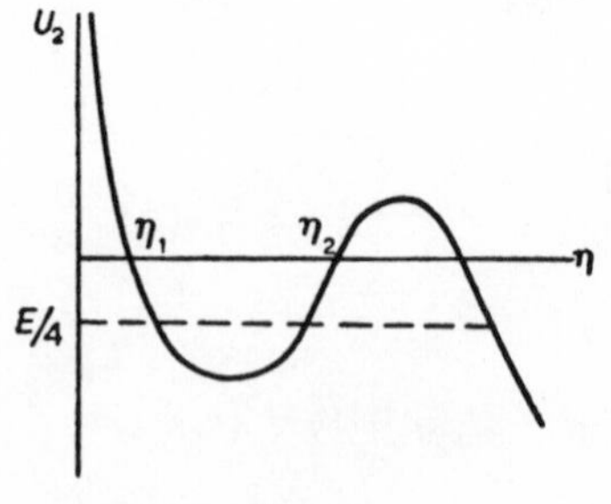

FIG. 26

† A detailed investigation shows that a more exact result is obtained by writing m^2 instead of m^2-1 in the expressions for U_1 and U_2. The integers n_1, n_2 are then equal to the parabolic quantum numbers.

vanishingly small. It increases exponentially with the field, however, and becomes considerable in fairly strong fields.†

PROBLEMS

PROBLEM 1. Determine the probability (per unit time) of the ionization of a hydrogen atom (in the ground state) in an electric field such that $\mathscr{E} \ll 1$ (in ordinary units, $\mathscr{E} \ll m^2|e|^5/\hbar^4$).

SOLUTION.‡ In parabolic coordinates there is a potential barrier "along the η coordinate" (Fig. 26); the "extraction" of the electron from the atom in the direction $z \to -\infty$ corresponds to its passage into the region of large η. To determine the ionization probability, it is necessary to investigate the form of the wave function for large η (and small ξ; we shall see below that small values of ξ are the important ones in the integral which determines the total probability current for the emerging electron). The wave function of the electron in the normal state (in the absence of the field) is

$$\psi = e^{-(\xi+\eta)/2}/\sqrt{\pi}. \tag{1}$$

When the field is present, the dependence of ψ on ξ in the region in which we are interested can be regarded as being the same as in (1), while to determine its dependence on η we have the equation

$$\frac{\partial^2\chi}{\partial\eta^2}+\left[-\tfrac{1}{4}+\frac{1}{2\eta}+\frac{1}{4\eta^2}+\tfrac{1}{4}\mathscr{E}\eta\right]\chi = 0, \tag{2}$$

where $\chi = \sqrt{\eta}\psi$ (the second equation (77.11) with $E = -\frac{1}{2}$, $m = 0$, $\beta_2 = \frac{1}{2}$). Let η_0 be some value of η (within the barrier) such that $1 \ll \eta_0 \ll 1/\mathscr{E}$. For $\eta \gtrsim \eta_0$, the wave function is quasi-classical. Since, on the other hand, equation (2) has the form of the one-dimensional Schrödinger's equation, we can use formulae (50.2). Using as the boundary condition that ψ must become the wave function (1) at $\eta = \eta_0$, we obtain in the region outside the barrier the expression

$$\chi = \left(\frac{\eta_0|p_0|}{\pi p}\right)^{1/2} \exp\left(\frac{\xi+\eta_0}{2} + i\int_{\eta_0}^{\eta_1} p\,d\eta+\tfrac{1}{4}i\pi\right),$$

where

$$p(\eta) = \sqrt{\left[-\tfrac{1}{4}+\frac{1}{2\eta}+\frac{1}{4\eta^2}+\tfrac{1}{4}\mathscr{E}\eta\right]}.$$

We shall be interested only in the square $|\chi|^2$. Hence the imaginary part of the exponent is unimportant. Denoting by η_1 the root of the equation $p(\eta) = 0$, we have

$$|\chi|^2 = \frac{\eta_0}{\pi}\frac{|p_0|}{p}e^{-\xi}\exp[-2\int_{\eta_0}^{\eta_1}|p|\,d\eta-\eta_0]. \tag{3}$$

---

† This phenomenon may serve to illustrate how a small perturbation may alter the nature of the energy spectrum. Even a weak field $\mathscr{E}$ is sufficient to create a potential barrier and produce a region, far from the nucleus, which is in principle accessible to the electron. As a result, the motion of the electron becomes, strictly speaking, infinite, and hence the energy spectrum becomes continuous instead of discrete. Nevertheless, the formal solution obtained by the methods of perturbation theory has a physical significance: it gives the energy levels of states which are not quite but "almost" stationary. An atom that is in such a state at some initial instant remains in it for a long period of time.

However, the series given by perturbation theory for the Stark splitting of the levels cannot be convergent in the strict sense, but is merely an asymptotic series: after a certain point in the series (which becomes later as the perturbation is reduced in magnitude) the terms increase, not decrease.

‡ In this problem we use atomic units.

In the coefficient of the exponential we put for $\eta \gg 1$

$$|p_0| \approx \tfrac{1}{2}, p \approx \tfrac{1}{2}\sqrt{(\mathscr{E}\eta-1)};$$

in the exponent we must keep also the next term of the expansion of $p(\eta)$:

$$|\chi|^2 = \frac{\eta_0}{\pi\sqrt{(\mathscr{E}\eta-1)}}e^{-\xi}\exp\left[-\int_{\eta_0}^{\eta_1}\sqrt{(1-\mathscr{E}\eta)}\,d\eta+\int_{\eta_0}^{\eta_1}\frac{d\eta}{\eta\sqrt{(1-\mathscr{E}\eta)}}-\eta_0\right],$$

where $\eta_1 \approx 1/\mathscr{E}$. Effecting the integration and neglecting $\eta_0\mathscr{E}$ compared with 1 wherever possible, we obtain

$$|\chi|^2 = \frac{4}{\pi\mathscr{E}}e^{-2/3\mathscr{E}}\frac{e^{-\xi}}{\sqrt{(\mathscr{E}\eta-1)}}. \tag{4}$$

The total probability current through a plane perpendicular to the z-axis (i.e. the required ionization probability w) is

$$w = \int_0^\infty |\psi|^2 v_z 2\pi\rho\,d\rho,$$

where ρ is the cylindrical radius in this plane. For large η (and small ξ) we can put

$$d\rho = d\sqrt{(\xi\eta)} \approx \tfrac{1}{2}\sqrt{(\eta/\xi)}\,d\xi.$$

Substituting also for the velocity of the electron

$$v_z \approx \sqrt{[2(-\tfrac{1}{2}+\tfrac{1}{2}\mathscr{E}\eta)]} = \sqrt{(\mathscr{E}\eta-1)},$$

we have

$$w = \int_0^\infty |\chi|^2\pi\sqrt{(\mathscr{E}\eta-1)}\,d\xi,$$

that is, finally,

$$w = (4/\mathscr{E})e^{-2/3\mathscr{E}}, \tag{5}$$

or, in ordinary units,

$$w = (4m^3|e|^9/\mathscr{E}\hbar^7)\exp(-2m^2|e|^5/3\mathscr{E}\hbar^4).$$

PROBLEM 2. Find the probability that an electron will be removed by an electric field from a potential well with short-range forces, in which the electron is in a bound s state. The electric field is assumed weak, in the sense that $|e|\mathscr{E} \ll \hbar^2\kappa^3/m$, where $\kappa = \sqrt{(2m|E|)}/\hbar$, E is the binding energy of the electron in the well and m is the electron mass (Yu. N. Demkov and G. F. Drukarev 1964).

SOLUTION. As in Problem 1, for a weak electric field, large distances from the centre ($\kappa r \gg 1$) are important. At these distances the wave function of the bound state of the electron in the well (without the field $\mathscr{E}$) has the asymptotic form

$$\psi = \frac{A\sqrt{\kappa}}{r}e^{-\kappa r},$$

where A is a dimensionless constant depending on the specific form of the well.† In parabolic coordinates we have $r = \tfrac{1}{2}(\xi+\eta)$, and in the region $\eta \gg \xi$ the wave function has the form

$$\psi \approx \frac{2A\sqrt{\kappa}}{\eta}\exp[-\tfrac{1}{2}\kappa(\xi+\eta)]. \tag{6}$$

† For example, if the radius a of the well is so small that $a\kappa \ll 1$, then $A = 1/\sqrt{(2\pi)}$; see §133.

In the rest of this solution the units of mass, length and time will be m, $1/\kappa$ and $m/\hbar\kappa^2$ respectively.

The function (6) is a product of functions of ξ and η. In the presence of an electric field, the dependence of ψ on ξ may be taken to be the same as in (6) (cf. Problem 1). To determine its dependence on η, we use Schrödinger's equation in parabolic coordinates. Unlike the case of the Coulomb field, the rapid decrease of the field of the well has the result that this field may be neglected at the large distances important in the problem. The separation of variables in Schrödinger's equation then gives again the equations (77.11), in which we must put $E = -\frac{1}{2}$, $m = 0$, and the separation parameters now satisfy the condition

$$\beta_1 + \beta_2 = 0.$$

The parameter β_1 must be taken as $\frac{1}{2}$ (so that the dependence $\psi \sim e^{-\xi/2}$ satisfies the first equation (77.11)—approximately for small $\xi\mathscr{E}$); then $\beta_2 = -\frac{1}{2}$, and the equation for ψ as a function of η is

$$\frac{\partial^2 \chi}{\partial \eta^2} + \left(-\tfrac{1}{4} - \frac{1}{2\eta} + \frac{1}{4\eta^2} + \tfrac{1}{4}\mathscr{E}\eta\right)\chi = 0, \qquad \chi = \psi\sqrt{\eta}.$$

Solving this in the same way as (2), we have in place of (3)

$$|\chi|^2 = \frac{4A^2|p_0|}{\eta_0 p} e^{-\xi} \exp\left(-2\int_{\eta_0}^{\eta_1} |p|\,d\eta - \eta_0\right),$$

with

$$p(\eta) = \sqrt{\left[-\tfrac{1}{4} - \frac{1}{2\eta} + \frac{1}{4\eta^2} + \tfrac{1}{4}\mathscr{E}\eta\right]}.$$

Next, instead of (4) we find

$$|\chi|^2 = \frac{A^2\mathscr{E}}{\sqrt{(\mathscr{E}\eta - 1)}} \exp\left(-\xi - \frac{2}{3\mathscr{E}}\right),$$

and finally, instead of (5),

$$w = \pi A^2 \mathscr{E} \exp(-2/3\mathscr{E}),$$

or in ordinary units,

$$w = \frac{\pi|e|\mathscr{E}A^2}{\hbar\kappa} \exp\left(-\frac{2\hbar^2\kappa^3}{3m|e|\mathscr{E}}\right).$$

PROBLEM 3. Find with exponential accuracy the probability that an electron will be removed from a potential well by a uniform variable electric field $\mathscr{E} = \mathscr{E}_0 \cos \omega t$; it is assumed that the field frequency and amplitude satisfy the conditions

$$\hbar\omega \ll |E|, \qquad |e|\mathscr{E}_0 \ll \hbar^2\kappa^3/m,$$

where $\kappa = \sqrt{(2m|E|)}/\hbar$, and $|E|$ is the binding energy of the electron in the well (L. V. Keldysh 1964).†

SOLUTION. With the conditions stated, the removal probability w is exponentially small. To calculate just the exponent (not the coefficient of the exponential), it is sufficient to regard the motion as one-dimensional in the direction of the field (the z-axis).

It will be convenient to describe the electric field by a vector (not scalar) potential,

† This may refer, for example, to the ionization of a singly charged negative ion by an intense light-wave; here the potential well is created by the interaction of the electron with the neutral remainder of the atom. The condition $\hbar\omega \ll |E|$ then ensures that the field of the electromagnetic wave may be treated classically.

$A_z = A = -(c\mathscr{E}_0\omega)\sin\omega t$. Then the Hamiltonian of the electron in the region outside the well is

$$\hat{H} = \frac{1}{2m}\left(-i\hbar\frac{\partial}{\partial z} + \frac{|e|\mathscr{E}_0}{\omega}\sin\omega t\right)^2;$$

see (111.3). It does not contain z. With the dimensionless variables and parameters

$$\tau = \frac{\hbar\kappa^2}{2m}t, \qquad \eta = 2\kappa z, \qquad \Omega = \frac{2m\omega}{\hbar\kappa^2} = \frac{\hbar\omega}{|E|}, \qquad F = \frac{|e|m\mathscr{E}_0}{\hbar^2\kappa^3},$$

we can write Schrödinger's equation in the form

$$\tfrac{1}{4}i\frac{\partial\Psi}{\partial t} = -\left(\frac{\partial}{\partial\eta} + \frac{iF}{\Omega}\sin\Omega\tau\right)^2\Psi.$$

The boundary condition is that the solution $\Psi(\eta, \tau)$ with $\eta \to 0$ should be the same as the electron wave function in the well (with energy $E = -|E|$), unperturbed by the wave:

$$\Psi \to e^{i\tau} \quad \text{as} \quad \eta \to 0. \tag{7}$$

Since the problem is quasi-classical, we seek the solution (with exponential accuracy) in the form $\Psi = \exp(iS)$, where $S(\eta, \tau)$ is the classical action. Since the Hamiltonian is independent of the coordinate η, the generalized momentum $p_\eta = p$ is conserved along the classical path, so that

$$S = -\int_{\tau_0}^{\tau} H(p, \tau')\,d\tau' + \eta p + A, \qquad H(p, \tau) = 4\left(p + \frac{F}{\Omega}\sin\Omega\tau\right)^2, \tag{8}$$

where A and τ_0 are constants. From the significance of the action as a function of the coordinates (see *Mechanics*, §43), we must take p as the value which brings the path to the specified point η at time τ, i.e. regard p as a function of η and τ determined by the equation of motion $\partial S/\partial p$ = constant:

$$\eta = \int_{\tau_0}^{\tau} \frac{\partial H(p, \tau')}{\partial p}\,d\tau'; \tag{9}$$

the constant is chosen so that $\eta = 0$ for $\tau = \tau_0$. Formulae (8) and (9) give the action as a function of the two constants τ_0 and A. In order to obtain a solution satisfying the condition (1) we must (as in finding the general integral of the Hamilton–Jacobi equation; see the footnote in *Mechanics*, §47) regard A as a function of τ_0, and τ_0 as a function of coordinate and time defined by

$$\partial S/\partial\tau_0 = 0. \tag{10}$$

It is evident that $A(\tau_0) = \tau_0$; then, for $\eta = 0$ and $\tau = \tau_0$, we have $S = \tau_0$, i.e. $S = \tau$ in accordance with the condition (7). Equation (10) then becomes

$$H(p, \tau_0) + 1 = 0. \tag{11}$$

Equations (9) and (11) together determine the functions $\tau_0(\eta, \tau)$ and $p(\eta, \tau)$, and hence (after substitution in (8)) the wave function $\Psi(\eta, \tau)$.

The required probability w is proportional to the current density along the z-axis. In the classically accessible region, this is $v_z|\Psi|^2$. The coordinate value at which this region begins is given by the point where im S ceases to increase. At that point $(\partial\,\text{im}\,S/\partial\eta)_\tau = 0$, and since $\partial S/\partial\eta = p$, im $p = 0$; from (9) and (11) it then follows that re $p = 0$ also. From this condition we find the value of τ_0, and substituting $p = 0$ in (11) gives

$$\frac{4F^2}{\Omega^2}\sin^2\Omega\tau_0 = -1,$$

whence

$$\Omega\tau_0 = i \sinh^{-1}\gamma, \quad \gamma = \frac{\Omega}{2F} = \frac{\sqrt{(2m|E|)}\omega}{|e|\mathscr{E}_0};$$

the fact that the "time" τ_0 is imaginary corresponds to the classical impossibility of the process. Finally

$$w \sim \exp\left\{-2\,\mathrm{im}\left[\int_{\tau}^{\tau_0} \frac{4F^2}{\Omega^2}\sin^2\Omega\tau'\,d\tau' + \tau_0\right]\right\},$$

and τ may be taken to have any real value; the imaginary part of the integral is unaffected. Calculation of the integral gives

$$w \sim \exp\left\{-\frac{2|E|}{\hbar\omega}f(\gamma)\right\}, \quad f(\gamma) = \left(1+\frac{1}{2\gamma^2}\right)\sinh^{-1}\gamma - \frac{\sqrt{(1+\gamma^2)}}{2\gamma}. \tag{12}$$

The limiting forms of the function $f(\gamma)$ are

$$\begin{aligned} f(\gamma) &\approx \tfrac{2}{3}\gamma \quad \text{for} \quad \gamma \ll 1, \\ &\approx \log 2\gamma - \tfrac{1}{2} \quad \text{for} \quad \gamma \gg 1. \end{aligned}$$

The limiting value of w as $\gamma \to 0$ corresponds to the probability of removal of the particle from the potential well by a constant field.

Formula (12) is applicable if the exponent is large. For this we must in any case have $\hbar\omega \ll |E|$.

CHAPTER XI

THE DIATOMIC MOLECULE

§78. Electron terms in the diatomic molecule

In the theory of molecules an important part is played by the fact that the masses of atomic nuclei are very large compared with those of the electrons. Because of this difference in mass, the rates of motion of the nuclei in the molecule are small in comparison with the velocities of the electrons. This makes it possible to regard the motion of the electrons as being about fixed nuclei placed at given distances from one another. On determining the energy levels U_n for such a system, we find what are called the *electron terms* for the molecule. Unlike those for atoms, where the energy levels were certain numbers, the electron terms here are not numbers but functions of parameters, the distances between the nuclei in the molecule. The energy U_n includes also the electrostatic energy of the mutual interaction of the nuclei, so that U_n is essentially the total energy of the molecule for a given arrangement of the fixed nuclei.

We shall begin the study of molecules by taking the simplest type, the diatomic molecules, which permit the most complete theoretical investigation. The electron terms of the diatomic molecule are functions of only one parameter, the distance r between the nuclei.

One of the chief principles in the classification of the atomic terms was the classification according to the values of the total orbital angular momentum L. In molecules, however, there is no law of conservation of the total orbital angular momentum of the electrons, since the electric field of several nuclei is not centrally symmetric.

In diatomic molecules, however, the field has axial symmetry about an axis passing through the two nuclei. Hence the projection of the orbital angular momentum on this axis is here conserved, and we can classify the electron terms of the molecules according to the values of this projection. The absolute value of the projected orbital angular momentum along the axis of the molecule is customarily denoted by the letter Λ; it takes the values 0, 1, 2, ... The terms with different values of Λ are denoted by the capital Greek letters corresponding to the Latin letters for the atomic terms with various L. Thus, for $\Lambda = 0, 1, 2$ we speak of Σ, Π and Δ terms respectively; higher values of Λ usually need not be considered.

Next, each electron state of the molecule is characterized by the total spin S of all the electrons in the molecule. If S is not zero, there is degeneracy of degree $2S+1$ with respect to the directions of the total spin.† The number

† We here neglect the fine structure due to relativistic interactions (see §§83 and 84 below).

$2S+1$ is, as in atoms, called the *multiplicity* of the term, and is written as an index before the letter for the term; thus $^3\Pi$ denotes a term with $\Lambda = 1$, $S = 1$.

Besides rotations through any angle about the axis, the symmetry of the molecule allows also a reflection in any plane passing through the axis. If we effect such a reflection, the energy of the molecule is unchanged. The state obtained from the reflection is, however, not completely identical with the initial state. For, on reflection in a plane passing through the axis of the molecule, the sign of the angular momentum (which is an axial vector) about this axis is changed. Thus we conclude that all electron terms with non-zero values of Λ are doubly degenerate: to each value of the energy, there correspond two states which differ in the direction of the projection of the orbital angular momentum on the axis of the molecule. In the case where $\Lambda = 0$ the state of the molecule is not changed at all on reflection, so that the Σ terms are not degenerate. The wave function of a Σ term can only be multiplied by a constant as a result of the reflection. Since a double reflection in the same plane is an identity transformation, this constant is ± 1. Thus we must distinguish Σ terms whose wave functions are unaltered on reflection and those whose wave functions change sign. The former are denoted by Σ^+, and the latter by Σ^-.

If the molecule consists of two similar atoms, a new symmetry appears, and with it an additional characteristic of the electron terms. A diatomic molecule with identical nuclei has a centre of symmetry at the point bisecting the line joining the nuclei.† (We shall take this point as the origin.) Hence the Hamiltonian is invariant with respect to a simultaneous change of sign of the coordinates of all the electrons in the molecule (the coordinates of the nuclei remaining unchanged). Since the operator of this transformation‡ also commutes with the orbital angular momentum operator, we have the possibility of classifying terms with a given value of Λ according to their parity: the wave functions of *even* (g) states are unchanged when the coordinates of the electrons change sign, while those of *odd* (u) states change sign. The suffixes u, g indicating the parity are customarily written with the letter for the term: Π_u, Π_g, and so on.

It is an empirical fact that the normal electron state in the great majority of chemically stable diatomic molecules is completely symmetrical: the electron wave function is invariant with respect to all symmetry transformations in the molecule. The total spin S is zero too, in the great majority of cases, in the normal state. In other words, the ground term of the molecule is $^1\Sigma^+$, and it is $^1\Sigma^+{}_g$ if the molecule consists of two similar atoms. Exceptions to these rules are formed by the molecules O_2 (whose normal term is $^3\Sigma^-{}_g$) and NO (normal term $^2\Pi$).

---

† It has also a plane of symmetry perpendicularly bisecting the axis of the molecule. This element of symmetry need not be considered separately, however, since the existence of such a plane follows automatically from the existence of a centre of symmetry and of an axis of symmetry.

‡ Not to be confused with that of inversion of the coordinates of all the particles in the molecule (cf. §86).

PROBLEM

Effect the separation of variables in Schrödinger's equation for the electron terms of the ion H_2^+, using elliptic coordinates.

SOLUTION. Schrödinger's equation for an electron in the field of two protons at rest is (using atomic units)

$$\triangle\psi+2\left(E+\frac{1}{r_1}+\frac{1}{r_2}\right)\psi = 0$$

The elliptic coordinates ξ, η are defined by

$$\xi = (r_1+r_2)/R, \qquad \eta = (r_2-r_1)/R; \qquad 1 \leqslant \xi \leqslant \infty, \ -1 \leqslant \eta \leqslant 1,$$

and the third coordinate ϕ is the angle of rotation about an axis passing through the two nuclei at a distance R apart. (See *Mechanics*, §48.) The Laplacian operator in these co-ordinates is

$$\triangle = \frac{4}{R^2(\xi^2-\eta^2)}\left[\frac{\partial}{\partial\xi}(\xi^2-1)\frac{\partial}{\partial\xi}+\frac{\partial}{\partial\eta}(1-\eta^2)\frac{\partial}{\partial\eta}\right]+\frac{1}{R^2(\xi^2-1)(1-\eta^2)}\frac{\partial^2}{\partial\phi^2}.$$

Putting

$$\psi = X(\xi)Y(\eta)e^{i\Lambda\phi},$$

we obtain for X and Y the equations

$$\frac{d}{d\xi}\left[(\xi^2-1)\frac{dX}{d\xi}\right]+\left(\tfrac{1}{2}ER^2\xi^2+2R\xi+A-\frac{\Lambda^2}{\xi^2-1}\right)X = 0,$$

$$\frac{d}{d\eta}\left[(1-\eta^2)\frac{dY}{d\eta}\right]+\left(-\tfrac{1}{2}ER^2\eta^2-A-\frac{\Lambda^2}{1-\eta^2}\right)Y = 0,$$

where A is the separation parameter.

Each electron term $E(R)$ is described by three quantum numbers: Λ, and two "elliptical quantum numbers" n_ξ, n_η which determine the number of zeros of the functions $X(\xi)$ and $Y(\eta)$.

§79. The intersection of electron terms

The electron terms in a diatomic molecule, as functions of the distance r between the nuclei, can be represented graphically by plotting the energy as a function of r. It is of considerable interest to examine the intersection of the curves representing the different terms.

Let $U_1(r)$, $U_2(r)$ be two different electron terms. If they intersect at some point, then the functions U_1 and U_2 will have neighbouring values near this point. To decide whether such an intersection can occur, it is convenient to put the problem as follows. Let us consider a point r_0 where the functions $U_1(r)$, $U_2(r)$ have very close but not equal values (which we denote by E_1, E_2), and examine whether or not we can make U_1 and U_2 equal by displacing the point a short distance δr. The energies E_1 and E_2 are eigenvalues of the Hamiltonian $\hat{H}_0$ of the system of electrons in the field of the nuclei, which are at a distance r_0 from each other. If we add to the distance r_0 an increment δr, the Hamiltonian becomes $\hat{H}_0+\hat{V}$, where $\hat{V} = \delta r \,.\, \partial\hat{H}_0/\partial r$ is a small correction; the values of the functions U_1, U_2 at the point $r_0+\delta r$ can be regarded as eigenvalues of the new Hamiltonian. This point of view enables us to determine the values of the terms $U_1(r)$, $U_2(r)$ at the point $r_0+\delta r$ by means of perturbation theory, $\hat{V}$ being regarded as a perturbation to the operator $\hat{H}_0$.

The ordinary method of perturbation theory is here inapplicable, however, since the eigenvalues E_1, E_2 of the energy in the unperturbed problem are very close to each other, and their difference is in general not large compared with the magnitude of the perturbation; the condition (38.9) is not fulfilled. Since, in the limit as the difference E_2-E_1 tends to zero, we have the case of degenerate eigenvalues, it is natural to attempt to apply to the case of close eigenvalues a method similar to that developed in §39.

Let ψ_1, ψ_2 be the eigenfunctions of the unperturbed operator $\hat{H}_0$ which correspond to the energies E_1, E_2. As an initial zero-order approximation we take, instead of ψ_1 and ψ_2 themselves, linear combinations of them of the form

$$\psi = c_1\psi_1+c_2\psi_2. \tag{79.1}$$

Substituting this expression in the perturbed equation

$$(\hat{H}_0+\hat{V})\psi = E\psi, \tag{79.2}$$

we obtain

$$c_1(E_1+\hat{V}-E)\psi_1+c_2(E_2+\hat{V}-E)\psi_2 = 0.$$

Multiplying this equation on the left by ψ_1^* and ψ_2^* in turn, and integrating, we have two algebraic equations:

$$\left.\begin{aligned} c_1(E_1+V_{11}-E)+c_2V_{12} &= 0,\\ c_1V_{21}+c_2(E_2+V_{22}-E) &= 0, \end{aligned}\right\} \tag{79.3}$$

Since the operator $\hat{V}$ is Hermitian, the matrix elements V_{11} and V_{22} are real, while $V_{12} = V_{21}^*$. The compatibility condition for these equations is

$$\begin{vmatrix} E_1+V_{11}-E & V_{12} \\ V_{21} & E_2+V_{22}-E \end{vmatrix} = 0,$$

whence

$$E = \tfrac{1}{2}(E_1+E_2+V_{11}+V_{22})\pm\sqrt{[\tfrac{1}{4}(E_1-E_2+V_{11}-V_{22})^2+|V_{12}|^2]}. \tag{79.4}$$

This formula gives the required eigenvalues of the energy in the first approximation.

If the energy values of the two terms become equal at the point $r_0+\delta r$ (i.e. the terms intersect), this means that the two values of E given by formula (79.4) are the same. For this to happen, the expression under the radical in (79.4) must vanish. Since it is the sum of two squares, we obtain, as the condition for there to be points of intersection of the terms, the equations

$$E_1-E_2+V_{11}-V_{22} = 0, \quad V_{12} = 0. \tag{79.5}$$

However, we have at our disposal only one arbitrary parameter giving the perturbation $\hat{V}$, namely the magnitude δr of the displacement. Hence the two equations (79.5) cannot in general be simultaneously satisfied (we suppose that the functions ψ_1, ψ_2 are chosen to be real, so that V_{12} also is real).

It may happen, however, that the matrix element V_{12} vanishes identically; there then remains only one equation (79.5), which can be satisfied by a suitable choice of δr. This happens in all cases where the two terms considered are of different symmetry. By *symmetry* we here understand all possible forms of symmetry: with respect to rotations about an axis, reflections in planes, inversion, and also with respect to interchanges of electrons. In the diatomic molecule this means that we may be dealing with terms of different Λ, different parity or multiplicity, or (for Σ terms) Σ^+ and Σ^- terms.

The validity of this statement depends on the fact that the perturbation operator (like the Hamiltonian itself) commutes with all the symmetry operators for the molecule: the operator of the angular momentum about the axis, the reflection and inversion operators, and the operators of interchanges of electrons. It has been shown in §§29 and 30 that, for a scalar quantity whose operator commutes with the angular momentum and inversion operators, only the matrix elements for transitions between states of the same angular momentum and parity are non-zero. This proof remains valid, in essentially the same form, for the general case of an arbitrary symmetry operator. We shall not pause to repeat it here, especially since in §97 we shall give another general proof, based on group theory.

Thus we reach the result that, in a diatomic molecule, only terms of different symmetry can intersect, while the intersection of terms of like symmetry is impossible (E. Wigner and J. von Neumann 1929). If, as a result of some approximate calculation, we obtain two intersecting terms of the same symmetry, they are found to move apart on calculating the next approximation, as shown by the continuous lines in Fig. 27.

We emphasize that this result not only is true for the diatomic molecule, but is a general theorem of quantum mechanics; it holds for any case where the Hamiltonian contains some parameter and its eigenvalues are consequently functions of that parameter.

In the terminology of group theory (see §96), the general condition for the

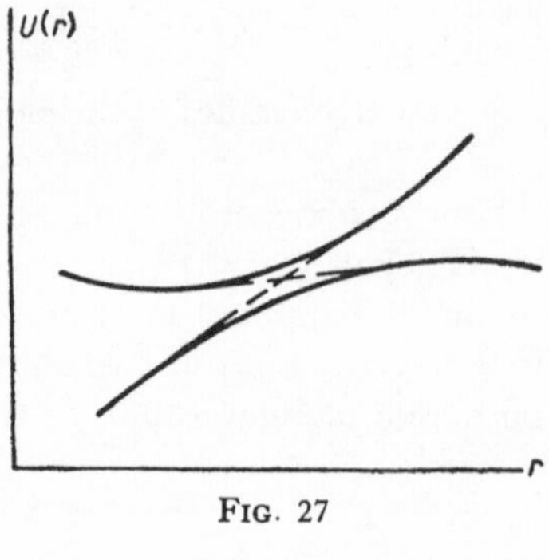

FIG. 27

possible intersection of terms is that the terms should belong to different irreducible representations of the symmetry group of the Hamiltonian of the system.†

In a polyatomic molecule, the electron terms are functions of not one but several parameters, the distances between the various nuclei. Let s be the number of independent distances between the nuclei; in a molecule of $N(>2)$ atoms, this number is $s = 3N-6$ for an arbitrary arrangement of the nuclei. Each term $U_n(r_1, \ldots, r_s)$ is, from the geometrical point of view, a surface in a space of $s+1$ dimensions, and we can speak of the intersections of these surfaces in manifolds of varying numbers of dimensions, from 0 (intersection in a point) to $s-1$. The derivation given above is wholly valid, except that the perturbation $\hat{V}$ is here determined not by one but by s parameters, the displacements $\delta r_1, \ldots, \delta r_s$. Even with two parameters, the two equations (79.5) can in general be satisfied. Thus we conclude that, in polyatomic molecules, any two terms may intersect. If the terms are of like symmetry, the intersection is given by the two conditions (79.5), from which it follows that the number of dimensions of the manifold in which the intersection occurs is $s-2$. If the terms are of different symmetry, on the other hand, there remains only one condition, and the intersection takes place in a manifold of $s-1$ dimensions.

Thus for $s = 2$ the terms are represented by surfaces in a three-dimensional system of coordinates. The intersection of these surfaces occurs in lines ($s-1 = 1$) when the symmetry of the terms is different, and in points ($s-2 = 0$) when it is the same. It is easy to ascertain the form of the surfaces near the point of intersection in the latter case. The value of the energy near the points of intersection of the terms is given by formula (79.4). In this expression the matrix elements V_{11}, V_{22}, V_{12} are linear functions of the displacements δr_1, δr_2, and hence are linear functions of the distances r_1, r_2 themselves. Such an equation determines an elliptic cone, as we know from analytical geometry. Thus, near the points of intersection, the terms are represented by the surface of an arbitrarily situated double elliptic cone (Fig. 28, p. 304).

§80. The relation between molecular and atomic terms

As we increase the distance between the nuclei in a diatomic molecule, we have in the limit two isolated atoms (or ions). The question thus arises of the correspondence between the electron terms of the molecule and the states of the atoms obtained by moving them apart (E. Wigner and E. Witmer 1928).

† An apparent exception to this rule occurs for the electron terms of the H_2^+ ion. These are described by the angular momentum component Λ and the two elliptical quantum numbers n_ξ and n_η (see §78, Problem). Since all these numbers are related to functions of different variables, there is in general nothing to prevent the intersection of terms $E(R)$ which differ in the values of n_ξ and n_η for the same Λ, even though such terms have the same symmetry with respect to rotations and reflections. In reality, however, the separability of variables in Schrödinger's equation for this system means that its Hamiltonian has a higher symmetry than follows from its geometrical properties; with respect to this complete group, the symmetries of states with different values of n_ξ and n_η are of different types.

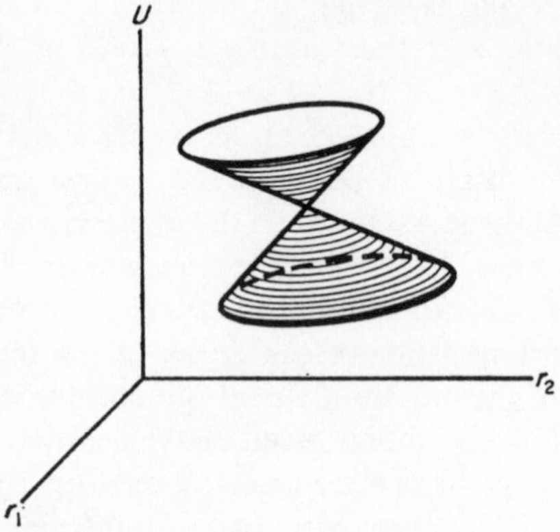

FIG. 28

This relation is not one-to-one; if we bring together two atoms in given states, we may obtain a molecule in various electron states.

Let us first suppose that the molecule consists of two different atoms. Let the isolated atoms be in states with orbital angular momenta L_1, L_2 and spins S_1, S_2, and let $L_1 \geqslant L_2$. The projections of the angular momenta on the line joining the nuclei take the values $M_1 = -L_1, -L_1+1, \dots, L_1$ and $M_2 = -L_2, -L_2+1, \dots, L_2$. The absolute value of the sum M_1+M_2 determines the angular momentum Λ obtained on bringing the atoms together. On combining all possible values of M_1 and M_2, we find the following values for the numbers of times that we obtain the various values of $\Lambda = |M_1+M_2|$:

$$\begin{array}{ll}
\Lambda = L_1+L_2 & \text{twice} \\
\quad L_1+L_2-1 & \text{four times} \\
\quad \dots & \dots \\
\quad L_1-L_2 & 2(2L_2+1)\ \text{times} \\
\quad L_1-L_2-1 & 2(2L_2+1)\ \text{times} \\
\quad \dots & \dots \\
\quad 1 & 2(2L_2+1)\ \text{times} \\
\quad 0 & 2L_2+1\ \text{times.}
\end{array}$$

Remembering that all terms with $\Lambda \neq 0$ are doubly degenerate, while those with $\Lambda = 0$ are not degenerate, we find that there will be

$$\left.\begin{array}{r}
1 \text{ term with } \Lambda = L_1+L_2, \\
2 \text{ terms with } \Lambda = L_1+L_2-1, \\
\dots\dots\dots\dots\dots\dots \\
2L_2+1 \text{ terms with } \Lambda = L_1-L_2 \\
2L_2+1 \text{ terms with } \Lambda = L_1-L_2-1, \\
\dots\dots\dots\dots\dots\dots \\
2L_2+1 \text{ terms with } \Lambda = 0;
\end{array}\right\} \quad (80.1)$$

in all, $(2L_2+1)(L_1+1)$ terms with values of Λ from 0 to L_1+L_2.

The spins S_1, S_2 of the two atoms combine to form the total spin of the molecule in accordance with the general rule for the addition of angular momenta, giving the following possible values of S:

$$S = S_1+S_2,\quad S_1+S_2-1,\quad \dots,\quad |S_1-S_2|. \tag{80.2}$$

On combining each of these values with each value of Λ in (80.1), we obtain the complete list of all possible terms in the molecule formed.

For Σ terms there is also the question of sign. This is easily resolved by noticing that the wave functions of the molecule can be written, as $r \to \infty$, in the form of products (or sums of products) of the wave functions of the two atoms. An angular momentum $\Lambda = 0$ can be obtained either by adding two non-zero angular momenta of the atoms such that $M_1 = -M_2$, or from $M_1 = M_2 = 0$. We denote the wave functions of the first and second atoms by $\psi^{(1)}{}_{M_1}$, $\psi^{(2)}{}_{M_2}$. For $M = |M_1| = |M_2| \neq 0$, we form the symmetrized and antisymmetrized products

$$\psi^+ = \psi^{(1)}{}_M\psi^{(2)}{}_{-M}+\psi^{(1)}{}_{-M}\psi^{(2)}{}_M,$$

$$\psi^- = \psi^{(1)}{}_M\psi^{(2)}{}_{-M}-\psi^{(1)}{}_{-M}\psi^{(2)}{}_M.$$

A reflection in a vertical plane (i.e. one passing through the axis of the molecule) changes the sign of the projection of the angular momentum on the axis, so that $\psi^{(1)}{}_M$, $\psi^{(2)}{}_M$ are changed into $\psi^{(1)}{}_{-M}$, $\psi^{(2)}{}_{-M}$ respectively, and *vice versa*. The function ψ^+ is thereby unchanged, while ψ^- changes sign; the former therefore corresponds to a Σ^+ term and the latter to a Σ^- term. Thus, for each value of M, we obtain one Σ^+ and one Σ^- term. Since M can take L_2 different values ($M = 1, \dots, L_2$), we have in all L_2 Σ^+ terms and L_2 Σ^- terms.

If, on the other hand, $M_1 = M_2 = 0$, the wave function of the molecule is of the form $\psi = \psi^{(1)}{}_0\psi^{(2)}{}_0$. In order to ascertain the behaviour of the function $\psi^{(1)}{}_0$ on reflection in a vertical plane, we take a coordinate system with its origin at the centre of the first atom, and the z-axis along the axis of the molecule, and we notice that a reflection in the vertical xz-plane is equivalent to an inversion with respect to the origin, followed by a rotation through 180° about the y-axis. On inversion, the function $\psi^{(1)}{}_0$ is multiplied by P_1, where $P_1 = \pm 1$ is the parity of the given state of the first atom. Next, the result of applying to the wave function the operation of an infinitely small rotation (and therefore that of any finite rotation) is entirely determined by the total orbital angular momentum of the atom. Hence it is sufficient to consider the particular case of an atom having one electron, with orbital angular momentum l (and a z-component of the angular momentum $m = 0$); on putting L in place of l in the result, we obtain the required solution for any atom. The angular part of the wave function of an electron with $m=0$ is, apart from a constant coefficient, $P_l(\cos\theta)$ (see (28.8)). A rotation through 180° about the y-axis is the transformation $x \to -x$, $y \to y$, $z \to -z$ or, in spherical polar coordinates, $r \to r$, $\theta \to \pi - \theta$, $\phi \to \pi - \phi$. Then $\cos\theta \to -\cos\theta$, and the function $P_l(\cos\theta)$ is multiplied by $(-1)^l$.

Thus we conclude that, as a result of reflection in a vertical plane, the function $\psi^{(1)}{}_0$ is multiplied by $(-1)^{L_1}P_1$. Similarly, $\psi^{(2)}{}_0$ is multiplied by $(-1)^{L_2}P_2$, so that the wave function $\psi = \psi^{(1)}{}_0\psi^{(2)}{}_0$ is multiplied by $(-1)^{L_1+L_2}P_1P_2$. The term is Σ^+ or Σ^- according as this factor is $+1$ or -1.

Summarizing the results obtained, we find that, of the total number $2L_2+1$ of Σ terms (with every possible multiplicity), L_2+1 terms are Σ^+ and L_2 are Σ^-, if $(-1)^{L_1+L_2}P_1P_2 = +1$, and *vice versa* if $(-1)^{L_1+L_2}P_1P_2 = -1$.

Let us now turn to a molecule consisting of similar atoms. The rules for the addition of the spins and orbital angular momenta of the atoms to form the total S and Λ for the molecule remain the same here as for a molecule composed of different atoms. The difference is that the terms may be even or odd. Here we must distinguish two cases, according as the combined atoms are in the same or different states.

If the atoms are in different states,† the total number of possible terms is doubled in comparison with the number when the atoms are different. For a reflection with respect to the origin (this being the point bisecting the axis of the molecule) results in an interchange of the states of the two atoms. Symmetrizing or antisymmetrizing the wave function of the molecule with respect to an interchange of the states of the atoms, we obtain two terms (with the same Λ and S), of which one is even and the other odd. Thus we have altogether the same number of even and odd terms.

If, on the other hand, both atoms are in the same state, the total number of states is the same as for a molecule with different atoms. An investigation which we shall not give here on account of its length‡ leads to the following results for the parity of these states. Let N_g, N_u be the numbers of even and odd terms with given values of Λ and S. Then

if Λ is odd, $N_g = N_u$;

if Λ is even and S is even ($S = 0, 2, 4, \ldots$), $N_g = N_u+1$;

if Λ is even and S is odd ($S = 1, 3, 5, \ldots$), $N_u = N_g+1$.

Finally, we must distinguish, among the Σ terms, between Σ^+ and Σ^-. Here,

if S is even, $N_g^+ = N_u^-+1 = L+1$;

if S is odd, $N_u^+ = N_g^-+1 = L+1$,

where $L_1 = L_2 \equiv L$. All the Σ^+ terms are of parity $(-1)^S$, and all Σ^- terms are of parity $(-1)^{S+1}$.

Besides the problem that we have examined of the relation between the molecular terms and those of the atoms obtained as $r \to \infty$, we may also propose the question of the relation between the molecular terms and those of the "composite atom" obtained as $r \to 0$, i.e. when both nuclei are brought to a single point (for example, between the terms of the H_2 molecule and those of the He atom). The following rules can be deduced without difficulty. From a term of the "composite" atom having spin S, orbital angular momen-

† In particular, we may be discussing the combination of a neutral and an ionized atom.

‡ See E. Wigner and E. Witmer, *Zeitschrift für Physik* **51**, 859, 1928.

tum L and parity P, we can obtain, on moving the constituent atoms apart, molecular terms with spin S and angular momentum about the axis $\Lambda = 0, 1, \dots, L$, with one term for each of these values of Λ. The parity of the molecular term is the same as the parity P of the atomic term (g for $P = +1$ and u for $P = -1$). The molecular term with $\Lambda = 0$ is a Σ^+ term if $(-1)^L P = +1$, and a Σ^- term if $(-1)^L P = -1$.

PROBLEMS

Problem 1. Determine the possible terms for the molecules H_2, N_2, O_2, Cl_2 which can be obtained by combining atoms in the normal state.

Solution. According to the rules given above, we find the following possible terms: H_2 molecule (atoms in the ^2S state):

$$^1\Sigma^+_g,\quad ^3\Sigma^+_u;$$

N_2 molecule (atoms in the ^4S state):

$$^1\Sigma^+_g,\quad ^3\Sigma^+_u,\quad ^5\Sigma^+_g,\quad ^7\Sigma^+_u;$$

Cl_2 molecule (atoms in the ^2P state):

$$2\,^1\Sigma^+_g,\quad ^1\Sigma^-_u,\quad ^1\Pi_g,\quad ^1\Pi_u,\quad ^1\Delta_g,\quad 2\,^3\Sigma^+_u,\quad ^3\Sigma^-_g,\quad ^3\Pi_g,\quad ^3\Pi_u,\quad ^3\Delta_u;$$

O_2 molecule (atoms in the ^3P state):

$$2\,^1\Sigma^+_g,\quad ^1\Sigma^-_u,\quad ^1\Pi_g,\quad ^1\Pi_u,\quad ^1\Delta_g,\quad 2\,^3\Sigma^+_u,\quad ^3\Sigma^-_g,\quad ^3\Pi_u,\quad ^3\Pi_g,\quad ^3\Delta_u,$$
$$2\,^5\Sigma^+_g,\quad ^5\Sigma^-_u,\quad ^5\Pi_g,\quad ^5\Pi_u,\quad ^5\Delta_g.$$

The figures in front of the symbols indicate the number of terms of the type concerned, if this number exceeds unity.

Problem 2. The same as Problem 1, but for the molecules HCl, CO.

Solution. When unlike atoms are combined, the parity of their states is important also. From formula (31.6) we find that the normal states of the H. O and C atoms are even, while that of the Cl atom is odd (see Table 3 for the electron configurations of these atoms). From the rules given above, we have

HCl molecule (atoms in the ^2S_g and ^2P_u states):

$$^{1,3}\Sigma^+,\quad ^{1,3}\Pi;$$

CO molecule (both atoms in the ^3P_g state):

$$2\,^{1,3,5}\Sigma^+,\quad ^{1,3,5}\Sigma^-,\quad 2\,^{1,3,5}\Pi,\quad ^{1,3,5}\Delta.$$

§81. Valency

The property of atoms of combining with one another to form molecules is described by means of the concept of *valency*. To each atom we ascribe a definite valency, and when atoms combine their valencies must be mutually satisfied, i.e. to each valency bond of an atom there must correspond a valency bond of another atom. For example, in the methane molecule CH_4, the four valency bonds of the quadrivalent carbon atom are satisfied by the four univalent hydrogen atoms. In going on to give a physical interpretation of valency, we shall begin with the simplest example, the combination of two hydrogen atoms to form the molecule H_2.

Let us consider two hydrogen atoms in the ground state (^2S). When they approach, the resulting system may be in the molecular state $^1\Sigma^+_g$ or $^3\Sigma^+_u$. The singlet term corresponds to an antisymmetrical spin wave function, and the triplet term to a symmetrical function. The coordinate wave function, on the other hand, is symmetrical for the $^1\Sigma$ term and antisymmetrical for the $^3\Sigma$ term. It is evident that the ground term of the H_2 molecule can only be the $^1\Sigma$ term. For an antisymmetrical wave function $\phi(\mathbf{r}_1, \mathbf{r}_2)$ (where $\mathbf{r}_1$ and $\mathbf{r}_2$ are the radius vectors of the two electrons) always has nodes (since it vanishes for $\mathbf{r}_1 = \mathbf{r}_2$), and hence cannot belong to the lowest state of the system.

A numerical calculation shows that the electron term $^1\Sigma$ in fact has a deep minimum corresponding to the formation of a stable H_2 molecule. In the $^3\Sigma$ state, the energy $U(r)$ decreases monotonically as the distance between the nuclei increases, corresponding to the mutual repulsion of the two hydrogen atoms† (Fig. 29).

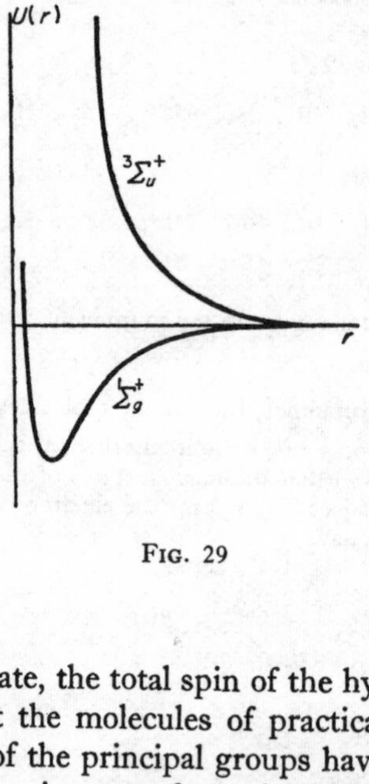

FIG. 29

Thus, in the ground state, the total spin of the hydrogen molecule is zero, $S = 0$. It is found that the molecules of practically all chemically stable compounds of elements of the principal groups have this property. Among inorganic molecules, exceptions are formed by the diatomic molecules O_2 (ground state $^3\Sigma$) and NO (ground state $^2\Pi$) and the triatomic molecules NO_2, ClO_2 (total spin $S = \frac{1}{2}$). Elements of the intermediate groups have special properties which we shall discuss below, after studying the valency properties of the elements of the principal groups.

The property of atoms of combining with one another is thus related to their spin (W. Heitler and H. London 1927). The combination occurs in

† Here we ignore the van der Waals attraction forces between the atoms (see §89). The existence of these forces causes a minimum (at a greater distance) on the $U(r)$ curve for the $^3\Sigma$ term also. This minimum, however, is very shallow in comparison with that on the $^1\Sigma$ curve, and would not be perceptible on the scale of Fig. 29.

such a way that the spins of the atoms compensate one another. As a quantitative characteristic of the mutual combining powers of atoms, it is convenient to use an integer, twice the spin of the atom. This is equal to the chemical valency of the atom. Here it must be borne in mind that the same atom may have different valencies according to the state it is in.

Let us examine, from this point of view, the elements of the principal groups in the periodic system. The elements of the first group (the first column in Table 3, the group of alkali metals) have a spin $S = \frac{1}{2}$ in the normal state, and accordingly their valencies are unity. An excited state with a higher spin can be attained only by exciting an electron from a completed shell. Accordingly, these states are so high that the excited atom cannot form a stable molecule.†

The atoms of elements in the second group (the second column in Table 3, the group of alkaline-earth metals) have a spin $S = 0$ in the normal state. Hence these atoms cannot enter into chemical compounds in the normal state. However, comparatively close to the ground state there is an excited state having a configuration sp instead of s^2 in the incomplete shell, and a total spin $S = 1$. The valency of an atom in this state is 2, and this is the principal valency of the elements in the second group.

The elements of the third group have an electron configuration s^2p in the normal state, with a spin $S = \frac{1}{2}$. However, by exciting an electron from the completed s-shell, an excited state is obtained having a configuration sp^2 and a spin $S = 3/2$, and this state lies close to the normal one. Accordingly, the elements of this group are both univalent and tervalent. The first two elements in the group (boron, aluminium) behave only as tervalent elements. The tendency to exhibit a valency 1 increases with the atomic number, and thallium behaves equally as a univalent and as a tervalent element (for example, in the compounds TlCl and $TlCl_3$). This is due to the fact that, in the first few elements, the binding energy in the tervalent compounds is greater than for the univalent compounds, and this difference exceeds the excitation energy of the atom.

In the elements of the fourth group, the ground state has the configuration s^2p^2 with a spin of 1, and the adjacent excited state has a configuration sp^3 with a spin 2. The valencies 2 and 4 correspond to these states. As in the third group, the first two elements (carbon, silicon) exhibit mainly the higher valency (though the compound CO, for example, forms an exception), and the tendency to exhibit the lower valency increases with the atomic number.

In the atoms of the elements of the fifth group, the ground state has the configuration s^2p^3 with a spin $S = 3/2$, so that the corresponding valency is three. An excited state of higher spin can be obtained only by the transition of one of the electrons into the shell with the next higher value of the principal quantum number. The nearest such state has the configuration sp^3s' and a spin $S = 5/2$ (by s' we conventionally denote here an s state of an electron with a principal quantum number one greater than in the state s).

† See the end of this section for the elements copper, silver and gold.

Although the excitation energy of this state is comparatively high, the excited atom can still form a stable compound. Accordingly, the elements of the fifth group behave as both tervalent and quinquevalent elements (thus, nitrogen is tervalent in NH_3 and quinquevalent in HNO_3).

In the sixth group of elements, the spin is 1 in the ground state (configuration s^2p^4), so that the atom is bivalent. The excitation of one of the p electrons leads to a state s^2p^3s' of spin 2, while the excitation of an s electron in addition gives a state $sp^3s'p'$ of spin 3. In both excited states the atom can enter into stable molecules, and accordingly exhibits valencies of 4 and 6. The first element of the sixth group (oxygen) shows only the valency 2, while the subsequent elements show higher valencies also (thus, sulphur in H_2S, SO_2, SO_3 is respectively bivalent, quadrivalent and sexivalent).

In the seventh group (the halogen group), the atoms are univalent in the ground state (configuration s^2p^5, spin $S = \frac{1}{2}$). They can, however, enter into stable compounds when they are in excited states having configurations s^2p^4s', $s^2p^3s'p'$, $sp^3s'p'^2$ with spins 3/2, 5/2, 7/2 and valencies 3, 5, 7 respectively. The first element in the group (fluorine) is always univalent, but the subsequent elements also exhibit the higher valencies (thus, chlorine in HCl, $HClO_2$, $HClO_3$, $HClO_4$ is respectively univalent, tervalent, quinquevalent and septivalent).

Finally, the atoms of the elements in the group of inert gases have completely filled shells in their ground states (so that the spin $S = 0$), and their excitation energies are high. Accordingly, the valency is zero, and these elements are chemically inactive.†

The following general remark should be made concerning all these discussions. The assertion that an atom enters into a molecule with a valency pertaining to an excited state does not mean that, on moving the atoms apart to large distances, we necessarily obtain an excited atom. It means only that the distribution of the electron density in the molecule is such that, near the nucleus of the atom in question, it is close to that in the isolated and excited atom; but the limit to which the electron distribution tends as the distance between the nuclei is increased may correspond to non-excited atoms.

When atoms combine to form a molecule, the completed electron shells in the atoms are not much changed. The distribution of the electron density in the incomplete shells, on the other hand, may be considerably altered. In the most clearly defined cases of what is called ***heteropolar binding***, all the valency electrons pass over from their own atoms to other atoms, so that we

---

† Some of them nevertheless form stable compounds with fluorine and oxygen. These valencies may be due to a transfer of electrons from the outermost complete shell to the incomplete f or d states, whose energies are comparatively near.

There is also an attraction which occurs in the interaction of an inert gas atom with an excited atom of the same element. This is due to the doubling in the number of possible states obtained on bringing together two atoms, if these atoms are of the same element but in different states (see §80). The transition of the excitation from one atom to the other here replaces the exchange interaction which brings about the ordinary valency. The molecule He_2 is an example of such a molecule. The same type of bond occurs in molecular ions composed of two similar atoms (for instance, $H_2{}^+$).

may say that the molecule consists of ions with charges equal (in units of *e*) to the valency. The elements of the first group are electropositive: in heteropolar compounds they lose electrons, forming positive ions. As we pass to the subsequent groups the electropositive character of the elements becomes gradually less marked and changes into electronegative character, which is present to the greatest extent in the elements of the seventh group. Regarding heteropolarity the same remark should be made as was made above concerning excited atoms in the molecule. If a molecule is heteropolar, this does not mean that, on moving the atoms apart, we necessarily obtain two ions. Thus, from the molecule CsF we should in fact obtain the ions Cs^+ and F^-, but the molecule NaF gives in the limit the neutral atoms Na and F (since the affinity of fluorine for an electron is greater than the ionization potential of caesium but less than that of sodium).

In the opposite limiting case of what is called ***homopolar binding***, the atoms in the molecule remain neutral on the average. Homopolar molecules, unlike heteropolar ones, have no appreciable dipole moment. The difference between the heteropolar and homopolar types is purely quantitative, and any intermediate case may occur.

Let us now turn to the elements of the intermediate groups. Those of the palladium and platinum groups are very similar to the elements of the principal groups as regards their valency properties. The only difference is that, owing to the comparatively deep position of the *d* electrons inside the atom, they interact only slightly with the other atoms in the molecule. As a result, "unsaturated" compounds, whose molecules have non-zero spin (though in practice not exceeding $\frac{1}{2}$), are often found among the compounds of these elements. Each of the elements can exhibit various valencies, and these may differ by unity, and not only by two as with the elements of the principal groups (where the change in valency is due to the excitation of some electron whose spin is compensated, so that the spins of two electrons are simultaneously released).

The elements of the rare-earth group are characterized by the presence of an incomplete *f* shell. The *f* electrons lie much deeper than the *d* electrons, and therefore take no part in the valency. Thus the valency of the rare-earth elements is determined only by the *s* and *p* electrons in the incomplete shells.† However, it must be borne in mind that, when the atom is excited, *f* electrons may pass into *s* and *p* states, thereby increasing the valency by one. Hence the rare-earth elements too exhibit valencies differing by unity (in practice they are all tervalent and quadrivalent).

The elements of the actinium group occupy a unique position. Actinium and thorium have no *f* electrons, and their valencies involve *d* electrons. In their chemical properties they are therefore analogous to elements of the palladium and platinum groups, not to the rare earths. The uranium atom in the normal state contains *f* electrons, but in its compounds it too has no *f*

---

† The *d* electrons which are found in the incomplete shells of the atoms of some rare-earth elements are unimportant, since these atoms in practice always form compounds in excited states where there are no *d* electrons.

electrons. Finally, the atoms of the elements neptunium, plutonium, americium and curium contain *f* electrons in compounds also, but the electrons which participate in their valencies are again *s* and *d* electrons. In this sense they are homologues of uranium. The maximum possible number of "unpaired" *s* and *d* electrons is one and five respectively, and so the maximum valency of elements in the actinium group is six, whereas the maximum valency of the rare-earth elements (with *s* and *p* electrons participating in the valency) is $1+3 = 4$.

The elements of the iron group occupy, as regards their valency properties, a position intermediate between the rare-earth elements and those of the palladium and platinum groups. In their atoms, the *d* electrons lie comparatively deep, and in many compounds take no part in the valency bonds. In these compounds, therefore, the elements of the iron group behave like rare-earth elements. Such compounds include those of ionic type (for instance $FeCl_2$, $FeCl_3$), in which the metal atom enters as a simple cation. Like the rare-earth elements, the elements of the iron group can show very various valencies in these compounds.

Another type of compound of the iron-group elements is formed by what are called *complex compounds*. These are characterized by the fact that the atom of the intermediate element enters into the molecule not as a simple ion, but as part of a complex ion (for instance the ion MnO_4^- in $KMnO_4$, or the ion $Fe(CN)_6^{4-}$ in $K_4Fe(CN)_6$). In these complex ions, the atoms are closer together than in simple ionic compounds, and in them the *d* electrons take part in the valency bond. Accordingly, the elements of the iron group behave in complex compounds like those of the palladium and platinum groups.

Finally, it must be mentioned that the elements copper, silver and gold, which in §73 we placed among the principal groups, behave as intermediate elements in some of their compounds. These elements can exhibit valencies of more than one, on account of a transition of an electron from a *d* shell to a *p* shell of nearly the same energy (for example, from 3*d* to 4*p* in copper). In such compounds the atoms have an incomplete *d* shell, and hence behave as intermediate elements: copper like the elements of the iron group, and silver and gold like those of the palladium and platinum groups.

PROBLEM

Determine the electron terms of the molecular ion H_2^+ obtained when a hydrogen atom in the normal state combines with an H^+ ion, for distances R between the nuclei large compared with the Bohr radius (L. Landau 1961; C. Herring 1961).†

SOLUTION. This problem is analogous in form to §50, Problem 3: instead of two one-dimensional potential wells we have here two three-dimensional wells (round the two nuclei) with axial symmetry about the line joining the nuclei. The level‡ $E_0 = -\frac{1}{2}$ (the ground level of the hydrogen atom) is split into two levels $U_g(R)$ and $U_u(R)$ (the terms ${}^2\Sigma_g^+$ and

† For the corresponding problem of the H_2 molecule, see L. P. Gor'kov and L. P. Pitaevskii, *Soviet Physics Doklady* **8**, 788, 1964; C. Herring and M. Flicker, *Physical Review* **134**, A362, 1964 (the second of these papers corrects an error of calculation in the first).

‡ Here we are using atomic units.

$^2\Sigma_u^+$), corresponding to the electron wave functions

$$\psi_{g,u}(x,y,z) = \frac{1}{\sqrt{2}}[\psi_0(x,y,z) \pm \psi_0(-x,y,z)],$$

which are symmetrical and antisymmetrical about the plane $x = 0$ which bisects the line joining the nuclei (which are at $(\pm\frac{1}{2}R, 0, 0)$). Here $\psi_0(x, y, z)$ is the wave function of the electron in one of the potential wells. Exactly as in §50, Problem 3, we find

$$U_{g,u}(R) - E_0 = \mp \int\int \psi_0 \frac{\partial \psi_0}{\partial x} \, dy \, dz, \tag{1}$$

where the integration is over the plane $x = 0$.†

The function ψ_0 (corresponding to motion around nucleus 1, say, at $x = \frac{1}{2}R$) is sought in the form

$$\psi_0 = \frac{a}{\sqrt{\pi}} e^{-r_1}, \tag{2}$$

where a is a slowly varying function (for a hydrogen atom, $a = 1$). The function ψ_0 must satisfy Schrödinger's equation

$$\tfrac{1}{2}\triangle\psi + \left(-\tfrac{1}{2} - \frac{1}{R} + \frac{1}{r_1} + \frac{1}{r_2}\right)\psi = 0, \tag{3}$$

where r_1, r_2 are the distances of the electron from nuclei 1 and 2. In this equation the total energy of the electron is $E_0 - 1/R$, since E_0 itself includes the energy $1/R$ of the Coulomb repulsion of the nuclei.

Since the function ψ_0 decreases rapidly away from the x-axis, only the region where y and z are small compared with R is important in the integral (1). For $y, z \ll R$, substitution of (2) in (3) gives

$$\frac{\partial a}{\partial x} + \frac{a}{\frac{1}{2}R + x} - \frac{a}{R} = 0;$$

here we have neglected the second derivatives of the slowly varying function a and put $r_2 \approx \frac{1}{2}R + x$. The solution of this equation which becomes unity as $x \to \frac{1}{2}R$ (i.e. in the neighbourhood of nucleus 1) is

$$a = \frac{2R}{R+2x} \exp\left(\frac{x}{R} - \tfrac{1}{2}\right).$$

Formula (1) now gives

$$U_{g,u} - E_0 = \mp \frac{4}{\pi e} \int_{R/2}^{\infty} e^{-2r_1} \, . \, 2\pi r_1 \, dr_1$$

$$= \mp 2Re^{-R-1}.$$

The amount of the splitting is‡

$$U_g - U = -4Re^{-R-1}. \tag{4}$$

At sufficiently large distances this expression decreases exponentially and becomes less than the effect in the second approximation with respect to the dipole interaction of the H atom and the H^+ ion. Since the polarizability of the hydrogen atom in the normal state is 9/2 (see (77.9)), and the field of the H^+ ion is $\mathscr{E} = 1/R^2$, the corresponding interaction energy

---

† Note that the effect sought is therefore determined by the range of distances at which the electron interacts in the same way with both nuclei.

‡ The corresponding result for the H_2 molecule, according to the papers quoted above, is

$$U_g - U_u = -1{\cdot}64 R^{5/2} e^{-2R}.$$

is $-9/4R^4$, and when this is taken into account we have

$$U_{g,u}(R)-E_0 = \mp\frac{2}{e}Re^{-R}-\frac{9}{4R^4}. \tag{5}$$

The second term becomes comparable with the first when $R = 10{\cdot}8$. It may also be noted that the term U_u has a minimum of $-5{\cdot}8\times 10^{-5}$ atomic unit ($-1{\cdot}6\times 10^{-3}$ eV) when $R = 12{\cdot}6$.†

§82. Vibrational and rotational structures of singlet terms in the diatomic molecule

As has been pointed out at the beginning of this chapter, the great difference in the masses of the nuclei and the electrons makes it possible to divide the problem of determining the energy levels of a molecule into two parts. We first determine the energy levels of the system of electrons, for nuclei at rest, as functions of the distance between the nuclei (the electron terms). We can then consider the motion of the nuclei for a given electron state; this amounts to regarding the nuclei as particles interacting with one another in accordance with the law $U_n(r)$, where U_n is the corresponding electron term. The motion of the molecule is composed of its translational displacement as a whole, together with the motion of the nuclei about their centre of mass. The translational motion is, of course, without interest, and we can regard the centre of mass as fixed.

For convenience of discussion, let us first consider the electron terms in which the total spin S of the molecule is zero (the singlet terms). The problem of the relative motion of two particles (the nuclei) which interact according to the law $U(r)$ reduces to that of the motion of a single particle of mass M (the reduced mass of the two particles) in a centrally symmetric field $U(r)$. By $U(r)$ we mean the energy of the electron term considered. The problem of motion in a centrally symmetric field $U(r)$, however, reduces in turn to that of a one-dimensional motion in a field where the effective energy is equal to the sum of $U(r)$ and the centrifugal energy.

We denote by $\mathbf{K}$ the total angular momentum of the molecule, composed of the orbital angular momentum $\mathbf{L}$ of the electrons and the angular momentum of the rotation of the nuclei. Then the operator of the centrifugal energy of the nuclei is

$$B(r)(\hat{\mathbf{K}}-\hat{\mathbf{L}})^2,$$

where we have introduced the notation

$$B(r) = \hbar^2/2Mr^2 \tag{82.1}$$

which is customary in the theory of diatomic molecules. Averaging this quantity over the electron state (for a given r), we obtain the centrifugal

† This minimum, which is due to van der Waals forces, is very shallow compared with that of the term $U_g(R)$ which corresponds to the normal state of the stable ion H_2^+: the latter minimum is $-0{\cdot}60$ atomic unit ($-16{\cdot}3$ eV), at $R = 2{\cdot}0$.

energy as a function of r, which must appear in the effective potential energy $U_K(r)$. Thus

$$U_K(r) = U(r)+B(r)\overline{(\mathbf{K}-\mathbf{L})^2}, \tag{82.2}$$

where the line denotes the average mentioned.

Let us carry out the averaging for a state in which the molecule has a definite value of the square of the total angular momentum $\mathbf{K}^2 = K(K+1)$ (where K is integral) and a definite value of the component of the electron angular momentum along the axis of the molecule (the z-axis) $L_z = \Lambda$. Expanding the parenthesis in (82.2), we have

$$U_K(r) = U(r)+B(r)K(K+1)-2B(r)\,\overline{\mathbf{L}}.\,\mathbf{K}+B(r)\overline{\mathbf{L}^2}. \tag{82.3}$$

The last term depends only on the electron state and does not contain the quantum number K; it may be simply included in the energy $U(r)$. We shall show that the same is true of the term preceding it.

If the component of the angular momentum along an axis has a definite value, the mean value of the angular momentum vector is also along that axis; see the end of §27. If $\mathbf{n}$ denotes a unit vector along the z-axis, we therefore have $\overline{\mathbf{L}} = \Lambda\mathbf{n}$. In classical mechanics, the angular momentum of rotation of a system of two particles, such as nuclei, is $\mathbf{r}\times\mathbf{p}$, where $\mathbf{r} = r\mathbf{n}$ is the radius vector between the two particles and $\mathbf{p}$ the momentum of their relative motion. This quantity is perpendicular to $\mathbf{n}$. In quantum mechanics, the same will be true of the operator of the angular momentum of rotation of the nuclei: $(\hat{\mathbf{K}}-\hat{\mathbf{L}}).\mathbf{n} = 0$, or $\hat{\mathbf{K}}.\mathbf{n} = \hat{\mathbf{L}}.\mathbf{n}$. Since the operators are equal, so of course are their eigenvalues, and, since $\mathbf{n}.\mathbf{L} = L_z = \Lambda$, we have

$$\mathbf{K}\,.\,\mathbf{n} = \Lambda. \tag{82.4}$$

Thus, in the term before the last in (82.3), $\overline{\mathbf{L}}.\mathbf{K} = \mathbf{n}.\mathbf{K}\Lambda = \Lambda^2$, and is independent of K. Redefining the function $U(r)$, we can finally write the effective potential energy as

$$U_K(r) = U(r)+B(r)K(K+1). \tag{82.5}$$

From the equation $K_z = \Lambda$ it follows that, for a given value of Λ, the quantum number K can take only values

$$K \geqslant \Lambda. \tag{82.6}$$

On solving the one-dimensional Schrödinger's equation with the potential energy (82.5), we obtain a series of energy levels. We arbitrarily number these levels (for each given K) in order of increasing energy, using a number $v = 0, 1, 2, \ldots$; $v = 0$ corresponds to the lowest level. Thus the motion of the nuclei causes a splitting of each electron term into a series of levels characterized by the values of the two quantum numbers K and v.

The number of these levels (for a given electron term) may be either finite or infinite. If the electron state is such that, as $r \to \infty$, the molecule becomes two isolated neutral atoms, then as $r \to \infty$ the potential energy $U(r)$ (and therefore $U_K(r)$) tends to a constant limiting value $U(\infty)$ (the sum of the energies of the two isolated atoms) more rapidly than $1/r$ tends to zero (see §89). The number of levels in such a field is finite (see §18), though in actual molecules it is very large. The levels are so distributed that, for any given value of K, there is a definite number of levels (with different values of v), while the number of levels with the same K diminishes as K increases, until a value of K is reached for which there are no levels at all.

If, on the other hand, as $r \to \infty$ the molecule disintegrates into two ions, at large distances $U(r)-U(\infty)$ becomes the energy of the attraction of the ions according to Coulomb's law ($\sim 1/r$). In such a field there is an infinite number of levels, which become closer and closer as we approach the limiting value $U(\infty)$. We may remark that, for the majority of molecules, the previous case is found in the normal state; only a comparatively small number of molecules become pairs of ions when their nuclei are moved apart.

The dependence of the energy levels on the quantum numbers cannot be completely calculated in a general form. Such a calculation is possible only for low excited levels which lie not too far above the ground level.† Small values of the quantum numbers K and v correspond to these levels. It is with such levels that we are in fact usually concerned in the study of molecular spectra, and hence they are of particular interest.

The motion of the nuclei in slightly excited states can be regarded as small vibrations about the equilibrium position. Accordingly we can expand $U(r)$ in a series of powers of $\xi = r - r_e$, where r_e is the value of r for which $U(r)$ has a minimum. Since $U'(r_e) = 0$, we have as far as terms of the second order

$$U(r) = U_e + \tfrac{1}{2}M\omega_e^2\xi^2,$$

where $U_e = U(r_e)$, and ω_e is the frequency of the vibrations.

In the second term in (82.5)—the centrifugal energy—it is sufficient to put $r = r_e$, since it already contains the small quantity $K(K+1)$. Thus we have

$$U_K(r) = U_e + B_e K(K+1) + \tfrac{1}{2}M\omega_e^2\xi^2, \tag{82.7}$$

where $B_e = \hbar^2/2Mr_e^2 = \hbar^2/2I$ is what is called the *rotational constant* ($I = Mr_e^2$ is the moment of inertia of the molecule).

The first two terms in (82.7) are constants, while the third corresponds to a one-dimensional harmonic oscillator. Hence we can at once write down the required energy levels:

$$E = U_e + B_e K(K+1) + \hbar\omega_e(v+\tfrac{1}{2}). \tag{82.8}$$

† We refer always to levels belonging to the same electron term.

Thus, in the approximation considered, the energy levels are composed of three independent parts:

$$E = E_{el}+E_r+E_v. \tag{82.9}$$

The first term, $E_{el} = U_e$, is the electron energy (including the energy of the Coulomb interaction of the nuclei for $r = r_e$); the second term is

$$E_r = B_e K(K+1) \tag{82.10}$$

the rotational energy from the rotation of the molecule,† and the third term

$$E_v = \hbar\omega_e(v+\tfrac{1}{2}) \tag{82.11}$$

is the energy of the vibrations of the nuclei within the molecule. The number v denumerates, by definition, the levels with a given K in order of increasing energy; it is called the *vibrational quantum number*.

For a given form of the potential energy curve $U(r)$, the frequency ω_e is inversely proportional to $\sqrt{M}$. Hence the intervals ΔE_v between the vibrational levels are proportional to $1/\sqrt{M}$. The intervals ΔE_r between the rotational levels contain in the denominator the moment of inertia I, and are therefore proportional to $1/M$. The intervals ΔE_{el} between the electron levels, however, are independent of M, like the levels themselves. Since m/M (m being the electron mass) is a small parameter in the theory of diatomic molecules, we see that

$$\Delta E_{el} \gg \Delta E_v \gg \Delta E_r. \tag{82.12}$$

This shows the rather unusual distribution of the energy levels of the molecule. The vibrational motion of the nuclei splits the electron terms into levels lying comparatively close together. These levels, in turn, exhibit a fine splitting due to the rotational motion of the molecule.‡

In subsequent approximations, the separation of the energy into independent vibrational and rotational parts is impossible; rotational-vibrational terms appear, which contain both K and v. On calculating the successive

† The wave function describing the rotation of a diatomic molecule (without spin) is essentially the same as that of a symmetrical top (§103). Unlike the top, the molecule has a rotation described by only two angles ($\alpha \equiv \phi$, $\beta \equiv \theta$), which define the direction of its axis. The rotational wave function differs from (103.8) by the absence of the factor $e^{ik\gamma}/\sqrt{(2\pi)}$ and in the notation for the quantum numbers. Since, by (82.4), the number Λ is equal to the component of the total angular momentum **K** along the axis of the molecule (the ζ-axis in §103), we must replace J, M and k by K, M and Λ (here $M = K_z$). Thus

$$\psi_{rot}(\phi, \theta) = i^K \sqrt{\frac{2K+1}{4\pi}}\, D_{\Lambda M}^{(K)}(\phi, \theta, 0).$$

‡ As an example, we give the values of U_e, $\hbar\omega_e$ and B_e (in electron-volts) for a few molecules:

	H_2	N_2	O_2
$-U_e$	4·7	7·5	5·2
$\hbar\omega_e$	0·54	0·29	0·20
$10^3\times B_e$	7·6	0·25	0·18

approximations, we should obtain the levels E as an expansion in powers of the quantum numbers K and v.

We shall calculate here the next approximation after (82.8). To do this, we must continue the expansion of $U(r)$ in powers of ξ up to terms of the fourth order (cf. the problem of an anharmonic oscillator in §38). Similarly, the expansion of the centrifugal energy is extended as far as the terms in ξ^2. We then obtain

$$U_K(r) = U_e+\tfrac{1}{2}M\omega_e^2\xi^2+(\hbar^2/2Mr_e^2)K(K+1)- \\ -a\xi^3+b\xi^4-(\hbar^2/Mr_e^3)K(K+1)\xi+(3\hbar^2/2Mr_e^4)K(K+1)\xi^2. \qquad (82.13)$$

Let us now calculate the correction to the eigenvalues (82.8), regarding the last four terms in (82.13) as the perturbation operator. Here it is sufficient, for the terms in ξ^2 and ξ^4, to take the first approximation of perturbation theory, but for those in ξ and ξ^3 we must calculate the second approximation, since the diagonal matrix elements of ξ and ξ^3 vanish identically. All the matrix elements needed for the calculation are derived in §23 and in §38, Problem 3. As a result, we obtain an expression which is usually written in the form

$$E = E_{el}+\hbar\omega_e(v+\tfrac{1}{2})-x_e\hbar\omega_e(v+\tfrac{1}{2})^2+B_vK(K+1)-D_eK^2(K+1)^2, \quad (82.14)$$

where

$$B_v = B_e-\alpha_e(v+\tfrac{1}{2}) \equiv B_0-\alpha_e v. \quad (82.15)$$

The constants x_e, B_e, α_e, D_e are related to the constants appearing in (82.13) by

$$B_e = \hbar^2/2I, \qquad D_e = 4B_e^3/\hbar^2\omega_e^2,$$

$$\alpha_e = \frac{6B_e^2}{\hbar\omega_e}\left(\frac{a\hbar}{M\omega_e^2}\sqrt{\frac{2}{MB_e}}-1\right), \qquad x_e = \frac{3}{2\hbar\omega_e}\left(\frac{\hbar}{M\omega_e}\right)^2\left[\frac{5}{2}\frac{a^2}{M\omega_e^2}-b\right]. \quad (82.16)$$

The terms independent of v and K are included in E_{el}.

PROBLEM

Determine the accuracy of the approximation which gives the separation of the electron and nuclear motions in a diatomic molecule.

SOLUTION. The total Hamiltonian of the molecule may be written $\hat{H} = \hat{T}_r+\hat{H}_{el}$, where $\hat{T}_r = \hat{\mathbf{p}}^2/2M$ is the operator of the kinetic energy of the relative motion of the nuclei ($\hat{\mathbf{p}} = -i\hbar\partial/\partial\mathbf{r}$; $\mathbf{r}$ is the radius vector joining the nuclei; M is the reduced mass). The Hamiltonian $\hat{H}_{el}$ includes the operators of the kinetic energy of the electrons, the potential energy of their Coulomb interaction with one another and with the nuclei, and the energy of the Coulomb interaction of the nuclei.† The solution of Schrödinger's equation

$$\hat{H}\psi = (\hat{T}_r+\hat{H}_{el})\psi = E\psi \quad (1)$$

is sought in the form

$$\psi = \sum_m \chi_m(\mathbf{r})\phi_m(q, r), \quad (2)$$

† The Hamiltonian $\hat{H}$ relates to a frame of reference in which the centre of mass of the whole molecule is at rest ($\mathbf{P}_n+\mathbf{P}_e = 0$, where $\mathbf{P}_n$ is the total momentum of the two nuclei, and $\mathbf{P}_e$ that of the electrons). We do not, however, include in it the term corresponding to the kinetic energy of the motion of the centre of mass $\hat{\mathbf{P}}_n^2/2(M_1+M_2) = \hat{\mathbf{P}}_e^2/2M_1+M_2)$. This term is certainly small, in the ratio m/M, compared with the kinetic energy of the electrons.

where the functions $\phi_m(q, r)$ are orthonormalized solutions of the equation

$$\hat{H}_{el}\phi_m(q, r) = U_m(r)\phi_m(q, r), \tag{3}$$

q denoting the set of electron coordinates; $U_m(r)$ are the eigenvalues of the Hamiltonian $\hat{H}_{el}$, which depend on r as a parameter. Substituting (2) in (1), multiplying by $\phi_n^*(q, r)$ and integrating over q, we obtain

$$\left[\frac{\hat{\mathbf{p}}^2}{2M} + V''_{nn} + U_n(r) - E\right]\chi_n(\mathbf{r}) = -\sum_m{}' (\hat{V}'_{nm} + V''_{nm})\chi_m(\mathbf{r}), \tag{4}$$

where

$$\hat{V}'_{nm} = \frac{1}{M}\mathbf{p}_{nm}\cdot\hat{\mathbf{p}}, \qquad V''_{nm} = \frac{1}{2M}(\mathbf{p}^2)_{nm},$$

and $\mathbf{p}_{nm} = \int \phi_n^*\hat{\mathbf{p}}\,\phi_m\,dq$; $(\mathbf{p}^2)_{nm}$ are the matrix elements with respect to the electron wave functions; the diagonal element $\mathbf{p}_{nn}$ is zero, by symmetry.

The electron functions ϕ_n vary appreciably only over atomic distances, and their differentiation with respect to $\mathbf{r}$ therefore does not introduce the large parameter M/m (where m is the electron mass). The quantity V''_{nn} is consequently small in comparison with $U_n(r)$, in the ratio m/M, and may be omitted. If the terms on the right of (4) are regarded as a small perturbation, then in the zero-order approximation the functions $\chi_n(\mathbf{R})$ are given by the solutions of the equation

$$\left[\frac{\hat{\mathbf{p}}^2}{2M} + U_n(r)\right]\chi_{nv} = E_{nv}\chi_{nv}, \tag{5}$$

which describes the motion of the nuclei in the field $U_n(r)$ (the v being the quantum numbers for this motion). The condition for perturbation theory to be applicable is

$$|\langle nv'|\hat{V}'_{nm} + V''_{nm}|mv\rangle| \ll |E_{nv'} - E_{mv}|.$$

The right-hand side of this inequality is the difference of the energies of different electron terms; these quantities are of order zero with respect to the smallness parameter m/M. The left-hand side contains the matrix elements with respect to the nuclear wave functions. The term in V''_{nm} contains the factor m/M and is certainly small. In the matrix element of $\hat{V}'_{nm}$, the operator $\hat{\mathbf{p}}$, acting on the function χ_{mv}, multiplies it by a quantity of the order of the momentum of the nuclei. If the nuclei execute small oscillations, their momentum $\sim \sqrt{(M\hbar\omega_e)}$; since the frequency ω_e is itself inversely proportional to $\sqrt{M}$, the matrix element $\langle nv'|V'_{nm}|mv\rangle$ is of the order $(m/M)^{3/4}$.

§83. Multiplet terms. Case *a*

Let us now turn to the question of the classification of molecular levels with non-zero spin S. In the zero-order approximation, when relativistic effects are entirely neglected, the energy of the molecule, like that of any system of particles, is independent of the direction of the spin (the spin is "free"), and this results in a $(2S+1)$-fold degeneracy of the levels. When relativistic effects are taken into account, however, the degenerate levels are split, and the energy consequently becomes a function of the projection of the spin on the axis of the molecule. We shall refer to relativistic interactions in molecules as the *spin–axis interaction*. The chief part in this is played (as in the case of atoms) by the interaction of the spins with the orbital motion of the electrons.†

† Besides the spin–orbit and spin–spin interactions there is also an interaction of the spin and orbital motion of the electrons with the rotation of the molecule. This part of the interaction is very small, however, and it is of possible interest only for terms with spin $S = \frac{1}{2}$ (see §84).

The nature and classification of molecular levels depend markedly on the relative parts played by the interaction of the spin with the orbital motion, on the one hand, and the rotation of the molecule, on the other. The part played by the latter is characterized by the distances between adjacent rotational levels. Accordingly, we have to consider two limiting cases. In one, the energy of the spin–axis interaction is large compared with the energy differences between the rotational levels, while in the other it is small. The first case is usually called *case* (or coupling type) *a*, and the second is called *case b* (F. Hund 1933).

Case *a* is the one most often found. An exception is formed by the Σ terms, where case *b* chiefly occurs, since the effect of the spin–axis interaction is very small for these terms† (see below). For other terms, case *b* is sometimes found in the lightest molecules, since the spin–axis interaction is here comparatively weak, while the distances between the rotational levels are large (the moment of inertia being small).

Of course, cases intermediate between *a* and *b* are also possible. It must also be borne in mind that the same electron state may pass continuously from case *a* to case *b* as the rotational quantum number changes. This is due to the fact that the distances between adjacent rotational levels increase with the rotational quantum number, and hence, when this is large, the distances may become large compared with the energy of the spin–axis coupling (case *b*), even if case *a* is found for the lower rotational levels.

In case *a*, the classification of the levels is in principle little different from that of the terms with zero spin. We first consider the electron terms for nuclei at rest, i.e. we neglect rotation entirely; besides the projection Λ of the orbital angular momentum of the electrons, we must now take into account the projection of the total spin on the axis of the molecule. This projection is denoted by‡ Σ; it takes the values $S, S-1, \ldots, -S$. We arbitrarily regard Σ as positive when the projection of the spin is in the same direction as that of the orbital angular momentum about the axis (we recall that Λ denotes the absolute value of the latter). The quantities Λ and Σ combine to give the total angular momentum of the electrons about the axis of the molecule:

$$\Omega = \Lambda + \Sigma; \tag{83.1}$$

this takes the values $\Lambda+S, \Lambda+S-1, \ldots, \Lambda-S$. Thus the electron term with orbital angular momentum Λ is split into $2S+1$ terms with different values of Ω; this splitting, as with atomic terms, is called the ***fine structure*** or ***multiplet splitting*** of the electron levels. The value of Ω is usually indicated as a suffix to the symbol for the term: thus, for $\Lambda = 1$, $S = \frac{1}{2}$ we obtain the terms $^2\Pi_{1/2}$, $^2\Pi_{3/2}$.

When the motion of the nuclei is taken into account, vibrational and rotational structures appear in each of these terms. The various rotational levels are characterized by the values of the quantum number *J*, which gives the

† A special case is the normal electron term of the molecule O_2 (the term $^3\Sigma$). For this we have a type of coupling intermediate between *a* and *b* (see §84, Problem 3).

‡ Not to be confused with the symbol for terms with $\Lambda = 0$.

total angular momentum of the molecule, including the orbital and spin angular momenta of the electrons and the angular momentum of the rotation of the nuclei.† This number takes all integral values from $|\Omega|$ upwards:

$$J \geqslant |\Omega|, \tag{83.2}$$

which is an obvious generalization of (82.6).

Let us now derive quantitative formulae to determine the molecular levels in case *a*. First of all, we consider the fine structure of an electron term. In discussing the fine structure of atomic terms in §72, we used formula (72.4), according to which the mean value of the spin–orbit interaction is proportional to the projection of the total spin of the atom on the orbital angular momentum vector. Similarly, the spin–axis interaction in a diatomic molecule (averaged over electron states for a given distance r between the nuclei) is proportional to the projection Σ of the total spin of the molecule on its axis, so that we can write the split electron term in the form

$$U(r)+A(r)\Sigma,$$

where $U(r)$ is the energy of the original (unsplit) term, and $A(r)$ is some function of r; this function depends on the original term (and in particular on Λ), but not on Σ. Since one usually uses the quantum number Ω and not Σ, it is more convenient to put $A\Omega$ in place of $A\Sigma$; these expressions differ by $A\Lambda$, which can be included in $U(r)$. Thus we have for an electron term the expression

$$U(r)+A(r)\Omega. \tag{83.3}$$

We may notice that the components of the split term are equidistant from one another: the distance between adjacent components (with values of Ω differing by unity) is $A(r)$, independent of Ω.

It is easy to see from general considerations that the value of A for Σ terms is zero. To show this, we perform the operation of changing the sign of the time. The energy must then remain unchanged, but the state of the molecule changes in that the direction of the orbital and spin angular momenta about the axis is reversed. In the energy $A(r)\Sigma$, the sign of Σ is changed, and if the energy remains unchanged $A(r)$ must change sign. If $\Lambda \neq 0$, we can draw no conclusions regarding the value of $A(r)$, since this depends on the orbital angular momentum, which itself changes sign. If $\Lambda = 0$, however, we can say that $A(r)$ is certainly unchanged, and consequently it must vanish identically. Thus, for the Σ terms, the spin–orbit interaction causes no splitting in the first approximation; splitting (proportional to Σ^2) would occur only on taking account of this interaction in the second approximation or the spin–spin interaction in the first approximation, and would be relatively small. This is the reason for the fact, already mentioned, that case *b* usually occurs for Σ terms.

---

† The notation **K** is, as usual, reserved for the total angular momentum of the molecule without allowance for its spin. In case *a* there is no quantum number K, since the angular momentum **K** is not even approximately conserved.

When the multiplet splitting has been determined, we can take account of the rotation of the molecule as a perturbation, just as in the derivation given at the beginning of §82. The angular momentum of the rotation of the nuclei is obtained from the total angular momentum by subtracting the orbital angular momentum and spin of the electrons. Hence the operator of the centrifugal energy now has the form

$$B(r)(\hat{\mathbf{J}}-\hat{\mathbf{L}}-\hat{\mathbf{S}})^2.$$

Averaging this quantity with respect to the electron state and adding to (83.3), we obtain the required effective potential energy $U_J(r)$:

$$\begin{aligned} U_J(r) &= U(r)+A(r)\Omega+B(r)\overline{(\mathbf{J}-\mathbf{L}-\mathbf{S})^2} \\ &= U(r)+A(r)\Omega+B(r)[\mathbf{J}^2-2\mathbf{J}\cdot\overline{(\mathbf{L}+\mathbf{S})}+\overline{\mathbf{L}^2}+2\overline{\mathbf{L}\cdot\mathbf{S}}+\overline{\mathbf{S}^2}]. \end{aligned}$$

The eigenvalue of $\mathbf{J}^2$ is $J(J+1)$. Next, by the same argument as in §82, we have

$$\overline{\mathbf{L}} = \mathbf{n}\Lambda, \qquad \overline{\mathbf{S}} = \mathbf{n}\Sigma, \tag{83.4}$$

and also $(\hat{\mathbf{J}}-\hat{\mathbf{L}}-\hat{\mathbf{S}})\cdot\mathbf{n} = 0$, whence we have for the eigenvalues

$$\mathbf{J}\cdot\mathbf{n} = (\mathbf{L}+\mathbf{S})\cdot\mathbf{n} = \Lambda+\Sigma = \Omega. \tag{83.5}$$

Substituting these values, we find

$$U_J(r) = U(r)+A(r)\Omega+B(r)[J(J+1)-2\Omega^2+\overline{\mathbf{L}^2}+2\overline{\mathbf{L}\cdot\mathbf{S}}+\overline{\mathbf{S}^2}].$$

The averaging with respect to the electron state is effected by means of the wave functions of the zero-order† approximation. In this approximation, however, the magnitude of the spin is conserved, and hence $\mathbf{S}^2 = S(S+1)$. The wave function is the product of the spin and coordinate functions; hence the averaging of the angular momenta **L** and **S** takes place independently, and we obtain

$$\overline{\mathbf{L}\cdot\mathbf{S}} = \Lambda\mathbf{n}\cdot\overline{\mathbf{S}} = \Lambda\Sigma.$$

Finally, the mean value of the squared orbital angular momentum $\mathbf{L}^2$ is independent of the spin, and is some function of r characterizing the given (unsplit) electron term. All the terms which are functions of r but independent of J and Σ can be included in $U(r)$, while the term proportional to Σ (or, what is the same thing, to Ω) can be included in the expression $A(r)\Omega$. Thus we have for the effective potential energy the formula

$$U_J(r) = U(r)+A(r)\Omega+B(r)[J(J+1)-2\Omega^2]. \tag{83.6}$$

† That is, the zero-order approximation with respect to both the effect of the rotation of the molecule and the spin–axis interaction.

The energy levels of the molecule can be obtained from this by the same method as in §82 when using the formula (82.5). Expanding $U(r)$ and $A(r)$ in series of powers of ξ, and retaining the terms up to and including the second order in the expansion of $U(r)$, but only the terms of zero order in the second and third terms, we obtain the energy levels in the form

$$E = U_e + A_e\Omega + \hbar\omega_e(v+\tfrac{1}{2}) + B_e[J(J+1) - 2\Omega^2], \tag{83.7}$$

where $A_e = A(r_e)$ and B_e are constants characterizing the given (unsplit) electron term. On continuing the expansion to higher terms, we obtain a series of terms in higher powers of the quantum numbers, but we shall not pause to write these out here.

§84. Multiplet terms. Case *b*

Let us now turn to case *b*. Here the effect of the rotation of the molecule predominates over the multiplet splitting. Hence we must first consider the effect of rotation, neglecting the spin–axis interaction, and then the latter must be taken into account as a perturbation.

In a molecule with "free" spin, not only the total angular momentum **J** but also the sum **K** of the orbital angular momentum of the electrons and the angular momentum of the nuclei are conserved; the latter is related to **J** by

$$\mathbf{J} = \mathbf{K} + \mathbf{S}. \tag{84.1}$$

The quantum number K distinguishes different states of a rotating molecule with free spin that are obtained from a given electron term. The effective potential energy $U_K(r)$ in a state with a given value of K is evidently determined by the same formula (82.5) as for terms with $S = 0$:

$$U_K(r) = U(r) + B(r)K(K+1), \tag{84.2}$$

where K takes the values $\Lambda, \Lambda+1, \ldots$.

When the spin–axis interaction is included, there is a splitting of each term into $2S+1$ terms in general (or $2K+1$ if $K < S$), which differ in the value of the total angular momentum† J. According to the general rule for the addition of angular momenta, the number J takes (for a given K) values from $K+S$ to $|K-S|$:

$$|K-S| \leqslant J \leqslant K+S. \tag{84.3}$$

To calculate the energy of the splitting (in the first approximation of perturbation theory), we must determine the mean value of the operator of the spin–axis interaction energy for the state in the zero-order approximation (with respect to this interaction). In the case considered, this means averaging with respect to both the electron state and the rotation of the molecule (for a given r). The result of the first averaging is an operator of the form

† In case *b*, the projection $\mathbf{n} \cdot \mathbf{S}$ of the spin on the axis of the molecule does not have definite values, so that there is no quantum number Σ (or Ω).

$A(r)\mathbf{n}\cdot\hat{\mathbf{S}}$, which is proportional to the projection $\mathbf{n}\cdot\hat{\mathbf{S}}$ of the spin operator on the axis of the molecule. Next we average this operator with respect to the rotation of the molecule, taking the direction of the spin vector to be arbitrary; then $\overline{\mathbf{n}\cdot\hat{\mathbf{S}}} = \bar{\mathbf{n}}\cdot\hat{\mathbf{S}}$. The mean value $\bar{\mathbf{n}}$ is a vector which, from considerations of symmetry, must have the same direction as the "vector" $\hat{\mathbf{K}}$, the only vector which characterizes the rotation of the molecule. Thus we can write

$$\bar{\mathbf{n}} = \text{constant} \times \hat{\mathbf{K}}.$$

The coefficient of proportionality is easily determined by multiplying both sides of this equation by $\hat{\mathbf{K}}$; noting that the eigenvalues of $\mathbf{n}\cdot\mathbf{K}$ and $\mathbf{K}^2$ are respectively Λ (see (82.4)) and $K(K+1)$, we find

$$\overline{\mathbf{n}\cdot\hat{\mathbf{S}}} = \Lambda\hat{\mathbf{K}}\cdot\hat{\mathbf{S}}/K(K+1).$$

Finally, the eigenvalue of the product $\mathbf{K}\cdot\mathbf{S}$, according to the general formula (31.3), is

$$\mathbf{K}\cdot\mathbf{S} = \tfrac{1}{2}[J(J+1)-K(K+1)-S(S+1)]. \tag{84.4}$$

As a result, we arrive at the following expression for the required mean value of the energy of the spin–axis interaction:

$$\begin{aligned} &A(r)\Lambda[J(J+1)-S(S+1)-K(K+1)]/2K(K+1) \\ &= A(r)\Lambda[(J-S)(J+S+1)]/2K(K+1)-\tfrac{1}{2}A(r)\Lambda. \end{aligned}$$

This expression must be added to the energy (84.2). The term $\frac{1}{2}A(r)\Lambda$, being independent of K and J, can be included in $U(r)$, so that we have finally for the effective potential energy the expression

$$U_K(r) = U(r)+B(r)K(K+1)+A(r)\Lambda(J-S)(J+S+1)/2K(K+1). \tag{84.5}$$

An expansion in powers of $\xi = r-r_e$ gives, in the usual manner, an expression for the energy levels of the molecule in case b:

$$E = U_e+\hbar\omega_e(v+\tfrac{1}{2})+B_eK(K+1)+A_e\Lambda(J-S)J+S+1)/2K(K+1). \tag{84.6}$$

As has been pointed out in the previous section, the spin–orbit interaction for Σ terms does not give a multiplet splitting in the first approximation, and to determine the fine structure we must take into account the spin–spin interaction, whose operator is quadratic with respect to the spins of the electrons. We are at present interested not in this operator itself, but in the result of averaging it with respect to the electron state of the molecule, as was done for the operator of the spin–orbit interaction. It is evident from considerations of symmetry that the required averaged operator must be proportional to the squared projection of the total spin of the molecule on the axis, i.e. it can be written in the form

$$\alpha(r)\,(\hat{\mathbf{S}}\cdot\mathbf{n})^2, \tag{84.7}$$

where $\alpha(r)$ is again some function of the distance r, characterizing the given electron state. Symmetry allows also a term proportional to $\hat{\mathbf{S}}^2$, but this is immaterial since the absolute value of the spin is just a constant. We shall not pause here to derive the lengthy general formula for the splitting due to the operator (84.7); in Problem 1 of this section we give the derivation of the formula for triplet Σ terms.

The doublet Σ terms form a special case. According to Kramers' theorem (§60), the double degeneracy in a system of particles with total spin $S = \frac{1}{2}$ certainly persists, even when the internal relativistic interactions in the system are fully allowed for. Hence the $^2\Sigma$ terms remain unsplit, even when we take account of both the spin–orbit and the spin–spin interaction, and in any approximation.

The splitting is obtained here only by taking into account the relativistic interaction of the spin with the rotation of the molecule; this effect is very small. The averaged operator of this interaction must evidently be of the form $\gamma\hat{\mathbf{K}}\cdot\hat{\mathbf{S}}$, and its eigenvalues are determined by the formula (84.4), in which we must put $S = \frac{1}{2}$, $J = K \pm \frac{1}{2}$. As a result, we obtain for the $^2\Sigma$ terms the formula

$$E = U_e + \hbar\omega_e(v+\tfrac{1}{2}) + B_e K(K+1) \pm \tfrac{1}{2}\gamma(K+\tfrac{1}{2}); \tag{84.8}$$

a constant $-\frac{1}{4}\gamma$ is included in U_e.

PROBLEMS

PROBLEM 1. Determine the multiplet splitting of a $^3\Sigma$ term in case b (H. A. Kramers 1929).

SOLUTION. The required splitting is determined by the operator (84.7), which must be averaged with respect to the rotation of the molecule. We write it in the form $\alpha_e n_i n_k \hat{S}_i \hat{S}_k$, where $\alpha_e = \alpha(r_0)$. Since the vector $\mathbf{S}$ is conserved, only the products $n_i n_k$ need be averaged. According to the formula derived in §29, Problem, we have

$$\overline{n_i n_k} = -\frac{\hat{K}_i\hat{K}_k + \hat{K}_k\hat{K}_i}{(2K-1)(2K+3)} + \dots;$$

here we have not written out the terms proportional to δ_{ik} whose contribution to the energy is independent of J and therefore does not cause any splitting of the type under consideration. Thus the splitting is given by the operator

$$-\frac{\alpha_e}{(2K-1)(2K+3)}\hat{S}_i\hat{S}_k(\hat{K}_i\hat{K}_k + \hat{K}_k\hat{K}_i).$$

Since $\hat{\mathbf{S}}$ commutes with $\hat{\mathbf{K}}$,

$$\hat{S}_i\hat{S}_k\hat{K}_i\hat{K}_k = \hat{S}_i\hat{K}_i\hat{S}_k\hat{K}_k = (\mathbf{S}\cdot\mathbf{K})^2,$$

where the eigenvalue $\mathbf{S}\cdot\mathbf{K}$ is given by (84.4). We also have

$$\begin{aligned}\hat{S}_i\hat{S}_k\hat{K}_k\hat{K}_i &= \hat{S}_i\hat{S}_k\hat{K}_i\hat{K}_k + i\hat{S}_i\hat{S}_k e_{kil}\hat{K}_l \\ &= (\mathbf{S}\cdot\mathbf{K})^2 - \tfrac{1}{2}(\hat{S}_i\hat{S}_k - \hat{S}_k\hat{S}_i) i e_{ikl}\hat{K}_l \\ &= (\mathbf{S}\cdot\mathbf{K})^2 + \tfrac{1}{2}e_{ikl}e_{ikm}\hat{S}_m\hat{K}_l \\ &= (\mathbf{S}\cdot\mathbf{K})^2 + \mathbf{S}\cdot\mathbf{K}.\end{aligned}$$

The values $J = K, K\pm 1$ correspond to the three components E_K of the triplet $^3\Sigma$ ($S = 1$). For the intervals between these components we find

$$E_{K+1} - E_K = -\alpha_e \frac{K+1}{2K+3}, \quad E_{K-1} - E_K = -\alpha_e \frac{K}{2K-1}.$$

PROBLEM 2. Determine the energy of a doublet term (with $\Lambda \neq 0$) for cases intermediate between a and b (E. Hill and J. H. van Vleck 1928).

SOLUTION. Since the rotational energy and the energy of the spin–axis interaction are supposed of the same order of magnitude, they must be considered together in perturbation theory, so that the perturbation operator is of the form†

$$\hat{V} = B_e \hat{\mathbf{K}}^2 + A_e \mathbf{n} . \hat{\mathbf{S}}.$$

As wave functions in the zero-order approximation it is convenient to use those of states in which the angular momenta K and J have definite values (i.e. those of case b). Since $S = \frac{1}{2}$ for a doublet term, the quantum number K, for a given J, can take the values $K = J\pm\frac{1}{2}$. To construct the secular equation, we must calculate the matrix elements $\langle nSKJ|V|nSK'J\rangle$ (n denoting the assembly of quantum numbers defining the electron term), where K,K' take the above values. The matrix of the operator $\hat{\mathbf{K}}^2$ is diagonal; the diagonal elements are $K(K+1)$. The matrix elements of $\mathbf{n} . \mathbf{S}$ are calculated from the general formula (109.5), in which we must put $j_1 = S, j_2 = K$; the reduced matrix elements of $\mathbf{n}$ are given by (87.4). Calculation gives the secular equation

$$\begin{vmatrix} B_e(J+\frac{1}{2})(J+\frac{3}{2}) - A_e\Lambda/(2J+1) - E^{(1)} & A_e\sqrt{[(J+\frac{1}{2})^2-\Lambda^2]}/(2J+1) \\ A_e\sqrt{[(J+\frac{1}{2})^2-\Lambda^2]}/(2J+1) & B_e(J+\frac{1}{2})(J-\frac{1}{2}) + A_e\Lambda/(2J+1) - E^{(1)} \end{vmatrix} = 0.$$

Solving this equation and adding $E^{(1)}$ to the unperturbed energy, we have

$$E = U_e + \hbar\omega_e(v+\tfrac{1}{2}) + B_e J(J+1) \pm \sqrt{[B_e^2(J+\tfrac{1}{2})^2 - A_e B_e \Lambda + \tfrac{1}{4}A_e^2]};$$

a constant $\frac{1}{4}B_e$ is included in U_e. The inequality $A_e \gg B_e J$ corresponds to case a, and the opposite one to case b.

PROBLEM 3. Determine the intervals between the components of a triplet level $^3\Sigma$ in a case intermediate between a and b.

SOLUTION. As in Problem 2, the rotational energy and the energy of the spin–spin interaction are considered together in the perturbation theory. The perturbation operator is of the form

$$\hat{V} = B_e \hat{\mathbf{K}}^2 + \alpha_e (\mathbf{n} . \hat{\mathbf{S}})^2.$$

As wave functions in the zero-order approximation we use those of case b. The matrix elements $\langle K|\mathbf{n} . \mathbf{S}|K'\rangle$ (we omit all suffixes with respect to which the matrix is diagonal) are again calculated from (109.5) and (87.4), this time with $\Lambda = 0$, $S = 1$. The non-zero elements are of the form

$$\langle J|\mathbf{n} . \mathbf{S}|J-1\rangle = \sqrt{[(J+1)/(2J+1)]}, \quad \langle J|\mathbf{n} . \mathbf{S}|J+1\rangle = \sqrt{[J/(2J+1)]}.$$

For a given J, the number K can take the values $K = J, J\pm 1$. For the matrix elements $\langle K|V|K'\rangle$ we find

$$\langle J|V|J\rangle = B_e J(J+1) + \alpha_e, \quad \langle J-1|V|J-1\rangle = B_e(J-1)J + \alpha_e(J+1)/(2J+1),$$

$$\langle J+1|V|J+1\rangle = B_e(J+1)(J+2) + \alpha_e J/(2J+1),$$

$$\langle J-1|V|J+1\rangle = \langle J+1|V|J-1\rangle = \alpha_e \sqrt{[J(J+1)/(2J+1)]}.$$

† The averaging with respect to vibrations must be done before that with respect to rotation. Hence, restricting ourselves to the first terms of the expansions in ξ, we have replaced the functions $B(r)$ and $A(r)$ by the values B_e and A_e, and the unperturbed energy levels are $E^{(0)} = U_e + \hbar\omega_e(v+\frac{1}{2})$.

We see that there are no transitions between states with $K = J$ and those with $K = J \pm 1$. Hence one of the levels is simply $E_1 = \langle J|V|J \rangle$. The other two ($E_2$, E_3) are obtained by solving the quadratic secular equation formed from the matrix elements for transitions between states $J \pm 1$. Since we are here interested only in the relative position of the components of the triplet, we subtract the constant α_e from all three energies E_1, E_2, E_3. As a result we obtain

$$E_1 = B_e J(J+1),$$

$$E_{2,3} = B_e(J^2+J+1) - \tfrac{1}{2}\alpha_e) \pm \sqrt{[B_e^2(2J+1)^2 - \alpha_e B_e + \tfrac{1}{4}\alpha_e^2]}.$$

In case *b* (α small), by considering three levels with the same K and different J ($J = K$, $K \pm 1$), we again obtain the formulae of Problem 1.

§85. Multiplet terms. Cases *c* and *d*

Besides cases of *a* and *b* coupling and those intermediate between them, there are also other types of coupling. These originate as follows. The occurrence of the quantum number Λ is due ultimately to the electric interaction of the two atoms in the molecule, which results in the axial symmetry of the problem of determining the electron terms (this interaction in the molecule is called the coupling between the orbital angular momentum and the axis). The distances between terms with different values of Λ give a measure of the magnitude of this interaction. Previously we have tacitly supposed this interaction so strong that these distances are large both compared with the intervals in the multiplet splitting and compared with those in the rotational structure of the terms. There are, however, opposite cases where the interaction of the orbital angular momentum with the axis is comparable with or even small compared with the other effects; in such cases, of course, we cannot in any approximation speak of a conservation of the projection of the orbital angular momentum on the axis, so that the number Λ is no longer meaningful.

If the coupling of the orbital angular momentum with the axis is small in comparison with the spin-orbit coupling, we say that we have case *c*. It is found in molecules which contain an atom of a rare-earth element. These atoms are characterized by the presence of *f* electrons with uncompensated angular momenta; their interaction with the axis of the molecule is weakened by the deep position of the *f* electrons in the atom. Cases intermediate between the *a* and *c* types of coupling are found in molecules consisting of heavy atoms.

If the coupling of the orbital angular momentum with the axis is small compared with the intervals in the rotational structure, we say that we have case *d*. This case is found for high rotational levels (with large J) in some electron terms of the lightest molecules (H_2, He_2). These terms are characterized by the presence in the molecule of a highly excited electron, whose interaction with the remaining electrons (or, as we say, with the "core" of the molecule) is so weak that its orbital angular momentum is not quantized along the axis of the molecule (whereas the "core" has a definite angular momentum Λ_{core} about the axis).

As the distance r between the nuclei increases, the interaction between the atoms is diminished, and finally becomes small compared with the spin–orbit interaction within the atoms. Hence, if we consider the electron terms for fairly large r, we shall have case c. This must be borne in mind when ascertaining the relation between the electron terms of the molecule and the states of the atoms obtained as $r \to \infty$. In §80 we have already discussed this relation, neglecting the spin–orbit interaction. When the fine structure of the terms is included, there arises also the question of the relation between the values J_1 and J_2 of the total angular momenta of the isolated atoms and the values of the quantum number Ω for the molecule. We shall give the results here, without reiterating arguments which are entirely similar to those of §80.

If the molecule consists of different atoms, the possible values of† $|\Omega|$ obtained on combining atoms with angular momenta J_1, J_2 ($J_1 \geqslant J_2$) are given by the same table (80.1), in which we must put J_1, J_2 in place of L_1, L_2, and $|\Omega|$ in place of Λ. The only difference is that, for half-integral J_1+J_2, the smallest value of $|\Omega|$ is not zero as shown in the table, but $\frac{1}{2}$. For integral J_1+J_2, on the other hand, there are $2J_2+1$ terms with $\Omega = 0$, for which (as for Σ terms when the fine structure is neglected) we have to decide the question of sign. If J_1 and J_2 are each half-integral, the number $2J_2+1$ is even, and there are equal numbers of terms, which we shall denote by 0^+ and 0^-. If J_1 and J_2 are both integral, however, then J_2+1 terms are 0^+ and J_2 are 0^- (if $(-1)^{J_1+J_2}P_1P_2 = 1$) or *vice versa* (if $(-1)^{J_1+J_2}P_1P_2 = -1$).

If the molecule consists of similar atoms in different states, the resulting molecular states are the same as in the case of different atoms, the only difference being that the total number of terms is doubled, with each term appearing once as an even and once as an odd term.

Finally, if the molecule consists of similar atoms in the same state (with angular momenta $J_1 = J_2 \equiv J$), the total number of states is the same as in the case of different atoms, while their distribution in parity is such that,

if J is integral and Ω is even, $N_g = N_u+1$;
if J is integral and Ω is odd, $N_g = N_u$;
if J is half-integral and Ω is even, $N_u = N_g$;
if J is half-integral and Ω is odd, $N_u = N_g+1$.

All the 0^+ terms are even and all the 0^- terms odd.

As the nuclei approach, a coupling of type c usually passes into one of type a‡. Here the following interesting circumstance may arise. As already mentioned, the term with $\Lambda = 0$ belongs to case b, and as regards the classification of case a this means that multiplet levels with different values of Ω (and the same $\Lambda = 0$) have the same energy; but such levels can occur on the approach of atoms which are in different fine-structure states.

---

† In adding the two total angular momenta J_1, J_2 of the atoms to form the resultant angular momentum Ω, the sign of Ω is clearly immaterial.

‡ The correspondence between the classification of terms of types a and c cannot be derived in a general form. Its derivation necessitates a consideration of the actual potential energy curves, taking into account the rule that levels of like symmetry cannot intersect (§79).

Thus it may happen that the same molecular term corresponds to different pairs of atomic fine-structure states. A similar situation may occur for terms with $\Omega = 0$ which, on the approach of the nuclei, become a molecular term with $\Lambda \neq 0$ (and therefore $\Sigma = -\Lambda$). Such levels are doubly degenerate, since in case *a* the same energy corresponds to the terms 0^+ and 0^- (which may arise from different pairs of atomic states).†

§86. Symmetry of molecular terms

In §78 we have already examined some symmetry properties of the terms of a diatomic molecule. These properties characterized the behaviour of the wave functions in transformations which leave the coordinates of the nuclei unaltered. Thus the symmetry of the molecule with respect to reflection in a plane passing through its axis brings about the difference between Σ^+ and Σ^- terms; the symmetry with respect to a change in sign of the coordinates‡ of all the electrons (for molecules composed of like atoms) gives rise to the classification of terms into even and odd. These symmetry properties characterize the electron terms, and are the same for all rotational levels belonging to the same electron term.

The states of the molecule, like those of any system of particles (see §30), are characterized by their behaviour with respect to inversion, i.e. a simultaneous change in sign of the coordinates of all the electrons and the nuclei. For this reason, all the terms for the molecule can be divided into *positive* (whose wave functions are unaltered when the sign of the coordinates of the electrons and nuclei is reversed) and *negative* (whose wave functions change sign on inversion).‖

For $\Lambda \neq 0$, each term is doubly degenerate, on account of the two possible directions of the angular momentum about the axis of the molecule. As a result of inversion, the angular momentum itself does not change sign, but the direction of the axis of the molecule is reversed (since the atoms change places), and hence the direction of the angular momentum Λ relative to the molecule is reversed. Hence two wave functions belonging to the same energy level are transformed into each other, and from them we can always form a linear combination that is invariant with respect to inversion and one that changes sign under this transformation. Thus we obtain for each term two states, of which one is positive and the other negative. In practice, every term with $\Lambda \neq 0$ is split, however (see §88), and so these two states correspond to different values of the energy.

The Σ terms require special consideration to determine their sign. First

---

† We here neglect what is called Λ-doubling (see §88).

‡ The origin is supposed to be taken on the axis of the molecule, and half-way between the two nuclei.

‖ We retain the customary terminology. It is unfortunate, however, since in the case of an atom the behaviour of the terms with respect to the operation of inversion is referred to as *parity* and not *sign*.

The sign of which we are here speaking must not be confused with the + and − which are added as indices to Σ terms.

of all, it is clear that the spin bears no relation to the sign of the term: the inversion operation changes only the coordinates of the particles, leaving the spin part of the wave function unaltered. Hence all the components of the multiplet structure of any given term have the same sign. In other words, the sign of the term depends only on K, and not on J.†

The wave function of the molecule is the product of the electron and nuclear wave functions. It has been shown in §82 that, in a Σ state, the motion of the nuclei is equivalent to that of a single particle, of orbital angular momentum K, in a centrally symmetric field $U(r)$. Hence we can say that, when the sign of the coordinates is changed, the nuclear wave function is multiplied by $(-1)^K$ (see (30.7)).

The electron wave function characterizes the electron term, and to ascertain its behaviour under inversion we must consider it in a system of coordinates rigidly connected to the nuclei and rotating with them. Let x, y, z be a system of coordinates fixed in space, and ξ, η, ζ a rotating system of coordinates in which the molecule is fixed. The direction of the axes of ξ, η, ζ is defined so that the ζ-axis coincides with the axis of the molecule from (say) nucleus 1 to nucleus 2, and the relative position of the positive directions of the axes of ξ, η, ζ is the same as in the system x, y, z (i.e. if the system x, y, z is left-handed, the system ξ, η, ζ is so too). As a result of the inversion operation, the direction of the axes of x, y, z is reversed, and the system changes from left-handed to right-handed. The system ξ, η, ζ must also become right-handed, but the ζ-axis, being rigidly connected to the nuclei, retains its former direction. Hence the direction of either one of the axes of ξ, η must be reversed. Thus the operation of inversion in the fixed system of coordinates is equivalent in the moving system to a reflection in a plane passing through the axis of the molecule. Under such a reflection, however, the electron wave function of a Σ^+ term is unaltered, while that of a Σ^- term changes sign.

Thus the sign of the rotational components of a Σ^+ term is determined by the factor $(-1)^K$; all the levels with even K are positive, while those with odd K are negative. For a Σ^- term, the sign of the rotational levels is determined by the factor $(-1)^{K+}$; all levels with even K are negative, while those with odd K are positive.

If the molecule consists of similar atoms,‡ its Hamiltonian is also invariant with respect to an interchange of the coordinates of the two nuclei. A term is said to be *symmetric* with respect to the nuclei if its wave function is unaltered when they are interchanged and *antisymmetric* if its wave function changes sign. The symmetry with respect to the nuclei is closely related to the parity and sign of the term. An interchange of the coordinates of the nuclei is equivalent to a change in sign of the coordinates of all the particles (electrons and nuclei), followed by a change in sign of the coordinates of the electrons only. Hence it follows that, if the term is even and positive (or

† We recall that case b usually holds for Σ terms, and so it is necessary to use the quantum numbers K and J.

‡ The two atoms must be not only of the same element, but also of the same isotope.

odd and negative), it is symmetric with respect to the nuclei. If, on the other hand, it is even and negative (or odd and positive), then it is antisymmetric with respect to the nuclei.

At the end of §62 we have established a general theorem that the coordinate wave function of a system of two identical particles is symmetrical when the total spin of the system is even, and antisymmetrical when it is odd. If we apply this result to the two nuclei of a molecule composed of similar atoms, we find that the symmetry of a term is related to the parity of the total spin I obtained by adding the spins i of the two nuclei. The term is symmetric when I is even, and antisymmetric when I is odd.† In particular, if the nuclei have no spin ($i = 0$), I is zero also; hence the molecule has no antisymmetric terms. We see that the nuclear spin has an important indirect influence on the molecular terms, although its direct influence (the hyperfine structure of the terms) is quite unimportant.

When the spin of the levels is taken into account, an additional degeneracy of the levels results. Again in §62, we have calculated the number of states with even and odd values of I that are obtained on adding two spins i. Thus, when i is half-integral, the number of states with even I is $i(2i+1)$, and with odd I is $(i+1)(2i+1)$. From what was said above, we conclude that the ratio of the degrees g_s, g_a of the degeneracy‡ of symmetric and antisymmetric terms for terms with half-integral i is

$$g_s/g_a = i/(i+1). \tag{86.1}$$

For integral i, we similarly find that this ratio is

$$g_s/g_a = (i+1)/i. \tag{86.2}$$

We have seen that the sign of the rotational components of a Σ^+ term is determined by the number $(-1)^K$. Hence, for example, the rotational components of a $\Sigma^+{}_g$ term for even K are positive, and therefore symmetric, while for odd K they are negative and consequently antisymmetric. Bearing in mind the results obtained above, we conclude that the nuclear statistical weights of the rotational components of a $\Sigma^+{}_g$ level with successive values of K take alternate values, in the ratios (86.1) or (86.2). A similar situation is found for $\Sigma^+{}_u$, $\Sigma^-{}_g$ and $\Sigma^-{}_u$ levels. In particular, for $i = 0$ the statistical weights of levels with even K for $\Sigma^+{}_u$ and $\Sigma^-{}_g$ terms, and of levels with odd K for $\Sigma^+{}_g$ and $\Sigma^-{}_u$ terms, are zero. In other words, in the electron states $\Sigma^+{}_u$, $\Sigma^-{}_g$ there are no rotational states with even K, and in $\Sigma^+{}_g$, $\Sigma^-{}_u$ states there are none with odd K.

Because of the extremely weak interaction of the nuclear spins with the electrons, the probability of a change in I is very small, even in collisions of

† Recalling the relation between the parity, sign and symmetry of terms, we conclude that, when the total spin I of the nuclei is even, the positive levels are even and the negative levels odd, and *vice versa* when I is odd.

‡ The degree of degeneracy of a level is often referred to in this connection as its *statistical weight*. Formulae (86.1), (86.2) determine the ratio of the nuclear statistical weights of symmetric and antisymmetric levels.

molecules. Hence molecules differing in the parity of I, and accordingly having only symmetric or only antisymmetric terms, behave almost as different forms of matter. Such, for instance, are orthohydrogen and parahydrogen; in the molecule of the former, the spins $i = \frac{1}{2}$ of the two nuclei are parallel ($I = 1$), while in that of the latter they are antiparallel ($I = 0$).

§87. Matrix elements for the diatomic molecule

In this section we shall give some general formulae for the matrix elements of physical quantities in a diatomic molecule. Let us first consider the matrix elements for transitions between states with zero spin.

Let **A** be some vector physical quantity for a molecule with fixed nuclei, such as its electric or magnetic dipole moment. Let us first consider this quantity in coordinates ξ, η, ζ that rotate with the molecule, the ζ-axis being along the axis of the molecule. The angular momentum of the molecule relative to this system (i.e. the electron angular momentum **L**) is not conserved completely, but its ζ-component is conserved. The selection rules for the quantum number $L_\zeta = \Lambda$ therefore remain valid (and they are the same as for M in §29). Thus the non-zero matrix elements of the vector are

$$\left.\begin{array}{c}\langle n'\Lambda|A_\zeta|n\Lambda\rangle, \quad \langle n'\Lambda|A_\xi + iA_\eta|n, \Lambda-1\rangle, \\ \langle n', \Lambda-1|A_\xi - iA_\eta|n\Lambda\rangle,\end{array}\right\} \tag{87.1}$$

where n numbers the electron terms for the given Λ.

If both the terms are Σ terms, we must also bear in mind the selection rule arising from the symmetry with respect to reflection in a plane passing through the axis of the molecule. In such a reflection, the ζ-component of an ordinary (polar) vector is unchanged, while that of an axial vector changes sign. Hence we conclude that, for a polar vector, A_ζ has non-zero matrix elements only for the transitions $\Sigma^+ \to \Sigma^+$ and $\Sigma^- \to \Sigma^-$, and for an axial vector only for $\Sigma^+ \to \Sigma^-$. We need not discuss the components A_ξ, A_η, since for these no transitions without change of Λ are possible.

If the molecule consists of similar atoms, there is also a selection rule regarding parity. The components of a polar vector change sign under inversion. Hence its matrix elements are non-zero only for transitions between states of different parity (the reverse is true for an axial vector). In particular, all the diagonal matrix elements of the components of a polar vector vanish identically.

The question of the relation between the matrix elements (87.1) and those of the same vector in a fixed coordinate system x, y, z is solved by the general formulae derived in §110 below for any axially symmetric physical system.

After separating the dependence (the same for any vector) on the quantum number M_K (the z-component of the total angular momentum **K** of the molecule), we are left with the reduced matrix elements $\langle n'K'\Lambda'||A||nK\Lambda\rangle$. Their relation to the matrix elements (87.1) is given by (110.7) with $k = k' = 1$

(corresponding to a vector) and the appropriate change in the quantum-number notation; from (82.4), Λ is equal to the ζ-component of the total angular momentum **K**. Using the relation (107.1) between the components of a spherical tensor of rank one and the Cartesian components of a vector, and the values of the $3j$-symbols from Table 9 (§106), we obtain the following expressions for the matrix elements diagonal with respect to Λ:

$$\left.\begin{aligned} \langle n'K\Lambda\|A\|nK\Lambda\rangle &= \Lambda\sqrt{\frac{2K+1}{K(K+1)}}\langle n'\Lambda|A_\zeta|n\Lambda\rangle, \\ \langle n', K-1, \Lambda\|A\|nK\Lambda\rangle &= i\sqrt{\frac{K^2-\Lambda^2}{K}}\langle n'\Lambda|A_\zeta|n\Lambda\rangle, \end{aligned}\right\} \tag{87.2}$$

and for those not diagonal with respect to Λ,

$$\left.\begin{aligned} &\langle n'K\Lambda\|A\|nK, \Lambda-1\rangle \\ &\quad = \left[\frac{(2K+1)(K+\Lambda)(K-\Lambda+1)}{4K(K+1)}\right]^{1/2}\langle n'\Lambda|A_\xi+iA_\eta|n, \Lambda-1\rangle, \\ &\langle n'K\Lambda\|A\|n, K-1, \Lambda-1\rangle \\ &\quad = i\left[\frac{(K+\Lambda)(K+\Lambda-1)}{4K}\right]^{1/2}\langle n'\Lambda|A_\xi+iA_\eta|n, \Lambda-1\rangle, \\ &\langle n', K-1, \Lambda\|A\|nK, \Lambda-1\rangle \\ &\quad = i\left[\frac{(K-\Lambda)(K-\Lambda+1)}{4K}\right]^{1/2}\langle n'\Lambda|A_\xi+iA_\eta|n, \Lambda-1\rangle. \end{aligned}\right\} \tag{87.3}$$

The remaining non-zero elements are found from these by means of the Hermitian property of the reduced matrix elements:

$$\langle nK\Lambda\|A\|n'K'\Lambda'\rangle = \langle n'K'\Lambda'\|A\|nK\Lambda\rangle^*,$$

and of the matrix elements in the coordinates ξ, η, ζ,

$$\begin{aligned} \langle n\Lambda|A_\xi-iA_\eta|n'\Lambda'\rangle &= \langle n'\Lambda'|A_\xi+iA_\eta|n\Lambda\rangle^*, \\ \langle n\Lambda|A_\zeta|n'\Lambda'\rangle &= \langle n'\Lambda'|A_\zeta|n\Lambda\rangle^*. \end{aligned}$$

The following are the particular formulae for the matrix elements of the vector $\mathbf{A} = \mathbf{n}$, a unit vector along the axis of the molecule. In this case, we have simply $A_\xi = A_\eta = 0$, $A_\zeta = 1$, and in the coordinates ξ, η, ζ only the diagonal elements non-zero, $\langle n\Lambda|A_\zeta|n\Lambda\rangle = 1$. The reduced matrix elements are diagonal in all suffixes except K, and if only this suffix is written we have

$$\langle K\|n\|K\rangle = \Lambda\sqrt{\frac{2K+1}{K(K+1)}}, \langle K-1\|n\|K\rangle = i\sqrt{\frac{K^2-\Lambda^2}{K}} \quad (87.4)$$

(H. Hönl and F. London 1925). For $\Lambda = 0$, these formulae give

$$\langle K\|n\|K\rangle = 0, \langle K-1\|n\|K\rangle = i\sqrt{K},$$

which agree, as we should expect, with the matrix elements for a unit vector in motion in a centrally symmetric field; see (29.14).

Let us now see how the formulae we have obtained should be modified for transitions between states with non-zero spin. Here it is important to know whether the states belong to case *a* or to case *b*.

If both states belong to case *a*, the formulae are changed essentially only as regards notation. The quantum numbers K and M_K do not exist, and instead we have the total angular momentum J and its projection M_J on the z-axis. There are also the additional numbers S and $\Omega = \Lambda+\Sigma$, so that the reduced matrix elements are

$$\langle n'J'S'\Omega'\Lambda'\|A\|nJS\Omega\Lambda\rangle.$$

Let $\mathbf{A}$ be any orbital vector (i.e. one which does not depend on the spin). Its operator commutes with the spin operator $\hat{\mathbf{S}}$, so that its matrix is diagonal with respect to the quantum numbers S and $S_\zeta = \Sigma$; the quantum number $\Omega = \Lambda+\Sigma$ therefore changes together with Λ (i.e. $\Omega'-\Omega = \Lambda'-\Lambda$). Formulae (87.2)–(87.4) are changed only in that the matrix elements have further suffixes, and in the other factors K and Λ are to be replaced by J and Ω. For example, the first formula (87.2) becomes

$$\langle n'J\Omega\Lambda\|A\|nJ\Omega\Lambda\rangle = \Omega\sqrt{\frac{2J+1}{J(J+1)}}\langle n'\Omega\Lambda|A_\zeta|n\Omega\Lambda\rangle$$

(the diagonal suffix S being omitted).

Now, let $\mathbf{A} = \mathbf{S}$. Since the spin operator commutes with the orbital angular momentum, and also with the Hamiltonian, its matrix is diagonal with respect to n and Λ, but not with respect to S and Σ (or Ω). The matrix elements of the components A_ξ, A_η, A_ζ for transitions $S, \Sigma \to S', \Sigma'$ are given by formulae (27.13), with S and Σ in place of L and M. The change to the coordinates x, y, z is then made by means of formulae (87.2) and (87.3), with J and Ω in place of K and Λ. Thus we have, for instance,

$$\langle J\Omega \| S \| J, \Omega-1\rangle = \left[\frac{(2J+1)(J+\Omega)(J-\Omega+1)}{4J(J+1)}\right]^{1/2} \times$$

$$\times \langle \Omega | S_\xi + iS_\eta | \Omega-1\rangle$$

$$= \left[\frac{(2J+1)(J+\Omega)(J-\Omega+1)(S+\Sigma)(S-\Sigma+1)}{4J(J+1)}\right]^{1/2}$$

(the diagonal suffixes n, S and Λ being omitted).

Next, let both states belong to case b, and let $\mathbf{A}$ be an orbital vector. The calculation of the matrix elements is performed in two stages. First we consider the rotating molecule without taking into account the addition of $\mathbf{S}$ and $\mathbf{K}$; the matrix elements are diagonal with respect to the number S, and are determined by the same formulae (87.2), (87.3). In the second stage, the angular momentum K is added to S to give the total angular momentum J, and the new matrix elements are obtained by the general formulae (109.3), with K, S, J in place of j_1, j_2, J. For example, the elements diagonal with respect to J, K and Λ are first

$$\langle n'JK\Lambda \| A \| nJK\Lambda\rangle$$
$$= (-1)^{K+J+S+1}(2J+1)\begin{Bmatrix} K & J & S \\ J & K & 1 \end{Bmatrix} \langle n'K\Lambda \| A \| nK\Lambda\rangle$$

and then, with the $6j$-symbol from Table 10 (§108) and the reduced matrix element from (87.2), finally

$$\langle n'JK\Lambda \| A \| nJK\Lambda\rangle$$
$$= \Lambda\left[\frac{2J+1}{J(J+1)}\right]^{1/2} \frac{J(J+1)+K(K+1)-S(S+1)}{2K(K+1)} \langle n'\Lambda | A_\zeta | n\Lambda\rangle.$$

The calculation of the matrix elements for transitions between states of which one belongs to case a and the other to case b is carried out similarly. We shall not pause to discuss it here.

PROBLEMS

PROBLEM 1. Determine the Stark splitting of the terms for a diatomic molecule having a constant dipole moment, in the case where the term belongs to case a.

SOLUTION. The energy of a dipole $\mathbf{d}$ in an electric field $\mathscr{E}$ is $-\mathbf{d}\cdot\mathscr{E}$. From considerations of symmetry, it is evident that the dipole moment of a diatomic molecule is directed along its axis; $\mathbf{d} = d\mathbf{n}$, where d is a constant. Taking the direction of the field as the z-axis, we obtain the perturbation operator in the form $-dn_z\mathscr{E}$.

Determining the diagonal matrix elements of n_z in accordance with the formulae derived above, we find that in case *a* the splitting of the levels is given by the formula†

$$\Delta E = -\mathscr{E} d M_J \Omega / J(J+1).$$

PROBLEM 2. The same as Problem 1, but for the case where the term belongs to case *b* (and $\Lambda \neq 0$).

SOLUTION. By the same method we have

$$\Delta E = -\mathscr{E} d M_J \Lambda \frac{J(J+1) - S(S+1) + K(K+1)}{2K(K+1)J(J+1)}.$$

PROBLEM 3. The same as Problem 2, but for a $^1\Sigma$ term.

SOLUTION. For $\Lambda = 0$ the linear effect is absent, and we must go to the second approximation of perturbation theory. In the summation in the general formula (38.10), it is sufficient to retain only those terms which correspond to transitions between rotational components of the electron term concerned; for other terms the energy differences in the denominators are large. Thus we find

$$\Delta E = d^2 \mathscr{E}^2 \left\{ \frac{|\langle K M_K | n_z | K-1, M_K \rangle|^2}{E_K - E_{K-1}} + \frac{|\langle K M_K | n_z | K+1, M_K \rangle|^2}{E_K - E_{K+1}} \right\},$$

where $E_K = BK(K+1)$. A simple calculation gives

$$\Delta E = \frac{d^2 \mathscr{E}^2}{B} \frac{K(K+1) - 3M_K^2}{2K(K+1)(2K-1)(2K+3)}.$$

§88. Λ-doubling

The double degeneracy of the terms with $\Lambda \neq 0$ (§78) is in fact only approximate. It occurs only so long as we neglect the effect of the rotation of the molecule on the electron state (and also the higher approximations with respect to the spin–orbit interaction), as we have done throughout the above theory. When the interaction between the electron state and the rotation is taken into account, a term with $\Lambda \neq 0$ is split into two levels close together. This phenomenon is called Λ-*doubling* (E. Hill and J. H. van Vleck, and R. de L. Kronig, 1928).

To consider this effect quantitatively, we again begin with the singlet terms ($S = 0$). We have calculated (in §82) the energy of the rotational levels in the first approximation of perturbation theory, determining the diagonal matrix elements (i.e. the mean value) of the operator

$$B(r)\,(\hat{\mathbf{K}} - \hat{\mathbf{L}})^2.$$

To calculate the subsequent approximations, we must consider the elements of this operator that are not diagonal with respect to Λ. The operators $\hat{\mathbf{K}}^2$ and $\hat{\mathbf{L}}^2$ are diagonal with respect to Λ, so that we need consider only the operator $2B\hat{\mathbf{K}} \cdot \hat{\mathbf{L}}$.

† It may seem that there is here a contradiction of the general assertion that there is no linear Stark effect (§76). In fact, of course, there is no contradiction, since the presence of a linear Stark effect is here due to the double degeneracy of the levels with $\Omega \neq 0$; the formula obtained is therefore applicable provided that the energy of the Stark splitting is large compared with that of what is called the Λ-doubling (§88).

The calculation of the matrix elements of $\hat{\mathbf{K}}\,.\,\hat{\mathbf{L}}$ is conveniently effected by means of the general formula (29.12), in which we must put $\mathbf{A} = \mathbf{K}$, $\mathbf{B} = \mathbf{L}$; the parts of L and M are taken by K and M_K, while in place of n we must put n, Λ, where n denotes the assembly of quantum numbers (other than Λ) which determine the electron term. Since the matrix of the vector $\mathbf{K}$, which is conserved, is diagonal with respect to n, Λ, while that of the vector $\mathbf{L}$ contains non-diagonal elements only for transitions in which Λ changes by unity (cf. what was said in §87 concerning an arbitrary vector $\mathbf{A}$), we find, using formulae (87.3),

$$\langle n'\Lambda K M_K|\mathbf{K}\,.\,\mathbf{L}|n, \Lambda-1, KM_K\rangle = \tfrac{1}{2}\langle n'\Lambda|L_\xi + iL_\eta|n, \Lambda-1\rangle \sqrt{[(K+\Lambda)(K+1-\Lambda)]}. \quad (88.1)$$

There are no non-zero matrix elements corresponding to any greater change in Λ.

The perturbing effect of the matrix elements with $\Lambda \to \Lambda-1$ can cause the appearance of an energy difference between states with $\pm\Lambda$ only in the 2Λth approximation of perturbation theory. Accordingly, the effect is proportional to $B^{2\Lambda}$, i.e. to $(m/M)^{2\Lambda}$, where M is the mass of the nuclei and m that of the electron. For $\Lambda > 1$, this quantity is so small that it is of no interest. Thus the Λ-doubling effect is of importance only for Π terms ($\Lambda = 1$), which are considered below.

For $\Lambda = 1$ we must go to the second approximation. The corrections to the eigenvalues of the energy can be determined from the general formula (38.10). In the denominators of the terms in the sum occurring in this equation we have energy differences, of the form $E_{n\Lambda K} - E_{n',\Lambda-1,K}$. In these differences, the terms containing K cancel, since, for a given distance r between the nuclei, the rotational energy is the same quantity, $B(r)K(K+1)$, for all the terms. Hence the dependence on K of the required splitting ΔE is entirely determined by the squared matrix elements in the numerators. Among these are the squared elements for transitions in which Λ changes from 1 to 0 and from 0 to -1; these both give, by (88.1), the same dependence on K, and we find that the splitting of the $^1\Pi$ term is of the form

$$\Delta E = \text{constant} \times K(K+1), \quad (88.2)$$

where the constant is of the order of magnitude of B^2/ϵ, ϵ being the order of magnitude of the differences between neighbouring electron terms.

Let us pass now to terms with non-zero spin ($^2\Pi$ and $^3\Pi$ terms; higher values of S are not found in practice). If the term belongs to case b, the multiplet splitting has no effect on the Λ-doubling of the rotational levels, which is determined as before by formula (88.2).

In case a, however, the effect of the spin is important. Here each electron term is characterized by the number Ω as well as Λ. If we simply replace Λ by $-\Lambda$, then $\Omega = \Lambda + \Sigma$ is changed, so that we obtain an entirely different term. The states with Λ, Ω and $-\Lambda$, $-\Omega$ are mutually degenerate. This degeneracy can here be removed not only by the effect, considered above, of the interaction between the orbital angular momentum and the rotation of the molecule, but also by the effect of the spin–orbit interaction. The conservation of the projection Ω of the total angular momentum on the axis of the molecule is (if the nuclei are fixed) an exact conservation law, and so cannot be destroyed by the spin–orbit interaction; the latter can, however, change Λ and Σ (i.e. there are matrix elements for the corresponding transitions) in such a way that Ω remains unchanged. This effect, alone or in combination with the orbit–rotation interaction (which alters Λ but not Σ), may cause Λ-doubling.

Let us first consider the $^2\Pi$ terms. For the $^2\Pi_{1/2}$ term ($\Lambda=1$, $\Sigma=-\frac{1}{2}$, $\Omega=\frac{1}{2}$), the splitting is obtained on taking into account simultaneously the spin–orbit and orbit–rotation interactions, each in the first approximation. For the former gives the transition $\Lambda=1$, $\Sigma=-\frac{1}{2}\rightarrow\Lambda=0$, $\Sigma=\frac{1}{2}$, and then the latter converts the state $\Lambda=0$, $\Sigma=\frac{1}{2}$ into $\Lambda=-1$, $\Sigma=\frac{1}{2}$, which differs from the initial state by the signs of Λ and Ω being reversed. The matrix elements of the spin–orbit interaction are independent of the rotational quantum number J, while the dependence of those for the orbit–rotation interaction is determined by formula (88.1), in which (under the radical) we must replace K and Λ by J and Ω. Thus we have for the Λ-doubling of a $^2\Pi_{1/2}$ term the expression

$$\Delta E_{1/2} = \text{constant} \times (J+\tfrac{1}{2}), \tag{88.3}$$

where the constant is of the order of AB/ϵ. For a $^2\Pi_{3/2}$ term, on the other hand, the splitting can be found only in higher approximations, so that in practice $\Delta E_{3/2}=0$.

Finally, let us consider $^3\Pi$ terms. For a $^3\Pi_0$ term ($\Lambda=1$, $\Sigma=-1$), the splitting is obtained on taking into account the spin–orbit interaction in the second approximation (because of the transitions $\Lambda=1$, $\Sigma=-1\rightarrow\Lambda=0$, $\Sigma=0\rightarrow\Lambda=-1$, $\Sigma=1$). Accordingly, the Λ-doubling in this case is entirely independent of J:

$$\Delta E_0 = \text{constant} \sim A^2/\epsilon. \tag{88.4}$$

For a $^3\Pi_1$ term, $\Sigma=0$, and so the spin has no effect on the splitting; hence we again have a formula like (88.2), but with K replaced by J:

$$\Delta E_1 = \text{constant} \times J(J+1). \tag{88.5}$$

For a $^3\Pi_2$ term, higher approximations are needed, so that we can suppose $\Delta E_2=0$.

One of the levels of the doublet resulting from Λ-doubling is always positive, and the other negative; we have already discussed this in §86. An investigation of the wave functions of the molecule enables us to establish the regularities of the alternation of positive and negative levels. Here we shall give only the results of the investigation.† It is found that if, for some

† This may be found in E. Wigner and E. Witmer, *Zeitschrift für Physik* **51**, 859, 1928.

value of J, the positive level is below the negative one, then in the doublet for $J+1$ the order is opposite, the positive level being above the negative one, and so on; the order varies alternately as the total angular momentum takes successive values. We are speaking here of case *a* terms; for case *b*, the same holds for successive values of the angular momentum K.

PROBLEM

Determine the Λ-splitting for a $^1\Delta$ term.

SOLUTION. Here the effect appears in the fourth approximation of perturbation theory. Its dependence on K is determined by the products of the four matrix elements (88.1) for transitions with change of $\Lambda : 2 \to 1, 1 \to 0, 0 \to -1, -1 \to -2$. This gives

$$\Delta E = \text{constant} \times (K-1)K(K+1)(K+2),$$

where the constant is of order of B^4/ϵ^3.

§89. The interaction of atoms at large distances

Let us consider two atoms which are at a great distance from each other (relative to their size), and determine the energy of their interaction. In other words, we shall discuss the determination of the form of the electron terms when the distance between the nuclei is large.

To solve this problem we apply perturbation theory, regarding the two isolated atoms as the unperturbed system, and the potential energy of their electrical interaction as the perturbation operator. As we know (see *Fields*, §§41, 42), the electrical interaction of two systems of charges at a large distance r apart can be expanded in powers of $1/r$, and successive terms of this expansion correspond to the interaction of the total charges, dipole moments, quadrupole moments, etc., of the two systems. For neutral atoms, the total charges are zero. The expansion here begins with the dipole–dipole interaction ($\sim 1/r^3$); then follow the dipole–quadrupole terms ($\sim 1/r^4$), the quadrupole–quadrupole (and dipole–octupole) terms ($\sim 1/r^5$), and so on.

Let us first suppose that both atoms are in the S state. Then it is easily seen that there is no interaction between the atoms in the first approximation of perturbation theory. The energy of the interaction of the atoms is there determined as the diagonal matrix element of the perturbation operator, calculated with respect to the unperturbed wave functions of the system (expressed in terms of products of the wave functions for the two atoms).† In S states, however, the diagonal matrix elements, i.e. the mean values of the dipole, quadrupole, etc., moments, are zero; this follows immediately, since the distribution of charge density in the atoms is spherically symmetrical. Hence each term of the expansion of the perturbation operator in powers of $1/r$ is zero in the first approximation of perturbation theory.‡

† Here we neglect the exchange effects, which decrease exponentially with increasing distance; cf. §62, Problem 1, and §81, Problem.

‡ This, of course, does not imply that the mean value of the interaction energy of the atoms is precisely zero. It diminishes exponentially with distance, i.e. more rapidly than every finite power of $1/r$, and hence each term of the expansion vanishes. This occurs because the expansion of the interaction operator in terms of the multipole moments involves the assumption that the charges of the two atoms are at a large distance r apart, whereas in quantum mechanics the electron density distribution has finite (though exponentially small) values even at large distances.

In the second approximation it is sufficient to restrict ourselves to the dipole interaction in the perturbation operator, since this decreases least rapidly as r increases, i.e. to the term

$$V = [-\mathbf{d}_1 \cdot \mathbf{d}_2 + 3(\mathbf{d}_1 \cdot \mathbf{n})(\mathbf{d}_2 \cdot \mathbf{n})]/r^3, \tag{89.1}$$

where $\mathbf{n}$ is a unit vector in the direction joining the two atoms. Since the non-diagonal matrix elements of the dipole moment are in general different from zero, we obtain in the second approximation of perturbation theory a non-vanishing result which, being quadratic in V, is proportional to $1/r^6$. The correction in the second approximation to the lowest eigenvalue is always negative (§38). Hence we obtain for the interaction energy of atoms in their normal states an expression of the form

$$U(r) = -\text{constant}/r^6, \tag{89.2}$$

where the constant is positive† (F. London 1928).

Thus two atoms in normal S states, at a great distance apart, attract each other with a force $(-dU/dr)$ which is inversely proportional to the seventh power of the distance. The attractive forces between atoms at large distances are usually called *van der Waals forces*. These forces cause the appearance of minima on the potential energy curves of the electron terms even for atoms which do not form a stable molecule. These depressions, however, are very shallow (being only tenths or even hundredths of an electron-volt in depth) and lie at distances several times greater than the distances between atoms in stable molecules.

If only one of the atoms is in the S state, the same result (89.2) is obtained for the interaction energy, since, for the first approximation to vanish, it is sufficient for the dipole (etc.) moment of only one atom to be zero. The constant in the numerator of (89.2) here depends, not only on the states of the two atoms, but also on their mutual orientation, i.e. on the value Ω of the projection of the angular momentum on the axis joining the atoms.

If both atoms have non-zero orbital and total angular momenta, however, the situation is changed. The mean value of the dipole moment is zero in every state of the atom (§75). The mean values of the quadrupole moment in states with $L \neq 0, J \neq 0$ or $\frac{1}{2}$ are not zero, however. Hence the quadrupole–quadrupole term in the perturbation operator gives a non-zero result even in the first approximation, and the interaction energy of the atoms diminishes as the fifth, not the sixth, power of the distance:

$$U(r) = \text{constant}/r^5. \tag{89.3}$$

Here the constant may be either positive or negative, i.e. we may have either attraction or repulsion. As in the previous case, this constant depends not

† As examples, the value of the constant (in atomic units) for two atoms of hydrogen is 6.5, of helium 1.5, of argon 68, of krypton 130.

only on the states of the atoms, but also on the state of the system formed by the two atoms.

A special case is the interaction of two similar atoms in different states. The unperturbed system (the two isolated atoms) has here an additional degeneracy due to the possibility of interchanging the states of the atoms. Accordingly, the correction in the first approximation will be given by the secular equation, in which the non-diagonal matrix elements of the perturbation appear as well as the diagonal ones. If the states of the two atoms have different parities, and angular momenta L differing by ± 1 or 0 but not both zero (the same restriction being placed on J), then the non-diagonal matrix elements of the dipole moment for transitions between these states are in general not zero. Hence an effect in the first approximation is obtained from the dipole term in the perturbation operator. Thus the interaction energy of the atoms is here proportional to $1/r^3$:

$$U(r) = \text{constant}/r^3, \tag{89.4}$$

where the constant may have either sign.

Usually, however, what is of interest is the interaction of the atoms averaged over all possible orientations of their angular momenta (this formulation corresponds, for instance, to the interaction of atoms in a gas). With this averaging, the mean values of all multipole moments vanish, and so do all effects, linear in these moments in the first approximation of perturbation theory, in the interaction of the atoms. Hence the averaged interaction forces between atoms at large distances always follow the law (89.2).†

Let us further consider the kindred question of the interaction between an atom and an ion. In the first approximation of perturbation theory, this interaction is given by the mean value of the operator (76.8), the energy of the quadrupole in the Coulomb field of the ion. Since the potential of this field is $\phi \sim 1/r$, the atom–ion interaction energy is proportional to $1/r^3$. This effect exists, however, only if the atom has a mean quadrupole moment. Even then, it vanishes on averaging over all directions of the angular momentum $\mathbf{J}$.

The next interaction in order of powers of $1/r$, which is always non-zero, is that in the second order of perturbation theory with respect to the dipole operator (76.1). Since the ion field strength is $\sim 1/r^2$, the energy of this interaction is proportional to $1/r^4$. It can be expressed in terms of the polarizability α of the atom (in the S state) by

$$U = -\alpha e^2/2r^4. \tag{89.5}$$

If the atom is in its ground state, this energy (like all corrections to the

† This law, derived on the basis of the non-relativistic theory, is valid only so long as the retardation of electromagnetic interactions is unimportant. For this to be so, the distance r between the atoms must be small compared with c/ω_{0n}, where ω_{0n} are the frequencies of transitions between the ground state and the excited states of the atom. See *RQT*, §85, for the interaction of atoms when the retardation is taken into account.

ground-state energy) is negative, i.e. there is an attraction between the atom and the ion.†

PROBLEM

Derive a formula giving the van der Waals forces in terms of the matrix elements of dipole moments for two like atoms in S states.

SOLUTION. The answer is obtained by applying the general formula (38.10) of perturbation theory to the operator (89.1). On account of the isotropy of the atoms in the S state it is evident *a priori* that, on summation over all intermediate states, the squared matrix elements of the three components of each of the vectors $\mathbf{d}_1$ and $\mathbf{d}_2$ give equal contributions, while the terms which contain products of different components give zero. The result is

$$U(r) = -\frac{6}{r^6}\sum_{n,n'}\frac{\langle n|d_z|0\rangle^2\langle n'|d_z|0\rangle^2}{E_n+E_{n'}-2E_0},$$

where E_0 and E_n are the unperturbed values of the energies of the ground state and excited states of the atom. Since by hypothesis $L = 0$ in the ground state, the matrix elements $(d_z)_{0n}$ are non-zero only for transitions to P states ($L = 1$). Using formulae (29.7), we bring $U(r)$ to the final form

$$U(r) = -\frac{2}{3r^6}\sum_{n,n'}\frac{\langle n1\|d\|00\rangle^2\langle n'1\|d\|00\rangle^2}{E_{n1}+E_{n'1}-2E_{00}},$$

where in the suffixes nL of the energy levels and the reduced matrix elements the second suffix gives the value of L and the first represents the assembly of the remaining quantum numbers which determine the energy level.

§90. Pre-dissociation

A basic premise of the theory of diatomic molecules as given in this chapter is the assumption that the wave function of the molecule falls into the product of an electron wave function (depending on the distance between the nuclei as a parameter) and a wave function for the motion of the nuclei. This supposition amounts to neglecting, in the exact Hamiltonian of the molecule, certain small terms corresponding to the interaction of the nuclear and electron motions.

When these terms are taken into account and perturbation theory is applied, transitions between different electron states appear. Physically, the transitions between states of which at least one belongs to the continuous spectrum are of particular importance.

Figure 30 shows curves for the potential energy of two electron terms (more precisely, the effective potential energy U_J in some given rotational states of the molecule). The energy E' (the lower dashed line in Fig. 30) is the energy of some vibrational level of a stable molecule in the electron state 2. In state 1, this energy lies in the range of the continuous spectrum. In other words, in passing from state 2 to state 1 the molecule automatically

† A similar attraction occurs between an atom and an electron at large distances. This attraction is the reason for the ability of the atom to form a negative ion by attachment of an electron (with binding energy from a fraction of an electron-volt to several electron-volts). Not all atoms possess this property, however, since, in a field which decreases at large distances as $1/r^4$ (or $1/r^3$), the number of levels (corresponding to bound states of the electron) is always finite and, in particular cases, may be zero.

disintegrates; this phenomenon is called *pre-dissociation.*† As a result of pre-dissociation, the state of the discrete spectrum corresponding to curve 2 has in reality a finite lifetime. This means that the discrete energy level is broadened, i.e. acquires a certain width (see the end of §44).

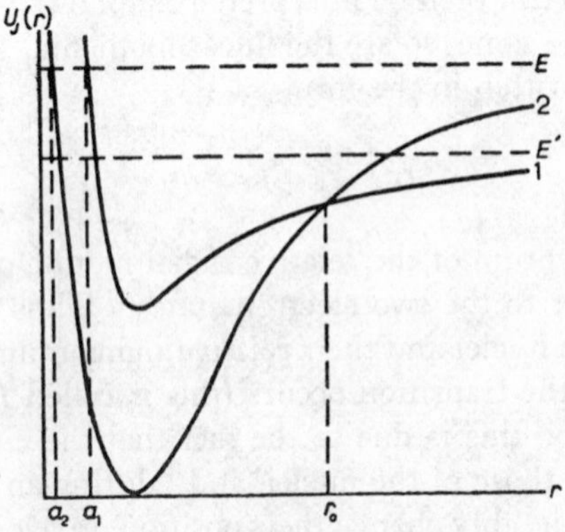

FIG. 30

If, on the other hand, the total energy E lies above the dissociation limit in both states (the upper dashed line in Fig. 30), the transition from one state to the other corresponds to what is called a *collision of the second kind.* Thus the transition $1 \to 2$ signifies the collision of two atoms, as a result of which the atoms are left in excited states, and separate with diminished kinetic energy (for $r \to \infty$, curve 1 passes below curve 2; the difference $U_2(\infty) - U_1(\infty)$ is the excitation energy of the atoms).

Because of the large masses of the nuclei their motion is quasi-classical. The problem of determining the probability of the transitions under consideration is therefore of the kind discussed in §52. From the general considerations given there we can say that the transition probability will be mainly determined by the point at which the transition could occur classically.‡ Since the total energy of the system of two atoms (the molecule) is conserved in the transition, the condition for it to be "classically possible" is that the effective potential energies should be equal: $U_{J1}(r) = U_{J2}(r)$. On account of the conservation of the total angular momentum of the molecule also, the centrifugal energies are the same in the two states, and so this condition means that the potential energies are equal:

$$U_1(r) = U_2(r), \tag{90.1}$$

the angular momentum not being involved at all.

If equation (90.1) has no real roots in the classically accessible region (where $E > U_{J1}, U_{J2}$), the transition probability according to §52 is expo-

† Curve 1 may have no minimum at all if it corresponds to purely repulsive forces between the atoms.

‡ Or else by the point $r = 0$ at which the potential energy becomes infinite.

nentially small.† Transitions occur with an appreciable probability only if the potential energy curves intersect in the classically accessible region (as shown in Fig. 30). Then the exponent in formula (52.1) is zero (and this formula is therefore, of course, invalid); accordingly, the transition probability is determined by a non-exponential expression which will be derived below. The condition (90.1) can then be interpreted as follows. If the potential (and total) energies are the same, so are the linear momenta. Hence the condition (90.1) may also be written in the form

$$r_1 = r_2, \quad p_1 = p_2, \tag{90.2}$$

where p is the momentum of the relative radial motion of the nuclei, and the suffixes 1 and 2 refer to the two electron states. Thus we can say that the distance between the nuclei and their relative momentum remain unchanged at the instant when the transition occurs (this is called *Franck and Condon's principle*). Physically, this is due to the fact that the electron velocities are large compared with those of the nuclei, and "during an electron transition" the nuclei cannot noticeably change their position or velocity.

It is not difficult to establish the selection rules for the transitions in question. First of all, there are two obvious exact rules. The total angular momentum J and the sign of the term (positive or negative; see §86) cannot change in a transition. This follows at once from the fact that the conservation of the total angular momentum and of the behaviour of the wave function under inversion of the coordinate system are exact laws for any (closed) system of particles.

Next, the rule which forbids (for molecules composed of similar atoms) transitions between states of unlike parity is very nearly accurate. For the parity of the state is uniquely determined by the nuclear spin and the sign of the term. The conservation of the sign of the term is an exact law, however, while the nuclear spin is very nearly conserved by virtue of the weakness of its interaction with the electrons.

The requirement that there should be a point of intersection of the potential energy curves means that the terms must be of different symmetry (see §79). Let us consider transitions occurring in the first approximation of perturbation theory; the probability of transitions which occur only in higher approximations is relatively small. First of all, we notice that the terms in the Hamiltonian which lead to the transitions in question are just those which cause the Λ-doubling of the levels. Among these terms are, firstly, terms representing the spin-orbit interaction. They are the product of two axial vectors, of which one is of spin character (i.e. is composed of the operators of the electron spins), and the other is of coordinate character; we emphasize,

† A peculiar situation must occur in the case of a transition involving a molecular term which can arise from two different pairs of atomic states (see the end of §85), i.e. when the potential energy curve is, as it were, split into two branches with increasing distance. In this case the transition probability is considerably greater; an example is given by A. I. Voronin and E. E. Nikitin, *Optics and Spectroscopy* 25, 450, 1969.

however, that these vectors are not simply the vectors $\hat{\mathbf{S}}$ and $\hat{\mathbf{L}}$. Hence they have non-zero matrix elements for transitions in which S and Λ change by $0, \pm 1$. The case where ΔS and $\Delta\Lambda$ are both zero (and $\Lambda \neq 0$) must be omitted, since the symmetry of the term would then be unchanged in the transition. The transition between two Σ terms is possible if one of them is a Σ^+ term and the other a Σ^- term; an axial vector has non-zero matrix elements only for transitions between Σ^+ and Σ^- (see §87).

The term in the Hamiltonian which corresponds to the interaction between the rotation of the molecule and its orbital angular momentum is proportional to $\hat{\mathbf{J}}\cdot\hat{\mathbf{L}}$. Its matrix elements are non-zero for transitions with $\Delta\Lambda = \pm 1$ without change of spin (only the ζ-component of the vector, i.e. L_ζ, has elements with $\Delta\Lambda = 0$, but L_ζ is diagonal with respect to the electron states).

As well as the terms we have considered, there is also a perturbation due to the fact that the operator of the kinetic energy of the nuclei (i.e. the operator of differentiation with respect to the coordinates of the nuclei) acts, not only on the wave function of the nuclei, but also on the electron function, which depends on r as a parameter. The corresponding terms in the Hamiltonian are of the same symmetry as the unperturbed Hamiltonian. Hence they can lead only to transitions between electron terms of like symmetry, the probability of which is negligible in view of the non-intersection of these terms.

Let us go on to the actual calculation of the transition probability. For definiteness, we shall consider a collision of the second kind. According to the general formula (43.1), the required probability is given by the expression

$$w = \frac{2\pi}{\hbar}\left|\int \chi_{\text{nuc},2}{}^* V(r)\chi_{\text{nuc},1}\,\mathrm{d}r\right|^2, \tag{90.3}$$

where $\chi_{\text{nuc}} = r\psi_{\text{nuc}}$ (ψ_{nuc} being the wave function of the radial motion of the nuclei) and $V(r)$ is the perturbing energy; we have taken, as the quantity ν_f in (43.1), the energy E and integrated with respect to it. The final wave function $\chi_{\text{nuc},2}$ must be normalized by the delta function of energy. The quasi-classical function (47.5), thus normalized, is

$$\chi_{\text{nuc},2} = \sqrt{\frac{2}{\pi\hbar v_2}}\cos\left\{\frac{1}{\hbar}\int_{a_2}^{r} p_2\,\mathrm{d}r - \tfrac{1}{4}\pi\right\}. \tag{90.4}$$

The normalizing factor is determined by the rule given at the end of §21. The wave function of the initial state can be written in the form

$$\chi_{\text{nuc},1} = \frac{2}{\sqrt{v_1}}\cos\left\{\frac{1}{\hbar}\int_{a_1}^{r} p_1\,\mathrm{d}r - \tfrac{1}{4}\pi\right\}. \tag{90.5}$$

It is normalized so that the current density is unity in each of the two travelling waves into which the stationary wave (90.5) can be resolved; v_1 and v_2 are the velocities of the relative radial motion of the nuclei. On substituting these functions in (90.3), we obtain the dimensionless transition probability w. It can be regarded as the transition probability for the nuclei to pass twice the point $r = r_0$ (the point of intersection of the levels). It must be borne in mind that the wave function (90.5) corresponds, in a certain sense, to a double passage through this point, since it contains both the incident and the reflected travelling waves.

The matrix element of $V(r)$, calculated with respect to the functions (90.4), (90.5), contains in the integrand a product of cosines, which can be written in terms of the cosines of the sum and difference of the arguments. On integrating near the point $r = r_0$, only the second cosine is important, so that

$$w = \frac{4}{\hbar^2}\left|\int \cos\left[\frac{1}{\hbar}\int_{a_1}^{r} p_1\,\mathrm{d}r - \frac{1}{\hbar}\int_{a_2}^{r} p_2\,\mathrm{d}r\right]\frac{V(r)\,\mathrm{d}r}{\sqrt{(v_1 v_2)}}\right|^2.$$

The integral rapidly converges as we move away from the point of intersection. Hence we can expand the argument of the cosine in powers of $\xi = r - r_0$ and integrate over ξ from $-\infty$ to $+\infty$ (replacing the slowly varying coefficient of the cosine by its value at $r = r_0$). Bearing in mind that, at the point of intersection, $p_1 = p_2$, we find

$$\int_{a_1}^{r} p_1\,\mathrm{d}r - \int_{a_2}^{r} p_2\,\mathrm{d}r \approx S_0 + \tfrac{1}{2}\left(\frac{\mathrm{d}p_1}{\mathrm{d}r_0} - \frac{\mathrm{d}p_2}{\mathrm{d}r_0}\right)\xi^2,$$

where S_0 is the value of the difference of the integrals at the point $r = r_0$. The derivative of the momentum can be expressed in terms of the force $F = -\mathrm{d}U/\mathrm{d}r$: differentiating the equation $p_1^2/2\mu + U_1 = p_2^2/2\mu + U_2$ (where μ is the reduced mass of the nuclei), we have $v_1\,\mathrm{d}p_1/\mathrm{d}r - v_2\,\mathrm{d}p_2/\mathrm{d}r = F_1 - F_2$. Thus

$$\int_{a_1}^{r} p_1\,\mathrm{d}r - \int_{a_2}^{r} p_2\,\mathrm{d}r \approx S_0 + \frac{F_1 - F_2}{2v}\xi^2,$$

where v is the common value of v_1 and v_2 at the point of intersection. The integration is effected by means of the well-known formula

$$\int_{-\infty}^{\infty} \cos(\alpha + \beta\xi^2)\,\mathrm{d}\xi = \sqrt{\frac{\pi}{\beta}}\cos(\alpha + \tfrac{1}{4}\pi),$$

and as a result we have

$$w = \frac{8\pi V^2}{\hbar v|F_2 - F_1|}\cos^2\left(\frac{S_0}{\hbar} + \tfrac{1}{4}\pi\right). \qquad (90.6)$$

The quantity $S_0/\hbar$ is large and varies rapidly with the energy E. Hence, on averaging over even a small interval of energy, the squared cosine can be replaced by its mean value. As a result we obtain the formula

$$w = 4\pi V^2/\hbar v|F_2-F_1| \qquad (90.7)$$

(L. Landau 1932). All the quantities on the right-hand side of the equation are taken at the point of intersection of the potential-energy curves.

In the application to pre-dissociation, we are interested in the probability of the disintegration of the molecule in unit time. In this time, the nuclei in their vibrations pass $2(\omega/2\pi)$ times through the point $r = r_0$. Hence the required pre-dissociation probability is obtained by multiplying w (the probability for a double passage) by $\omega/2\pi$, i.e. it is

$$2V^2\omega/\hbar v|F_2-F_1|. \qquad (90.8)$$

The following remark must be made concerning these calculations. In speaking of the intersection of terms, we have had in mind the eigenvalues of the "unperturbed" Hamiltonian $\hat{H}_0$ of the electron motion in the molecule; in this, the terms $\hat{V}$ which lead to the transitions concerned are not taken into account. If we include these terms in the Hamiltonian, the intersection of the terms becomes impossible, and the curves move apart slightly, as shown in Fig. 31. This follows from the results of §79 when regarded from a slightly different point of view.

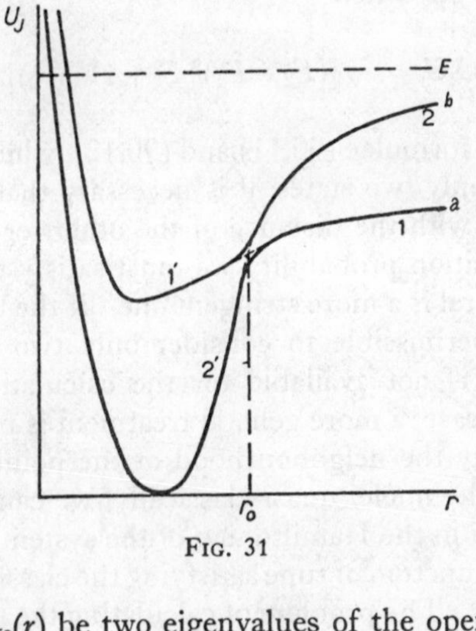

FIG. 31

Let $U_{J1}(r)$ and $U_{J2}(r)$ be two eigenvalues of the operator $\hat{H}_0$ (in which r is regarded as a parameter). In the region near the point r_0 where the curves $U_{J1}(r)$ and $U_{J2}(r)$ intersect, to determine the eigenvalues $U(r)$ of the perturbed operator $\hat{H}_0+\hat{V}$ we must use the method given in §79, as a result of which we obtain the formula

$$U_{b,a}(r) = \tfrac{1}{2}(U_{J1}+U_{J2}+V_{11}+V_{22}) \pm \sqrt{[\tfrac{1}{4}(U_{J1}-U_{J2}+V_{11}-V_{22})^2+V_{12}^2]},$$

where all quantities are functions of r; the function $U_b(r)$ (with the upper sign in the formula) corresponds to the upper continuous curve $(1'-2)$ in Fig. 31, and $U_a(r)$ to the lower one $(2'-1)$. The matrix elements V_{11} and V_{22} may be included in the definition of the functions U_{J1} and U_{J2} respectively; V_{12} will be denoted by $V(r)$ simply. Then the above formula becomes

$$U_{b,a}(r) = \tfrac{1}{2}(U_{J1}+U_{J2}) \pm \tfrac{1}{2}\sqrt{[(U_{J1}-U_{J2})^2+4V^2]}. \qquad (90.9)$$

The interval between the two levels is now

$$\Delta U = \sqrt{[(U_{J1}-U_{J2})^2+4V^2]}. \qquad (90.10)$$

Thus, if there are transitions between the two states ($V \neq 0$), the intersection of the levels disappears. The minimum distance between the curves occurs at $r = r_0$, where $U_{J1} = U_{J2}$:

$$(\Delta U)_{\min} = 2|V(r_0)|. \qquad (90.11)$$

Near this point, we can expand the difference $U_{J1}-U_{J2}$ in powers of the small difference $\xi = r-r_0$, putting

$$U_{J1}-U_{J2} = U_1-U_2 \approx \xi(F_2-F_1),$$

where $F = -(\mathrm{d}V/\mathrm{d}r)_{r_0}$. Then

$$\Delta U = \sqrt{[(F_2-F_1)^2\,\xi^2+4V^2(r_0)]}. \qquad (90.12)$$

For the validity of formulae (90.11) and (90.12), which have been derived by consideration of only two states, it is necessary that $(\Delta U)_{\min}$ should be small in comparison with the distance of the other terms. For the validity of (90.7) as the transition probability we must satisfy the condition (90.19) below, which in general is a more stringent one. If the latter condition is not satisfied, it is still permissible to consider only two terms, but ordinary perturbation theory is not available for the calculation of the transition probability. In that case, a more general treatment is needed.

If we consider only the neighbourhood of the point of intersection and treat the motion of the nuclei quasi-classically, we can replace the velocity operator of the nuclei in the Hamiltonian of the system by a constant v, and the coordinate r by a function of time satisfying the classical equation $\mathrm{d}r/\mathrm{d}t = v$, i.e. $\xi = r-r_0 = vt$. The problem of calculating the transition probability then reduces to that of solving the wave equation for the electron wave functions with a Hamiltonian explicitly dependent on the time:

$$i\hbar\partial\Psi/\partial t = [\hat{H}_0(t)+\hat{V}(t)]\Psi. \qquad (90.13)$$

Let ψ_a and ψ_b be the wave functions of the electron states corresponding to the curves a and b. They are solutions of the equations

$$(\hat{H}_0+\hat{V})\psi_{a,b} = U_{a,b}(t)\psi_{a,b},$$

in which t is a parameter. The solution of equation (90.13) is sought in the form

$$\Psi = a(t)\psi_a + b(t)\psi_b. \tag{90.14}$$

If the equation is solved with the boundary condition $a = 1$, $b = 0$ as $t \to -\infty$, then $|b(\infty)|^2$ gives the probability that the molecule enters the state ψ_b, representing a transition from curve a to curve b, as the nuclei pass through the point $r = r_0$. Similarly, $|a(\infty)|^2 = 1-|b(\infty)|^2$ is the probability that the molecule remains on the curve a. A transition from curve a to curve b in a twofold passage through the point r_0 (as the nuclei approach and then recede) can occur in two ways: $a \to b \to b$ (with the transition $1 \to 1'$ as they approach and the molecule remaining on the curve $1'2$ as they recede), or $a \to a \to b$ (with $1 \to 2'$ as they approach and $2' \to 2$ as they recede). Hence the required probability for such a transition is

$$w = 2|b(\infty)|^2[1-|b(\infty)|^2], \tag{90.15}$$

where we have used the fact that the transition probability in a passage through the point $r = r_0$ is of course independent of the direction of motion.

The value of $b(\infty)$ can be found by the method described in §53, without making direct use of (90.13).† To do so, we note that the curves of $U_a(t)$ and $U_b(t)$ intersect at the imaginary points

$$t_0^{(\pm)} = \pm i\frac{2|V|}{|F_2-F_1|v} \equiv \pm i\tau_0. \tag{90.16}$$

For large negative t, the coefficient $a(t)$ in (90.14) has a form that is "quasi-classical with respect to time":

$$a(t) = \exp\left\{-\frac{1}{\hbar}\int_{-\infty}^{t} U_a(t)\,dt\right\}.$$

Let us now move from the left half of the real axis in the plane of the complex variable t to the right half along a contour on which the "quasi-classicality"

† In §53 the process was assumed to be entirely adiabatic, and accordingly its probability was found to be exponentially small. In the present case, however, this condition may be violated when the nuclei are in the immediate neighbourhood of the point r_0, if their velocity v is not sufficiently small. However, it is clear from the analysis in §§52 and 53 that only the adiabaticity for large $|t|$, and the possibility of considering just two levels in the system, are important as regards the applicability of the method itself.

condition is always satisfied; since $U_a < U_b$, the path used must be in the upper half-plane, passing round the point $t_0^{(+)}$ (cf. §53). The function $a(t)$ then becomes $b(t)$, with

$$|b(\infty)|^2 = \exp\left\{\frac{2}{\hbar}\,\mathrm{im}\left[\int_{t_1}^{i\tau_0} U_a(t)\,\mathrm{d}t + \int_{i\tau_0}^{t_1} U_b(t)\,\mathrm{d}t\right]\right\}$$

$$= \exp\left\{-\frac{2}{\hbar}\,\mathrm{im}\int_{t_1}^{i\tau_0} \Delta U\,\mathrm{d}t\right\},$$

where t_1 may be taken to be any point on the real axis, for instance $t_1 = 0$. According to (90.12) we have

$$\Delta U = \sqrt{[(F_2-F_1)^2 v^2 t^2 + 4V^2]}, \tag{90.17}$$

and the required integral, with the substitution $t = i\tau$, becomes

$$i\int_0^{\tau_0} \sqrt{[4V^2-(F_2-F_1)^2 v^2\tau^2]}\,\mathrm{d}\tau = i\frac{\pi V^2}{v|F_2-F_1|}.$$

Thus we have the following final expression for the transition probability:

$$w = 2\exp\left(-\frac{2\pi V^2}{\hbar v|F_2-F_1|}\right)\left[1-\exp\left(-\frac{2\pi V^2}{\hbar v|F_2-F_1|}\right)\right] \tag{90.18}$$

(C. Zener 1932). We see that the transition probability becomes small in two limiting cases. For $V^2 \gg \hbar v|F_2-F_1|$ it is exponentially small (the adiabatic case), and for

$$V^2 \ll \hbar v|F_2-F_1| \tag{90.19}$$

formula (90.18) becomes (90.7). From (90.17) we see that $\tau \sim |V|/|F_2-F_1|v$ is the "passage time" for the nuclei through the point of intersection; the corresponding frequency $\omega_\tau \sim 1/\tau$. Hence the attainment of the two limiting cases mentioned is determined by the relation between $\hbar\omega_\tau$ and the characteristic energy $|V|$ of the problem.

Finally, let us consider the phenomenon, akin to pre-dissociation, of what are called ***perturbations*** in the spectra of diatomic molecules. If two discrete molecular levels E_1 and E_2 corresponding to two intersecting electron terms are close together, the possibility of a transition between the two electron states results in a displacement of the levels. According to the general formula (79.4) of perturbation theory, we have for the displaced levels the expression

$$\tfrac{1}{2}(E_1+E_2)\pm\sqrt{[\tfrac{1}{4}(E_1-E_2)^2+|V_{12,\text{nuc}}|^2]}, \tag{90.20}$$

where $V_{12,\text{nuc}}$ is the matrix element of the perturbation for the transition between the molecular states 1 and 2; the matrix elements $V_{11,\text{nuc}}$ and $V_{22,\text{nuc}}$ must, of course, be included in E_1 and E_2. From this formula we see that the two levels are moved apart, being displaced in opposite directions (the higher level is raised and the other lowered). The amount of the displacement is the greater, the smaller the difference $|E_1-E_2|$.

The matrix element $V_{12,\text{nuc}}$ is calculated in exactly the same way as for determining the probability of a collision of the second kind. The only difference is that the wave functions $\chi_{\text{nuc},1}$ and $\chi_{\text{nuc},2}$ belong to the discrete spectrum, and hence must be normalized to unity. According to (48.3) we have

$$\chi_{\text{nuc},1} = \sqrt{\frac{2\omega_1}{\pi v_1}}\cos\left\{\frac{1}{\hbar}\int_{a_1}^{r} p_1\,\mathrm{d}r-\tfrac{1}{4}\pi\right\},$$

and similarly for $\chi_{\text{nuc},2}$. A comparison with formulae (90.3) to (90.5) shows that the matrix element $V_{12,\text{nuc}}$ here considered is related to the transition probability w for a twofold passage through the point of intersection by

$$|V_{12,\text{nuc}}|^2 = w(\hbar\omega_1/2\pi)(\hbar\omega_2/2\pi). \tag{90.21}$$

PROBLEMS

PROBLEM 1. Determine the total cross-section for collisions of the second kind, as a function of the kinetic energy E of the colliding atoms, for transitions pertaining to the spin–orbit interaction (L. D. Landau 1932).

SOLUTION. On account of the quasi-classical motion of the nuclei, we can introduce the concept of the *impact parameter* ρ (the distance at which the nuclei would pass if there were no interaction between them) and define the cross-section $\mathrm{d}\sigma$ as the product of the "target area" $2\pi\rho\,\mathrm{d}\rho$ and the transition probability $w(\rho)$ per collision (cf. *Mechanics*, §18). The total cross-section σ is obtained by integrating with respect to ρ.

For spin–orbit interaction, the matrix element $V(r)$ is independent of the angular momentum M of the colliding atoms. We write the velocity v at the point $r = r_0$, where the curves intersect, in the form

$$v = \sqrt{[(2/\mu)(E-U-M^2/2\mu r_0^2)]} = \sqrt{[(2/\mu)(E-U-\rho^2E/r_0^2)]}.$$

Here U is the common value of U_1 and U_2 at the point of intersection, μ is the reduced mass of the atoms, and the angular momentum $M = \mu\rho v_\infty$, where v_∞ is the relative velocity of the atoms at infinity. The zero of energy is chosen so that the interaction energy of the atoms in the initial state is zero at infinity; then $E = \tfrac{1}{2}\mu v_\infty^2$. Substituting this expression in (90.7), we find

$$\mathrm{d}\sigma = 2\pi\rho\,\mathrm{d}\rho\,.\,w = \frac{8\pi^2V^2}{\hbar|F_2-F_1|}\frac{\rho\mathrm{d}\rho}{\sqrt{[2(E-U-\rho^2E/r_0^2)/\mu]}}.$$

The integration with respect to ρ must be taken from zero up to the value for which the velocity v vanishes. As a result we have

$$\sigma = \frac{4\sqrt{(2\mu)}\pi^2 V^2 r_0^2}{\hbar|F_2-F_1|}\frac{\sqrt{(E-U)}}{E}.$$

PROBLEM 2. The same as Problem 1, but for transitions pertaining to the interaction between the rotation of the molecule and its orbital angular momentum (L. D. Landau 1932).

SOLUTION. The matrix element V is of the form $V(r) = MD/\mu r^2$, where $D(r)$ is the matrix element of the electron orbital angular momentum. By the same method as in Problem 1 we obtain

$$\sigma = \frac{16\sqrt{2}\pi^2 D^2}{3\hbar\sqrt{\mu}|F_2-F_1|}\frac{(E-U)^{3/2}}{E}.$$

PROBLEM 3. Determine the transition probability for energies E close to the value U_J of the potential energy at the point of intersection.

SOLUTION. For small values of $E-U_J$, formula (90.7) is inapplicable, since the velocity v of the nuclei cannot be regarded as constant near the point of intersection, and hence it cannot be taken outside the integral as it was in deriving (90.7).

Near the point of intersection we replace the curves of U_{J1}, U_{J2} by the straight lines

$$U_{J1} = U_J - F_{J1}\xi, \quad U_{J2} = U_J - F_{J2}\xi, \quad \xi = r - r_0.$$

The wave functions $\chi_{\mathrm{nuc},1}$ and $\chi_{\mathrm{nuc},2}$ in this region are wave functions of one-dimensional motion in a homogeneous field (§24). The calculations are conveniently effected by means of wave functions in the momentum representation. The wave function normalized by the delta function of energy is of the form (see §24, Problem)

$$a_2 = \frac{1}{\sqrt{(2\pi\hbar|F_{J2}|)}}\exp\left\{\frac{i}{\hbar F_{J2}}[(E-U_J)p-p^3/6\mu]\right\},$$

while the wave function normalized to unit current density in the incident and reflected waves is obtained by multiplying by $\sqrt{(2\pi\hbar)}$:

$$a_1 = \frac{1}{\sqrt{|F_{J1}|}}\exp\left\{\frac{i}{\hbar F_{J1}}[(E-U_J)p-p^3/6\mu]\right\}.$$

On integrating, the perturbing energy (matrix element) V may again be taken outside the integral, replacing it by its value at the point of intersection;

$$w = \frac{2\pi}{\hbar}\left|V\int_{-\infty}^{\infty} a_1 a_2^* \,dp\right|^2.$$

As a result we obtain

$$w = \frac{4\pi V^2(2\mu)^{2/3}}{\hbar^{4/3}(F_{J1}F_{J2})^{1/3}(F_{J2}-F_{J1})^{2/3}}\Phi^2\left[-(E-U_J)\left(\frac{2\mu}{\hbar^2}\right)^{1/3}\left(\frac{1}{F_{J2}}-\frac{1}{F_{J1}}\right)^{2/3}\right],$$

where $\Phi(\xi)$ is the Airy function (see §b of the Mathematical Appendices). For large $E-U_J$, this formula reduces to (90.7).

PROBLEM 4. Determine the probability of charge exchange in a distant slow (relative velocity $v \ll 1$) collision between a hydrogen atom and a hydrogen ion (proton) (O. B. Firsov 1951).†

SOLUTION. We shall regard the system $H+H^+$ as a molecular hydrogen ion (see §81, Problem). Charge exchange consists in the transfer of the electron from a state ψ_1 localized

† In this problem, atomic units are used.

at the first nucleus to a state ψ_2 near the second nucleus. These are not stationary states, even when the nuclei are at rest. The stationary states are

$$\psi_{g,u} = \frac{1}{\sqrt{2}}(\psi_1 \pm \psi_2).$$

Their energies are $U_{g,u}(R)$ as functions of the distance R between the nuclei. When the nuclei are in a given slow motion (which we regard as classical), these energies are slowly varying functions of the time, and the time dependence of the wave functions is given by the factors "quasi-classical with respect to time"

$$\exp\left(-i \int U_{g,u}(t)\, dt\right);$$

cf. §53. The superposition of the two states that is equal to ψ_1 when $t = -\infty$ is

$$\Psi = \frac{1}{\sqrt{2}}\left[\psi_g \exp\left(-i \int_{-\infty}^{t} U_g\, dt\right) + \psi_u \exp\left(-i \int_{-\infty}^{t} U_u\, dt\right)\right].$$

When $t \to \infty$, this function is a linear combination, of the form $c_1\psi_1 + c_2\psi_2$, and the probability of charge exchange is $w = |c_2|^2$. A simple calculation gives

$$w = \sin^2 \eta, \qquad \eta = \tfrac{1}{2}\int_{-\infty}^{\infty} (U_u - U_g)\, dt.$$

In a collision with a large impact parameter ρ (and a sufficiently low velocity v), the motion of the nuclei may be assumed to take place in a straight line, with $R = \sqrt{(\rho^2 + v^2 t^2)}$. The difference $U_u - U_g$ for $R \gg 1$ is given by formula (4) in §81, Problem. Then

$$\eta = \frac{4}{v}\int_{\rho}^{\infty} \frac{R^2 e^{-R-1}}{\sqrt{(R^2 - \rho^2)}}\, dR.$$

For $\rho \gg 1$, the important range of values of R in the integral is that near the lower limit; putting $R = \rho(1+x)$, we obtain

$$\eta \approx \frac{2\sqrt{2}}{ev}\rho^2 e^{-\rho}\int_0^{\infty} \frac{e^{-\rho x}}{\sqrt{x}}\, dx$$

$$= \frac{2\sqrt{(2\pi)}}{ev}\rho^{3/2} e^{-\rho}.$$

CHAPTER XII

THE THEORY OF SYMMETRY

§91. Symmetry transformations

THE classification of terms in the polyatomic molecule is fundamentally related to its symmetry, as in the diatomic molecule. Hence we shall begin by examining the types of symmetry which a molecule can have.

The symmetry of a body is determined by the assembly of all those rearrangements after which the body is unaltered; these rearrangements are called *symmetry transformations*. Any possible symmetry transformation can be represented as a combination of one or more of the three fundamental types of transformation. These three essentially different types are: the *rotation* of the body through a definite angle about some axis, the *reflection* of it in some plane, and the *parallel displacement* of the body over some distance. Of these, the last evidently is applicable only to an infinite medium (a crystal lattice). A body of finite dimensions (in particular, a molecule) can be symmetrical only with respect to rotations and reflections.

If the body is unaltered on rotation through an angle $2\pi/n$ about some axis, then that axis is said to be an *axis of symmetry of the* n*th order*. The number n can take any integral value: $n = 2, 3, \ldots$. The value $n = 1$ corresponds to a rotation through an angle of 2π or, what is the same thing, of 0, i.e. it corresponds to an identical transformation. We shall symbolically denote by C_n the operation of rotation through an angle $2\pi/n$ about a given axis. Repeating this operation two, three, ... times, we obtain rotations through angles $2(2\pi/n)$, $3(2\pi/n)$, ... , which also leave the body unaltered; these rotations may be denoted by C_n^2, C_n^3, ... It is obvious that, if p divides n,

$$C_n^p = C_{n/p}. \tag{91.1}$$

In particular, performing the rotation n times, we return to the initial position, i.e. we effect an identical transformation. The latter is customarily denoted by E, so that we can write

$$C_n^n = E. \tag{91.2}$$

If the body is left unaltered by a reflection in some plane, this plane is said to be a *plane of symmetry*. We shall denote by the symbol σ the operation of reflection in a plane. It is evident that a double reflection in the same plane is the identical transformation:

$$\sigma^2 = E. \tag{91.3}$$

A simultaneous application of the two transformations (rotation and reflection) gives what are called the *rotary–reflection axes*. A body has a rotary–reflection axis of the nth order if it is left unaltered by a rotation through an angle $2\pi/n$ about this axis, followed by a reflection in a plane perpendicular to the axis (Fig. 32). It is easy to see that this is a new form

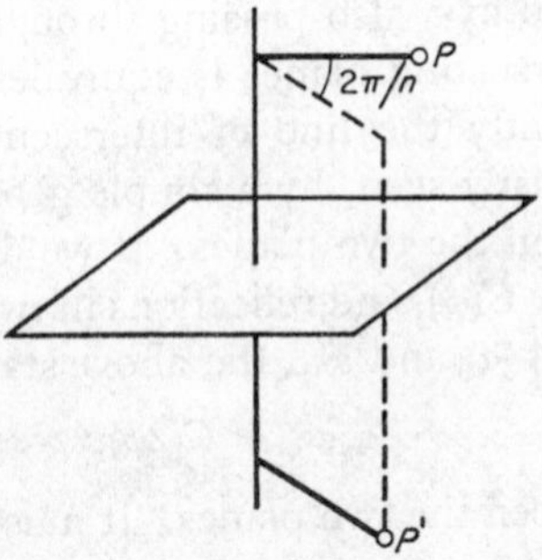

FIG. 32

of symmetry only when n is even. For, if n is odd, an n-fold repetition of the rotary–reflection transformation would be equivalent to a simple reflection in a plane perpendicular to the axis (since the angle of rotation is 2π, while an odd number of reflections in the same plane amounts to a simple reflection). Repeating this transformation a further n times, we have as a result that the rotary–reflection axis reduces to the simultaneous presence of an axis of symmetry of the nth order and an independent plane of symmetry perpendicular to this axis. If, however, n is even, an n-fold repetition of the rotary–reflection transformation returns the body to its initial position.

We denote the rotary–reflection transformation by the symbol S_n. Denoting by σ_h a reflection in a plane perpendicular to a given axis, we can put, by definition,

$$S_n = C_n\sigma_h = \sigma_h C_n; \tag{91.4}$$

the order in which the operations C_n and σ_h are performed clearly does not affect the result.

An important particular case is a rotary–reflection axis of the second order. It is easy to see that a rotation through an angle π, followed by a reflection in a plane perpendicular to the axis of rotation, is the inversion transformation, whereby a point P of the body is carried into another point P', lying on the continuation of the line which joins P to the intersection O of the axis and the plane, and such that the distances OP and OP' are the same. A body symmetrical with respect to this transformation is said to have a *centre of symmetry*. We shall denote the operation of inversion by I, so that we have

$$I \equiv S_2 = C_2\sigma_h. \tag{91.5}$$

It is also evident that $I\sigma_h = C_2$, $IC_2 = \sigma_h$; in other words, an axis of the second order, a plane of symmetry perpendicular to it and a centre of symmetry at their point of intersection are mutually dependent: if any two of these elements are present, the third is automatically present also.

We shall now give various purely geometrical properties of rotations and reflections which it is useful to bear in mind in studying the symmetry of bodies.

A product of two rotations about axes intersecting at some point is a rotation about some third axis also passing through that point. A product of two reflections in intersecting planes is equivalent to a rotation; the axis of this rotation is evidently the line of intersection of the planes, while the angle of rotation is easily seen, by a simple geometrical construction, to be twice the angle between the two planes. If we denote a rotation through an angle ϕ about an axis by $C(\phi)$, and reflections in two planes passing through that axis by the symbols† σ_v and σ'_v, the above statement can be written as

$$\sigma_v\sigma'_v = C(2\phi), \tag{91.6}$$

where ϕ is the angle between the two planes. It must be noted that the order in which the two reflections are performed is not immaterial. The transformation $\sigma_v\sigma'_v$ gives a rotation in the direction from the plane of σ'_v to that of σ_v; on interchanging the factors we have a rotation in the opposite direction. Multiplying equation (91.6) on the left by σ_v, we obtain

$$\sigma'_v = \sigma_v C(2\phi); \tag{91.7}$$

in other words, the operation of rotation, followed by reflection in a plane passing through the axis, is equivalent to a reflection in another plane intersecting the first at half the angle of rotation. In particular, it follows from this that an axis of symmetry of the second order and two mutually perpendicular planes of symmetry passing through it are mutually dependent; if two of them are present, so is the third.

We shall show that the product of rotations through an angle π about two axes intersecting at an angle ϕ (Oa and Ob in Fig. 33) is a rotation through an angle 2ϕ about an axis perpendicular to the first two (PP' in Fig. 33).

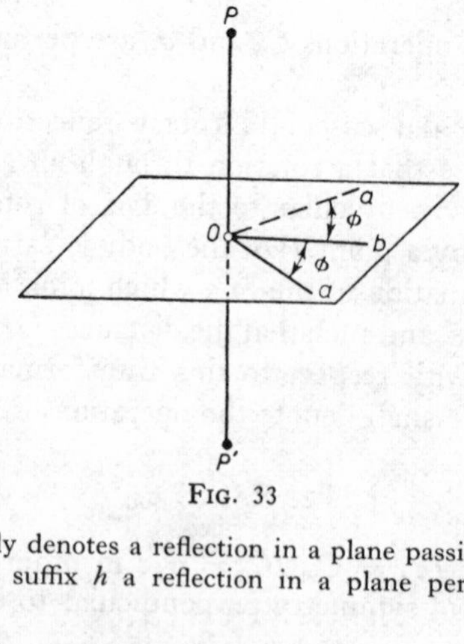

FIG. 33

† The suffix v customarily denotes a reflection in a plane passing through a given axis (a "vertical" plane), and the suffix h a reflection in a plane perpendicular to the axis (a "horizontal" plane).

For it is obvious that the resulting transformation is also a rotation; after the first rotation (about Oa) the point P is carried into P', and after the second (about Ob) it returns to its original position. This means that the line PP' remains fixed, and is therefore an axis of rotation. To determine the angle of rotation, it is sufficient to note that, in the first rotation, the axis Oa remains fixed, while after the second it takes the position Oa', which makes an angle 2ϕ with Oa. In the same way we can see that, when the order of the two transformations is reversed, we obtain a rotation in the opposite direction.

Although the result of two successive transformations in general depends on the order in which they are performed, in some cases the order of operations is immaterial: the transformations commute. This is so for the following transformations:

(1) Two rotations about the same axis.

(2) Two reflections in mutually perpendicular planes (equivalent to a rotation through π about their line of intersection).

(3) Two rotations through π about mutually perpendicular axes (equivalent to a rotation through π about the third perpendicular axis).

(4) A rotation and a reflection in a plane perpendicular to the axis of rotation.

(5) Any rotation or reflection and an inversion with respect to a point lying on the axis of rotation or in the plane of reflection; this follows from (1) and (4).

§92. Transformation groups

The set of all the symmetry transformations for a given body is called its *symmetry transformation group* (or simply its *symmetry group*). Hitherto we have spoken of these transformations as geometrical rearrangements of the body. However, in quantum-mechanical applications it is more convenient to regard symmetry transformations as transformations of the coordinates which leave the Hamiltonian of the system in question invariant. It is obvious that, if the system is left unaltered by some rotation or reflection, the corresponding transformation of the coordinates does not change its Schrödinger's equation. Thus we shall speak of a transformation group with respect to which a given Schrödinger's equation is invariant.†

† This point of view enables us to include in our considerations not only the rotation and reflection groups discussed here, but also other types of transformation which leave Schrödinger's equation unaltered. These include the interchange of the coordinates of identical particles forming part of the system considered (a molecule or atom). The set of all possible permutations of identical particles in a given system is called its *permutation group* (we have already met these permutations in §63). The general properties of groups given below apply to permutation groups also; we shall not pause to study this type of group in more detail here.

The following remark should be made concerning the notation which we use in this chapter. Symmetry transformations are essentially operators just like those which we consider all through the book. They ought, therefore, to be denoted by letters with circumflexes. We do not do this, in view of the generally accepted notation, and because this omission cannot lead to misunderstandings in the present chapter. For the same reason we denote the identical transformation by the customary symbol E, and not by 1, which would correspond to the notation in the other chapters. Lastly, the inversion operator is denoted in this chapter by I, instead of P as in §30, although the latter is customary in recent literature on quantum mechanics.

Symmetry groups are conveniently studied with the help of the general mathematical techniques of what is called *group theory*, the fundamentals of which we shall explain below. At first we shall consider groups, each of which contains a finite number of transformations (known as *finite groups*). Each of the transformations forming a group is said to be an *element* of the group.

Symmetry groups have the following important properties. Each group contains the identical transformation E (called the *unit element* of the group). The elements of a group can be *multiplied* by one another; by the *product* of two (or more) transformations we mean the result of applying them in succession. It is obvious that the product of any two elements of a group is also an element of that group. For the multiplication of elements we have the associative law $(AB)C = A(BC)$, where A, B, C are elements of a group. There is evidently no general commutative law; in general, $AB \neq BA$. For each element A of a group there is in the same group an *inverse* element A^{-1} (the *inverse transformation*), such that $AA^{-1} = E$. In some cases an element may be its own inverse; in particular, $E^{-1} = E$. It is evident that mutually inverse elements A and A^{-1} commute.

The element inverse to the product AB of two elements is

$$(AB)^{-1} = B^{-1}A^{-1},$$

and similarly for the product of a greater number of elements; this is easily seen by effecting the multiplication and using the associative law.

If all the elements of a group commute, the group is said to be *Abelian*. A particular case of Abelian groups is formed by what are called *cyclic* groups. By a cyclic group we mean a group, all of whose elements can be obtained by raising one of them to successive powers, i.e. a group consisting of the elements

$$A, A^2, A^3, \dots, A^n = E,$$

where n is some integer.

Let $\boldsymbol{G}$ be some group.† If we can separate from it some set of elements $\boldsymbol{H}$ such that the latter is itself a group, then the group $\boldsymbol{H}$ is called a *sub-group* of the group $\boldsymbol{G}$. A given element of a group may appear in several of its sub-groups.

By taking any element A of a group and raising it to successive powers, we finally obtain the unit element (since the total number of elements in the group is finite). If n is the smallest number for which $A^n = E$, then n is called the *order* of the element A, and the set of elements $A, A^2, \dots, A^n = E$ is called the *period* of A. The period is denoted by $\{A\}$; it is itself a group, i.e. it is a sub-group of the original group, and is cyclic.

In order to find whether a given set of elements of a group is a sub-group of it, it is sufficient to find whether, on multiplying any two of its elements, we obtain another element of the set. For in that case we have, together with each element A, all its powers, including A^{n-1} (where n is the order of A),

† We shall denote groups by bold italic letters.

which is the inverse of A (since $A^{n-1} A = A^n = E$); and there will obviously be a unit element.

The total number of elements in a group is called its *order*. It is easy to see that the order of a sub-group is a factor of the order of the whole group. To show this, let us consider a sub-group $\boldsymbol{H}$ of a group $\boldsymbol{G}$, and let G_1 be some element of $\boldsymbol{G}$ which does not belong to $\boldsymbol{H}$. Multiplying all the elements of $\boldsymbol{H}$ (on the right, say) by G_1, we obtain a set (or *complex*, as it is called) of elements, denoted by $\boldsymbol{H}G_1$. All the elements of this complex clearly belong to the group $\boldsymbol{G}$. However, none of them belongs to $\boldsymbol{H}$; for, if for any two elements H_a, H_b belonging to $\boldsymbol{H}$ we had $H_a G_1 = H_b$, it would follow that $G_1 = H_a^{-1} H_b$, i.e. G_1 would also belong to the sub-group $\boldsymbol{H}$, which is contrary to hypothesis. Similarly we can show that, if G_2 is an element of $\boldsymbol{G}$ not belonging to $\boldsymbol{H}$ or to $\boldsymbol{H}G_1$, none of the elements of the complex $\boldsymbol{H}G_2$ will belong to $\boldsymbol{H}$ or to $\boldsymbol{H}G_1$. Continuing this process, we finally exhaust all the elements contained in the finite group $\boldsymbol{G}$. Thus all the elements are divided among the complexes (called the *cosets* of $\boldsymbol{H}$ in $\boldsymbol{G}$)

$$\boldsymbol{H}, \boldsymbol{H}G_1, \boldsymbol{H}G_2, \dots, \boldsymbol{H}G_m$$

each of which contains h elements, h being the order of the sub-group $\boldsymbol{H}$. Hence it follows that the order g of the group $\boldsymbol{G}$ is $g = hm$, and this proves the theorem. The integer $m = g/h$ is called the *index* of the sub-group $\boldsymbol{H}$ in the group $\boldsymbol{G}$.

If the order of a group is a prime number, it follows at once from the above that the group has no sub-groups (except itself and E). The converse theorem is also valid: a group having no sub-groups is of prime order and in addition must be cyclic (since otherwise it would contain elements whose period would form a sub-group).

We shall now introduce the important concept of *conjugate* elements. Two elements A and B are said to be conjugate if

$$A = CBC^{-1},$$

where C is also an element of the group; multiplying this equation on the right by C and on the left by C^{-1}, we have the converse equation $B = C^{-1}AC$. An important property of conjugate elements is that, if A is conjugate to B, and B to C, then A is conjugate to C; for, if $B = P^{-1}AP$, $C = Q^{-1}BQ$ (P and Q being elements of the group), it follows that $C = (PQ)^{-1}A(PQ)$. For this reason we can speak of sets of conjugate elements of a group. Such sets are called *classes of conjugate elements*, or simply *classes*, of the group. Each class is completely determined by any one element A of it; for, given A, we obtain the whole class by forming the products GAG^{-1}, where G is successively every element of the group (of course, this may give each element of the class several times). Thus we can divide the whole group into classes; each element of the group can clearly appear in only one class. The unit element of the group is a class by itself, since for every element of the group $GEG^{-1} = E$. If the group is Abelian, each of its elements is a class by itself; since all the elements, by definition, commute, each element is conjugate only to itself. We emphasize that a class of a group (not being E) is not a sub-group of it; this is evident from the fact that it does not contain a unit element.

All the elements of a given class are of the same order. For, if n is the order of the element A (so that $A^n = E$), then for a conjugate element $B = CAC^{-1}$ we have $(CAC^{-1})^n = CA^nC^{-1} = E$.

Let $\boldsymbol{H}$ be a sub-group of $\boldsymbol{G}$, and G_1 an element of $\boldsymbol{G}$ not belonging to $\boldsymbol{H}$. It is easy to see that the set of elements $G_1\boldsymbol{H}G_1^{-1}$ has all the properties of a group, i.e. it also is a sub-group of the group $\boldsymbol{G}$. The sub-groups $\boldsymbol{H}$ and $G_1\boldsymbol{H}G_1^{-1}$ are said to be *conjugate*; each element of one is conjugate to one element of the other. By giving G_1 various values, we obtain a series of conjugate sub-groups, which may partly coincide. It may happen that all the sub-groups conjugate to $\boldsymbol{H}$ are $\boldsymbol{H}$ itself. In this case $\boldsymbol{H}$ is called a *normal divisor* or *invariant sub-group* of the group $\boldsymbol{G}$. Thus, for example, every sub-group of an Abelian group is clearly a normal divisor of it.

Let us consider a group $\boldsymbol{A}$ with n elements $A, A', A'', \ldots$, and a group $\boldsymbol{B}$ with m elements $B, B', B'', \ldots$, and suppose that all the elements of $\boldsymbol{A}$ (apart from the unit E) are different from those of $\boldsymbol{B}$ but commute with them. If we multiply every element of group $\boldsymbol{A}$ by every element of group $\boldsymbol{B}$, we obtain a set of nm elements, which also form a group. For, for any two elements of this set we have $AB \,.\, A'B' = AA' \,.\, BB' = A''B''$, i.e. another element of the set. The group of order nm thus obtained is denoted by $\boldsymbol{A}\times\boldsymbol{B}$, and is called the *direct product* of the groups $\boldsymbol{A}$ and $\boldsymbol{B}$.

Finally, we shall introduce the concept of the *isomorphism* of groups. Two groups $\boldsymbol{A}$ and $\boldsymbol{B}$ of the same order are said to be *isomorphous* if we can establish a one-to-one correspondence between their elements, such that, if the element B corresponds to the element A, and B' to A', then $B'' = BB'$ corresponds to $A'' = AA'$. Two such groups, considered in the abstract, clearly have identical properties, though the actual meaning of their elements may be different.

§93. Point groups

Transformations which appear in the symmetry group of a body of finite dimensions (in particular, a molecule) must be such that at least one point of the body remains fixed when any of these transformations is applied. In other words, all axes and planes of symmetry of a molecule must have at least one common point of intersection. For a successive rotation of the body about two non-intersecting axes or a reflection in two non-intersecting planes results in a translation of the body, which obviously cannot then be left unaltered. Symmetry groups having the above property are called *point groups*.

Before going on to construct the possible types of point group, we shall explain a simple geometrical procedure whereby the elements of a group may be easily divided into classes. Let Oa be some axis, and let the element A of the group be a rotation through a definite angle about this axis. Next, let G be a transformation (rotation or reflection) in the same group, which on being applied to the same axis Oa carries it to the position Ob. We shall show that the element $B = GAG^{-1}$ then corresponds to a rotation about the axis Ob through the same angle as that of the rotation about Oa to which the element A corresponds. For, let us consider the effect of the transformation GAG^{-1} on the axis Ob itself. The transformation G^{-1} inverse to G carries the axis Ob to the position Oa, so that the subsequent rotation A leaves it in this position; finally, G carries it back to its initial position. Thus the axis Ob remains fixed, so that B is a rotation about this axis. Since A and B belong to the same class, their orders are the same; this means that they effect rotations through the same angle.

Thus we reach the result that two rotations through the same angle belong to the same class if there is, among the elements of the group, a transformation whereby one axis of rotation can be carried into the other. In exactly the same way, we can show that two reflections in different planes belong to the same class if some transformation in the group carries one plane into the other. The axes or planes of symmetry whose directions can be carried into each other are said to be *equivalent*.

Some additional comments are necessary in the case where both rotations are about the same axis. The element inverse to the rotation $C_n{}^k$ ($k = 1, 2, \ldots, n-1$) about an axis of symmetry of the nth order is the element $C_n{}^{-k} = C_n{}^{n-k}$, i.e. a rotation through an angle $(n-k)2\pi/n$ in the same direction or, what is the same thing, a rotation through an angle $2k\pi/n$ in the opposite direction. If, among the transformations in the group, there is a rotation through an angle π about a perpendicular axis (this rotation reverses the direction of the axis under consideration), then, by the general rule proved above, the rotations $C_n{}^k$ and $C_n{}^{-k}$ belong to the same class. A reflection σ_h in a plane perpendicular to the axis also reverses its direction; however, it must be borne in mind that the reflection also changes the direction of rotation. Hence the existence of σ_h does not render $C_n{}^k$ and $C_n{}^{-k}$ conjugate. A reflection σ_v in a plane passing through the axis, on the other hand, does not change the direction of the axis, but changes the direction of rotation, and therefore $C_n{}^{-k} = \sigma_v C_n{}^k \sigma_v$, so that $C_n{}^k$ and $C_n{}^{-k}$ belong to the same class if such a plane of symmetry exists. If rotations about an axis through the same angle in opposite directions are conjugate, we shall call it *bilateral*.

The determination of the classes of a point group is often facilitated by the following rule. Let $\mathbf{G}$ be some group not containing the inversion I, and $\mathbf{C}_i$ a group consisting of the two elements I and E. Then the direct product $\mathbf{G}\times\mathbf{C}_i$ is a group containing twice as many elements as $\mathbf{G}$; half of them are the same as the elements of the group $\mathbf{G}$, while the remainder are obtained by multiplying the latter by I. Since I commutes with any other transformation of a point group, it is clear that the group $\mathbf{G}\times\mathbf{C}_i$ contains twice as many classes as $\mathbf{G}$; to each class A of the group $\mathbf{G}$ there correspond the two classes A and AI in the group $\mathbf{G}\times\mathbf{C}_i$. In particular, the inversion I always forms a class by itself.

Let us now go on to enumerate all possible point groups. We shall construct these by starting from the simplest ones and adding new elements of symmetry. We shall denote point groups by bold italic Latin letters with appropriate suffixes.

I. C_n *groups*

The simplest type of symmetry has a single axis of symmetry of the nth order. The group $\boldsymbol{C}_n$ is the group of rotations about an axis of the nth order. This group is evidently cyclic. Each of its n elements forms a class by itself. The group $\boldsymbol{C}_1$ contains only the identical transformation E, and corresponds to the absence of any symmetry.

II. S_{2n} *groups*

The group $\boldsymbol{S}_{2n}$ is the group of rotary reflections about a rotary–reflection axis of even order $2n$. It contains $2n$ elements and is evidently cyclic. In particular, the group $\boldsymbol{S}_2$ contains only two elements, E and I; it is also denoted by $\boldsymbol{C}_i$. We may note also that, if the order of a group is a number of the form $2n = 4p+2$, inversion is among its elements; it is clear that $(S_{4p+2})^{2p+1} = C_2 \sigma_h = I$. Such a group can be written as a direct product $\boldsymbol{S}_{4p+2} = \boldsymbol{C}_{2p+1} \times \boldsymbol{C}_i$; it is also denoted by $\boldsymbol{C}_{2p+1,i}$.

III. C_{nh} *groups*

These groups are obtained by adding to an axis of symmetry of the nth order a plane of symmetry perpendicular to it. The group $\boldsymbol{C}_{nh}$ contains $2n$ elements: n rotations of the group $\boldsymbol{C}_n$ and n rotary–reflection transformations $C_n{}^k\sigma_h$, $k = 1, 2, \ldots, n$ (including the reflection $C_n{}^n\sigma_h = \sigma_h$). All the elements of the group commute, i.e. it is Abelian; the number of classes is the same as the number of elements. If n is even ($n = 2p$), the group contains a centre of symmetry (since $C_{2p}{}^p\sigma_h = C_2\sigma_h = I$). The simplest group, $\boldsymbol{C}_{1h}$, contains only two elements, E and σ_h; it is also denoted by $\boldsymbol{C}_s$.

IV. C_{nv} *groups*

If we add to an axis of symmetry of the nth order a plane of symmetry passing through it, this automatically gives another $n-1$ planes intersecting along the axis at angles of π/n, as follows at once from the geometrical theorem† (91.7). The group $\boldsymbol{C}_{nv}$ thus obtained therefore contains $2n$ elements: n rotations about the axis of the nth order, and n reflections σ_v in vertical planes. Figure 34 shows, as an example, the systems of axes and planes of symmetry for the groups $\boldsymbol{C}_{3v}$ and $\boldsymbol{C}_{4v}$.

To determine the classes, we notice that, because of the presence of planes of symmetry passing through the axis, the latter is bilateral. The actual distribution of the elements among the classes depends on whether n is even or odd.

If n is odd ($n = 2p+1$), successive rotations C_{2p+1} carry each of the planes successively into each of the other $2p$ planes, so that all the planes of

† In a finite group, there cannot be two planes of symmetry intersecting at an angle which is not a rational fraction of 2π. If there were two such planes, it would follow that there were an infinite number of other planes of symmetry, intersecting along the same line and obtained by reflecting one plane in the other *ad infinitum*. In other words, if there are two such planes, there must be complete axial symmetry.

symmetry are equivalent, and the reflections in them belong to a single class. Among rotations about the axis there are $2p$ operations apart from the identity, and these are conjugate in pairs, forming p classes each of two elements ($C_{2p+1}{}^k$ and $C_{2p+1}{}^{-k}$, $k = 1, 2, \dots, p$); moreover, E forms an extra class. Thus there are $p+2$ classes altogether.

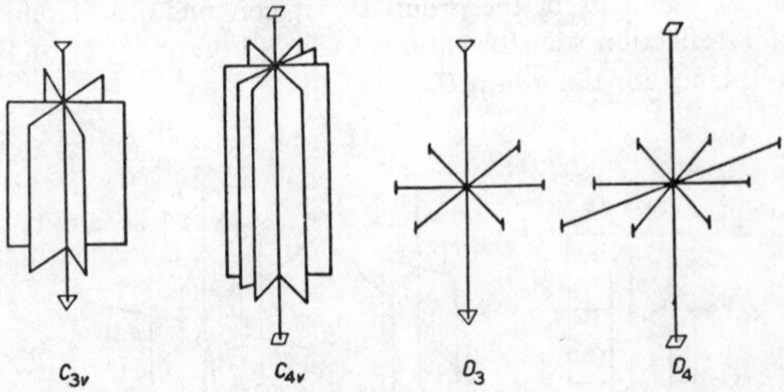

FIG. 34

If, on the other hand, n is even ($n = 2p$), only every alternate plane can be interchanged by successive rotations C_{2p}; two adjacent planes cannot be carried into each other. Thus there are two sets of p equivalent planes, and accordingly two classes of p elements (reflections) each. Of the rotations about the axis, $C_{2p}{}^{2p} = E$ and $C_{2p}{}^{p} = C_2$ each form a class by themselves, while the remaining $2p-2$ rotations are conjugate in pairs and give another $p-1$ classes, each of two elements. The group $C_{2p,v}$ thus has $p+3$ classes altogether.

V. *D_n groups*

If we add to an axis of symmetry of the nth order an axis of the second order perpendicular to it, this involves the appearance of a further $n-1$ such axes, so that there are altogether n horizontal axes of the second order, intersecting at angles π/n. The resulting group D_n contains $2n$ elements: n rotations about an axis of the nth order, and n rotations through an angle π about horizontal axes (we shall denote the latter by U_2, reserving the notation C_2 for a rotation through an angle π about a vertical axis). Fig. 34 shows, as an example, the systems of axes for the groups D_3 and D_4.

In an exactly similar manner to case IV, we may verify that the axis of the nth order is bilateral, while the horizontal axes of the second order are all equivalent if n is odd, or form two non-equivalent sets if n is even. Consequently, the group D_{2p} has the following $p+3$ classes: E, 2 classes each of p rotations U_2, the rotation C_2, and $p-1$ classes each of two rotations about the vertical axis. The group D_{2p+1}, on the other hand, has $p+2$ classes: E, $2p+1$ rotations U_2, and p classes each of two rotations about the vertical axis.

An important particular case is the group D_2. Its system of axes is composed of three mutually perpendicular axes of the second order. This group is also denoted by V.

VI. D_{nh} *groups*

If we add to the system of axes of a group D_n a horizontal plane of symmetry passing through the n axes of the second order, n vertical planes automatically appear, each of which passes through the vertical axis and one of the horizontal axes. The group D_{nh} thus obtained contains $4n$ elements; besides the $2n$ elements of the group D_n, it contains also n reflections σ_v and n rotary–reflection transformations $C_n^k\sigma_h$. Figure 35 shows the system of axes and planes for the group D_{3h}.

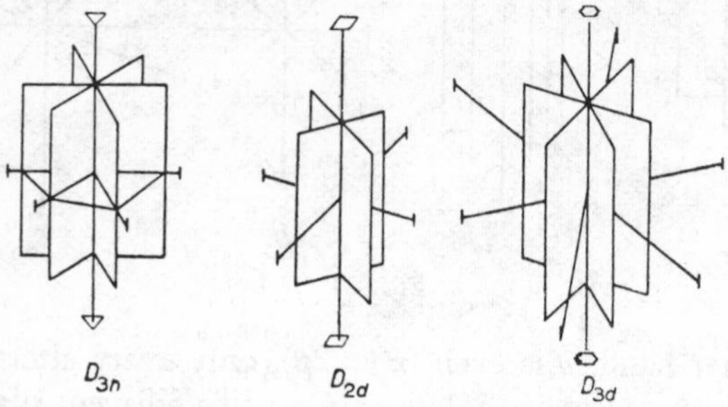

FIG. 35

The reflection σ_h commutes with all the other elements of the group; hence we can write D_{nh} as the direct product $D_{nh} = D_n \times C_s$, where C_s is the group consisting of the two elements E and σ_h. For even n the inversion operation is among the elements of the group, and we can also write $D_{2p.h} = D_{2p} \times C_i$.

Hence it follows that the number of classes in the group D_{nh} is twice the number in the group D_n. Half of them are the same as those of the group D_n (rotations about axes), while the remainder are obtained by multiplying these by σ_h. The reflections σ_v in vertical planes all belong to a single class (if n is odd) or form two classes (if n is even). The rotary–reflection transformations $\sigma_h C_n^k$ and $\sigma_h C_n^{-k}$ are conjugate in pairs.

VII. D_{nd} *groups*

There is another way of adding planes of symmetry to the system of axes of the group D_n. This is to draw vertical planes through the axis of the nth order, midway between each adjacent pair of horizontal axes of the second order. The adding of one such plane again involves the appearance of another $(n-1)$ planes. The system of axes and planes of symmetry thus obtained determines the group D_{nd}. Figure 35 shows the axes and planes for the groups D_{2d} and D_{3d}.

The group D_{nd} contains $4n$ elements. To the $2n$ elements of the group D_n are added n reflections in the vertical planes (denoted by σ_d—the "diagonal" planes) and n transformations of the form $G = U_2\sigma_d$. In order to ascertain the nature of these latter, we notice that the rotation U_2 can, by (91.6), be written in the form $U_2 = \sigma_h\sigma_v$, where σ_v is a reflection in the vertical plane passing through the corresponding axis of the second order. Then $G = \sigma_h \sigma_v \sigma_d$ (the transformations σ_v, σ_h alone are not, of course, among the elements of the group). Since the planes of the reflections σ_v and σ_d intersect along an axis of the nth order, forming an angle $(2k+1)\pi/2n$, where $k = 1, \ldots, (n-1)$ (since here the angle between adjacent planes is $\pi/2n$), it follows that, by (91.6), we have $\sigma_v \sigma_d = C_{2n}^{2k+1}$. Thus we find that $G = \sigma_h C_{2n}^{2k+1} = S_{2n}^{2k+1}$, i.e. these elements are rotary–reflection transformations about the vertical axis, which is consequently not a simple axis of symmetry of the nth order, but a rotary–reflection axis of the $2n$th order.

The diagonal planes reflect two adjacent horizontal axes of the second order into each other; hence, in the groups under consideration, all axes of the second order are equivalent (for both even and odd n). Similarly, all diagonal planes are equivalent. The rotary–reflection transformations S_{2n}^{2k+1} and S_{2n}^{-2k-1} are conjugate in pairs.†

Applying these considerations to the group $\boldsymbol{D}_{2p,\,d}$, we find that it contains the following $2p+3$ classes: E, the rotation C_2 about the axis of the nth order, $(p-1)$ classes each of two conjugate rotations about the same axis, one class of the $2p$ rotations U_2, one class of $2p$ reflections σ_d, and p classes each of two rotary–reflection transformations.

For odd n $(= 2p+1)$, inversion is among the elements of the group; this is seen from the fact that, in this case, one of the horizontal axes is perpendicular to a vertical plane. Hence we can write $\boldsymbol{D}_{2p+1,\,d} = \boldsymbol{D}_{2p+1} \times \boldsymbol{C}_i$, so that the group $\boldsymbol{D}_{2p+1,\,d}$ contains $2p+4$ classes, which are obtained at once from the $p+2$ classes of the group $\boldsymbol{D}_{2p+1}$.

VIII. *The group* $\boldsymbol{T}$ *(the tetrahedron group)*

The system of axes of this group is the system of axes of symmetry of a tetrahedron. It can be obtained by adding to the system of axes of the group $\boldsymbol{V}$ four oblique axes of the third order, rotations about which carry the three axes of the second order into one another. This system of axes is conveniently represented by showing the three axes of the second order as passing through the centres of opposite faces of a cube, and those of the third order as the spatial diagonals of the cube. Figure 36 shows the position of these axes in a cube and in a tetrahedron (one axis of each type is shown).

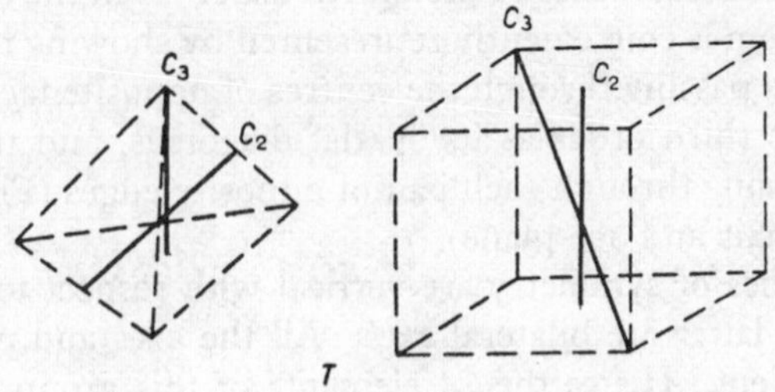

FIG. 36

† For we have

$$\sigma_d S_{2n}^{2k+1} \sigma_d = \sigma_d \sigma_h C_{2n}^{2k+1} \sigma_d = \sigma_h \sigma_d C_{2n}^{2k+1} \sigma_d = \sigma_h C_{2n}^{-2k-1} = S_{2n}^{-2k-1}$$

The three axes of the second order are mutually equivalent. The axes of the third order are also equivalent, since they are carried into one another by the rotations C_2, but they are not bilateral axes. Hence it follows that the twelve elements in the group T are divided into four classes: E, the three rotations C_2, the four rotations C_3 and the four rotations C_3^2.

IX. *The group* T_d

This group contains all the symmetry transformations of the tetrahedron. Its system of axes and planes can be obtained by adding to the axes of the group T planes of symmetry, each of which passes through one axis of the

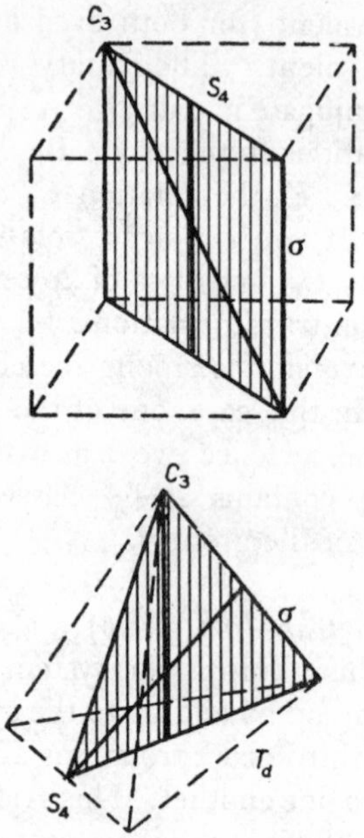

FIG. 37

second order and two of the third order. The axes of the second order thereby become rotary–reflection axes of the fourth order (as in the case of the group D_{2d}). This system is conveniently represented by showing the three rotary–reflection axes as passing through the centres of opposite faces of a cube, the four axes of the third order as its spatial diagonals, and the six planes of symmetry as passing through each pair of opposite edges (Fig. 37 shows one of each kind of axis and one plane).

Since the planes of symmetry are vertical with respect to the axes of the third order, the latter are bilateral axes. All the axes and planes of a given kind are equivalent. Hence the 24 elements of this group are divided into the following five classes: E, eight rotations C_3 and C_3^2, six reflections in planes, six rotary–reflection transformations S_4 and S_4^3, and three rotations $C_2 = S_4^2$.

X. *The group* T_h

This group is obtained from T by adding a centre of symmetry: $T_h = T \times C_i$. As a result, three mutually perpendicular planes of symmetry appear, passing through each pair of axes of the second order, and the axes of the third order become rotary–reflection axes of the sixth order (Fig. 38 shows one of each kind of axis and one plane).

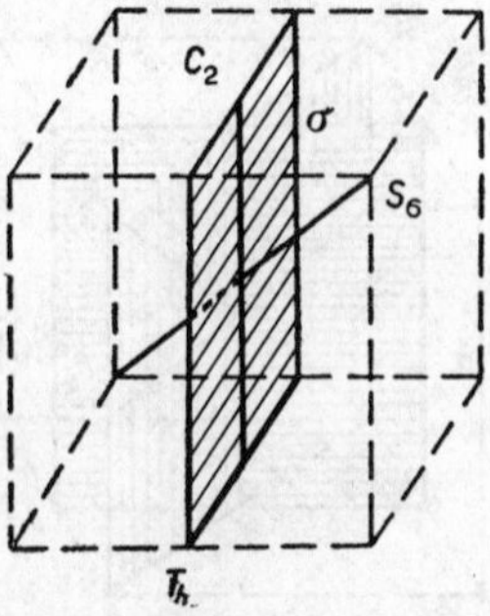

FIG. 38

The group contains 24 elements divided among eight classes, which are obtained at once from those of the group T.

XI. *The group* O *(the octahedron group)*

The system of axes of this group is the system of axes of symmetry of a cube: three axes of the fourth order pass through the centres of opposite

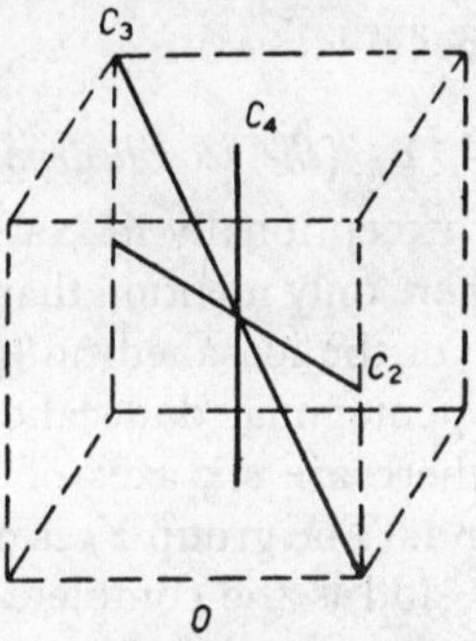

FIG. 39

faces, four axes of the third order through opposite corners, and six axes of the second order through the midpoints of opposite edges (Fig. 39).

It is easy to see that all the axes of a given order are equivalent, and each of them is bilateral. Hence the 24 elements are divided among the following five classes: E, eight rotations C_3 and C_3^2, six rotations C_4 and C_4^3, three rotations C_4^2 and six rotations C_2.

XII. *The group* O_h

This is the group of all symmetry transformations of the cube.† It is

† The groups T, T_d, T_h, O, O_h are called *cubic groups.*

obtained by adding to the group $\boldsymbol{O}$ a centre of symmetry: $\boldsymbol{O}_h = \boldsymbol{O} \times \boldsymbol{C}_i$. The axes of the third order in the group $\boldsymbol{O}$ are thereby converted into rotary-reflection axes of the sixth order (the spatial diagonals of the cube); in addition, another six planes of symmetry appear, passing through each pair of opposite edges, and three planes parallel to the faces of the cube (Fig. 40). The group contains 48 elements divided among ten classes, which

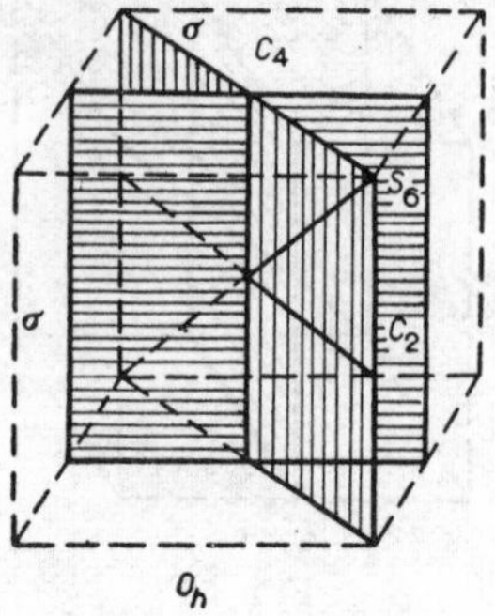

FIG. 40

can be at once obtained from those of the group $\boldsymbol{O}$; five classes are the same as those of the group $\boldsymbol{O}$, while the remainder are: I, eight rotary-reflection transformations S_6 and S_6^5, six rotary-reflection transformations $C_4\sigma_h$, $C_4^3\sigma_h$ about axes of the fourth order, three reflections σ_h in planes horizontal with respect to the axes of the fourth order, and six reflections σ_v in planes vertical with respect to these axes.

XIII, XIV. *The groups* $\boldsymbol{Y}$, $\boldsymbol{Y}_h$ *(the icosahedron groups)*

These groups occur only exceptionally in Nature as symmetry groups of molecules. Hence we shall here only mention that $\boldsymbol{Y}$ is a group of 60 rotations about the axes of symmetry of the icosahedron (a regular solid with twenty triangular faces) or of the pentagonal dodecahedron (a regular solid with twelve pentagonal faces); there are six axes of the fifth order, ten of the third and fifteen of the second. The group $\boldsymbol{Y}_h$ is obtained by adding a centre of symmetry: $\boldsymbol{Y}_h = \boldsymbol{Y} \times \boldsymbol{C}_i$, and is the complete group of symmetry transformations of the above-mentioned polyhedra.

This exhausts all possible types of point group containing a finite number of elements. In addition, we must consider what are called *continuous point groups*, which contain an infinite number of elements. This we shall do in §98.

§94. Representations of groups

Let us consider any symmetry group, and let ψ_1 be some one-valued function of the coordinates in the configuration space of the physical system concerned. Under the transformation of the coordinate system which corresponds to an element G of the group, this function is changed into some other function. On performing in turn all the g transformations in the group (g being the order of the group), we in general obtain g different functions from ψ_1. For certain ψ_1, however, some of these functions may be linearly dependent. As a result we obtain some number f ($\leqslant g$) of linearly independent functions $\psi_1, \psi_2, \dots, \psi_f$, which are transformed into linear combinations of one another under the transformations belonging to the group in question. In other words, as a result of the transformation G, each of the functions ψ_i ($i = 1, 2, 3, \dots, f$) is changed into a linear combination of the form

$$\sum_{k=1}^{f} G_{ki}\psi_k,$$

where the G_{ik} are constants depending on the transformation G. The array of these constants is called the *matrix* of the transformation.†

In this connection it is convenient to regard the elements G of the group as operators acting on the functions ψ_i, so that we can write

$$\hat{G}\psi_i = \sum_k G_{ki}\psi_k; \tag{94.1}$$

the functions ψ_i can always be chosen so as to be orthonormal. Then the concept of the matrix of the transformation is the same as that of the matrix of the operator, in the form defined in §11:

$$G_{ik} = \int \psi_i^* \hat{G}\psi_k \, dq. \tag{94.2}$$

To the product of two elements G and H of the group there corresponds the matrix obtained from the matrices of G and H by the ordinary rule of matrix multiplication (11.12):

$$(GH)_{ik} = \sum_l G_{il}H_{lk}. \tag{94.3}$$

The set of matrices of all the elements in a group is called a ***representation*** of the group. The functions $\psi_1, \dots, \psi_f$ with respect to which these matrices are defined are called the *basis* of the representation. The number f of these functions gives what is called the *dimension* of the representation.

Let us consider the integrals $\int\psi_i^*\psi_k \, dq$. Since the integration is taken over all space, it is evident that the values of the integrals are unchanged by any rotation or reflection of the coordinate system. That is, the symmetry transformations do not destroy the orthonormality of the base functions, and therefore the operators $\hat{G}$ are unitary (see §12).‡ Accordingly, the matrices which represent the elements of a group in a representation with an orthonormalized basis are also unitary.

---

† Since the functions ψ_i are assumed one-valued, a definite matrix corresponds to each element of the group.

‡ In this argument it is important that the integrals are either equal to zero (for $i \neq k$), or definitely not zero (for $i = k$) because the integrand $|\psi_i|^2$ is positive.

Suppose that we perform on the system of functions $\psi_1, \ldots, \psi_f$ the linear unitary transformation

$$\psi'_i = \hat{S}\psi_i. \tag{94.4}$$

This gives a new system of functions $\psi'_1, \ldots, \psi'_f$, which are also orthonormal (see §12).† If we now take, as the basis of the representation, the functions ψ'_i, we obtain a new representation of the same dimension. Such representations, obtained from one another by a linear transformation of their base functions, are said to be *equivalent*; it is evident that they are not essentially different.

The matrices of equivalent representations can be simply expressed in terms of one another. According to (12.7), the matrix of the operator $\hat{G}$ in the new representation is the matrix of the operator

$$\hat{G}' = \hat{S}^{-1}\hat{G}\hat{S} \tag{94.5}$$

in the old representation.

The sum of the diagonal elements (i.e. the trace) of the matrix representing an element G of a group is called its *character*; we shall denote it by $\chi(G)$. It is a very important result that the characters of the matrices of equivalent representations are the same (see (12.11)). This circumstance gives particular importance to the description of group representations by stating their characters: it enables us to distinguish at once the fundamentally different representations from those which are equivalent. Henceforward we shall regard as different representations only those which are not equivalent.

If we take S in (94.5) to be that element of the group which relates the conjugate elements G and G', we have the result that, in any given representation of a group, the characters of the matrices representing elements of the same class are the same.

The identical transformation corresponds to the unit element E of the group. Hence the matrix representing the latter is diagonal in every representation, and the diagonal elements are unity. The character $\chi(E)$ is consequently just the dimension of the representation:

$$\chi(E) = f. \tag{94.6}$$

Let us consider some representation of dimension f. It may happen that, as a result of a suitable linear transformation (94.4), the base functions divide into sets of $f_1, f_2, \ldots$ functions ($f_1+f_2+\ldots = f$), in such a way that, when any element of the group acts on them, the functions in each set are transformed only into combinations of themselves, and do not involve functions from other sets. In such a case the representation in question is said to be *reducible*.

If, on the other hand, the number of base functions that are transformed only into combinations of themselves cannot be reduced by any linear transformation of them, the representation which they give is said to be *irreducible*.

† From (12.12), the unitarity of the transformations implies that the sum of the squared moduli of the base functions is invariant.

Any reducible representation can, as we say, be *decomposed* into irreducible ones. This means that, by the appropriate linear transformation, the base functions divide into several sets, of which each is transformed by some irreducible representation when the elements of the group act on it. Here it may be found that several different sets transform by the same irreducible representation; in such a case this irreducible representation is said to be contained so many times in the reducible one.

Irreducible representations are an important characteristic of a group, and play a fundamental part in all quantum-mechanical applications of group theory. We shall give the chief properties of irreducible representations.†

It may be shown that the number of different irreducible representations of a group is equal to the number r of classes in the group. We shall distinguish the characters of the various irreducible representations by indices; the characters of the matrices of the element G in the representations are $\chi^{(1)}(G), \chi^{(2)}(G), \ldots, \chi^{(r)}(G)$.

The matrix elements of irreducible representations satisfy a number of orthogonality relations. First of all, for two different irreducible representations the relations

$$\sum_G G^{(\alpha)}{}_{ik} G^{(\beta)}{}_{lm}{}^* = 0 \tag{94.7}$$

hold, where α and β ($\alpha \neq \beta$) refer to the two irreducible representations, and the summation is taken over all the elements of the group. For any irreducible representation the relations

$$\sum_G G^{(\alpha)}{}_{ik} G^{(\alpha)}{}_{lm}{}^* = \frac{g}{f_\alpha}\delta_{il}\delta_{km} \tag{94.8}$$

hold, i.e. only the sums of the squared moduli of the matrix elements are not zero:

$$\sum_G |G^{(\alpha)}{}_{ik}|^2 = g/f_\alpha.$$

The relations (94.7), (94.8) can be combined in the form

$$\sum_G G^{(\alpha)}{}_{ik} G^{(\beta)}{}_{lm}{}^* = \frac{g}{f_\alpha}\delta_{\alpha\beta}\delta_{il}\delta_{km}. \tag{94.9}$$

In particular, we can obtain from this an important orthogonality relation for the characters of the representations. Summing both sides of equation (94.9) over equal values of the suffixes i, k and l, m, we have

$$\sum_G \chi^{(\alpha)}(G)\chi^{(\beta)}(G)^* = g\delta_{\alpha\beta}. \tag{94.10}$$

For $\alpha = \beta$ we have

$$\sum_G |\chi^{(\alpha)}(G)|^2 = g,$$

† The proof of these properties may be found in any textbook on group theory.

i.e. the sum of the squared moduli of the characters of an irreducible representation is equal to the order of the group. We may notice that this relation can be used as a criterion of the irreducibility of a representation; for a reducible representation, this sum is always greater than g (for instance, it is ng if the representation contains n different irreducible parts).

It also follows from (94.10) that the equality of the characters of two irreducible representations is not only a necessary but also a sufficient condition for them to be equivalent.

Since the characters of elements of the same class are equal, the sum (94.10) actually contains only r independent terms, and can be written in the form

$$\sum_C g_C \chi^{(\alpha)}(C)\chi^{(\beta)}(C)^* = g\delta_{\alpha\beta}, \tag{94.11}$$

where the summation is over the r classes of the group (arbitrarily denoted by C) and g_C is the number of elements in class C.

Since the number of irreducible representations is equal to the number of classes, the quantities $f_{\alpha C} = \sqrt{(g_C/g)}\chi^{(\alpha)}(C)$ form a square matrix of r^2 quantities.

The orthogonality relations for the first suffix,

$$\sum_C f_{\alpha C} f_{\beta C}^* = \delta_{\alpha\beta},$$

then automatically give those for the second suffix,

$$\sum_\alpha f_{\alpha C} f_{\alpha C'}^* = \delta_{CC'}.$$

Hence, besides (94.11), we have

$$\sum_\alpha \chi^{(\alpha)}(C)\chi^{(\alpha)}(C')^* = (g/g_C)\delta_{CC'}. \tag{94.12}$$

Among the irreducible representations of any group there is always a trivial one, given by a single base function invariant under all the transformations in the group. This one-dimensional representation is called the *unit representation*; in it, all characters are unity. If one of the representations in the orthogonality relation (94.10) or (94.11) is the unit representation, the other is such that

$$\sum_G \chi^{(\alpha)}(G) = \sum_C g_C \chi^{(\alpha)}(C) = 0, \tag{94.13}$$

i.e. the sum of the characters of all the elements of the group is zero for every irreducible representation.

The relation (94.10) enables any reducible representation to be very easily decomposed into irreducible ones if the characters of both are known. Let $\chi(G)$ be the characters of some reducible representation of dimension f, and let the numbers $a^{(1)}, a^{(2)}, \ldots, a^{(r)}$ indicate how many times the corresponding irreducible representations are contained in it, so that

$$\sum_{\beta=1}^{r} a^{(\beta)} f_\beta = f, \tag{94.14}$$

where f_β are the dimensions of the irreducible representations. Then the characters $\chi(G)$ can be written

$$\chi(G) = \sum_{\beta=1}^{r} a^{(\beta)} \chi^{(\beta)}(G). \tag{94.15}$$

Multiplying this equation by $\chi^{(\alpha)}(G)^*$ and summing over all G, we have by (94.10)

$$a^{(\alpha)} = \frac{1}{g} \sum_G \chi(G) \chi^{(\alpha)}(G)^*. \tag{94.16}$$

Let us consider a representation of dimension $f = g$, given by the g functions $\hat{G}\psi$, ψ being some general function of the coordinates (so that all the g functions $\hat{G}\psi$ obtained from it are linearly independent); such a representation is said to be *regular*. It is clear that none of the matrices of this representation will contain any diagonal elements, with the exception of the matrix corresponding to the unit element; hence $\chi(G) = 0$ for $G \neq E$, while $\chi(E) = g$. Decomposing this representation into irreducible ones, we have for the numbers $a^{(\alpha)}$, by (94.16), the values $a^{(\alpha)} = (1/g) g f^{(\alpha)} = f^{(\alpha)}$, i.e. each irreducible representation is contained in the reducible one under consideration as many times as its dimension. Substituting this in (94.14), we find the relation

$$f_1^2 + f_2^2 + \ldots + f_r^2 = g; \tag{94.17}$$

the sum of the squared dimensions of the irreducible representations of a group is equal to its order.† Hence it follows, in particular, that for Abelian groups (where $r = g$) all the irreducible representations are of dimension one ($f_1 = f_2 = \ldots = f_r = 1$).

We may also remark, without proof, that the dimensions of the irreducible representations of a group divide its order.

In practice, the decomposition of a regular representation into irreducible parts is made by means of the formula

$$\psi_i^{(\alpha)} = \frac{f_\alpha}{g} \sum_G G_{ik}^{(\alpha)*} \hat{G}\psi. \tag{94.18}$$

It is easy to verify that the functions $\psi_i^{(\alpha)}$ ($i = 1, 2, \ldots, f_\alpha$) represented by this formula with a given value of k are transformed according to

$$\hat{G}\psi_i^{(\alpha)} = \sum_l G_{li}^{(\alpha)} \psi_l^{(\alpha)},$$

† It may be mentioned that, for point groups, equation (94.17) for given r and g can in practice be satisfied in only one way by a set of integers $f_1, \ldots, f_r$.

i.e. they are a basis of the αth irreducible representation. By giving various values to k we obtain in this way f_α different sets of base functions $\psi_i^{(\alpha)}$ for one irreducible representation, in accordance with the fact that each irreducible representation appears f_α times in the regular representation.

Any function ψ may be written as a sum of functions transformed by the irreducible representations of the group. This problem is solved by the formulae

$$\psi = \sum_\alpha \sum_i \psi_i^{(\alpha)}, \quad \psi_i^{(\alpha)} = \frac{f_\alpha}{g} \sum_G G_{ii}^{(\alpha)*} \hat{G}\psi. \tag{94.19}$$

To prove this, we substitute the second formula in the first and calculate the sum over i, obtaining

$$\psi = \frac{1}{g} \sum_\alpha f_\alpha \chi^{(\alpha)*}(G) \,.\, \hat{G}\psi. \tag{94.20}$$

Since the dimensions f_α coincide with the characters $\chi^{(\alpha)}(E)$ of the unit element of the group, we can use the orthogonality relation (94.12) to show that the sum in (94.20) is non-zero (and equal to g) only if G is the unit element of the group. Hence the right-hand side of (94.20) is identically equal to ψ.

Let us consider two different systems of functions $\psi_1^{(\alpha)}, \dots, \psi_{f_\alpha}^{(\alpha)}$ and $\psi_1^{(\beta)}, \dots, \psi_{f_\beta}^{(\beta)}$, which form two irreducible representations of a group. By forming the products $\psi_i^{(\alpha)}\psi_k^{(\beta)}$ we obtain a system of $f_\alpha f_\beta$ new functions, which can serve as the basis for a new representation of dimension $f_\alpha f_\beta$. This representation is called the *direct product* or *Kronecker product* of the other two; it is irreducible only if f_α or f_β is unity. It is easy to see that the characters of the direct product are equal to the products of the characters of the two component representations. For, if

$$\hat{G}\psi_i^{(\alpha)} = \sum_l G_{li}^{(\alpha)}\psi_l^{(\alpha)}, \qquad \hat{G}\psi_k^{(\beta)} = \sum_m G_{mk}^{(\beta)}\psi_m^{(\beta)},$$

then

$$\hat{G}\psi_i^{(\alpha)}\psi_k^{(\beta)} = \sum_{l,m} G_{li}^{(\alpha)}G_{mk}^{(\beta)}\psi_l^{(\alpha)}\psi_m^{(\beta)};$$

hence we have for the characters, which we denote by $(\chi^{(\alpha)}\times\chi^{(\beta)})(G)$,

$$(\chi^{(\alpha)}\times\chi^{(\beta)})(G) = \sum_{i,k} G_{ii}^{(\alpha)}G_{kk}^{(\beta)} = \sum_i G_{ii}^{(\alpha)} \sum_k G_{kk}^{(\beta)},$$

i.e.

$$(\chi^{(\alpha)}\times\chi^{(\beta)})(G) = \chi^{(\alpha)}(G)\chi^{(\beta)}(G). \tag{94.21}$$

The two irreducible representations so multiplied may, in particular, be the same; in this case we have two different sets of functions $\psi_1, \dots, \psi_f$ and $\phi_1, \dots, \phi_f$ giving the same representation, while the direct product of the representation with itself is given by the f^2 functions $\psi_i\phi_k$, and has the characters

$$(\chi\times\chi)(G) = [\chi(G)]^2.$$

This reducible representation can be at once decomposed into two representations of smaller dimension (although these are, in general, themselves reducible). One of them is given by the $\frac{1}{2}f(f+1)$ functions $\psi_i\phi_k+\psi_k\phi_i$, the other by the $\frac{1}{2}f(f-1)$ functions $\psi_i\phi_k-\psi_k\phi_i$, $i \neq k$; it is evident that the functions in each of these sets are transformed only into combinations of themselves. The former is called the *symmetric product* of the representation with itself, and its characters are denoted by the symbol $[\chi^2](G)$; the latter is called the *antisymmetric product*, and its characters are denoted by $\{\chi^2\}(G)$. To determine the characters of the symmetric product, we write

$$\begin{aligned}\hat{G}(\psi_i\phi_k+\psi_k\phi_i) &= \sum_{l,m} G_{li}G_{mk}(\psi_l\phi_m+\psi_m\phi_l)\\ &= \tfrac{1}{2}\sum_{l,m}(G_{li}G_{mk}+G_{mi}G_{lk})(\psi_l\phi_m+\psi_m\phi_l).\end{aligned}$$

Hence we have for the character

$$[\chi^2](G) = \tfrac{1}{2}\sum_{i,k}(G_{ii}G_{kk}+G_{ik}G_{ki}).$$

But $\sum_i G_{ii} = \chi(G)$, and $\sum_{i,k} G_{ik}G_{ki} = \chi(G^2)$; thus we finally obtain the formula

$$[\chi^2](G) = \tfrac{1}{2}\{[\chi(G)]^2+\chi(G^2)\}, \tag{94.22}$$

which enables us to determine the characters of the symmetric product of a representation with itself from the characters of the representation. In an exactly similar manner, we find for the characters of the antisymmetric product the formula†

$$\{\chi^2\}(G) = \tfrac{1}{2}\{[\chi(G)]^2-\chi(G^2)\}. \tag{94.23}$$

If the functions ψ_i and ϕ_i are the same, we can evidently construct from them only the symmetric product, formed by the squares $\psi_i{}^2$ and the products $\psi_i\psi_k$, $i \neq k$. In applications, symmetric products of higher orders are also encountered; their characters may be obtained in a similar manner.

An important property of direct products is the following. The decomposition of the direct product of two different irreducible representations into irreducible parts contains the unit representation (and only once) only if the representations multiplied together are complex conjugates. For real representations, the unit representation is present only in the direct product of an irreducible representation with itself, and is of course in the symmetric part. In order to know whether the unit representation is present in the

† It is useful to note that, for representations of dimension 2, the characters $\{\chi^2\}(G)$ are equal to the determinants of the linear transformations G, as can easily be shown by direct calculation.

representation (94.21), we simply sum its characters with respect to G and divide the result by the order g of the group, in accordance with (94.16). The conclusion stated then follows at once from the orthogonality relations (94.10).

Finally, we shall make a few remarks regarding the irreducible representations of a group which is the direct product of two other groups (not to be confused with the direct product of two representations of the same group). If the functions $\psi_i^{(\alpha)}$ give an irreducible representation of the group $\boldsymbol{A}$, and the functions $\phi_k^{(\beta)}$ give one of the group $\boldsymbol{B}$, the products $\phi_k^{(\beta)}\psi_i^{(\alpha)}$ are the basis of an $f_\alpha f_\beta$-dimensional representation of the group $\boldsymbol{A}\times\boldsymbol{B}$, and this representation is irreducible. The characters of this representation are obtained by multiplying the corresponding characters of the original representations (cf. the derivation of formula (94.21)); to an element $C = AB$ of the group $\boldsymbol{A}\times\boldsymbol{B}$ there corresponds the character

$$\chi(C) = \chi^{(\alpha)}(A)\chi^{(\beta)}(B). \tag{94.24}$$

Multiplying together in this way all the irreducible representations of the groups $\boldsymbol{A}$ and $\boldsymbol{B}$, we obtain all the irreducible representations of the group $\boldsymbol{A}\times\boldsymbol{B}$.

§95. Irreducible representations of point groups

Let us pass now to the actual determination of the irreducible representations of point groups. The great majority of molecules have axes of symmetry only of the second, third, fourth or sixth order. Hence we shall not consider the icosahedron groups $\boldsymbol{Y}$, $\boldsymbol{Y}_h$; we shall examine the groups $\boldsymbol{C}_n$, $\boldsymbol{C}_{nh}$, $\boldsymbol{C}_{nv}$, $\boldsymbol{D}_n$, $\boldsymbol{D}_{nh}$ only for the values $n = 1, 2, 3, 4, 6$, and the groups $\boldsymbol{S}_{2n}$, $\boldsymbol{D}_{nd}$ only for $n = 1, 2, 3$.

The characters of the representations of these groups are shown in Table 7. Isomorphous groups have the same representations and are given together. The numbers in front of the symbols for the elements of a group in the upper rows show the numbers of elements in the corresponding classes (see §93). The left-hand columns show the conventional names usually given to the representations. The one-dimensional representations are denoted by the letters A, B, the two-dimensional ones by E, and the three-dimensional ones by F; the notation E for a two-dimensional irreducible representation should not be confused with the unit element of a group.† The base functions of A representations are symmetric, and those of B representations antisymmetric, with respect to rotations about a principal axis of the nth order. The functions of different symmetry with respect to a reflection σ_h are distinguished by the number of primes (one or two), while the suffixes g and u show the symmetry with respect to inversion. Beside the symbols for the representations are placed the letters x, y, z; these show the repre-

† The reason why two complex conjugate one-dimensional representations are shown as one two-dimensional one is explained in §96.

sentations by which the coordinates themselves are transformed. The z-axis is always taken along the principal axis of symmetry. The letters ϵ and ω denote

$$\epsilon = e^{2\pi i/3}, \qquad \omega = e^{2\pi i/6} = -\omega^4;$$
$$\epsilon + \epsilon^2 = -1, \qquad \omega^2 - \omega = -1.$$

The simplest problem is to determine the irreducible representations for

TABLE 7

Characters of irreducible representations of point groups

C_i			E	I
	C_2		E	C_2
		C_s	E	σ
A_g	$A; z$	$A'; x, y$	1	1
$A_u; x, y, z$	$B; x, y$	$A''; z$	1	-1

C_3	E	C_3	C_3^2
$A; z$	1	1	1
$E; x\pm iy$	1	ϵ	ϵ^2
	1	ϵ^2	ϵ

C_{2h}			E	C_2	σ_h	I
	C_{2v}		E	C_2	σ_v	σ'_v
		D_2	E	C_2^z	C_2^y	C_2^x
A_g	$A_1; z$	A	1	1	1	1
B_g	$B_2; y$	$B_3; x$	1	-1	-1	1
$A_u; z$	A_2	$B_1; z$	1	1	-1	-1
$B_u; x, y$	$B_1; x$	$B_2; y$	1	-1	1	-1

C_{3v}		E	$2C_3$	$3\sigma_v$
	D_3	E	$2C_3$	$3U_2$
$A_1; z$	A_1	1	1	1
A_2	$A_2; z$	1	1	-1
$E; x, y$	$E; x, y$	2	-1	0

C_4		E	C_4	C_2	C_4^3
	S_4	E	S_4	C_2	S_4^3
$A; z$	A	1	1	1	1
B	$B; z$	1	-1	1	-1
$E; x\pm iy$	$E; x\pm iy$	1	i	-1	$-i$
		1	$-i$	-1	i

C_6	E	C_6	C_3	C_2	C_3^2	C_6^5
$A; z$	1	1	1	1	1	1
B	1	-1	1	-1	1	-1
E_1	1	ω^2	$-\omega$	1	ω^2	$-\omega$
	1	$-\omega$	ω^2	1	$-\omega$	ω^2
$E_2; x\pm iy$	1	ω	ω^2	-1	$-\omega$	$-\omega^2$
	1	$-\omega^2$	$-\omega$	-1	ω^2	ω

TABLE 7—*continued*

C_{4v}	D_4	D_{2d}					
			E	C_2	$2C_4$	$2\sigma_v$	$2\sigma'_v$
			E	C_2	$2C_4$	$2U_2$	$2U'_2$
			E	C_2	$2S_4$	$2U_2$	$2\sigma_d$
$A_1; z$	A_1	A_1	1	1	1	1	1
A_2	$A_2; z$	A_2	1	1	1	−1	−1
B_1	B_1	B_1	1	1	−1	1	−1
B_2	B_2	$B_2; z$	1	1	−1	−1	1
$E; x, y$	$E; x, y$	$E; x, y$	2	−2	0	0	0

D_6	C_{6v}	D_{3h}						
			E	C_2	$2C_3$	$2C_6$	$3U_2$	$3U'_2$
			E	C_2	$2C_3$	$2C_6$	$3\sigma_v$	$3\sigma'_v$
			E	σ_h	$2C_3$	$2S_3$	$3U_2$	$3\sigma'_v$
A_1	$A_1; z$	A_1'	1	1	1	1	1	1
$A_2; z$	A_2	A_2'	1	1	1	1	−1	−1
B_1	B_2	A_1''	1	−1	1	−1	1	−1
B_2	B_1	$A_2''; z$	1	−1	1	−1	−1	1
E_2	E_2	$E'; x, y$	2	2	−1	−1	0	0
$E_1; x, y$	$E_1; x, y$	E''	2	−2	−1	1	0	0

T	E	$3C_2$	$4C_3$	$4C_3^2$
A	1	1	1	1
E	1	1	ε	ε^2
	1	1	ε^2	ε
$F; x, y, z$	3	−1	0	0

O	T_d					
		E	$8C_3$	$3C_2$	$6C_2$	$6C_4$
		E	$8C_3$	$3C_2$	$6\sigma_d$	$6S_4$
A_1	A_1	1	1	1	1	1
A_2	A_2	1	1	1	−1	−1
E	E	2	−1	2	0	0
F_2	$F_2; x, y, z$	3	0	−1	1	−1
$F_1; x, y, z$	F_1	3	0	−1	−1	1

the cyclic groups (C_n, S_n). A cyclic group, like any Abelian group, has only one-dimensional representations. Let G be a *generating element* of the group (i.e. one which, on being raised to successive powers, gives all the elements of the group). Since $G^g = E$ (where g is the order of the group), it is clear that, when the operator $\hat{G}$ acts on a base function ψ, the latter can be multiplied only by $1^{1/g}$, i.e.†

$$\hat{G}\psi = e^{2\pi ik/g}\psi \qquad (k = 1, 2, \ldots, g).$$

The group C_{2h} (and the isomorphous groups C_{2v} and D_2) is Abelian, so that all its irreducible representations are one-dimensional, and the characters can only be $\pm$ 1 (since the square of every element is E).

Next we consider the group C_{3v}. As compared with the group C_3, the reflections σ_v in vertical planes (all belonging to one class) are here added. A function invariant with respect to rotation about the axis (a base function of the representation A of the group C_3) may be either symmetric or antisymmetric with respect to the reflections σ_v. Functions multiplied by ϵ and ϵ^2 under the rotation C_3, on the other hand (base functions of the complex conjugate representations E), change into each other on reflection.‡ It follows from these considerations that the group C_{3v} (and D_3, which is isomorphous with it) has two one-dimensional irreducible representations and one two-dimensional, with the characters shown in the table. The fact that we have indeed found all the irreducible representations may be seen from the result $1^2+1^2+2^2 = 6$, which is the order of the group.

Similar considerations give the characters of the representations of other groups of the same type (C_{4v}, C_{6v}).

The group T is obtained from the group $D_2 \equiv V$ by adding rotations about four oblique axes of the third order. A function invariant with respect to transformations of the group V (a basis of the representation A) can be multiplied, under the rotation C_3, by 1, ϵ or ϵ^2. The base functions of the three one-dimensional representations B_1, B_2, B_3 of the group V change into one another under rotations about the axes of the third order (this is seen. for example, if we take as these functions the coordinates x, y, z themselves). Thus we obtain three one-dimensional irreducible representations and one three-dimensional ($1^2+1^2+1^2+3^2 = 12$).

Finally, let us consider the isomorphous groups O and T_d. The group T_d is obtained from the group T by adding reflections σ_d in planes each of which passes through two axes of the third order. A base function of the unit representation A of the group T may be symmetric or antisymmetric with respect to these reflections (which all belong to one class), and this gives two one-dimensional representations of the group T_d. Functions multiplied by ϵ or ϵ^2 under a rotation about an axis of the third order (the

† For the point group C_n we can, for example, take as the functions ψ the functions $e^{ik\phi}$, $k = 1, 2, \ldots, n$, where ϕ is the angle of rotation about the axis, measured from some fixed direction.

‡ These functions may, for example, be taken as $\psi_1 = e^{i\phi}$, $\psi_2 = e^{-i\phi}$. On reflection in a vertical plane, ϕ changes sign.

basis of the complex conjugate representations E of the group T) change into each other on reflection in a plane passing through this axis, so that one two-dimensional representation is obtained. Finally, of three base functions of the representation F of the group T, one is transformed into itself on reflection (and can either remain unaltered or change sign), while the other two change into each other. Thus we have altogether two one-dimensional representations, one two-dimensional and two three-dimensional.†

The representations of the remaining point groups in which we are interested can be obtained immediately from those already given, if we notice that the remaining groups are direct products of those already considered with the group C_i (or C_s):

$$C_{3h} = C_3 \times C_s \qquad D_{2h} = D_2 \times C_i \qquad D_{3d} = D_3 \times C_i$$
$$C_{4h} = C_4 \times C_i \qquad D_{4h} = D_4 \times C_i \qquad D_{6h} = D_6 \times C_i$$
$$C_{6h} = C_6 \times C_i \qquad S_6 = C_3 \times C_i \qquad T_h = T \times C_i$$
$$O_h = O \times C_i$$

Each of these direct products has twice as many irreducible representations as the original group, half of them being symmetric (denoted by the suffix g) and the other half antisymmetric (suffix u) with respect to inversion. The characters of these representations are obtained from those of the representations of the original group by multiplying by ± 1 (in accordance with the rule (94.24)). Thus, for instance, we have for the group D_{3d} the representations:

D_{3d}	E	$2C_3$	$3U_2$	I	$2S_6$	$3\sigma_d$
A_{1g}	1	1	1	1	1	1
A_{2g}	1	1	−1	1	1	−1
E_g	2	−1	0	2	−1	0
A_{1u}	1	1	1	−1	−1	−1
A_{2u}	1	1	−1	−1	−1	1
E_u	2	−1	0	−2	1	0

§96. Irreducible representations and the classification of terms

The quantum-mechanical applications of group theory are based on the fact that the Schrödinger's equation for a physical system (an atom or molecule) is invariant with respect to symmetry transformations of the

† Irreducible representations of higher dimension (4 and 5) occur in the icosahedron groups.

system.† It follows at once from this that, on applying the elements of a group to a function satisfying Schrödinger's equation for some value of the energy (an eigenvalue), we must again obtain solutions of the same equation for the same value of the energy. In other words, under a symmetry transformation the wave functions of the stationary states of the system belonging to a given energy level transform into linear combinations of one another, i.e. they give some representation of the group. An important fact is that this representation is irreducible. For functions which are invariably transformed into linear combinations of themselves under symmetry transformations must belong to the same energy level; the equality of the eigenvalues of the energy corresponding to several groups of functions (into which the basis of a reducible representation can be divided), which are not transformed into combinations of one another, would be an improbable coincidence.‡

Thus, to each energy level of the system, there corresponds some irreducible representation of its symmetry group. The dimension of this representation determines the degree of degeneracy of the level concerned, i.e. the number of different states with the energy in question. The fixing of the irreducible representation determines all the symmetry properties of the given state, i.e. its behaviour with respect to the various symmetry transformations.

Irreducible representations of dimension greater than one are found only in groups containing non-commuting elements; Abelian groups have only one-dimensional irreducible representations. It is apposite to recall here that the relation between degeneracy and the presence of operators which do not commute with one another (but do commute with the Hamiltonian) has already been found above from considerations unrelated to group theory (§10).

The following important reservation should be made regarding all these statements. As has already been pointed out (§18), the symmetry (valid in the absence of a magnetic field) with respect to a change in the sign of the time has, in quantum mechanics, the result that complex conjugate wave functions must belong to the same eigenvalue of the energy. Hence it follows that, if some set of functions and the set of complex conjugate functions give different (non-equivalent) irreducible representations of a group, these two complex conjugate representations must be regarded as forming together a single "physically irreducible" representation of twice the dimension. This will be assumed below. In the preceding section we had examples of such representations. Thus the group $\boldsymbol{C}_3$ has only one-dimensional representations; however, two of these are complex conjugates, and correspond physically to doubly degenerate energy levels. (In the presence of a magnetic field there is no symmetry with respect to a change in the sign of the time,

---

† The methods of group theory were first applied in quantum mechanics by E. P. Wigner (1926).

‡ Provided that there is no special reason for this. Reference may be made here to the "accidental" degeneracy that arises because the Hamiltonian of a system can have a higher symmetry than the purely geometrical symmetry considered in the present chapter (see the end of §36).

and hence complex conjugate representations correspond to different energy levels.)†

Let us suppose that a physical system is subjected to the action of some perturbation (i.e. the system is placed in an external field). The question arises to what extent the perturbation can result in a splitting of the degenerate levels. The external field has itself a certain symmetry.‡ If this symmetry is the same as or higher‖ than that of the unperturbed system, the symmetry of the perturbed Hamiltonian $\hat{H} = \hat{H}_0 + \hat{V}$ is the same as the symmetry of the unperturbed operator $\hat{H}_0$. It is clear that, in this case, no splitting of the degenerate levels occurs. If, however, the symmetry of the perturbation is lower than that of the unperturbed system, then the symmetry of the Hamiltonian $\hat{H}$ is the same as that of the perturbation $\hat{V}$. The wave functions which gave an irreducible representation of the symmetry group of the operator $\hat{H}_0$ will also give a representation of the symmetry group of the perturbed operator $\hat{H}$, but this representation may be reducible, and this means that the degenerate level is split.

We shall show by means of an example how the mathematical techniques of group theory enable us to solve the problem of the splitting of any given level.

Let the unperturbed system have symmetry T_d, and let us consider a triply degenerate level corresponding to the irreducible representation F_2 of this group. The characters of this representation are

E	$8C_3$	$3C_2$	$6\sigma_d$	$6S_4$
3	0	−1	1	−1

Let us assume that the system is subjected to the action of a perturbation with symmetry C_{3v} (with the third-order axis coinciding with one of those of the group T_d). The three wave functions of the degenerate level give a representation of the group C_{3v} (which is a sub-group of the group T_d), and the characters of this representation are equal to those of the same elements in the original representation of the group T_d, i.e.

E	$2C_3$	$3\sigma_v$
3	0	1

This representation, however, is reducible. Knowing the characters of the

† Strictly speaking, the fact that the characters are real (i.e. that the complex conjugate representations are equivalent) is not a sufficient condition for the possibility of choosing real base functions of the representation of the group. For irreducible representations of point groups, however, it is sufficient (though not for the "double" point groups; see §99).

‡ For example, in the case of the energy levels of the *d* and *f* shells of ions in a crystal lattice which interact slightly with the surrounding atoms, the perturbation (the external field) is the field acting on an ion due to the other atoms.

‖ If a symmetry group $\boldsymbol{H}$ is a sub-group of the group $\boldsymbol{G}$, we say that $\boldsymbol{H}$ corresponds to a *lower symmetry* and $\boldsymbol{G}$ to a *higher symmetry*. It is evident that the symmetry of the sum of two expressions, one of which has the symmetry of $\boldsymbol{G}$ and the other that of $\boldsymbol{H}$, is the lower symmetry, that of $\boldsymbol{H}$.

irreducible representations of the group C_{3v}, it is easy to decompose it into irreducible parts, using the general rule (94.16). Thus we find that it consists of the representations A_1 and E of the group C_{3v}. The triply degenerate level F_2 is therefore split into one non-degenerate level A_1 and one doubly degenerate level E. If the same system is subjected to the action of a perturbation of symmetry C_{2v}, which is also a sub-group of the group T_d, then the wave functions of the same level F_2 give a representation with characters

E	C_2	σ_v	σ'_v
3	−1	1	1

Decomposing this into irreducible parts, we find that it contains the representations A_1, B_1, B_2. Thus in this case the level is completely split into three non-degenerate levels.

§97. Selection rules for matrix elements

Group theory not only enables us to carry out a classification of the terms of any symmetrical physical system, but also gives us a simple method of finding the selection rules for the matrix elements of the various quantities which characterize the system.

This method is based on the following general theorem. Let $\psi_i^{(\alpha)}$ be one of the base functions of an irreducible (non-unit) representation of a symmetry group. Then the integral of this function over all space† vanishes identically:

$$\int \psi_i^{(\alpha)}\, dq = 0. \tag{97.1}$$

The proof is based on the evident fact that the integral over all space is invariant with respect to any transformation of the coordinate system, including any symmetry transformation. Hence

$$\int \psi_i^{(\alpha)}\, dq = \int \hat{G}\psi_i^{(\alpha)}\, dq = \int \sum_k G_{ki}^{(\alpha)}\psi_k^{(\alpha)}\, dq.$$

We sum this equation over all the elements of the group. The integral on the left is simply multiplied by g, the order of the group, and we have

$$g\int \psi_i^{(\alpha)}\, dq = \sum_k \int \psi_k^{(\alpha)} \sum_G G_{ki}^{(\alpha)}\, dq.$$

However, for any non-unit irreducible representation we have identically

$$\sum_G G_{ki}^{(\alpha)} = 0;$$

this is a particular case of the orthogonality relations (94.7), when one of the irreducible representations is the unit representation. This proves the theorem.

† That is, the configuration space of the physical system concerned.

If ψ is a function belonging to the basis of some reducible representation of a group, the integral $\int \psi \, dq$ will be zero except when this representation contains the unit representation. This theorem is a direct consequence of the previous one.

The matrix elements of a physical quantity f are given by the integrals

$$\langle \beta k | f | \alpha i \rangle = \int \psi_k^{(\beta)} \hat{f} \psi_i^{(\alpha)} \, dq, \qquad (97.2)$$

where the indices α and β distinguish different energy levels of the system, and the suffixes i, k denumerate wave functions belonging to the same degenerate level.† We denote the irreducible representations of the symmetry group of the system concerned that are given by the functions $\psi_i^{(\alpha)}$ and $\psi_k^{(\beta)}$ by the symbols $D^{(\alpha)}$ and $D^{(\beta)}$, and by D_f the representation of the same group that corresponds to the symmetry of the quantity f; this representation depends on the tensor character of f. For example, if f is a true scalar, then its operator $\hat{f}$ is invariant under all the symmetry transformations, and D_f is the unit representation. The same occurs for a pseudoscalar quantity if the group contains only axes of symmetry, but if there are also reflections, D_f is not the unit representation, though its dimension is unity. If f is a vector, then D_f is a representation given by the three vector components that are transformed into combinations of each other; this representation is in general different for polar and axial vectors.

The products $\psi_k^{(\beta)} \hat{f} \psi_i^{(\alpha)}$ give the representation that is the direct product $D^{(\beta)} \times D_f \times D^{(\alpha)}$. The matrix elements are non-zero if this representation contains the unit representation or, equivalently, if the direct product $D^{(\beta)} \times D^{(\alpha)}$ contains D_f. In practice, it is more convenient to decompose into irreducible parts the product $D^{(\alpha)} \times D_f$; this gives us immediately all the types $D^{(\beta)}$ of states for transitions into which (from a state of the type $D^{(\alpha)}$) the matrix elements are not zero.

In the simplest case of a scalar quantity, for which D_f is the unit representation, it then follows immediately that the matrix elements are non-zero only for transitions between states of the same type: the direct product $D^{(\alpha)} \times D^{(\beta)}$ of two different irreducible representations does not contain the unit representation, but the latter is always present in the direct product of an irreducible representation with itself. This is most general statement of a theorem of which particular cases have already been met with.

The matrix elements diagonal with respect to energy, i.e. those for transitions between states belonging to the same term (as opposed to transitions between states belonging to two different terms of the same type), need special treatment. In this case we have only one set of functions $\psi_1^{(\alpha)}$, $\psi_2^{(\alpha)}$, ..., not two different ones. The selection rules here are found by different methods, depending on the behaviour of the quantity f under time reversal.

† Since the base functions can be taken as real when "physically irreducible" representations are used, we do not distinguish in (97.2) between the wave functions and their complex conjugates.

Let us consider a state described by a wave function of the form $\psi = \Sigma\, c_i\psi_i^{(\alpha)}$. The mean value of f in this state is given by the sum

$$\bar{f} = \sum_{i,k} c_k^* c_i \langle \alpha k|\, f\,|\alpha i\rangle.$$

In the state with the complex conjugate wave function $\psi^* = \Sigma\, c_i^*\psi_i^{(\alpha)}$, we have

$$\bar{f} = \sum_{i,k} c_k c_i^* \langle \alpha k|\, f\,|\alpha i\rangle$$
$$= \sum_{i,k} c_i c_k^* \langle \alpha i|\, f\,|\alpha k\rangle.$$

If f is invariant under time reversal, the two states not only belong to the same energy level but must also have the same value of $\bar{f}$. Since the coefficients c_i are arbitrary, this means that

$$\langle \alpha k|\, f\,|\alpha i\rangle = \langle \alpha i|\, f\,|\alpha k\rangle.$$

Hence, in order to find the selection rules, we must consider not the direct product $D^{(\alpha)}\times D^{(\alpha)}$ as a whole, but only its symmetric part $[D^{(\alpha)2}]$; there are non-zero matrix elements if $[D^{(\alpha)2}]$ contains D_f.†

If, however, f changes sign under time reversal, the change from ψ to ψ^* has to be accompanied by a change in the sign of $\bar{f}$. Hence we find by the same method that

$$\langle \alpha k|\, f\,|\alpha i\rangle = -\langle \alpha i|\, f\,|\alpha k\rangle.$$

In this case, therefore, the selection rules are determined by the decomposition of the antisymmetric part of the direct product, $\{D^{(\alpha)2}\}$.

PROBLEMS

PROBLEM 1. Find the selection rules for the matrix elements of the electric and magnetic dipole moments $\mathbf{d}$ and $\boldsymbol{\mu}$ when symmetry $\boldsymbol{O}$ is present.

SOLUTION. The group $\boldsymbol{O}$ includes no reflections; the polar vector $\mathbf{d}$ and the axial vector $\boldsymbol{\mu}$ are therefore transformed by the same irreducible representation, F_1. The decompositions of the direct products of F_1 with the other representations of the group $\boldsymbol{O}$ are

$$\left.\begin{aligned} &F_1\times A_1 = F_1,\quad F_1\times A_2 = F_2,\quad F_1\times E = F_1+F_2,\\ &F_1\times F_1 = A_1+E+F_1+F_2,\quad F_1\times F_2 = A_2+E+F_1+F_2.\end{aligned}\right\} \qquad (1)$$

Hence the non-zero non-diagonal (with respect to energy) matrix elements are those for the transitions

$$F_1 \leftrightarrow A_1, E, F_1, F_2; \qquad F_2 \leftrightarrow A_2, E, F_2.$$

---

† The product $[D^{(\alpha)2}]$ always contains the unit representation, so that the diagonal elements (and non-diagonal elements between states of the same type) are non-zero for a scalar quantity.

The symmetric and antisymmetric products of the irreducible representations of the group $\boldsymbol{O}$ are

$$\left.\begin{aligned} &[A_1^2] = [A_2^2] = A_1, \quad [E^2] = A_1+E, \quad [F_1^2] = [F_2^2] = A_1+E+F_2, \\ &\{E^2\} = A_2, \quad \{F_1^2\} = \{F_2^2\} = F_1. \end{aligned}\right\} \quad (2)$$

The symmetric products do not contain F_1; hence there are no diagonal (with respect to energy) matrix elements of the vector **d** (which is invariant under time reversal). The magnetic moment, which changes sign under time reversal, has diagonal matrix elements for the states F_1 and F_2.

PROBLEM 2. The same as Problem 1, but for symmetry $\boldsymbol{D}_{3d}$.

SOLUTION. The vectors **d** and $\boldsymbol{\mu}$ have different transformation laws in the group $\boldsymbol{D}_{3d}$:

$$\begin{aligned} &d_x, d_y \sim E_u, \quad d_z \sim A_{2u}, \\ &\mu_x, \mu_y \sim E_g, \quad \mu_z \sim A_{2g}; \end{aligned}$$

here and in the Problems below, the symbol $\sim$ stands for the words "is transformed by the representation". We have

$$\left.\begin{aligned} &E_u \times A_{1g} = E_u \times A_{2g} = E_u, \quad E_u \times A_{1u} = E_u \times A_{2u} = E_g, \\ &E_u \times E_u = A_{1g}+A_{2g}+E_g, \quad E_u \times E_g = A_{1u}+A_{2u}+E_u. \end{aligned}\right\} \quad (3)$$

Hence the non-diagonal matrix elements of d_x, d_y are non-zero for the transitions $E_u \leftrightarrow A_{1g}$, A_{2g}, E_g; $E_g \leftrightarrow A_{1u}$, A_{2u}. In the same way we find the selection rules

for d_z: $A_{1g} \leftrightarrow A_{2u}$; $A_{2g} \leftrightarrow A_{1u}$; $E_g \leftrightarrow E_u$;
for μ_x, μ_y: $E_g \leftrightarrow A_{1g}$, A_{2g}, E_g; $E_u \leftrightarrow A_{1u}$, A_{2u}, E_u;
for μ_z: $A_{1g} \leftrightarrow A_{2g}$; $A_{1u} \leftrightarrow A_{2u}$; $E_g \leftrightarrow E_g$; $E_u \leftrightarrow E_u$.

The symmetric and antisymmetric products of the irreducible representations of the group $\boldsymbol{D}_{3d}$ are

$$\left.\begin{aligned} &[A_{1g}^2] = [A_{1u}^2] = [A_{2g}^2] = [A_{2u}^2] = A_{1g}, \\ &[E_g^2] = [E_u^2] = E_g + A_{1g}, \quad \{E_g^2\} = \{E_u^2\} = A_{2g}. \end{aligned}\right\} \quad (4)$$

Hence we see that there are no diagonal (with respect to energy) matrix elements for any of the components **d**; for the vector $\boldsymbol{\mu}$, there are diagonal matrix elements of μ_z for transitions between states belonging to a degenerate level of the type E_g or E_u.

PROBLEM 3. Find the selection rules for the matrix elements of the electric quadrupole moment tensor Q_{ik} when symmetry $\boldsymbol{O}$ is present.

SOLUTION. The components of the tensor Q_{ik} (a symmetrical tensor with the sum Q_{ii} equal to zero) with respect to group $\boldsymbol{O}$ are transformed by the laws

$$Q_{xy}, Q_{xz}, Q_{yz} \sim F_2, \quad Q_{xx}+\epsilon Q_{yy}+\epsilon^2 Q_{zz}, \; Q_{xx}+\epsilon^2 Q_{yy}+\epsilon Q_{zz} \sim E$$
$$(\epsilon = e^{2\pi i/3}).$$

Decomposing the direct products of F_2 and E with all the representations of the group, we find the selection rules for the non-diagonal matrix elements:

for Q_{xy}, Q_{xz}, Q_{yz}: $F_1 \leftrightarrow A_2$, E, F_1, F_2; $F_2 \leftrightarrow A_1$, E, F_1, F_2;
for Q_{xx}, Q_{yy}, Q_{zz}: $E \leftrightarrow A_1$, A_2, E; $F_1 \leftrightarrow F_1$, F_2; $F_2 \leftrightarrow F_2$.

The diagonal matrix elements exist (as we see from (2)) in the following states:

for Q_{xy}, Q_{xz}, Q_{yz}: F_1, F_2,
for Q_{xx}, Q_{yy}, Q_{zz}: E, F_1, F_2.

PROBLEM 4. The same as Problem 3, but for symmetry $\boldsymbol{D}_{3d}$.

SOLUTION. The transformation laws of the components Q_{ik} with respect to the group $\boldsymbol{D}_{3d}$ are

$$Q_{zz} \sim A_{1g}; \quad Q_{xx}-Q_{yy}, Q_{xy} \sim E_g; \quad Q_{xz}, Q_{yz} \sim E_g.$$

Q_{zz} behaves as a scalar. Decomposing the direct products of E_g with all the representations of the group, we find the selection rules for the non-diagonal matrix elements of the remaining components Q_{ik}:

$$E_g \leftrightarrow A_{1g}, A_{2g}, E_g; \quad E_u \leftrightarrow A_{1u}, A_{2u}, E_u.$$

The diagonal elements are non-zero (as we see from (4)) only for the states E_g and E_u.

§98. Continuous groups

As well as the finite point groups enumerated in §93, there exist also what are called *continuous point groups*, having an infinite number of elements. These are the groups of axial and spherical symmetry.

The simplest axial symmetry group is the group C_∞, which contains rotations $C(\phi)$ through any angle ϕ about the axis of symmetry; this is called the *two-dimensional rotation group*. It may be regarded as the limiting case of the groups C_n as $n \to \infty$. Similarly, as limiting cases of the groups C_{nh}, C_{nv}, D_n, D_{nh} we obtain the continuous groups $C_{\infty h}$, $C_{\infty v}$, D_∞, $D_{\infty h}$.

A molecule has axial symmetry only if it consists of atoms lying in a straight line. If it meets this condition, but is asymmetric about its midpoint, its point group will be the group $C_{\infty v}$, which, besides rotations about the axis, contains also reflections σ_v in any plane passing through the axis. If, on the other hand, the molecule is symmetrical about its midpoint, its point group will be $D_{\infty h} = C_{\infty v} \times C_i$. The groups C_∞, $C_{\infty h}$, D_∞ cannot appear as the symmetry groups of a molecule.

The group of complete spherical symmetry contains rotations through any angle about any axis passing through the centre, and reflections in any plane passing through the centre; this group, which we shall denote by K_h, is the symmetry group of a single atom. It contains as a sub-group the group K of all spatial rotations (called the *three-dimensional rotation group*, or simply the *rotation group*). The group K_h can be obtained from the group K by adding a centre of symmetry ($K_h = K \times C_i$).

The elements of a continuous point group may be distinguished by one or more parameters which take a continuous range of values. Thus, in the rotation group, the parameters might be the three Eulerian angles, which define a rotation of the coordinates.

The general properties of finite groups described in §92, and the concepts appertaining to them (sub-groups, conjugate elements, classes, etc.), can be at once generalized to continuous groups. Of course, the statements which directly concern the order of the group (for instance, that the order of a sub-group divides the order of the group) are no longer meaningful.

In the group $C_{\infty v}$ all planes of symmetry are equivalent, so that all reflections σ_v form a single class with a continuous series of elements; the axis of symmetry is bilateral, so that there is a continuous series of classes, each containing two elements $C(\pm\phi)$. The classes of the group $D_{\infty h}$ are obtained at once from those of the group $C_{\infty v}$, since $D_{\infty h} = C_{\infty v} \times C_i$.

In the rotation group K, all axes are equivalent and bilateral; hence the classes of this group are rotations through an angle of fixed absolute magnitude $|\phi|$ about any axis. The classes of the group $\boldsymbol{K}_h$ are obtained at once from those of the group $\boldsymbol{K}$.

The concept of representations, reducible and irreducible, can also be immediately generalized to continuous groups. Each irreducible representation contains an infinite sequence of matrices, but the number of base functions transformed into combinations of one another (the dimension of the representation) is finite. These functions may always be chosen so as to make the representation unitary. The number of different irreducible representations of a continuous group is infinite, but they form a discrete sequence, i.e. they can be numbered successively. For the matrix elements and characters of these representations there are orthogonality relations which generalize the corresponding ones for finite groups. Instead of (94.9), we now have

$$\int G_{ik}^{(\alpha)} G_{lm}^{(\beta)*} \mathrm{d}\tau_G = \frac{1}{f_\alpha} \delta_{\alpha\beta}\delta_{il}\delta_{km} \int \mathrm{d}\tau_G, \tag{98.1}$$

and instead of (94.10)

$$\int \chi^{(\alpha)}(G)\chi^{(\beta)}(G)^* \,\mathrm{d}\tau_G = \delta_{\alpha\beta} \int \mathrm{d}\tau_G. \tag{98.2}$$

The integration in these formulae is what is called an *invariant integration* over the group; the element $\mathrm{d}\tau_G$ is expressed in terms of the parameters of the group and their differentials in such a way as to remain an element when subjected to any transformation in the group.† For example, in the rotation group we can take $\mathrm{d}\tau_G = \sin\beta \,\mathrm{d}\alpha\,\mathrm{d}\beta\,\mathrm{d}\gamma$, where α, β and γ are the Eulerian angles, which define a rotation of the system of coordinates (§58); in this case, $\int \mathrm{d}\tau_G = 8\pi^2$.

We have already found, in essence, the irreducible representations of the three-dimensional rotation group (without using the terminology of group theory), when determining the eigenvalues and eigenfunctions of the total angular momentum. For the angular momentum component operators are (apart from a constant factor) the operators of infinitely small rotations,‡ and the eigenvalues of the angular momentum characterize the behaviour of the wave functions with respect to spatial rotations. To a value j of the angular momentum there correspond $2j+1$ different eigenfunctions ψ_{jm}, differing in the values of the component m of the angular momentum and all belonging to one $(2j+1)$-fold degenerate energy level. Under rotations of the coordinate system, these functions are transformed into linear combinations of themselves, and thus give irreducible representations of the

† The statements made here about the properties of irreducible representations of continuous groups are valid only if the integrals (98.1) and (98.2) converge; in particular, the "volume of the group" $\int \mathrm{d}\tau_G$ must be finite. This condition is satisfied for continuous point groups (but not, for instance, for the Lorentz group which occurs in the relativistic theory).

‡ In mathematical terms, these operators are the *generators* of the rotation group.

rotation group. Thus, from the group-theory point of view, the numbers j number the irreducible representations of the rotation group, and one $(2j+1)$-dimensional representation corresponds to each j. The number j takes integral and half-integral values, so that the dimension $2j+1$ of the representations takes all the integral values 1, 2, 3, ...

The base functions of these representations have been, in essence, investigated in §§56 and 57, and the matrices of the representations have been found in §58. The basis of a representation of given j is formed by the $2j+1$ independent components of a symmetrical spinor of rank $2j$ (which are equivalent to the set of $2j+1$ functions ψ_{jm}).

The irreducible representations of the rotation group which correspond to half-integral values of j are distinguished by an important property. Under a rotation through 2π, the base functions of the representations change sign (being components of a spinor of odd rank). Since, however, a rotation through 2π is the same as the unit element of the group, we reach the result that representations with half-integral j are, as we say, *two-valued*; to each element of the group (a rotation through an angle ϕ, $0 \leqslant \phi \leqslant 2\pi$, about some axis) there correspond in such a representation not one but two matrices, with characters differing in sign.†

An isolated atom has, as we have already remarked, the symmetry $\boldsymbol{K_h} = \boldsymbol{K} \times \boldsymbol{C_i}$. Hence, from the group-theory point of view, there corresponds to each term of the atom some irreducible representation of the rotation group $\boldsymbol{K}$ (determining the value of the total angular momentum J of the atom) and an irreducible representation of the group $\boldsymbol{C_i}$ (determining the parity of the state).‡

When the atom is placed in an external electric field, its energy levels are split. The number of different levels resulting and the symmetry of the corresponding states can be determined by the method described in §96. It is necessary to decompose the $(2J+1)$-dimensional representation of the symmetry group of the external field (given by the functions ψ_{JM}) into irreducible representations of this group. This requires a knowledge of the characters of the representation given by the functions ψ_{JM}.

Since the characters of the irreducible representations of elements of one class are the same, it is sufficient to consider rotations about the z-axis. By a rotation through an angle ϕ about this axis the wave functions ψ_{JM}

† It must be mentioned that two-valued representations of a group are not representations in the true sense of the word, since they are not given by one-valued base functions; see also §99.

‡ Moreover, the Hamiltonian of the atom is invariant with respect to interchanges of the electrons. In the non-relativistic approximation, the coordinate and spin wave functions are separable, and we can speak of representations of the permutation group that are given by the coordinate functions. If the irreducible representation of the permutation group is given, the total spin S of the atom is determined (§63). When the relativistic interactions are taken into account, however, the separation of the wave functions into coordinate and spin parts is not possible. The symmetry with respect to simultaneous interchange of the coordinates and spins of the particles does not characterize the term, since Pauli's principle admits only those total wave functions which are antisymmetric with respect to all the electrons. This is in accordance with the fact that, when the relativistic interactions are taken into account, the spin is not, strictly speaking, conserved; only the total angular momentum J is conserved.

are, as we know, multiplied by $e^{iM\phi}$, where M is the component of the angular momentum along this axis. The transformation matrix for the functions ψ_{JM} will therefore be diagonal, with character

$$\chi^{(J)}(\phi) = \sum_{m=-J}^{J} e^{iM\phi} = \frac{e^{i(J+1)\phi} - e^{-iJ\phi}}{e^{i\phi}-1},$$

or†

$$\chi^{(J)}(\phi) = \frac{\sin(J+\frac{1}{2})\phi}{\sin\frac{1}{2}\phi}. \tag{98.3}$$

With respect to inversion I, all the functions ψ_{JM} with different M behave in the same way, being multiplied by $+1$ or -1 according as the state of the atom is even or odd. Hence the character

$$\chi^{(J)}(I) = \pm(2J+1). \tag{98.4}$$

Finally, the characters corresponding to reflection in a plane σ and rotary reflection through an angle ϕ are found by writing these symmetry transformations as

$$\sigma = IC_2, \qquad S(\phi) = IC(\pi+\phi).$$

Let us pause to consider also the irreducible representations of the axial symmetry group $\boldsymbol{C}_{\infty v}$. This problem has, in essence, been solved when we ascertained the classification of the electron terms of a diatomic molecule having this symmetry $\boldsymbol{C}_{\infty v}$ (i.e. when the two atoms are different). To the terms 0^+ and 0^- (with $\Omega = 0$) there correspond two one-dimensional irreducible representations: the unit representation A_1 and the representation A_2, in which the base function is invariant under all rotations and changes sign under reflections in planes σ_v, while to the doubly degenerate terms with $\Omega = 1, 2, \ldots$ there correspond two-dimensional representations denoted by $E_1, E_2, \ldots$. Under a rotation through an angle ϕ about the axis, the base functions are multiplied by $e^{\pm i\Omega\phi}$, while on reflection in planes σ_v they change into each other. The characters of these representations are

$\boldsymbol{C}_{\infty v}$	E	$2C(\phi)$	$\infty\sigma_v$
A_1	1	1	1
A_2	1	1	-1
E_k	2	$2\cos k\phi$	0

(98.5)

† To avoid misunderstanding, it should be emphasized that this formula corresponds to a parametrization of the group elements other than that by the Eulerian angles: the transformation is specified by the direction of the axis of rotation and the angle ϕ of the rotation about the axis. It can be shown that, with this parametrization, the integration in (98.2), for example, is to be taken over $2(1-\cos\phi)d\phi\, do$, where do is the element of solid angle for the direction of the axis of rotation.

The irreducible representations of the group $D_{\infty h} = C_{\infty v} \times C_i$ are obtained at once from those of the group $C_{\infty v}$ (and correspond to the classification of the terms of a diatomic molecule composed of like nuclei).

If we take half-integral values for Ω, the functions $e^{\pm i\Omega\phi}$ give two-valued irreducible representations of the group $C_{\infty v}$, corresponding to the terms of the molecule having half-integral spin.†

§99. Two-valued representations of finite point groups

To the states of a system with half-integral spin (and therefore half-integral total angular momentum) there correspond two-valued representations of the point symmetry group of the system. This is a general property of spinors, and therefore holds for both continuous and finite point groups. The necessity thus arises of finding the two-valued irreducible representations of finite point groups.

As we have already remarked, the two-valued representations are not really true representations of a group. In particular, the relations discussed in §94 do not apply to them, and where all irreducible representations were considered in these relations (for example, in the relation (94.17) for the sum of the squared dimensions of the irreducible representations), only the true one-valued representations were meant.

To find the two-valued representations, it is convenient to employ the following artifice (H. A. Bethe 1929). We introduce, in a purely formal manner, the concept of a new element of the group (denoted by Q); this is a rotation through an angle of 2π about an arbitrary axis, and is not the unit element, but gives the latter when applied twice: $Q^2 = E$. Accordingly, rotations C_n about the axes of symmetry of the nth order will give identical transformations only after being applied $2n$ times (and not n times):

$$C_n^{\,n} = Q, \qquad C_n^{\,2n} = E. \tag{99.1}$$

The inversion I, being an element which commutes with all rotations, must give E as before on being applied twice. A twofold reflection in a plane, however, gives Q, not E:

$$\sigma^2 = Q, \qquad \sigma^4 = E; \tag{99.2}$$

this follows, since the reflection can be written in the form $\sigma_h = IC_2$. As a

† Contrary to the result for the three-dimensional rotation group, it would here be possible, by a suitable choice of fractional values of Ω, to obtain not only one-valued and two-valued representations, but also those of three or more values. However, the physically possible eigenvalues of the angular momentum, which is the operator of an infinitely small rotation, are determined by the representations of the aforementioned three-dimensional rotation group. Hence the three (or more)-valued representations of the two-dimensional rotation group (and of any finite symmetry group), though mathematically determinate, are without physical significance.

result we obtain a set of elements forming some fictitious point symmetry group, whose order is twice that of the original group; such groups we shall call *double* point groups. The two-valued representations of the actual point group will clearly be one-valued (i.e. true) representations of the corresponding double group, so that they can be found by the usual methods.

The number of classes in the double group is greater than in the original group (but not, in general, twice as great). The element Q commutes with all the other elements of the group,† and hence always forms a class by itself. If the axis of symmetry is bilateral, the elements $C_n^{\ k}$ and $C_n^{\ 2n-k} = QC_n^{\ n-k}$ are conjugate in the double group. Hence, when axes of the second order are present, the distribution of the elements among classes depends also on whether these axes are bilateral (in ordinary point groups this is unimportant, since C_2 is the same as the opposite rotation C_2^{-1}).

Thus, for instance, in the group T the axes of the second order are equivalent, and each of them is bilateral, while the axes of the third order are equivalent but not bilateral. Hence the 24 elements of the double group‡ T' are distributed in seven classes: E, Q, the class of three rotations C_2 and three C_2Q, and the classes $4C_3$, $4C_3^{\ 2}$, $4C_3Q$, $4C_3^{\ 2}Q$.

The irreducible representations of a double point group include, firstly, representations which are the same as the one-valued representations of the simple group (a unit matrix corresponding to both Q and E); secondly, the two-valued representations of the simple group, a negative unit matrix corresponding to Q. It is these latter representations in which we are now interested.

The double groups C_n' ($n = 1, 2, 3, 4, 6$) and S_4', like the corresponding simple groups, are cyclic.|| All their irreducible representations are one-dimensional, and can be found without difficulty as shown in §95.

The irreducible representations of the groups D_n' (or C_{nv}', which are isomorphous with them) can be found by the same method as for the corresponding simple groups. These representations are given by functions of the form $e^{\pm ik\phi}$, where ϕ is the angle of rotation about an axis of the nth order, and k is given half-integral values (the integral values correspond to the ordinary one-valued representations). Rotations about horizontal axes of the second order change these functions into one another, while the rotation C_n multiplies them by $e^{\pm 2\pi ik/n}$.

It is a little less easy to find the representations of the double cubic groups. The 24 elements of the group T' are divided among seven classes. Hence there are altogether seven irreducible representations, of which four are the same as those of the simple group T. The sum of the squared dimensions of the remaining three representations must be 12, and hence we find that they are all two-dimensional. Since the elements C_2 and C_2Q belong to the

† This is obvious for rotations and inversion; for a reflection in a plane, it follows since the reflection can be represented as the product of an inversion and a rotation.

‡ We distinguish the double groups by primes to the symbols for the ordinary groups.

|| The groups $S_2' \equiv C_i'$, $S_6' \equiv C_{3i}'$, however, which contain the inversion I, are Abelian but not cyclic.

TABLE 8

Two-valued representations of point groups

D_2'	E	Q	$C_2^{(x)}$ $C_2^{(x)}Q$	$C_2^{(y)}$ $C_2^{(y)}Q$	$C_2^{(z)}$ $C_2^{(z)}Q$
E'	2	−2	0	0	0

D_3'	E	Q	C_3 C_3^2Q	C_3^2 C_3Q	$3U_2$	$3U_2Q$
E_1'	1	−1	−1	1	i	$-i$
	1	−1	−1	1	$-i$	i
E_2'	2	−2	1	−1	0	0

D_6'	E	Q	C_2 C_2Q	C_3 C_3^2Q	C_3^2 C_3Q	C_6 C_6^5Q	C_6^5 C_6Q	$3U_2$ $3U_2Q$	$3U'_2$ $3U'_2Q$
E_1'	2	−2	0	1	−1	$\sqrt{3}$	$-\sqrt{3}$	0	0
E_2'	2	−2	0	1	−1	$-\sqrt{3}$	$\sqrt{3}$	0	0
E_3'	2	−2	0	−2	2	0	0	0	0

D_4'	E	Q	C_2 C_2Q	C_4 C_4^3Q	C_4^3 C_4Q	$2U_2$ $2U_2Q$	$2U'_2$ $2U'_2Q$
E_1'	2	−2	0	$\sqrt{2}$	$-\sqrt{2}$	0	0
E_2'	2	−2	0	$-\sqrt{2}$	$\sqrt{2}$	0	0

T'	E	Q	$4C_3$	$4C_3^2$	$4C_3Q$	$4C_3^2Q$	$3C_2$ $3C_2Q$
E'	2	−2	1	−1	−1	1	0
G'	2	−2	ε	$-\varepsilon^2$	$-\varepsilon$	ε^2	0
	2	−2	ε^2	$-\varepsilon$	$-\varepsilon^2$	ε	0

O'	E	Q	$4C_3$ $4C_3^2Q$	$4C_3^2$ $4C_3Q$	$3C_4^2$ $3C_4^2Q$	$3C_4$ $3C_4^3Q$	$3C_4^3$ $3C_4Q$	$6C_2$ $6C_2Q$
E_1'	2	−2	1	−1	0	$\sqrt{2}$	$-\sqrt{2}$	0
E_2'	2	−2	1	−1	0	$-\sqrt{2}$	$\sqrt{2}$	0
G'	4	−4	−1	1	0	0	0	0

same class, $\chi(C_2) = \chi(C_2 Q) = -\chi(C_2)$, whence we conclude that $\chi(C_2) = 0$ in all three representations. Next, at least one of the three representations must be real, since complex representations can occur only in conjugate pairs. Let us consider this representation, and suppose that the matrix of the element C_3 is brought to diagonal form, with diagonal elements a_1, a_2. Since $C_3{}^3 = Q$, $a_1{}^3 = a_2{}^3 = -1$. In order that $\chi(C_3) = a_1 + a_2$ may be real, we must take $a_1 = e^{\pi i/3}$, $a_2 = e^{-\pi i/3}$. Hence we find that $\chi(C_3) = 1$, $\chi(C_3{}^2) = a_1{}^2 + a_2{}^2 = -1$. Thus one of the required representations is obtained. By comparing its direct products with the two complex conjugate one-dimensional representations of the group $\boldsymbol{T}$, we find the other two representations.

By means of similar arguments, which we shall not pause to give here, we may find the representations of the group $\boldsymbol{O}'$. Table 8 gives the characters of the representations of the double groups mentioned above. Only those representations are shown which correspond to two-valued representations of the ordinary groups. The isomorphous double groups have the same representations.

The remaining point groups are isomorphous with those we have considered, or else are obtained by direct multiplication of the latter by the group $\boldsymbol{C}_i$, so that their representations do not need to be specially calculated.

For the same reasons as for ordinary representations, two complex conjugate two-valued representations must be regarded as one physically irreducible representation of twice the dimension. It is necessary to pair one-dimensional two-valued representations even when they have real characters. For (see §60) in systems with half-integral spin, complex conjugate wave functions are linearly independent. Hence, if we have a two-valued one-dimensional representation† with real characters (given by some function ψ), then, although the complex conjugate function ψ^* is transformed by an equivalent representation, we can nevertheless see that ψ and ψ^* are linearly independent. Since, on the other hand, the complex conjugate wave functions must belong to the same energy level, we see that in physical applications this representation must be doubled.

The whole of the discussion in §97 concerning the method of finding the selection rules for the matrix elements of various physical quantities f remains valid for states of a system with half-integral spin, except as regards the matrix elements diagonal with respect to energy. On repeating the analysis at the end of §97 but with formulae (60.2) and (60.3), we find that, if the quantity f is even or odd under time reversal, we must use, in finding the selection rules, respectively the antisymmetric $\{D^{(\alpha)2}\}$ and symmetric $[D^{(\alpha)2}]$ products of the representation $D^{(\alpha)}$ with itself; this is the opposite of the rule stated in §97 for systems with integral spin.‡

† Such representations are found in the group $\boldsymbol{C}_n{}'$ for odd n; the characters are $\chi(C_n{}^k) = (-1)^k$.

‡ In connection with the application of these rules, it may be noted that for two-valued representations the unit representation is in the antisymmetric, not the symmetric, product of the representation with itself. For a two-valued representation with dimension 2, the product $\{D^{(\alpha)2}\}$ is just the unit representation.

PROBLEM

Determine how the levels of an atom (with given values of the total angular momentum J) are split when it is placed in a field having the cubic symmetry† $\boldsymbol{O}$.

SOLUTION. The wave functions of the states of an atom with angular momentum J and various values M_J give a $(2J+1)$-dimensional reducible representation of the group O, with characters determined by the formula (98.3). Decomposing this representation into irreducible parts (one-valued for integral J and two-valued for half-integral J), we at once find the required splitting (cf. §96). We shall list the irreducible parts of the representations corresponding to the first few values of J:

$J = 0$	A_1
$1/2$	E_1'
1	F_1
$3/2$	G'
2	$E+F_2$
$5/2$	$E_2'+G'$
3	$A_2+F_1+F_2$
...	...

† For example, an atom in a crystal lattice. The presence or absence of a centre of symmetry in the symmetry group of the external field is immaterial to this problem, since the behaviour of the wave function on inversion (the parity of the level) is unrelated to the angular momentum J.

CHAPTER XIII

POLYATOMIC MOLECULES

§100. The classification of molecular vibrations

In its applications to polyatomic molecules, group theory first of all resolves the problem of the classification of their electron terms, i.e. of the energy levels for a given situation of the nuclei. They are classified according to the irreducible representations of the point symmetry group appropriate to the configuration of the nuclei. Here, however, we must emphasize what is really obvious, that the classification thus obtained belongs to the definite nuclear configuration considered, since the symmetry is in general destroyed when the nuclei are displaced. We usually discuss the configuration corresponding to the equilibrium position of the nuclei. In this case the classification continues to possess a certain amount of meaning even when the nuclei execute small vibrations, but of course becomes meaningless when the vibrations can no longer be regarded as small.

In the diatomic molecule this question did not arise, since its axial symmetry is of course preserved under any displacement of the nuclei. A similar situation occurs for triatomic molecules also. The three nuclei always lie in a plane, which is a plane of symmetry of the molecule. Hence the classification of the electron terms of the triatomic molecule with respect to this plane (wave functions symmetric or antisymmetric with respect to reflection in the plane) is always possible.

For the normal electron terms of polyatomic molecules there is an empirical rule according to which, in the great majority of molecules, the wave function of the normal electron state is completely symmetrical (this rule, for diatomic molecules, has already been mentioned in §78). Thus, the wave function is invariant with respect to all the elements of the symmetry group of the molecule, i.e. it belongs to the unit irreducible representation of the group.

The application of the methods of group theory is particularly significant in the investigation of molecular vibrations (E. P. Wigner 1930). Before beginning a quantum-mechanical investigation of this problem, a purely classical discussion of the vibrations of the molecule is necessary, in which it is regarded as a system of several interacting particles (the nuclei).

A system of N particles (not lying in a straight line) has $3N-6$ vibrational degrees of freedom; of the total number of degrees of freedom $3N$, three correspond to translational and three to rotational motion of the system as a whole (see *Mechanics*, §§23, 24).† The energy of a system of particles

† If all the particles lie in a straight line, the number of vibrational degrees of freedom is $3N-5$; in this case, only two coordinates correspond to rotation, since it is meaningless to speak of the rotation of a linear molecule about its axis.

executing small vibrations can be written

$$E = \tfrac{1}{2} \sum_{i,k} m_{ik}\dot{u}_i\dot{u}_k + \tfrac{1}{2} \sum_{i,k} k_{ik}u_iu_k, \qquad (100.1)$$

where m_{ik}, k_{ik} are constant coefficients, and the u_i are the components of the vector displacements of the particles from their equilibrium positions (the suffixes i, k denumerate both the components of the vector and the particles). By a suitable linear transformation of the quantities u_i, we can eliminate from (100.1) the coordinates corresponding to translational motion and rotation of the system, and take the vibrational coordinates in such a way that both the quadratic forms in (100.1) are transformed into sums of squares. Normalizing these coordinates so as to make all the coefficients in the expression for the kinetic energy unity, we obtain the vibrational energy in the form

$$E = \tfrac{1}{2} \sum_{i,\alpha} \dot{Q}_{\alpha i}^2 + \tfrac{1}{2} \sum_{\alpha} \omega_\alpha^2 \sum_i Q_{\alpha i}^2. \qquad (100.2)$$

The vibrational coordinates $Q_{\alpha i}$ are said to be *normal*; the ω_α are the frequencies of the corresponding independent vibrations. It may happen that the same frequency (which is then said to be *multiple*) corresponds to several normal coordinates; the suffix α to the normal coordinate gives the number of the frequency, and the suffix $i = 1, 2, \ldots, f_\alpha$ numbers the coordinates belonging to a given frequency (f_α being the *multiplicity* of the frequency).

The expression (100.2) for the energy of the molecule must be invariant with respect to symmetry transformations. This means that, under any transformation belonging to the point symmetry group of the molecule, the normal coordinates $Q_{\alpha i}$, $i = 1, 2, \ldots, f_\alpha$ (for any given α) are transformed into linear combinations of themselves, in such a way that the sum of the squares $\sum_i Q_{\alpha i}^2$ remains unchanged. In other words, the normal coordinates belonging to any particular eigenfrequency of the vibrations of the molecule give some irreducible representation of its symmetry group; the multiplicity of the frequency determines the dimension of the representation. The irreducibility follows from the same considerations as were given in §96 for the solutions of Schrödinger's equation. The equality of the frequencies corresponding to two different irreducible representations would be an improbable coincidence. A reservation is again necessary: since the normal coordinates are by their physical nature real quantities, two complex conjugate representations correspond physically to one eigenfrequency of twice the multiplicity.

These considerations enable us to carry out a classification of the eigenvibrations of a molecule without solving the complex problem of actually determining its normal coordinates. To do so, we must first find (by the method described below) the representation given by all the vibrational coordinates together, which we shall call the *total vibrational* representation; this representation is reducible, and on decomposing it into irreducible parts we determine the multiplicities of the eigenfrequencies and the symmetry properties of the corresponding vibrations. Here it may happen that the same irreducible representation appears several times in the total representation; this means that there are several different frequencies of the same multiplicity and with vibrations of the same symmetry.

To find the total vibrational representation, we start from the fact that the characters of a representation are invariant with respect to a linear transformation of the base functions. Hence they can be calculated by using as base functions not the normal coordinates, but simply the components u_i of the vectors of the displacements of the nuclei from their equilibrium positions.

First of all, it is evident that, to calculate the character of some element G of a point group, we need consider only those nuclei which (or, more exactly, whose equilibrium positions) remain fixed under the given symmetry transformation. For if, under the rotation or reflection G in question, nucleus 1 is moved to a new position, previously occupied by a similar nucleus 2, this means that under the operation G a displacement of nucleus 1 is transformed into a displacement of nucleus 2. In other words, there will be no diagonal elements in the rows of the matrix G_{ik} which correspond to this nucleus (i.e. to its displacement u_i). The components of the displacement vector of a nucleus whose equilibrium position is not affected by the operation G, on the other hand, are evidently transformed into combinations of themselves, so that they may be considered independently of the displacement vectors of the remaining nuclei.

Let us first consider a rotation $C(\phi)$ through an angle ϕ about some symmetry axis. Let u_x, u_y, u_z be the components of the displacement vector of some nucleus, whose equilibrium position is on the axis, and hence is unaffected by the rotation. Under the rotation these components are transformed, like those of any ordinary (polar) vector, according to the formulae (the z-axis being the axis of symmetry)

$$u'_x = u_x \cos\phi + u_y \sin\phi,$$
$$u'_y = -u_x \sin\phi + u_y \cos\phi,$$
$$u'_z = u_z.$$

The character, i.e. the sum of the diagonal terms of the transformation matrix, is $1+2\cos\phi$. If altogether N_C nuclei lie on the axis in question, the total character is

$$N_C(1+2\cos\phi). \tag{100.3}$$

However, this character corresponds to the transformation of all the $3N$ displacements u_i; hence it is necessary to separate the part corresponding to the transformations of translation and (small) rotation of the molecule as a whole. The translation is determined by the displacement vector $\mathbf{U}$ of the centre of mass of the molecule; the corresponding part of the character is therefore $1+2\cos\phi$. The rotation of the molecule as a whole is determined by the vector $\delta\mathbf{\Omega}$ of the angle of rotation.† The vector $\delta\mathbf{\Omega}$ is axial, but with respect to rotations of the coordinate system an axial vector behaves like a polar vector. Hence a character of $1+2\cos\phi$ also corresponds to the vector $\delta\mathbf{\Omega}$. Altogether, therefore, we must subtract from (100.3) a quantity $2(1+2\cos\phi)$. Thus we finally have the character $\chi(C)$ of the rotation $C(\phi)$ in the total vibrational representation:

$$\chi(C) = (N_C-2)(1+2\cos\phi). \tag{100.4}$$

The character of the unit element is evidently just the total number of vibrational degrees of freedom: $\chi(E) = 3N-6$ (as is obtained from (100.4) when $N_C = N$, $\phi = 0$).

In a similar manner, we calculate the character of the rotary–reflection transformation $S(\phi)$ (a rotation through an angle ϕ about the z-axis and a reflection in the xy-plane). Here a vector is transformed according to the formulae

$$u'_x = u_x\cos\phi+u_y\sin\phi,$$
$$u'_y = -u_x\sin\phi+u_y\cos\phi,$$
$$u'_z = -u_z,$$

to which there corresponds a character $-1+2\cos\phi$. Hence the character of the representation given by all the $3N$ displacements u_i is

$$N_S(-1+2\cos\phi), \tag{100.5}$$

where N_S is the number of nuclei left unmoved by the operation $S(\phi)$; this number is evidently either none or one. To the vector $\mathbf{U}$ of the displacement of the centre of mass there corresponds a character $-1+2\cos\phi$. The vector $\delta\mathbf{\Omega}$ being an axial vector, is unchanged by an inversion of the coordinate system; on the other hand, the rotary–reflection transformation $S(\phi)$ can be represented in the form

$$S(\phi) = C(\phi)\sigma_h = C(\phi)C_2I = C(\pi+\phi)I,$$

i.e. as a rotation through an angle $\pi+\phi$, followed by an inversion. Hence the character of the transformation $S(\phi)$ applied to the vector $\delta\mathbf{\Omega}$ is equal to the character of the transformation $C(\pi+\phi)$ applied to an ordinary vector, i.e. it is $1+2\cos(\pi+\phi) = 1-2\cos\phi$. The sum $(-1+2\cos\phi)+(1-2\cos\phi) = 0$, so that we reach the conclusion that the expression (100.5) is equal to the required character $\chi(S)$ of the rotary–reflection transformation $S(\phi)$ in the total vibrational representation:

$$\chi(S) = N_S(-1+2\cos\phi). \tag{100.6}$$

In particular, the character of reflection in a plane ($\phi = 0$) is $\chi(\sigma) = N_\sigma$, while that of an inversion ($\phi = \pi$) is $\chi(I) = -3N_I$.

---

† As is well known, the angle of a small rotation can be regarded as a vector $\delta\mathbf{\Omega}$, whose modulus is equal to the angle of rotation and which is directed along the axis of rotation in the direction determined by the corkscrew rule. The vector $\delta\mathbf{\Omega}$ so defined is clearly axial.

Having thus determined the characters χ of the total vibrational representation, we have only to decompose it into irreducible representations, which is done by formula (94.16) and the character tables given in §95 (see the Problems at the end of the present section).

To classify the vibrations of a linear molecule there is no need to have recourse to group theory. The total number of vibrational degrees of freedom is $3N-5$. Among the vibrations, we must distinguish those in which the atoms remain in a straight line, and those where this does not happen.† The number of degrees of freedom in the motion of N particles in a straight line is N; of these, one corresponds to the translational motion of the molecule as a whole. Hence the number of normal coordinates of the vibrations which leave the atoms in a straight line is $N-1$; in general, $N-1$ different eigenfrequencies correspond to them. The remaining $(3N-5)-(N-1) = 2N-4$ normal coordinates relate to vibrations which destroy the collinearity of the molecule; to these, there correspond $N-2$ different double frequencies (two normal coordinates, corresponding to the same vibrations in two mutually perpendicular planes, belong to each frequency).‡

PROBLEMS

PROBLEM 1. Classify the normal vibrations of the molecule NH_3 (an equilateral triangular pyramid, with the N atom at the vertex and the H atoms at the corners of the base; Fig. 41).

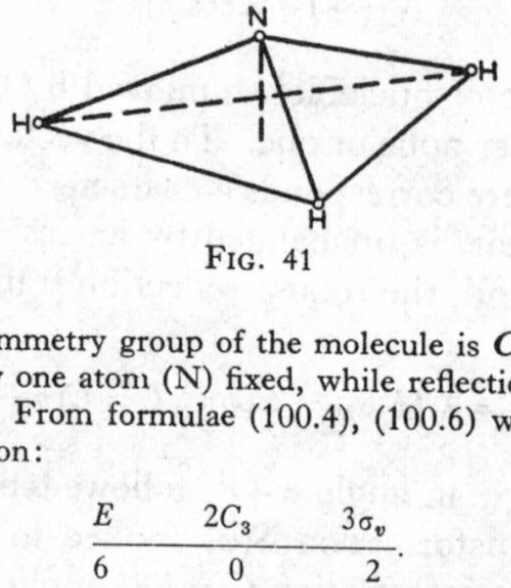

FIG. 41

SOLUTION. The point symmetry group of the molecule is C_{3v}. Rotations about an axis of the third order leave only one atom (N) fixed, while reflections in planes each leave two atoms fixed (N and one H). From formulae (100.4), (100.6) we find the characters of the total vibrational representation:

E	$2C_3$	$3\sigma_v$
6	0	2

Decomposing this representation into irreducible parts, we find that it contains the representations A_1 and E twice each. Thus there are two simple frequencies corresponding to vibrations of the type A_1, which conserve the complete symmetry of the molecule (what are called *totally symmetric* vibrations), and two double frequencies with corresponding normal coordinates which are transformed into combinations of each other by the representation E.

PROBLEM 2. The same as Problem 1, but for the molecule H_2O (Fig. 42).

SOLUTION. The symmetry group is C_{2v}. The transformation C_2 leaves the O atom fixed; the transformation σ_v (a reflection in the plane of the molecule) leaves all three atoms fixed;

† If the molecule is symmetrical about its centre, a further characteristic of the vibrations appears; see Problem 10 at the end of this section.

‡ Using the notation for the irreducible representations of the group $C_{\infty v}$ (see §98), we can say that there are $N-1$ vibrations of the type A_1, and $N-2$ of the type E_1.

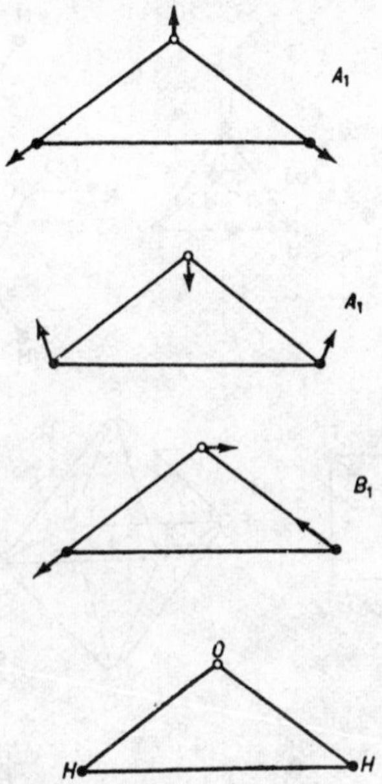

FIG. 42

the reflection σ'_v leaves only the O atom fixed. The characters of the total vibrational representation are

E	C_2	σ_v	σ'_v
3	1	3	1

This representation divides into the irreducible representations $2A_1$, $1B_1$, i.e. there are two totally symmetric vibrations and one with the symmetry given by the representation B_1; all the frequencies are simple. Fig. 42 shows the corresponding normal vibrations.

PROBLEM 3. The same as Problem 1, but for the molecule $CHCl_3$ (Fig. 43a).

SOLUTION. The symmetry group of the molecule is $\boldsymbol{C_{3v}}$. By the same method we find that there are three totally symmetric vibrations A_1 and three double vibrations of the type E.

PROBLEM 4. The same as Problem 1, but for the molecule CH_4 (the C atom is at the centre of a tetrahedron with the H atoms at the vertices; Fig. 43b).

SOLUTION. The symmetry of the molecule is $\boldsymbol{T_d}$. The vibrations are $1A_1$, $1E$, $2F_2$.

PROBLEM 5. The same as Problem 1, but for the molecule C_6H_6 (Fig. 43c).

SOLUTION. The symmetry of the molecule is $\boldsymbol{D_{6h}}$. The vibrations are $2A_{1g}$, $1A_{2g}$, $1A_{2u}$, $1B_{1g}$, $1B_{1u}$, $1B_{2g}$, $3B_{2u}$, $1E_{1g}$, $3E_{1u}$, $4E_{2g}$, $2E_{2u}$.

PROBLEM 6. The same as Problem 1, but for the molecule OsF_8 (the Os atom is at the centre of a cube with the F atoms at the vertices; Fig. 43d).

SOLUTION. The symmetry of the molecule is $\boldsymbol{O_h}$. The vibrations are $1A_{1g}$, $1A_{2u}$, $1E_g$, $1E_u$, $2F_{1u}$, $2F_{2g}$, $1F_{2u}$.

PROBLEM 7. The same as Problem 1, but for the molecule UF_6 (the U atom is at the centre of an octahedron with the F atoms at the vertices; Fig. 43e).

SOLUTION. The symmetry of the molecule is $\boldsymbol{O_h}$. The vibrations are $1A_{1g}$, $1E_g$, $2F_{1u}$, $1F_{2g}$, $1F_{2u}$.

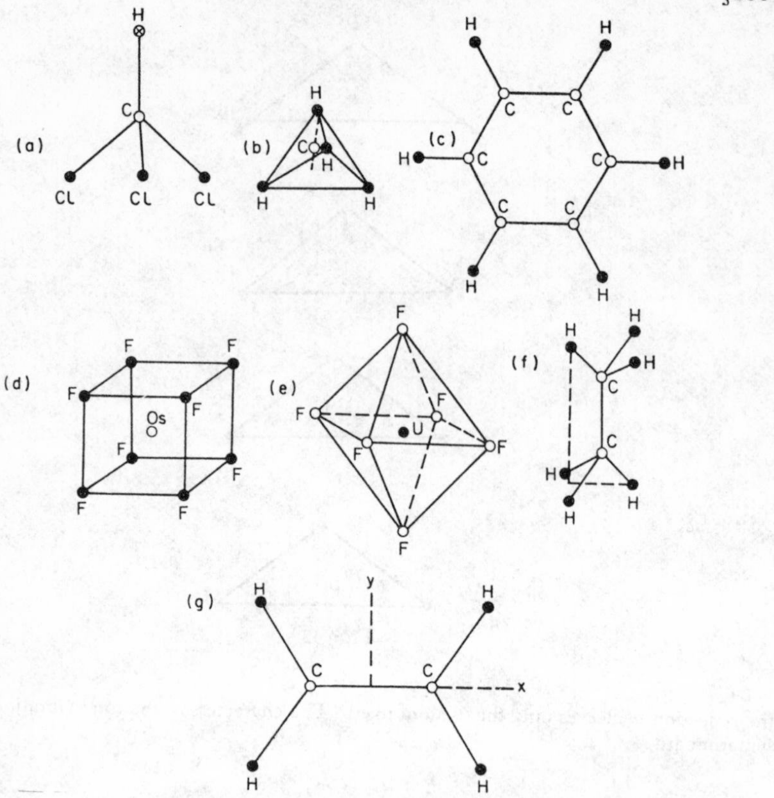

FIG. 43

PROBLEM 8. The same as Problem 1, but for the molecule C_2H_6 (Fig. 43f).

SOLUTION. The symmetry of the molecule is $\boldsymbol{D}_{3d}$. The vibrations are $3A_{1g}$, $1A_{1u}$, $2A_{2u}$, $3E_g$, $3E_u$.

PROBLEM 9. The same as Problem 1, but for the molecule C_2H_4 (Fig. 43g; all the atoms are coplanar).

SOLUTION. The symmetry of the molecule is $\boldsymbol{D}_{2h}$. The vibrations are $3A_{1g}$, $1A_{1u}$, $2B_{1g}$, $1B_{1u}$, $2B_{3u}$, $1B_{2g}$, $2B_{2u}$; the axes of coordinates are taken as shown in the figure.

PROBLEM 10. The same as Problem 1, but for a linear molecule of N atoms symmetrical about its centre.

SOLUTION. To the classification of the vibrations of a linear molecule considered in the text, we must add the classification from the behaviour with respect to inversion in the centre. There are two distinct cases, according as N is even or odd.

If N is even ($N = 2p$), there is no atom at the centre of the molecule. On giving to the p atoms in one half of the molecule independent displacements along the line, and to the remaining p atoms equal and opposite displacements, we find that p of the vibrations leaving the atoms in line are symmetrical with respect to the centre, while the remaining $(2p-1)-p = p-1$ vibrations of this type are antisymmetrical. Next, p atoms have $2p$ degrees of freedom for motions in which the atoms do not remain in line. On giving equal and opposite displacements to symmetrically placed atoms, we should obtain $2p$ symmetrical vibrations; of these, however, the two corresponding to a rotation of the molecule must be removed. Thus there are $p-1$ double frequencies of vibrations which bring the atoms out of line and are symmetrical about the centre, and the same number $[(2p-2)-(p-1) = p-1]$ which are antisymmetrical. Using the notation for the irreducible representations of the group $\boldsymbol{D}_{\infty h}$ (see the end of §98), we can say that there are p vibrations of the type A_{1g} and $p-1$ of the types A_{1u}, E_{1g}, E_{1u}.

If N is odd ($N = 2p+1$), similar arguments show that there are p vibrations of each of the types A_{1g}, A_{1u}, E_{1u} and $p-1$ of the type E_{1g}.

§101. Vibrational energy levels

From the viewpoint of quantum mechanics, the vibrational energy of a molecule is determined by the eigenvalues of the Hamiltonian

$$\hat{H}^{(v)} = \tfrac{1}{2} \sum_{\alpha} \sum_{i=1}^{f_\alpha} (\hat{P}_{\alpha i}{}^2 + \omega_\alpha{}^2 Q_{\alpha i}{}^2), \tag{101.1}$$

where $\hat{P}_{\alpha i} = -i\hbar\partial/\partial Q_{\alpha i}$ are the momentum operators corresponding to the normal coordinates $Q_{\alpha i}$. Since this Hamiltonian falls into the sum of independent terms $(\hat{P}_{\alpha i}{}^2 + \omega_\alpha{}^2 Q_{\alpha i}{}^2)$, the energy levels are given by the sums

$$E^{(v)} = \hbar \sum_{\alpha} \omega_\alpha \sum_{i} (v_{\alpha i} + \tfrac{1}{2}) = \sum_{\alpha} \hbar\omega_\alpha (v_\alpha + \tfrac{1}{2} f_\alpha), \tag{101.2}$$

where $v_\alpha = \Sigma\, v_{\alpha i}$, and f_α is the multiplicity of the frequency ω_α. The wave functions are given by the products of the corresponding wave functions for linear harmonic oscillators:

$$\psi = \prod_{\alpha} \psi_\alpha, \tag{101.3}$$

$$\text{where } \psi_\alpha = \text{constant} \times \exp\{-\tfrac{1}{2} c_\alpha{}^2 \sum_{i} Q_{\alpha i}{}^2\} \prod_{i} H_{v_{\alpha i}}(c_\alpha Q_{\alpha i}), \tag{101.4}$$

where H_v denotes the Hermite polynomial of order v, and $c_\alpha = \sqrt{(\omega_\alpha/\hbar)}$. If there are multiple frequencies among the ω_α, the vibrational energy levels are in general degenerate. The energy (101.2) depends only on the sums $v_\alpha = \Sigma\, v_{\alpha i}$. Hence the degree of degeneracy of the level is equal to the number of ways of forming the given set of numbers v_α from the $v_{\alpha i}$. For a single number v_α it is†

$$(v_\alpha + f_\alpha - 1)!/v_\alpha!(f_\alpha - 1)!.$$

Hence the total degree of degeneracy is

$$\prod_{\alpha} \frac{(v_\alpha + f_\alpha - 1)!}{v_\alpha!(f_\alpha - 1)!}. \tag{101.5}$$

For double frequencies, the factors in this product are $v_\alpha + 1$, while for triple frequencies they are $\tfrac{1}{2}(v_\alpha + 1)(v_\alpha + 2)$.

It must be borne in mind that this degeneracy occurs only so long as we consider purely harmonic vibrations. When terms of higher order in the normal coordinates are taken into account in the Hamiltonian (*anharmonic* vibrations), the degeneracy is in general removed, though not completely (see §104 for a further discussion of this point).

---

† This is the number of ways in which v_α balls can be distributed among f_α urns.

The wave functions (101.3) belonging to the same degenerate vibrational term give some representation (in general reducible) of the symmetry group of the molecule. The functions belonging to different frequencies are transformed independently of one another. Hence the representation given by all the functions (101.3) is the product of the representations given by the functions (101.4), so that we need consider only the latter.

The exponential factor in (101.4) is invariant with respect to all the symmetry transformations. In the Hermite polynomials, the terms of any given degree are transformed only into similar terms; a symmetry transformation evidently does not change the degree of any term. Since, on the other hand, each Hermite polynomial is completely determined by its highest term, it follows that it is sufficient to consider only the highest term, writing

$$\prod_{i=1}^{f_\alpha} H_{v_{\alpha i}}(c_\alpha Q_{\alpha i}) = \text{constant} \times Q_{\alpha 1}{}^{v_{\alpha 1}} Q_{\alpha 2}{}^{v_{\alpha 2}} \dots Q_{\alpha f_\alpha}{}^{v_{\alpha f_\alpha}}$$
$$+ \text{ terms of lower degree.}$$

The functions for which the sum $v_\alpha = \Sigma\, v_{\alpha i}$ has the same value belong to the same term. Thus we have a representation given by the products of v_α quantities $Q_{\alpha i}$; this is just the symmetric product (see §94) of the irreducible representation given by the $Q_{\alpha i}$ with itself v_α times (L. Tisza 1933).

For one-dimensional representations, the finding of the characters of their symmetric products with themselves v times is trivial:†

$$\chi_v(G) = [\chi(G)]^v.$$

For two- and three-dimensional representations it is convenient to use the following mathematical device.‡ The sum of the squared base functions of an irreducible representation is invariant with respect to all symmetry transformations. Hence we can formally regard these functions as the components of a vector in two or three dimensions, and the symmetry transformations as some rotations (or reflections) applied to these vectors. We emphasize that there is in general no relation between these rotations and reflections and the actual symmetry transformations, the former depending (for any given element G of the group) also on the particular representation considered.

Let us consider two-dimensional representations more closely. Let $\chi(G)$ be the character of some element of the group in the two-dimensional representation concerned, with $\chi(G) \neq 0$. The sum of the diagonal elements of the transformation matrix for the components x, y of a two-dimensional vector on rotation through an angle ϕ in a plane is $2\cos\phi$. Putting

$$2\cos\phi = \chi(G), \tag{101.6}$$

we find the angle of the rotation which formally corresponds to the element

† We use the notation $\chi_v(G)$ in place of the cumbersome $[\chi^v](G)$.

‡ It was applied to this problem by A. S. Kompaneets (1940).

G in the irreducible representation considered. The symmetric product of the representation with itself v times is the representation whose basis is formed by the $v+1$ quantities $x^v, x^{v-1}y, \ldots, y^v$. The characters of this representation are†

$$\chi_v(G) = \sin(v+1)\phi/\sin\phi. \tag{101.7}$$

The case where $\chi(G) = 0$ requires special consideration, since a zero character corresponds both to a rotation through $\frac{1}{2}\pi$ and to a reflection. If $\chi(G^2) = -2$, we have a rotation through $\frac{1}{2}\pi$, and for $\chi_v(G)$ we obtain

$$\chi_v(G) = -\tfrac{1}{2}[1+(-1)^v]. \tag{101.8}$$

If $\chi(G^2) = 2$, on the other hand, $\chi(G)$ must be regarded as the character of a reflection (i.e. a transformation $x \to x, y \to -y$); then

$$\chi_v(G) = \tfrac{1}{2}[1+(-1)^v]. \tag{101.9}$$

We can similarly obtain the formulae for the symmetric products of three-dimensional representations. The finding of the rotation or reflection which formally corresponds to an element of the group in a given representation is easily accomplished with the aid of Table 7 (§95). This is the transformation which corresponds to the given $\chi(G)$ in that isomorphous group in which the coordinates are transformed by the representation in question. Thus, for the representation F_1 of the groups $\boldsymbol{O}$ and $\boldsymbol{T}_d$ we must take a transformation from the group $\boldsymbol{O}$, but for the representation F_2 we must take one from the group $\boldsymbol{T}_d$. We shall not pause here to derive the corresponding formulae for the characters $\chi_v(G)$.

§102. Stability of symmetrical configurations of the molecule

For a symmetrical position of the nuclei, an electron term of the molecule may be degenerate, if there are among the irreducible representations of the symmetry group one or more whose dimensions exceed unity. We may ask whether such a symmetrical configuration is a stable equilibrium configuration of the molecule. Here we shall neglect the effect of spin (if any), which is usually insignificant in polyatomic molecules. The degeneracy of the electron terms of which we shall speak is therefore only the orbital degeneracy, and is unrelated to the spin.

If the configuration in question is stable, the energy of the molecule as a function of the distances between the nuclei must be a minimum for the given position of the nuclei. This means that the change in the energy due to a small displacement of the nuclei must contain no terms linear in the displacements.

† For purposes of calculation it is convenient to take the base functions in the form

$$(x+iy)^v, (x+iy)^{v-1}(x-iy), \ldots, (x-iy)^v;$$

the matrix of the rotation is then diagonal, and the sum of the diagonal elements takes the form

$$e^{iv\phi}+e^{i(v-2)\phi}+\ldots+e^{-iv\phi}.$$

Let $\hat{H}$ be the Hamiltonian of the electron state of the molecule, the distances between the nuclei being regarded as parameters. We denote by $\hat{H}_0$ the value of this Hamiltonian for the symmetrical configuration considered. The quantities defining the small displacements of the nuclei can be taken as the normal vibrational coordinates $Q_{\alpha i}$. The expansion of $\hat{H}$ in powers of the $Q_{\alpha i}$ is of the form

$$\hat{H} = \hat{H}_0 + \sum_{\alpha,i} V_{\alpha i} Q_{\alpha i} + \sum_{\alpha,\beta,i,k} W_{\alpha i,\beta k} Q_{\alpha i} Q_{\beta k} + \dots . \tag{102.1}$$

The expansion coefficients V, W, ... are functions only of the coordinates of the electrons. Under a symmetry transformation, the quantities $Q_{\alpha i}$ are transformed into combinations of one another, and the sums in (102.1) are changed into other sums of the same form. Hence we can formally regard the symmetry transformation as a transformation of the coefficients in these sums, the $Q_{\alpha i}$ remaining unchanged. Here, in particular, the coefficients $V_{\alpha i}$ (for any given α) will be transformed by the same representation of the symmetry group as the corresponding coordinates $Q_{\alpha i}$. This follows at once from the fact that, by virtue of the invariance of the Hamiltonian under all symmetry transformations, the group of terms of any given order in its expansion must be invariant also, and in particular the linear terms must be invariant.†

Let us consider some electron term E_0 which is degenerate in the symmetrical configuration. A displacement of the nuclei which destroys the symmetry of the molecule generally results in a splitting of the term. The amount of the splitting is determined, as far as terms of the first order in the displacements of the nuclei, by the secular equation formed from the matrix elements of the linear term in the expansion (102.1),

$$V_{\rho\sigma} = \sum_{\alpha,i} Q_{\alpha i} \int \psi_\rho V_{\alpha i} \psi_\sigma \, dq, \tag{102.2}$$

where ψ_ρ, ψ_σ are the wave functions of electron states belonging to the degenerate term in question (and are chosen to be real). The stability of the symmetrical configuration requires that the splitting linear in Q should be zero, i.e. all the roots of the secular equation must vanish identically. This means that the matrix $V_{\rho\sigma}$ must itself be zero. Here, of course, we must consider only those normal vibrations which destroy the symmetry of the molecule, i.e. we must omit the totally symmetric vibrations (which correspond to the unit representation of the group).

Since the $Q_{\alpha i}$ are arbitrary, the matrix elements (102.2) vanish only if all the integrals

$$\int \psi_\rho V_{\alpha i} \psi_\sigma \, dq \tag{102.3}$$

† Strictly speaking, the quantities $V_{\alpha i}$ must be transformed by the representation which is the complex conjugate of that by which the $Q_{\alpha i}$ are transformed. However, as we have already pointed out, if two complex conjugate representations are not the same, they must physically be considered together as one representation of twice the dimension. The above remark is therefore unimportant.

vanish. Let D^{el} be the irreducible representation by which the electron wave functions ψ_ρ are transformed, and D_α the same for the quantities $V_{\alpha i}$; as we have already remarked, the representations D_α are those by which the corresponding normal coordinates $Q_{\alpha i}$ are transformed. According to the results of §97, the integrals (102.3) will be non-zero if the product $[D^{(el)2}] \times D_\alpha$ contains the unit representation or, what is the same thing, if $[D^{(el)2}]$ contains D_α. Otherwise all the integrals vanish.

Thus a symmetrical configuration is stable if the representation $[D^{(el)2}]$ does not contain any (except the unit representation) of the irreducible representations D_α which characterize the vibrations of the molecule. This condition is always satisfied for non-degenerate electron states, since the symmetric product of a one-dimensional representation with itself is the unit representation.

Let us consider, for instance, a molecule of the type CH_4, in which one atom (C) is at the centre of a tetrahedron, with four atoms (H) at the vertices. This configuration has the symmetry T_d. The degenerate electron terms correspond to the representations E, F_1, F_2 of this group. The molecule has one normal vibration A_1 (a totally symmetric vibration), one double vibration E, and two triple vibrations F_2 (see §100, Problem 4). The symmetric products of the representations E, F_1, F_2 with themselves are

$$[E^2] = A_1 + E, \qquad [F_1^2] = [F_2^2] = A_1 + E + F_2.$$

We see that each of these contains at least one of the representations E, F_2, and hence the tetrahedral configuration considered is unstable when there are degenerate electron states.

This result constitutes a general rule, the *Jahn–Teller theorem* (H. A. Jahn and E. Teller 1937): when there is a degenerate electron state, any symmetrical position of the nuclei (except when they are collinear) is unstable. As a result of this instability, the nuclei move in such a way that the symmetry of their configuration is destroyed, the degeneracy of the term being completely removed. In particular, we can say that the normal electron term of a symmetrical (non-linear) molecule can only be non-degenerate.†

As we have just mentioned, the linear molecules alone form an exception. This is easily seen, without using group theory. A displacement of a nucleus whereby it moves off the axis of the molecule is an ordinary vector with ξ and η components (the ζ-axis being along the axis of the molecule). We have seen in §87 that such vectors have matrix elements only for transitions in which the angular momentum Λ about the axis changes by unity. On the other hand, to a degenerate term of a linear molecule there correspond states with angular momenta Λ and $-\Lambda$ about the axis ($\Lambda \geqslant 1$). A transition between them changes the angular momentum by at least 2, and therefore the

† The physical idea of destruction of symmetry in an electron state that is degenerate because of the same symmetry is due to L. D. Landau (1934). Jahn and Teller proved the theorem by considering all possible types of symmetrical configuration of the nuclei in the molecule and examining each one by the method given above.

matrix elements always vanish. Thus the linear position of the nuclei in the molecule may be stable, even if the electron state is degenerate.

A constructive general proof of the theorem is based on the following consideration (E. Ruch 1957). The degeneracy of electron states due to the symmetry of the configuration of the nuclei can exist only in point symmetry groups of the molecule which include at least one rotation (C_n) or rotary-reflection (S_n) axis of order $n > 2$. Then the wave functions of mutually degenerate states (i.e. the base functions of the corresponding representation $D^{(el)}$) include at least one for which the electron density $\rho = |\psi|^2 = \psi^2$ is not invariant under rotations about this axis; the electric field due to the electrons, like the electron density, will not be symmetrical about the axis. In a (non-linear) molecule, there are equivalent nuclei, not on the axis, which move to one another's positions in the rotation C_n or S_n. Thus equivalent nuclei lie at points that are not equivalent as regards the electric field. But an equivalence of the equilibrium positions of charged particles in a field that is not required by the symmetry of the field is impossible, in the sense that it can only result from an unlikely coincidence.

The systematic proof is a mathematical embodiment of this physical situation. We shall indicate the structure of the proof (E. Ruch and A. Schönhofer 1965).

Let us consider (in a non-linear molecule) some nucleus a that is not at the "centre" of the molecule (i.e. not at the fixed point for the transformations in its symmetry group) and not on the principal axis of symmetry, if any.† Let $\boldsymbol{H}$ be the set of all symmetry transformations of the molecule that leave the nucleus a fixed; $\boldsymbol{H}$ is a sub-group of the total symmetry group $\boldsymbol{G}$ of the molecule, and may be one of the point groups $\boldsymbol{C}_1$, $\boldsymbol{C}_s$, $\boldsymbol{C}_n$, $\boldsymbol{C}_{nv}$. The transformations in $\boldsymbol{G}$ that are not in $\boldsymbol{H}$ move the nucleus a to the positions of other equivalent nuclei a', a'', ... ; let s be the number of nuclei in this set. It is evident that the order of the sub-group $\boldsymbol{H}$ is g/s, where g is the order of the whole group $\boldsymbol{G}$ (i.e. s is the index of the sub-group $\boldsymbol{H}$ in the group $\boldsymbol{G}$).‡

The number s is certainly at least 3, since for the existence of an irreducible representation $D^{(el)}$ with dimension exceeding unity it is necessary (as already mentioned) that there should be at least one axis of symmetry of higher than the second order, and the nucleus a is not on this axis, by the condition stated.

The representation $D^{(el)}$ of the group $\boldsymbol{G}$ is in general reducible with respect to the group $\boldsymbol{H}$, which has a lower symmetry. Let us suppose that its decomposition into irreducible representations of $\boldsymbol{H}$ includes a representation $d^{(el)}$ of dimension unity. This is given by an electron wave function ψ, one of the base functions of the representation $D^{(el)}$. Since the dimension of $d^{(el)}$ is unity, $\rho = \psi^2$ is invariant under all transformations in $\boldsymbol{H}$, i.e. it gives a unit irreducible representation of this group.

† By the principal axis we mean (in symmetry groups other than cubic and icosahedron) the axis C_n or S_n with $n > 2$.

‡ All the elements of the group $\boldsymbol{G}$ can be divided into s cosets $\boldsymbol{H}$, $G'\boldsymbol{H}$, $G''\boldsymbol{H}$, ..., where G', G'', ..., are the elements that move the nucleus a to a', a'', ...

Such a unit representation of $\boldsymbol{H}$ can also be obtained by taking as the basis one of the displacements Q_a of the atom a: the displacement along the position vector of the nucleus a from the centre of the molecule.

By applying to this displacement all the operations in the group $\boldsymbol{G}$, we obtain the basis of a representation D_Q (in general reducible) of this group. Since every transformation in $\boldsymbol{G}$ that is not in $\boldsymbol{H}$ changes the displacement Q_a into a displacement of one of the other $s-1$ equivalent nuclei a', a'', ..., and the displacements of different nuclei are of course linearly independent, the dimension of D_Q is s. The displacements Q_a, $Q_{a'}$, ... which form the basis of D_Q certainly cannot correspond to either translation or rotation of the molecule as a whole: if there are three or more equivalent nuclei, radial displacements of them cannot be combined to form such displacements of the molecule.

In a similar way, we can find a representation D_ρ of the group $\boldsymbol{G}$ by applying all its transformations to the function $\rho = \psi^2$. The dimension of D_ρ may be s, but may also be less, since there is no reason to suppose that all the s functions ρ, $G'\rho$, $G''\rho$, ... are linearly independent. We can, however, say that the representation D_ρ, if not the same as D_Q, will always be entirely contained in D_Q.† Moreover, it is not the unit representation, since ψ^2 is certainly not invariant with respect to the whole group $\boldsymbol{G}$; only the sum of the squares of the base functions of the irreducible representation $D^{(el)}$, with dimension exceeding unity, is invariant.

These properties of the representations D_Q and D_ρ give the required result immediately: D_Q is part of the total vibrational representation, and D_ρ is part of the representation $[D^{(el)2}]$, not containing the unit representation. The fact that D_Q contains D_ρ therefore means that $[D^{(el)2}]$ contains at least one of the non-unit vibrational representations D_α, as was to be proved.

In these arguments, however, it has also been assumed that the decomposition of $D^{(el)}$ into irreducible representations of the sub-group $\boldsymbol{H}$ includes a representation of dimension unity. This assumption is correct in the great majority of cases. For example, it is certainly correct if $\boldsymbol{H} = \boldsymbol{C}_1, \boldsymbol{C}_s, \boldsymbol{C}_2, \boldsymbol{C}_{2v}$ (since all irreducible representations of these groups have dimension unity). It is certainly correct also if $\boldsymbol{H} = \boldsymbol{C}_n, \boldsymbol{C}_{nv}$ with $n > 2$, provided that the dimension of $D^{(el)}$ is odd (since the groups $\boldsymbol{C}_n$ and $\boldsymbol{C}_{nv}$ have only irreducible representations of dimension 1 or 2). An examination of the character tables of the irreducible representations of point groups shows that an exception occurs for the two-dimensional representations of the cubic groups $\boldsymbol{G} = \boldsymbol{O}, \boldsymbol{T}_d, \boldsymbol{O}_h$, with respect to the sub-groups $\boldsymbol{H} = \boldsymbol{C}_4, \boldsymbol{C}_{4v}$.

Let us take the particular case of $\boldsymbol{G} = \boldsymbol{O}$ and $\boldsymbol{H} = \boldsymbol{C}_4$; this affects only

† The significance of this statement is as follows. Let one representation (dimension f) of the sub-group $\boldsymbol{H}$ be given by different sets of base functions, and let one set, when all the transformations in $\boldsymbol{G}$ are applied to it, give a representation of $\boldsymbol{G}$ with dimension sf, where s is the index of the sub-group $\boldsymbol{H}$ in the group $\boldsymbol{G}$. Then we can say that the representation of the group $\boldsymbol{G}$ obtained by the same method from any other such set of functions is either the same as the first or entirely contained in it. A rigorous proof is given by E. Ruch and A. Schönhofer, *Theoretica Chimica Acta* **3**, 291, 1965.

the naming of the representations. The two electron functions ψ_1, ψ_2 give the representation $D^{(el)} = E$ of the group $\boldsymbol{O}$, and the representation $d^{(el)} = E$ of the sub-group $\boldsymbol{C}_4$; the representation of the latter given by the products $\psi_1{}^2$, $\psi_2{}^2$, $\psi_1\psi_2$, is $[E^2] = A+E$. A similar representation of $\boldsymbol{C}_4$ is obtained with the three components of the vectors of any displacement Q_a of the nucleus a as basis. The representation D_ρ of $\boldsymbol{O}$ is in this case $[D^{(el)2}] = A_1+E$, and does not contain the representation F_2 corresponding to the vector of translation or rotation of the molecule as a whole; it contains both the unit and a non-unit representation. Hence the fact that D_ρ is (for the same reasons as previously) contained in D_Q (whose dimension here is $3s$) proves that in this case also the molecule is unstable.†

In accordance with the remark at the beginning of this section, the whole of the above discussion has regarded the degeneracy of the electron states as being purely orbital in origin. It may be mentioned, however, that the Jahn–Teller theorem remains valid even when the spin–orbit and spin–spin interactions are taken into account, the only difference being that in (non-linear) molecules with half-integral spin the Kramers double degeneracy does not cause instability, in accordance with the general theorem proved in §60. The latter case corresponds to the two-dimensional two-valued irreducible representations of the double point groups. The absence of instability in this case may be seen by the following formal argument. To determine the selection rules for matrix elements (102.3) in the case of two-valued representations $D^{(el)}$, we must consider the antisymmetric products $\{D^{(el)2}\}$, not the symmetric ones (see §99). But, for every two-valued irreducible representation with dimension 2, these products are the unit representation, i.e. they certainly do not contain representations corresponding to any vibrations of the molecule that are not totally symmetrical.

§103. Quantization of the rotation of a top

The investigation of the rotational levels of a polyatomic molecule is often hampered by the necessity of considering the rotation simultaneously with the vibrations. As a preliminary example, let us consider the rotation of a molecule as a solid body, i.e. with the atoms "rigidly fixed" (a *top*).

Let ξ, η, ζ be a system of coordinates with axes along the three principal axes of inertia of a top, and rotating with it. The corresponding Hamiltonian is obtained by replacing the components J_ξ, J_η, J_ζ of the angular momentum of the rotation, in the classical expression for the energy, by the corresponding operators:

$$\hat{H} = \tfrac{1}{2}\hbar^2\left(\frac{\hat{J}_\xi{}^2}{I_A}+\frac{\hat{J}_\eta{}^2}{I_B}+\frac{\hat{J}_\zeta{}^2}{I_C}\right), \tag{103.1}$$

where I_A, I_B, I_C are the principal moments of inertia of the top.

† One further exceptional case comprises the four-dimensional representations of the icosahedron groups. This is treated similarly, and gives the same result.

The commutation rules for the operators $\hat{J}_\xi$, $\hat{J}_\eta$, $\hat{J}_\zeta$ of the angular momentum components in a rotating system of coordinates are not obvious, since the usual derivation of the commutation rules relates to the components $\hat{J}_x$, $\hat{J}_y$, $\hat{J}_z$ in a fixed system of coordinates. They are, however, easily obtained by using the formula

$$(\hat{\mathbf{J}}\cdot\mathbf{a})(\hat{\mathbf{J}}\cdot\mathbf{b})-(\hat{\mathbf{J}}\cdot\mathbf{b})(\hat{\mathbf{J}}\cdot\mathbf{a}) = -i\hat{\mathbf{J}}\cdot\mathbf{a}\times\mathbf{b}, \tag{103.2}$$

where **a**, **b** are any two commuting vectors which characterize the body in question. This formula is easily verified by calculating the left-hand side of the equation in the fixed system of coordinates x, y, z, using the general rules for the commutation of angular momentum components with one another and with the components of an arbitrary vector.

Let **a** and **b** be unit vectors along the ξ and η axes. Then $\mathbf{a}\times\mathbf{b}$ is a unit vector along the ζ-axis, and (103.2) gives

$$\hat{J}_\xi\hat{J}_\eta-\hat{J}_\eta\hat{J}_\xi = -i\hat{J}_\zeta. \tag{103.3}$$

Two other relations are obtained similarly. Thus the commutation rules for the operators of the angular momentum components in the rotating system of coordinates differ from those in the fixed system only in the sign on the right-hand side of the equation.† Hence it follows that all the results which we have previously obtained from the commutation rules, relating to the eigenvalues and matrix elements, hold for J_ξ, J_η, J_ζ also, with the difference that all expressions must be replaced by their complex conjugates. In particular, the eigenvalues of J_ζ (which will be denoted in this section by k, whereas the eigenvalues of J_z are denoted by M) take the values $k = -J$, ..., $+J$, where J, an integer, is the magnitude of the angular momentum of the top.

The spherical top

The finding of the eigenvalues of the energy of a rotating top is simplest for the case where all three principal moments of inertia of the body are equal: $I_A = I_B = I_C \equiv I$. This holds for a molecule in cases where it has the symmetry of one of the cubic point groups. The Hamiltonian (103.1) takes the form

$$\hat{H} = \hbar^2\hat{\mathbf{J}}^2/2I,$$

and its eigenvalues are

$$E = \hbar^2 J(J+1)/2I. \tag{103.4}$$

Each of these energy levels is degenerate with respect to the $2J+1$ directions of the angular momentum relative to the body itself (i.e. with respect to the values of $J_\zeta = k$).‡

† This expresses the fact that, as regards its effect on the wave function of the top, a rotation of the system x, y, z is equivalent to an opposite rotation of the system ξ, η, ζ.

‡ Here and subsequently we ignore the $(2J+1)$-fold degeneracy with respect to the directions of the angular momentum relative to a fixed coordinate system. This degeneracy always occurs, and is not physically important. If it is included, the total degree of degeneracy of the energy levels of a spherical top is $(2J+1)^2$.

THE SYMMETRICAL TOP

There is also no difficulty in calculating the energy levels in the case where only two of the moments of inertia of the top are the same: $I_A = I_B \neq I_C$. This holds for molecules having one axis of symmetry of order above the second. The Hamiltonian (103.1) takes the form

$$\hat{H} = \hbar^2(\hat{J}_\xi^2+\hat{J}_\eta^2)/2I_A+\hbar^2\hat{J}_\zeta^2/2I_C$$

$$= \hbar^2\hat{\mathbf{J}}^2/2I_A+\tfrac{1}{2}\hbar^2\left(\frac{1}{I_C}-\frac{1}{I_A}\right)\hat{J}_\zeta^2. \tag{103.5}$$

Hence we see that, in a state with given values of J and k, the energy is

$$E = \frac{\hbar^2}{2I_A}J(J+1)+\tfrac{1}{2}\hbar^2\left(\frac{1}{I_C}-\frac{1}{I_A}\right)k^2, \tag{103.6}$$

which determines the energy levels of a symmetrical top.

The degeneracy with respect to values of k which occurred for a spherical top is here partly removed. The values of the energy are the same only for values of k differing in sign alone, corresponding to opposite directions of the angular momentum relative to the axis of the top. Thus the energy levels of a symmetrical top are (for $k \neq 0$) doubly degenerate.

The stationary states of a symmetrical top are thus described by three quantum numbers: the angular momentum J and its components along the axis of the top ($J_\zeta = k$) and along the z-axis fixed in space ($J_z = M$); the energy of the top is independent of the last of these. The simultaneous measurability of the angular momentum and its components along an axis fixed in space and along an axis rigidly attached to a physical system† follows since the operators $\hat{\mathbf{J}}^2$ and $\hat{J}_z$ commute not only with each other but also with the operator $\hat{J}_\zeta = \hat{\mathbf{J}}\cdot\mathbf{n}$, where $\mathbf{n}$ is a unit vector along the ζ-axis. This is easily shown by direct calculation, but is also obvious *a priori*: the angular momentum operator is that of an infinitesimal rotation, and the scalar product $\mathbf{J}\cdot\mathbf{n}$ of two vectors fixed to the top is invariant under any rotation of the coordinate axes.

The determination of the wave functions of the stationary states of a symmetrical top therefore amounts to finding the common eigenfunctions of the operators $\hat{\mathbf{J}}^2$, $\hat{J}_z$ and $\hat{J}_\zeta$. This is in turn mathematically dependent on the law of transformation of the angular momentum eigenfunctions under finite rotations. With a change in the notation for the quantum numbers, we can write this law (58.7) as

$$\psi_{JM} = \sum_k D_{kM}^{(J)}(\alpha, \beta, \gamma)\psi_{Jk}. \tag{103.7}$$

We shall take ψ_{JM} to be the wave function of a state of the top described in terms of the fixed coordinates x, y, z, and the ψ_{Jk} to be the wave functions of states described in terms of the axes ξ, η, ζ attached to the top. In coordinates

† Not to be confused with the components (not simultaneously measurable) along two axes fixed in space.

rigidly attached to a physical system (such as the top), the ψ_{Jk} have definite values $\psi_{Jk}^{(0)}$ independent of the spatial orientation of the system. Formula (103.7) gives the angle dependence of the ψ_{JM}. Let the state $|JM\rangle$ also have a definite value k of the angular-momentum component along the ζ-axis. This means that only the $\psi_{Jk}^{(0)}$ with that k will be non-zero. Then the sum in (103.7) reduces to one term:

$$\psi_{JMk} = \psi_{Jk}^{(0)} D_{kM}^{(J)}(\alpha, \beta, \gamma).$$

This gives the dependence of the wave functions of the states $|JMk\rangle$ on the Eulerian angles, which define the rotation of the axes of the top with respect to the fixed axes. Normalizing the wave function by the condition

$$\int |\psi_{JMk}|^2 \sin\beta \, d\alpha \, d\beta \, d\gamma = 1,$$

we have

$$\psi_{JMk} = i^J \sqrt{\frac{2J+1}{8\pi^2}} D_{kM}^{(J)}(\alpha, \beta, \gamma); \qquad (103.8)$$

the phase factor is chosen so that, for $k = 0$, the function (103.8) becomes the eigenfunction of the free (not attached to the ζ-axis) integral angular momentum J with component M, i.e. the ordinary (spherical harmonic) function; cf. (58.25).†

The asymmetrical top

For $I_A \neq I_B \neq I_C$, the calculation of the energy levels in a general form is impossible. The degeneracy with respect to the directions of the angular momentum relative to the top is here removed completely, so that $2J+1$ different non-degenerate levels correspond to any given J. To calculate these levels (for a given J) we have to start from Schrödinger's equation written in matrix form (O. Klein 1929). This is done as follows.

The wave functions ψ_{Jk} of states of the top with definite values of J and the ζ-component of the angular momentum are the functions (103.8) derived above (we shall omit for brevity the suffix M giving the z-component of the angular momentum, since the energy does not depend on this); in these states the energy of the asymmetrical top does not have definite values. In the stationary states, on the other hand, the component J_ζ does not have definite values, i.e. no definite values of k can be assigned to the energy levels. The wave functions of these states are sought as linear combinations

$$\psi_J = \sum_k c_k \psi_{Jk}; \qquad (103.9)$$

† The direct derivation of (103.8) without the use of the theory of finite rotations is given in Problem 1. See §§110 and 87 for the calculation of the matrix elements of various quantities with respect to the wave functions (103.8); the corresponding formulae differ from those for a diatomic molecule (without spin) only in the nomenclature of the quantum numbers (see the second footnote to §82).

it is assumed that all functions have some common value of M. Substitution in Schrödinger's equation $\hat{H}\psi_J = E_J\psi_J$ leads to the equations

$$\sum_{k'} (\langle Jk|H|Jk'\rangle - E\delta_{kk'})c_{k'} = 0, \tag{103.10}$$

and the condition for these to have a solution leads to the secular equation

$$|\langle Jk|H|Jk'\rangle - E\delta_{kk'}| = 0. \tag{103.11}$$

The roots of this equation give the energy levels of the top, and then the equations (103.10) give the linear combinations (103.9) which diagonalize the Hamiltonian, i.e. the wave functions of the stationary states of the top with a given value of J (and of M). The calculation of the matrix elements of any physical quantity with respect to these wave functions thus reduces to that of the matrix elements of the symmetrical top.

The operators $\hat{J}_\xi$, $\hat{J}_\eta$ have matrix elements only for transitions in which k changes by unity, while $\hat{J}_\zeta$ has only diagonal elements (see formulae (27.13) in which we must write J, k instead of L, M). Hence the operators $\hat{J}_\xi^2$, $\hat{J}_\eta^2$, $\hat{J}_\zeta^2$, and therefore $\hat{H}$, have matrix elements only for transitions with $k \to k$ or $k \pm 2$. The absence of matrix elements for transitions between states with even and odd k has the result that the secular equation of degree $2J+1$ immediately falls into two independent equations of degrees J and $J+1$. One of these contains matrix elements for transitions between states with even k, and the other contains those for transitions between states with odd k. Each of these equations, in turn, can be reduced to two equations of lower degree. To do this, we must use the matrix elements defined, not with respect to the functions ψ_{Jk}, but with respect to the functions

$$\left.\begin{aligned} \psi_{Jk}^{+} &= (\psi_{Jk}+\psi_{J,-k})/\sqrt{2}, \qquad \psi_{J0}^{+} = \psi_{J0}, \\ \psi_{Jk}^{-} &= (\psi_{Jk}-\psi_{J,-k})/\sqrt{2} \qquad (k \neq 0). \end{aligned}\right\} \tag{103.12}$$

Functions differing in the index + and − are of different symmetry (with respect to a reflection in a plane passing through the ζ-axis, which changes the sign of k), and hence the matrix elements for transitions between them vanish. Consequently we can form the secular equations separately for the + and − states.

The Hamiltonian (103.1), with the commutation rules (103.3), has a particular symmetry: it is invariant with respect to a simultaneous change in sign of any two of the operators $\hat{J}_\xi$, $\hat{J}_\eta$, $\hat{J}_\zeta$. This symmetry formally corresponds to the group $\boldsymbol{D}_2$. Hence the levels of an asymmetrical top can be classified in accordance with the irreducible representations of this group. Thus there are four types of non-degenerate level, corresponding to the representations A, B_1, B_2, B_3 (see Table 7, §95).

It is easy to establish which states of the asymmetrical top belong to each of these types. To do so, we must find the symmetry properties of the ψ_{Jk} and the functions (103.12). This could be done directly from (103.8), but it is simpler to begin from the more usual spherical harmonic functions, noting that, as regards their symmetry properties, the wave functions of states with definite values of the ζ-component of the angular momentum are the same as the angular-momentum eigenfunctions

$$\psi_{Jk} \sim Y_{Jk}^*(\theta, \phi) \sim e^{-ik\phi}\Theta_{Jk}(\theta), \tag{103.13}$$

where θ and ϕ are the spherical angles in the axes ξ, η, ζ, and the symbol $\sim$ stands for the words "is transformed as"; the complex conjugate is taken in (103.13) because of the change in sign on the right of the commutation relations (103.3).

A rotation through an angle π about the ζ-axis (i.e. the symmetry operation $C_2^{(\zeta)}$) multiplies the function (103.13) by $(-1)^k$:

$$C_2^{(\zeta)}:\ \psi_{Jk} \rightarrow (-1)^k \psi_{Jk}.$$

The operation $C_2^{(\eta)}$ may be regarded as the result of successively performing an inversion and a reflection in the $\xi\zeta$-plane; the first operation multiplies ψ_{Jk} by $(-1)^J$, and the second (a change in sign of ϕ) is equivalent to changing the sign of k. Using the definition (28.6) of the functions $\Theta_{J,-k}$, we therefore have

$$C_2^{(\eta)}:\ \psi_{Jk} \rightarrow (-1)^{J+k} \psi_{J,-k}.$$

Finally, the operation $C_2^{(\xi)} = C_2^{(\eta)} C_2^{(\zeta)}$ gives

$$C_2^{(\xi)}:\ \psi_{Jk} \rightarrow (-1)^J \psi_{J,-k}.$$

Using these transformation rules, we find that the states corresponding to the functions (103.12) belong to the following types of symmetry:

$$\left.\begin{array}{l} \psi_{Jk}^{+}\left\{\begin{array}{lll} J \text{ even}, & k \text{ even} & A \\ J \text{ even}, & k \text{ odd} & B_3 \\ J \text{ odd}, & k \text{ even} & B_1 \\ J \text{ odd}, & k \text{ odd} & B_2 \end{array}\right. \\ \psi_{Jk}^{-}\left\{\begin{array}{lll} J \text{ even}, & k \text{ even} & B_1 \\ J \text{ even}, & k \text{ odd} & B_2 \\ J \text{ odd}, & k \text{ even} & A \\ J \text{ odd}, & k \text{ odd} & B_3 \end{array}\right. \end{array}\right\} \tag{103.14}$$

By simple counting it is easy to find the number of states of each type for a given value of J. The following numbers of states correspond to the types A and each of B_1, B_2, B_3:

	A	B_1, B_2, B_3	
J even	$\frac{1}{2}J+1$	$\frac{1}{2}J$	(103.15)
J odd	$\frac{1}{2}J-\frac{1}{2}$	$\frac{1}{2}J+\frac{1}{2}$	

For the asymmetrical top there are selection rules for the matrix elements of transitions between states of the types A, B_1, B_2, B_3; these rules are easily obtained from symmetry considerations in the usual way. Thus, for the components of a vector physical quantity **A** we have the selection rules

$$\left.\begin{array}{lll} \text{for } A_\xi: & A \leftrightarrow B_3^{(\xi)}, & B_1^{(\zeta)} \leftrightarrow B_2^{(\eta)}, \\ \text{for } A_\eta: & A \leftrightarrow B_2^{(\eta)}, & B_1^{(\zeta)} \leftrightarrow B_3^{(\xi)}, \\ \text{for } A_\zeta: & A \leftrightarrow B_1^{(\zeta)}, & B_2^{(\eta)} \leftrightarrow B_3^{(\xi)}. \end{array}\right\} \quad (103.16)$$

For clarity we show, as an index to the symbol for the representation, the axis about which a rotation has the character +1 in the representation concerned.

PROBLEMS

PROBLEM 1. Find the wave functions of the states $|JMk\rangle$ of a symmetrical top by direct calculation as the eigenfunctions of the operators $\hat{\mathbf{J}}^2$, $\hat{J}_z$, $\hat{J}_\zeta$ (F. Reiche and H. Rademacher 1926).

SOLUTION. In order to obtain the ψ_{JMk} as functions of the Eulerian angles α, β, γ, we must express in terms of them the operators of the angular momentum components along fixed axes x, y, z. Since the operator of such a component along any axis is $-i\partial/\partial\phi$, where ϕ is the angle of rotation about this axis, we can write

$$\hat{J}_x = -i\partial/\partial\phi_x, \quad \hat{J}_y = -i\partial/\partial\phi_y, \quad \hat{J}_z = -i\partial/\partial\phi_z,$$

where ϕ_x, ϕ_y, ϕ_z are the angles of rotation about the corresponding axes. The derivatives with respect to these angles can be expressed in terms of those with respect to α, β, γ by noting that infinitesimal rotations are added like vectors along the axes of rotation. Figure 20 (§58) shows the directions of the vectors $\delta\alpha$, $\delta\beta$, $\delta\gamma$ of the infinitesimal rotations in terms of Eulerian angles. Taking components along the fixed axes x, y, z, we obtain for the angles of rotation about these axes

$$\begin{aligned} \delta\phi_x &= -\sin\alpha\,\delta\beta + \cos\alpha\sin\beta\,\delta\gamma, \\ \delta\phi_y &= \cos\alpha\,\delta\beta + \sin\alpha\sin\beta\,\delta\gamma, \\ \delta\phi_z &= \delta\alpha + \cos\beta\,\delta\gamma. \end{aligned}$$

Hence, conversely,

$$\begin{aligned} \delta\alpha &= -\cot\beta\cos\alpha\,\delta\phi_x - \cot\beta\sin\alpha\,\delta\phi_y + \delta\phi_z, \\ \delta\beta &= -\sin\alpha\,\delta\phi_x + \cos\alpha\,\delta\phi_y, \\ \delta\gamma &= \frac{\cos\alpha}{\sin\beta}\,\delta\phi_x + \frac{\sin\alpha}{\sin\beta}\,\delta\phi_y. \end{aligned}$$

From these expressions we find

$$\begin{aligned} \hat{J}_x &= -i\left(-\cos\alpha\cot\beta\frac{\partial}{\partial\alpha} - \sin\alpha\frac{\partial}{\partial\beta} + \frac{\cos\alpha}{\sin\beta}\frac{\partial}{\partial\gamma}\right), \\ \hat{J}_y &= -i\left(-\sin\alpha\cot\beta\frac{\partial}{\partial\alpha} + \cos\alpha\frac{\partial}{\partial\beta} + \frac{\sin\alpha}{\sin\beta}\frac{\partial}{\partial\gamma}\right), \\ \hat{J}_z &= -i\frac{\partial}{\partial\alpha}. \end{aligned}$$

When the operators $\hat{J}_z = -i\partial/\partial\alpha$ and $\hat{J}_\zeta = -i\partial/\partial\gamma$ (γ being the angle of rotation about the ζ-axis) act on the function ψ_{JMk}, they are replaced by M and k (the corresponding dependence of the wave function on the angles α and γ is given by the factor $\exp(i\alpha M+i\gamma k)$). Then

$$\hat{J}_+ = \hat{J}_x+i\hat{J}_y = e^{i\alpha}\left(\frac{\partial}{\partial\beta}-M\cot\beta+\frac{k}{\sin\beta}\right),$$

$$\hat{J}_- = \hat{J}_x-i\hat{J}_y = e^{-i\alpha}\left(-\frac{\partial}{\partial\beta}-M\cot\beta+\frac{k}{\sin\beta}\right).$$

The rest of the derivation is exactly as at the end of §28. We start from the equation $\hat{J}_+\psi_{JJk} = 0$, which is valid for the wave function with $M = J$. Hence

$$\left(\frac{\partial}{\partial\beta}-J\cot\beta+\frac{k}{\sin\beta}\right)\psi_{JJk} = 0.$$

The normalized solution of this equation is

$$\psi_{JJk} = i^J(-1)^{J-k}\left[\frac{(2J+1)!}{2(J+k)!(J-k)!}\right]^{1/2}(\cos\tfrac{1}{2}\beta)^{J+k}(\sin\tfrac{1}{2}\beta)^{J-k}\times$$
$$\times\frac{e^{i(J\alpha+k\gamma)}}{2\pi};$$

the normalization integral is an Euler beta function. This expression is in fact the same, apart from a phase factor, as

$$\sqrt{\frac{2J+1}{8\pi^2}}D^{(J)}_{kJ}(\alpha,\beta,\gamma);$$

cf. (58.26). The phase factor is chosen in accordance with the definition in (103.7).

The wave functions with $M<J$ are then calculated by repeated application to ψ_{JJk} of the formula

$$\hat{J}_-\psi_{J,M+1,k} = \sqrt{[J-M)(J+M+1)]}\psi_{JMk}.$$

The final result is the same as (103.8), where the functions $D^{(J)}_{kM}$ are (58.10) and (58.11), and the symmetry properties (58.18) of these functions are to be taken into account.

PROBLEM 2. Calculate the matrix elements $\langle Jk'|H|Jk\rangle$ for an asymmetrical top.

SOLUTION. From formulae (27.13) we find

$$\langle k|J_\xi^2|k\rangle = \langle k|J_\eta^2|k\rangle = \tfrac{1}{2}[J(J+1)-k^2],$$
$$\langle k|J_\xi^2|k+2\rangle = \langle k+2|J_\xi^2|k\rangle = -\langle k|J_\eta^2|k+2\rangle = -\langle k+2|J_\eta^2|k\rangle$$
$$= \tfrac{1}{4}\sqrt{[(J-k)(J-k-1)(J+k+1)(J+k+2)]};$$

for brevity, we everywhere omit the diagonal suffixes J, J of the matrix elements. Hence we have for the required matrix elements of the Hamiltonian†

$$\left.\begin{aligned}\langle k|H|k\rangle &= \tfrac{1}{4}\hbar^2(a+b)[J(J+1)-k^2]+\tfrac{1}{2}\hbar^2ck^2,\\ \langle k|H|k+2\rangle &= \langle k+2|H|k\rangle\\ &= \tfrac{1}{8}\hbar^2(a-b)\sqrt{[(J-k)(J-k-1)(J+k+1)(J+k+2)]}.\end{aligned}\right\}\quad(1)$$

The matrix elements with respect to the functions (103.12) are expressed in terms of the elements in formulae (1) by

$$\left.\begin{aligned}&\langle k\pm|H|k\pm\rangle = \langle k|H|k\rangle\ (k\neq 1),\quad \langle 1\pm|H|1\pm\rangle = \langle 1|H|1\rangle\pm\langle 1|H|-1\rangle,\\ &\langle k\pm|H|k+2,\pm\rangle = \langle k|H|k+2\rangle\ (k\neq 0),\quad \langle 0+|H|2+\rangle = \sqrt{2}\langle 0|H|2\rangle.\end{aligned}\right\}\quad(2)$$

† In Problems 2–5, the formulae are simplified by using the notation

$$a = 1/I_A,\quad b = 1/I_B,\quad c = 1/I_C.$$

PROBLEM 3. Determine the energy levels for an asymmetrical top with $J = 1$.

SOLUTION. The secular equation, of the third degree, falls into three linear equations. One of these gives

$$E_1 = \langle 0+|H|0+\rangle = \tfrac{1}{2}\hbar^2(a+b). \tag{3}$$

From this we can at once write down the other two energy levels, since it is obvious that the three parameters a, b, c enter the problem in a symmetrical manner. Hence

$$E_2 = \tfrac{1}{2}\hbar^2(a+c), \quad E_3 = \tfrac{1}{2}\hbar^2(b+c). \tag{4}$$

The levels E_1, E_2, E_3 belong† to the symmetry types B_1, B_2, B_3 respectively. The wave functions of these states are $\psi_1 = \psi_{10}^+$, $\psi_2 = \psi_{11}^+$, $\psi_3 = \psi_{11}^-$.

PROBLEM 4. The same as Problem 2, but for $J = 2$.

SOLUTION. The secular equation, of the fifth degree, falls into three linear equations and one quadratic. One of the linear equations gives

$$E_1 = \langle 2-|H|2-\rangle = 2\hbar^2 c + \tfrac{1}{2}\hbar^2(a+b), \tag{5}$$

a level of the type B_1. Hence we at once conclude that there must be two other levels, of the types B_2 and B_3:

$$E_2 = 2\hbar^2 b + \tfrac{1}{2}\hbar^2(a+c), \quad E_3 = 2\hbar^2 a + \tfrac{1}{2}\hbar^2(b+c).$$

These three levels have the wave functions $\psi_1 = \psi_{22}^-$, $\psi_2 = \psi_{21}^-$, $\psi_3 = \psi_{21}^+$.

The equation of the second degree is

$$\begin{vmatrix} \langle 0+|H|0+\rangle - E & \langle 2+|H|0+\rangle \\ \langle 2+|H|0+\rangle & \langle 2+|H|2+\rangle - E \end{vmatrix} = 0. \tag{6}$$

Solving this, we obtain

$$E_{4,5} = \hbar^2(a+b+c) \pm \hbar^2[(a+b+c)^2 - 3(ab+bc+ac)]^{1/2}. \tag{7}$$

These levels belong to the type A. The corresponding wave functions are linear combinations of ψ_{20}^+ and ψ_{22}^+.

PROBLEM 5. The same as Problem 2, but for $J = 3$.

SOLUTION. The secular equation, of the seventh degree, falls into one linear equation and three quadratic. The linear equation gives

$$E_1 = \langle 2-|H|2-\rangle = 2\hbar^2(a+b+c), \tag{8}$$

a level of the type A. One of the quadratic equations is equation (6) of Problem 3, with a different value of J. Its roots are

$$E_{2,3} = \frac{5\hbar^2}{2}(a+b) + \hbar^2 c \pm \hbar^2[4(a-b)^2 + c^2 + ab - ac - bc]^{1/2}, \tag{9}$$

levels of the type B_1. The remaining levels are obtained from these by permuting a, b and c.

PROBLEM 6. Determine the splitting of the levels of a system having a quadrupole moment, in an arbitrary external electric field.

SOLUTION. Taking as coordinate axes the principal axes of the tensor $\partial^2\phi/\partial x_i \partial x_k$ (see §76, Problem 3), we bring the quadrupole part of the Hamiltonian of the system to the form

$$\hat{H} = A\hat{J}_x^2 + B\hat{J}_y^2 + C\hat{J}_z^2, \quad A+B+C = 0.$$

---

† This follows at once from considerations of symmetry. The energy E_1, for instance, is symmetrical with respect to the parameters a and b, and this property belongs to the energy of a state whose symmetry about the ξ and η axes is the same, i.e. a state of the type B_1.

Owing to the complete formal analogy between this expression and the Hamiltonian (103.1), the problem under consideration is equivalent to that of finding the energy levels of an asymmetrical top, the only difference being that here the sum of the coefficients $A+B+C=0$, and the angular momentum can have half-integral values also. For these the calculations must be done afresh by the same method, but for integral J we can use the results of Problems 3 to 5, obtaining the following values for the energy displacement ΔE for the first few values of J:

$$J=1: \quad \Delta E = -A, -B, -C;$$
$$J=3/2: \quad \Delta E = \pm\sqrt{[3(A^2+B^2+C^2)/2]};$$
$$J=2: \quad \Delta E = 3A, 3B, 3C, \pm\sqrt{[6(A^2+B^2+C^2)]}.$$

For $J = 3/2$ the energy levels remain doubly degenerate, in accordance with Kramers' theorem (§60).

§104. The interaction between the vibrations and the rotation of the molecule

Hitherto we have regarded the rotation and the vibrations as independent motions of the molecule. In reality, however, the simultaneous presence of both motions results in a peculiar interaction between them (E. Teller, L. Tisza and G. Placzek 1932–33).

Let us start by considering linear polyatomic molecules. A linear molecule can execute vibrations of two types (see the end of §100): longitudinal vibrations with simple frequencies and transverse ones with double frequencies. We shall here be interested in the latter. A molecule executing transverse vibrations has in general some angular momentum. This is evident from simple mechanical considerations,† but it can also be shown by a quantum-mechanical discussion. The latter also enables us to determine the possible values of this angular momentum in a given vibrational state.

Let us suppose that some one double frequency ω_α has been excited in the molecule. The energy level with the vibrational quantum number v_α is $(v_\alpha+1)$-fold degenerate. To this level there correspond the $v_\alpha+1$ wave functions

$$\psi_{v_{\alpha1}v_{\alpha2}} = \text{constant} \times e^{-c_\alpha^2(Q_{\alpha1}^2+Q_{\alpha2}^2)/2} H_{v_{\alpha1}}(c_\alpha Q_{\alpha1}) H_{v_{\alpha2}}(c_\alpha Q_{\alpha2}),$$

where $v_{\alpha1}+v_{\alpha2}=v_\alpha$, or any independent linear combinations of them. The total degree (in $Q_{\alpha1}$ and $Q_{\alpha2}$ together) of the polynomial by which the exponential factor is multiplied is the same in all these functions, and is equal to v_α. It is evident that we can always take, as the fundamental functions, linear combinations of the functions $\psi_{v_{\alpha1}v_{\alpha2}}$ of the form

$$\psi_{v_\alpha l_\alpha} = \text{constant} \times e^{-c_\alpha^2(Q_{\alpha1}^2+Q_{\alpha2}^2)/2} [(Q_{\alpha1}+iQ_{\alpha2})^{(v_\alpha+l_\alpha)/2}(Q_{\alpha1}-iQ_{\alpha2})^{(v_\alpha-l_\alpha)/2}+\dots]. \tag{104.1}$$

The square brackets contain a determinate polynomial, of which we have

† For example, two mutually perpendicular transverse vibrations with a phase difference of $\frac{1}{2}\pi$ can be regarded as a pure rotation of a bent molecule about a longitudinal axis.

written out only the highest term. l_α is an integer, which can take the $v_\alpha+1$ different values $v_\alpha, v_\alpha-2, v_\alpha-4, \ldots, -v_\alpha$.

The normal coordinates $Q_{\alpha 1}$, $Q_{\alpha 2}$ of the transverse vibration are two mutually perpendicular displacements off the axis of the molecule. Under a rotation through an angle ϕ about this axis, the highest term of the polynomial (and therefore the whole function $\psi_{v_\alpha l_\alpha}$) is multiplied by

$$e^{i\phi(v_\alpha+l_\alpha)/2}e^{-i\phi(v_\alpha-l_\alpha)/2} = e^{il_\alpha\phi}.$$

Hence we see that the function (104.1) corresponds to a state with angular momentum l_α about the axis.

Thus we reach the result that, in a state where the double frequency ω_α is excited (with quantum number v_α), the molecule has an angular momentum (about its axis) which takes the values

$$l_\alpha = v_\alpha, v_\alpha-2, v_\alpha-4, \ldots, -v_\alpha. \tag{104.2}$$

This is called the *vibrational angular momentum* of the molecule. If several transverse vibrations are excited simultaneously, the total vibrational angular momentum is equal to the sum Σl_α. On being added to the electron orbital angular momentum, it gives the total angular momentum l of the molecule about its axis.

The total angular momentum J of the molecule cannot be less than the angular momentum about the axis (just as in a diatomic molecule), i.e. J takes the values

$$J = |l|, \quad |l|+1, \ldots .$$

In other words, there are no states with $J = 0, 1, \ldots, |l|-1$.

For harmonic vibrations, the energy depends only on the numbers v_α, and not on l_α. The degeneracy of the vibrational levels (with respect to the values of l_α) is removed by the presence of anharmonic vibrations. The removal is not complete, however: the levels remain doubly degenerate, the same energy belonging to states differing by a simultaneous change of sign of all the l_α and of l. In the next approximation (after that of harmonic motion), a term quadratic in the angular momenta l_α, of the form $\sum_{\alpha\beta} g_{\alpha\beta} l_\alpha l_\beta$ (the $g_{\alpha\beta}$ being constants), appears in the energy. This remaining double degeneracy is removed by an effect similar to the Λ-doubling in diatomic molecules.

When we turn to non-linear molecules, we must first of all make the following remark, which has a purely mechanical significance. For an arbitrary (non-linear) system of particles, the question arises how we can at all separate the vibrational motion from the rotation; in other words, what we are to understand by a "non-rotating system". At first sight it might be thought that the vanishing of the angular momentum,

$$\Sigma\, m\mathbf{r}\times\mathbf{v} = 0 \tag{104.3}$$

(the summation being over the particles in the system), could serve as a criterion of the absence of rotation. However, the expression on the left-hand side is not the complete derivative, with respect to time, of any function of the coordinates. Hence the above equation cannot be integrated with respect to time in such a way as to be formulated as the vanishing of some function of the coordinates. This, however, is necessary if a reasonable definition of the concepts of "pure vibrations" and "pure rotation" is to be possible.

As a definition of the absence of rotation, we must therefore use the condition

$$\Sigma\, m\mathbf{r}_0 \times \mathbf{v} = 0, \qquad (104.4)$$

where $\mathbf{r}_0$ are the radius vectors of the equilibrium positions of the particles. Putting $\mathbf{r} = \mathbf{r}_0 + \mathbf{u}$, where $\mathbf{u}$ are the displacements in small vibrations, we have $\mathbf{v} = \dot{\mathbf{r}} = \dot{\mathbf{u}}$. The equation (104.4) can be integrated with respect to time, giving

$$\Sigma\, m\mathbf{r}_0 \times \mathbf{u} = 0. \qquad (104.5)$$

The motion of the molecule will be regarded as a combination of the purely vibrational motion, in which the condition (104.5) is satisfied, and the rotation of the molecule as a whole.†

Writing the angular momentum in the form

$$\Sigma\, m\mathbf{r} \times \mathbf{v} = \Sigma\, m\mathbf{r}_0 \times \mathbf{v} + \Sigma\, m\mathbf{u} \times \mathbf{v},$$

we see that, in accordance with the definition (104.4) of the absence of rotation, the vibrational angular momentum must be understood as the sum $\Sigma\, m\mathbf{u} \times \mathbf{v}$. However, it must be borne in mind that this angular momentum, being only a part of the total angular momentum of the system, is not conserved. Hence only a mean value of the vibrational angular momentum can be ascribed to each vibrational state.

Molecules having no axis of symmetry of order above the second belong to the asymmetrical-top type. In a molecule of this type, all the frequencies are simple (their symmetry groups have only one-dimensional irreducible representations). Hence none of the vibrational levels is degenerate. In any non-degenerate state, however, the mean angular momentum vanishes (see §26). Thus, in a molecule of the asymmetrical-top type, the mean vibrational angular momentum vanishes in every state.

If, among the symmetry elements of the molecule, there is one axis of order higher than the second, the molecule is of the symmetrical-top type. Such a molecule has vibrations with both simple and double frequencies. The mean vibrational angular momentum of the former again vanishes. To the double frequencies, however, there corresponds a non-zero mean angular momentum component along the axis of the molecule.

It is easy to find an expression for the energy of the rotational motion of the molecule (of the symmetrical-top type), taking into account the vibrational angular momentum. The operator of this energy differs from (103.5) in that

† The translational motion is supposed removed from the start, by choosing a system of coordinates in which the centre of mass of the molecule is at rest.

the rotational angular momentum of the top is replaced by the difference between the total (conserved) angular momentum $\mathbf{J}$ of the molecule and its vibrational angular momentum $\mathbf{J}^{(v)}$:

$$\hat{H}_{\text{rot}} = \frac{\hbar^2}{2I_A}(\hat{\mathbf{J}}-\hat{\mathbf{J}}^{(v)})^2+\tfrac{1}{2}\hbar^2\left(\frac{1}{I_C}-\frac{1}{I_A}\right)(\hat{J}_\zeta-\hat{J}_\zeta^{(v)})^2. \qquad (104.6)$$

The required energy is the mean value $\bar{H}_{\text{rot}}$. The terms in (104.6) which contain the squared components of $\mathbf{J}$ give a purely rotational energy which is the same as (103.6). The terms which contain the squared components of $\mathbf{J}^{(v)}$ give constants independent of the rotational quantum numbers and may be omitted; the terms which contain products of components of $\mathbf{J}$ and $\mathbf{J}^{(v)}$ constitute the interaction here considered between the vibrations of the molecule and its rotation. This is called the *Coriolis interaction* (since it corresponds to the Coriolis forces in classical mechanics). In averaging these terms it must be borne in mind that the mean values of the transverse (ξ,η) components of the vibrational angular momentum are zero. The mean energy of the Coriolis interaction is therefore

$$E_{\text{Cor}} = -\hbar^2 k k_v / I_C, \qquad (104.7)$$

where the integer k is, as in §103, the component of the total angular momentum along the axis of the molecule, and $k_v = \overline{J_\zeta^{(v)}}$ is the mean value of the component of the vibrational angular momentum for the vibrational state concerned; k_v, unlike k, is not an integer.

Finally, let us consider molecules of the spherical-top type. These include molecules whose symmetry is that of any of the cubic groups. Such molecules have simple, double and triple frequencies (there being one-, two- and three-dimensional irreducible representations of the cubic groups). The degeneracy of the vibrational levels is, as usual, partly removed by the presence of anharmonic motion; when these effects have been taken into account there remain, apart from the non-degenerate levels, only doubly and triply degenerate levels. Here we shall discuss these levels that are split by the presence of anharmonic motion.

It is easy to see that, for molecules of the spherical-top type, the mean vibrational angular momentum is zero not only in the non-degenerate vibrational states but also in the doubly degenerate ones. This follows from simple considerations based on symmetry properties. The mean angular momentum vectors in two states belonging to the same degenerate energy level must be transformed into each other in all possible symmetry transformations of the molecule. None of the cubic symmetry groups, however, allows the existence of two directions transformed only into each other; only sets of three or more directions are so transformed.

From these arguments it follows that, in states corresponding to triply degenerate vibrational levels, the mean vibrational angular momentum is non-zero. After averaging over the vibrational state, this angular momentum is represented by an operator whose matrix elements correspond to transitions between three mutually degenerate states. In accordance with the number of these states, this operator must have the form $\zeta\hat{\mathbf{l}}$, where $\hat{\mathbf{l}}$ is the operator of an angular momentum of unity (for which $2l+1 = 3$) and ζ is a constant characterizing the vibrational level in question. The Hamiltonian of the rotational motion of the molecule is

$$\hat{H}_{\text{rot}} = (\hbar^2/2I)(\hat{\mathbf{J}} - \hat{\mathbf{J}}^{(v)})^2$$

and, after averaging, becomes the operator

$$\hat{H}_{\text{rot}} = \frac{\hbar^2}{2I}\hat{\mathbf{J}}^2 + \frac{\hbar^2}{2I}\overline{\hat{\mathbf{J}}^{(v)2}} - \frac{\hbar^2}{I}\zeta^2\hat{\mathbf{J}}.\hat{\mathbf{l}}. \tag{104.8}$$

The eigenvalues of the first term give the ordinary rotational energy (103.4); the second term gives an unimportant constant, which does not depend on the rotational quantum number. The last term in (104.8) gives the desired energy of the Coriolis splitting of the vibrational level. The eigenvalues of the quantity $\mathbf{J}.\mathbf{l}$ are calculated in the usual way; it has (for a given $\mathbf{J}$) three different values corresponding to the values $J+1$, $J-1$, J of the vector $\mathbf{J}+\mathbf{l}$. The result is

$$\left.\begin{aligned} E_{\text{Cor}}^{(J+1)} &= -\hbar^2\zeta J/I, \\ E_{\text{Cor}}^{(J-1)} &= \hbar^2\zeta(J+1)/I, \\ E_{\text{Cor}}^{(J)} &= \hbar^2\zeta/I. \end{aligned}\right\} \tag{104.9}$$

§105. The classification of molecular terms

The wave function of a molecule is the product of the electron wave function, the wave function of the vibrational motion of the nuclei, and the rotational wave function. We have already discussed the classification and types of symmetry of these functions separately. It now remains to examine the question of the classification of molecular terms as a whole, i.e. of the possible symmetry of the total wave function.

It is clear that, if the symmetry of all three factors with respect to some transformation is given, the symmetry of the product with respect to that transformation is determined. For a complete description of the symmetry of the state, we must also specify the behaviour of the total wave function when the coordinates of all the particles in the molecule (electrons and nuclei) are inverted simultaneously. The state is said to be *negative* or *positive*, according as the wave function does or does not change sign under this transformation.†

It must be remembered, however, that the characterization of the state with respect to inversion is significant only for molecules which do not

† We use the same customary, though unfortunate, terminology as for diatomic molecules (§86).

possess stereoisomers. If stereoisomerism is present, the molecule assumes on inversion a configuration which can by no rotation in space be made to coincide with the original configuration; these are the "right-hand" and "left-hand" modifications of the substance.† Hence, when stereoisomerism is present, the wave functions obtained from each other on inversion belong essentially to different molecules, and it is meaningless to compare them.‡

We have seen in §86 that, for diatomic molecules, the spin of the nuclei exerts an important indirect effect on the arrangement of the molecular terms by determining their degree of degeneracy, and in some cases entirely forbidding levels of a certain symmetry. The same is true for polyatomic molecules. Here, however, the investigation of the problem is considerably more complex, and requires the application of the methods of group theory to each particular case.

The idea of the method is as follows. The total wave function must contain, besides the coordinate part (the only part we have considered so far), a spin factor, which is a function of the projections of the spins of all the nuclei on some chosen direction in space. The projection σ of the spin of a nucleus takes $2i+1$ values (where i is the spin of the nucleus); by giving to all the $\sigma_1, \sigma_2, \ldots, \sigma_N$ (where N is the number of atoms in the molecule) all possible values, we obtain altogether $(2i_1+1)(2i_2+1)\ldots(2i_N+1)$ different values of the spin factor. In each symmetry transformation, certain nuclei (of the same kind) change places, and if we imagine the spin values to "remain fixed", the transformation is equivalent to an interchange of spin values among the nuclei. Accordingly, the various spin factors will be transformed into linear combinations of one another, thus giving some representation (in general reducible) of the symmetry group of the molecule. Decomposing this into irreducible parts, we find the possible types of symmetry for the spin wave function.

A general formula can easily be written down for the characters $\chi_{sp}(G)$ of the representation given by the spin factors. To do this, it is sufficient to notice that, in a transformation, only those spin factors are unchanged in which the nuclei changing places have the same σ_a; otherwise, one spin factor changes into another and contributes nothing to the character. Bearing in mind that σ_a takes $2i_a+1$ values, we find that

$$\chi_{sp}(G) = \Pi\,(2i_a+1), \tag{105.1}$$

where the product is taken over the groups of atoms which change places under the transformation G considered (there being one factor in the product from each group).

We are, however, interested not so much in the symmetry of the spin function as in that of the coordinate wave function (by which we mean its

† For stereoisomerism to be possible, the molecule must have no symmetry element pertaining to reflection (i.e. no centre of symmetry, plane of symmetry, or rotary–reflection axis).

‡ Strictly speaking, quantum mechanics always gives a non-zero probability for the transition from one modification to the other. This probability, however, which relates to the passage of nuclei through a barrier, is extremely small.

symmetry with respect to interchanges of the coordinates of the nuclei, the coordinates of the electrons remaining unchanged). These two symmetries are directly related, however, since the total wave function must remain unchanged or change sign when any pair of nuclei are interchanged, according as they obey Bose statistics or Fermi statistics (in other words, it must be multiplied by $(-1)^{2i}$, where i is the spin of the nuclei that are interchanged). Introducing the appropriate factor in the characters (105.1), we obtain the system of characters $\chi(G)$ for the representation containing all the irreducible representations by which the coordinate wave functions are transformed:

$$\chi(G) = \Pi(2i_a+1)(-1)^{2i_a(n_a-1)}, \tag{105.2}$$

where n_a is the number of nuclei in each group which change places under the transformation in question. Decomposing this representation into irreducible parts, we obtain the possible types of symmetry of the coordinate wave functions of the molecule, together with the degrees of degeneracy of the corresponding energy levels (here and later we mean the degeneracy with respect to the different spin states of the system of nuclei).†

Each type of symmetry of the states is related to definite values of the total spins of the groups of equivalent nuclei in the molecule (i.e. groups of nuclei which change places under the various symmetry transformations of the molecule). This relation is not one-to-one; each type of symmetry of states can, in general, be brought about with various values of the spins of equivalent groups. The relation can also be established, in any particular case, by means of group theory.

As an example, let us consider a molecule of the asymmetrical-top type, the ethylene molecule $C^{12}{}_2H^1{}_4$ (Fig. 43g), with the symmetry group $\boldsymbol{D}_{2h}$. The index to the chemical symbol indicates the isotope to which the nucleus belongs; this indication is necessary, since the nuclei of different isotopes may have different spins. In this case, the spin of the H^1 nucleus is $\frac{1}{2}$, while the C^{12} nucleus has no spin. Hence we need consider only the hydrogen atoms.

We take the system of coordinates shown in Fig. 43g; the z-axis is perpendicular to the plane of the molecule, while the x-axis is along the axis of the molecule. A reflection in the xy-plane leaves all the atoms fixed, while other reflections and rotations interchange the hydrogen atoms in pairs. From formula (105.2) we have the following characters of the representation:

E	$\sigma(xy)$	$\sigma(xz)$	$\sigma(yz)$	I	$C_2(x)$	$C_2(y)$	$C_2(z)$
16	16	4	4	4	4	4	4

Decomposing this representation into irreducible parts, we find that it contains the following irreducible representations of the group $\boldsymbol{D}_{2h}$: $7A_g$, $3B_{1g}$, $3B_{2u}$, $3B_{3u}$. The figures show the number of times each irreducible represen-

† The degree of degeneracy of the level in this respect is often called its *nuclear statistical weight*; see the last footnote to §86.

tation appears in the reducible one; these numbers are also the nuclear statistical weights of the levels with the corresponding symmetry.†

The classification of the states of the ethylene molecule thus obtained relates to the symmetry of the total (coordinate) wave function, including the electron, vibrational and rotational parts. Usually, however, it is of interest to arrive at these results from a different point of view. Knowing the possible symmetries of the total wave function, we can find at once which rotational levels are possible (and with what statistical weights) for any prescribed electron and vibrational state.

Let us consider, for instance, the rotational structure of the lowest vibrational level (that for which the vibrations are not excited at all) of the normal electron term, assuming that the electron wave function of the normal state is completely symmetrical (as is the case for practically all polyatomic molecules). Then the symmetry of the total wave function with respect to rotations about the axes of symmetry is the same as the symmetry of the rotational wave function. Comparing this with the results obtained above, we therefore conclude that in the ethylene molecule the rotational levels of the types A and B_1 (see §103) are positive with statistical weights 7 and 3, while those of the types B_2 and B_3 are negative with statistical weight 3.

As with diatomic molecules (see the end of §86), owing to the extreme weakness of the interaction between the nuclear spins and the electrons, transitions between states of different nuclear symmetry in the ethylene molecule do not occur in practice. Hence molecules in such states behave like different modifications of the substance. Thus ethylene $C^{12}{}_2H^1{}_4$ has four modifications, with nuclear statistical weights 7, 3, 3, 3.

In reaching this conclusion it is important that states with different symmetry belong to different energy levels (the intervals between which are large compared with the interaction energy of nuclear spins). The conclusion is therefore invalid for molecules in which there exist states of different nuclear symmetry belonging to the same degenerate energy level.

Let us consider another example, the ammonia molecule $N^{14}H^1{}_3$, of the symmetrical-top type (Fig. 41), whose symmetry group is C_{3v}. The spin of the nucleus N^{14} is 1, and that of H^1 is $\frac{1}{2}$. Using formula (105.2), we find the characters of the representation of the group C_{3v} in which we are interested:

E	$2C_3$	$3\sigma_v$
24	6	-12.

It contains the following irreducible representations of the group C_{3v}: $12A_2$, $6E$. Thus two types of level are possible; their nuclear statistical weights are‡ 12 and 6.

The rotational levels of a symmetrical top are classified (for a given J) according to the values of the quantum number k. Let us consider, as in the

† The relation between the symmetry of states and the values of the total spin of the four hydrogen nuclei in the ethylene molecule is derived in Problem 1.

‡ A total spin of the hydrogen nuclei of 3/2 corresponds to the terms of symmetry A_2, and one of 1/2 to those of symmetry E.

previous example, the rotational structure of the normal electron and vibrational state of the NH_3 molecule (i.e. we suppose the electron and vibrational wave functions to be completely symmetrical). In determining the symmetry of the rotational wave function, we must bear in mind that it is meaningful to speak of its behaviour only with respect to rotations about axes. Hence we replace the planes of symmetry by axes of symmetry of the second order perpendicular to them, a reflection in a plane being equivalent to a rotation about such an axis, followed by an inversion. In the present case, therefore, instead of the group C_{3v} we have to consider the isomorphous point group D_3.

The rotational wave functions with $k = \pm|k|$ are multiplied by $e^{\pm 2\pi i|k|/3}$ respectively under a rotation C_3 about a vertical axis of the third order, while under a rotation U_2 about a horizontal axis of the second order they change into each other, thus giving a two-dimensional representation of the group D_3. If $|k|$ is not a multiple of three, this representation is irreducible; it is E. The representation of the group C_{3v} corresponding to the total wave function is obtained by multiplying the character $\chi(U_2)$ by 1 or -1, according as the term is positive or negative. Since, however, in the representation E we have $\chi(U_2) = 0$, we obtain the same representation E in either case (but this time as a representation of the group C_{3v}, and not D_3). Bearing in mind the results obtained above, we thus conclude that, when $|k|$ is not a multiple of three, both positive and negative levels are possible, with nuclear statistical weights of 6 (the symmetry of the total coordinate wave function being of the type E).

When $|k|$ is a multiple of three (but not zero), the rotational functions give a representation (of the group D_3) with characters

E	$2C_3$	$3U_2$
2	2	0

This representation is reducible, and divides into the representations A_1, A_2. In order that the total wave function should belong to the representation A_2 of the group C_{3v}, the rotational level A_1 must be negative and A_2 positive. Thus, when $|k|$ is a multiple of three and not zero, both positive and negative levels are possible, with nuclear statistical weights of 12 (levels of the type A_2).

Finally, only one rotational function corresponds to an angular momentum component $k = 0$; it gives a representation with characters†

E	$2C_3$	$3U_2$
1	1	$(-1)^J$

If the total wave function has the symmetry A_2, its behaviour with respect to inversion must therefore be given by the factor $(-1)^{J+1}$. Thus, for $k = 0$,

† On rotation through an angle π, the eigenfunction of the angular momentum with magnitude J and component zero is multiplied by $(-1)^J$.

levels with even and odd J can only be negative and positive respectively; the statistical weight is 6 in either case (levels of the type A_2).

Summarizing these results, we have the following table of possible states for various values of the quantum number k for the normal electron and vibrational term of the molecule $N^{14}H^1{}_3$ (the symbols + and − denoting positive and negative states).

		+	−
$\lvert k \rvert$ not a multiple of 3		$6E$	$6E$
$\lvert k \rvert$ a multiple of 3		$12A_2$	$12A_2$
$k = 0$	J even	−	$6A_2$
	J odd	$6A_2$	−

For given J and k, the energy levels of the NH_3 molecule are in general degenerate (see also the table for ND_3 in Problem 3). This degeneracy is partly removed by a peculiar effect due to the flat shape of the ammonia molecule and the small mass of the hydrogen atoms. By a fairly small vertical displacement of the atoms in this molecule a transition can be brought about between two configurations obtained from one another by a reflection in a plane parallel to the base of the pyramid (Fig. 44). These transitions cause a splitting of the levels, separating positive and negative levels (an effect similar to the one-dimensional case considered in §50, Problem 3). The magnitude of the splitting is proportional to the probability of passage of the atoms through the "potential barrier" separating the two configurations of the molecule. Although this probability is comparatively high in the ammonia molecule, owing to the above-mentioned properties, the splitting is still small (1×10^{-4} eV).

An example of a molecule of the spherical-top type is discussed in Problem 5.

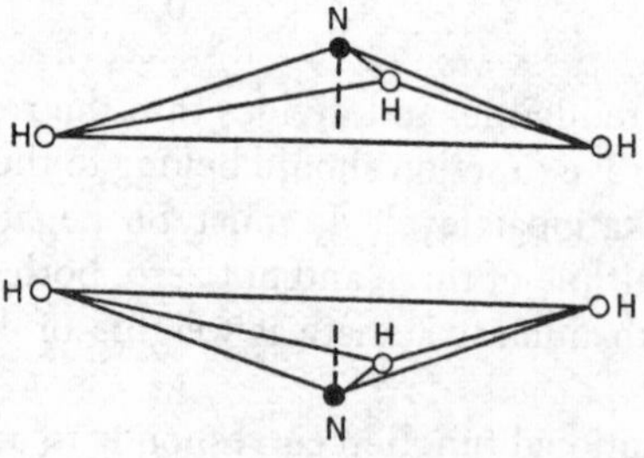

FIG. 44

PROBLEMS

PROBLEM 1. Find the relation between the symmetry of the state of the $C^{12}{}_2H^1{}_4$ molecule and the total spin of the hydrogen nuclei in the molecule.

SOLUTION.† The total spin of the four H^1 nuclei can take the values $I = 2, 1, 0$, and its component M_I takes values from 2 to −2. Let us consider the representations given by the spin factors for each value of M_I, beginning with the largest.

† A method of solving problems of this kind, based on the theory of permutation groups, is given in Kaplan's book (quoted in §63), ch. VI, §2.

The value $M_I = 2$ corresponds to only one spin factor, in which all the nuclei have a spin component $+\frac{1}{2}$. The value $M_I = 1$ corresponds to four different spin factors differing as regards the nucleus which has spin component $-\frac{1}{2}$. Finally, the value $M_I = 0$ is given by six spin factors, depending on the pair of nuclei which have spin components $-\frac{1}{2}$. The characters of the three corresponding representations are as follows:

	E	$\sigma(xy)$	$\sigma(xz)$	$\sigma(yz)$	I	$C_2(z)$	$C_2(y)$	$C_2(x)$
$M_I = 2$	1	1	1	1	1	1	1	1
$M_I = 1$	4	4	0	0	0	0	0	0
$M_I = 0$	6	6	2	2	2	2	2	2

The first of these representations is the unit representation A_g; since the value $M_I = 2$ can occur only for $I = 2$, we conclude that a state with symmetry A_g corresponds to spin $I = 2$.

The value $M_I = 1$ can occur for both $I = 1$ and $I = 2$. Subtracting the first representation from the second and decomposing the result into irreducible parts, we find that states B_{1g}, B_{2u}, B_{3u} correspond to spin $I = 1$.

Finally, the value $M_I = 0$ can occur in all cases where $M_I = 1$ is possible, and also for $I = 0$. Subtracting the second representation from the third, we find two states A_g corresponding to spin $I = 0$.

PROBLEM 2. Determine the types of symmetry of the total (coordinate) wave functions, and the statistical weights of the corresponding levels, for the molecules $C^{12}{}_2H^2{}_4$, $C^{13}{}_2H^1{}_4$, $N^{14}{}_2O^{16}{}_4$ (all these molecules are of the same form; the spins are $i(H^2) = 1$, $i(C^{13}) = \frac{1}{2}$, $i(N^{14}) = 1$).

SOLUTION. By the method shown in the text for the molecule $C^{12}{}_2H^1{}_4$, we find the following states (the axes of coordinates being taken the same as above):

Molecule	+	−
$C^{12}{}_2H^2{}_4$	$27A_g$, $18B_{1g}$	$18B_{2u}$, $18B_{3u}$
$C^{13}{}_2H^1{}_4$	$16A_g$, $12B_{1g}$	$12B_{2u}$, $24B_{3u}$
$N^{14}{}_2O^{16}{}_4$	$6A_g$	$3B_{3u}$

PROBLEM 3. The same as Problem 2, but for the molecule $N^{14}H^2{}_3$.

SOLUTION. In the way shown in the text for the molecule $N^{14}H^1{}_3$, we find the states $30A_1$, $3A_2$, $24E$.

In the normal electron and vibrational term, the following states are possible for various values of the quantum number k:

	+	−
$\lvert k\rvert$ not a multiple of 3	$24E$	$24E$
$\lvert k\rvert$ a multiple of 3	$30A_1$, $3A_2$	$30A_1$, $3A_2$
$k = 0$, J even	$30A_1$	$3A_2$
$k = 0$, J odd	$3A_2$	$30A_1$

PROBLEM 4. The same as Problem 2, but for the molecule $C^{12}{}_2H^1{}_6$ (see Fig. 43f; the symmetry is $\boldsymbol{D}_{3d}$).

SOLUTION. The possible states are of the types $7A_{1g}$, $1A_{1u}$, $3A_{2g}$, $13A_{2u}$, $9E_g$, $11E_u$.

In the normal electron and vibrational term, the following states are obtained:

	+	−
$\lvert k\rvert$ not a multiple of 3	$9E_g$	$11E_u$
$\lvert k\rvert$ a multiple of 3	$7A_{1g}$, $3A_{2g}$	$1A_{1u}$, $13A_{2u}$
$k = 0$, J even	$7A_{1g}$	$1A_{1u}$
$k = 0$, J odd	$3A_{2g}$	$13A_{2u}$

PROBLEM 5. The same as Problem 2, but for the methane molecule $C^{12}H^1{}_4$ (the C atom is at the centre of a tetrahedron with the H atoms at the vertices).

SOLUTION. The molecule is of the spherical-top type, and has the symmetry $\boldsymbol{T}_d$. Following the same method, we find that the possible states are of the types $5A_2$, $1E$, $3F_1$ (corresponding to a total spin of the molecule of 2, 0, 1 respectively).

The rotational states of a spherical top are classified according to the values of the total angular momentum J. The $2J+1$ rotational functions belonging to a particular value of J give a $(2J+1)$-dimensional representation of the group $\boldsymbol{O}$, which is isomorphous with the group T_d; it is obtained from the latter by replacing all planes of symmetry by axes of the second order perpendicular to them. The characters in this representation are given by formula (98.3). Thus for example, for $J = 3$ we obtain a representation with characters

E	$8C_3$	$6C_2$	$6C_4$	$3C_4^2$
7	1	−1	−1	−1.

This contains the following irreducible representations of the group O: A_2, F_1, F_2. Again considering the rotational structure of the normal electron and vibrational term, we therefore conclude that, for $J = 3$, the states with a symmetry A_2 of the total wave function can only be positive, while those of type F_1 can be either positive or negative. For the first few values of J we thus obtain the following states (which we write together with their statistical weights):

	+	−
$J = 0$	—	$5A_2$
$J = 1$	$3F_1$	—
$J = 2$	$1E$	$1E$, $3F_1$
$J = 3$	$5A_2$, $3F_1$	$3F_1$
$J = 4$	$1E$, $3F_1$	$5A_2$, $1E$, $3F_1$

CHAPTER XIV

ADDITION OF ANGULAR MOMENTA

§106. 3j-symbols

THE rule of addition of angular momenta deduced in §31 gives the possible values of the total angular momentum of a system consisting of two particles (or more complex components) with angular momenta j_1 and j_2.† This rule is in fact closely related to the properties of wave functions with respect to spatial rotations, and follows immediately from the properties of spinors.

The wave functions of particles with angular momenta j_1 and j_2 are symmetrical spinors of ranks $2j_1$ and $2j_2$, and the wave function of the system is their product,

$$\psi^{(1)} \underbrace{{}^{\lambda\mu\ldots}}_{2j_1} \psi^{(2)} \underbrace{{}^{\rho\sigma\ldots}}_{2j_2}. \tag{106.1}$$

Symmetrizing this product with respect to all the indices, we obtain a symmetrical spinor of rank $2(j_1+j_2)$, corresponding to a state with total angular momentum j_1+j_2. If we contract the product (106.1) with respect to one pair of indices, of which one must belong to $\psi^{(1)}$ and the other to $\psi^{(2)}$ (since otherwise the result is zero), the symmetry of each of the spinors $\psi^{(1)}$ and $\psi^{(2)}$ shows that it does not matter which indices are taken from $\lambda, \mu, \ldots$ and $\rho, \sigma, \ldots$. After symmetrization we obtain a symmetrical spinor of rank $2(j_1+j_2-1)$, corresponding to a state with angular momentum j_1+j_2-1.‡ Continuing this process, we find, in agreement with the rule already known, that j takes values from j_1+j_2 to $|j_1-j_2|$, each occurring once.

Mathematically, this involves the decomposition of the direct product $D^{(j_1)} \times D^{(j_2)}$ of two irreducible representations of the rotation group (with

† Strictly speaking, we shall always be considering (without explicitly mentioning the fact each time) a system whose parts interact so weakly that their angular momenta may be regarded as conserved in a first approximation.

All the results given below apply, of course, not only to the addition of the total angular momenta of two particles (or systems) but also to the addition of the orbital angular momentum and spin of the same system, assuming that the spin–orbit coupling is sufficiently weak.

‡ To avoid misunderstanding, the following comment is useful. The wave function of a system of two particles is always a spinor of rank $2(j_1+j_2)$, and this is in general not equal to $2j$, where j is the total angular momentum of the system. Such a spinor may, however, be equivalent to a spinor of lower rank. For example, the wave function of a system of two particles with angular momenta $j_1 = j_2 = \frac{1}{2}$ is a spinor of rank two; but if the total angular momentum $j = 0$, this spinor is antisymmetrical, and therefore reduces to a scalar. In general, the total angular momentum j determines the symmetry of the spinor wave function of the system: this is symmetrical with respect to $2j$ indices and antisymmetrical with respect to the remainder.

dimensions $2j_1+1$ and $2j_2+1$) into irreducible parts. The addition rule for angular momenta may then be written as

$$D^{(j_1)} \times D^{(j_2)} = D^{(j_1+j_2)} + D^{(j_1+j_2-1)} + \dots + D^{(|j_1-j_2|)}.$$

For a complete solution of the problem of the addition of angular momenta, we must also consider the problem of constructing the wave function of a system with a given total angular momentum from those of its two component particles.

Let us begin with the simple case of the addition of two angular momenta to give a zero total angular momentum. Here we must evidently have $j_1 = j_2$ and angular momentum components $m_1 = -m_2$. Let ψ_{jm} be the normalized wave functions of the states of one particle with angular momentum j and component thereof m (in the non-spinor representation). The required wave function ψ_0 of the system is the sum of the products of the wave functions of the two particles with opposite values of m:

$$\psi_0 = \frac{1}{\sqrt{(2j+1)}} \sum_{m=-j}^{j} (-1)^{j-m} \psi^{(1)}{}_{jm} \psi^{(2)}{}_{j,-m}, \tag{106.2}$$

where j is the common value of j_1 and j_2. The factor preceding the sum is due to the normalization. The coefficients in the sum must all have the same absolute value, since all values of the components m of the angular momenta of the particles are equally probable. The sequence of signs in (106.2) is easily found by means of the spinor representation of the wave functions. In spinor notation the sum in (106.2) is a scalar (the total angular momentum of the system being zero)

$$\psi^{(1)\lambda\mu\dots} \psi^{(2)}{}_{\lambda\mu\dots}, \tag{106.3}$$

formed from two spinors of rank $2j$. Using this, we find the signs in (106.2) directly from (57.3).

It should be borne in mind, however, that in general only the relative signs of the terms in the sum (106.2) are determinate, while the sign of the whole sum may depend on the "order of addition" of the angular momenta. For, if we lower all spinor indices ($j+m$ ones and $j-m$ twos) in $\psi^{(1)}$ and raise them in $\psi^{(2)}$, the scalar (106.3) is multiplied by $(-1)^{2j}$, and therefore changes sign when j is half-integral.

Next we consider a system with zero total angular momentum consisting of three particles with angular momenta j_1, j_2, j_3 and components thereof m_1, m_2, m_3. The condition for the total angular momentum to be zero is that $m_1+m_2+m_3 = 0$ and j_1, j_2, j_3 have values such that each of them can be obtained by vector addition of the other two, i.e. geometrically j_1, j_2, j_3 must be the sides of a closed triangle. In other words, each of them lies between the difference and the sum of the other two:

$$|j_1-j_2| \leqslant j_3 \leqslant j_1+j_2, \text{ etc.}$$

It is evident that the algebraic sum $j_1+j_2+j_3$ is an integer.

The wave function of the system under consideration is the sum

$$\psi_0 = \sum_{m_1, m_2, m_3} \begin{pmatrix} j_1 & j_2 & j_3 \\ m_1 & m_2 & m_3 \end{pmatrix} \psi^{(1)}{}_{j_1 m_1} \psi^{(2)}{}_{j_2 m_2} \psi^{(3)}{}_{j_3 m_3}, \qquad (106.4)$$

taken over the values of each m_i from $-j_i$ to j_i. The coefficients in this formula are termed *Wigner 3j-symbols*. By definition they are non-zero only if $m_1+m_2+m_3 = 0$.

When the suffixes 1, 2, 3 are permuted, the wave function (106.4) can change only by an unimportant phase factor. The 3j-symbols can in fact be defined as purely real quantities (see below), and then the indeterminacy of ψ_0 can consist only in its sign as a whole being indefinite (as is true of the function (106.2) also). This means that interchanging the columns of a 3j-symbol can either leave it unchanged or change its sign.

The most symmetrical way of defining the coefficients in the sum (106.4), which is the definition generally used for the 3j-symbols, is as follows. In spinor notation, ψ_0 is a scalar formed by contracting the product of the three spinors $\psi^{(1)\lambda\mu\ldots}$, $\psi^{(2)\lambda\mu\ldots}$, $\psi^{(3)\lambda\mu\ldots}$ with respect to all pairs of indices belonging to two different spinors. In each pair belonging to particles 1 and 2 the spinor index will be written superior with $\psi^{(1)}$ and inferior with $\psi^{(2)}$; in a pair belonging to particles 2 and 3, superior with $\psi^{(2)}$ and inferior with $\psi^{(3)}$; and in a pair belonging to particles 3 and 1, superior with $\psi^{(3)}$ and inferior with $\psi^{(1)}$. It is easily seen that the total number of pairs of each kind is $j_1+j_2-j_3$, $j_2+j_3-j_1$, $j_1+j_3-j_2$ respectively. This rule determines uniquely the sign of ψ_0.

It is evident that, with this definition, cyclic interchange of the indices 1, 2 and 3 leaves ψ_0 unchanged. This means that the 3j-symbol is unchanged when its columns are cyclically permuted. Interchange of any two indices is easily seen to require the raising of the lower indices and lowering of the upper indices in all $j_1+j_2+j_3$ pairs. This means that ψ_0 is multiplied by $(-1)^{j_1+j_2+j_3}$; in other words, the 3j-symbols have the property

$$\begin{pmatrix} j_2 & j_1 & j_3 \\ m_2 & m_1 & m_3 \end{pmatrix} = (-1)^{j_1+j_2+j_3} \begin{pmatrix} j_1 & j_2 & j_3 \\ m_1 & m_2 & m_3 \end{pmatrix} \text{ etc.}, \qquad (106.5)$$

i.e. they change sign when two columns are interchanged if $j_1+j_2+j_3$ is odd.

Finally, we easily see that

$$\begin{pmatrix} j_1 & j_2 & j_3 \\ -m_1 & -m_2 & -m_3 \end{pmatrix} = (-1)^{j_1+j_2+j_3} \begin{pmatrix} j_1 & j_2 & j_3 \\ m_1 & m_2 & m_3 \end{pmatrix}: \qquad (106.6)$$

a change in the sign of the z-component of each angular momentum can be regarded as the result of a rotation through an angle π about the y-axis, and this is equivalent to raising all the lower spinor indices and lowering all the upper ones (see (58.5)).

From (106.4) we can derive an important formula which gives the wave function ψ_{jm} of a system consisting of two particles and having given values of j and m. To do so, we consider the particles 1 and 2 together as one system. Since the angular momentum $\mathbf{j}$ of this system together with the angular momentum $\mathbf{j}_3$ of particle 3 gives a total angular momentum of zero, we must have $j = j_3$, $m = -m_3$. According to (106.2) we can then write

$$\psi_0 = \frac{1}{\sqrt{(2j+1)}} \sum_m (-1)^{j-m} \psi_m \psi^{(3)}{}_{j,-m}. \tag{106.7}$$

This formula is to be compared with (106.4) (in which we replace j_3, m_3 by j, $-m$). Here, however, we must first take into account the fact that the rule for constructing the sum in (106.7) according to (106.3) does not correspond to the rule for constructing the sum (106.4): to bring (106.7) to the form (106.4) we must, as is easily seen, interchange pairs of upper and lower indices corresponding to particles 1 and 3. This leads to an additional factor $(-1)^{j_1-j_2+j_3}$. The result is†

$$\psi_{jm} = (-1)^{j_1-j_2+m} \sqrt{(2j+1)} \sum_{m_1, m_2} \begin{pmatrix} j_1 & j_2 & j \\ m_1 & m_2 & -m \end{pmatrix} \psi^{(1)}{}_{j_1 m_1} \psi^{(2)}{}_{j_2 m_2}, \tag{106.8}$$

where the summation over m_1 and m_2 is subject to the condition $m_1 + m_2 = m$.

Formula (106.8) gives the required expression for obtaining the wave function of a system from those of its two particles, which have definite angular momenta j_1 and j_2. It can be written in the form

$$\psi_{jm} = \sum_{m_1, m_2} \langle m_1 m_2 | jm \rangle \psi^{(1)}{}_{j_1 m_1} \psi^{(2)}{}_{j_2 m_2} \qquad (m_2 = m - m_1). \tag{106.9}$$

The coefficients

$$\langle m_1 m_2 | jm \rangle = (-1)^{j_1-j_2+m} \sqrt{(2j+1)} \begin{pmatrix} j_1 & j_2 & j \\ m_1 & m_2 & -m \end{pmatrix} \tag{106.10}$$

form the matrix of the transformation from the complete orthonormal set of $(2j_1+1)(2j_2+1)$ wave functions of states $|m_1\, m_2\rangle$ to the similar set of wave functions of states $|jm\rangle$ (for given values of j_1, j_2). They are called *vector addition coefficients* or *Clebsch–Gordan coefficients*. The notation $\langle m_1 m_2 | jm\rangle$ corresponds to the general notation for the coefficients in the expansion of one set of functions in terms of another (11.18). To simplify, we have omitted the quantum numbers j_1 and j_2, which are the same in both sets of functions. When necessary, these are included, in the form $\langle j_1 m_1 j_2 m_2 | j_1 j_2 jm\rangle$.‡

† Under time reversal, the wave functions change in accordance with (60.2):

$$\psi_{jm} \to (-1)^{j-m} \psi_{j,-m}.$$

It is easily verified that the function ψ_{jm} on the left of (106.8) is transformed in this way if the functions $\psi_{j_1 m_1}$ and $\psi_{j_2 m_2}$ on the right are.

‡ The Clebsch–Gordan coefficients are also denoted in the literature by

$$C^{jm}_{m_1 m_2} \text{ or } C^{jm}_{j_1 m_1 j_2 m_2}.$$

The matrix of the transformation (106.9) is unitary (see §12). The coefficients of the inverse transformation

$$\psi^{(1)}{}_{j_1 m_1}\psi^{(2)}{}_{j_2 m_2} = \sum_{j=|j_2-j_1|}^{j_1+j_2} \langle j, m_1+m_2|m_1 m_2\rangle \psi_{j,m_1+m_2} \tag{106.11}$$

are therefore the complex conjugates of those in the transformation (106.9). We shall see later that these coefficients are real, so that we have simply

$$\langle m_1 m_2|jm\rangle = \langle jm|m_1 m_2\rangle.$$

According to the general rules of quantum mechanics, the squares of the coefficients in the expansion (106.11) give the probability for the system to have any particular values of j and m (for given j_1, m_1 and j_2, m_2).

The unitarity of the transformation (106.9) means that its coefficients satisfy certain orthogonality conditions. According to formulae (12.5) and (12.6)

$$\begin{aligned}\sum_{m_1,m_2} \langle m_1 m_2|jm\rangle\langle m_1 m_2|j'm'\rangle &= (2j+1)\sum_{m_1,m_2}\begin{pmatrix} j_1 & j_2 & j \\ m_1 & m_2 & -m\end{pmatrix}\begin{pmatrix} j_1 & j_2 & j' \\ m_1 & m_2 & -m'\end{pmatrix} \\ &= \delta_{jj'}\delta_{mm'},\end{aligned} \tag{106.12}$$

$$\begin{aligned}\sum_{j} \langle m_1 m_2|jm\rangle\langle m_1' m_2'|jm\rangle &= \sum_j (2j+1)\begin{pmatrix} j_1 & j_2 & j \\ m_1 & m_2 & -m\end{pmatrix}\begin{pmatrix} j_1 & j_2 & j \\ m_1' & m_2' & -m\end{pmatrix} \\ &= \delta_{m_1 m_1'}\delta_{m_2 m_2'}.\end{aligned} \tag{106.13}$$

The explicit general form of the 3j-symbols is quite lengthy. It can be written as†

$$\begin{pmatrix} j_1 & j_2 & j_3 \\ m_1 & m_2 & m_3\end{pmatrix} = \left[\frac{(j_1+j_2-j_3)!(j_1-j_2+j_3)!(-j_1+j_2+j_3)!}{(j_1+j_2+j_3+1)!}\right]^{1/2} \times$$

† The coefficients in (106.9) were first calculated by E. P. Wigner (1931). Their symmetry properties and the symmetrical expression (106.14) were first derived by G. Racah (1942). The most direct method of calculation is probably to go immediately from the spinor representation of ψ_0 (appropriately normalized) to the representation in the form of the sum (106.4) by means of the correspondence formula (57.6); it may be noted that, since the coefficient in this formula is real, so also must be the 3j-symbols. Another derivation is given by A. R. Edmonds, *Angular Momentum in Quantum Mechanics*, Princeton, 1957. The table of 3j-symbols given below is also taken from Edmonds's book.

$$\times [(j_1+m_1)!(j_1-m_1)!(j_2+m_2)!(j_2-m_2)!(j_3+m_3)!(j_3-m_3)!]^{1/2} \times$$

$$\times \sum_z \frac{(-1)^{z+j_1-j_2-m_3}}{z!(j_1+j_2-j_3-z)!(j_1-m_1-z)!(j_2+m_2-z)!(j_3-j_2+m_1+z)!(j_3-j_1-m_2+z)!}. \tag{106.14}$$

The summation is over all integers z but, since the factorial of a negative number is infinite, the sum contains only a finite number of terms. The coefficient of the sum is obviously symmetrical in the suffixes 1, 2, 3; the symmetry of the sum itself appears if the values of the summation variable z are interchanged.

Besides the symmetry properties (106.5) and (106.6), which follow immediately from the definition of the $3j$-symbols, the latter also have other symmetry properties, though the derivation of these is more complex and will not be given here. The properties in question can be conveniently formulated in terms of a three-by-three array of numbers derived from the parameters of the $3j$-symbol as follows:

$$\begin{pmatrix} j_1 & j_2 & j_3 \\ m_1 & m_2 & m_3 \end{pmatrix} = \begin{bmatrix} j_2+j_3-j_1 & j_3+j_1-j_2 & j_1+j_2-j_3 \\ j_1-m_1 & j_2-m_2 & j_3-m_3 \\ j_1+m_1 & j_2+m_2 & j_3+m_3 \end{bmatrix}; \tag{106.15}$$

the sum of the numbers in each row and each column of this array is $j_1+j_2+j_3$. Then (1) interchange of any two columns of the array multiplies the $3j$-symbol by $(-1)^{j_1+j_2+j_3}$ (the same property as that given by (106.5)); (2) the same is true for interchange of any two rows (for the two lower rows, the same property as that given by (106.6)); (3) the $3j$-symbol is unchanged when the rows and columns of the array are interchanged.†

Some of the simpler formulae for particular cases will be given here. The value

$$\begin{pmatrix} j & j & 0 \\ m & -m & 0 \end{pmatrix} = (-1)^{j-m} \frac{1}{\sqrt{(2j+1)}} \tag{106.16}$$

corresponds to formula (106.2). The formulae

$$\begin{pmatrix} j_1 & j_2 & j_1+j_2 \\ m_1 & m_2 & -m_1-m_2 \end{pmatrix} = (-1)^{j_1-j_2+m_1+m_2} \times$$

$$\times \left[\frac{(2j_1)!(2j_2)!(j_1+j_2+m_1+m_2)!(j_1+j_2-m_1-m_2)!}{(2j_1+2j_2+1)!(j_1+m_1)!(j_1-m_1)!(j_2+m_2)!(j_2-m_2)!} \right]^{1/2}, \tag{106.17}$$

$$\begin{pmatrix} j_1 & j_2 & j_3 \\ j_1 & -j_1-m_3 & m_3 \end{pmatrix} = (-1)^{-j_1+j_2+m_3} \times$$

† See T. Regge, *Il nuovo cimento* [10] **10**, 544, 1958; **11**, 116, 1959. The more profound mathematical features of the symmetry property (106.15) (and of the property (108.3) of the $6j$-symbols) are discussed in the review article by Ya. A. Smorodinskii and L. A. Shelepin, *Soviet Physics Uspekhi* **15**, 1, 1972.

$$\times\left[\frac{(2j_1)!(-j_1+j_2+j_3)!(j_1+j_2+m_3)!(j_3-m_3)!}{(j_1+j_2+j_3+1)!(j_1-j_2+j_3)!(j_1+j_2-j_3)!(-j_1+j_2-m_3)!(j_3+m_3)!}\right]^{1/2}$$

are obtained directly from (106.14). The derivation of the formula

$$\begin{pmatrix} j_1 & j_2 & j_3 \\ 0 & 0 & 0 \end{pmatrix} = (-1)^p\left[\frac{(j_1+j_2-j_3)!(j_1-j_2+j_3)!(-j_1+j_2+j_3)!}{(2p+1)!}\right]^{1/2}\times$$

$$\times\frac{p!}{(p-j_1)!(p-j_2)!(p-j_3)!}, \tag{106.18}$$

where $2p = j_1+j_2+j_3$ is even, requires a number of additional calculations;† when $2p$ is odd, this 3j-symbol is zero owing to the symmetry property (106.6).

Table 9 gives for reference the values of the 3j-symbols for $j_3 = \frac{1}{2}, 1, \frac{3}{2}, 2$. For each j_3 the minimum number of 3j-symbols are shown from which the remainder may be obtained by means of the relations (106.5), (106.6).

TABLE 9

Formulae for 3j-symbols

$$\begin{pmatrix} j+\frac{1}{2} & j & \frac{1}{2} \\ m & -m-\frac{1}{2} & \frac{1}{2} \end{pmatrix} = (-1)^{j-m-1/2}\left[\frac{j-m+\frac{1}{2}}{(2j+1)(2j+2)}\right]^{1/2}$$

$$(-1)^{j-m}\begin{pmatrix} j_1 & j & 1 \\ m & -m-m_3 & m_3 \end{pmatrix}$$

j_1 \ m_3	0
j	$\dfrac{2m}{[2j(2j+1)(2j+2)]^{1/2}}$
$j+1$	$-\left[\dfrac{2(j+m+1)(j-m+1)}{(2j+1)(2j+2)(2j+3)}\right]^{1/2}$

j_1 \ m_3	1
j	$\left[\dfrac{2(j-m)(j+m+1)}{2j(2j+1)(2j+2)}\right]^{1/2}$
$j+1$	$-\left[\dfrac{(j-m)(j-m+1)}{(2j+1)(2j+2)(2j+3)}\right]^{1/2}$

$$(-1)^{j-m+1/2}\begin{pmatrix} j_1 & j & \frac{3}{2} \\ m & -m-m_3 & m_3 \end{pmatrix}$$

j_1 \ m_3	$\frac{1}{2}$
$j+\frac{1}{2}$	$-(j+3m+\frac{3}{2})\left[\dfrac{j-m+\frac{1}{2}}{2j(2j+1)(2j+2)(2j+3)}\right]^{1/2}$
$j+\frac{3}{2}$	$\left[\dfrac{3(j-m+\frac{1}{2})(j-m+\frac{3}{2})(j+m+\frac{3}{2})}{(2j+1)(2j+2)(2j+3)(2j+4)}\right]^{1/2}$

† See Edmonds's book already quoted.

$j_1 \backslash m_3$	$\frac{3}{2}$
$j+\frac{1}{2}$	$-\left[\frac{3(j-m-\frac{1}{2})(j-m+\frac{1}{2})(j+m+\frac{3}{2})}{2j(2j+1)(2j+2)(2j+3)}\right]^{1/2}$
$j+\frac{3}{2}$	$\left[\frac{(j-m-\frac{1}{2})(j-m+\frac{1}{2})(j-m+\frac{3}{2})}{(2j+1)(2j+2)(2j+3)(2j+4)}\right]^{1/2}$

$$(-1)^{j-m}\begin{pmatrix} j_1 & j & 2 \\ m & -m-m_3 & m_3 \end{pmatrix}$$

$j_1 \backslash m_3$	0
j	$\frac{2[3m^2-j(j+1)]}{[(2j-1)2j(2j+1)(2j+2)(2j+3)]^{1/2}}$
$j+1$	$-2m\left[\frac{6(j+m+1)(j-m+1)}{2j(2j+1)(2j+2)(2j+3)(2j+4)}\right]^{1/2}$
$j+2$	$\left[\frac{6(j+m+2)(j+m+1)(j-m+2)(j-m+1)}{(2j+1)(2j+2)(2j+3)(2j+4)(2j+5)}\right]^{1/2}$

$j_1 \backslash m_3$	1
j	$(1+2m)\left[\frac{6(j+m+1)(j-m)}{(2j-1)2j(2j+1)(2j+2)(2j+3)}\right]^{1/2}$
$j+1$	$-2(j+2m+2)\left[\frac{(j-m+1)(j-m)}{2j(2j+1)(2j+2)(2j+3)(2j+4)}\right]^{1/2}$
$j+2$	$2\left[\frac{(j+m+2)(j-m+2)(j-m+1)(j-m)}{(2j+1)(2j+2)(2j+3)(2j+4)(2j+5)}\right]^{1/2}$

$j_1 \backslash m_3$	2
j	$\left[\frac{6(j-m-1)(j-m)(j+m+1)(j+m+2)}{(2j-1)2j(2j+1)(2j+2)(2j+3)}\right]^{1/2}$
$j+1$	$-2\left[\frac{(j-m-1)(j-m)(j-m+1)(j+m+2)}{2j(2j+1)(2j+2)(2j+3)(2j+4)}\right]^{1/2}$
$j+2$	$\left[\frac{(j-m-1)(j-m)(j-m+1)(j-m+2)}{(2j+1)(2j+2)(2j+3)(2j+4)(2j+5)}\right]^{1/2}$

PROBLEM

Determine the angle dependence of the wave functions of a particle with spin $\frac{1}{2}$ in states with given values of the orbital angular momentum l, the total angular momentum j and component thereof m.

SOLUTION. The problem is solved by the general formula (106.8), in which $\psi^{(1)}$ must be taken as the eigenfunctions of the orbital angular momentum (i.e. the spherical harmonic functions Y_{lm_l}), and $\psi^{(2)}$ as the spin wave function $\chi(\sigma)$ (where $\sigma = \pm\frac{1}{2}$):

$$\psi_{jm} = (-1)^{l+m-1/2}\sqrt{(2j+1)}\sum_{\sigma}\begin{pmatrix} l & \frac{1}{2} & j \\ m-\sigma & \sigma & -m \end{pmatrix} Y_{l,m-\sigma}\chi(\sigma).$$

Substituting the values of the $3j$-symbols, we obtain

$$\psi_{l+1/2,m} = \sqrt{\frac{j+m}{2j}}\chi(\tfrac{1}{2})Y_{l,m-1/2} + \sqrt{\frac{j-m}{2j}}\chi(-\tfrac{1}{2})Y_{l,m+1/2},$$

$$\psi_{l-1/2,m} = -\sqrt{\frac{j-m+1}{2j+2}}\chi(\tfrac{1}{2})Y_{l,m-1/2} + \sqrt{\frac{j+m+1}{2j+2}}\chi(-\tfrac{1}{2})Y_{l,m+1/2}.$$

§107. Matrix elements of tensors

In §29 formulae have been obtained which give the matrix elements of a vector physical quantity in terms of the value of the angular momentum component. These formulae are really a particular case of the corresponding general formulae for an irreducible (see §57) tensor of any rank.†

The set of $2k+1$ components of an irreducible tensor of rank k (an integer) are equivalent, as regards their transformation properties, to a set of $2k+1$ spherical harmonic functions Y_{kq}, $q = -k, \ldots, k$ (see the last footnote to §57). This means that, by means of appropriate linear combinations of the components of the tensor, we can obtain a set of quantities which are transformed under rotations as the functions Y_{kq}. A set of such quantities, which will be denoted here by f_{kq}, is called a *spherical tensor* of rank k.

For example, $k = 1$ for a vector, and the quantities f_{1q} are related to the components of the vector by the formulae

$$f_{10} = ia_z, \qquad f_{1,\pm 1} = \mp \frac{i}{\sqrt{2}}(a_x \pm ia_y); \tag{107.1}$$

cf. (57.7). The corresponding formulae for a tensor of rank two are

$$\left.\begin{aligned} f_{20} = -\sqrt{\tfrac{3}{2}}a_{zz}, \qquad f_{2,\pm 1} = \pm(a_{xz} \pm ia_{yz}), \\ f_{2,\pm 2} = -\tfrac{1}{2}(a_{xx} - a_{yy} \pm 2ia_{xy}), \end{aligned}\right\} \tag{107.2}$$

with $a_{xx} + a_{yy} + a_{zz} = 0$.‡

The construction of tensor products from two (or more) spherical tensors $f_{k_1q_1}$, $f_{k_2q_2}$ is effected in accordance with the rules for addition of angular momenta, with k_1, k_2 formally representing the "angular momenta" corresponding to these tensors. Thus from two spherical tensors of ranks k_1 and k_2, one can form spherical tensors of ranks $K = k_1 + k_2, \ldots, |k_1 - k_2|$ by means of the formulae

$$\begin{aligned} (f_{k_1}g_{k_2})_{KQ} &= \sum_{q_1, q_2} \langle q_1 q_2 | KQ \rangle f_{k_1q_1} g_{k_2q_2} \\ &= (-1)^{k_1-k_2+Q}\sqrt{(2K+1)} \sum_{q_1, q_2} \begin{pmatrix} k_1 & k_2 & K \\ q_1 & q_2 & -Q \end{pmatrix} f_{k_1q_1} g_{k_2q_2}; \end{aligned} \tag{107.3}$$

cf. (106.9). The scalar product of two spherical tensors of the same rank k is, however, usually defined as

$$(f_k g_k)_{00} = \sum_q (-1)^{k-q} f_{kq} g_{k,-q}, \tag{107.4}$$

which differs from the definition according to (107.3) with $K = Q = 0$ by

† The analysis of the problems discussed in §§107–109, and most of the results given, are due to G. Racah (1942–1943).

‡ It is assumed that the quantities f_{kq} are complex only because of the use of spherical components, i.e. that the original Cartesian components of the tensor are real.

the factor $\sqrt{(2k+1)}$; cf. (106.2).† This definition can also be written in the form

$$(f_k g_k)_{00} = \sum_q f_{kq} g_{kq}^*$$

if we note that the complex conjugate of a spherical tensor is given by

$$f_{kq}^* = (-1)^{k-q} f_{k,-q};$$

cf. (28.9).‡

The representation of physical quantities in the form of spherical tensors is particularly convenient for the calculation of their matrix elements, since it allows the direct application of the results of the theory of addition of angular momenta.

By the definition of the matrix elements we have

$$\hat{f}_{kq}\psi_{njm} = \sum_{n'j'm'} \langle n'j'm'| f_{kq}|njm\rangle \psi_{n'j'm'}, \tag{107.5}$$

where the ψ_{njm} are the wave functions of the stationary states of the system, described by its angular momentum j, component thereof m and the set of the remaining quantum numbers n. As regards transformation properties, the functions on the right-hand and left-hand sides of equation (107.5) correspond to the respective sides of equation (106.11). Hence we immediately obtain the selection rules: the matrix elements of the components f_{kq} of an irreducible tensor of rank k are zero except for transitions $jm \to j'm'$ which satisfy the "angular momentum addition rule" $\mathbf{j}' = \mathbf{j}+\mathbf{k}$; the numbers j', j, k must satisfy the "triangle rule", i.e. must be able to form the sides of a closed triangle, and $m' = m+q$. In particular, the diagonal matrix elements can be different from zero only if $2j \geqslant k$.

Next, it follows from the same transformation correspondence that the coefficients in the sum (107.5) must be proportional to the coefficients in (106.11) (the *Wigner–Eckart theorem*). This determines the dependence of the coefficients on the numbers m and m', and the matrix elements are therefore written in the form

$$\langle n'j'm'| f_{kq}|njm\rangle = i^k(-1)^{j_{\max}-m'}\begin{pmatrix} j' & k & j \\ -m' & q & m \end{pmatrix}\langle n'j'\| f_k\|nj\rangle, \tag{107.6}$$

where $j_{\max}$ is the greater of j and j', and the $\langle n'j'\| f_k\|nj\rangle$ are quantities independent of m, m' and q, called the *reduced matrix elements*. This formula gives the solution to the problem of determining the dependence of the matrix elements on the angular momentum components. The dependence

† If **A** and **B** are two vectors corresponding to the spherical tensors f_{1q} and g_{1q} according to formulae (107.1), then $(f_1 g_1)_{00} = \mathbf{A}\,.\,\mathbf{B}$.

‡ Here we may repeat the comment made concerning formula (106.8): with this rule, taking the complex conjugate of the tensors of ranks k_1 and k_2 on the right of (107.3) leads to a similar result for the tensor of rank K on the left.

is entirely governed by the symmetry properties with respect to the rotation group, whereas the dependence on the other quantum numbers is determined by the physical nature of the f_{kq} themselves.†

The operators $\hat{f}_{kq}$ are related by

$$\hat{f}_{kq}{}^{+} = (-1)^{k-q}\hat{f}_{k,-q}. \tag{107.7}$$

The equation

$$\langle n'j'm'|f_{kq}|njm\rangle^* = (-1)^{k-q}\langle njm|f_{k,-q}|n'j'm'\rangle \tag{107.8}$$

therefore holds for their matrix elements. Substituting (107.6) and using the properties (106.5) and (106.6) of the 3*j*-symbols, we obtain for the reduced matrix elements the "Hermitian" relation‡

$$\langle n'j'\| f_k \| nj\rangle = \langle nj\| f_k \| n'j'\rangle^*. \tag{107.9}$$

The matrix elements of the scalar (107.4) are diagonal in j and m. According to rule of matrix multiplication

$$\langle n'jm|(f_k g_k)_{00}|njm\rangle = \sum_q (-1)^{k-q} \sum_{n'',j'',m''} \langle n'jm|f_{kq}|n''j''m''\rangle \times \\ \times \langle n''j''m''|g_{k,-q}|njm\rangle.$$

Substituting here the expressions (107.6) and effecting the summation over q and m'' by means of the orthogonality relation for the 3*j*-symbols, we obtain

$$\langle n'jm|(f_k g_k)_{00}|njm\rangle = \frac{1}{2j+1}\sum_{n'',j''} \langle n'j\| f_k \| n''j''\rangle\langle n''j''\| g_k \| nj\rangle. \tag{107.10}$$

Similarly, we easily obtain the following formulae for the sums of the squared matrix elements:

$$\sum_{q,m'} |\langle n'j'm'|f_{kq}|njm\rangle|^2 = \frac{1}{2j+1}|\langle n'j'\| f_k \| nj\rangle|^2, \tag{107.11}$$

$$\sum_{m,m'} |\langle n'j'm'|f_{kq}|njm\rangle|^2 = \frac{1}{2k+1}|\langle n'j'\| f_k \| nj\rangle|^2. \tag{107.12}$$

In the first of these the summation is over q and m' for a given value of m, and in the second it is over m and m' for a given value of q (in every case, $m' = m+q$).

---

† From these results, in particular, there follow at once the selection rules given in §29 for the matrix elements of a vector and formulae (29.7)–(29.9) for these.

‡ The phase factor in the definition (107.6) is in fact chosen so as to make this relation valid.

For reference purposes we may consider the case where the quantities f_{kq} are the spherical harmonic functions Y_{kq} themselves, and give their matrix elements for transitions between the states of one particle with integral orbital angular momenta l_1 and l_2, i.e. the integrals

$$\langle l_1m_1|Y_{lm}|l_2m_2\rangle = \int Y_{l_1m_1}{}^* Y_{lm} Y_{l_2m_2}\, do. \tag{107.13}$$

Besides the selection rule corresponding to the rule of addition of angular momenta ($\mathbf{l}+\mathbf{l}_2 = \mathbf{l}_1$), there is also a rule for these matrix elements whereby the sum $l+l_1+l_2$ must be even. This is due to the conservation of parity, according to which the product $(-1)^{l_1+l_2}$ of the parities of the two states must be the same as the parity $(-1)^l$ of the physical quantity considered (see §30).

The matrix elements (107.13) are a particular case of a more general integral to be calculated in §110 (see the footnote to that section). They are given by

$$\begin{aligned}\langle l_1m_1|Y_{lm}|l_2m_2\rangle \\ = (-1)^{m_1} i^{-l_1+l_2+l}\begin{pmatrix} l_1 & l & l_2 \\ -m_1 & m & m_2\end{pmatrix}\begin{pmatrix} l_1 & l & l_2 \\ 0 & 0 & 0\end{pmatrix} \times \\ \times \left[\frac{(2l+1)(2l_1+1)(2l_2+1)}{4\pi}\right]^{1/2}.\end{aligned} \tag{107.14}$$

In particular, for $m_1 = m_2 = m = 0$ we find the integral of the product of three Legendre polynomials:

$$\int_{-1}^{1} P_l(\mu)P_{l_1}(\mu)P_{l_2}(\mu)\, d\mu = 2\begin{pmatrix} l_1 & l & l_2 \\ 0 & 0 & 0\end{pmatrix}^2. \tag{107.15}$$

§108. 6*j*-symbols

In §106 we have defined 3*j*-symbols as the coefficients in the sum (106.4) which represents the wave function of a system of three particles with zero total angular momentum. As regards the transformation properties under rotations, this sum is a scalar. Hence it follows that the set of 3*j*-symbols with given values of j_1, j_2, j_3 (and all possible m_1, m_2, m_3) may be regarded as a set of quantities which are transformed under rotations according to a law contragredient to that for the products $\psi_{j_1m_1}\psi_{j_2m_2}\psi_{j_3m_3}$, so that the sum as a whole is invariant.

From this viewpoint we may put the problem of constructing a scalar consisting of 3*j*-symbols only. This scalar must depend only on the numbers j, and not on the numbers m, which are altered by rotations. In other words, it must be expressible in terms of sums over all the numbers m. Each such sum consists of a "contraction" of the product of two 3*j*-symbols according to the formula

$$\sum_m (-1)^{j-m}\begin{pmatrix} j & \cdot & \cdot \\ m & \cdot & \cdot\end{pmatrix}\begin{pmatrix} j & \cdot & \cdot \\ -m & \cdot & \cdot\end{pmatrix}; \tag{108.1}$$

cf. the method of constructing the scalar (106.2).

Since in each "contraction" a pair of numbers m is involved, we must consider products of an even number of $3j$-symbols in constructing the complete scalar. The contraction of the product of two $3j$-symbols gives, owing to their orthogonality, the trivial result

$$\sum_{m_1,m_2,m_3} \begin{pmatrix} j_1 & j_2 & j_3 \\ m_1 & m_2 & m_3 \end{pmatrix} \begin{pmatrix} j_1 & j_2 & j_3 \\ -m_1 & -m_2 & -m_3 \end{pmatrix} (-1)^{j_1+j_2+j_3-m_1-m_2-m_3}$$

$$= \sum_{m_1,m_2,m_3} \begin{pmatrix} j_1 & j_2 & j_3 \\ m_1 & m_2 & m_3 \end{pmatrix}^2 = 1;$$

here we have used the equation $m_1+m_2+m_3 = 0$ and formulae (106.6) and (106.12). The smallest number of factors needed to form a non-trivial scalar is therefore four.

In each $3j$-symbol the three numbers j form a closed triangle. Since each number j must appear in the "contraction" in two $3j$-symbols, it is clear

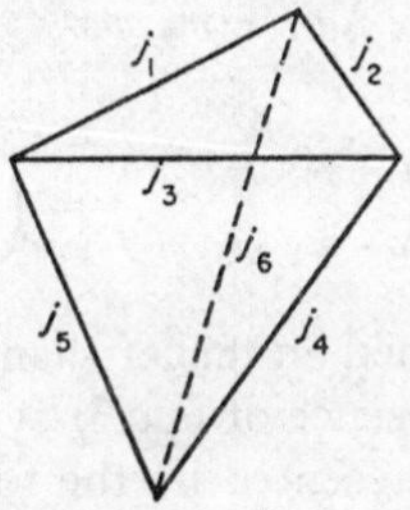

FIG. 45

that in the construction of a scalar from the products of four $3j$-symbols there are six numbers j forming the edges of an irregular tetrahedron (Fig. 45), one face of which corresponds to each $3j$-symbol. In defining the required scalar it is customary to use a certain condition as regards the contraction process, given by the formula

$$\begin{Bmatrix} j_1 & j_2 & j_3 \\ j_4 & j_5 & j_6 \end{Bmatrix} = \sum_{\text{all } m} (-1)^{\Sigma_i(j_i-m_i)} \begin{pmatrix} j_1 & j_2 & j_3 \\ -m_1 & -m_2 & -m_3 \end{pmatrix} \times$$

$$\times \begin{pmatrix} j_1 & j_5 & j_6 \\ m_1 & -m_5 & m_6 \end{pmatrix} \begin{pmatrix} j_4 & j_2 & j_6 \\ m_4 & m_2 & -m_6 \end{pmatrix} \begin{pmatrix} j_4 & j_5 & j_3 \\ -m_4 & m_5 & m_3 \end{pmatrix}. \quad (108.2)$$

The summation here is over all possible values of all the numbers m; however, since the sum of the three m in every $3j$-symbol must be zero, only three of the six m are in fact independent. The quantities defined by the formula (108.2) are called *6j-symbols* or *Racah coefficients*.†

† The notation

$$W(j_1j_2j_4j_5; j_3j_6) = (-1)^{j_1+j_2+j_4+j_5} \begin{Bmatrix} j_1 & j_2 & j_3 \\ j_4 & j_5 & j_6 \end{Bmatrix}$$

is also used in the literature.

From the definition (108.2), using the symmetry properties of the $3j$-symbols, we easily see that a $6j$-symbol is unchanged by any permutation of its three columns, and in any pair of columns the two numbers can be simultaneously interchanged. Owing to these symmetry properties, the sequence of numbers $j_1, \ldots, j_6$ in the $6j$-symbol can be put in 24 equivalent forms.† In addition, the $6j$-symbols have another, less evident, symmetry property which states an equality between symbols with different sets of numbers j:‡

$$\begin{Bmatrix} j_1 & j_2 & j_3 \\ j_4 & j_5 & j_6 \end{Bmatrix} = \begin{Bmatrix} j_1 & \frac{1}{2}(j_2+j_5+j_3-j_6) & \frac{1}{2}(j_3+j_6+j_2-j_5) \\ j_4 & \frac{1}{2}(j_2+j_5+j_6-j_3) & \frac{1}{2}(j_3+j_6+j_5-j_2) \end{Bmatrix}. \tag{108.3}$$

We may mention a useful relation between the $6j$- and $3j$-symbols which can be derived from the definition (108.2):

$$\sum_{m_4,m_5,m_6} (-1)^{j_4+j_5+j_6-m_4-m_5-m_6} \begin{pmatrix} j_1 & j_5 & j_6 \\ m_1 & -m_5 & m_6 \end{pmatrix} \times$$

$$\times \begin{pmatrix} j_4 & j_2 & j_6 \\ m_4 & m_2 & -m_6 \end{pmatrix} \begin{pmatrix} j_4 & j_5 & j_3 \\ -m_4 & m_5 & m_3 \end{pmatrix} = \begin{pmatrix} j_1 & j_2 & j_3 \\ m_1 & m_2 & m_3 \end{pmatrix} \begin{Bmatrix} j_1 & j_2 & j_3 \\ j_4 & j_5 & j_6 \end{Bmatrix}. \tag{108.4}$$

The expression which is summed on the left-hand side of the equation differs from that in (108.2) by the absence of one $3j$-symbol. We can therefore say that the sum in (108.4) is represented by the tetrahedron (Fig. 45) without one of its faces; this determines the difference of the sum from a scalar. In other words, as regards transformation properties it corresponds to one $3j$-symbol, the one on the right-hand side of equation (108.4), to which it must be proportional. The proportionality coefficient (the $6j$-symbol on the right-hand side of the equation) is easily found by multiplying both sides by

$$\begin{pmatrix} j_1 & j_2 & j_3 \\ m_1 & m_2 & m_3 \end{pmatrix}$$

and summing over the remaining numbers m_1, m_2, m_3.

The $6j$-symbols arise naturally in connection with the following problem concerning the addition of three angular momenta.

Let three angular momenta j_1, j_2, j_3 be added to give a resultant angular momentum J. If the value of J (and of its component M) is given, the state of the system is not yet uniquely determined, but depends also on the manner of addition of the angular momenta (or, as we say, on their coupling scheme).

For example, let us consider two such coupling schemes: (1) first the angular momenta j_1 and j_2 are added to give a total angular momentum j_{12}, and then j_{12} and j_3 are added to give the final angular momentum J; (2) the

† If we regard the tetrahedron in Fig. 45 as being regular, the 24 equivalent permutations of the numbers j can be obtained by means of the 24 symmetry transformations (rotations and reflections) of the tetrahedron.

‡ See T. Regge, *Il nuovo cimento* [10] **10**, 544, 1958; **11**, 116, 1959.

angular momenta j_2 and j_3 are added to give j_{23}, and then j_{23} and j_1 to give J. The former scheme corresponds to states in which the quantity j_{12} (as well as j_1, j_2, j_3, J, M) has a definite value; their wave functions will be denoted by $\psi_{j_{12}JM}$ (omitting for brevity the repeated suffixes j_1, j_2, j_3). Similarly, the wave functions of the second coupling scheme are denoted by $\psi_{j_{23}JM}$. In both cases the values of the "intermediate" angular momentum (j_{12} or j_{23}) are in general not unique, so that (for given J and M) we have two different sets of states differing in the values of j_{12} or j_{23}. According to the general rules, functions of these two sets are related by a certain unitary transformation:

$$\psi_{j_{23}JM} = \sum_{j_{12}} \langle j_{12}|j_{23}\rangle \psi_{j_{12}JM}. \tag{108.5}$$

It is evident from physical considerations that the coefficients in this transformation are independent of the number M: they must be independent of the orientation of the whole system in space. Thus they depend only on the values of the six angular momenta $j_1, j_2, j_3, j_{12}, j_{23}, J$, not on their components, i.e. are scalar quantities (in the sense defined above). The actual calculation of these coefficients is easily effected as follows.

By a repeated application of formula (106.9) we find

$$\begin{aligned}\psi_{j_{23}JM} &= \sum_{(m)} \langle m_1 m_{23}|JM\rangle \psi_{j_1m_1}\psi_{j_{23}m_{23}}\\ &= \sum_{(m)} \langle m_1 m_{23}|JM\rangle\langle m_2 m_3|j_{23}m_{23}\rangle \psi_{j_1m_1}\psi_{j_2m_2}\psi_{j_3m_3},\end{aligned}$$

$$\psi_{j_{12}JM} = \sum_{(m)} \langle m_3 m_{12}|JM\rangle\langle m_1 m_2|j_{12}m_{12}\rangle \psi_{j_1m_1}\psi_{j_2m_2}\psi_{j_3m_3},$$

where (m) denotes that the summation is over all numbers $m_1, m_2, \ldots$ that appear in the expression. From the orthonormality of the functions ψ_{jm} we have

$$\begin{aligned}\langle j_{12}|j_{23}\rangle &\equiv \int \psi_{j_{12}JM}{}^*\psi_{j_{23}JM}\,dq\\ &= \sum_{(m)} \langle m_3 m_{12}|JM\rangle\langle m_1 m_{23}|JM\rangle\langle m_1 m_2|j_{12}m_{12}\rangle\langle m_2 m_3|j_{23}m_{23}\rangle.\end{aligned}$$

The sum on the right-hand side is taken for a fixed M, but the result is actually independent of M (for the reason already mentioned). The summation can therefore be extended over the values of M if the sum is multiplied by a factor $1/(2j+1)$. Expressing the coefficients $\langle m_1 m_2|jm\rangle$ in terms of $3j$-symbols by (106.10), we obtain the following expression:

$$\langle j_{12}|j_{23}\rangle = (-1)^{j_1+j_2+j_3+J}\sqrt{[(2j_{12}+1)(2j_{23}+1)]}\begin{Bmatrix} j_1 & j_2 & j_{12}\\ j_3 & J & j_{23}\end{Bmatrix}. \tag{108.6}$$

The relation between the $6j$-symbols and the transformation coefficients in (108.5) makes it easy to derive some useful formulae for the sums of products of $6j$-symbols.

First of all, since the transformation (108.5) is unitary and its coefficients are real, the relation

$$\sum_{j}(2j+1)(2j''+1)\begin{Bmatrix} j_1 & j_2 & j' \\ j_3 & j_4 & j \end{Bmatrix}\begin{Bmatrix} j_3 & j_2 & j \\ j_1 & j_4 & j'' \end{Bmatrix} = \delta_{j'j''} \tag{108.7}$$

holds.

Next, let us consider the three coupling schemes of three angular momenta, with intermediate sums j_{12}, j_{23} and j_{31} respectively. The coefficients (108.6) of the corresponding transformations are related, according to the matrix multiplication rule, by

$$\sum_{j_{23}} \langle j_{12}|j_{23}\rangle\langle j_{23}|j_{31}\rangle = \langle j_{12}|j_{31}\rangle.$$

Substituting here (108.6) and renumbering the suffixes we have

$$\sum_{j}(-1)^{j+j_3+j_6}(2j+1)\begin{Bmatrix} j_2 & j_4 & j_6 \\ j_1 & j_5 & j \end{Bmatrix}\begin{Bmatrix} j_4 & j_1 & j \\ j_2 & j_5 & j_3 \end{Bmatrix} = \begin{Bmatrix} j_1 & j_2 & j_3 \\ j_4 & j_5 & j_6 \end{Bmatrix}. \tag{108.8}$$

Finally, by considering the various coupling schemes of four angular momenta, we can derive† the following addition formula for the products of three $6j$-symbols:

$$\sum_{j}(-1)^{j+\sum_1^9 j_i}(2j+1)\begin{Bmatrix} j_4 & j_2 & j_6 \\ j_9 & j_8 & j \end{Bmatrix}\begin{Bmatrix} j_2 & j_1 & j_3 \\ j_7 & j & j_9 \end{Bmatrix}\begin{Bmatrix} j_4 & j_3 & j_5 \\ j_7 & j_8 & j \end{Bmatrix}$$

$$= \begin{Bmatrix} j_1 & j_2 & j_3 \\ j_4 & j_5 & j_6 \end{Bmatrix}\begin{Bmatrix} j_6 & j_1 & j_5 \\ j_7 & j_8 & j_9 \end{Bmatrix} \tag{108.9}$$

(L. C. Biedenharn, and J. P. Elliott, 1953).

For reference we shall give some explicit formulae for the $6j$-symbols. In the general case, a $6j$-symbol can be written as the following sum:

$$\begin{Bmatrix} j_1 & j_2 & j_3 \\ j_4 & j_5 & j_6 \end{Bmatrix} = \triangle(j_1j_2j_3)\triangle(j_1j_5j_6)\triangle(j_4j_2j_6)\triangle(j_4j_5j_3)\times$$

$$\times\sum_{z}\frac{(-1)^z(z+1)!}{(z-j_1-j_2-j_3)!(z-j_1-j_5-j_6)!(z-j_4-j_2-j_6)!(z-j_4-j_5-j_3)!}\times$$

$$\times\frac{1}{(j_1+j_2+j_4+j_5-z)!(j_2+j_3+j_5+j_6-z)!(j_3+j_1+j_6+j_4-z)!}, \tag{108.10}$$

where

$$\triangle(abc) = \left[\frac{(a+b-c)!(a-b+c)!(-a+b+c)!}{(a+b+c+1)!}\right]^{1/2},$$

and the sum is taken over all positive integers z for which none of the factorials in the denominator has a negative argument.

† See Edmonds's book quoted in §106.

Table 10 gives the values of the 6j-symbols for cases where one of the parameters is 0, $\frac{1}{2}$ or 1.

Finally, we shall make a few remarks concerning the higher-order scalars constructed from the 3j-symbols.

The next in complexity after the 6j-symbols is a scalar formed by contracting products of six 3j-symbols. These 3j-symbols contain 18 numbers j equal in pairs, and so the resulting scalar depends on 9 parameters j. It is called a *9j-symbol* and is defined as follows† (E. P. Wigner 1951):

TABLE 10

Formulae for 6j-symbols

$$\begin{Bmatrix} a & b & c \\ 0 & c & b \end{Bmatrix} = \frac{(-1)^s}{\sqrt{[(2b+1)(2c+1)]}}, \qquad s = a+b+c$$

$$\begin{Bmatrix} a & b & c \\ \frac{1}{2} & c-\frac{1}{2} & b+\frac{1}{2} \end{Bmatrix} = (-1)^s\left[\frac{(s-2b)(s-2c+1)}{(2b+1)(2b+2)2c(2c+1)}\right]^{1/2}$$

$$\begin{Bmatrix} a & b & c \\ \frac{1}{2} & c-\frac{1}{2} & b-\frac{1}{2} \end{Bmatrix} = (-1)^s\left[\frac{(s+1)(s-2a)}{2b(2b+1)2c(2c+1)}\right]^{1/2}$$

$$\begin{Bmatrix} a & b & c \\ 1 & c-1 & b-1 \end{Bmatrix} = (-1)^s\left[\frac{s(s+1)(s-2a-1)(s-2a)}{(2b-1)2b(2b+1)(2c-1)2c(2c+1)}\right]^{1/2}$$

$$\begin{Bmatrix} a & b & c \\ 1 & c-1 & b \end{Bmatrix} = (-1)^s\left[\frac{2(s+1)(s-2a)(s-2b)(s-2c+1)}{2b(2b+1)(2b+2)(2c-1)2c(2c+1)}\right]^{1/2}$$

$$\begin{Bmatrix} a & b & c \\ 1 & c-1 & b+1 \end{Bmatrix} = (-1)^s\left[\frac{(s-2b-1)(s-2b)(s-2c+1)(s-2c+2)}{(2b+1)(2b+2)(2b+3)(2c-1)2c(2c+1)}\right]^{1/2}$$

$$\begin{Bmatrix} a & b & c \\ 1 & c & b \end{Bmatrix} = (-1)^{s+1}\frac{2[b(b+1)+c(c+1)-a(a+1)]}{[2b(2b+1)(2b+2)2c(2c+1)(2c+2)]^{1/2}}$$

$$\begin{Bmatrix} j_{11} & j_{12} & j_{13} \\ j_{21} & j_{22} & j_{23} \\ j_{31} & j_{32} & j_{33} \end{Bmatrix} = \sum_{\text{all } m} \begin{pmatrix} j_{11} & j_{12} & j_{13} \\ m_{11} & m_{12} & m_{13} \end{pmatrix}\begin{pmatrix} j_{21} & j_{22} & j_{23} \\ m_{21} & m_{22} & m_{23} \end{pmatrix}\begin{pmatrix} j_{31} & j_{32} & j_{33} \\ m_{31} & m_{32} & m_{33} \end{pmatrix}\times$$
$$\times\begin{pmatrix} j_{11} & j_{21} & j_{31} \\ m_{11} & m_{21} & m_{31} \end{pmatrix}\begin{pmatrix} j_{12} & j_{22} & j_{32} \\ m_{12} & m_{22} & m_{32} \end{pmatrix}\begin{pmatrix} j_{13} & j_{23} & j_{33} \\ m_{13} & m_{23} & m_{33} \end{pmatrix}. \quad \textbf{(108.11)}$$

This quantity can also be written as the sum of products of three 6j-symbols:

$$\begin{Bmatrix} j_{11} & j_{12} & j_{13} \\ j_{21} & j_{22} & j_{23} \\ j_{31} & j_{32} & j_{33} \end{Bmatrix} = \sum_j (-1)^{2j}(2j+1)\begin{Bmatrix} j_{11} & j_{21} & j_{31} \\ j_{32} & j_{33} & j \end{Bmatrix}\begin{Bmatrix} j_{12} & j_{22} & j_{32} \\ j_{21} & j & j_{23} \end{Bmatrix}\begin{Bmatrix} j_{13} & j_{23} & j_{33} \\ j & j_{11} & j_{12} \end{Bmatrix}. \quad (108.12)$$

† According to the general rule of contraction (108.1) it would be necessary to write the arguments m in the last three 3j-symbols with the minus sign and to include in the summand a factor $(-1)^{\Sigma(j-m)}$. However, by using the property (106.6) of the 3j-symbols and the fact that in this case, as is easily seen, the sum $\sum m$ of all the nine numbers m is zero, we have the definition (108.11).

The equivalence of (108.11) and (108.12) can be seen by substituting in (108.12) the definition (108.2) and using the orthogonality properties of the $3j$-symbols.

The $9j$-symbol has a high degree of symmetry, which follows directly from the definition (108.11) and the symmetry properties of the $3j$-symbols. It is easily seen that when any two rows or columns are interchanged the $9j$-symbol is multiplied by $(-1)^{\Sigma j}$. Moreover, the $9j$-symbol is unaltered by transposition, i.e. interchange of rows and columns.

Scalars of still higher orders depend on a still larger number of parameters j. It is evident that this number will always be a multiple of three ($3nj$-symbols). We shall not pause to discuss their properties, but merely mention that for every $n > 3$ there is more than one type of $3nj$-symbol, and these do not reduce to one another. For example, there are two different types of $12j$-symbol.†

§109. Matrix elements for addition of angular momenta

Let us again consider a system consisting of two parts (referred to as sub-systems 1 and 2), and let $f_{kq}^{(1)}$ be a spherical tensor pertaining to sub-system 1. Its matrix elements with respect to the wave functions of this sub-system are given, according to (107.6), by the formula

$$\langle n_1'j_1'm_1'|f_{kq}^{(1)}|n_1j_1m_1\rangle = i^k(-1)^{j_{1,\max}-m_1'}\begin{pmatrix} j_1' & k & j_1 \\ -m_1' & q & m_1 \end{pmatrix}\langle n_1'j_1'\| f_k^{(1)}\| n_1j_1\rangle. \tag{109.1}$$

The question arises of calculating the matrix elements of these quantities with respect to the wave functions of the system as a whole. We shall show how they may be expressed in terms of the same reduced matrix elements as appear in the expression (109.1).

The states of the system as a whole are defined by the quantum numbers j_1, j_2, J, M, n_1, n_2 (where J and M are the angular momentum and its component for the whole system). Since $f_{kq}^{(1)}$ refers to sub-system 1, its operator commutes with the angular momentum operator of sub-system 2. Its matrix is therefore diagonal with respect to j_2; it is also diagonal with respect to the remaining quantum numbers n_2 of this sub-system. These indices j_2, n_2 will be omitted, for brevity, and the required matrix elements will be written as

$$\langle n_1'j_1'J'M'|f_{kq}^{(1)}|n_1j_1JM\rangle.$$

According to (107.6), their dependence on the number M is given by

$$\langle n_1'j_1'J'M'|f_{kq}^{(1)}|n_1j_1JM\rangle = i^k(-1)^{J_{\max}-M'}\begin{pmatrix} J' & k & J \\ -M' & q & M \end{pmatrix}\langle n_1'j_1'J'\| f_k^{(1)}\| n_1j_1J\rangle. \tag{109.2}$$

† A more detailed account of the theory of $9j$-symbols and of the properties of $3nj$-symbols is given in Edmonds's book quoted in §106 and by A. P. Yutsis, I. B. Levinson and V. V. Vanagas, *Mathematical Apparatus of the Theory of Angular Momentum*, Oldbourne Press, London 1963 (*Matematicheskii apparat teorii momenta kolichestva dvizheniya*, Vilnius 1960).

To establish the relation between the reduced matrix elements on the right-hand sides of (109.1) and (109.2) we write, using the definition of the matrix elements,

$$\langle n_1'j_1'J'M'|f_{kq}^{(1)}|n_1j_1JM\rangle = \int \psi_{J'M'}^{*}\hat{f}_{kq}^{(1)}\psi_{JM}\,dq$$

$$= \sum_{m_1,m_1'} (-1)^{j_1'-j_1+M'-M}\sqrt{[(2J'+1)(2J+1)]}\begin{pmatrix} j_1' & j_2 & J' \\ m_1' & m_2 & -M' \end{pmatrix}\times$$

$$\times\begin{pmatrix} j_1 & j_2 & J \\ m_1 & m_2 & -M \end{pmatrix}\langle n_1'j_1'm_1'|f_{kq}^{(1)}|n_1j_1m_1\rangle.$$

Substituting here (109.1) and (109.2) and comparing the resulting relation with formula (108.4), we see that the ratio of the reduced matrix elements in (109.1) and (109.2) must be proportional to a certain 6j-symbol. A careful comparison of the two relations mentioned leads to the final formula

$$\langle n_1'j_1'J'\| f_k^{(1)}\|n_1j_1J\rangle = (-1)^{j_{1,\max}+j_2+J_{\min}+k}\sqrt{[(2J+1)(2J'+1)]}\times$$

$$\times\begin{Bmatrix} j_1' & J' & j_2 \\ J & j_1 & k \end{Bmatrix}\langle n_1'j_1'\| f_k^{(1)}\|n_1j_1\rangle, \quad (109.3)$$

where $j_{1,\max}$ is the greater of j_1, j_1', and $J_{\min}$ the smaller of J, J'. A similar formula for the reduced matrix elements of the spherical tensor pertaining to the second sub-system is

$$\langle n_2'j_2'J'\| f_k^{(2)}\|n_2j_2J\rangle = (-1)^{j_1+j_{2,\min}+J_{\max}+k}\sqrt{[(2J+1)(2J'+1)]}\times$$

$$\times\begin{Bmatrix} j_2' & J' & j_1 \\ J & j_2 & k \end{Bmatrix}\langle n_2'j_2'\| f_k^{(2)}\|n_2j_2\rangle. \quad (109.4)$$

The lack of complete symmetry between the expressions (109.3) and (109.4) (in the exponent of -1) is due to the dependence of the phase of the wave functions on the order of addition of the angular momenta. The difference must be borne in mind when calculating matrix elements for both sub-systems simultaneously.

We shall also derive a useful formula for the matrix elements, with respect to the wave functions of the whole system, of a scalar product (see the definition (107.4)) of two spherical tensors of the same rank k pertaining to different sub-systems (and therefore commuting). According to (107.10), these matrix elements are given in terms of the reduced matrix elements of each tensor (with respect to the wave functions of the whole system) by

$$\langle n_1'n_2'j_1'j_2'JM|(f_k^{(1)}f_k^{(2)})_{00}|n_1n_2j_1j_2JM\rangle$$

$$= \frac{1}{2J+1}\sum_{J''}\langle n_1'j_1'J\| f_k^{(1)}\|n_1j_1J''\rangle\langle n_2'j_2'J''\| f_k^{(2)}\|n_2j_2J\rangle,$$

where we have used the fact that the matrix of a quantity pertaining to one sub-system is diagonal with respect to the quantum numbers of the other sub-system. Substituting (109.3) and (109.4) and using the summation formula (108.8), we obtain the desired formula expressing the matrix elements of the scalar product in terms of the reduced matrix elements of each tensor with respect to the wave functions of the corresponding sub-systems:

$$\langle n_1'n_2'j_1'j_2'JM|(f_k^{(1)}f_k^{(2)})_{00}|n_1n_2j_1j_2JM\rangle$$
$$= (-1)^{j_{1,\min}+j_{2,\max}+J}\begin{Bmatrix} J & j_2' & j_1' \\ k & j_1 & j_2 \end{Bmatrix}\langle n_1'j_1'\| f_k^{(1)}\|n_1j_1\rangle\langle n_2'j_2'\| f_k^{(2)}\|n_2j_2\rangle. \quad (109.5)$$

§110. Matrix elements for axially symmetric systems

The basis for calculating the matrix elements of quantities pertaining to systems of the symmetrical-top type is the expression for the integral of the product of three D functions.

To derive this, we return to the expansion (106.11):

$$\psi_{j_1m_1}\psi_{j_2m_2} = \sum_j \langle jm|m_1m_2\rangle\psi_{jm}, \qquad m = m_1+m_2,$$

and transform both sides by a finite rotation of the coordinates. Each of the functions ψ is transformed according to (58.7), so that we have

$$\sum_{m_1',m_2'} D^{(j_1)}_{m_1'm_1}D^{(j_2)}_{m_2'm_2}\psi_{j_1m_1'}\psi_{j_2m_2'} = \sum_j\sum_m \langle jm|m_1m_2\rangle D^{(j)}_{m'm}\psi_{jm'}.$$

Now, expressing the functions $\psi_{jm'}$ on the right by means of the expansion (106.9) and comparing the coefficients of the respective products $\psi_{j_1m_1}\psi_{j_2m_2}$, we find the relations

$$D^{(j_1)}_{m_1'm_1}(\omega)D^{(j_2)}_{m_2'm_2}(\omega) = \sum_j \langle m_1'm_2'|jm'\rangle D^{(j)}_{m'm}(\omega)\langle m_1m_2|jm\rangle, \quad (110.1)$$

where $m = m_1+m_2$, $m' = m_1'+m_2'$, and ω denotes the set of three Eulerian angles α, β, γ. Expressed in terms of $3j$-symbols, this formula becomes

$$D^{(j_1)}_{m_1'm_1}(\omega)D^{(j_2)}_{m_2'm_2}(\omega)$$
$$= \sum_j (2j+1)\begin{pmatrix} j_1 & j_2 & j \\ m_1' & m_2' & -m' \end{pmatrix}\begin{pmatrix} j_1 & j_2 & j \\ m_1 & m_2 & -m \end{pmatrix}D^{(j)*}_{-m',-m}(\omega); \quad (110.2)$$

here we have also used the property (58.19) of the D functions.

Multiplying both sides of equation (110.2) by $D^{(j)}_{-m',-m}(\omega)$ and integrating with respect to ω by means of the orthogonality relation (58.20), we have

$$\int D^{(j_1)}_{m_1'm_1}(\omega)D^{(j_2)}_{m_2'm_2}(\omega)\,D^{(j_3)}_{m_3'm_3}(\omega)\frac{d\omega}{8\pi^2} = \begin{pmatrix} j_1 & j_2 & j_3 \\ m_1' & m_2' & m_3' \end{pmatrix}\begin{pmatrix} j_1 & j_2 & j_3 \\ m_1 & m_2 & m_3 \end{pmatrix}; \quad (110.3)$$

here the indices have been renamed in an obvious manner, in order to make the result more symmetrical. This is the required formula.†

Let $f_{kq'}$ be a spherical tensor of rank k pertaining to a top, in coordinates $x', y', z' \equiv \xi, \eta, \zeta$ fixed to the top (with ζ along the axis): for example, the electric or magnetic multipole moment tensor. Let f_{kq} be the components of the same tensor relative to the fixed coordinates x, y, z. The relation between these is given by the matrix of finite rotations:

$$f_{kq} = \sum_{q'} D^{(k)}_{q'q}(\omega) f_{kq'}. \tag{110.4}$$

The wave functions describing the rotation of the system as a whole differ from the D functions only as regards normalization:

$$\psi_{jm\mu} = i^j \sqrt{\frac{2j+1}{8\pi^2}} D^{(j)}_{\mu m}(\omega), \tag{110.5}$$

where j is the total angular momentum of the system, m its component along the fixed z-axis, μ its component along the axis of the system; the phase factor is chosen so that, for integral j and $\mu = 0$, the function (110.5) becomes the eigenfunction of the free angular momentum (cf. (103.8)). On calculating with respect to these functions the matrix element of the quantity (110.4) by means of formula (110.3), and expressing the complex conjugate D function by means of (58.19), we find

$$\begin{aligned}&\langle j'\mu'm'|f_{kq}|j\mu m\rangle\\ &= i^{j-j'}(-1)^{\mu'-m'}\sqrt{[(2j+1)(2j'+1)]}\times\\ &\times\begin{pmatrix} j' & k & j \\ -\mu' & q' & \mu\end{pmatrix}\begin{pmatrix} j' & k & j \\ -m' & q & m\end{pmatrix}\langle\mu'|f_{kq'}|\mu^*\rangle;\end{aligned} \tag{110.6}$$

here $q' = \mu'-\mu$, $q = m'-m$.

This formula gives the solution of the problem proposed, expressing the dependence of the matrix elements on the angular momenta j, j' and their components m, m'. The dependence on the quantum numbers μ, μ' is, of course, indeterminate; their values are related to the "internal" states of the system, between which the "internal" matrix element $\langle\mu'|f_{kq'}|\mu\rangle$ is taken.

The dependence of the matrix elements (110.6) on the numbers m, m' is, of course, of the same kind as for any system with a given total angular momentum. Separating out this dependence by using the reduced matrix elements according to (107.6), we obtain for the latter the expression

$$\begin{aligned}&\langle j'\mu'\|f_k\|j\mu\rangle\\ &= i^{j-j'-k}(-1)^{j_{\max}-\mu'}\sqrt{[(2j+1)(2j'+1)]} = \begin{pmatrix} j' & k & j \\ -\mu' & q' & \mu\end{pmatrix}\langle\mu'|f_{kq'}|\mu\rangle.\end{aligned} \tag{110.7}$$

† For integral values of $j_1 = l_1, j_2 = l_2$ and $j_3 = l_3$, and $m_1' = m_2' = m_3' = 0$, the functions $D^{(l)}_{0m}$ reduce, according to (58.25), to spherical harmonic functions, and formula (110.3) gives an expression for the integral of a product of three such functions (107.14).

The squared modulus of the matrix element (110.6), summed over all values of the final number m' (and over $q = m'-m$) for a given m, is independent of the value of m and equal, by the general rule (107.11), to

$$\sum_{q,m'} |\langle j'\mu'm'|f_{kq}|j\mu m\rangle|^2 = \frac{1}{2j+1}|\langle j'\mu'\|f_k\|j\mu\rangle|^2$$

$$= (2j'+1)\begin{pmatrix} j' & k & j \\ -\mu' & q' & \mu \end{pmatrix}^2 |\langle\mu'|f_{kq'}|\mu\rangle|^2. \qquad (110.8)$$

The Hermitian relations (107.9) for the reduced matrix elements in the coordinates x, y, z (110.7) are, as we should expect, in agreement with the relations (107.8)

$$\langle\mu'|f_{kq'}|\mu\rangle = (-1)^{k-q'}\langle\mu|f_{k,-q'}|\mu'\rangle^*$$

for the matrix elements in the coordinates ξ, η, ζ.

The rotation of such axially symmetric systems as a diatomic molecule (or an axial nucleus) is described by only two angles ($\alpha \equiv \phi$, $\beta \equiv \theta$), which define the direction of the axis of the system. The rotational wave function differs in this case from (110.5) by the absence of the factor $e^{i\mu\gamma}/\sqrt{(2\pi)}$; cf. the second footnote to §82. This difference, however, does not affect the matrix elements: since the dependence of the functions $D^{(j)}_{m'm}(\alpha, \beta, \gamma)$ on γ is represented by the factor $e^{im'\gamma}$, formula (110.3) can be written as

$$\delta_{m'0}\int D^{(j_1)}_{m_1'm_1}(\alpha, \beta, 0)D^{(j_2)}_{m_2'm_2}(\alpha, \beta, 0)D^{(j_3)}_{m_3'm_3}(\alpha, \beta, 0)\frac{\sin\alpha\, d\alpha d\beta}{4\pi}$$

$$= \begin{pmatrix} j_1 & j_2 & j_3 \\ m_1 & m_2 & m_3 \end{pmatrix}\begin{pmatrix} j_1 & j_2 & j_3 \\ m_1' & m_2' & m_3' \end{pmatrix},$$

where $m' = m_1'+m_2'+m_3'$; the result of calculating the integral is unchanged. The selection rule for the axial component of the angular momentum is the same as before ($\mu'-\mu = q'$), resulting from the orthogonality of the electron wave functions (because of the symmetry of the molecule about the ζ-axis). In formulae (110.6) and (110.7), $\langle\mu'|f_{kq'}|\mu\rangle$ must now be understood as the matrix elements with respect to the electron states for nuclei at rest.

CHAPTER XV

MOTION IN A MAGNETIC FIELD

§111. Schrödinger's equation in a magnetic field

A PARTICLE that has a spin also has a certain "intrinsic" magnetic moment $\boldsymbol{\mu}$. The corresponding quantum-mechanical operator is proportional to the spin operator $\hat{\mathbf{s}}$, and can therefore be written as

$$\hat{\boldsymbol{\mu}} = \mu \hat{\mathbf{s}}/s, \tag{111.1}$$

where s is the magnitude of the particle spin and μ a constant characterizing the particle. The eigenvalues of the magnetic moment component are $\mu_z = \mu\sigma/s$. Hence we see that the coefficient μ (which is usually called just the magnitude of the magnetic moment) is the maximum possible value of μ_z, reached when the spin component $\sigma = s$.

The ratio $\mu/\hbar s$ gives the ratio of the intrinsic magnetic moment and the intrinsic angular momentum of the particle (when both are along the z-axis). For the ordinary (orbital) angular momentum, this ratio is $e/2mc$ (see *Fields*, §44). The coefficient of proportionality between the intrinsic magnetic moment and the spin of the particle is not the same. For an electron it is $-|e|/mc$, i.e. twice the usual value, as is found theoretically from Dirac's relativistic wave equation (see *RQT*, §33). The intrinsic magnetic moment of the electron (spin $\frac{1}{2}$) is consequently $-\mu_B$, where

$$\mu_B = |e|\hbar/2mc = 0{\cdot}927 \times 10^{-20} \text{ erg/gauss}. \tag{111.2}$$

This quantity is called the *Bohr magneton*.

The magnetic moment of heavy particles is customarily measured in *nuclear magnetons*, defined as $e\hbar/2m_pc$, with m_p the mass of the proton. The intrinsic magnetic moment of the proton is found by experiment to be 2·79 nuclear magnetons, the moment being parallel to the spin. The magnetic moment of the neutron is opposite to the spin, and is 1·91 nuclear magneton.

It should be noted that the quantities $\boldsymbol{\mu}$ and $\mathbf{s}$ on the two sides of (111.1) are the same type of vector, as they should be: both are axial vectors. A similar equation for the electric dipole moment $\mathbf{d}$ (= constant × $\mathbf{s}$) would contradict the symmetry under inversion of the coordinates: the relative sign of the two sides would be changed by inversion.†

† Such an equation (and therefore the existence of an electric moment of an elementary particle) would also contradict the symmetry under time reversal: a change in the sign of the time does not alter $\mathbf{d}$, but does change the sign of the spin (as is evident, for example, from the definitions of these quantities in orbital motion, that of $\mathbf{d}$ involving only the coordinates, whereas that of the angular momentum also involves the velocity of the particle).

In non-relativistic quantum mechanics, the magnetic field may be regarded as an external field only. The magnetic interaction between the particles is a relativistic effect, and a consistent relativistic theory is needed to take it into account.

In the classical theory, the Hamilton's function of a charged particle in an electromagnetic field is

$$H = \frac{1}{2m}\left(\mathbf{p} - \frac{e}{c}\mathbf{A}\right)^2 + e\phi,$$

where ϕ is the scalar and $\mathbf{A}$ the vector potential of the field, and $\mathbf{p}$ the generalized momentum of the particle; see *Fields*, §16. If the particle has no spin, the transition to quantum mechanics can be made in the usual manner: the generalized momentum must be replaced by the operator $\hat{\mathbf{p}} = -i\hbar\nabla$, and we obtain the Hamiltonian†

$$\hat{H} = \frac{1}{2m}\left(\hat{\mathbf{p}} - \frac{e}{c}\mathbf{A}\right)^2 + e\phi. \tag{111.3}$$

If, on the other hand, the particle has a spin, this procedure does not suffice. This is because the intrinsic magnetic moment of the particle interacts directly with the magnetic field. In the classical Hamilton's function, this interaction does not appear, since the spin, which is a purely quantum effect, vanishes in the limit of classical mechanics. The correct expression for the Hamiltonian is obtained by including in (111.3) an extra term $-\hat{\boldsymbol{\mu}}\cdot\mathbf{H}$ corresponding to the energy of the magnetic moment $\boldsymbol{\mu}$ in the field $\mathbf{H}$. Thus the Hamiltonian of a particle having a spin is‡

$$\hat{H} = \frac{1}{2m}\left(\hat{\mathbf{p}} - \frac{e}{c}\mathbf{A}\right)^2 - \hat{\boldsymbol{\mu}}\cdot\mathbf{H} + e\phi. \tag{111.4}$$

In expanding the square $(\hat{\mathbf{p}} - e\mathbf{A}/c)^2$, we must bear in mind that $\hat{\mathbf{p}}$ does not in general commute with the vector $\mathbf{A}$, which is a function of the coordinates. Hence we must write

$$\hat{H} = \hat{\mathbf{p}}^2/2m - (e/2mc)(\mathbf{A}\cdot\hat{\mathbf{p}} + \hat{\mathbf{p}}\cdot\mathbf{A}) + e^2\mathbf{A}^2/2mc^2 - (\mu/s)\hat{\mathbf{s}}\cdot\mathbf{H} + U. \tag{111.5}$$

According to the rule (16.4) for the commutation of the momentum operator with any function of the coordinates, we have

$$\hat{\mathbf{p}}\cdot\mathbf{A} - \mathbf{A}\cdot\hat{\mathbf{p}} = -i\hbar\,\mathrm{div}\,\mathbf{A}. \tag{111.6}$$

† The generalized momentum is here denoted by the same letter $\mathbf{p}$ as the ordinary momentum (and not by $\mathbf{P}$ as in *Fields*, §16), in order to emphasize that it corresponds to the same operator.

‡ There should be no misunderstanding here caused by the use of the same letter for the field and the Hamiltonian, since the latter always has a circumflex over it.

Thus $\hat{\mathbf{p}}$ and $\mathbf{A}$ commute if div $\mathbf{A} \equiv 0$. This holds, in particular, for a uniform field, if its vector potential is expressed in the form

$$\mathbf{A} = \tfrac{1}{2}\mathbf{H} \times \mathbf{r}. \tag{111.7}$$

The equation $i\hbar\partial\Psi/\partial t = \hat{H}\Psi$ with the Hamiltonian (111.4) is a generalization of Schrödinger's equation to the case where a magnetic field is present. The wave functions on which the Hamiltonian acts in this equation are symmetrical spinors of rank $2s$.

The wave functions of a particle in an electromagnetic field are not uniquely defined, because the choice of the field potentials is not unique: they are defined (see *Fields*, §18) only to within a gauge transformation

$$\mathbf{A} \to \mathbf{A} + \nabla f, \; \phi \to \phi - \frac{1}{c}\frac{\partial f}{\partial t}, \tag{111.8}$$

where f is an arbitrary function of the coordinates and the time. This transformation does not affect the values of the field strengths, and it is therefore clear that it cannot essentially alter the solutions of the wave equation; in particular, it must leave $|\Psi|^2$ unchanged, since it is easy to see that the original equation is restored if we make the changes (111.8) in the Hamiltonian and at the same time change the wave function according to

$$\Psi \to \Psi \exp(ief/\hbar c). \tag{111.9}$$

This non-uniqueness of the wave function does not affect any quantity having a physical significance (in whose definition the potentials do not appear explicitly).

In classical mechanics, the generalized momentum of a particle is related to its velocity by

$$m\mathbf{v} = \mathbf{p} - e\mathbf{A}/c.$$

In order to find the operator $\hat{\mathbf{v}}$ in quantum mechanics, we have to commute the vector $\mathbf{r}$ with the Hamiltonian. A simple calculation gives the result

$$m\hat{\mathbf{v}} = \hat{\mathbf{p}} - e\mathbf{A}/c, \tag{111.10}$$

which is exactly analogous to the classical expression. For the operators of the velocity components we have the commutation rules

$$\left.\begin{aligned} \{\hat{v}_x, \hat{v}_y\} &= i(e\hbar/m^2c)H_z, \\ \{\hat{v}_y, \hat{v}_z\} &= i(e\hbar/m^2c)H_x, \\ \{\hat{v}_z, \hat{v}_x\} &= i(e\hbar/m^2c)H_y, \end{aligned}\right\} \tag{111.11}$$

which are easily verified directly. We see that, in a magnetic field, the operators of the three velocity components of a (charged) particle do not commute. This means that the particle cannot simultaneously have definite values of the velocity components in all three directions.

In motion in a magnetic field, the symmetry with respect to time reversal occurs only if the sign of the field **H** (and of the vector potential **A**) is changed. This means (see §§18 and 60) that Schrödinger's equation $\hat{H}\psi = E\psi$ must keep the same form when we take complex conjugates and change the sign of **H**. This is immediately evident for all terms in the Hamiltonian (111.4) except $-\hat{\mathbf{s}}\,.\,\mathbf{H}$. The term $-\hat{\mathbf{s}}\,.\,\mathbf{H}\psi$ in Schrödinger's equation becomes $\mathbf{s}^*\,.\,\mathbf{H}\psi^*$ under the transformation in question, and at first sight this destroys the required invariance, since the operator $\hat{\mathbf{s}}^*$ is not the same as $-\hat{\mathbf{s}}$. It must be remembered, however, that the wave function is in reality a spinor $\psi^{\lambda\mu\cdots}$, and on time reversal a contravariant spinor must be replaced by a covariant one (see §60), so that in Schrödinger's equation the term $-\hat{\mathbf{s}}\,.\,\mathbf{H}\psi^{\lambda\mu\cdots}$ is replaced by $\hat{\mathbf{s}}^*\,.\,\mathbf{H}\psi_{\lambda\mu\cdots}$. It is easily seen by means of the definitions (57.4), (57.5) that the result of the action of the operator $\hat{\mathbf{s}}^*$ on the components of the covariant spinor has the opposite sign to that of the operator $\hat{\mathbf{s}}$ on the components of the contravariant spinor. The operation of time reversal therefore leads to a Schrödinger's equation for the components $\psi_{\lambda\mu\ldots}$ which is of the same form as the original equation for the components $\psi^{\lambda\mu\cdots}$.

§112. **Motion in a uniform magnetic field**

Let us determine the energy levels of a particle in a constant uniform magnetic field (L. D. Landau 1930). The vector potential of the uniform field is conveniently taken here not in the form (111.7), but as

$$A_x = -Hy, \qquad A_y = A_z = 0 \tag{112.1}$$

(the z-axis being taken in the direction of the field).

The Hamiltonian then becomes

$$\hat{H} = \frac{1}{2m}(\hat{p}_x+eHy/c)^2+\frac{\hat{p}_y^2}{2m}+\frac{\hat{p}_z^2}{2m}-(\mu/s)\hat{s}_z H. \tag{112.2}$$

First of all, we notice that the operator $\hat{s}_z$ commutes with the Hamiltonian, since the latter does not contain the operators of the other components of the spin. This means that the z-component of the spin is conserved, and therefore that $\hat{s}_z$ can be replaced by the eigenvalue $s_z = \sigma$. Then the spin dependence of the wave function becomes unimportant, and ψ in Schrödinger's equation can be taken as the ordinary coordinate function. For this function we have the equation

$$\frac{1}{2m}\left[\left(\hat{p}_x+\frac{eH}{c}y\right)^2+\hat{p}_y^2+\hat{p}_z^2\right]\psi-\frac{\mu}{s}\sigma H\psi = E\psi. \tag{112.3}$$

The Hamiltonian of this equation does not contain the coordinates x and z explicitly. The operators $\hat{p}_x$ and $\hat{p}_z$ (of differentiation with respect to x and z) therefore also commute with the Hamiltonian, i.e. the x and z components of the generalized momentum are conserved. We accordingly seek ψ in the form

$$\psi = e^{(i/\hbar)(p_x x+p_z z)}\chi(y). \tag{112.4}$$

The eigenvalues p_x and p_z take all values from $-\infty$ to $+\infty$. Since $A_z = 0$, the z-component of the generalized momentum is equal to the ordinary momentum component mv_z. Thus the velocity of the particle in the direction of the field can take any value; we can say that the motion along the field is "not quantized".

Substituting (112.4) in (112.3), we obtain the following equation for the function $\chi(y)$:

$$\chi'' + \frac{2m}{\hbar^2}\left[\left(E + \frac{\mu\sigma}{s}H - \frac{p_z{}^2}{2m}\right) - \tfrac{1}{2}m\omega_H{}^2(y-y_0)^2\right]\chi = 0, \qquad (112.5)$$

with the notation $y_0 = -cp_x/eH$ and

$$\omega_H = |e|H/mc. \qquad (112.6)$$

Equation (112.5) is formally identical with Schrödinger's equation (23.6) for a linear oscillator, oscillating with frequency ω_H. Hence we can conclude immediately that the expression in round brackets in (112.5), which takes the part of the oscillator energy, can have the values $(n+\frac{1}{2})\hbar\omega_H$, where $n = 0, 1, 2, \ldots$.

Thus we obtain the following expression for the energy levels of a particle in a uniform magnetic field:

$$E = (n+\tfrac{1}{2})\hbar\omega_H + p_z{}^2/2m - \mu\sigma H/s. \qquad (112.7)$$

The first term here gives the discrete energy values corresponding to motion in a plane perpendicular to the field; they are called *Landau levels*. For an electron, $\mu/s = -|e|\hbar/mc$, and formula (112.7) becomes

$$E = (n+\tfrac{1}{2}+\sigma)\hbar\omega_H + p_z{}^2/2m. \qquad (112.8)$$

The eigenfunctions $\chi_n(y)$ corresponding to the energy levels (112.7) are given by (23.12) with the appropriate changes of notation:

$$\chi_n(y) = \frac{1}{\pi^{1/4}a_H{}^{1/2}\sqrt{(2^n n!)}}\exp\left[-\frac{(y-y_0)^2}{2a_H{}^2}\right]H_n\left(\frac{y-y_0}{a_H}\right), \qquad (112.9)$$

where $a_H = \sqrt{(\hbar/m\omega_H)}$.

In classical mechanics, the motion of particles in a plane perpendicular to the field **H** (the xy-plane) takes place in a circle about a fixed centre. The quantity y_0, which is conserved in the quantum case, corresponds to the classical y coordinate of the centre of the circle. The quantity $x_0 = cp_y eH + x$ is also conserved; it is easy to see that its operator commutes with the Hamiltonian (112.2). This quantity x_0 corresponds to the classical x coordinate of the centre of the circle.† The operators $\hat{x}_0$ and $\hat{y}_0$, however, do not commute. In other words, the coordinates x_0 and y_0 cannot take definite values simultaneously.

Since (112.7) does not contain the quantity p_x, which assumes a continuous sequence of values, the energy levels are continuously degenerate. However, the degree of degeneracy becomes finite if the motion in the xy-plane is restricted to a large, but finite, area $S = L_x L_y$. The number of (now discrete) possible values of p_x in an interval Δp_x is $(L_x/2\pi\hbar)\Delta p_x$. All values of p_x are admissible for which the orbit centre is inside S (we neglect the radius of the orbit in comparison with the large quantity L_y). From the condition $0 < y_0 < L_y$ we have $\Delta p_x = eHL_y/c$. Hence the number of states (for given n, p_z) is $eHS/2\pi\hbar c$. If the region of motion is bounded in the z-direction also (dimension L_z), the number of possible values of p_z in an interval Δp_z is $(L_z/2\pi\hbar)\Delta p_z$ and the number of states in this interval is

$$\frac{eHS}{2\pi\hbar c}\frac{L_z}{2\pi\hbar}\Delta p_z = \frac{eHV\Delta p_z}{4\pi^2\hbar^2 c}. \tag{112.10}$$

For an electron there is an additional degeneracy: the levels (112.8) with n, $\sigma = \frac{1}{2}$ and $n+1$, $\sigma = -\frac{1}{2}$ are the same.

PROBLEMS

PROBLEM 1. Find the wave functions of an electron in a uniform magnetic field in states in which it has definite values of the momentum and angular momentum in the direction of the field.

SOLUTION. In cylindrical polar coordinates ρ, ϕ, z with the z-axis in the direction of the field, the vector potential has components $A_\phi = \frac{1}{2}H\rho$, $A_z = A_\rho = 0$, and Schrödinger's equation is‡

$$-\frac{\hbar^2}{2M}\left[\frac{1}{\rho}\frac{\partial}{\partial\rho}\left(\rho\frac{\partial\psi}{\partial\rho}\right)+\frac{\partial^2\psi}{\partial z^2}+\frac{1}{\rho^2}\frac{\partial^2\psi}{\partial\phi^2}\right]-\tfrac{1}{2}i\hbar\omega_H\frac{\partial\psi}{\partial\phi}+\tfrac{1}{8}M\omega_H{}^2\rho^2\psi = E\psi. \tag{1}$$

We seek a solution in the form

$$\psi = \frac{1}{\sqrt{(2\pi)}}R(\rho)e^{im\phi}e^{ip_z z/\hbar},$$

obtaining for the radial function the equation

$$\frac{\hbar^2}{2M}\left(R''+\frac{R'}{\rho}-\frac{m^2R}{\rho^2}\right)+\left[E-\frac{p_z{}^2}{2M}-\tfrac{1}{8}M\omega_H{}^2\rho^2-\tfrac{1}{2}\hbar\omega_H m\right]R = 0.$$

---

† For, in classical motion in a circle of radius cmv_t/eH (where v_t is the projection of the velocity on the xy-plane; see *Fields*, §21), we have

$$y_0 = -cp_x/eH = -cmv_x/eH+y.$$

It is evident from this that y_0 is the y coordinate of the centre of the circle. The other coordinate is

$$x_0 = cmv_y/eH+x = cp_y/eH+x.$$

‡ The electron charge is written as $e = -|e|$, and the electron mass is denoted by M to distinguish it from the angular momentum m. The spin term is unimportant in this problem, and is omitted.

Defining a new independent variable $\xi = (M\omega_H/2\hbar)\rho^2$, we can write this equation in the form

$$\xi R''+R'+\left(-\tfrac{1}{4}\xi+\beta-\frac{m^2}{4\xi}\right)R = 0,$$

$$\beta = \frac{1}{\hbar\omega_H}\left(E-\frac{p_z{}^2}{2M}\right)-\tfrac{1}{2}m.$$

As $\xi \to \infty$ the required function behaves as $e^{-\frac{1}{2}\xi}$, and for $\xi \to 0$ as $\xi^{|m|/2}$. Accordingly we seek a solution in the form

$$R(\xi) = e^{-\xi/2}\xi^{|m|/2}w(\xi);$$

the equation for $w(\xi)$ is satisfied by the confluent hypergeometric function

$$w = F\{-(\beta-\tfrac{1}{2}|m|-\tfrac{1}{2}),\ |m|+1,\ \xi\}.$$

If the wave function is everywhere finite, the quantity $\beta-\frac{1}{2}|m|-\frac{1}{2}$ must be a non-negative integer n_ρ. The energy levels are then given by the formula

$$E = \hbar\omega_H(n_\rho+\tfrac{1}{2}|m|+\tfrac{1}{2}m+\tfrac{1}{2})+\frac{p_z{}^2}{2M},$$

which is equivalent to (112.7). The corresponding radial wave functions are

$$R_{n_\rho m}(\rho) = \frac{1}{a_H{}^{1+|m|}|m|!}\left[\frac{(|m|+n_\rho)!}{2^{|m|}n_\rho!}\right]^{1/2}\exp\left(-\frac{\rho^2}{4a_H{}^2}\right)\rho^{|m|}\times$$
$$\times F(-n_\rho,\ |m|+1,\ \rho^2/2a_H{}^2), \qquad (2)$$

where $a_H = \sqrt{(\hbar/M\omega_H)}$; these are normalized by the condition

$$\int_0^\infty R^2\rho\, d\rho = 1.$$

The hypergeometric function is here a generalized Laguerre polynomial.

PROBLEM 2. Find the lowest energy level corresponding to a bound state of an electron in a potential well $U(r)$ of small depth ($|U| \ll \hbar^2/ma^2$, where a is the range of the forces in the well), with a uniform magnetic field superimposed (Yu. A. Bychkov 1960).

SOLUTION. The condition stated for the field $U(r)$ ensures (in the absence of the magnetic field) that perturbation theory is applicable to it, and there are no bound states in the well (§45). When the magnetic field also is present, $U(r)$ can be regarded as a perturbation only for the motion in the plane transverse to **H**; the (discrete) nature of the energy spectrum of this motion is unaltered when U is applied. The nature of the motion in the direction of **H** is altered, however, becoming (as we shall see) finite instead of infinite, i.e. the spectrum becomes discrete instead of continuous. Thus for this motion the field of the well cannot be treated by perturbation theory.

Accordingly, when the variables are separated in Schrödinger's equation (equation (1) of Problem 1, with the added term $U\psi$ on the left), the radial functions $R(\rho)$ are taken in the previous form (2); the lowest level corresponds to the values $n_p = m = 0$ of the quantum numbers. Substituting in Schrödinger's equation $\psi = R_{00}(\rho)\chi(z)$, multiplying by $R_{00}(\rho)$ and integrating over $\rho\, d\rho$, we obtain for $\chi(z)$ the equation

$$-\frac{\hbar^2}{2m}\chi''+\bar{U}(z)\chi = \epsilon\chi, \qquad (3)$$

where $\epsilon = E-\frac{1}{2}\hbar\omega$,

$$\bar{U}(z) = \int_0^\infty U(\sqrt{(z^2+\rho^2)})R_{00}^2(\rho)\rho\, d\rho,$$

and m is again the mass of the particle. This equation has the same form as Schrödinger's equation for one-dimensional motion in a potential well $U(z)$, with ϵ the energy of this motion. We can therefore simply use the result of §45, Problem 1, according to which the discrete level is

$$\epsilon = -\frac{m}{2\hbar^2}\left[\int_{-\infty}^{\infty} \bar{U}(z)\, dz\right]^2$$

$$= -\frac{m}{2\hbar^2}\left[\int_{-\infty}^{\infty}\int_{0}^{\infty} U(\sqrt{(z^2+\rho^2)})R_{00}^2(\rho)\rho\, d\rho\, dz\right]^2. \tag{4}$$

The wave function $R_{00}(\rho)$ is damped at distances $\rho \sim a_H$. If the magnetic field is so weak that $a_H \gg a$, the integral over ρ is governed by the range $\rho \lesssim a$, in which we can take $R_{00}(\rho) \approx R_{00}(0) = 1/a_H$. Then

$$\epsilon = -\frac{me^2H^2}{8\pi^2\hbar^4c^2}\left(\int U(r)\, dV\right)^2; \tag{5}$$

here $dV = 2\pi\rho\, d\rho\, dz \to 4\pi r^2 dr$. In the opposite case of a strong magnetic field, when $a_H \ll a$, the integral in (4) is governed by the range $\rho \lesssim a_H$, in which we can put $U(\sqrt{(z^2+\rho^2)}) \approx U(z)$. Then the integral over ρ reduces to the normalization integral of the function R_{00} and is equal to unity, so that

$$\epsilon = -\frac{2m}{\hbar^2}\left(\int_0^{\infty} U(z)\, dz\right)^2. \tag{6}$$

In either case, an estimate of the integral shows that $\epsilon \ll \hbar\omega_H$.

PROBLEM 3. Find the energy levels of a hydrogen atom in a magnetic field that is so strong that $a_H \ll a_B$, where a_B is the Bohr radius (R. J. Elliott and R. Loudon 1960).

SOLUTION. With the condition stated, $\hbar\omega_H \gg me^4/\hbar^2$, the influence of the Coulomb field of the nucleus on the motion of the electron in the plane transverse to $\mathbf{H}$ may be regarded as a small perturbation. We thus return to the situation considered in Problem 2, and equation (3) may be used, with

$$\bar{U}(z) = -e^2\int_0^{\infty} \frac{R_{00}^2(\rho)\rho}{\sqrt{(\rho^2+z^2)}}\, d\rho. \tag{7}$$

By writing the radial function R_{00} in this expression, we take only the energy levels of the longitudinal motion, pertaining to the zero Landau level ($\frac{1}{2}\hbar\omega_H$) of the transverse motion.

The ground-state wave function $\chi_0(z)$ extends to a distance $|z| \lesssim a_B$ and varies slowly over this distance (without zeros, so that it does not vanish at $z = 0$). Hence the ground level satisfies the conditions used in §45, Problem 1, and we can use formula (6), which is based on the solution of that problem. The logarithmically divergent integral is "cut off" at an upper limit of distance $|z| \sim a_B$ and at a lower limit $|z| \sim a_H$ (where $|z| \sim \rho$ and it is not permissible to replace $\sqrt{(\rho^2+z^2)}$ by $|z|$ in (7)). The result is

$$\epsilon_0 = -\frac{2me^4}{\hbar^2}\log^2\frac{a_B}{a_H}$$

$$= -\frac{me^4}{2\hbar^2}\log^2\frac{\hbar^3 H}{m^2c|e|^3}. \tag{8}$$

This formula has what is called logarithmic accuracy. The assumption made is that not only the ratio a_B/a_H but also its logarithm is large. A numerical factor in the argument of the logarithm remains undetermined.

The excited states of the discrete spectrum are obtained as solutions of Schrödinger's equation (3) with the field $\bar{U}(z) \approx -e^2/z$ (obtained from (7) with $z \sim a_B \gg \rho$). But this equation can be brought, by the substitution $\chi = z\phi(z)$, to the form

$$-\frac{\hbar^2}{2m}\frac{1}{z^2}\frac{d}{dz}\left(z^2\frac{d\phi}{dz}\right)-\frac{e^2\phi}{z} = \epsilon\phi, \qquad (9)$$

which is the same as the equation for the radial wave functions of s states in the three-dimensional Coulomb problem. The required levels are therefore given by (36.10):

$$\epsilon_n = -me^4/2\hbar^2n^2, \qquad (10)$$

with $n = 1, 2, 3, \ldots$. This expression too has only logarithmic accuracy; the next correction term would be small in comparison with the principal term, but only in the ratio 1/log (a_B/a_H).

Equation (9) gives the wave function only for $z > 0$. It can be continued into the region $z < 0$ as $\chi(-z) = \chi(z)$ or as $\chi(-z) = -\chi(z)$. Accordingly, in this approximation the levels (10) are doubly degenerate. This degeneracy is, however, removed in higher approximations with respect to a_H/a_B.

§113. An atom in a magnetic field

Let us consider an atom in a uniform magnetic field **H**. Its Hamiltonian is

$$\hat{H} = \frac{1}{2m}\sum_a\left[\mathbf{p}_a+\frac{|e|}{c}\mathbf{A}(\mathbf{r}_a)\right]^2+U+\frac{|e|\hbar}{mc}\mathbf{H}\cdot\hat{\mathbf{S}}, \qquad (113.1)$$

where the summation is taken over all the electrons (the electron charge being written as $-|e|$); U is the energy of interaction of the electrons with the nucleus and with one another, and $\hat{\mathbf{S}} = \Sigma\hat{\mathbf{s}}_a$ is the operator of the total (electron) spin of the atom.

If the vector potential of the field is taken in the form (111.7), then, as already noted, the operator $\hat{\mathbf{p}}$ commutes with **A**. Expanding the bracket in (113.1) and denoting by $\hat{H}_0$ the Hamiltonian of the atom in the absence of the field, we find

$$\hat{H} = \hat{H}_0+\frac{|e|}{mc}\sum_a\mathbf{A}_a\cdot\hat{\mathbf{p}}_a+\frac{e^2}{2mc^2}\sum_a\mathbf{A}_a{}^2+\frac{|e|\hbar}{mc}\mathbf{H}\cdot\hat{\mathbf{S}}.$$

Substituting **A** from (111.7), we obtain

$$\hat{H} = \hat{H}_0+\frac{|e|}{2mc}\mathbf{H}\cdot\sum_a\mathbf{r}_a\times\hat{\mathbf{p}}_a+\frac{e^2}{8mc^2}\sum_a(\mathbf{H}\times\mathbf{r}_a)^2+\frac{|e|\hbar}{mc}\mathbf{H}\cdot\hat{\mathbf{S}}.$$

The vector $\mathbf{r}_a\times\hat{\mathbf{p}}_a$, however, is the operator of the orbital angular momentum of the electron, and the summation over all the electrons gives the operator $\hbar\hat{\mathbf{L}}$ of the total orbital angular momentum of the atom. Thus

$$\hat{H} = \hat{H}_0+\mu_B(\hat{\mathbf{L}}+2\hat{\mathbf{S}})\cdot\mathbf{H}+(e^2/8mc^2)\sum_a(\mathbf{H}\times\mathbf{r}_a)^2, \qquad (113.2)$$

where μ_B is the Bohr magneton. The operator

$$\hat{\boldsymbol{\mu}}_{\text{at}} = -\mu_B(\hat{\mathbf{L}}+2\hat{\mathbf{S}}) \qquad (113.3)$$

may be regarded as the operator of the "intrinsic" magnetic moment of the atom, which it possesses in the absence of the field.

The external magnetic field splits the atomic levels and removes the degeneracy with respect to the directions of the total angular momentum (the *Zeeman effect*). Let us determine the amount of this splitting for atomic levels having definite values of the quantum numbers J, L and S (i.e. assuming the case of LS coupling for the levels; see §72).

We shall assume that the magnetic field is so weak that $\mu_B H$ is small compared with the distances between the energy levels of the atom, including the fine-structure intervals. Then the second and third terms in (113.2) can be regarded as a perturbation, the unperturbed levels being the separate components of the multiplets. In the first approximation we can neglect the third term, which is quadratic with respect to the field, in comparison with the second term, which is linear.

In this approximation, the energy ΔE of the splitting is determined by the mean values of the perturbation in the (unperturbed) states which have different values of the projection of the total angular momentum on the direction of the field. Taking this direction as the z-axis, we have

$$\Delta E = \mu_B H(\bar{L}_z + 2\bar{S}_z) = \mu_B H(\bar{J}_z + \bar{S}_z). \tag{113.4}$$

The mean value $\bar{J}_z$ is just the given eigenvalue of $J_z = M_J$. The mean value $\bar{S}_z$ can be found as follows, using stepwise averaging (cf. §72).

We first average the operator $\hat{\mathbf{S}}$ over a state of the atom with fixed values of S, L and J, but not of M_J. The operator $\bar{\hat{\mathbf{S}}}$ thus averaged must be "parallel" to $\hat{\mathbf{J}}$, the only conserved "vector" characterizing a free atom. We can therefore write

$$\bar{\mathbf{S}} = \text{constant} \times \mathbf{J}.$$

In this form, however, the equation is purely conventional, since the three components of the vector $\mathbf{J}$ cannot simultaneously have definite values. Its z-component can be taken literally:

$$\bar{S}_z = \text{constant} \times J_z = \text{constant} \times M_J,$$

as can the equation

$$\bar{\mathbf{S}} \cdot \mathbf{J} = \text{constant} \times \mathbf{J}^2 = \text{constant} \times J(J+1),$$

which is obtained on multiplying both sides by $\mathbf{J}$. Taking the conserved vector $\mathbf{J}$ under the averaging sign gives $\bar{\mathbf{S}} \cdot \mathbf{J} = \overline{\mathbf{S} \cdot \mathbf{J}}$. The mean value $\overline{\mathbf{S} \cdot \mathbf{J}}$ is the same as the eigenvalue

$$\mathbf{S} \cdot \mathbf{J} = \tfrac{1}{2}[J(J+1) - L(L+1) + S(S+1)],$$

to which it is equal in a state having definite values of $\mathbf{L}^2$, $\mathbf{S}^2$ and $\mathbf{J}^2$; cf.

formula (31.4). Determining the constant from the second equation and substituting in the first equation, we therefore have

$$\bar{S}_z = M_J \mathbf{J} \cdot \mathbf{S}/\mathbf{J}^2. \tag{113.5}$$

Collecting the above expressions and substituting in (113.4), we find the following final expression for the amount of the splitting:

$$\Delta E = \mu_B g M_J H, \tag{113.6}$$

where

$$g = 1 + \frac{J(J+1) - L(L+1) + S(S+1)}{2J(J+1)} \tag{113.7}$$

is called the *Landé factor* or the *gyromagnetic factor*. If there is no spin ($S = 0$, and so $J = L$), $g = 1$; if $L = 0$ (and so $J = S$), $g = 2$.†

Formulae (113.6) gives different values of the energy for all the $2J+1$ values $M_J = -J, -J+1, \ldots, J$. Thus the magnetic field completely removes the degeneracy of the levels with respect to directions of the angular momentum, unlike the electric field, which leaves the levels with $M_J = \pm|M_J|$ unsplit (§76).‡ However, the linear splitting described by (113.6) does not occur if $g = 0$; this can refer even to states for which $J \neq 0$, such as ${}^4D_{1/2}$.

We have seen in §76 that there is a relation between the displacement of an energy level of an atom in an electric field and its mean electric dipole moment. A similar relation exists in the magnetic case. The potential energy of a system of charges is given, in classical theory, by $-\boldsymbol{\mu} \cdot \mathbf{H}$, where $\boldsymbol{\mu}$ is the magnetic moment of the system. In the quantum theory, it is replaced by the corresponding operator, so that the Hamiltonian of the system is

$$\hat{H} = \hat{H}_0 - \hat{\boldsymbol{\mu}} \cdot \mathbf{H} = \hat{H}_0 - \hat{\mu}_z H.$$

Now, applying (11.16), with the field H as the parameter λ, we find that the mean value of the magnetic moment is

$$\bar{\mu}_z = -\partial \Delta E/\partial H, \tag{113.8}$$

where ΔE is the displacement of the energy level for the given state of the atom. Substituting (113.6), we see that an atom in a state with a definite

† The splitting is described by the general formulae (113.6) and (113.7) is often called the *anomalous Zeeman effect*. This unfortunate name arose because, before the spin of the electron was discovered, the effect described by (113.6) with $g = 1$ was regarded as normal.

‡ The arguments applied to the electric-field case in §76 are not valid for a magnetic field. The reason is that $\mathbf{H}$ is an axial vector and therefore changes sign on reflection in any plane containing it. Hence the states obtained from each other by this operation belong to atoms in different fields, not in the same field.

value M_J of the projection of the angular momentum on some direction z has a mean magnetic moment in that direction

$$\bar{\mu}_z = -\mu_B g M_J. \tag{113.9}$$

If the atom has neither spin nor orbital angular momentum ($S = L = 0$), the second term in (113.2) gives no displacement of the level, either in the first approximation or in any higher one (since the matrix elements of **L** and **S** vanish). Hence, in this case, the whole effect arises from the third term in (113.2), and in the first approximation of perturbation theory the displacement of the level is equal to the mean value

$$\Delta E = \frac{e^2}{8mc^2}\sum_a \overline{(\mathbf{H}\times\mathbf{r}_a)^2}. \tag{113.10}$$

Putting $(\mathbf{H}\times\mathbf{r}_a)^2 = H^2 r_a^2 \sin^2\theta$, where θ is the angle between $\mathbf{r}_a$ and **H**, and averaging with respect to the directions of $\mathbf{r}_a$, we have $\overline{\sin^2\theta} = 1-\overline{\cos^2\theta} = 2/3$ (the wave function of a state with $L = S = 0$ is spherically symmetrical, and so the averaging over directions is independent of that over distances r_a). Thus

$$\Delta E = \frac{e^2}{12mc^2}H^2\sum_a \overline{r_a^2}. \tag{113.11}$$

The magnetic moment calculated from (113.8) is then proportional to the magnitude of the field (an atom with $L = S = 0$ has, of course, no magnetic moment in the absence of a field). Writing it in the form χH, we can regard the coefficient χ as the magnetic susceptibility of the atom, given by *Langevin's formula* (P. Langevin 1905):

$$\chi = -\frac{e^2}{6mc^2}\sum_a \overline{r_a^2}. \tag{113.12}$$

It is negative, i.e. the atom is diamagnetic.†

If $J = 0$, but $S = L \neq 0$, the displacement linear with respect to the field again vanishes, but the quadratic effect from the perturbation $-\boldsymbol{\mu}_{at}\cdot\mathbf{H}$ in the second approximation exceeds the effect (113.11).‡ This is because, according to the general formula (38.10), the correction to the eigenvalue of the energy in the second approximation is given by a sum of expressions whose denominators contain the differences between the unperturbed

† The Thomas–Fermi model cannot be used to calculate the mean square distance of the electrons from the nucleus. Although the integral $\int nr^2\, dr$ with the Thomas–Fermi density $n(r)$ converges, it does so too slowly, and therefore the results are quite different from those given by experiment.

‡ For $S = L \neq 0$, the non-diagonal matrix elements of L_z, S_z for the transitions $S, L, J \to S, L, J\pm 1$ are not in general zero.

energy levels, in this case the fine-structure intervals of the level, which are small quantities. We have remarked in §38 that the correction to the normal level in the second approximation is always negative. Hence the magnetic moment in the normal state is positive, i.e. an atom in the normal state with $J = 0$, $L = S \neq 0$ is paramagnetic.

In strong magnetic fields, where $\mu_0 H$ is comparable with or greater than the intervals in the fine structure, the splitting of the levels differs from that predicted by formulae (113.6), (113.7); this phenomenon is called the *Paschen–Back effect*.

The calculation of the energy of the splitting is very simple in the case where the Zeeman splitting is large in comparison with the intervals in the fine structure but still, of course, small compared with the distances between the different multiplets (when it may be shown that we can, as before, neglect the third term of the Hamiltonian (113.2) in comparison with the second). In other words, the energy in the magnetic field considerably exceeds the spin–orbit interaction.† Hence we can neglect this interaction in the first approximation. The projections M_L and M_S of the orbital angular momentum and spin are then conserved, as well as the projection of the total angular momentum, so that the splitting is given by the formula

$$\Delta E = \mu_B H(M_L + 2M_S). \tag{113.13}$$

The multiplet splitting is superposed on the splitting in the magnetic field. It is determined by the mean value of the operator $A\hat{\mathbf{L}} \cdot \hat{\mathbf{S}}$ (72.4) with respect to the state with the given M_L, M_S (we are considering the multiplet splitting due to the spin–orbit interaction). For a given value of one of the angular momentum components, the mean values of the other two are zero. Hence $\overline{\mathbf{L} \cdot \mathbf{S}} = M_L M_S$, so that the energy of the levels is given in the next approximation by the formula

$$\Delta E = \mu_B H(M_L + 2M_S) + AM_L M_S. \tag{113.14}$$

The calculation of the Zeeman effect in the general case of any type of coupling (not LS) is not possible. We can say only that the splitting (in a weak field) is linear with respect to the field and proportional to the projection M_J of the total angular momentum, i.e. it has the form

$$\Delta E = \mu_B g_{nJ} H M_J, \tag{113.15}$$

where the g_{nJ} are some coefficients characterizing the term in question; n denotes the assembly of all the quantum numbers, except J, which characterize the term. Though these coefficients cannot be calculated separately, it is possible to obtain a formula, useful in applications, which gives their sum taken over all possible states of the atom with the given electron configuration and total angular momentum.

† For intermediate cases, where the effect of the magnetic field is comparable with the spin–orbit interaction, the splitting cannot be calculated in a general form; the calculation for $S = \frac{1}{2}$ is given in Problem 1.

By definition,

$$g_{nJ}M_J = \langle nJM_J|L_z+2S_z|nJM_J\rangle.$$

The quantities $g_{SLJ}M_J$, where g_{SLJ} is the Landé factor (113.7) corresponding to LS coupling, are the diagonal matrix elements

$$g_{SLJ}M_J = \langle SLJM_J|L_z+2S_z|SLJM_J\rangle,$$

calculated with a different complete set of wave functions. The functions in each set are obtained from the other set by a linear unitary transformation. But this transformation leaves unchanged the sum of the diagonal matrix elements (§12). Hence we conclude that

$$\sum_n g_{nJ}M_J = \sum_{S,L} g_{SLJ}M_J,$$

or, since g_{nJ} and g_{SLJ} do not depend on M,

$$\sum_n g_{nJ} = \sum_{S,L} g_{SLJ}. \qquad (113.16)$$

The summation is taken over all states with the given value of J which are possible for the given electron configuration. This is the required relation.

PROBLEMS

PROBLEM 1. Determine the splitting of a term with $S = \frac{1}{2}$ by the Paschen–Back effect.

SOLUTION. The magnetic field and the spin–orbit interaction have to be taken into account simultaneously by perturbation theory, i.e. the perturbation operator is†

$$\hat{V} = A\hat{\mathbf{L}}\cdot\hat{\mathbf{S}}+\mu_B(\hat{L}_z+2\hat{S}_z)H.$$

As the initial wave functions for the zero-order approximation, we take functions corresponding to states with definite values of L, $S = \frac{1}{2}$, M_L, M_S (L given; $M_L = -L, \ldots, L$; $M_S = \pm\frac{1}{2}$). In the perturbed states, only the sum $M \equiv M_J = M_L+M_S$ is conserved ($\hat{V}$ commutes with $\hat{J}_z$), so that we can ascribe definite values of M to the components of the split term.

The values $M = \pm(L+\frac{1}{2})$ can occur in only one way each: with $|M_LM_S\rangle = |L\frac{1}{2}\rangle$ and $|-L, -\frac{1}{2}\rangle$. Hence the corrections to the energy of the states with these M are simply equal to the diagonal matrix elements $\langle M_LM_S|V|M_LM_S\rangle$ with the indicated values of $|M_LM_S\rangle$. The remaining values of M can occur in two ways each: $|M-\frac{1}{2}, \frac{1}{2}\rangle$ and $|M+\frac{1}{2}, -\frac{1}{2}\rangle$. Here two different values of the energy correspond to each M; they are determined from the secular equation formed from the matrix elements for transitions between these two states. The matrix elements of $\mathbf{L}\,.\,\mathbf{S}$ are calculated by directly multiplying the matrices $\langle M_L|\mathbf{L}|M'_L\rangle$ and $\langle M_S|\mathbf{S}|M'_S\rangle$, and are

$$\langle M_LM_S|\mathbf{L}\,.\,\mathbf{S}|M_LM_S\rangle = M_LM_S,$$

$$\langle M+\tfrac{1}{2}, -\tfrac{1}{2}|\mathbf{L}\,.\,\mathbf{S}|M-\tfrac{1}{2}, \tfrac{1}{2}\rangle = \langle M-\tfrac{1}{2}, \tfrac{1}{2}|\mathbf{L}\,.\,\mathbf{S}|M+\tfrac{1}{2}, -\tfrac{1}{2}\rangle = \tfrac{1}{2}\sqrt{[(L+M+\tfrac{1}{2})(L-M+\tfrac{1}{2})]}.$$

† We do not include in $\hat{V}$ the term proportional to $(\hat{\mathbf{L}}\,.\,\hat{\mathbf{S}})^2$ (the spin–spin interaction). It must be borne in mind, however, that, for a spin $S = \frac{1}{2}$, the expression $(\hat{\mathbf{L}}\,.\,\hat{\mathbf{S}})^2$ reduces by virtue of the properties of the Pauli matrices (see §55) to $\hat{\mathbf{L}}\,.\,\hat{\mathbf{S}}$, and is therefore included in the formula as written here.

In the absence of a magnetic field, the term is a doublet, the distance between the components being $\epsilon = A(L+\frac{1}{2})$; see (72.6). We take the lower of these levels as the origin of energy. Then the final formulae for the levels in the magnetic field are

$$E = \epsilon \pm \mu_B H(L+1) \text{ for } M = \pm(L+\tfrac{1}{2}),$$

$$E^{\pm} = \tfrac{1}{2}\epsilon + \mu_B HM \pm \sqrt{[\tfrac{1}{4}(\epsilon^2+\mu_B{}^2H^2)+\mu_B HM\epsilon/(2L+1)]},$$

$$M = L-\tfrac{1}{2}, \ldots, -L+\tfrac{1}{2}.$$

For $\mu_B H/\epsilon \ll 1$ we have

$$E^{+} = \epsilon+\mu_B HM \,.\, 2(L+1)/(2L+1), \quad E^{-} = \mu_B HM \,.\, 2L/(2L+1),$$

in accordance with formulae (113.6), (113.7) (in which we must put $S = \frac{1}{2}$, $J = L\pm\frac{1}{2}$). For $\mu_B H/\epsilon \gg 1$ we have

$$E^{\pm} = \mu_B H(M\pm\tfrac{1}{2})+\tfrac{1}{2}\epsilon \pm \frac{M\epsilon}{2L+1},$$

in accordance with (113.14).

Problem 2. Determine the Zeeman splitting for the terms of a diatomic molecule in case *a*.

Solution. The magnetic moment arising from the motion of the nuclei is very small in comparison with the magnetic moment of the electrons. Hence the perturbation due to the magnetic field can be written for the molecule as for a system of electrons, i.e. in the form used previously: $\hat{V} = \mu_B \mathbf{H}\,.\,(\hat{\mathbf{L}}+2\hat{\mathbf{S}})$, where **L**, **S** are the electron orbital and spin angular momenta.

Averaging the perturbation with respect to the electron state, we have in case *a*

$$\mu_B Hn_z(\Lambda+2\Sigma) = \mu_B Hn_z(2\Omega-\Lambda).$$

The mean value of n_z with respect to the rotation of the molecule is the diagonal matrix element

$$\langle JM|n_z|JM\rangle = \Omega M/J(J+1),$$

where $M \equiv M_J$; the matrix element is calculated from the reduced matrix element given by (87.4) with J and Ω in place of K and Λ. Thus the required splitting is

$$\Delta E = \mu_B HM\frac{\Omega(2\Omega-\Lambda)}{J(J+1)}.$$

Problem 3. The same as Problem 2, but for case *b*.

Solution. The diagonal matrix elements $\langle \Lambda KJ|V|\Lambda KJ\rangle$ which determine the required splitting could be calculated from the general rules given in §87. However, it is simpler and more comprehensible to perform the calculation as follows. Averaging the perturbation operator with respect to the orbital and electron states, we obtain

$$\mu_B H(\Lambda n_z+2\hat{S}_z)$$

(the spin operator is unaffected by this averaging). Next, we average with respect to rotation of the molecule; the mean value of n_z is given by formula (87.4), and so we have

$$\mu_B H[\{\Lambda^2/K(K+1)\}\hat{K}_z+2\hat{S}_z].$$

Lastly, we average with respect to the spin wave function; after the whole averaging, the mean values of the vectors must be directed parallel to the total angular momentum **J**, which is the only conserved vector. Hence we have (cf. (113.5))

$$\frac{\mu_B H}{J(J+1)}\left[\frac{\Lambda^2}{K(K+1)}\mathbf{K}\,.\,\mathbf{J}+2\mathbf{S}\,.\,\mathbf{J}\right]M$$

($M \equiv M_J$), or finally

$$\Delta E = \frac{\mu_B}{J(J+1)}\left\{\frac{\Lambda^2}{2K(K+1)}[J(J+1)+K(K+1)-S(S+1)]+ \right.$$

$$\left. +[J(J+1)-K(K+1)+S(S+1)]\right\}HM.$$

PROBLEM 4. A diamagnetic atom is in an external magnetic field. Determine the strength of the induced magnetic field at the centre of the atom.

SOLUTION. For $S = L = 0$ the Hamiltonian contains no perturbation linear in the field, and so the wave function of the atom involves no correction of the first order with respect to the magnetic field. The change $\mathbf{j}'$ in the electron current in the atom induced by the external magnetic field is due (again in the first approximation with respect to H) only to the addition of the term $(|e|/mc)\mathbf{A}$ to the electron velocity operator. We therefore have†

$$\mathbf{j}' = -\rho(e^2/mc)\mathbf{A} = -\rho(e^2/2mc)\mathbf{H}\times\mathbf{r}, \tag{1}$$

where ρ is the electron density in the atom. The magnetic field produced at the centre of the atom by this additional current is

$$\mathbf{H}_{\text{ind}} = -\frac{1}{c}\int\frac{\mathbf{j}'\times\mathbf{r}}{r^3}\,dV;$$

cf. (121.8). Substituting (1) and averaging in the integrand over the directions of $\mathbf{r}$, we obtain

$$\mathbf{H}_{\text{ind}} = -\frac{e^2}{3mc^2}\mathbf{H}\int\frac{\rho}{r}\,dV$$

$$= \frac{e}{3mc^2}\phi_e(0)\mathbf{H}, \tag{2}$$

where $\phi_e(0)$ is the potential of the field at the centre of the atom due to its electron envelope.

In the Thomas–Fermi model $\phi_e(0) = -1{\cdot}80Z^{4/3}me^3/\hbar^2$ (see (70.8)), so that

$$\mathbf{H}_{\text{ind}} = -0{\cdot}60(e^2/\hbar c)^2 Z^{4/3}\mathbf{H}$$

$$= -3{\cdot}2\times10^{-5}Z^{4/3}\mathbf{H}.$$

§114. Spin in a variable magnetic field

Let us consider an electrically neutral particle having a magnetic moment, and situated in a magnetic field which is uniform but varies with time. We may have in mind either an elementary particle (a neutron) or a complex one (an atom). The magnetic field is supposed so weak that the magnetic energy of the particle in the field is small compared with the intervals between its energy levels. Then we can consider the motion of the particle as a whole, its internal state being given.

Let $\hat{\mathbf{s}}$ be the operator of the "intrinsic" angular momentum of the particle—the spin of an elementary particle, or the total angular momentum $\mathbf{J}$ for an atom. The magnetic moment operator can be represented in the form (111.1). The Hamiltonian for the motion of a neutral particle as a whole can be written

$$\hat{H} = -(\mu/s)\hat{\mathbf{s}}\cdot\mathbf{H}; \tag{114.1}$$

† This expression corresponds to the Larmor precession of the electron envelope of the atom round the direction of the external magnetic field; see *Fields*, §45.

here we write out only the part of the Hamiltonian that depends on the spin.

In a uniform field, this operator does not contain the coordinates explicitly†. Hence the wave function of the particle falls into a product of a coordinate and a spin function. Of these, the former is simply the wave function of free motion; in what follows, we shall be interested only in the spin part. We shall show that the problem of a particle with any angular momentum s can be reduced to the simpler problem of the motion of a particle of spin $\frac{1}{2}$ (E. Majorana). To do this, it is sufficient to use the method which we have already employed in §57. That is, instead of one particle of spin s, we can formally introduce a system of $2s$ "particles" of spin $\frac{1}{2}$. The operator $\hat{\mathbf{s}}$ is then represented as a sum $\Sigma\, \hat{\mathbf{s}}_a$ of the spin operators of these "particles", and the wave function as a product of $2s$ spinors of rank one. The Hamiltonian (114.1) then falls into the sum of $2s$ independent Hamiltonians:

$$\hat{H} = \sum_a \hat{H}_a, \qquad \hat{H}_a = -(\mu/s)\mathbf{H}\,.\,\hat{\mathbf{s}}_a, \tag{114.2}$$

so that the motion of each of the $2s$ "particles" is determined independently of the others. When this has been done, we need only reintroduce the components of an arbitrary symmetrical spinor of rank $2s$ in place of the products of components of $2s$ spinors of rank one.

PROBLEMS

PROBLEM 1. Determine the spin wave function for a neutral particle of spin $\frac{1}{2}$, in a uniform magnetic field which is constant in direction but varies in absolute magnitude according to an arbitrary law $H = H(t)$.

SOLUTION. The wave function is a spinor ψ^ν satisfying the wave equation

$$i\hbar\, \partial\psi^\nu/\partial t = -2\mu\mathbf{H}\,.\,\hat{\mathbf{s}}\psi^\nu. \tag{1}$$

Taking the direction of the field as the z-axis, we can write this equation in spinor components:

$$i\hbar\, \partial\psi^1/\partial t = -\mu H\psi^1, \qquad i\hbar\, \partial\psi^2/\partial t = \mu H\psi^2.$$

Hence

$$\psi^1 = c_1 e^{(i\mu/\hbar)\int H\,dt}, \qquad \psi^2 = c_2 e^{-(i\mu/\hbar)\int H\,dt}.$$

The constants c_1, c_2 must be determined from the initial conditions and from the normalization condition $|\psi^1|^2 + |\psi^2|^2 = 1$.

PROBLEM 2. The same as Problem 1, but for a uniform magnetic field constant in absolute magnitude, whose direction rotates uniformly, with angular velocity ω, around the z-axis and at an angle θ to it.

SOLUTION. The magnetic field has the components

$$H_x = H \sin\theta \cos\omega t, \qquad H_y = H \sin\theta \sin\omega t, \qquad H_z = H\cos\theta,$$

† These arguments can also be applied to the case where any particle (charged or not) moves in a non-uniform magnetic field, if its motion can be regarded as quasi-classical. The magnetic field, which varies as the particle moves along its path, can then be regarded simply as a function of time, and we can apply the same equations to the variation of the spin wave function.

and from (1) we obtain the equations

$$\dot{\psi}^1 = i\omega_H(\psi^1 \cos\theta + \psi^2 e^{-i\omega t}\sin\theta),$$
$$\dot{\psi}^2 = i\omega_H(\psi^1 e^{i\omega t}\sin\theta - \psi^2\cos\theta),$$

where $\omega_H = \mu H/\hbar$. The substitution $\psi^1 = e^{-\frac{1}{2}i\omega t}\phi^1$, $\psi^2 = e^{\frac{1}{2}i\omega t}\phi^2$ converts these equations into linear equations with constant coefficients, whose solution gives

$$\psi^1 = e^{-i\omega t/2}(c_1 e^{i\Omega t/2} + c_2 e^{-i\Omega t/2}),$$

$$\psi^2 = 2\omega_H e^{i\omega t/2}\sin\theta\left[\frac{c_1}{\Omega+\omega+2\omega_H\cos\theta}e^{i\Omega t/2} - \frac{c_2}{\Omega-\omega-2\omega_H\cos\theta}e^{-i\Omega t/2}\right],$$

where

$$\Omega = \sqrt{[(\omega+2\omega_H\cos\theta)^2 + 4\omega_H^2\sin^2\theta]}.$$

§115. The current density in a magnetic field

We shall now derive the quantum-mechanical expression for the current density when a charged particle moves in a magnetic field.

We start from the formula†

$$\delta H = -(1/c)\int \mathbf{j}.\delta\mathbf{A}\,dV; \qquad (115.1)$$

this determines the change in the Hamilton's function of charges distributed in space when the vector potential is varied.‡ In quantum mechanics this formula must be applied to the mean value of the Hamiltonian of the charged particle:

$$\bar{H} = \int \Psi^*[(\hat{\mathbf{p}} - e\mathbf{A}/c)^2/2m - (\mu/s)\mathbf{H}.\mathbf{s}]\Psi\,dV. \qquad (115.2)$$

Effecting the variation and bearing in mind that $\delta\mathbf{H} = \mathbf{curl}\,\delta\mathbf{A}$, we find

$$\delta\bar{H} = \int \Psi^*\left[-\frac{e}{2mc}(\hat{\mathbf{p}}.\delta\mathbf{A} + \delta\mathbf{A}.\hat{\mathbf{p}}) + \frac{e^2}{mc^2}\mathbf{A}.\delta\mathbf{A}\right]\Psi\,dV - (\mu/s)\int \mathbf{curl}\,\delta\mathbf{A}.\Psi^*\hat{\mathbf{s}}\Psi\,dV. \qquad (115.3)$$

The term in $\hat{\mathbf{p}}.\delta\mathbf{A}$ is transformed by integration by parts:

$$\int \Psi^*\hat{\mathbf{p}}.\delta\mathbf{A}\Psi\,dV = -i\hbar\int \Psi^*\nabla(\delta\mathbf{A}.\Psi)\,dV$$
$$= i\hbar\int \delta\mathbf{A}.\Psi\nabla\Psi^*\,dV$$

† In this section, $\mathbf{j}$ denotes the electric current density, i.e. the particle flux density multiplied by the particle charge e.

‡ Lagrange's function for a charge in a magnetic field contains a term $e\mathbf{v}.\mathbf{A}/c$, or, if the charge is distributed in space, $(1/c)\int \mathbf{j}.\mathbf{A}\,dV$.

The change in the Lagrange's function when $\mathbf{A}$ is varied is therefore

$$\delta L = (1/c)\int \mathbf{j}.\delta\mathbf{A}\,dV.$$

An infinitely small change in Hamilton's function is, however, equal to the change in Lagrange's function, taken with the opposite sign (see *Mechanics*, §40).

(the integral over an infinitely distant surface vanishing in the usual way). The integration by parts is also used in the last term in (115.3), together with the well-known formula of vector analysis

$$\mathbf{a}\,.\,\mathbf{curl}\,\mathbf{b} = -\mathrm{div}(\mathbf{a}\times\mathbf{b})+\mathbf{b}\,.\,\mathbf{curl}\,\mathbf{a}.$$

The integral of the div term vanishes, so that we have

$$\int \Psi^*\hat{\mathbf{s}}\Psi\,.\,\mathbf{curl}\,\delta\mathbf{A}\,\mathrm{d}V = \int \delta\mathbf{A}\,.\,\mathbf{curl}(\Psi^*\hat{\mathbf{s}}\Psi)\,\mathrm{d}V.$$

The final result is

$$\delta\bar{H} = -\frac{ie\hbar}{2mc}\int \delta\mathbf{A}\,.(\Psi\nabla\Psi^*-\Psi^*\nabla\Psi)\,\mathrm{d}V+\frac{e^2}{mc^2}\int \mathbf{A}\,.\,\delta\mathbf{A}\Psi\Psi^*\,\mathrm{d}V-$$
$$-(\mu/s)\int \delta\mathbf{A}\,.\,\mathbf{curl}(\Psi^*\hat{\mathbf{s}}\Psi)\,\mathrm{d}V.$$

Comparing this expression with (115.1), we find the following expression for the current density:

$$\mathbf{j} = \frac{ie\hbar}{2m}[(\nabla\Psi^*)\Psi-\Psi^*\nabla\Psi]-\frac{e^2}{mc}\mathbf{A}\Psi^*\Psi+(\mu/s)c\,\mathbf{curl}(\Psi^*\hat{\mathbf{s}}\Psi). \quad (115.4)$$

We emphasize that, though this expression contains the vector potential explicitly, it is nevertheless one-valued, as it should be. This is easily seen by direct calculation, recalling that the transformation (111.8) of the vector potential must be accompanied by the transformation (111.9) of the wave function.

It is also easy to verify that the current (115.4) and the charge density $\rho = e|\Psi|^2$ satisfy, as they should, the continuity equation

$$\partial\rho/\partial t+\mathrm{div}\,\mathbf{j} = 0.$$

The last term in (115.4) gives the contribution of the magnetic moment of the particle to the current density. It is c **curl m**, where

$$\mathbf{m} = (\mu/s)\Psi^*\hat{\mathbf{s}}\Psi = \Psi^*\hat{\boldsymbol{\mu}}\Psi \quad (115.5)$$

is the spatial density of the magnetic moment.

The expression (115.4) is the mean value of the current. It may be regarded as a diagonal matrix element of the current density operator $\hat{\mathbf{j}}$. This operator is most simply written in the second quantization form, with Ψ and Ψ^* replaced by operators $\hat{\Psi}$ and $\hat{\Psi}^+$ (and, according to the general rule, with Ψ^+ on the left of Ψ in each term). The non-diagonal matrix elements of this operator can be determined also:

$$\mathbf{j}_{nm} = \frac{ie\hbar}{2m}[(\nabla\Psi_n^*)\Psi_m-\Psi_n^*\nabla\Psi_m]-\frac{e^2}{mc}\mathbf{A}\Psi_n^*\Psi_m+$$
$$+(\mu/s)c\,\mathbf{curl}(\Psi_n^*\hat{\mathbf{s}}\Psi_m). \quad (115.6)$$

CHAPTER XVI

NUCLEAR STRUCTURE

§116. Isotopic invariance

THERE is as yet no complete theory of *nuclear forces*—that is, the forces which act between nuclear particles or *nucleons* and hold them together in the nucleus of an atom. In consequence, to describe nuclear forces it is still necessary to rely on experiment to a much greater extent than would be needed if a consistent theory were available.

The two types of particle which are nucleons differ mainly in their electrical properties, the proton (p) having a positive charge, while the neutron (n) is neutral. They have the same spin $\frac{1}{2}$, and their masses are almost equal (1836·1 and 1838·6 electron masses respectively). This similarity is no accident. Despite the difference in electrical properties, the proton and the neutron are very similar particles, and this similarity is of fundamental importance.

It is found that, apart from the relatively weak electric forces, the forces of interaction between two protons are very similar to those between two neutrons. This is called the *charge symmetry* of nuclear forces.†

In so far as this symmetry is maintained we can, in particular, say that systems of two protons (pp) and two neutrons (nn) have states whose properties are the same. Here, of course, it is important that protons and neutrons obey the same statistics (namely Fermi statistics) and so only states with the same symmetry of the wave functions $\psi(\mathbf{r}_1, \sigma_1; \mathbf{r}_2, \sigma_2)$ are permissible for the pp and nn systems, namely those antisymmetrical with respect to a simultaneous interchange of the coordinates and spins of the particles.

Charge symmetry is, however, only one of the manifestations of a still more far-reaching physical similarity between protons and neutrons, known as *isotopic invariance*.‡ This leads to the existence of an analogy not only between pp and nn systems (obtained from one other by interchanging all protons and neutrons), but also between these and the pn system, which consists of different particles. There cannot be a complete analogy here, of course, since the possible states of the pn system, in which the particles are non-identical, are certainly not restricted to those with antisymmetrical wave functions. It is found, however, that among the possible states of the pn system there are some whose properties are almost exactly the same as

† It appears, in particular, in the similarity of the properties (binding energy, energy spectrum, etc.) of what are called *mirror nuclei*, i.e. those which differ in that the numbers of protons and neutrons are interchanged.

‡ Also called *isobaric invariance* in the literature.

those of systems of two identical nucleons†; these states are, of course, described by antisymmetrical wave functions (the remaining states of the *pn* system are described by symmetrical wave functions and do not occur in the *pp* and *nn* systems).

Isotopic invariance, like charge symmetry, is valid only if electromagnetic interactions are neglected. Another reason why isotopic invariance is only approximately true is the slight mass difference between the neutron and the proton; if there were exact symmetry between neutrons and protons, their masses would of course be identical also.‡

A convenient formalism may be used to describe the isotopic invariance. It follows naturally from the fact that isotopic invariance is equivalent to the possibility of classifying the states of a system of nucleons with respect to the symmetry of its coordinate–spin wave functions ψ, independent of the types of nucleons concerned. The required formalism must therefore enable us to define for the description of the states of the system a new quantum number which uniquely determines the symmetry of the functions ψ. A similar situation has already been encountered in connection with the properties of a system of particles with spin $\frac{1}{2}$. We have seen in §63 that, if the total spin S of such a system is specified, then the symmetry of its coordinate wave function ϕ is uniquely determined, regardless of which of the two possible values $(\pm\frac{1}{2})$ is taken by the component σ of the spin of each particle.

It is therefore reasonable that, for a formal description of isotopic invariance, the neutron and the proton should be regarded as two different "charge states" of one particle, the nucleon, differing in the value of the component of a new vector $\boldsymbol{\tau}$, whose formal properties are analogous to those of the vector of spin $\frac{1}{2}$. This new quantity, which is usually called the *isotopic spin* or *isospin*,|| is a vector in "isotopic space" ξ, η, ζ (which, of course, is not related in any way to real space).

The component of the isotopic spin of a nucleon along the ζ-axis can take only the two values $\tau_\zeta = \pm\frac{1}{2}$. The value $+\frac{1}{2}$ is arbitrarily assigned to the proton and $-\frac{1}{2}$ to the neutron.†† The isotopic spins of several nucleons add to give the total isotopic spin of the system in accordance with the same rules as for the addition of ordinary spins. The ζ-component of the total isotopic spin of the system is equal to the sum of the values of τ_ζ for the component particles. For a nucleus in which the number of protons (i.e. the atomic number) is Z, the number of neutrons N and the mass number $A = Z+N$ we have

$$T_\zeta = \Sigma\tau_\zeta = \tfrac{1}{2}(Z-N) = Z-\tfrac{1}{2}A, \qquad (116.1)$$

i.e. T_ζ gives the total charge of the system if the number of nucleons is fixed.

---

† This was shown from an analysis of experimental data on the scattering of neutrons and protons by protons (G. Breit, E. U. Condon and R. D. Present 1936).

‡ In reality this mass difference between the neutron and the proton is probably also electromagnetic in origin.

|| First used by W. Heisenberg (1932), and applied to the description of isotopic invariance by B. Cassen and E. U. Condon (1936).

†† The opposite assignment is also found in the literature.

It is therefore clear that there is a strict conservation of the quantity T_ζ, which simply expresses the conservation of charge.

The absolute magnitude T of the isotopic spin of the system determines the symmetry of the "charge part" ω of the wave function of the system, just as the total spin S determines the symmetry of the spin wave function. It therefore determines also the symmetry of the coordinate–spin (i.e. the ordinary) wave function ψ, since the total wave function of a system of nucleons (i.e. the product $\psi\omega$) must have a definite symmetry; as for all fermions, it must be antisymmetrical with respect to simultaneous interchange of the coordinates, spins and "charge variables" τ_ζ of the particles. The existence of a definite symmetry of the wave functions ψ of any system of nucleons is therefore expressed, in this treatment, by the conservation of the quantity T.

We can say, in other words, that isotopic invariance signifies the invariance of the properties of the system with respect to rotations in isotopic space. States differing only in the value of T_ζ (with T and the remaining quantum numbers having given values) have identical properties. In particular, charge symmetry—the invariance of the properties of the system with respect to the replacement of neutrons by protons and *vice versa*, being a particular case of isotopic invariance, is described as invariance with respect to a simultaneous change of sign of all the τ_ζ, i.e. with respect to rotation in isotopic space through an angle of 180° about an axis lying in the $\xi\eta$-plane.

It may be noted that the obvious violation of isotopic invariance by the Coulomb interaction is also formally evident from this treatment. The Coulomb interaction depends on the charge, i.e. on the ζ-components of the isotopic spin, which are not invariant with respect to rotations in $\xi\eta\zeta$-space.

Let us consider, for example, a system of two nucleons. Its total isotopic spin can take the values $T = 1$ and $T = 0$. For $T = 1$, the possible values of the component T_ζ are 1, 0, -1. According to (116.1), the corresponding charge values are 2, 1, 0, i.e. a system with $T = 1$ may be *pp*, *pn* or *nn*. The charge part ω of the wave function with $T = 1$ is symmetrical (just as a symmetrical spin function corresponds to a spin $S = 1$; cf. §62). Hence states with antisymmetrical ordinary wave functions ψ correspond to the value $T = 1$. For $T = 0$ we can only have $T_\zeta = 0$, and the corresponding function ω is antisymmetrical; this therefore relates to states of the *pn* system with symmetrical wave functions ψ.

The isotopic spin corresponds to an operator $\hat{\boldsymbol{\tau}}$ which acts on the charge variable τ_ζ in the wave function, just as the spin operator $\hat{\mathbf{s}}$ acts on the spin variable σ. By virtue of the complete formal analogy between the two, the operators $\hat{\tau}_\xi$, $\hat{\tau}_\eta$, $\hat{\tau}_\zeta$ are given by the same Pauli matrices (55.7) as the operators $\hat{s}_x$, $\hat{s}_y$, $\hat{s}_z$.

Here we may note some combinations of these operators which have a simple and evident meaning. The sum

$$\hat{\tau}_+ = \hat{\tau}_\xi + i\hat{\tau}_\eta = \begin{pmatrix} 0 & 1 \\ 0 & 0 \end{pmatrix}$$

is an operator which, acting on a neutron wave function, converts it into a proton wave function, and acting on a proton wave function gives zero. Similarly, the operator

$$\hat{\tau}_- = \hat{\tau}_\xi - i\hat{\tau}_\eta = \begin{pmatrix} 0 & 0 \\ 1 & 0 \end{pmatrix}$$

converts a proton into a neutron and "annihilates" a neutron. Finally, the operator

$$\tfrac{1}{2} + \hat{\tau}_\zeta = \begin{pmatrix} 1 & 0 \\ 0 & 0 \end{pmatrix}$$

leaves a proton wave function unchanged and annihilates a neutron; on multiplication by e, it may be called the nucleon charge operator.

We shall also show that the operator $\hat{P}$ of the interchange of two particles may be expressed in terms of the operators $\hat{\boldsymbol{\tau}}_1$, $\hat{\boldsymbol{\tau}}_2$ of their isotopic spins. By definition, the result of the action of the interchange operator on the wave function $\psi(\mathbf{r}_1, \sigma_1; \mathbf{r}_2, \sigma_2)$ of the system of two particles consists in interchanging their coordinates and spins, i.e. interchanging the variables $\mathbf{r}_1$, σ_1 and $\mathbf{r}_2$, σ_2. The eigenvalues of this operator are ± 1, and occur when it acts on a symmetrical or antisymmetrical function ψ:

$$\hat{P}\psi_{\text{sym}} = \psi_{\text{sym}};\ \hat{P}\psi_{\text{ant}} = -\psi_{\text{ant}}. \tag{116.2}$$

We have seen above that the functions ψ_{sym} and ψ_{ant} correspond to charge functions ω_T with values of the total isotopic spin $T = 0$ and $T = 1$. Hence, in order to put the operator $\hat{P}$ in a form in which it acts on charge variables, it must have the properties

$$\hat{P}\omega_0 = \omega_0,\ \hat{P}\omega_1 = -\omega_1. \tag{116.3}$$

These conditions are satisfied by the operator $1-\hat{\mathbf{T}}^2$, as is easily seen by noting that ω_T is the eigenfunction of the operator $\hat{\mathbf{T}}^2$ corresponding to the eigenvalue $T(T+1)$. Finally, writing $\mathbf{T} = \boldsymbol{\tau}_1 + \boldsymbol{\tau}_2$ and using the fact that $\boldsymbol{\tau}_1^2$ and $\boldsymbol{\tau}_2^2$ have the same definite values $\tau(\tau+1) = \frac{3}{4}$, we find the required expression†

$$\hat{P} = 1 - \hat{\mathbf{T}}^2 = -\tfrac{1}{2} - 2\hat{\boldsymbol{\tau}}_1 \cdot \hat{\boldsymbol{\tau}}_2. \tag{116.4}$$

For the matrix elements of different physical quantities in a system of nucleons there are certain selection rules for the isotopic spin (L.A. Radicati 1952). Let F be some quantity (of any tensor rank) having the property of additivity, in the sense that its value for the system is equal to the sum of its values for the individual nucleons. We write the operator of such a quantity as

$$\hat{F} = \sum_p \hat{f}_p + \sum_n \hat{f}_n,$$

where the summations are over all protons and all neutrons in the system.

---

† An operator of this form derived from the ordinary spins of the particles has already been met in §62, Problems.

This expression can be written in the identical form

$$\hat{F} = \Sigma(\tfrac{1}{2}+\hat{\tau}_\zeta)\hat{f}_p+\Sigma(\tfrac{1}{2}-\hat{\tau}_\zeta)\hat{f}_n$$
$$= \tfrac{1}{2}\Sigma(\hat{f}_p+\hat{f}_n)+\Sigma(\hat{f}_p-\hat{f}_n)\hat{\tau}_\zeta, \qquad (116.5)$$

where the summation in each term is over all nucleons (both protons and neutrons). The first term in (116.5) is a scalar; the second is the ζ-component of a vector in isotopic space. The same selection rules therefore apply to them, with respect to the isotopic spin, as to scalars and vectors in ordinary space with respect to the orbital angular momentum (see §29): the isotopic scalar allows only transitions without change of T; the ζ-component of the isotopic vector has matrix elements only for transitions in which $\Delta T = 0$ or ± 1, and in addition transitions with $\Delta T = 0$ are forbidden between states with $T_\zeta = 0$, i.e. systems with the same number of neutrons and protons; the latter rule follows from the fact that the matrix element of a transition with $\Delta T = 0$ is proportional to T_ζ (see (29.7)).

For example, for the dipole moment of the nucleus the quantities f_p are the products $e\mathbf{r}$, and $f_n = 0$. The first term in (116.5) is then

$$\tfrac{1}{2}e\Sigma\mathbf{r} = (e/2m)\,\Sigma m\mathbf{r},$$

is therefore proportional to the radius vector of the centre of mass, and can be made to vanish by a suitable choice of the origin. Thus the dipole moment of the nucleus reduces to the ζ-component of the isotopic vector.

§117. Nuclear forces

The principal characteristic of the specifically nuclear forces which act between nucleons is their short range of action: they decrease exponentially at distances of the order of 10^{-13} cm.

In the non-relativistic limit we can say that nuclear forces are independent of the velocities of the nucleons and have a potential; the velocities of the nucleons in the nucleus are about one-quarter of the velocity of light (see below). The potential energy U of the interaction of two nucleons depends not only on the distance r between them but also quite strongly on their spins.† The precise dependence on r could, of course, be established only by a consistent theory of nuclear forces. The nature of the spin dependence, however, can be found from simple considerations based on the properties of spin operators.

We have at our disposal only three vectors on which the interaction energy U can depend: the unit vector $\mathbf{n}$ in the direction of the radius vector between the two nucleons, and their spins $\mathbf{s}_1$ and $\mathbf{s}_2$. According to the general properties of an operator of spin $\frac{1}{2}$, any function of it reduces to a linear function (§55). It must also be taken into account that the product $\mathbf{n.s}$ is not a true scalar but a pseudoscalar (since $\mathbf{n}$ is a polar vector and $\mathbf{s}$ an axial vector).

† In this respect the interaction of nucleons differs considerably from the interaction of electrons, for which the spin–spin interaction is purely relativistic and is small (in atoms).

Thus it is evident that only two independent scalar quantities linear in each of the spins can be constructed from the three vectors $\mathbf{n}$, $\mathbf{s}_1$, $\mathbf{s}_2$, namely $\mathbf{s}_1.\mathbf{s}_2$ and $(\mathbf{n}.\mathbf{s}_1)(\mathbf{n}.\mathbf{s}_2)$.†

Consequently, the operator of the interaction of two nucleons, as regards its dependence on the spins, can be written as the sum of three independent terms:

$$\hat{U}_{\text{ord}} = U_1(r) + U_2(r)(\hat{\mathbf{s}}_1.\hat{\mathbf{s}}_2) + U_3(r)[3(\hat{\mathbf{s}}_1.\mathbf{n})(\hat{\mathbf{s}}_2.\mathbf{n}) - \hat{\mathbf{s}}_1.\hat{\mathbf{s}}_2], \tag{117.1}$$

of which two depend on the spins and one does not. The third term is here written in a form which gives zero on averaging over the directions of $\mathbf{n}$. The forces described by this term are usually called *tensor forces*.

In (117.1) we have used the suffix ord (for "ordinary") in order to emphasize the fact that this operator does not affect the charge state of the nucleons. There is another possible interaction which converts a proton into a neutron and *vice versa*. The operator of this "exchange" interaction differs in form from (117.1) by the presence of the particle interchange operator (116.4):

$$\hat{U}_{\text{exch}} = \{U_4(r) + U_5(r)(\hat{\mathbf{s}}_1.\hat{\mathbf{s}}_2) + U_6(r)[3(\hat{\mathbf{s}}_1.\mathbf{n})(\hat{\mathbf{s}}_2.\mathbf{n}) - \hat{\mathbf{s}}_1.\hat{\mathbf{s}}_2]\}\hat{P}. \tag{117.2}$$

The total interaction operator is the sum

$$\hat{U} = \hat{U}_{\text{ord}} + \hat{U}_{\text{exch}}. \tag{117.3}$$

Thus the interaction of two nucleons is described by six different functions of the distance between them. All these terms are in general of the same order of magnitude.‡

The spin operators appearing in (117.1) and (117.2) can be expressed in terms of the total-spin operator $\hat{\mathbf{S}}$. By squaring the equations $\hat{\mathbf{S}} = \hat{\mathbf{s}}_1 + \hat{\mathbf{s}}_2$ and $\hat{\mathbf{S}}.\mathbf{n} = \hat{\mathbf{s}}_1.\mathbf{n} + \hat{\mathbf{s}}_2.\mathbf{n}$ and using the results $\hat{\mathbf{s}}_1^2 = \hat{\mathbf{s}}_2^2 = \frac{3}{4}$, $(\hat{\mathbf{s}}_1.\mathbf{n})^2 = (\hat{\mathbf{s}}_2.\mathbf{n})^2 = \frac{1}{4}$ (see (55.10)), we find

$$\hat{\mathbf{s}}_1.\hat{\mathbf{s}}_2 = \tfrac{1}{2}(\hat{\mathbf{S}}^2 - \tfrac{3}{2}), \quad (\hat{\mathbf{s}}_1.\mathbf{n})(\hat{\mathbf{s}}_2.\mathbf{n}) = \tfrac{1}{2}[(\hat{\mathbf{S}}.\mathbf{n})^2 - \tfrac{1}{2}]. \tag{117.4}$$

The operator $\hat{\mathbf{S}}^2$ commutes with the operator $\hat{\mathbf{S}}$, and so the interactions described by the first two terms in (117.1) and (117.2) conserve the total spin vector of the system. The tensor interaction contains the operator $(\hat{\mathbf{S}}.\mathbf{n})^2$, which commutes with the square $\hat{\mathbf{S}}^2$ but not with the vector $\hat{\mathbf{S}}$ itself. In consequence, only the magnitude of the total spin is conserved, not its direction.

The total spin S of a system of two nucleons can take the values 0 and 1, as can the total isotopic spin T. Hence all possible states of this system fall into four groups with various pairs of values of S and T. For states in each

† Here it is assumed that the nuclear forces are invariant under spatial inversion, i.e. that they cannot include pseudoscalar quantities. There are at present no experimental results to disprove this assumption.

‡ It may also be mentioned that the interaction dependent on the velocities of the nucleons, in an approximation linear with respect to these velocities, is described by an operator of the form $[\phi_1(r) + \phi_2(r)\hat{P}]\hat{\mathbf{L}}\,.\,\hat{\mathbf{S}}$, where $\mathbf{L} = \mathbf{r}\times\mathbf{p}$ is the orbital angular momentum of the relative motion of the nucleons, $\mathbf{p}$ the linear momentum of this motion, and $\mathbf{S} = \mathbf{s}_1 + \mathbf{s}_2$; this operator contains two functions of r. Terms of the form $\mathbf{p}\,.\,\mathbf{n}$ and $\mathbf{S}\,.\,\mathbf{n}$ are excluded by the requirements of invariance under inversion and under time reversal.

of these groups there is an interaction operator of the form $A(r)$ (for $S = 0$) or $A(r)+B(r)\,[(\hat{\mathbf{S}}.\mathbf{n})^2-\frac{2}{3}]$ (for $S = 1$), to which the general operator (117.3) reduces in these cases (see Problem 1).†

For given values of S and T the states of the system are classified with respect to the values of the total angular momentum J and the parity. As we know, the values $T = 0$ and $T = 1$ correspond to the states with symmetrical and antisymmetrical wave functions ψ respectively. Since, on the other hand, the value of S determines the symmetry of the wave function with respect to the spin variables (symmetrical for $S = 1$ and antisymmetrical for $S = 0$), it is clear that, if the two numbers S and T are specified, the symmetry of the wave function with respect to the space variables (i.e. the parity of the state) is also determined. Evidently the states of the system with isotopic spin $T = 0$ can only be even triplets ($S = 1$) or odd singlets ($S = 0$), while those with isotopic spin $T = 1$ are odd triplets or even singlets.

Since the spin, as a vector, is not conserved, the orbital angular momentum also need not in general be conserved; only the sum $\mathbf{J} = \mathbf{L}+\mathbf{S}$ is conserved. Nevertheless, the magnitude L may be conserved simply because specified values of J, S and the parity (or J, S and T) may be compatible with only one particular value of L (the parity of a system of two particles, it will be remembered, is $(-1)^L$). For example, an odd state with $S = 1$, $J = 1$ can only have $L = 1$, i.e. it is $^3P_1$. In other cases two different values of L may correspond to given values of J, S and the parity, so that L is not conserved. For example, in an odd state with $S = 1$, $J = 2$ we can have $L = 1$ or $L = 3$, i.e. it is a superposition $^3P_2+{}^3F_2$.

Thus we arrive at the following possible states of a system of two nucleons (the signs $\pm$ indicating the parity):

$$\text{for } T=1:\ {}^3P_0^-, {}^3P_1^-, ({}^3P_2+{}^3F_2)^-, {}^3F_3^-, \ldots;$$
$$\qquad {}^1S_0^+, {}^1D_2^+, {}^1G_4^+, \ldots;$$
$$\text{for } T=0:\ ({}^3S_1+{}^3D_1)^+, {}^3D_2^+, ({}^3D_3+{}^3G_3)^+, \ldots;$$
$$\qquad {}^1P_1^-, {}^1F_3^-, \ldots.$$

Nuclear forces are not in general additive. This means that the interaction in a system of more than two nucleons does not reduce to a sum of interactions between each pair of particles. It seems, however, that ternary and higher interactions are relatively unimportant in comparison with binary interactions, and so, in discussing the properties of complex nuclei, we can to a considerable extent take as basis the properties of binary interactions.

Experimental results concerning nuclei show that, as the number A of

† The experimental results concerning the properties of the deuteron show that for $T = 0$, $S = 1$ the nucleon interaction involves a strong attraction with a deep "potential well" (the presence of tensor forces makes it difficult to formulate this fact in terms of properties of the functions $A(r)$ and $B(r)$); in addition, it follows from the sign of the observed quadrupole moment of the deuteron that in this state the coefficient $B(r)$ in the tensor forces is negative. From nucleon scattering results it follows that for $T = 1$, $S = 0$ there is also an attraction, but one which is weaker and, in particular, does not lead to the formation of a stable system of two particles.

particles increases, the system of nucleons begins to behave like a macroscopic "nuclear matter", whose volume and energy increase in proportion to A (apart from effects due to the Coulomb interaction of protons and the existence of a free surface of the nucleus). The property of nuclear forces which gives rise to this phenomenon is called *saturation*.

The existence of this property imposes certain restrictions on the functions $U_1, \ldots, U_6$ which determine the binary interactions of nucleons. Let us suppose that all the particles are concentrated in a volume whose dimensions are of the order of the radius of action of nuclear forces. Then every pair of particles interact. If there is a configuration of certain nucleons (and an orientation of their spins) for which attractive forces act between every pair, then the potential energy of such a system is negative and proportional to A^2; the kinetic energy is positive and proportional to $A^{5/3}$, a smaller power of A.† It is clear that under such conditions a sufficiently large number of nucleons will in fact be concentrated in a small volume independent of A, i.e. will not form nuclear matter. The condition for saturation of nuclear forces must therefore be expressed as the conditions for the absence of configurations leading to a negative interaction energy proportional to A^2 (see Problem 2).

The proportionality between the volume of nuclear matter and the number of particles is expressed by a relation of the form

$$R = r_0 A^{1/3}, \tag{117.5}$$

which connects the radius R of the nucleus and the number A of particles in it. Experimental results (on the scattering of electrons by nuclei) lead to the value $r_0 = 1{\cdot}1 \times 10^{-13}$ cm.

We may determine the limiting momentum of nucleons in nuclear matter (cf. §70). The volume of phase space corresponding to particles in unit volume of physical space and with momenta $p \leqslant p_0$ is $4\pi p_0^3/3$. Dividing by $(2\pi\hbar)^3$, we obtain the number of "cells" in each of which two protons and two neutrons can be simultaneously. Putting the number of protons equal to the number of neutrons, we obtain $4(4\pi/3)(p_0/2\pi\hbar)^3 = A/V$, where V is the volume of the nucleus. Substitution of (117.5) gives

$$p_0 = (3\pi^2 A/2V)^{1/3}\hbar = (9\pi)^{1/3}\hbar/2r_0$$
$$= 1{\cdot}4 \times 10^{-14} \text{ g.cm/sec.}$$

The corresponding energy $p_0^2/2m_p$, where m_p is the nucleon mass, is ~ 40 MeV, and the velocity $p_0/m_p \approx \frac{1}{4}c$.

PROBLEMS

PROBLEM 1. Find the operators of the interaction of two nucleons in states with definite values of S and T.

† The density n at which the particles are concentrated in a given volume is proportional to their number A, and the kinetic energy of each particle is proportional to $n^{2/3}$ (cf. (70.1)). The total kinetic energy is therefore $\sim A.A^{2/3}$.

SOLUTION. The required operators $\hat{U}_{ST}$ are obtained from the general expression (117.1)–(117.3), using (116.3) and (117.4):

$$\hat{U}_{00} = U_1 - \tfrac{3}{4}U_2 + U_4 - \tfrac{3}{4}U_5,$$
$$\hat{U}_{01} = U_1 - \tfrac{3}{4}U_2 - U_4 + \tfrac{3}{4}U_5,$$
$$\hat{U}_{10} = U_1 + \tfrac{1}{4}U_2 + U_4 + \tfrac{1}{4}U_5 + \tfrac{1}{2}(U_3 + U_6)[3(\hat{\mathbf{S}}.\mathbf{n})^2 - 2],$$
$$\hat{U}_{11} = U_1 + \tfrac{1}{4}U_2 - U_4 - \tfrac{1}{4}U_5 + \tfrac{1}{2}(U_3 - U_6)[3(\hat{\mathbf{S}}.\mathbf{n})^2 - 2].$$

PROBLEM 2. Find the conditions for the saturation of nuclear forces, assuming tensor forces absent. The radii of action of forces of all other types are supposed equal.

SOLUTION. Let us consider some extreme cases (between which lie all other possible cases) for the state of a system of A nucleons, and write down the conditions for the interaction energy of an "average" pair of nucleons in this system to be positive.

Let the total spin and the isotopic spin of the nucleus have the greatest possible values: $S_{\text{nuc}} = T_{\text{nuc}} = \frac{1}{2}A$ (when all the particles in the system are protons with their spins parallel). Then for each pair of nucleons we have $S = T = 1$, and the condition is

$$U_{11} > 0. \tag{1}$$

Next, let $T_{\text{nuc}} = \frac{1}{2}A$, $S_{\text{nuc}} = 0$. Then for each pair of nucleons $T = 1$, and the mean value of s_z for an individual nucleon is zero. The latter result means that the nucleon can have $s_z = \frac{1}{2}$ and $s_z = -\frac{1}{2}$ with equal probability; under these conditions the probabilities that a pair of nucleons are in states with $S = 0$ or 1 are respectively $\frac{1}{4}$ and $\frac{3}{4}$ (being proportional to the number $2S+1$ of possible values of S_z). The condition for the mean energy of the pair to be positive is therefore

$$\tfrac{1}{4}U_{01} + \tfrac{3}{4}U_{11} > 0. \tag{2}$$

Similarly, a discussion of the state with $T_{\text{nuc}} = 0$, $S_{\text{nuc}} = \frac{1}{2}A$ gives the condition

$$\tfrac{1}{4}U_{10} + \tfrac{3}{4}U_{11} > 0. \tag{3}$$

In a state with $T_{\text{nuc}} = S_{\text{nuc}} = 0$, the probability for a pair of nucleons to have $S = T = 1$ is $\frac{3}{4} \cdot \frac{3}{4}$, that for $T = 1$, $S = 0$ is $\frac{3}{4} \cdot \frac{1}{4}$, and so on. Hence we find the condition

$$\frac{9}{16}U_{11} + \frac{3}{16}(U_{10} + U_{01}) + \frac{1}{16}U_{00} > 0. \tag{4}$$

Finally, let the system consist of $\frac{1}{2}A$ protons and $\frac{1}{2}A$ neutrons, with the spins of all the protons in one direction and the spins of all the neutrons in the other direction. An individual nucleon can with equal probability be p or n, i.e. have $\tau_\zeta = \frac{1}{2}$ or $\tau_\zeta = -\frac{1}{2}$; the probability for a pair of nucleons to have $T = 0$ is $\frac{1}{4}$. Here one of the pair of nucleons is p and the other n, and hence $S_z = 0$. This value of S_z can occur with equal probability from states with $S = 0$ and $S = 1$. Consequently the probabilities for the pair to be in the states with $T = 0$, $S = 0$ and $T = 0$, $S = 1$ are each $\frac{1}{4} \cdot \frac{1}{2} = \frac{1}{8}$. The probability of the state with $T = 1$, $S = 0$ is the same, and the remaining $\frac{5}{8}$ relates to the state with $T = S = 1$. Thus we have the condition

$$\tfrac{1}{8}(U_{00} + U_{01} + U_{10}) + \tfrac{5}{8}U_{11} > 0. \tag{5}$$

The inequalities (1)–(5) form the required set of conditions for the saturation of nuclear forces.

§118. The shell model

Many properties of nuclei can be well described by means of the *shell model*, which is basically similar to the structure of the electron shells of an atom. In this model each nucleon in the nucleus is regarded as moving in a self-consistent field due to all the other nucleons; owing to the small range of action of nuclear forces, this field decreases rapidly outside the volume bounded by the "surface" of the nucleus. Accordingly the state of the nucleus as a whole is described by specifying the states of the individual nucleons.

The self-consistent field is spherically symmetrical, and the centre of symmetry is, of course, the centre of mass of the nucleus. The following difficulty arises here, however. In the self-consistent field method, the wave function of the system is constructed as the product (or the appropriately symmetrized sum of the products) of the wave functions of the individual particles. Such a function, however, does not keep the centre of mass fixed: although the mean velocity of the centre of mass calculated from this function is zero, it gives a finite probability of non-zero values of the velocity.†

The difficulty can be avoided by first eliminating the motion of the centre of mass in calculating any physical quantity by means of the wave functions $\psi(\mathbf{r}_1, \ldots, \mathbf{r}_A)$ of the self-consistent field method. Let $f(\mathbf{r}_i,\mathbf{p}_i)$ be some physical quantity, a function of the coordinates and momenta of the nucleons. Then, in calculating its matrix elements by means of the functions ψ, we must, without changing $\psi(\mathbf{r}_i)$, alter the arguments of the function f as follows:

$$\mathbf{r}_i \to \mathbf{r}_i - \mathbf{R}, \quad \mathbf{p}_i \to \mathbf{p}_i - \mathbf{P}/A, \tag{118.1}$$

where $\mathbf{R}$ is the radius vector of the centre of mass of the nucleus, A the number of particles in it, $\mathbf{P}$ the momentum of its motion as a whole; the second change in (118.1) corresponds to subtracting the velocity $\mathbf{V}$ of the centre of mass from the velocities $\mathbf{v}_i$ of the nucleons, the momentum $\mathbf{P}$ being related to $\mathbf{V}$ by $\mathbf{P} = Am_p\mathbf{V}$ (S. Gartenhaus and C. Schwartz 1957).

For example, the dipole moment operator of the nucleus is $\mathbf{d} = e\Sigma\mathbf{r}_p$, where the summation is over all protons in the nucleus. To calculate the matrix elements in the self-consistent field method, this operator must be replaced by $e\Sigma(\mathbf{r}_p - \mathbf{R})$. The coordinates of the centre of mass of the nucleus are

$$\mathbf{R} = \frac{1}{A}\Big(\sum_p \mathbf{r}_p + \sum_n \mathbf{r}_n\Big),$$

where the summation is over all protons and neutrons. Since the number of protons in the nucleus is Z, the dipole moment operator must finally be changed thus:

$$e\sum_p \mathbf{r}_p \to e\Big(1-\frac{Z}{A}\Big)\sum_p \mathbf{r}_p - e\frac{Z}{A}\sum_n \mathbf{r}_n. \tag{118.2}$$

The protons appear here with an "effective charge" $e(1-Z/A)$ and the neutrons with a "charge" $-eZ/A$. It may be noted that the relative order of magnitude of the resulting correction terms in the calculation of the dipole moment is seen from (118.2) to be of the order of unity. The corrections in

† For electrons in an atom this difficulty did not arise, because the centre of mass coincided in position with the fixed heavy nucleus, and was therefore necessarily at rest.

the calculation of the magnetic and higher electric multipole moments are easily found to be of relative order $1/A$.

In the non-relativistic approximation the interaction of a nucleon with the self-consistent field is independent of the spin of the nucleon: such a dependence can be given only by a term proportional to $\hat{\mathbf{s}}.\mathbf{n}$, where $\mathbf{n}$ is a unit vector in the direction of the radius vector $\mathbf{r}$ of the nucleon, and this product is a pseudoscalar, not a true scalar.

A dependence of the nucleon energy on the spin appears, however, when relativistic terms depending on the velocity of the particle are taken into account. The largest of these is the term linear in the velocity. From the three vectors $\mathbf{s}$, $\mathbf{n}$ and $\mathbf{v}$ a true scalar $\mathbf{n}\times\mathbf{v}.\mathbf{s}$ can be formed. The spin–orbit coupling operator of the nucleon in the nucleus is therefore

$$\hat{V}_{sl} = -\phi(r)\mathbf{n}\times\hat{\mathbf{v}}.\hat{\mathbf{s}}, \tag{118.3}$$

where $\phi(r)$ is some function of r; see also the third footnote to §117. Since $m_p\mathbf{r}\times\mathbf{v}$ is the orbital angular momentum $\hbar\mathbf{l}$ of the particle, the expression (118.3) can also be written as

$$\hat{V}_{sl} = -f(r)\hat{\mathbf{l}}.\hat{\mathbf{s}}, \tag{118.4}$$

where $f = \hbar\phi/rm_p$. It should be emphasized that this interaction is of the first order in v/c, whereas the spin–orbit coupling of an electron in an atom is a second-order effect (§72). This difference is due to the fact that nuclear forces depend on the spin even in the non-relativistic approximation, whereas the non-relativistic interaction of electrons (Coulomb forces) is not spin-dependent.

The energy of the spin–orbit interaction is mainly concentrated near the surface of the nucleus, i.e. the function $f(r)$ decreases inside the nucleus. This is because, in infinite nuclear matter, there would be no such interaction at all, as is clear from the fact that, the system being homogeneous, there is no preferred direction in it which could be that of the vector $\mathbf{n}$.

The interaction (118.4) brings about a splitting of the nucleon level with orbital angular momentum l into two levels with angular momenta $j = l\pm\frac{1}{2}$. Since

$$\left.\begin{aligned}\mathbf{l}.\mathbf{s} &= \tfrac{1}{2}l \text{ for } j = l+\tfrac{1}{2},\\ &= -\tfrac{1}{2}(l+1) \text{ for } j = l-\tfrac{1}{2}\end{aligned}\right\} \tag{118.5}$$

(according to formula (31.3)), the amount of this splitting is

$$\begin{aligned}\Delta E &= E_{l-\frac{1}{2}} - E_{l+\frac{1}{2}}\\ &= \overline{f(r)}(l+\tfrac{1}{2}).\end{aligned} \tag{118.6}$$

Experiment shows that the level with $j = l+\frac{1}{2}$ (the vectors **l** and **s** parallel) is below the level with $j = l-\frac{1}{2}$; this means that $f(r) > 0$.

The spin–orbit coupling of a nucleon in the nucleus is relatively weak in comparison with its interaction in the self-consistent field. It is, nevertheless, in general large compared with the energy of the direct interaction of two nucleons in the nucleus, on account of the more rapid decrease of the latter with increasing atomic weight.

This relation between the energies of the various interactions has the result that the classification of the nuclear levels must be of the *jj* coupling type: the spins and orbital angular momenta of the various nucleons are added to give the total angular momenta $\mathbf{j} = \mathbf{l}+\mathbf{s}$, which are definite quantities, since the relation between **l** and **s** is not affected by the direct interaction between the particles (M. Göppert-Mayer 1949; O. Haxel, J. H. D. Jensen and H. E. Suess 1949)†. The vectors **j** of the individual nucleons are then added to give the total angular momentum **J** of the nucleus (usually called simply the *nuclear spin*, as if the nucleus were an elementary particle). In this respect the classification of nuclear levels differs essentially from that of atomic levels: in the electron shells of the atom, the relativistic spin–orbit coupling is in general small in comparison with the direct electric and exchange interactions, and so the level classification is usually based on *LS* coupling.

The state of each nucleon in a nucleus, however, is described by its angular momentum j and its parity. Although the vectors **l** and **s** are not separately conserved, the absolute magnitude of the orbital angular momentum of the nucleon is nevertheless definite. For the angular momentum j can arise either from a state with $l = j-\frac{1}{2}$ or from one with $l = j+\frac{1}{2}$. For a given (half-integral) j, these two states have different parities $(-1)^l$, and so, if j and the parity are specified, the quantum number l is determined also.

The states of nucleons with given l and j are customarily numbered (in order of increasing energy) by the "principal quantum number" n, which takes integral values starting from 1.‡ The various states are denoted by the symbols $1s_{1/2}$, $1p_{1/2}$, $1p_{3/2}$, etc., where the figure before the letter is the principal quantum number, the letters $s, p, d, \ldots$ indicate as usual the value of l, and the suffix is the value of j. Not more than $2j+1$ neutrons and the same number of protons can simultaneously be in a state with given values of n, l and j.

The states of the nucleus as a whole (in a given configuration) are customarily described by a figure giving the value of J and the sign + or − indicating the parity of the state (the latter being determined in the shell model by the parity of the algebraic sum of the values of l for all the nucleons).

From an analysis of experimental results concerning the properties of nuclei it is possible to derive a number of regularities in the positions of the nuclear levels. First of all, it is found that the energy of the nucleon increases with the orbital angular momentum l. This rule arises because, when l

---

† The coupling is closer to *LS* only for the lightest nuclei.

‡ Unlike the usual procedure for electron levels in an atom, where the number n takes values starting from $l+1$.

increases, so does the centrifugal energy of the particle, and its binding energy is therefore reduced.

Next, for a given value of l the level with $j = l+\frac{1}{2}$ (i.e. that which corresponds to parallel vectors **l** and **s**) lies below the level with $j = l-\frac{1}{2}$. This rule has already been mentioned in connection with the properties of the spin–orbit coupling of the nucleon in the nucleus.

The following rule relates to the isotopic spin of nuclei. The component T_ζ of the isotopic spin is known to be determined by the atomic weight and atomic number of the nucleus (see 116.1)). For a given value of T_ζ, the absolute magnitude of the isotopic spin can take any value such that $T \geqslant |T_\zeta|$. Usually, the ground state of the nucleus has the smallest of these possible values of the isotopic spin, i.e.

$$T_{gr} = |T_\zeta| = \tfrac{1}{2}(N-Z). \tag{118.7}$$

This rule is due to a property of the neutron–proton interaction, namely that in the *np* system the state with isotopic spin $T = 0$ (the deuteron state) has a greater binding energy than the state with $T = 1$; see the fourth footnote to §117.

We can also formulate certain rules relating to the spins of the ground states of nuclei. These rules determine the way in which the angular momenta **j** of the individual nucleons add to give the total spin of the nucleus. They represent the tendency of protons or neutrons in like states in the nucleus to "pair off" with opposite angular momenta; the binding energy of such *pp* and *nn* pairs is of the order of 1 or 2 MeV.

This phenomenon has, in particular, the result that, if the nucleus contains even numbers of both protons and neutrons (an *even–even* nucleus), then the angular momenta of all the nucleons balance in pairs, so that the total angular momentum of the nucleus is zero.

If the nucleus contains an odd number of protons or neutrons, however, with all nucleons outside closed shells being in like states, the total angular momentum of the nucleus is usually equal to that of one nucleon, as if a single nucleon were left over after the pairing of all possible pairs of protons and of neutrons (the total angular momenta of complete shells being necessarily zero).

For *odd–odd* nuclei (Z odd and N odd) there is no sufficiently general rule to determine the spin of the ground state.

A discussion of the actual manner in which the shells are filled in nuclei would require a detailed analysis of the available experimental results, and is outside the scope of this book. Here we shall add only some general remarks.

In studying the properties of atoms we have seen that their electron states can be divided into groups such that the binding energy of the electron decreases as each group is completed and the next is begun. A similar situation occurs for nuclei, the nucleon states being distributed among the following groups:

	Nucleons	
$1s_{1/2}$	2	
$1p_{3/2}$, $1p_{1/2}$	6	
$1d_{5/2}$, $1d_{3/2}$, $2s_{1/2}$	12	(118.8)
$1f_{7/2}$, $2p_{3/2}$, $1f_{5/2}$, $2p_{1/2}$, $1g_{9/2}$	30	
$2d_{5/2}$, $1g_{7/2}$, $1h_{11/2}$, $2d_{3/2}$, $3s_{1/2}$	32	
$2f_{7/2}$, $1h_{9/2}$, $1i_{13/2}$, $2f_{5/2}$, $3p_{3/2}$, $3p_{1/2}$	44	

For each group the total number of proton or neutron vacancies is shown. According to these numbers the occupation of a group is completed when the total number Z of protons or N of neutrons in the nucleus is equal to one of the numbers

$$2,\ 8,\ 20,\ 50,\ 82,\ 126.$$

These are commonly called *magic numbers*.†

The "doubly magic" nuclei, in which both Z and N are magic numbers, are particularly stable. In comparison with adjacent nuclei they have an unusually small affinity for a further nucleon, and their first excited states are unusually high.‡

The various states in each of the groups (118.8) are listed in approximate order of successive occupation in the series of nuclei. In reality, however, considerable irregularities are observed in the occupation process. Moreover, it must be borne in mind that, in heavy nuclei not close to the magic numbers, the distances between the various levels may be comparable with the "pairing energy", and the concept of individual states of components of a pair is then itself largely meaningless.

We may make some comments regarding the calculation of the magnetic moment of the nucleus in the shell model. By this we mean, of course, the magnetic moment averaged with respect to the motion of the particles in the nucleus. This mean magnetic moment $\bar{\boldsymbol{\mu}}$ is evidently in the direction of the nuclear spin $\mathbf{J}$, which is the only preferred direction in the nucleus; its operator is therefore

$$\hat{\bar{\boldsymbol{\mu}}} = \mu_0 g \hat{\mathbf{J}}, \tag{118.9}$$

where μ_0 is the nuclear magneton and g the gyromagnetic factor. The eigenvalue of the projection of this moment is $\bar{\mu}_z = \mu_0 g M_J$. Usually (cf. (111.1)) the magnetic moment μ of the nucleus is taken to be simply the maximum value of its projection, i.e. $\mu = \mu_0 g J$. In this notation

$$\hat{\bar{\boldsymbol{\mu}}} = \mu \hat{\mathbf{J}}/J. \tag{118.10}$$

---

† The states $1f_{7/2}$ (8 vacancies) are sometimes placed in a separate group, since the number 28 also has some magic properties.

‡ These nuclei include ${}^4_2He_2$, ${}^{16}_8O_8$, ${}^{40}_{20}Ca_{20}$, ${}^{208}_{82}Pb_{126}$; the ${}^4He$ nucleus is incapable of adding another nucleon.

The magnetic moment of the nucleus is composed of the magnetic moments of the nucleons outside closed shells, since the moments of nucleons in completed shells cancel out. Each nucleon produces in the nucleus a magnetic moment which consists of two parts: a spin part and (in the case of the proton) an orbital part, i.e. is represented by the sum $g_s\hat{\mathbf{s}}+g_l\hat{\mathbf{l}}$. (Here and henceforward we omit the factor μ_0, assuming, as is usual, that magnetic moments are measured in units of the nuclear magneton.) The spin and orbital gyromagnetic factors are $g_l = 1, g_s = 5{\cdot}585$ for the proton and $g_l = 0$, $g_s = -3{\cdot}826$ for the neutron.

After averaging with respect to the motion of the nucleon in the nucleus, its magnetic moment becomes proportional to $\mathbf{j}$; writing it in the form $g_j\mathbf{j}$, we have

$$g_j\hat{\mathbf{j}} = g_s\bar{\hat{\mathbf{s}}}+g_l\bar{\hat{\mathbf{l}}}$$

$$= \tfrac{1}{2}(g_l+g_s)\hat{\mathbf{j}}+\tfrac{1}{2}(g_l-g_s)\overline{(\hat{\mathbf{l}}-\hat{\mathbf{s}})}.$$

Multiplying both sides of this equation by $\hat{\mathbf{j}} = \hat{\mathbf{l}}+\hat{\mathbf{s}}$ and taking eigenvalues, we obtain

$$g_jj(j+1) = \tfrac{1}{2}(g_l+g_s)j(j+1)+\tfrac{1}{2}(g_l-g_s)[l(l+1)-s(s+1)],$$

and, putting $s = \tfrac{1}{2}, j = l\pm\tfrac{1}{2}$,

$$g_j = g_l\pm\frac{g_s-g_l}{2l+1} \text{ for } j = l\pm\tfrac{1}{2}. \tag{118.11}$$

With the above values of the gyromagnetic factors, this gives for the magnetic moment of the proton $\mu_p = g_jj$

$$\left.\begin{aligned} \mu_p &= \left(1-\frac{2{\cdot}29}{j+1}\right)j && \text{for } j = l-\tfrac{1}{2},\\ \mu_p &= j+2{\cdot}29 && \text{for } j = l+\tfrac{1}{2}, \end{aligned}\right\} \tag{118.12}$$

and for that of the neutron

$$\left.\begin{aligned} \mu_n &= \frac{1{\cdot}91}{j+1}j && \text{for } j = l-\tfrac{1}{2},\\ \mu_n &= -1{\cdot}91 && \text{for } j = l+\tfrac{1}{2} \end{aligned}\right\} \tag{118.13}$$

(T. Schmidt 1937).

If there is only one nucleon outside the closed shells, formulae (118.12) and (118.13) give directly the magnetic moment of the nucleus. For two nucleons, the addition of their magnetic moments is also elementary (see Problem 1). When the number of nucleons exceeds two, the averaging of the magnetic moment must be effected by means of the wave function of the system, constructed in the appropriate manner from the wave functions of the individual nucleons. If the nucleon configuration and the state of the nucleus as a whole are given, the wave function can be constructed uniquely in cases where only one state of the system with the given values of J and T can correspond to the given configuration (see, for example, Problem 3); otherwise, the state of the nucleus is a mixture of several independent states (with the same J and T), and in general the coefficients in the linear combination which gives the wave function of the nucleus remain unknown.†

Finally, we may mention that the existence of spin–orbit coupling of nucleons in the nucleus leads to the appearance of a certain magnetic moment of the protons in the nucleus, additional to (118.9) (M. Göppert-Mayer and J. H. D. Jensen 1952). The reason is that, when the interaction operator depends explicitly on the velocity of the particle, the case where an external field is present is obtained by replacing the momentum operator $\hat{\mathbf{p}}$ by $\hat{\mathbf{p}}-e\mathbf{A}/c$. Carrying out this replacement in (118.3) and using the expression (111.7) for the vector potential, we find that the Hamiltonian of the proton contains an additional term

$$\begin{aligned}\phi(r)\frac{e}{cm_p}\mathbf{n}\times\mathbf{A}.\hat{\mathbf{s}} &= f(r)\frac{e}{2c\hbar}\mathbf{r}\times(\mathbf{H}\times\mathbf{r}).\hat{\mathbf{s}}\\ &= f(r)\frac{e}{2c\hbar}\mathbf{r}\times(\hat{\mathbf{s}}\times\mathbf{r}).\mathbf{H}.\end{aligned}$$

This term is equivalent to the appearance of an additional magnetic moment whose operator is

$$\begin{aligned}\hat{\boldsymbol{\mu}}_{\text{add}} &= -\frac{e}{2c\hbar}f(r)\mathbf{r}\times(\hat{\mathbf{s}}\times\mathbf{r})\\ &= -\frac{e}{2c\hbar}r^2f(r)\{\hat{\mathbf{s}}-(\hat{\mathbf{s}}.\mathbf{n})\mathbf{n}\}.\end{aligned} \tag{118.14}$$

PROBLEMS

Problem 1. Determine the magnetic moment of a system of two nucleons (with total angular momentum $\mathbf{J} = \mathbf{j}_1+\mathbf{j}_2$), expressing it in terms of the magnetic moments μ_1 and μ_2 of the two nucleons.

Solution. Similarly to the derivation of formula (118.11) we obtain

$$\frac{\mu}{J} = \tfrac{1}{2}\left(\frac{\mu_1}{j_1}+\frac{\mu_2}{j_2}\right)+\tfrac{1}{2}\left(\frac{\mu_1}{j_1}-\frac{\mu_2}{j_2}\right)\frac{(j_1-j_2)(j_1+j_2+1)}{J(J+1)}.$$

Problem 2. Find the possible states of a system of three nucleons with angular momenta $j = 3/2$ (and the same principal quantum numbers).

Solution. We proceed as in §67 when finding the possible states of a system of equivalent electrons. Each nucleon can be in one of eight states with the following pairs of values of (m_j, τ_ζ):

(3/2, 1/2), (1/2, 1/2), (−1/2, 1/2), (−3/2, 1/2),
(3/2, −1/2), (1/2, −1/2), (−1/2, −1/2), (−3/2, −1/2).

† Note, however, that the "single-particle" calculation of nuclear magnetic moments is in practice fairly inaccurate. The pairs of values (118.12) and (118.13) are upper and lower limits rather than exact values of the moments.

Combining these states in groups of three different ones, we find the following pairs of values of (M_J, T_ζ) for the system of three nucleons:

(7/2, 1/2), 2(5/2, 1/2), (3/2, 3/2), 4(3/2, 1/2), (1/2, 3/2), 5(1/2, 1/2).

(The number before the parenthesis indicates the number of such states; states with negative values of M_J and T_ζ need not be written out.) These correspond to states of the system with the following values of (J,T):

(7/2, 1/2), (5/2, 1/2), (3/2, 3/2), (3/2, 1/2), (1/2, 1/2).

PROBLEM 3. Determine the magnetic moment of the ground state of a configuration of two neutrons and one proton in $p_{3/2}$ states (with the same n), taking account of isotopic invariance.†

SOLUTION. The ground state of such a configuration has $J = 3/2$, and from the rule given in the text its isotopic spin has the minimum possible value $T = |T_\zeta| = \frac{1}{2}$.

Let us determine the wave function of the system corresponding to the greatest possible value $M_J = 3/2$. This value can occur (when Pauli's principle for two like nucleons is applied) for the following sets of values of m_j for the nucleons p,n,n respectively:

(3/2, 3/2, −3/2), (3/2, 1/2, −1/2), (1/2, 3/2, −1/2), (−1/2, 3/2, 1/2).

Hence the required wave function $\psi_{TT_\zeta}^{JM_J}$ is a linear combination of the form

$$\Psi_{1/2,-1/2}^{3/2,3/2} = a[\psi_{1/2}^{3/2}\psi_{-1/2}^{3/2}\psi_{-1/2}^{-3/2}] + b[\psi_{1/2}^{3/2}\psi_{-1/2}^{1/2}\psi_{-1/2}^{-1/2}] +$$

$$+ c[\psi_{-1/2}^{3/2}\psi_{1/2}^{1/2}\psi_{-1/2}^{-1/2}] + d[\psi_{-1/2}^{3/2}\psi_{-1/2}^{1/2}\psi_{1/2}^{-1/2}], \tag{1}$$

where [...] denotes the normalized antisymmetrized product (i.e. a determinant of the form (61.5)) of the wave functions $\psi_{\tau_\zeta}^{m_j}$ of the individual nucleons.

The function (1) must vanish under the action of the operators

$$\hat{T}_- = \sum_{i=1}^{3} \hat{\tau}_-^{(i)} \quad \text{and} \quad \hat{J}_+ = \sum_{i=1}^{3} \hat{j}_+^{(i)};$$

see §67, Problem. The operators $\hat{\tau}_-^{(i)}$ convert the proton function of the ith nucleon to the neutron function, and the latter to zero. It is therefore easily seen that the operator $\hat{T}_-$ reduces the first term in (1) to a determinant with two identical rows, i.e. to zero, while the determinants in the three remaining terms become equal; thus we have the condition $b+c+d = 0$.

Next, for a single nucleon with $j = 3/2$ and various values of m_j we have, according to (27.12),

$$\hat{j}_+\psi^{3/2} = 0, \quad \hat{j}_+\psi^{1/2} = \sqrt{3}\psi^{3/2}, \quad \hat{j}_+\psi^{-1/2} = 2\psi^{1/2}, \quad \hat{j}_+\psi^{-3/2} = \sqrt{3}\psi^{-1/2}.$$

Hence we see that the action of the operator $\hat{J}_+$ on the function (1) gives

$$\hat{J}_+\Psi_{1/2,-1/2}^{3/2,3/2} = \sqrt{3}(a+b-c)[\psi_{1/2}^{3/2}\psi_{-1/2}^{3/2}\psi_{-1/2}^{-1/2}] +$$

$$+ 2(c-d)[\psi_{-1/2}^{3/2}\psi_{1/2}^{1/2}\psi_{-1/2}^{1/2}];$$

the change in sign of some terms is due to interchanging the rows of the determinant. The conditions for this expression to vanish are

$$a+b-c = 0, \quad c-d = 0.$$

Together with the normalization condition for the function (1), these relations give

$$a = 3/\sqrt{15}, \quad b = -2/\sqrt{15}, \quad c = d = 1/\sqrt{15}.$$

Since the mean value of the magnetic moment component of the proton (or neutron) in a state with given m_j is $\mu_p m_j/j$ (or $\mu_n m_j/j$), we find that the mean value of the angular momen-

† The nucleus Li^7 has this configuration (outside the closed shell $(1s_{1/2})^4$).

tum of the system calculated by means of the wave function (1) is

$$\mu = \bar{\mu}_z = \frac{9}{15}\mu_p + \frac{4}{15}\mu_p + \frac{1}{15}(\tfrac{1}{3}\mu_p + \tfrac{2}{3}\mu_n) +$$

$$+\frac{1}{15}(-\tfrac{1}{3}\mu_p + \frac{4}{3}\mu_n) = \frac{1}{15}(13\mu_p + 2\mu_n).$$

From formulae (118.12), (118.13) it follows that, for a nucleon in the $p_{3/2}$ state, $\mu_n = -1{\cdot}91$ and $\mu_p = 3{\cdot}79$. Thus $\mu = 3{\cdot}03$.

PROBLEM 4. Determine the magnetic moment of a nucleus in which all nucleons outside closed shells are in like states and the number of protons is equal to the number of neutrons.

SOLUTION. Since, for $N = Z$, the component T_ζ of the isotopic spin is zero, diagonal matrix elements occur only for the isotopic-scalar part of the operator

$$\hat{\mu} = \sum_n g_n \hat{\mathbf{j}}_n + \sum_p g_p \hat{\mathbf{j}}_p;$$

see the end of §116. Separating this part in accordance with formula (116.5), we find that it is

$$\tfrac{1}{2}(g_n + g_p)\sum_{n,p} \hat{\mathbf{j}} = \tfrac{1}{2}(g_n + g_p)\hat{\mathbf{J}}.$$

The total mean magnetic moment of the nucleus is therefore $\frac{1}{2}(g_n + g_p)J$.

PROBLEM 5. Calculate the additional magnetic moment of a nucleon with angular momentum j, expressing it in terms of the spin–orbit splitting (118.6) (M. Göppert-Mayer and J. H. D. Jensen 1952).

SOLUTION. The averaging of the angular part of the operator (118.14) (the expression in braces in that formula, which we denote by $\hat{\boldsymbol{\sigma}}$) is effected by means of the formula derived in §29, Problem. The result is

$$\bar{\hat{\boldsymbol{\sigma}}} \equiv \overline{\hat{\mathbf{s}} - (\hat{\mathbf{s}}.\mathbf{n})\mathbf{n}}$$

$$= \tfrac{2}{3}\hat{\mathbf{s}} + \frac{(\hat{\mathbf{s}}.\hat{\mathbf{l}})\hat{\mathbf{l}} - \hat{\mathbf{l}}(\hat{\mathbf{s}}.\hat{\mathbf{l}}) - \frac{2}{3}l(l+1)\hat{\mathbf{s}}}{(2l-1)(2l+3)}. \quad (2)$$

After complete averaging with respect to the motion of the nucleon, the mean value of $\boldsymbol{\sigma}$ can only be in the direction of $\mathbf{j}$, i.e. $\bar{\bar{\hat{\boldsymbol{\sigma}}}} = a\bar{\hat{\mathbf{j}}}$, whence $a = \bar{\hat{\boldsymbol{\sigma}}}\cdot\hat{\mathbf{j}}/\mathbf{j}^2$. Taking the component of the vector (2) in the direction of $\mathbf{j}$ (noting that the operator $\hat{\mathbf{j}}$ commutes with $\hat{\mathbf{l}}\cdot\hat{\mathbf{s}}$), and taking eigenvalues of the quantities $\mathbf{l}\cdot\mathbf{s}$, $\mathbf{l}^2$ etc., we easily find the following expression for the additional magnetic moment of the nucleon (in units of the nuclear magneton):

$$\mu_{\text{add}} = \mp\overline{f(r)}\frac{m_p R^2}{\hbar^2}\frac{2j+1}{4(j+1)} \quad \text{for } j = l \pm \tfrac{1}{2}, \quad (3)$$

where m_p is the nucleon mass and R the radius of the nucleus. In the averaging of $r^2 f$, the factor r^2 is replaced by R^2 owing to the rapid decrease of $f(r)$ inside the nucleus. The mean value $\bar{f}$ in (3) can be expressed in terms of the spin–orbit splitting by means of (118.6).

§119. Non-spherical nuclei

A system of particles in a spherically symmetric field cannot have a rotational energy spectrum; in quantum mechanics, the concept of rotation has no meaning for such a system. This applies to the shell model of the nucleus with a spherically symmetric self-consistent field considered in §117.

The division of the energy of the system into "internal" and "rotational" parts has no precise meaning in quantum mechanics. It can only be approximate and is possible where, for physical reasons, the consideration of the system as an assembly of particles, moving in a given field which is not spherically symmetric, is a good approximation. The rotational structure of the levels is then a consequence of taking into account the possibility of rotating this field with respect to a fixed system of coordinates. Such a case occurred, for example, in molecules, whose electron terms can be determined as the energy levels of a system of electrons moving in a given field of fixed nuclei.

Experiment shows that the majority of nuclei in fact have no rotational structure. This means that the spherically symmetric self-consistent field is a good approximation for such nuclei, i.e. they are spherical in shape apart from quantum fluctuations.

There exists also, however, a class of nuclei which have an energy spectrum of the rotational type; they lie approximately in the ranges of atomic weight $150 < A < 190$ and $A > 220$. This property means that the approximation of the spherically symmetric self-consistent field is entirely inapplicable to such nuclei, and for them the self-consistent field must in principle be sought without any initial assumptions regarding its symmetry, in order that the shape of the nucleus should also be "self-consistently" determined. Experiment shows that a correct model for nuclei of this type is given by a self-consistent field having an axis of symmetry and a plane of symmetry perpendicular to it (i.e. having the symmetry of a spheroid). The concept of non-spherical nuclei has been most extensively developed in the work of A. Bohr and B. R. Mottelson (1952–3).

It should be emphasized that we are concerned here with two qualitatively different classes of nuclei. This is seen, in particular, from the fact that nuclei are either spherical or else non-spherical with a "degree of deformation" that is not small.

The occurrence of non-sphericity is favoured by the presence of incomplete shells in the nucleus, and the phenomenon of nucleon pairing also appears to be of considerable importance here. Closed shells, on the other hand, tend to give spherical nuclei. A characteristic example is the doubly magic nucleus $^{208}_{82}Pb$: owing to the marked completeness of its nucleon configuration, this nucleus (and also those adjoining it) is spherical, and this brings about a gap in the sequence of non-spherical heavy nuclei.

The energy levels of a non-spherical nucleus consist of two parts: the levels of the "fixed" nucleus and the energy of its rotation as a whole. In even–even nuclei the intervals of the rotational structure of the levels are small in comparison with the distances between the levels of the "fixed" nucleus.

The classification of the levels of a non-spherical nucleus is in many ways similar to that for a diatomic molecule consisting of like atoms, since the symmetry of the field in which the particles (nucleons or electrons) move is the same in each case. We can therefore apply directly a number of the results obtained in Chapter XI.†

† It must be emphasized that we are referring to the analogy with the classification of levels of the diatomic molecule, not of the symmetrical top. For a system of particles moving in an axially symmetric field, the concept of rotation about the field axis has no meaning, like that of rotation about any axis for a system in a centrally symmetric field.

Let us first consider the classification of states of the "fixed" nucleus. In a field with axial symmetry, only the component of the angular momentum along the axis of symmetry is conserved. Each state of the nucleus is therefore described first of all by the value Ω of the component of its total angular momentum,† which can be either integral or half-integral. The levels are described as even (g) or odd (u) according to the behaviour of the wave function when the coordinates of all the nucleons (with respect to the centre of the nucleus) change sign.

In addition, for $\Omega = 0$ positive and negative states are distinguished, according to the behaviour of the wave function on reflection in a plane passing through the axis of the nucleus (see §78).

The ground states of even–even non-spherical nuclei are 0_g^+ (the zero indicating the value of Ω), corresponding to zero angular momentum and the highest symmetry of the wave function. This is a result of the pairing of all the neutrons and protons. If the nucleus contains an odd number of protons or neutrons, however, we can consider the state of the "odd" nucleon in the self-consistent field of the even–even remainder of the nucleus. Here the value of Ω is determined by the component ω of the angular momentum of this nucleon. Similarly, in an odd–odd nucleus the value of Ω is obtained from the angular momentum components of the odd neutron and proton:

$$\Omega = |\omega_p \pm \omega_n|.$$

It should be emphasized at the same time that we cannot speak of definite values of the components of the orbital angular momentum and spin of the nucleon. The reason is that, although the spin–orbit coupling of the nucleon is small in comparison with the energy of its interaction with the self-consistent field of the remainder of the nucleus, it is not small compared with the distances between adjoining energy levels of the nucleon in that field, as it would have to be for perturbation theory to be applicable, so that the orbital angular momentum and the spin of the nucleon could, to a good approximation, be considered separately.‡

Let us now consider the rotational structure of a non-spherical nucleus. The intervals in this structure are small compared with the spin–orbit interaction of the nucleons in the nucleus. This corresponds to case *a* in the theory of diatomic molecules (§83).

The total angular momentum $\mathbf{J}$ of a rotating nucleus is, of course, conserved. For given Ω its magnitude J takes values from Ω upwards:

$$J = \Omega,\ \Omega+1, \Omega+2, \dots; \tag{119.1}$$

see (83.2). An additional restriction on the possible values of J occurs for

---

† By definition, $\Omega \geqslant 0$ (just as the quantum number Λ is positive for diatomic molecules). It may be recalled that negative values of Ω for diatomic molecules could arise only because Ω was defined as the sum $\Lambda+\Sigma$, and Σ can be either positive or negative, depending on the relative directions of the orbital angular momentum and the spin.

‡ In spherical nuclei this was still possible, owing to the simultaneous conservation of parity and angular momentum.

nuclei with $\Omega = 0$: in states 0_g^+ and 0_u^- the number J takes only even values, and in states 0_g^- and 0_u^+ only odd values (see §86). In particular, in the rotational levels of the ground term for even–even nuclei (0_g^+) the number J takes the values 0, 2, 4, ...

The rotational energy of the nucleus is given by the formula

$$E_{\text{rot}} = \frac{\hbar^2}{2I} J(J+1), \tag{119.2}$$

where I is the moment of inertia of the nucleus (about an axis perpendicular to its axis of symmetry); this formula corresponds to the similar expression in the theory of diatomic molecules (the term depending on J in (83.6)). The lowest level corresponds to the least possible value of J, i.e. $J = \Omega$.

On account of (119.2) the rotational structure of the levels is described by certain interval rules which do not depend on the other characteristics of the level (for given Ω). For instance, the components of the rotational structure of the ground term of an even–even nucleus (with $J = 2, 4, 6, 8, \ldots$) are at distances in the ratio 1:3·3:7:12 ... from the lowest level ($J = 0$).

Formula (119.2), however, is insufficient for states with $\Omega = \frac{1}{2}$, which can occur in nuclei with an odd number of nucleons. In this case there is a contribution to the energy, comparable with (119.2), due to the interaction of the odd nucleon with the centrifugal field of the rotating nucleus. Its dependence on J can be found as follows.

It is known from mechanics (*Mechanics*, §39) that the energy of a particle in a rotating coordinate system contains an additional term equal to the product of the angular velocity of rotation and the angular momentum of the particle. The corresponding term in the Hamiltonian of the nucleus can be written in the form $2b\hat{\mathbf{K}}.\hat{\boldsymbol{\sigma}}$, where b is some constant, $\mathbf{K}$ the angular momentum of the remainder of the nucleus (excluding the last nucleon), and $\boldsymbol{\sigma}$ the angular momentum of that nucleon. Here the latter must be understood in a purely formal sense; in reality, the angular momentum vector of the nucleon does not exist in the axial field of the nucleus. This sense is that of an operator analogous to the operator of spin $\frac{1}{2}$, which gives transitions between states with values of the angular momentum component $\pm\frac{1}{2}$, in accordance with the value $\Omega = \frac{1}{2}$.† Since $\mathbf{K} = \mathbf{J} - \boldsymbol{\sigma}$, the eigenvalues of this operator are

$$2b\mathbf{K}\cdot\boldsymbol{\sigma} = b[J(J+1) - K(K+1) - \tfrac{3}{4}].$$

Adding for convenience the constant $\frac{1}{2}b$, which is independent of J, we find that this quantity equals $\pm b(J+\frac{1}{2})$ when $J = K \pm \frac{1}{2}$.

This expression can be written $(-1)^{J-\frac{1}{2}}\, b(J+\frac{1}{2})$ if we use the fact that the angular momentum K of the even–even remainder of the nucleus is even.

---

† The specific property of the case $\Omega = \frac{1}{2}$ consists precisely in the existence of matrix elements of the energy perturbation for transitions between states differing only in the sign of the angular momentum component and therefore belonging to the same energy. This brings about a shift in energy even in the first approximation of perturbation theory.

The phenomenon concerned is analogous to the Λ-doubling of the levels of a diatomic molecule with $\Omega = \frac{1}{2}$ (§88).

Thus we have finally the following expression for the rotational energy of the nucleus with $\Omega = \frac{1}{2}$:

$$E_{\text{rot}} = \frac{\hbar^2}{2I}J(J+1)+(-1)^{J-1/2}b(J+\tfrac{1}{2}) \qquad (119.3)$$

(A. Bohr and B. R. Mottelson 1953). Note that, if the constant b is positive and sufficiently large, the level with $J = 3/2$ may lie below that with $J = \frac{1}{2}$, i.e. the normal order of rotational levels (where the lowest level corresponds to the smallest possible value of J) will be altered.

The moment of inertia of a non-spherical nucleus cannot be calculated as that of a solid of given shape. Such a calculation would be possible only if the nucleons moving in the self-consistent field of the nucleus could be regarded as not directly interacting. In reality, the pairing effect leads to a reduction in the moment of inertia, in comparison with the value for a rigid body.

The magnetic moment $\boldsymbol{\mu}$ of a non-spherical nucleus consists of the magnetic moment of the "fixed" nucleus and that due to the rotation of the nucleus. The former (after averaging over the motion of the nucleons in the nucleus) is along the axis of the nucleus; denoting its value by μ', and the unit vector along the axis of the nucleus by $\mathbf{n}$, we can write it in the form $\mu'\mathbf{n}$. The magnetic moment due to the rotation is (after the same averaging) along the vector $\mathbf{J}-\Omega\mathbf{n}$, the total angular momentum of the nucleus minus that of the nucleons in the "fixed nucleus".† Thus

$$\boldsymbol{\mu} = \mu'\mathbf{n}+g_r(\mathbf{J}-\Omega\mathbf{n}). \qquad (119.4)$$

Here g_r is the gyromagnetic factor for the rotation of the nucleus. Since the contribution to the magnetic moment in rotation comes only from the protons, we have

$$g_r = I_p/(I_p+I_n), \qquad (119.5)$$

where I_n and I_p are the neutron and proton parts of the moment of inertia of the nucleus; for a system of protons only, $g_r = 1$ simply. The ratio (119.5) is in general not the same as the ratio Z/A of the number of protons to the total mass of the nucleus.

After averaging over the rotation of the nucleus, the magnetic moment is in the direction of the conserved vector $\mathbf{J}$:

$$\bar{\hat{\boldsymbol{\mu}}} = \mu\hat{\mathbf{J}}/J = (\mu'-\Omega g_r)\bar{\hat{\mathbf{n}}}+g_r\hat{\mathbf{J}}.$$

As usual, we multiply both sides of this equation by $\hat{\mathbf{J}}$ and take eigenvalues. In the ground state of the nucleus $\Omega = J$, and so

$$\mu = (\mu'+g_r)J/(J+1). \qquad (119.6)$$

† This formulation can be used only if $\Omega\neq\frac{1}{2}$ (see Problem 2).

PROBLEMS

PROBLEM 1. Express the quadrupole moment Q of a rotating nucleus in terms of the quadrupole moment Q_0 relative to axes fixed to the nucleus (A. Bohr 1951).

SOLUTION. The operator of the quadrupole moment tensor of a rotating nucleus is given in terms of Q_0 by

$$Q_{ik} = \tfrac{3}{2}Q_0(n_i n_k - \tfrac{1}{3}\delta_{ik});$$

this is a symmetrical tensor with zero trace, formed from the components of the unit vector **n** along the axis of the nucleus, and $Q_{zz} = Q_0$. The averaging with respect to the rotational state of the nucleus is effected similarly to the solution in §29, Problem (with the difference that $n_i \hat{J}_i = \Omega$, not zero), and leads to an expression of the form (75.2) with

$$Q = Q_0 \frac{3\Omega^2 - J(J+1)}{(2J+3)(J+1)}.$$

For the ground state of a nucleus with $\Omega = J$ we obtain

$$Q = Q_0 \frac{(2J-1)J}{(2J+3)(J+1)}.$$

As J increases, the ratio Q/Q_0 tends to 1, but only slowly.

PROBLEM 2. Determine the magnetic moment in the ground state of a nucleus with $\Omega = \frac{1}{2}$.

SOLUTION. In this case the magnetic moment operator can be written by means of the operator $\hat{\boldsymbol{\sigma}}$ introduced in the text, in the form

$$\hat{\boldsymbol{\mu}} = 2\mu'\hat{\boldsymbol{\sigma}} + g_r\hat{\mathbf{K}}, \quad \hat{\mathbf{K}} = \hat{\mathbf{J}} - \hat{\boldsymbol{\sigma}}.$$

The subsequent calculation is similar to that in the text. If the value $J = \frac{1}{2}$ corresponds to the ground level of the nucleus (and $K = J - \frac{1}{2} = 0$), we have $\mu = \mu'$; if in the ground state $J = 3/2$ (and $K = J + \frac{1}{2} = 2$), then $\mu = (9g_r - 3\mu')/5$.

PROBLEM 3. Determine the energies of the first few levels of the rotational structure of the ground state of an even–even nucleus having ellipsoidal symmetry.

SOLUTION. The ground state of an even–even nucleus corresponds to the most symmetrical wave function of the "fixed" nucleus, i.e. the function whose symmetry corresponds to the representation A of the group $\boldsymbol{D}_2$. There are therefore altogether $\frac{1}{2}J+1$ (for even J) or $\frac{1}{2}(J-1)$ (for odd J) different levels for a given value of J. For $J = 2$ they are given by formula (7) in §103, Problem 3, and for $J = 3$ by formula (8) in §103, Problem 4.

§120. Isotopic shift

The specific properties of the nucleus (finite mass, dimensions, spin) which distinguish it from a fixed point centre of a Coulomb field have a certain influence on the electron energy levels of the atom.

One such effect is called the *isotopic shift* of levels, that is, a change in the energy of a level from one isotope of an element to another. In practice, of course, what is of interest is not the change in energy of one level but the change in the distance between two levels observed as a spectral line. For this reason we must in practice consider not the energy of the entire electron envelope of the atom but only the part due to the electron involved in the transition in question.

In light atoms the isotopic shift is due mainly to the finite mass of the nucleus. When the motion of the nucleus is taken into account a term

$$\frac{1}{2M}\Big(\sum_i \hat{\mathbf{p}}_i\Big)^2$$

appears in the Hamiltonian, where M is the mass of the nucleus and the $\mathbf{p}_i$ are the momenta of the electrons.† The isotopic shift due to this effect is therefore given by the mean value

$$\frac{1}{2}\left(\frac{1}{M_1}-\frac{1}{M_2}\right)\overline{\left(\sum_i \mathbf{p}_i\right)^2}, \tag{120.1}$$

calculated from the wave function of the relevant state of the atom (M_1 and M_2 being the masses of the nuclei of the isotopes).

In heavy atoms the main contribution to the isotopic shift comes from the finite size of the nucleus. This effect is in practice appreciable only for the levels of an outer electron in the *s* state, since the wave function of the *s* state (unlike those of states with $l \neq 0$) does not vanish as $r \to 0$, and so the probability of finding the electron "within the nucleus" is comparatively large. We shall calculate the isotopic shift for this case.‡

Let $\phi(r)$ be the true electrostatic potential of the field of the nucleus, as opposed to the potential Ze/r of the Coulomb field of a point charge Ze. Then the change in the electron energy in comparison with its value in a purely Coulomb field Ze/r is given by the integral

$$\Delta E = -e\int(\phi - Ze/r)\psi^2(r)\,\mathrm{d}V, \tag{120.2}$$

where $\psi(r)$ is the electron wave function; in the *s* state this function is spherically symmetric and real. Although the integration here is formally extended to all space, in practice the difference $\phi - Ze/r$ in the integrand is zero except within the nucleus. The wave function of the *s* state tends to a constant limit as $r \to 0$ (see §32), and this constant value is practically reached even outside the nucleus. We can therefore take ψ^2 outside the integral and replace $\psi(r)$ by its value at $r = 0$, calculated for the Coulomb field of a point charge.

For a further transformation of the integral we use the identity $\triangle r^2 = 6$ and write (120.2) as

$$\Delta E = -\tfrac{1}{6}e\psi^2(0)\int(\phi - Ze/r)\triangle r^2.\mathrm{d}V$$

$$= -\tfrac{1}{6}e\psi^2(0)\int r^2\triangle(\phi - Ze/r)\mathrm{d}V;$$

in transforming the volume integral we have used the fact that the resulting

† In the centre-of-mass system of the atom, the sum of the momenta of the nucleus and the electrons is zero: $\mathbf{p}_{\mathrm{nuc}} + \Sigma\mathbf{p}_i = 0$. Their total kinetic energy is therefore

$$\frac{\mathbf{p}_{\mathrm{nuc}}^2}{2M}+\frac{1}{2m}\sum_i \mathbf{p}_i^2 = \frac{1}{2M}\left(\sum_i \mathbf{p}_i\right)^2+\frac{1}{2m}\sum \mathbf{p}_i^2.$$

‡ The calculation given below does not take account of relativistic effects in the motion of the electron near the nucleus, and is valid only if the condition $Ze^2/\hbar c \ll 1$ holds.

integral over an infinitely remote surface is zero. But $\triangle(1/r) = -4\pi\delta(\mathbf{r})$, and $r^2\delta(\mathbf{r}) = 0$ for all r. According to the electrostatic Poisson's equation, $\triangle\phi = -4\pi\rho$, and in this case ρ is the density of the electric charge distribution in the nucleus. The final result is

$$\Delta E = \tfrac{2}{3}\pi\psi^2(0)Ze^2\overline{r^2}, \tag{120.3}$$

where

$$\overline{r^2} = (1/Ze)\int \rho r^2 \mathrm{d}V$$

is the proton mean square radius of the nucleus; for a uniform distribution of protons in the nucleus, $\overline{r^2} = 3R^2/5$, where R is the geometrical radius of the nucleus. The isotopic shift of the level is given by the difference of the expressions (120.3) for the two isotopes.

In §71 an estimate has been given of $\psi(0)$, and it was shown to depend on the atomic number (assumed large) as $\sqrt{Z}$. Hence the splitting (120.3) is proportional to R^2Z^2.

§121. Hyperfine structure of atomic levels

Another effect in atoms due to the properties of the nucleus is the splitting of atomic energy levels as a result of the interaction of electrons with the spin of the nucleus. This is called the *hyperfine structure* of the levels. On account of the weakness of this interaction the intervals in the hyperfine structure are very small, even in comparison with those in the fine structure. Hence the hyperfine structure must be considered separately for each component of the fine structure.

The spin of the nucleus will be denoted in this section (in accordance with the notation usual in atomic spectroscopy) by i, the notation J being retained for the total angular momentum of the electron envelope of the atom. The total angular momentum of the atom (including the nucleus) is denoted by $\mathbf{F} = \mathbf{J}+\mathbf{i}$. Each component of the hyperfine structure is described by a definite value of this angular momentum. According to the general rules for addition of angular momenta, the quantum number F takes the values

$$F = J+i,\ J+i-1, \ldots, |J-i|, \tag{121.1}$$

so that each level with given J is split into $2i+1$ components if $i < J$, or $2J+1$ if $i > J$.

Since the mean distances r between the electrons in the atom are large compared with the radius R of the nucleus, an important part in the hyperfine splitting is played by the interaction of the electrons with the lowest-order multipole moments of the nucleus. These are the magnetic dipole and electric quadrupole moments; the mean dipole moment is zero (see §75).

The magnetic moment of the nucleus is of the order of $\mu_{\text{nuc}} \sim eRv_{\text{nuc}}/c$, where v_{nuc} are the velocities of the nucleons in the nucleus. The energy of its interaction with the magnetic moment of the electron ($\mu_{el} \sim e\hbar/mc$) is of the order of

$$\frac{\mu_{nuc}\mu_{el}}{r} \sim \frac{e^2\hbar}{mc^2}\frac{Rv_{nuc}}{r^3}. \qquad (121.2)$$

The quadrupole moment $Q \sim eR^2$; the energy of interaction of the field which it produces with the charge on the electron is of the order of

$$eQ/r^3 \sim e^2R^2/r^3. \qquad (121.3)$$

Comparison of (121.2) and (121.3) shows that the magnetic interaction (and therefore the resulting splitting of the levels) is $(v_{nuc}/c)(\hbar/mcR) \sim 15$ times greater than the quadrupole interaction; although the ratio v_{nuc}/c is relatively small, the ratio $\hbar/mcR$ is large.

The operator of the magnetic interaction of the electrons with the nucleus is of the form

$$\hat{V}_{iJ} = a\hat{\mathbf{i}}\cdot\hat{\mathbf{J}} \qquad (121.4)$$

(similarly to the spin–orbit interaction of the electrons (72.4)). The dependence of the resulting splitting of the levels on F is therefore given by

$$\tfrac{1}{2}aF(F+1); \qquad (121.5)$$

cf. (72.5).

The operator of the quadrupole interaction of the electrons with the nucleus is constructed from the operator $\hat{Q}_{ik}$ of the quadrupole moment tensor of the nucleus and the components of the angular momentum vector $\hat{\mathbf{J}}$ of the electrons. It is proportional to the scalar $\hat{Q}_{ik}\hat{J}_i\hat{J}_k$ formed from these operators, i.e. has the form

$$b[\hat{i}_i\hat{i}_k + \hat{i}_k\hat{i}_i - \tfrac{2}{3}i(i+1)\delta_{ik}]\hat{J}_i\hat{J}_k; \qquad (121.6)$$

here we have used the fact that Q_{ik} is given in terms of the nuclear spin operator by a formula of the type (75.2). On calculating the eigenvalues of the operator (121.6) (in a manner entirely similar to the calculations in §84, Problem 1), we find that the dependence of the quadrupole hyperfine splitting of the levels on the quantum number F is given by the expression

$$\tfrac{1}{2}bF^2(F+1)^2 + \tfrac{1}{2}bF(F+1)[1-2J(J+1)-2i(i+1)]. \qquad (121.7)$$

The magnetic hyperfine splitting effect is especially noticeable for levels due to an outer electron in the s state, owing to the comparatively high probability that such an electron will be near the nucleus.

Let us calculate the hyperfine splitting for an atom containing one outer s electron (E. Fermi 1930). This electron is described by the spherically symmetric wave function $\psi(r)$ of its motion in the self-consistent field of the other electrons and the nucleus.†

† The following calculation assumes that the condition $Ze^2/\hbar c \ll 1$ is satisfied (cf. the second footnote to §120).

We shall seek the operator of the interaction with the nucleus as the operator $-\hat{\boldsymbol{\mu}}.\hat{\mathbf{H}}$ of the energy of the magnetic moment $\hat{\boldsymbol{\mu}} = \mu\hat{\mathbf{i}}/i$ of the nucleus in the magnetic field $\mathbf{H}$ created (at the origin) by the electron. According to a well-known formula of electrodynamics, this field is

$$\hat{\mathbf{H}} = \frac{1}{c}\int \frac{\mathbf{n}\times\hat{\mathbf{j}}}{r^2}\,\mathrm{d}V, \tag{121.8}$$

where $\hat{\mathbf{j}}$ is the operator of the current density due to the moving electron spin, and $\mathbf{r} = \mathbf{n}r$ the radius vector from the centre to the element $\mathrm{d}V$.† According to (115.4),

$$\begin{aligned}\hat{\mathbf{j}} &= -2\mu_B c\,\mathbf{curl}\,(\psi^2\hat{\mathbf{s}})\\ &= -2\mu_B c\frac{\mathrm{d}\psi^2(r)}{\mathrm{d}r}\mathbf{n}\times\hat{\mathbf{s}},\end{aligned}$$

where μ_B is the Bohr magneton. Writing $\mathrm{d}V = r^2\,\mathrm{d}r\mathrm{d}o$ and carrying out the integration, we find

$$\begin{aligned}\hat{\mathbf{H}} &= -2\mu_B\int_0^\infty \frac{\mathrm{d}\psi^2}{\mathrm{d}r}\,\mathrm{d}r\int \mathbf{n}\times(\mathbf{n}\times\hat{\mathbf{s}})\,\mathrm{d}o\\ &= -2\mu_B\psi^2(0)\frac{8\pi}{3}\hat{\mathbf{s}}.\end{aligned}$$

The interaction operator is, finally,

$$\hat{V}_{is} = -\hat{\boldsymbol{\mu}}\,.\,\hat{\mathbf{H}} = \frac{16\pi}{3i}\mu\mu_B\psi^2(0)\hat{\mathbf{i}}\,.\,\hat{\mathbf{s}}. \tag{121.9}$$

If the total angular momentum of the atom $J = S = \frac{1}{2}$, the hyperfine splitting leads to the appearance of a doublet ($F = i\pm\frac{1}{2}$); according to (121.5) and (121.9) we find for the distance between the two levels

$$E_{i+\frac{1}{2}} - E_{i-\frac{1}{2}} = (8\pi/3i)\mu\mu_B(2i+1)\psi^2(0). \tag{121.10}$$

Since the value of $\psi(0)$ is proportional to $\sqrt{Z}$ (see §71), the magnitude of this splitting is proportional to the atomic number.

PROBLEMS

Problem 1. Calculate the hyperfine splitting (due to magnetic interaction) for an atom containing (outside closed shells) one electron with orbital angular momentum i (E. Fermi 1930).

---

† See *Fields*, (43.7). In that formula, the vector $\mathbf{R}$ is in the opposite direction, from $\mathrm{d}V$ to the centre (the point at which the field is observed).

SOLUTION. The vector potential and the magnetic field strength due to the magnetic moment μ of the nucleus are

$$\mathbf{A} = \frac{\boldsymbol{\mu}\times\mathbf{n}}{r^2}, \quad \mathbf{H} = \frac{3\mathbf{n}(\boldsymbol{\mu}.\mathbf{n})-\boldsymbol{\mu}}{r^3}$$

(div $\mathbf{A} = 0$). Using these expressions, we can write the interaction operator in the form

$$\frac{|e|}{mc}\mathbf{A}.\hat{\mathbf{p}}+\frac{|e|\hbar}{mc}\hat{\mathbf{H}}.\hat{\mathbf{s}} = \frac{2\mu_B}{r^3}\hat{\boldsymbol{\mu}}.[\hat{\mathbf{l}}+3(\hat{\mathbf{s}}.\mathbf{n})\mathbf{n}-\hat{\mathbf{s}}].$$

After averaging over a state with a given value of j, the expression in the brackets is in the direction of $\mathbf{j}$. We can therefore write

$$\hat{V}_{ij} = 2\mu_B\hat{\boldsymbol{\mu}}.\hat{\mathbf{j}}[\hat{\mathbf{l}}.\hat{\mathbf{j}}+3\overline{(\hat{\mathbf{s}}.\mathbf{n})(\mathbf{n}.\hat{\mathbf{j}})}-\hat{\mathbf{s}}.\hat{\mathbf{j}}]\overline{r^{-3}}/j(j+1).$$

The mean value of $n_i n_k$ has been calculated in §29, Problem. Using this and taking eigenvalues, we find

$$\frac{2\mu_B\mu}{i}\mathbf{i}.\mathbf{j}\left[\mathbf{l}.\mathbf{j}+\frac{2l(l+1)\mathbf{s}.\mathbf{j}-6(\mathbf{s}.\mathbf{l})(\mathbf{j}.\mathbf{l})}{(2l-1)(2l+3)}\right]\overline{r^{-3}}/j(j+1),$$

whence, after a simple calculation, we have finally

$$\frac{\mu_B\mu}{i}\frac{l(l+1)}{j(j+1)}F(F+1)\overline{r^{-3}},$$

where $\mathbf{F} = \mathbf{j}+\mathbf{i}$, and $j = l\pm\frac{1}{2}$. The averaging of r^{-3} is with respect to the radial part of the electron wave function.

PROBLEM 2. Determine the Zeeman splitting of the components of the hyperfine structure of an atomic level (S. A. Goudsmit and R. F. Bacher 1930).

SOLUTION. In formula (113.4) (the field being assumed so weak that the splitting which it causes is small in comparison with the hyperfine structure intervals), the averaging must be effected not only with respect to the electron state but also with respect to the directions of the nuclear spin. From the first averaging we get $\Delta E = \mu_B g_J J_z H$, with the same g_J (113.7). The second averaging gives, analogously to (113.5),

$$\bar{J}_z = (\mathbf{J}\,.\,\mathbf{F})M_F/\mathbf{F}^2.$$

Thus we have finally

$$\Delta E = \mu_B g_F H M_F, \quad g_F = g_J\frac{F(F+1)+J(J+1)-i(i+1)}{2F(F+1)}.$$

§122. Hyperfine structure of molecular levels

The hyperfine structure of the energy levels of molecules is similar to that of the atomic levels.

In the great majority of molecules the total electron spin is zero. The main source of the hyperfine splitting of the levels is then the quadrupole interaction of the nuclei and the electrons; here, of course, only those nuclei participate in the interaction whose spin i is neither 0 nor $\frac{1}{2}$, since otherwise their quadrupole moment is zero.

On account of the comparative slowness of the motion of the nuclei in the molecule, the averaging of the quadrupole interaction operator with respect to the state of the molecule is effected in two stages: first we must average with respect to the electron state for fixed nuclei, and then with respect to the rotation of the molecule.

Let us first consider the diatomic molecule. The first averaging gives an interaction of each nucleus with the electrons that is expressed by an operator proportional to the scalar $\hat{Q}_{ik}n_i n_k$ formed from the operator of the quadrupole moment tensor of the nucleus and the unit vector **n** along the axis of the molecule—the only quantity which determines the orientation of the molecule with respect to the direction of the nuclear spin. Since $\hat{Q}_{ii} = 0$, this operator can be written in the form

$$b\hat{\imath}_i\hat{\imath}_k(n_i n_k - \tfrac{1}{3}\delta_{ik}); \tag{122.1}$$

for a given value of the component i_ζ of the nuclear spin along the axis of the molecule, this quantity is $b[i_\zeta^2 - \frac{1}{3}i(i+1)]$.

When the operator (122.1) is averaged with respect to the rotation of the molecule, it is expressed in terms of the operator $\hat{\mathbf{K}}$ of the conserved rotational angular momentum. The averaging of the product $n_i n_k$ is effected by means of the formula derived in §29, Problem (with the vector **K** in place of **l**), and the result is

$$-\frac{b}{(2K-1)(2K+3)}\hat{\imath}_i\hat{\imath}_k[\hat{K}_i\hat{K}_k + \hat{K}_k\hat{K}_i - \tfrac{2}{3}\delta_{ik}K(K+1)]. \tag{122.2}$$

The eigenvalues of this operator are found in the same way as for (121.6).

For a polyatomic molecule we obtain in general, instead of (122.1), an operator of the form

$$b_{ik}\hat{\imath}_i\hat{\imath}_k, \tag{122.3}$$

where b_{ik} is a tensor with zero trace which is a certain characteristic of the electron state of the molecule. After averaging with respect to the rotation of the molecule, this tensor is given in terms of the total rotational angular momentum **J** by a formula of the type

$$\bar{b}_{ik} = b[\hat{J}_i\hat{J}_k + \hat{J}_k\hat{J}_i - \tfrac{2}{3}J(J+1)\delta_{ik}]. \tag{122.4}$$

The coefficient b can in principle be expressed in terms of the components of the tensor b_{ik} relative to the principal axes of inertia of the molecule ξ,η,ζ; since these axes are fixed in the molecule, the components $b_{\xi\xi}$ etc. are a property of the molecule and unaffected by the averaging. Let us consider the scalar $\overline{b_{ik}J_iJ_k}$. A calculation using (122.4) gives

$$\overline{b_{ik}J_iJ_k} = bJ(J+1)[\tfrac{4}{3}J(J+1) - 1]; \tag{122.5}$$

the method is similar to that used in §29, Problem. Expanding the tensor product in components along the axes ξ,η,ζ, we obtain

$$\overline{b_{ik}J_iJ_k} = b_{\xi\xi}\overline{J_\xi^2} + b_{\eta\eta}\overline{J_\eta^2} + b_{\zeta\zeta}\overline{J_\zeta^2}, \tag{122.6}$$

where we have used the fact that the mean values of the products $J_\xi J_\zeta$ etc.

are zero.† The mean values of the squares $\overline{J_\xi^2}$ etc. are found, in principle, from the wave functions of the corresponding rotational states of the top. In particular, for a symmetrical top we have simply

$$\overline{J_\zeta^2} = k^2;\ \overline{J_\xi^2}, \overline{J_\eta^2} = \tfrac{1}{2}[J(J+1)-k^2].$$

If the spins of the nuclei are $\frac{1}{2}$, the quadrupole interaction is absent. In this case one of the main sources of hyperfine splitting is the direct magnetic interaction between the nuclear magnetic moments. The operator of the interaction of two magnetic moments $\boldsymbol{\mu}_1 = \mu_1\mathbf{i}_1/i_1$, $\boldsymbol{\mu}_2 = \mu_2\mathbf{i}_2/i_2$ is given by

$$\frac{\mu_1\mu_2}{i_1i_2r^3}[\mathbf{i}_1.\mathbf{i}_2 - 3(\mathbf{i}_1.\mathbf{n})(\mathbf{i}_2.\mathbf{n})].$$

To calculate the splitting energy, this must be averaged with respect to the state of the molecule, as described above.

When the molecule contains heavy atoms, comparable contributions to the hyperfine splitting are given by the direct interaction and by the indirect interaction of the nuclear moments through the electron envelope. Formally, this interaction is an effect in the second approximation of perturbation theory with respect to the interaction of the nuclear spin with the electrons. By means of the results of §121 we easily find that the ratio of this effect to the direct interaction of the nuclear moments is of the order $(Ze^2/\hbar c)^2$, and is comparable with unity for large Z.

Finally, some contribution to the hyperfine splitting of molecular levels comes from the interaction of the nuclear magnetic moment with the rotation of the molecule. The rotating molecule, being a moving system of charges, creates a certain magnetic field, which may be calculated, using the formulae of electrodynamics, from the given current density $\mathbf{j} = \rho\boldsymbol{\Omega}\times\mathbf{r}$, where ρ is the charge density (of electrons and nuclei) in the molecule at rest, and $\boldsymbol{\Omega}$ its angular velocity of rotation. The magnitude of the level splitting is found as the energy of the magnetic moment of the nucleus in this field; the components of the angular velocity of the molecule must be expressed in terms of those of its angular momentum (cf. §103).

† For in a representation where the matrix of one component of $\mathbf{J}$ (J_ζ, say) is diagonal, the matrices of the products $J_\xi J_\zeta$, $J_\eta J_\zeta$ have non-zero elements only when the quantum number k changes by 1, whereas the wave functions of stationary states of an asymmetrical top include functions ψ_{Jk} with values of k differing by an even number (see §103).

CHAPTER XVII

ELASTIC COLLISIONS

§123. The general theory of scattering

In classical mechanics, collisions of two particles are entirely determined by their velocities and impact parameter (the distance at which they would pass if they did not interact). In quantum mechanics the very wording of the problem must be changed, since in motion with definite velocities the concept of the path is meaningless, and therefore so is the impact parameter. The purpose of the theory is here only to calculate the probability that, as a result of the collision, the particles will deviate (or, as we say, be *scattered*) through any given angle. We are speaking here of what are called *elastic collisions*, in which the particles, or the internal state of the colliding particles if these are complex, are left unchanged.

The problem of an elastic collision, like any problem of two bodies, amounts to a problem of the scattering of a single particle, with the reduced mass, in the field $U(r)$ of a fixed centre of force.† This simplification is effected by changing to a system of coordinates in which the centre of mass of the two particles is at rest. The scattering angle in this system we denote by θ. It is simply related to the angles ϑ_1 and ϑ_2 giving the deviations of the two particles in the *laboratory system* of coordinates, in which the second particle (say) was at rest before the collision:

$$\tan\vartheta_1 = m_2 \sin\theta/(m_1 + m_2\cos\theta), \qquad \vartheta_2 = \tfrac{1}{2}(\pi-\theta), \tag{123.1}$$

where m_1, m_2 are the masses of the particles (see *Mechanics*, §17). In particular, if the masses of the two particles are the same ($m_1 = m_2$), we have simply

$$\vartheta_1 = \tfrac{1}{2}\theta, \qquad \vartheta_2 = \tfrac{1}{2}(\pi-\theta); \tag{123.2}$$

the sum $\vartheta_1 + \vartheta_2 = \frac{1}{2}\pi$, i.e. the particles diverge at right angles.

In this chapter we shall always use (unless the contrary is specifically stated) a system of coordinates in which the centre of mass is at rest, and m will denote the reduced mass of the colliding particles.

A free particle moving in the positive direction of the z-axis is described by a plane wave, which we take in the form $\psi = e^{ikz}$, i.e. we normalize so that the current density in the wave is equal to the particle velocity v. The scattered particles are described, at a great distance from the scattering

† Here we neglect the spin–orbit interaction of the particles (if they have spin). By assuming the field to be centrally symmetric, we exclude from consideration also processes such as the scattering of electrons by molecules.

centre, by an outgoing spherical wave of the form $f(\theta)e^{ikr}/r$, where $f(\theta)$ is some function of the scattering angle θ (the angle between the z-axis and the direction of the scattered particle). This function is called the *scattering amplitude*. Thus the exact wave function, which is a solution of Schrödinger's equation with potential energy $U(r)$, must have at large distances the asymptotic form

$$\psi \approx e^{ikz} + f(\theta)e^{ikr}/r. \tag{123.3}$$

The probability per unit time that the scattered particle will pass through a surface element $\mathrm{d}S = r^2\mathrm{d}o$ (where $\mathrm{d}o$ is an element of solid angle) is $(v/r^2)\,|f|^2\,\mathrm{d}S = v|f|^2\,\mathrm{d}o$.† Its ratio to the current density in the incident wave is

$$\mathrm{d}\sigma = |f(\theta)|^2\,\mathrm{d}o. \tag{123.4}$$

This quantity has the dimensions of area, and is called the *effective cross-section*, or simply the *cross-section*, for scattering into the solid angle $\mathrm{d}o$. If we put $\mathrm{d}o = 2\pi\sin\theta\,\mathrm{d}\theta$, we obtain for the cross-section

$$\mathrm{d}\sigma = 2\pi\sin\theta\,|f(\theta)|^2\,\mathrm{d}\theta \tag{123.5}$$

for scattering through angles in the range from θ to $\theta+\mathrm{d}\theta$.

A solution of Schrödinger's equation for scattering in a central field $U(r)$ must evidently be axially symmetric about the z-axis, the direction of the incident particles. Any such solution can be represented as a superposition of wave functions of the continuous spectrum, corresponding to motion, in the field concerned, of particles with given energy $\hbar^2k^2/2m$ and orbital angular momenta having various magnitudes l and zero z-components; these functions are independent of the azimuthal angle ϕ round the z-axis, i.e. they are axially symmetric. Thus the required wave function has the form

$$\psi = \sum_{l=0}^{\infty} A_l P_l(\cos\theta) R_{kl}(r), \tag{123.6}$$

where the A_l are constants and the R_{kl} are radial functions satisfying the equation

$$\frac{1}{r^2}\frac{\mathrm{d}}{\mathrm{d}r}\left(r^2\frac{\mathrm{d}R_{kl}}{\mathrm{d}r}\right) + \left[k^2 - \frac{l(l+1)}{r^2} - \frac{2m}{\hbar^2}U(r)\right]R_{kl} = 0. \tag{123.7}$$

The coefficients A_l must be chosen so that at large distances the function (123.6) has the asymptotic form (123.3). We shall show that this implies that

$$A_l = \frac{1}{2k}(2l+1)i^l \exp(i\delta_l), \tag{123.8}$$

† It is supposed that the incident beam of particles is defined by a wide (to avoid diffraction effects) but finite diaphragm, as happens in actual experiments on scattering. There is therefore no interference between the two terms of the expression (123.3); the squared modulus $|\psi|^2$ is taken at points where there is no incident wave.

where the δ_l are the phase shifts of the functions R_{kl}. This will also solve the problem of expressing the scattering amplitude in terms of these phases.

The asymptotic form of the function R_{kl} is given by (33.20):

$$R_{kl} \approx \frac{2}{r}\sin\left(kr - \tfrac{1}{2}l\pi + \delta_l\right)$$

$$= \frac{1}{ir}\{(-i)^l \exp[i(kr+\delta_l)] - i^l \exp[-i(kr+\delta_l)]\}.$$

Substituting this and (123.8) in (123.6), we obtain the asymptotic expression for the wave function in the form

$$\psi \approx \frac{1}{2ikr}\sum_{l=0}^{\infty}(2l+1)P_l(\cos\theta)[(-1)^{l+1}e^{-ikr} + S_l e^{ikr}], \qquad (123.9)$$

with the notation

$$S_l = \exp(2i\delta_l). \qquad (123.10)$$

The expansion of the plane wave (34.2), with the same transformation, is

$$e^{ikz} \approx \frac{1}{2ikr}\sum_{l=0}^{\infty}(2l+1)P_l(\cos\theta)[(-1)^{l+1}e^{-ikr} + e^{ikr}].$$

We see that, in the difference $\psi - e^{ikz}$, all terms containing the factors e^{-ikr} disappear, as they should. For the coefficient of e^{ikr}/r in this difference, i.e. the scattering amplitude, we obtain

$$f(\theta) = \frac{1}{2ik}\sum_{l=0}^{\infty}(2l+1)[S_l - 1]P_l(\cos\theta). \qquad (123.11)$$

The formula solves the problem of expressing the scattering amplitude in terms of the δ_l (H. Faxén and J. Holtsmark 1927).†

If we integrate $d\sigma$ over all angles, we obtain the total scattering cross-section σ, which is the ratio of the total probability (per unit time) that the particle will be scattered to the probability current density in the incident

† The problem of recovering the form of the scattering potential from the phases δ_l (assumed known) is of fundamental interest. This has been solved by I. M. Gel'fand, B. M. Levitan and V. A. Marchenko. It is found that, to determine $U(r)$, it is in principle sufficient to know $\delta_0(k)$ as a function of the wave number throughout the range from $k = 0$ to $k = \infty$, together with the coefficients a_n in the asymptotic expressions (for $r \to \infty$) $R_{n0} \approx (a_n/r)e^{-\kappa_n r}$ ($\kappa_n = \sqrt{(2m|E_n|)}/\hbar$) of the wave functions of states corresponding to the discrete (negative) energy levels E_n (if any). The determination of $U(r)$ from these data requires the solution of a certain linear integral equation. This topic is fully discussed by V. de Alfaro and T. Regge, *Potential Scattering*, North-Holland, Amsterdam, 1965.

wave. Substituting (123.11) in the integral

$$\sigma = 2\pi \int_0^\pi |f(\theta)|^2 \sin\theta \, d\theta,$$

and recalling that the Legendre polynomials with different l are orthogonal, while

$$\int_0^\pi P_l^2(\cos\theta) \sin\theta \, d\theta = 2/(2l+1),$$

we have for the total cross-section

$$\sigma = \frac{4\pi}{k^2} \sum_{l=0}^{\infty} (2l+1) \sin^2 \delta_l. \tag{123.12}$$

Each of the terms in this sum is a *partial cross-section* σ_l for the scattering of particles with given orbital angular momentum l. It may be noted that the maximum possible value of this cross-section is

$$\sigma_{l,\max} = (4\pi/k^2)(2l+1). \tag{123.13}$$

Comparing this with formula (34.5), we see that the number of particles scattered with angular momentum l may be four times the number of such particles in the incident flux. This is a purely quantum effect due to interference between the scattered and unscattered particles.

It will be useful later to employ also the *partial scattering amplitudes* f_l, which we define as the coefficients in the expansion

$$f(\theta) = \sum_l (2l+1) f_l P_l(\cos\theta). \tag{123.14}$$

According to (123.11) these are related to the phases δ_l by

$$f_l = \frac{1}{2ik}(S_l - 1) = \frac{1}{2ik}(e^{2i\delta_l} - 1), \tag{123.15}$$

and the partial cross-sections are

$$\sigma_l = 4\pi(2l+1)|f_l|^2. \tag{123.16}$$

§124. An investigation of the general formula

The formulae which we have obtained are in principle applicable to scattering in any field $U(r)$ which vanishes at infinity. The use of these formulae involves only an examination of the properties of the phases δ_l which appear in them.

To estimate the order of magnitude of the phases δ_l for large values of l, we use the fact that the motion is quasi-classical for large l (see §49). Hence the phase of the wave function is determined by the integral

$$\int_{r_0}^{r} \sqrt{\left[k^2 - \frac{(l+\frac{1}{2})^2}{r^2} - \frac{2mU(r)}{\hbar^2}\right]} \, dr + \tfrac{1}{4}\pi,$$

where r_0 is a zero of the expression under the radical ($r > r_0$ being the classically accessible region of motion). Subtracting from this the phase

$$\int_{r_0}^{r} \sqrt{\left[k^2 - \frac{(l+\frac{1}{2})^2}{r^2}\right]} \, dr + \tfrac{1}{4}\pi$$

of the wave function of free motion, and letting $r \to \infty$, we obtain, by definition, the quantity δ_l. For large values of l, the value of r_0 also becomes large; $U(r)$ is therefore small throughout the range of integration, and we have approximately

$$\delta_l = -\int_{r_0}^{\infty} mU(r) \, dr \Big/ \hbar^2 \sqrt{\left[k^2 - \frac{(l+\frac{1}{2})^2}{r^2}\right]}. \tag{124.1}$$

In order of magnitude this integral (if convergent) is

$$\delta_l \sim mU(r_0) r_0 / k\hbar^2. \tag{124.2}$$

The order of magnitude of r_0 is $r_0 \sim l/k$.

If $U(r)$ vanishes at infinity as $1/r^n$ with $n > 1$, the integral (124.1) converges, and the phases δ_l are finite. On the other hand, for $n \leqslant 1$ the integral diverges, so that the phases δ_l are infinite. This holds for any l, since the convergence or divergence of the integral (124.1) depends on the behaviour of $U(r)$ for large r, while at large distances (where the field $U(r)$ is weak) the radial motion is quasi-classical for all l. We shall show below how the formulae (123.11), (123.12) are to be interpreted when δ_l is infinite.

Let us first consider the convergence of the series (123.12) which gives the total scattering cross-section. For large l, the phases $\delta_l \ll 1$, as is seen from (124.1) if we take into account the fact that $U(r)$ decreases more rapidly than $1/r$. Hence we can put $\sin^2\delta_l \approx \delta_l^2$, and so the sum of the high terms in the series (123.12) will be of the order of $\sum_{l \gg 1} l\delta_l^2$. From the well-known integral test for the convergence of series, we conclude that the series in question converges if the integral $\int^{\infty} l\delta_l^2 dl$ does so. Substituting here (124.2) and replacing l by kr_0, we obtain the integral

$$\int^{\infty} U^2(r_0) r_0^3 \, dr_0.$$

If $U(r)$ decreases at infinity as $1/r^n$ with $n > 2$, this integral converges, and the total cross-section is finite. If, on the other hand, the field $U(r)$ decreases not more rapidly than $1/r^2$, the total cross-section appears to be infinite. The physical reason for this is that, when the field falls off only slowly with distance, the probability of scattering through small angles becomes extremely large. In this connection we may recall that, in classical mechanics, in any field which vanishes only as $r \to \infty$, a particle passing at any finite impact parameter ρ, however large, always undergoes a deviation through some angle which, though small, is not zero; hence the total scattering cross-section is infinite for any law of decrease of $U(r)$.† In quantum mechanics, this argument is invalid, since we can speak of scattering through a certain angle only if this angle is large compared with the indeterminacy in the direction of motion of the particle. If the impact parameter is known to within $\Delta\rho$, an indeterminacy $\hbar/\Delta\rho$ is caused in the transverse component of momentum, i.e. an indeterminacy $\sim \hbar/mv\Delta\rho$ in the angle.

In view of the important part played by small-angle scattering when $U(r)$ decreases only slowly, the question naturally arises whether the scattering amplitude $f(\theta)$ diverges for $\theta = 0$, even when $U(r)$ decreases more rapidly than $1/r^2$. Putting $\theta = 0$ in (123.11), we obtain for the high terms in the sum an expression proportional to $\sum_{l \gg 1} l\delta_l$. Arguing as in the previous case, our search for the criterion of the convergence of the sum leads us to the integral

$$\int^{\infty} U(r_0)r_0^2 \, dr_0,$$

which diverges for $U(r) \sim 1/r^n$ with $n \leq 3$. Thus the scattering amplitude becomes infinite at $\theta = 0$ for fields which decrease not more rapidly than $1/r^3$.

Finally, let us consider the case where the phase δ_l itself is infinite, as happens when $U(r) \sim 1/r^n$ with $n \leq 1$. It is evident from the results obtained above that, when the field decreases so slowly, both the total cross-section and the scattering amplitude for $\theta = 0$ will be infinite. There remains, however, the problem of calculating $f(\theta)$ for $\theta \neq 0$. First of all, we notice that the formula‡

$$\sum_{l=0}^{\infty} (2l+1)P_l(\cos\theta) = 4\delta(1-\cos\theta) \qquad (124.3)$$

---

† This is seen from the divergence of the integral $\int 2\pi\rho \, d\rho$ which gives the total cross-section in classical mechanics.

‡ This formula is the expansion of the delta function in Legendre polynomials, and can be immediately verified by multiplying both sides by $\sin\theta \, P_l(\cos\theta)$ and integrating over θ. Here the integral

$$\int_0^{\infty} \delta(x) \, dx$$

of the even function $\delta(x)$ is taken to be $\frac{1}{2}$.

holds. In other words, the sum vanishes for all $\theta \neq 0$. Hence, in the expression (123.11) for the scattering amplitude, we can omit unity in the square brackets in each term of the sum when $\theta \neq 0$, leaving

$$f(\theta) = \frac{1}{2ik}\sum_{l=0}^{\infty}(2l+1)P_l(\cos\theta)e^{2i\delta_l}. \tag{124.4}$$

If we multiply the right-hand side of the equation by the constant factor $e^{-2i\delta_0}$, the cross-section will be unchanged, since it is determined by the squared modulus $|f(\theta)|^2$, while the phase of the complex function $f(\theta)$ is changed only by an unimportant constant. On the other hand, the divergent integral of $U(r)$ cancels in the difference $\delta_l - \delta_0$ of expressions such as (124.1), and a finite quantity remains. Thus, to calculate the scattering amplitude in the case considered, we can use the formula

$$f(\theta) = \frac{1}{2ik}\sum_{l=0}^{\infty}(2l+1)P_l(\cos\theta)e^{2i(\delta_l-\delta_0)}. \tag{124.5}$$

§125. The unitary condition for scattering

The scattering amplitude in an arbitrary (not necessarily central) field satisfies certain relations which follow from general physical requirements.

The asymptotic form of the wave function at large distances for elastic scattering in an arbitrary field is

$$\psi \approx e^{ik\mathbf{rn}.\mathbf{n}'} + \frac{1}{r}f(\mathbf{n},\mathbf{n}')e^{ikr}. \tag{125.1}$$

This expression differs from (123.3) in that the scattering amplitude depends on the directions of two unit vectors, one ($\mathbf{n}$) in the direction of incidence of the particles and the other ($\mathbf{n}'$) along the direction of scattering, and not only on the angle between them.

Any linear combination of functions of the form (125.1) with different directions of incidence $\mathbf{n}$ also represents a possible scattering process. Multiplying the functions (125.1) by an arbitrary coefficient $F(\mathbf{n})$ and integrating over all directions $\mathbf{n}$ (solid angle element do), we can write such a linear combination as the integral

$$\int F(\mathbf{n})e^{ik\mathbf{rn}.\mathbf{n}'}\,do + \frac{e^{ikr}}{r}\int F(\mathbf{n})f(\mathbf{n},\mathbf{n}')\,do. \tag{125.2}$$

Since the distance r is arbitrarily large, the factor $e^{ik\mathbf{rn}.\mathbf{n}'}$ in the first integral is a rapidly oscillating function of the direction of the variable vector $\mathbf{n}$. The value of the integral is therefore determined mainly by the regions near those values of $\mathbf{n}$ for which the exponent has an extremum ($\mathbf{n} = \pm\mathbf{n}'$). In each of these regions the factor $F(\mathbf{n}) \approx F(\pm\mathbf{n}')$ can be taken outside the integral, and the integration then gives†

$$2\pi i F(-\mathbf{n}')\frac{e^{-ikr}}{kr} - 2\pi i F(\mathbf{n}')\frac{e^{ikr}}{kr} + \frac{e^{ikr}}{r}\int f(\mathbf{n},\mathbf{n}')F(\mathbf{n})\,do.$$

This expression can be written in a concise operator form, omitting the common factor $2\pi i/k$:

$$\frac{e^{-ikr}}{r}F(-\mathbf{n}') - \frac{e^{ikr}}{r}\hat{S}F(\mathbf{n}'), \tag{125.3}$$

where

$$\hat{S} = 1 + 2ik\hat{f} \tag{125.4}$$

and $\hat{f}$ is the integral operator defined by

$$\hat{f}F(\mathbf{n}') = \frac{1}{4\pi}\int f(\mathbf{n},\mathbf{n}')F(\mathbf{n})\,do. \tag{125.5}$$

The operator $\hat{S}$ is called the *scattering operator*, the *scattering matrix*, or simply the *S-matrix*; it was first used by W. Heisenberg (1943).

The first term in (125.3) represents a wave going in to the centre, and the second a wave going out from the centre. The conservation of the number of particles in elastic scattering is expressed by the equality of the total fluxes of particles in the ingoing and outgoing waves. In other words, these two waves must have the same normalization. To achieve this, the scattering operator must be unitary (§12), i.e. we must have

$$\hat{S}\hat{S}^{+} = 1, \tag{125.6}$$

or, substituting (125.4) and carrying out the multiplication,

$$\hat{f} - \hat{f}^{+} = 2ik\hat{f}\hat{f}^{+}. \tag{125.7}$$

Finally, using the definition (125.5), we can write the *unitarity condition* for scattering in the form

$$f(\mathbf{n},\mathbf{n}') - f^{*}(\mathbf{n}',\mathbf{n}) = \frac{ik}{2\pi}\int f(\mathbf{n},\mathbf{n}'')f^{*}(\mathbf{n}',\mathbf{n}'')\,do''. \tag{125.8}$$

For $\mathbf{n} = \mathbf{n}'$ the integral on the right-hand side of the equation is just the total scattering cross-section $\sigma = \int |f(\mathbf{n},\mathbf{n}'')|^2\,do''$. The difference on the left-hand side of the equation reduces in this case to the imaginary part of the amplitude $f(\mathbf{n},\mathbf{n})$. Thus we obtain the following general relation between the total elastic scattering cross-section and the imaginary part of the amplitude of scattering through an angle zero:

$$\operatorname{im} f(\mathbf{n},\mathbf{n}) = k\sigma/4\pi. \tag{125.9}$$

This is called the *optical theorem* for scattering.

† To calculate the integral, we displace the path of integration with respect to the variable $\mu = \cos\theta$ (θ being the angle between $\mathbf{n}$ and $\mathbf{n}'$) in the complex μ-plane in such a way that it bends into the upper half-plane, the end points $\mu = \pm 1$ being kept fixed. Then the function $e^{ikr\mu}$ decreases rapidly as we move away from these end points.

Another general property of the scattering amplitude can be derived from the requirement of symmetry with respect to time reversal. In quantum mechanics this symmetry is expressed by the fact that, if a function ψ describes any possible state, then the complex conjugate function ψ^* also corresponds to a possible state (§18). Hence the wave function

$$\frac{e^{ikr}}{r}F^*(-\mathbf{n}')-\frac{e^{-ikr}}{r}\hat{S}^*F^*(\mathbf{n}'),$$

which is the complex conjugate of (125.3), also describes some possible scattering process. We define a new arbitrary function by putting $-\hat{S}^*F^*(\mathbf{n}') = \Phi(-\mathbf{n}')$. Using the unitarity of the operator $\hat{S}$, we then have

$$F^*(\mathbf{n}') = -(\hat{S}^*)^{-1}\Phi(-\mathbf{n}') = -\tilde{\hat{S}}\Phi(-\mathbf{n}');$$

using the operator $\hat{P}$ of inversion of the coordinates, which changes the sign of the vectors $\mathbf{n}$ and $\mathbf{n}'$, we can write

$$F^*(-\mathbf{n}') = \hat{P}F^*(\mathbf{n}') = -\hat{P}\tilde{\hat{S}}\hat{P}\Phi(\mathbf{n}').$$

Thus we obtain the time-reversed wave function in the form

$$\frac{e^{-ikr}}{r}\Phi(-\mathbf{n}')-\frac{e^{ikr}}{r}\hat{P}\tilde{\hat{S}}\hat{P}\Phi(\mathbf{n}').$$

This must be essentially the same as the original wave function (125.3). Comparison shows that this implies the condition

$$\hat{P}\tilde{\hat{S}}\hat{P} = \hat{S}; \qquad (125.10)$$

then the two functions differ only in the notation for the arbitrary function.

The corresponding relation for the scattering amplitude is found by changing from the operator equation (125.10) to a matrix equation. Transposition interchanges the initial and final vectors $\mathbf{n}$ and $\mathbf{n}'$, while inversion changes their sign. Hence we have

$$S(\mathbf{n},\mathbf{n}') = S(-\mathbf{n}', -\mathbf{n}), \qquad (125.11)$$

or, what is the same thing,

$$f(\mathbf{n},\mathbf{n}') = f(-\mathbf{n}', -\mathbf{n}). \qquad (125.12)$$

This relation (called the *reciprocity theorem*) expresses the obvious result that the amplitudes are the same for two scattering processes such that each is the time reversal of the other. Time reversal interchanges the initial and final states and reverses the direction of motion of the particles in those states.

For scattering in a central field, the general relations obtained above can be simplified. In this case the amplitude $f(\mathbf{n},\mathbf{n}')$ depends only on the angle θ between $\mathbf{n}$ and $\mathbf{n}'$. The equation (125.12) therefore becomes an identity.

The unitary condition (125.8) becomes

$$\operatorname{im} f(\theta) = \frac{k}{4\pi} \int f(\gamma) f^*(\gamma') \mathrm{d}o'', \tag{125.13}$$

where γ, γ' are the angles between $\mathbf{n}$, $\mathbf{n}'$ and some direction $\mathbf{n}''$ fixed in space. If we use the expansion (123.14) for $f(\theta)$, the addition theorem (c.10) for spherical harmonics gives from (125.13) the following relation for the partial amplitudes:

$$\operatorname{im} f_l = k |f_l|^2. \tag{125.14}$$

This formula can also be derived directly from the expression (123.15), according to which $|2ikf_l + 1|^2 = 1$. The optical theorem (125.9) is also easily deduced directly from formulae (123.11) and (123.12) for the case of scattering in a central field.

Rewriting (125.14) as $\operatorname{im}(1/f_l) = -k$, we see that the amplitude f_l must have the form

$$f_l = 1/(g_l - ik), \tag{125.15}$$

where $g_l = g_l(k)$ is a real quantity; it is related to the phase δ_l by

$$g_l = k \cot \delta_l. \tag{125.16}$$

We shall several times make use of this formula for the amplitude.

Let us examine (for scattering in a central field) the relation between the scattering operator defined above and the quantities which appear in the theory given in §123.

Since the orbital angular momentum is conserved in a central field, the scattering operator commutes with the angular momentum operator. In other words, the S matrix is diagonal in the l representation, and since the operator $\hat{S}$ is unitary its eigenvalues must have unit modulus, i.e. must be of the form $e^{2i\delta_l}$ with real δ_l. It is easy to see that these quantities are the same as the phase shifts of the wave functions, so that the eigenvalues of the S-matrix are the quantities S_l defined in (123.10). The eigenvalues of the operator $\hat{f} = (\hat{S}-1)/2ik$ are the partial amplitudes (123.15). For, if we take $P_l(\cos\theta)$ as the function $F(\mathbf{n})$ (so that $F(-\mathbf{n}) = P_l(-\cos\theta) = (-1)^l P_l(\cos\theta)$), the wave function (125.3) must be the solution of Schrödinger's equation represented by a term in the sum (123.9). Thus $\hat{S}P_l(\cos\theta) = S_l P_l(\cos\theta)$.

For a plane wave incident along the z-axis, the function $F(\mathbf{n})$ in (125.3) is the delta function $F = 4\delta(1-\cos\theta)$, where θ is the angle between $\mathbf{n}$ and the z-axis, the delta function is defined as indicated in the second footnote to §124, and the coefficient of it is so chosen as to give simply $f(\theta)$ on substitution on the right-hand side of the definition (125.5); θ is now the angle between $\mathbf{n}'$ and the z-axis. Writing the delta function in the form (124.3):

$$F = 4\delta(1-\cos\theta) = \sum_{l=0}^{\infty} (2l+1) P_l(\cos\theta), \tag{125.17}$$

and applying the operator $\hat{f}$ to it, we find that the scattering amplitude has the form (123.14), as it should.

Finally, we may add the following remark. Mathematically, the unitary condition (125.8) signifies that not every specified function $f(\mathbf{n},\mathbf{n}')$ can be the scattering amplitude in some field. In particular, not every function $f(\theta)$ can be the scattering amplitude in some central field. From (125.13), a certain relation must hold between its real and imaginary parts. If we write $f(\theta) = |f|e^{i\alpha}$, then, when the modulus $|f|$ is given for all angles, (125.13) gives an integral equation from which the unknown phase $\alpha(\theta)$ can in principle be determined. In other words, from a scattering cross-section (i.e. $|f|^2$) known for all angles we can in principle recover the amplitude. This process is, however, not completely unique and determines the amplitude only to within the alternative

$$f(\theta) \to -f^*(\theta), \tag{125.18}$$

which leaves the equation (125.13) unchanged (and of course does not alter the cross-section $|f|^2$; the transformation (125.18) is equivalent to a simultaneous change of sign of all the phases δ_l in (123.11)). This non-uniqueness is, however, removed if the scattering amplitude is regarded as a function of energy as well as angle. We shall see below (§§128, 129) that the analytical properties of the amplitude as a function of energy are not invariant under the transformation (125.18).

§126. Born's formula

The scattering cross-section can be calculated in a general form in a very important case, namely that where the scattering field may be regarded as a perturbation.† It has been shown in §45 that this is possible when either of the two conditions

$$|U| \ll \hbar^2/ma^2 \tag{126.1}$$

and

$$|U| \ll \hbar v/a = (\hbar^2/ma^2)ka \tag{126.2}$$

holds, a being the range of action of the field $U(r)$ and U the order of magnitude of the field in the range where it is significant. When the first condition is satisfied, the approximation is valid for all velocities; the second condition shows that it is always applicable for sufficiently fast particles.

In accordance with §45, we seek the wave function in the form $\psi = \psi^{(0)}+\psi^{(1)}$, where $\psi^{(0)} = e^{i\mathbf{k}.\mathbf{r}}$ corresponds to an incident particle having wave vector $\mathbf{k}=\mathbf{p}/\hbar$. From formula (45.3) we then have

$$\psi^{(1)}(x,y,z) = -\frac{m}{2\pi\hbar^2}\int U(x',y',z')e^{i(\mathbf{k}.\mathbf{r}'+kR)}\frac{dV'}{R}. \tag{126.3}$$

† In the general theory derived in §123 this approximation corresponds to the case where all the phases δ_l are small; it is also necessary that these phases can be calculated from Schrödinger's equation with the potential energy regarded as a perturbation (see Problem 4).

Taking the origin at the scattering centre, we introduce the radius vector $\mathbf{R}_0$ from the origin to the point where the value of $\psi^{(1)}$ is required, and denote by $\mathbf{n}'$ a unit vector along $\mathbf{R}_0$. Let the radius vector of a volume element dV' be $\mathbf{r}'$; then $\mathbf{R} = \mathbf{R}_0 - \mathbf{r}'$. At large distances from the centre, $R_0 \gg r'$, so that

$$R = |\mathbf{R}_0 - \mathbf{r}'| \approx R_0 - \mathbf{r}' \cdot \mathbf{n}'.$$

Substituting this in (126.3), we have the following asymptotic expression for $\psi^{(1)}$:

$$\psi^{(1)} \approx -\frac{m}{2\pi\hbar^2}\frac{e^{ikR_0}}{R_0}\int U(\mathbf{r}')e^{i(\mathbf{k}-\mathbf{k}')\cdot\mathbf{r}'}\,dV'$$

(where $\mathbf{k}' = k\mathbf{n}'$ is the wave vector of the particle after scattering). Comparing this with the scattering amplitude given by formula (123.3), we find for the latter the expression

$$f = -\frac{m}{2\pi\hbar^2}\int Ue^{-i\mathbf{q}\cdot\mathbf{r}}\,dV, \tag{126.4}$$

where we have renamed the variable of integration and introduced the vector

$$\mathbf{q} = \mathbf{k}' - \mathbf{k}, \tag{126.5}$$

whose absolute magnitude is

$$q = 2k\sin\tfrac{1}{2}\theta, \tag{126.6}$$

θ being the angle between $\mathbf{k}$ and $\mathbf{k}'$, i.e. the scattering angle.

Finally, squaring the modulus of the scattering amplitude, we have the following expression for the cross-section for scattering into the solid angle element do:

$$d\sigma = \frac{m^2}{4\pi^2\hbar^4}\left|\int Ue^{-i\mathbf{q}\cdot\mathbf{r}}dV\right|^2 do. \tag{126.7}$$

We see that the scattering with a momentum change $\hbar\mathbf{q}$ is determined by the squared modulus of the corresponding Fourier component of the field U. Formula (126.7) was first obtained by M. Born (1926). In the theory of collisions, the approximation considered here is often called the *Born approximation*.

It may be noted that, in this approximation, the relation

$$f(\mathbf{k},\mathbf{k}') = f^*(\mathbf{k}',\mathbf{k}) \tag{126.8}$$

holds between the amplitudes of the direct and inverse scattering processes, i.e. processes differing by the interchange of the initial and final momenta, without the change of sign such as occurs in time reversal. Thus another symmetry property, in addition to the reciprocity theorem (125.12), appears in scattering. This property is closely related to the smallness of the scattering amplitudes in perturbation theory, and follows immediately from the unitarity condition (125.8) if we neglect the integral term quadratic in f.†

Formula (126.7) can also be obtained by another method (which, however, does not determine the phase of the scattering amplitude). We can start from the general formula (43.1), according to which the transition probability between states of the continuous spectrum is given by the expression

$$dw_{fi} = (2\pi/\hbar)|U_{fi}|^2\,\delta(E_f - E_i)d\nu_f.$$

In the case under consideration, we have to apply this formula to a transition from the state of the incident particle with momentum $\mathbf{p}$ to the state of the particle, with momentum $\mathbf{p}'$, scattered into the element of solid angle do'. As the interval of states $d\nu_f$ we can take $d^3p'/(2\pi\hbar)^3$. Substituting for the difference of the final and initial energies

$$E_f - E_i = (p'^2 - p^2)/2m,$$

we obtain

$$dw_{\mathbf{p}'\mathbf{p}} = (4\pi m/\hbar)|U_{\mathbf{p}'\mathbf{p}}|^2\delta(p'^2 - p^2)\,d^3p'/(2\pi\hbar)^3. \tag{126.9}$$

The wave functions of the incident and scattered particles are plane waves. Since we have taken as the interval of states $d\nu_f$ an element of $\mathbf{p}/2\pi\hbar$ space, the final wave function must be normalized by the delta function of $\mathbf{p}/2\pi\hbar$:

$$\psi_{\mathbf{p}'} = e^{(i/\hbar)\mathbf{p}'.\mathbf{r}}. \tag{126.10}$$

We normalize the initial wave function to unit current density:

$$\psi_{\mathbf{p}} = \sqrt{(m/p)}e^{(i/\hbar)\mathbf{p}.\mathbf{r}}. \tag{126.11}$$

Then (126.9) will have the dimensions of area, and is the differential scattering cross-section.

The presence of the delta function in formula (126.9) means that $p' = p$, i.e. the absolute magnitude of the momentum is unchanged, as it should be in elastic scattering. We can remove the delta function by changing to spherical coordinates in momentum space (i.e. by replacing d^3p' by $p'^2dp'\,do' = \frac{1}{2}p'd(p'^2)do'$) and integrating over p'^2. The integration amounts to replacing p' by p in the integrand, and we obtain

$$d\sigma = (mp/4\pi^2\hbar^4)\left|\int\psi_{\mathbf{p}'}{}^*U\psi_{\mathbf{p}}dV\right|^2do'.$$

Substituting the functions (126.10), (126.11), we reach once more the final expression (126.7).

In the form (126.7), this formula is applicable to scattering in a field

† Hence it is clear that this property no longer holds in even the second approximation of perturbation theory. This will be proved directly in §130; cf. (130.13).

$U(x, y, z)$ which is any function of the coordinates, and not only a function of r. In the case of a central field $U(r)$, however, this formula can be further transformed. In the integral

$$\int U(r)e^{-i\mathbf{q}\cdot\mathbf{r}}\,dV$$

we use spherical space coordinates r, ϑ, ϕ, with the polar axis in the direction of the vector $\mathbf{q}$, denoting the polar angle by ϑ to distinguish it from the scattering angle θ. The integration over ϑ and ϕ can be effected, and we obtain

$$\int_0^\infty\int_0^{2\pi}\int_0^\pi U(r)e^{iqr\cos\vartheta}r^2\sin\vartheta\,d\vartheta\,d\phi\,dr = 4\pi\int_0^\infty U(r)\frac{\sin qr}{q}r\,dr.$$

Substituting this expression in (126.4), we obtain the following formula for the scattering amplitude in a centrally symmetric field:

$$f = -\frac{2m}{\hbar^2}\int_0^\infty U(r)\frac{\sin qr}{q}r\,dr. \tag{126.12}$$

For $\theta = 0$ (i.e. $q = 0$), the integral diverges when $U(r)$ decreases at infinity not more rapidly than $1/r^3$ (in acordance with the general results of §124).

We may call attention to the following interesting fact. The momentum p of the particle and the scattering angle θ enter (126.12) only through q. Thus, in the Born approximation, the scattering cross-section depends on p and θ only in the combination $p\sin\frac{1}{2}\theta$.

Returning to the case of arbitrary fields $U(x, y, z)$, let us consider the limiting cases of small velocities ($ka \ll 1$) and large velocities ($ka \gg 1$). For small velocities, we can put $e^{-i\mathbf{q}\cdot\mathbf{r}} \approx 1$ in (126.4), so that

$$f = -\frac{m}{2\pi\hbar^2}\int U\,dV, \tag{126.13}$$

while if $U = U(r)$,

$$f = -\frac{2m}{\hbar^2}\int_0^\infty U(r)r^2\,dr. \tag{126.14}$$

Here the scattering is isotropic and independent of the velocity, in accordance with the general results of §132.

In the opposite limiting case of high velocities, the scattering is markedly anisotropic and mainly forward in a narrow cone of angle $\Delta\theta \sim 1/ka$; since outside this cone the quantity q is large, the factor $e^{-i\mathbf{q}\cdot\mathbf{r}}$ is a rapidly oscillating function, and the integral of its product with the slowly varying function U is almost zero.

The law of decrease for large q is not universal and depends on the specific form of the field. If the field $U(r)$ has a singularity at $r = 0$ or at any other real value of r, the integral (126.12) is mainly determined by the range near the singular point, and the cross-section decreases according to a power law. The same applies to the case where the function $U(r)$ has no singularity but is not an even function; here the region near $r = 0$ is the most important in the integral. If $U(r)$ is an even function of r, however, the integration may be formally extended to negative values of r, i.e. taken along the whole of the real axis of the variable r, after which (if $U(r)$ has no singularity on the real axis) the path of integration may be moved into the complex plane until it meets the nearest complex singularity. Then, for large q, the integral will decrease exponentially. It should be borne in mind, however, that the Born approximation is in general inadequate to calculate this exponentially small quantity (see also §131).

Although the value of the differential scattering cross-section within the cone $\Delta\theta \sim 1/ka$ does not depend greatly on the velocity, the total scattering cross-section (assuming that the integral $\int d\sigma$ does converge) decreases at high energies owing to the decreasing angle of the cone, in proportion to the solid angle of the cone, i.e. as $(\Delta\theta)^2 \sim 1/k^2a^2$, or inversely as the energy.

In many physical applications of collision theory the quantity which describes the scattering is the integral

$$\sigma_{tr} = \int (1-\cos\theta)\, d\sigma, \tag{126.15}$$

often called the *transport cross-section*. Arguments similar to those given above show that at high velocities this quantity is inversely proportional to the square of the energy.

PROBLEMS

PROBLEM 1. Determine, in the Born approximation, the scattering cross-section for a spherical potential well: $U = -U_0$ for $r < a$, $U = 0$ for $r > a$.

SOLUTION. The calculation of the integral in (126.12) gives

$$d\sigma = 4a^2\left(\frac{mU_0a^2}{\hbar^2}\right)^2 \frac{(\sin qa - qa\cos qa)^2}{(qa)^6}\, do.$$

The integration over all angles (which is conveniently effected by using the variable $q = 2k\sin\frac{1}{2}\theta$ and replacing do by $2\pi q\, dq/k^2$) gives the total scattering cross-section

$$\sigma = \frac{2\pi}{k^2}\left(\frac{mU_0a^2}{\hbar^2}\right)^2 \left[1-\frac{1}{(2ka)^2}+\frac{\sin 4ka}{(2ka)^3}-\frac{\sin^2 2ka}{(2ka)^4}\right].$$

In the limiting cases this formula gives

$$\sigma = \frac{16\pi a^2}{9}\left(\frac{mU_0a^2}{\hbar^2}\right)^2 \quad \text{for} \quad ka \ll 1,$$

$$\sigma = \frac{2\pi}{k^2}\left(\frac{mU_0a^2}{\hbar^2}\right)^2 \quad \text{for} \quad ka \gg 1.$$

PROBLEM 2. The same as Problem 1, but in a field $U = U_0 e^{-r^2/a^2}$.

SOLUTION. The calculation is conveniently effected from formula (126.7), taking the direction of **q** along one of the coordinate axes. The result is

$$d\sigma = \tfrac{1}{4}\pi a^2 \left(\frac{mU_0 a^2}{\hbar^2}\right)^2 e^{-q^2a^2/2}\, do,$$

and the total cross-section is

$$\sigma = \frac{\pi^2}{2k^2}\left(\frac{mU_0 a^2}{\hbar^2}\right)^2 (1 - e^{-2k^2a^2}).$$

The condition for these formulae to be applicable is given by the inequalities (126.1), (126.2) with U_0 in place of U. The formula for $d\sigma$ is also inapplicable if the exponent is large in absolute magnitude.†

PROBLEM 3. The same as Problem 1, but in a field $U = (\alpha/r)e^{-r/a}$.

SOLUTION. The calculation of the integral in (126.12) gives

$$d\sigma = 4a^2\left(\frac{\alpha m a}{\hbar^2}\right)^2 \frac{do}{(q^2a^2+1)^2}.$$

The total cross-section is

$$\sigma = 16\pi a^2 \left(\frac{\alpha m a}{\hbar^2}\right)^2 \frac{1}{4k^2a^2+1}.$$

The condition for these formulae to be applicable is found from (126.1) and (126.2) with α/a instead of U: $\alpha m a/\hbar^2 \ll 1$ or $\alpha/\hbar v \ll 1$.

PROBLEM 4. Determine the phases δ_l for scattering in a centrally symmetric field for the case corresponding to the Born approximation.

SOLUTION. For the radial wave function $\chi = rR$ for motion in the field $U(r)$, and for the function $\chi^{(0)}$ for free motion, we have the equations (see (32.10))

$$\chi'' + \left[k^2 - \frac{l(l+1)}{r^2} - \frac{2m}{\hbar^2}U\right]\chi = 0,$$

$$\chi^{(0)''} + \left[k^2 - \frac{l(l+1)}{r^2}\right]\chi^{(0)} = 0.$$

Multiplying the first equation by $\chi^{(0)}$, the second by χ, and subtracting, followed by integration with respect to r (using the boundary condition $\chi = 0$ at $r = 0$), we obtain

$$\chi'(r)\chi^{(0)}(r) - \chi(r)\chi^{(0)\prime}(r) = \frac{2m}{\hbar^2}\int_0^r U\chi\chi^{(0)}\, dr.$$

Regarding U as a perturbation, we can put $\chi \approx \chi^{(0)}$ on the right-hand side. For $r \to \infty$ the asymptotic expressions (33.12), (33.20) can be used on the left-hand side, while in the integral we substitute the exact expression (33.10). The result is

$$\sin\delta_l \approx \delta_l = -\frac{\pi m}{\hbar^2}\int_0^\infty U(r)[J_{l+1/2}(kr)]^2 r\, dr.$$

This formula could also be derived by a direct expansion of the Born scattering amplitude (126.4) in Legendre polynomials in accordance with (123.11) (for small δ_l).

---

† The inapplicability of perturbation theory in this case is easily seen by calculating the scattering amplitude in the second approximation (see (130.13)); although the coefficient of the exponential is small in comparison with the coefficient in the first-approximation term, the magnitude of the negative exponent is only half as great.

PROBLEM 5. Determine in the Born approximation the total scattering cross-section in a field $U = \alpha/(r^2+a^2)^{n/2}$ with $n>2$, for fast particles ($ka \gg 1$).

SOLUTION. We shall see that in this case the partial amplitudes with large angular momenta l are predominant in this scattering. The cross-section may therefore be calculated from formula (123.11), replacing the summation over l by integration; in the Born approximation, all $\delta_l \ll 1$, so that

$$\sigma \approx \frac{4\pi}{k^2}\int_0^\infty 2l\delta_l^2 \, dl. \tag{1}$$

The phases δ_l with large l are calculated from (124.1):

$$\delta_l = -\frac{\alpha m}{\hbar^2}\int_{l/k}^\infty \frac{dr}{(r^2+a^2)^{n/2}(k^2-l^2/r^2)^{1/2}}.$$

By the substitution $r^2+a^2 = (a^2+l^2/k^2)/\xi$, the integral is brought to the familiar Euler form, and the result is

$$\delta_l = -\frac{m\alpha k^{n-2}}{2\hbar^2(a^2k^2+l^2)^{(n-1)/2}}\frac{\Gamma(\frac{1}{2})\Gamma(\frac{1}{2}n-\frac{1}{2})}{\Gamma(\frac{1}{2}n)}. \tag{2}$$

The integral (1) is determined by the range $l \sim ak \gg 1$, and this justifies the assumption made. A calculation of the integral gives the result

$$\sigma = \frac{\pi^2}{n-2}\left[\frac{\Gamma(\frac{1}{2}n-\frac{1}{2})}{\Gamma(\frac{1}{2}n)}\right]^2\left(\frac{m\alpha}{k\hbar^2a^{n-2}}\right)^2. \tag{3}$$

According to (126.2), the condition for the Born approximation to be valid in this case is $m\alpha/\hbar^2ka^{n-1} \ll 1$. Note the dependence $\sigma \sim k^{-2}$, which is in accordance with the general statement made above.

PROBLEM 6. Determine in the Born approximation the scattering amplitude in the two-dimensional case of a field $U = U(x, z)$ with the particle flux incident along the z-axis.

SOLUTION. Using the second footnote to §45 and the known asymptotic expression of the Hankel function

$$H_0^{(1)}(u) \approx \sqrt{\frac{2}{\pi u}}\, e^{i(u-\pi/4)} \text{ as } u \to \infty,$$

we find for the correction to the wave function at large distances R_0 from the field axis (the y-axis) the expression

$$\psi^{(1)} \approx \frac{f(\theta)}{\sqrt{R_0}}\, e^{ikR_0},$$

where the scattering amplitude is

$$f(\theta) = -\frac{m}{\hbar^2\sqrt{(2\pi k)}}\, e^{i\pi/4}\int U(\boldsymbol{\rho})\, e^{-i\mathbf{q}\cdot\boldsymbol{\rho}}\, d^2\rho,$$

with $\boldsymbol{\rho} = (x, z)$ the two-dimensional radius vector, $d^2\rho = dx\,dz$, and θ the scattering angle in the xz-plane. In the two-dimensional case, the scattering amplitude has the dimensions of square root of length, and the scattering cross-section $d\sigma = |f|^2\, d\theta$ those of length.

§127. The quasi-classical case

Let us investigate the manner in which the passage occurs from the quantum-mechanical theory of scattering to the limit of the classical theory.

Omitting from consideration a scattering angle θ of zero, we can write the scattering amplitude given by the exact theory in the form (124.4):

$$f(\theta) = (1/2ik)\sum_{l=0}^{\infty}(2l+1)P_l(\cos\theta)e^{2i\delta_l}. \tag{127.1}$$

We know that the quasi-classical wave functions are characterized by having large phases. It is therefore natural to suppose that large phases δ_l correspond to the passage to the limit of the classical theory of scattering. The value of the sum (127.1) is mainly determined by the terms with large l. Hence we can replace $P_l(\cos\theta)$ by the asymptotic expression (49.7), which we write in the form

$$P_l(\cos\theta) \approx -\frac{i}{\sqrt{(2\pi l\sin\theta)}}[e^{i(l+1/2)\theta+i\pi/4}-e^{-i(l+1/2)\theta-i\pi/4}].$$

Substituting this expression in (127.1), we obtain

$$f(\theta) = \frac{1}{k}\sum_l\sqrt{\frac{l}{2\pi\sin\theta}}\{e^{i[2\delta_l-(l+1/2)\theta-\pi/4]}-e^{i[2\delta_l+(l+1/2)\theta+\pi/4]}\}. \tag{127.2}$$

The exponential factors, regarded as functions of l, are rapidly oscillating functions, since their phases are large. The majority of the terms in the sum in (127.2) therefore cancel. The sum is mainly determined by the range of values of l near that for which one of the exponents has an extremum, i.e. near the root of the equation

$$2\,\mathrm{d}\delta_l/\mathrm{d}l\pm\theta = 0. \tag{127.3}$$

In this region there are a large number of terms in the series for which the exponential factors have almost the same value (since the exponents vary slowly near the extremum), and which therefore will not cancel.

The phases δ_l in the quasi-classical case can be written (see §124) as the limit to which the difference between the phase

$$\tfrac{1}{4}\pi+\frac{1}{\hbar}\int_{r_0}^{r}\sqrt{\{2m[E-U(r)]-\hbar^2(l+\tfrac{1}{2})^2/r^2\}}\,\mathrm{d}r$$

of the quasi-classical wave function in the field $U(r)$ and the phase

$$kr-\tfrac{1}{2}l\pi$$

(see §33) of the wave function of free motion tends as $r\to\infty$. Thus

$$\delta_l = \int_{r_0}^{\infty}\left\{\frac{1}{\hbar}\sqrt{[2m(E-U)-\hbar^2(l+\tfrac{1}{2})^2/r^2]}-k\right\}\mathrm{d}r+\tfrac{1}{2}\pi(l+\tfrac{1}{2})-kr_0. \tag{127.4}$$

This expression is to be substituted in equation (127.3). In finding the derivative of the integral, it must be remembered that the limit of integration

r_0 also depends on l; the term $k\,dr_0/dl$ arising from this, however, cancels with the derivative of the term $-kr_0$ in δ_l.

$\hbar(l+\frac{1}{2})$ is the angular momentum of the particle. In classical mechanics, it can be written in the form $m\rho v$, where ρ is the impact parameter, and v is the velocity of the particle at infinity. We make this substitution; equation (127.3) then takes the final form

$$\int_{r_0}^{\infty} \frac{mv\rho\, dr}{r^2\sqrt{[2m(E-U)-(mv\rho/r)^2]}} = \tfrac{1}{2}(\pi\mp\theta). \tag{127.5}$$

In a repulsive field this equation has a root (for ρ) only for a minus sign in front of θ on the right-hand side, and in an attractive field only for a plus sign.

Equation (127.5) is exactly the same as the classical equation which determines the scattering angle from the impact parameter (see *Mechanics*, §18). It is easy to see that the classical expression for the cross-section is in fact obtained.

To prove this, we expand the exponent in (127.2) in powers of $l' = l-l_0(\theta)$, where $l_0(\theta)$ is determined by equations (127.3)–(127.5). We shall take in particular the first term in (127.2), and accordingly the lower sign in (127.3) (attraction). According to (127.3),

$$[d^2\delta_l/dl^2]_{l=l_0} = \tfrac{1}{2}d\theta/dl_0,$$

and so

$$i[2\delta_l-(l+\tfrac{1}{2})\theta-\tfrac{1}{4}\pi] \approx i[2\delta_{l_0}-(l_0+\tfrac{1}{2})\theta-\tfrac{1}{4}\pi]+\tfrac{1}{2}i(d\theta/dl_0)l'^2.$$

The summation over l in (127.2) is now replaced by integration over l' near the point $l' = 0$. Regarding l' as a complex variable, we take the path of integration near this point in the direction of steepest decrease of the exponent, i.e. at an angle of $\frac{1}{4}\pi$ or $-\frac{1}{4}\pi$ to the real axis, according to the sign of $d\theta/dl_0$. In other words, we put $l' = \xi \exp(\pm\frac{1}{4}i\pi)$ and integrate over real values of ξ; owing to the rapid convergence of the integral, it can be extended from $-\infty$ to ∞:

$$\int_{-\infty}^{\infty} \exp(-\tfrac{1}{2}\xi^2|d\theta/dl_0|)d\xi = (2\pi|dl_0/d\theta|)^{1/2}.$$

The result is

$$f(\theta) = \frac{1}{k}\left(\frac{l_0}{\sin\theta}\left|\frac{dl_0}{d\theta}\right|\right)^{1/2} \exp\{i[2\delta_{l_0}-(l_0+\tfrac{1}{2})\theta-\tfrac{1}{4}\pi]\}. \tag{127.6}$$

Hence

$$d\sigma = |f|^2 \,.\, 2\pi \sin\theta \, d\theta$$
$$= 2\pi(l_0/k^2)|dl_0/d\theta|d\theta, \qquad (127.7)$$

and with the impact parameter $\rho = l_0/k$ we arrive at the classical formula $d\sigma = 2\pi\rho\, d\rho$.

Thus the conditions for classical scattering through a given angle θ are that the value of l for which (127.3) holds should be large, and that δ_l should also be large for this value of l.† This latter condition has a simple interpretation. If we can speak of classical scattering through an angle θ when the particle is incident at an impact parameter ρ, it is necessary that the quantum-mechanical indeterminacies of these two quantities should be relatively small: $\Delta\rho \ll \rho$, $\Delta\theta \ll \theta$. The indeterminacy in the scattering angle is of the order of magnitude $\Delta\theta \sim \Delta p/p$, where p is the momentum of the particle and Δp is the indeterminacy in its transverse component. Since $\Delta p \sim \hbar/\Delta\rho \gg \hbar/\rho$, we have $\Delta\theta \gg \hbar/p\rho$, and thus

$$\theta \gg \hbar/\rho m v. \qquad (127.8)$$

Replacing the angular momentum $m\rho v$ by $\hbar l$, we obtain $\theta l \gg 1$, which is the same as $\delta_l \gg 1$ (since $\delta_l \sim l\theta$, as we see from (127.3)).

The classical angle of deviation of the particle can be estimated as the ratio of the transverse momentum increment Δp during the "collision time" $\tau \sim \rho/v$ and the original momentum mv. The force acting on the particle at a distance ρ is $F = -dU(\rho)/d\rho$; hence $\Delta p \sim F\rho/v$, so that $\theta \sim F\rho/mv^2$. This estimate is strictly valid only if $\theta \ll 1$, but it can be applied to give an order of magnitude even if $\theta \sim 1$. Substitution in (127.8) gives the condition for quasi-classical scattering in the form

$$|F|\rho^2 \gg \hbar v. \qquad (127.9)$$

This inequality must hold for all values of ρ such that $|U(\rho)| \lesssim E$.

If the field $U(r)$ decreases more rapidly than $1/r$, the condition (127.9) always ceases to be satisfied for sufficiently large ρ. Small θ, however, correspond to large ρ; thus scattering through sufficiently small angles is never classical. If, on the other hand, the field decreases less rapidly than $1/r$, the scattering through small angles is classical; whether the scattering through large angles is classical in this case depends on the behaviour of the field at small distances.

For a Coulomb field, $U = \alpha/r$, the condition (127.9) is satisfied if $\alpha \gg \hbar v$. This is the opposite condition to that for which the Coulomb field can be

† The relation between θ and ρ given by (127.5) may not be one-to-one; more than one value of ρ may correspond to the same value of θ. In such a case, the amplitude $f(\theta)$ is given by the sum of (127.6) with the appropriate values of l_0. At extrema of $\theta(\rho)$ the derivative $d\rho/d\theta$ and therefore the classical differential cross-section $d\sigma/do$ become infinite; near such an angle, the classical approximation is of course invalid (see Problem 2).

regarded as a perturbation. We shall see, however, that the quantum theory of scattering in a Coulomb field leads to a result which, as it happens, is always in agreement with the classical result.

PROBLEMS

PROBLEM 1. Find the total cross-section for quasi-classical scattering in a field which has the form $U = \alpha/r^n$ $(n>2)$ at sufficiently large distances.

SOLUTION. Bearing in mind that the phases δ_l with large l are the most important, we calculate them from (124.1):

$$\delta_l = -\frac{m\alpha}{\hbar^2}\int\limits_{l/k}^{\infty}\frac{\mathrm{d}r}{r^n\sqrt{(k^2-l^2/r^2)}}$$

$$= -\frac{m\alpha k^{n-2}}{2\hbar^2 l^{n-1}}\frac{\Gamma(\frac{1}{2})\Gamma(\frac{1}{2}n-\frac{1}{2})}{\Gamma(\frac{1}{2}n)}; \quad (1)$$

see §126, Problem 5, for the calculation of the integral.

Replacing the summation in (123.12) by an integration, we write

$$\sigma = \frac{4\pi}{k^2}\int\limits_0^{\infty} 2l\sin^2\delta_l\,\mathrm{d}l.$$

We substitute $\delta_l = u$ and integrate by parts with respect to u, reducing the integral to a gamma function. The result is

$$\sigma = 2\pi^{n/(n-1)}\sin\left[\tfrac{1}{2}\pi.\frac{n-3}{n-1}\right]\Gamma\left(\frac{n-3}{n-1}\right)\left[\frac{\Gamma(\frac{1}{2}n-\frac{1}{2})}{\Gamma(\frac{1}{2}n)}\right]^{2/(n-1)}\left(\frac{\alpha}{\hbar v}\right)^{2/(n-1)} \quad (2)$$

For $n = 3$, the indeterminacy can be resolved to give $\sigma = 2\pi^2\alpha/\hbar v$.

The chief condition for the applicability of this result is that $l \gg 1$ for $\delta_l \sim 1$; this gives the inequality

$$m\alpha k^{n-2}/\hbar^2 \gg 1.$$

A further condition arises from the requirement that the field $U(r)$ should have the form in question from distances

$$r \sim l/k \sim (m\alpha/\hbar^2 k)^{1/(n-1)}$$

outwards (l being obtained from $\delta_l \sim 1$), these distances playing the principal part in the integral (1). If this form is reached only at distances $r \gg a$, where a represents the characteristic dimensions of the field, we have the condition

$$m\alpha/\hbar^2 k a^{n-1} \gg 1,$$

which places an upper limit on the permissible velocities. In this case, for sufficiently high velocities ($m\alpha/\hbar^2 k a^{n-1} \ll 1$), $\sigma \sim k^{-2}$ (cf. §126, Problem 5).

PROBLEM 2. Find the angular distribution of scattering near an extremum of the classical scattering angle $\theta(\rho)$ as a function of the impact parameter $\rho = l/k$.

SOLUTION. The presence of an extremum of $\theta(l)$ for some $l = l_0$ implies, according to (127.3), that the phase δ_l near this point has the form

$$2\delta_l \approx 2\delta_{l_0} + \theta_0 l' + \tfrac{1}{3}\alpha l'^3,$$

where $\theta_0 = \theta(l_0)$, $l' = l = l_0$; here we again take the particular case of the lower sign in (127.3). The constant α is negative and positive respectively for a maximum or minimum of the function $\theta(l)$. For the scattering amplitude we have instead of (127.6)

$$|f(\theta)| = \frac{1}{k}\left(\frac{l_0}{2\pi \sin \theta_0}\right)^{1/2} \left| \int_{-\infty}^{\infty} \exp \{i(-l'\theta' + \tfrac{1}{3}\alpha l'^3)\}\, dl' \right|,$$

where $\theta' = \theta - \theta_0$. Expressing the integral in terms of the Airy function by means of (b.3), we finally have for the scattering cross-section†

$$d\sigma = \frac{4\pi l_0}{\alpha^{2/3} k^2} \Phi^2\left(-\frac{\theta'}{\alpha^{1/3}}\right) d\theta'$$

The differential cross-section $d\sigma/d\theta'$ decreases with increasing distance into the classically inaccessible region of scattering ($\theta' > 0$ for $\alpha < 0$, or $\theta' < 0$ for $\alpha > 0$); on the other side of the point $\theta' > 0$, it oscillates between zero and a gradually decreasing amplitude. Its maximum value occurs for $\theta'\alpha^{-1/3} = 1{\cdot}02$, where $\Phi^2 = 0{\cdot}90$.

PROBLEM 3. Find the angular distribution of quasi-classical scattering at small angles, if the classical angle of deviation θ is zero for some finite value of $\rho = l_0/k$.

SOLUTION. The assumption of quasi-classical scattering means here that $l_0 \gg 1$ and $\delta_{l_0} \ll 1$. Then the values of l close to l_0 are important in scattering. For small $l' = l - l_0$ we have

$$\delta_l \approx \delta_{l_0} + \tfrac{1}{2}\beta l'^2;$$

then, according to (127.3), $\theta = 0$ for $l' = 0$. This expression is to be substituted in (127.1), and $P_l(\cos\theta)$ can be written in the form (49.6). The summation over l is again replaced by integration over l' near the point $l' = 0$:‡

$$f = (l_0/ik) \exp(2i\delta_{l_0}) \int J_0(l\theta) \exp(i\beta l'^2)\, dl'.$$

The integral is determined by the range $l' \sim \beta^{-1/2}$. For angles $\theta \ll \sqrt{\beta}$, we can take the function $J_0(l\theta)$ outside the integral and use its value at $l = l_0$. The remaining integral is calculated as shown above. The result for the cross-section is‖

$$d\sigma = (\pi l_0^2/\beta k^2) J_0^2(l_0\theta)\, do.$$

A similar result is found for the cross-section for scattering at angles close to π if the classical scattering angle tends to π for some finite (non-zero) value of ρ.

§128. Analytical properties of the scattering amplitude

A number of important properties of the scattering amplitude can be established by considering it as a function of the energy E of the particle undergoing scattering, this energy being formally regarded as a complex variable.

Let us consider the motion of a particle in a field $U(r)$ which vanishes

† This type of scattering occurs in the theory of the rainbow, and it is therefore called *rainbow scattering*.

‡ Strictly speaking, the amplitude should include a term representing the contribution to small-angle scattering from impact parameters $\rho \to \infty$. This contribution, however, is in general small in comparison with that shown.

‖ This type of scattering is called *luminescence*, from its occurrence in the theory of certain meteorological phenomena.

sufficiently rapidly at infinity (the necessary degree of rapidity will be specified later). To simplify the discussion we shall first suppose that the orbital angular momentum l of the particle is zero. We can write down the asymptotic form of the wave function (the solution of Schrödinger's equation with $l = 0$ for any given value of E) as

$$\chi \equiv r\psi = A(E) \exp\left(-\frac{\sqrt{(-2mE)}}{\hbar}r\right) + B(E) \exp\left(\frac{\sqrt{(-2mE)}}{\hbar}r\right), \quad (128.1)$$

and regard E as a complex variable, defining $\sqrt{-E}$ as being positive when E is real and negative. The wave function is assumed normalized by some definite condition, say $\psi(0) = 1$.

On the left half of the real axis ($E < 0$) the exponential factors in the two terms in (128.1) are real; one decreases and the other increases as $r \to \infty$. From the condition that χ is real it follows that the functions $A(E)$ and $B(E)$ are real for $E < 0$, and from this in turn it follows that these functions have complex conjugate values at any two points lying symmetrically about the real axis:

$$A(E^*) = A^*(E), \qquad B(E^*) = B^*(E). \quad (128.2)$$

On going from the left half to the right half of the real axis through the upper half-plane we obtain an asymptotic expression for the wave function for $E > 0$ in the form

$$\chi = A(E)e^{ikr} + B(E)e^{-ikr}, \qquad k = \sqrt{(2mE)}/\hbar. \quad (128.3)$$

If a path through the lower half-plane is used, however, the result is

$$\chi = A^*(E)e^{-ikr} + B^*(E)e^{ikr}.$$

Since χ must be a single-valued function of E, this means that

$$A(E) = B^*(E) \quad \text{for} \quad E > 0; \quad (128.4)$$

this relation also follows directly from the fact that χ is real for $E > 0$. Nevertheless, because the root $\sqrt{-E}$ in (128.1) is not single-valued, the coefficients $A(E)$ and $B(E)$ themselves are not single-valued. To avoid this, we cut the complex plane along the right half of the real axis. The cut makes $\sqrt{-E}$ single-valued, and so the functions $A(E)$ and $B(E)$ are uniquely determined. They have complex conjugate values on the upper and lower edges of the cut (in (128.3), $A(E)$ and $B(E)$ are taken on the upper edge).

The complex plane cut in the manner described above will be called a *physical sheet* of the Riemann surface. According to our definition we have everywhere on this sheet

$$\operatorname{re}\sqrt{-E} > 0. \quad (128.5)$$

In particular, on the upper edge of the cut $\sqrt{-E}$ thus defined becomes $-i\sqrt{E}$.†

In (128.3) the factors e^{ikr} and e^{-ikr}, and so also the two terms in χ, are quantities of the same order of magnitude; an asymptotic expression of the form (128.3) is therefore always legitimate. Everywhere else on the physical sheet the first term in (128.1) decreases exponentially, and the second term increases exponentially, as $r \to \infty$ (because of (128.5)). Hence the two terms in (128.1) are of different orders of magnitude, and this expression may not be legitimate as the asymptotic form of the wave function: the small term compared with the large one may represent an unjustified exaggeration of accuracy. For the expression (128.1) to be legitimate the ratio of the small and large terms must not be less than the relative order of magnitude of the potential energy U/E, which is neglected in Schrödinger's equation on going to the asymptotic region. In other words, the field $U(r)$ must be such that

$$U(r) \text{ decreases more rapidly than } \exp\left(-\frac{2\sqrt{(2m)}}{\hbar} r \operatorname{re}\sqrt{-E}\right) \text{ as } r \to \infty. \tag{128.6}$$

When this condition is satisfied, the asymptotic expression of the form (128.1) is valid everywhere on the physical sheet. Being a solution of an equation with finite coefficients, it has no singularity with respect to E. This means that the functions $A(E)$ and $B(E)$ are regular everywhere on the physical sheet except the point $E = 0$, which, being the point where the cut begins, is a branch point of these functions.

The bound states of a particle in the field $U(r)$ correspond to wave functions which vanish as $r \to \infty$. This means that the second term in (128.1) cannot appear, i.e. the discrete energy levels correspond to zeros of the function $B(E)$. Since Schrödinger's equation has only real eigenvalues, all the zeros of $B(E)$ on the physical sheet are real (and lie on the left half of the real axis).

The functions $A(E)$ and $B(E)$ for $E > 0$ are directly related to the scattering amplitude in the field $U(r)$: comparing (128.3) with the asymptotic expression for χ written in the form (33.20),

$$\chi = \text{constant} \times [e^{i(kr+\delta_0)} - e^{-i(kr+\delta_0)}], \tag{128.7}$$

we see that

$$-A(E)/B(E) = e^{2i\delta_0(E)}. \tag{128.8}$$

† In the rest of this section we shall be considering the properties of the scattering amplitude on the physical sheet. Later, however, it will sometimes be necessary to consider another "non-physical" sheet of the Riemann surface (see §134). On this sheet

$$\operatorname{re}\sqrt{-E} < 0. \tag{128.5a}$$

The passage from the right half of the axis to the non-physical sheet is made directly down through the cut.

The scattering amplitude with angular momentum $l = 0$ is, according to (123.15),

$$f_0 = \frac{1}{2ik}(e^{2i\delta_0}-1) = \frac{\hbar}{2\sqrt{(-2mE)}}\left(\frac{A}{B}+1\right); \tag{128.9}$$

here A and B are taken on the upper edge of the cut.

Considering now the scattering amplitude as a function of E over the whole physical sheet, we see that the discrete energy levels are simple poles of this function. If the field $U(r)$ satisfies the condition (128.6), the above discussion shows that the scattering amplitude has no other singular points.†

Let us calculate the residue of the scattering amplitude at its pole for some discrete level $E = E_0 < 0$. To do so, we write down the equations satisfied by the function χ and its derivative with respect to energy:

$$\chi''+\frac{2m}{\hbar^2}(E-U)\chi = 0, \qquad \left(\frac{\partial\chi}{\partial E}\right)''+\frac{2m}{\hbar^2}(E-U)\frac{\partial\chi}{\partial E} = -\frac{2m}{\hbar^2}\chi.$$

Multiplying the first by $\partial\chi/\partial E$, the second by χ, subtracting, and integrating with respect to r, we obtain

$$\chi'\frac{\partial\chi}{\partial E}-\chi\left(\frac{\partial\chi}{\partial E}\right)' = \frac{2m}{\hbar^2}\int_0^r \chi^2\,dr. \tag{128.10}$$

We apply this relation for $E = E_0$ and $r \to \infty$. The integral on the right-hand side becomes unity for $r \to \infty$ if the wave function of the bound state is normalized by the usual condition $\int \chi^2\,dr = 1$. On the left-hand side we substitute χ from (128.1), using the fact that, near the point $E = E_0$,

$$A(E) \approx A(E_0) \equiv A_0, \qquad B(E) \approx (E+|E_0|)[dB/dE]_{E=E_0} \equiv \beta(E+|E_0|).$$

The result is

$$\beta = -\frac{1}{A_0\hbar}\sqrt{\frac{m}{2|E_0|}}.$$

By means of these expressions we find that, near the point $E = E_0$, the principal term in the scattering amplitude (i.e. the amplitude for $l = 0$) has the form

$$f = -\frac{\hbar^2 A_0^2}{2m}\frac{1}{E+|E_0|}. \tag{128.11}$$

Thus the residue of the scattering amplitude at the discrete level is determined by the coefficient A_0 in the asymptotic expression

$$\chi = A_0 \exp\left(-\frac{\sqrt{(2m|E_0|)}}{\hbar}r\right) \tag{128.12}$$

† Except the point $E = 0$, which is singular, because of the singularity of $A(E)$ and $B(E)$ previously mentioned. The scattering amplitude, however, remains finite as $E \to 0$ (see §132). In future we shall, for brevity, omit this qualification.

of the normalized wave function of the corresponding stationary state.

Returning to the examination of the analytical properties of the scattering amplitude, let us consider cases where the condition (128.6) is not satisfied. In such fields only the increasing term in (128.1) is the correct part of the asymptotic form of the solution of Schrödinger's equation over the whole of the physical sheet. Accordingly, we can, as before, assert that the function $B(E)$ has no singularity.

The function $A(E)$ under these conditions can be determined in the complex plane only as an analytical continuation of the function which is the coefficient in the asymptotic expression for χ on the right half of the real axis, where the two terms in χ are both legitimate. In general, however, such a continuation now gives different results according as it is carried out from the upper or the lower side of the cut. To obtain a single-valued function, we shall agree to define $A(E)$ in the upper and lower half-planes as the analytical continuation from the upper and lower sides of the right half of the real axis respectively; the cut must then in general be extended to the whole of the real axis. The function thus defined has as before the property $A(E^*) = A^*(E)$, but in general is not real either on the right or on the left half of the real axis. It may also, in principle, possess singularities.

We shall show, however, that there is nevertheless a class of fields for which the function $A(E)$ has no singularity on the physical sheet, although the condition (128.6) is not satisfied.

To do so, we regard χ as a function of a complex variable r for a given (complex) value of E. Here we need only consider values of E in the upper half-plane, since the values of the function $A(E)$ in the two half-planes are complex conjugates. For values of r such that Er^2 is real and positive, the two terms in the wave function (128.1) are of the same order, i.e. we return to the situation which occurs when $E > 0$ and r is real, when both terms in the asymptotic expression for χ are legitimate for any field $U(r)$ which tends to zero at infinity. We can therefore say that $A(E)$ cannot have singular points for values of E such that $U(r) \to 0$ when $r \to \infty$ along a line on which $Er^2 > 0$. When E takes all values in the upper half-plane, the condition $Er^2 > 0$ selects the lower right quadrant in the complex r-plane. Thus we conclude that $A(E)$ also has no singularity on the physical sheet when $U(r)$ satisfies the condition†

$$U(r) \to 0 \text{ when } r \to \infty \text{ in the right half-plane} \tag{128.13}$$

(L. D. Landau 1961).

The conditions (128.6) and (128.13) cover a very wide class of fields. We can therefore say that the scattering amplitude usually has no singularity in the two half-planes. On the left half of the axis (which is part of the physical sheet if not cut) the scattering amplitude has poles corresponding to the energies of the bound states; when the cut exists, there may be other singularities also.

† Since $U(r)$ is real on the real axis $U(r^*) = U^*(r)$; thus the condition (128.13) is satisfied throughout the right half-plane if it is satisfied in the lower right quadrant.

This happens, in particular, in fields of the form

$$U = \text{constant} \times r^n e^{-r/a} \tag{128.14}$$

with any n. On the segment $0 < -E < \hbar^2/8ma^2$ of the left half of the axis, the condition (128.6) holds, and so there need not be a cut; the scattering amplitude has only poles corresponding to the bound states. On the remainder of the left half of the axis there may be *redundant* poles and other singularities (S. T. Ma 1946). The appearance of these is due to the fact that the function (128.14) no longer tends to zero when $r \to \infty$ along a line on which $Er^2 > 0$, as soon as E moves below the left half of the axis (i.e. this line falls to the left of the imaginary axis in the complex r-plane).

Next, let us consider the analytical properties of the scattering amplitude as $|E| \to \infty$. When $E \to +\infty$ along the real axis, the Born approximation is valid and the scattering amplitude tends to zero. According to the above discussion, this situation also occurs when E tends to infinity in the complex plane along any line $\arg E$ = constant, if we consider complex values of r for which $Er^2 > 0$. If $U \to 0$ when $r \to \infty$ along a line $\arg r = -\frac{1}{2}\arg E$, and $U(r)$ has no singular point on this line, then the condition for the Born approximation to be valid is satisfied and the scattering amplitude again tends to zero. When $\arg E$ takes all values from 0 to π, $\arg r$ takes values from 0 to $-\frac{1}{2}\pi$.

We therefore conclude that the scattering amplitude tends to zero at infinity in all directions in the E-plane if the function $U(r)$ has no singular point in the right half-plane of r and tends to zero at infinity.

Although we have have spoken throughout of scattering with angular momentum $l = 0$, all the above results are in fact valid for the partial scattering amplitudes with any non-zero angular momentum. The only difference in the derivations is that, instead of the factors $e^{\pm ikr}$ in the asymptotic expressions for χ, we should have to use the exact radial wave functions for free motion (33.16).†

Some changes are needed in formulae (128.9) and (128.11) when $l \neq 0$. Instead of (128.7) we now have

$$\chi_l = rR_l = \text{constant} \times \{\exp[i(kr - \tfrac{1}{2}l\pi + \delta_l)] - \exp[-i(kr - \tfrac{1}{2}l\pi + \delta_l)]\}, \tag{128.15}$$

and for the partial amplitude f_l (defined according to (123.15)) we obtain

$$f_l = \frac{\hbar}{2\sqrt{(-2mE)}}\left((-1)^l\frac{A}{B} + 1\right). \tag{128.16}$$

The principal term in the scattering amplitude near the level $E = E_0$ with

† The limiting form (33.17) of these functions can be used only for $E > 0$; in the rest of the E-plane, where the two terms in χ are of different orders of magnitude, the use of these limiting expressions would involve an error in χ which is in general greater than that which arises from neglecting U in Schrödinger's equation.

angular momentum l is given by the formula

$$f \approx (2l+1)f_l P_l(\cos\theta)$$
$$= (-1)^{l+1}\frac{\hbar^2 A_0^2}{2m}\frac{1}{E+|E_0|}(2l+1)P_l(\cos\theta) \qquad (128.17)$$

instead of (128.11).

§129. The dispersion relation

In the previous section we have studied the analytical properties of the partial scattering amplitudes with given values of l, and have seen that these properties are complicated by the possible appearance of "redundant" singularities and non-regularity at infinity. The total amplitude, regarded as a function of energy for given values of the scattering angle, evidently has similar properties. The scattering amplitude for scattering angle zero forms an exception, however: as we shall now show, its analytical properties are considerably simpler.

Writing Schrödinger's equation for the wave function of the particle undergoing scattering as

$$\triangle\psi + k^2\psi = (2mU/\hbar^2)\psi, \qquad (129.1)$$

we may formally regard it as a wave equation with a non-zero right-hand side, i.e. as the equation of retarded potentials well known in electrodynamics.

The solution of this equation which describes the "emission" in some direction $\mathbf{k}'$ at large distances R_0 from the centre has the form (see *Fields*, §66)

$$\psi_{sc} = -\frac{1}{4\pi}\frac{e^{ikR_0}}{R_0}\int \frac{2mU}{\hbar^2}\psi e^{-i\mathbf{k}'.\mathbf{r}}\,dV. \qquad (129.2)$$

In the present case this represents the wave function of the scattered particle, and the coefficient of e^{ikR_0}/R_0 gives the scattering amplitude $f(\theta,E)$. In particular, putting $\mathbf{k}' = \mathbf{k}$ (where $\mathbf{k}$ is the wave vector of the incident particle), we obtain the scattering amplitude for scattering angle zero:

$$f(0,E) = -\frac{m}{2\pi\hbar^2}\int U\psi e^{-ikz}\,dV \qquad (129.3)$$

(the z-axis being taken in the direction of $\mathbf{k}$). This expression has, of course, only formal significance, since the integrand again involves the unknown wave function. However, it allows certain conclusions to be drawn concerning the analytical properties of the quantity $f(0,E)$ as a function of the energy E.†

The function ψ in the integrand consists of two parts when r is large, the

† It is assumed, of course, that the field $U(r)$ decreases, as $r\to\infty$, sufficiently rapidly for $f(0,E)$ to exist (when $E>0$); see §124.

incident wave and the outgoing wave. The latter is proportional to e^{ikr}, so that the corresponding part of the integral contains $e^{ik(r-z)}$ in the integrand. On the other hand, in the complex plane (going from the upper edge of the cut along the right half of the real axis) ik is replaced by $-\sqrt{(-2mE)}/\hbar$, and $\mathrm{re}\sqrt{-E} > 0$ everywhere on the physical sheet. Since $r \geqslant z$, $\mathrm{re}[ik(r-z)] < 0$, and the integral converges for any complex E. For the incident wave in ψ, proportional to e^{ikz}, the exponential factors cancel in the corresponding part of the integral, so that this part also converges.

The function ψ in the integral (129.3) is uniquely defined for all complex E as the solution of Schrödinger's equation which contains, in addition to the plane wave, only a part which is damped as $r \to \infty$. The whole of the convergent integral (129.2) is therefore uniquely determined also, and so its singularities can arise only through ψ's becoming infinite. This occurs at discrete energy levels.†

It is also easy to see that $f(0,E)$ remains finite as $|E| \to \infty$. For large $|E|$ the term in U can be neglected in Schrödinger's equation (129.1), so that only the plane wave remains in ψ: $\psi \sim e^{ikz}$. Thus the integral (129.2) becomes

$$f(0,\infty) = -\frac{m}{2\pi\hbar^2}\int U\,\mathrm{d}V,$$

which agrees, as it should, with the Born amplitude (126.4) for scattering through zero angle ($q = 0$); we denote it by $f_B(0)$.

Thus we conclude that the scattering amplitude for scattering angle zero is regular over the whole physical sheet (including infinity), except for the necessary poles on the left half of the real axis at the discrete energy levels.‡

Let us consider the integral

$$\frac{1}{2\pi i}\int_C \frac{f(0,E')-f_B}{E'-E}\,\mathrm{d}E', \tag{129.4}$$

taken along the contour shown in Fig. 46, which consists of an infinitely distant circle and an indentation round the cut along the right half of the

† To avoid misunderstanding, we should emphasize that here we are discussing the complete wave function ψ of the system, normalized by the condition that the coefficient of the plane wave in its asymptotic expression should be equal to unity (cf. 123.3)). In the previous section we were considering the parts ψ_l of the wave function which correspond to definite values of l, and ψ_l was assumed to be normalized in some arbitrary manner. If we expand the complete function ψ in terms of the functions ψ_l, the latter will appear in ψ with coefficients proportional to $1/B_l$. For example, the function (128.3) with $l = 0$ must appear in ψ in the form

$$\frac{\text{constant}}{r}\cdot\frac{1}{B}[(A+B)e^{ikr}-2iB\sin kr].$$

Hence ψ becomes infinite at the zeros of the functions $B_l(E)$, i.e. at the discrete energy levels.

‡ The idea of the foregoing proof is due to L. D. Faddeev (1958).

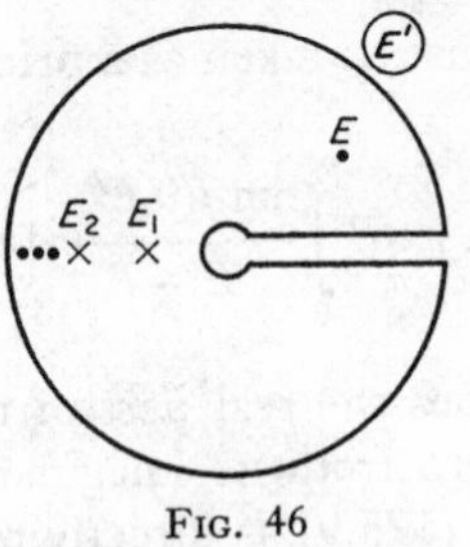

FIG. 46

real axis. The integral along the circle is zero, since $f(0,\infty)-f_B = 0$. The integration along the two sides of the cut gives

$$\frac{1}{\pi}\int_0^\infty \frac{\operatorname{im} f(0,E')}{E'-E}\,dE';$$

here we have used the fact that, according to the definition adopted in §128, the physical scattering amplitude for real positive values of E is given on the upper side of the cut, and has the complex conjugate value on the lower side.

According to Cauchy's theorem, the integral (129.4) is equal to the sum of $f(0,E)-f_B$ and the residues R_n of the integrand at all the poles $E' = E_n$ of the function $f(0,E')/(E'-E)$, where E_n are the discrete energy levels. These residues are determined by formula (128.17), and are

$$R_n = \frac{d_n}{E_n-E}, \qquad d_n = -(-1)^{l_n}(2l_n+1)\frac{\hbar^2 A_{0n}{}^2}{2m}, \tag{129.5}$$

where l_n is the angular momentum of the state with energy E_n. Thus we find

$$f(0,E) = f_B+\frac{1}{\pi}\int_0^\infty \frac{\operatorname{im} f(0,E')}{E'-E}\,dE'+\sum_n \frac{d_n}{E-E_n}. \tag{129.6}$$

This *dispersion relation* determines $f(0,E)$ at any point on the physical sheet from the values of its imaginary part for $E > 0$ (D. Y. Wong 1957, N. Khuri 1957).

When the point E tends to the upper side of the cut, the integral along the real axis in (129.6) must be taken by passing below the pole $E' = E$; if this is done along an infinitesimal semicircle (Fig. 47), the corresponding part of the

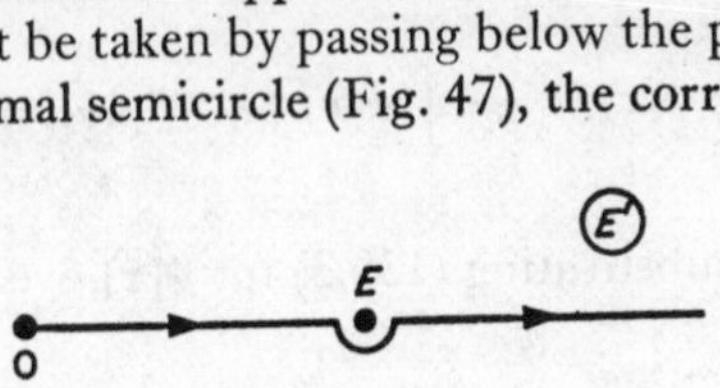

FIG. 47

integral gives $i \operatorname{im} f(0,E)$ on the right-hand side of (129.6), while the remaining integral from 0 to ∞ must be taken as a principal value. The result is

$$\operatorname{re} f(0,E) = f_B + \frac{1}{\pi} P \int_0^\infty \frac{\operatorname{im} f(0,E')}{E'-E} \mathrm{d}E' + \sum_n \frac{d_n}{E-E_n}, \tag{129.7}$$

which, for $E > 0$, determines the real part of the scattering amplitude for scattering through angle zero from its imaginary part. It may be recalled that the latter, according to (125.9), is directly related to the total scattering cross-section.

§130. The scattering amplitude in the momentum representation

The concept of the scattering amplitude involves only the directions of the initial and final momenta of the particle undergoing scattering. It can therefore naturally be approached also by formulating the scattering problem in the momentum representation, where there is no question of the spatial distribution of the process. We shall now show how this may be done.

First of all, let us transform to the momentum representation the original Schrödinger's equation

$$-\frac{\hbar^2}{2m} \Delta\psi(\mathbf{r}) + [U(\mathbf{r}) - E]\psi(\mathbf{r}) = 0, \tag{130.1}$$

changing from coordinate to momentum wave functions, i.e. to the Fourier components

$$a(\mathbf{q}) = \int \psi(\mathbf{r}) e^{-i\mathbf{q}\cdot\mathbf{r}} \, \mathrm{d}V. \tag{130.2}$$

Conversely,

$$\psi(\mathbf{r}) = \int a(\mathbf{q}) e^{i\mathbf{q}\cdot\mathbf{r}} \, \mathrm{d}^3q/(2\pi)^3. \tag{130.3}$$

We multiply equation (130.1) by $e^{-i\mathbf{q}\cdot\mathbf{r}}$ and integrate over $\mathrm{d}V$. In the first term, a repeated integration by parts gives

$$\int e^{-i\mathbf{q}\cdot\mathbf{r}} \Delta\psi(\mathbf{r}) \mathrm{d}V = \int \psi(\mathbf{r}) \Delta e^{-i\mathbf{q}\cdot\mathbf{r}} \, \mathrm{d}V = -\mathbf{q}^2 a(\mathbf{q}).$$

In the second term, substituting (130.3) for $\psi(\mathbf{r})$, we obtain

$$\begin{aligned}\int U(\mathbf{r})\psi(\mathbf{r}) e^{-i\mathbf{q}\cdot\mathbf{r}} \, \mathrm{d}V &= \int\int U(\mathbf{r}) e^{-i\mathbf{q}\cdot\mathbf{r}} a(\mathbf{q}') e^{i\mathbf{q}'\cdot\mathbf{r}} \, \mathrm{d}V \, \mathrm{d}^3q'/(2\pi)^3 \\ &= \int U(\mathbf{q}-\mathbf{q}') a(\mathbf{q}') \mathrm{d}^3q'/(2\pi)^3,\end{aligned}$$

where $U(\mathbf{q})$ is the Fourier component of the field $U(\mathbf{r})$:†

$$U(\mathbf{q}) = \int U(\mathbf{r})e^{-i\mathbf{q}.\mathbf{r}}\,dV.$$

Thus Schrödinger's equation in the momentum representation becomes

$$\left(\frac{\hbar^2 q^2}{2m} - E\right)a(\mathbf{q}) + \int U(\mathbf{q}-\mathbf{q}')a(\mathbf{q}')\frac{d^3q'}{(2\pi)^3} = 0. \qquad (130.4)$$

Note that this is an integral, not a differential, equation.

The wave function describing the scattering of particles with momentum $\hbar\mathbf{k}$ has the form

$$\psi_{\mathbf{k}}(\mathbf{r}) = e^{i\mathbf{k}.\mathbf{r}} + \chi_{\mathbf{k}}(\mathbf{r}), \qquad (130.5)$$

where $\chi_{\mathbf{k}}(\mathbf{r})$ is a function whose asymptotic form (as $r \to \infty$) is that of an outgoing spherical wave. Its Fourier component is

$$a_{\mathbf{k}}(\mathbf{q}) = (2\pi)^3\delta(\mathbf{q}-\mathbf{k}) + \chi_{\mathbf{k}}(\mathbf{q}), \qquad (130.6)$$

and substitution in (130.4) gives the following equation for the function $\chi_{\mathbf{k}}(\mathbf{q})$:‡

$$\frac{\hbar^2}{2m}(k^2 - q^2)\chi_{\mathbf{k}}(\mathbf{q}) = U(\mathbf{q}-\mathbf{k}) + \int U(\mathbf{q}-\mathbf{q}')\chi_{\mathbf{k}}(\mathbf{q}')d^3q'/(2\pi)^3. \qquad (130.7)$$

This equation may conveniently be transformed by using instead of $\chi_{\mathbf{k}}(\mathbf{q})$ another unknown function defined by

$$\chi_{\mathbf{k}}(\mathbf{q}) = \frac{2m}{\hbar^2}\frac{F(\mathbf{k}, \mathbf{q})}{q^2 - k^2 - i0}. \qquad (130.8)$$

This eliminates the singularity at $q^2 = k^2$ in the coefficients of equation (130.7), which becomes

$$F(\mathbf{k}, \mathbf{q}) = -U(\mathbf{q}-\mathbf{k}) - \frac{2m}{\hbar^2}\int \frac{U(\mathbf{q}-\mathbf{q}')F(\mathbf{k}, \mathbf{q}')}{q'^2 - k^2 - i0}\frac{d^3q'}{(2\pi)^3}. \qquad (130.9)$$

The term $i0$, which denotes the limit of $i\delta$ as $\delta \to +0$, is included in the definition (130.8) in order to give a definite sense to the integral in (130.9),

---

† For convenience of notation, we write $\mathbf{q}$ as an argument of the Fourier component instead of as a subscript.

‡ According to the properties of the delta function, the product $(q^2 - k^2)\delta(\mathbf{q}-\mathbf{k})$ gives zero when multiplied by any function $f(\mathbf{q})$ (not having a singularity at $\mathbf{q} = \mathbf{k}$) and integrated over d^3q. In this sense, $(q^2 - k^2)\delta(\mathbf{q}-\mathbf{k}) \equiv 0$.

since it establishes the manner of passage round the pole $q'^2 = k^2$ (cf. §43). We shall show that this manner of passage in fact corresponds to the required asymptotic form of the function

$$\chi_{\mathbf{k}}(\mathbf{r}) = \frac{2m}{\hbar^2}\int \frac{F(\mathbf{k},\ \mathbf{q})e^{i\mathbf{q}.\mathbf{r}}}{q^2-k^2-i0}\frac{d^3q}{(2\pi)^3}. \tag{130.10}$$

To do so, we write $d^3q = q^2\,dq\,do_{\mathbf{q}}$ and first integrate over $do_{\mathbf{q}}$, i.e. over the directions of the vector $\mathbf{q}$ relative to $\mathbf{r}$. An integration of this type has already been effected in transforming the first term in (125.2); in the region of large r, the result is

$$\chi_{\mathbf{k}}(\mathbf{r}) = -\frac{2m}{\hbar^2}\frac{2\pi i}{r}\int_0^\infty \frac{F(\mathbf{k},\ q\mathbf{n}')e^{iqr}-F(\mathbf{k},-q\mathbf{n}')e^{-iqr}}{q^2-k^2-i0}\frac{q\,dq}{(2\pi)^3},$$

where $\mathbf{n}' = \mathbf{r}/r$, or

$$\chi_{\mathbf{k}}(\mathbf{r}) = -\frac{im}{2\pi^2\hbar^2 r}\int_{-\infty}^{\infty}\frac{F(\mathbf{k},\ q\mathbf{n}')e^{iqr}\,q\,dq}{q^2-k^2-i0}.$$

The integrand has poles at the points $q = k+i0$ and $q = -k-i0$; the path of integration in the complex q-plane passes respectively below and above these (Fig. 48a). The path can be moved slightly into the upper half-plane

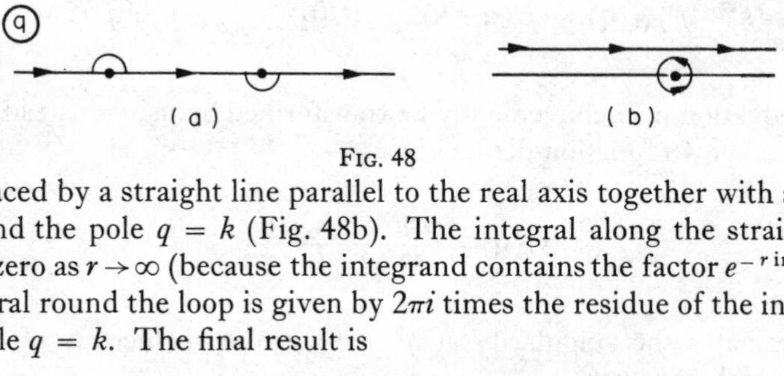

FIG. 48

and replaced by a straight line parallel to the real axis together with a closed loop round the pole $q = k$ (Fig. 48b). The integral along the straight line tends to zero as $r\to\infty$ (because the integrand contains the factor $e^{-r\,\mathrm{im}\,q}$), and the integral round the loop is given by $2\pi i$ times the residue of the integrand at the pole $q = k$. The final result is

$$\chi_{\mathbf{k}}(\mathbf{r}) = \frac{m}{2\pi\hbar^2}\frac{e^{ikr}}{r}F(k\mathbf{n},\ k\mathbf{n}'), \tag{130.11}$$

where $\mathbf{n}$ is a unit vector in the direction of $\mathbf{k}$. We have derived the required asymptotic form of the wave function, and the scattering amplitude is

$$f(\mathbf{n},\mathbf{n}') = \frac{m}{2\pi\hbar^2}F(k\mathbf{n},\ k\mathbf{n}'). \tag{130.12}$$

Thus the scattering amplitude is determined by the value at $q = k$ of the function $F(\mathbf{k},\ \mathbf{q})$ which satisfies the integral equation (130.9).

When perturbation theory is applicable, equation (130.9) is easily solved by iteration. In the first approximation, omitting the integral term, we have $F(\mathbf{k},\mathbf{q}) = -U(\mathbf{q}-\mathbf{k})$. In the next approximation, we substitute this in the integral term; the scattering amplitude (130.12) is then found to be (with a slight change of notation)

$$f(\mathbf{n}, \mathbf{n}') = -\frac{m}{2\pi\hbar^2}\left\{U(\mathbf{k}'-\mathbf{k})+\frac{2m}{\hbar^2}\int\frac{U(\mathbf{k}'-\mathbf{k}'')U(\mathbf{k}''-\mathbf{k})}{k^2-k''^2+i0}\frac{\mathrm{d}^3k''}{(2\pi)^3}\right\}, \quad (130.13)$$

with $\mathbf{k} = k\mathbf{n}$, $\mathbf{k}' = k\mathbf{n}'$. The first term is the same as (126.4) in the first Born approximation; the second term shows the contribution of the second approximation to the scattering amplitude.†

From (130.13) we have the result already mentioned in §126: even in the second approximation, the scattering amplitude does not have the symmetry property (126.8). At first sight it may seem that the integral term in (130.13) is also symmetrical with respect to interchange of the initial and final states. Actually, however, this symmetry does not exist, because the path of integration and the direction of passage round the pole are altered when the complex conjugate expression is taken.

§131. Scattering at high energies

If the potential energy is not small compared with $\hbar^2/ma^2$ (a being as usual the range of action of the field), a situation can occur where the energy of the particles undergoing scattering is so large that

$$|U| \ll E \sim (\hbar^2/ma^2)(ka)^2, \quad (131.1)$$

yet the condition

$$|U| \gtrsim (\hbar^2/ma^2)ka = \hbar v/a \quad (131.2)$$

holds; here it is, of course, assumed that

$$ka \gg 1. \quad (131.3)$$

In such a case we have scattering of fast particles to which the Born approximation is not applicable; neither of the conditions (126.1) and (126.2) is satisfied.

To examine this case, we can use the expression for the wave function in the form (45.9):

$$\psi = e^{ikz}F(\mathbf{r}),\ F(\mathbf{r}) = \exp\left(-\frac{i}{\hbar v}\int_{-\infty}^{z} U\,\mathrm{d}z\right); \quad (131.4)$$

† This result can, of course, also be easily obtained without recourse to the momentum representation: the fact that the second-approximation formula differs from the first-approximation formula by the replacement of $U(\mathbf{k}'-\mathbf{k})$ by the expression in the braces in (130.13) is evident from a comparison of (43.1) and (43.6).

for this to be applicable, the energy need only satisfy the condition $|U| \ll E$. In §45 it has been noted that this expression is valid only for $z \ll ka^2$, and so it cannot be immediately extended to distances where the asymptotic expression (123.3) holds. This is not necessary, however; to calculate the scattering amplitude, it is sufficient to know the wave function at distances z such that $a \ll z \ll ka^2$, and the integral in the exponent of $F(\mathbf{r})$ can be extended to infinity:

$$\psi = e^{ikz}S(\rho), \tag{131.5}$$

with the notation

$$S(\rho) = \exp[2i\delta(\rho)], \quad \delta(\rho) = -\frac{1}{2\hbar v}\int_{-\infty}^{\infty} U\,dz, \tag{131.6}$$

ρ being the radius vector in the xy-plane.

The scattering of fast particles takes place mainly through small angles, and these will be considered here. The change in the momentum $\hbar\mathbf{q}$ is relatively small ($q \ll k$), and hence the vector $\mathbf{q}$ may be regarded as being perpendicular to the wave vector $\mathbf{k}$ of the incident particle, i.e. as lying in the xy-plane. The scattered wave is obtained by subtracting from (131.5) the incident wave e^{ikz} (the function (131.4) for $z = -\infty$). The amplitude for scattering with wave vector $\mathbf{k}' = \mathbf{k}+\mathbf{q}$ is proportional to the corresponding Fourier component of the scattered wave:†

$$f \sim \int [S(\rho)-1]e^{-i\mathbf{q}\cdot\boldsymbol{\rho}}\,d^2\rho$$

($d^2\rho = dx\,dy$). The proportionality coefficient in this expression can then be derived by comparison with the limiting case of the Born approximation (see below).

The calculation can also be made by a different method, which leads directly to a completely definite expression. For this, we use (129.2) and substitute ψ from (131.4). Since, according to (45.8),

$$(2m/\hbar^2)UF = 2ik\partial F/\partial z,$$

we obtain for the scattering amplitude (the coefficient of e^{ikR_0}/R_0)

$$f = \frac{k}{2\pi i}\int \frac{\partial F}{\partial z} e^{-i\mathbf{q}\cdot\boldsymbol{\rho}}\,dx\,dy\,dz$$

$$= \frac{k}{2\pi i}\int [F(z=\infty)-F(z=-\infty)]e^{-i\mathbf{q}\cdot\boldsymbol{\rho}}\,dx\,dy.$$

† This method of determining the scattering amplitude is analogous to the one used in the discussion of Fraunhofer diffraction (*Fields*, §61). Diffraction effects make formula (131.4) inapplicable for $z \gtrsim ka^2$.

Substituting the expression for F, we have finally†

$$f = \frac{k}{2\pi i}\int [S(\rho)-1]e^{-i\mathbf{q}\cdot\boldsymbol{\rho}}\,d^2\rho. \tag{131.7}$$

If the energy is so high that $\delta \sim |U|a/\hbar v \ll 1$, the Born approximation is valid: expanding $S-1 \approx 2i\delta$, we obtain from (131.7)

$$f = -\frac{m}{2\pi\hbar^2}\int Ue^{-i\mathbf{q}\cdot\boldsymbol{\rho}}\,d^2\rho\,dz,$$

in accordance with (126.4).

Using the optical theorem (125.9), we can derive the total scattering cross-section from (131.7). The amplitude for scattering at zero angle is the value of f for $\mathbf{q} = 0$. We thus have

$$\sigma = \int 2\,\mathrm{re}\,(1-S)d^2\rho = \int 4\sin^2\delta(\rho)\,d^2\rho. \tag{131.8}$$

The integrand may be regarded as the scattering cross-section for particles with impact parameter in the range $d^2\rho$.‡

Formula (131.7) does not presuppose that the field has central symmetry. It is instructive to see how this formula can be derived, for a centrally symmetric field, directly from the exact general formula (123.11).

Under the conditions (131.1)–(131.3), the main role in the scattering is played by the partial amplitudes with large angular momenta l. The quasi-classical condition is therefore satisfied for the wave functions, and we can use formula (124.1) for δ_l. Putting there $r_0 \approx l/k$, $r^2 = z^2+l^2/k^2$, we find

$$\delta_l \approx -\frac{m}{\hbar^2}\int\limits_{l/k}^{\infty}\frac{U(r)\,dr}{\sqrt{(k^2-l^2/r^2)}} = -\frac{m}{\hbar^2 k}\int\limits_0^{\infty} U[\sqrt{(z^2+l^2/k^2)}]dz,$$

---

† In the two-dimensional case, the scattering amplitude in the field $U(x, z)$ is determined by the analogous formula

$$f = \sqrt{\frac{k}{2\pi i}}\int [S(x)-1]\,e^{-iqx}\,dx. \tag{131.7a}$$

The square $|f|^2 d\theta$ is the scattering cross-section per unit length along the y-axis, and θ is the angle of scattering in the xz-plane; cf. also §126, Problem 6.

‡ In §152 a generalization of formulae (131.7) and (131.8) to the case of scattering by a system of particles will be given.

in accordance with the value of $\delta(\rho)$ (131.6) when $\rho = l/k$.† Next, for small angles ($\theta \ll 1$) the Legendre polynomials with large l can be put in the form (49.6):

$$P_l(\cos\theta) \approx J_0(\theta l) = \frac{1}{2\pi}\int_0^{2\pi} e^{-i\theta l \cos\phi}\, \mathrm{d}\phi.$$

Substituting this in (123.11) and changing from summation (over large l) to integration, we obtain

$$f = \frac{1}{\pi}\int\int_0^{2\pi} f_l e^{-i\theta l \cos\phi}\, \mathrm{d}\phi\, l\, \mathrm{d}l$$

$$= \frac{k^2}{\pi}\int f_l e^{-i\mathbf{q}\cdot\boldsymbol{\rho}}\, \mathrm{d}^2\rho, \tag{131.9}$$

where $\mathbf{q}$ and $\boldsymbol{\rho}$ are two-dimensional vectors with magnitudes $q = k\theta$, $\rho = l/k$. Lastly, substituting for f_l in the form (123.15) with $\delta_l = \delta(l/k)$, we return to (131.7).

For scattering in a centrally symmetric field, formula (131.7), after the integration over the polar angle ϕ in the xy-plane ($\mathrm{d}^2\rho = \rho\, \mathrm{d}\rho\, \mathrm{d}\phi$) has been carried out, becomes

$$f = -ik\int \{\exp[2i\delta(\rho)] - 1\} J_0(q\rho)\rho\, \mathrm{d}\rho. \tag{131.10}$$

It has already been mentioned in §126 that the Born approximation is not applicable to the scattering of fast particles through large angles if the cross-section is exponentially small. The method given here is also inapplicable under these conditions. Such cases are actually quasi-classical, and perturbation theory cannot be used.

In accordance with the general rules of the quasi-classical approximation (cf. §§52, 53), the exponent in the exponential law of decrease of the scattering

---

† The quasi-classical function $2\hbar\delta(\rho)$ is the change in the action, caused by the field U, when the particle traverses a classical path. For a fast particle, this path may be taken as a straight line, and $2\delta(\rho)$ is then the difference of the classical action integrals

$$\int_{-\infty}^{\infty}\sqrt{\left(k^2 - \frac{2mU}{\hbar^2}\right)}\, \mathrm{d}z - \int_{-\infty}^{\infty} k\, \mathrm{d}z \approx -\frac{m}{\hbar^2 k}\int_{-\infty}^{\infty} U\, \mathrm{d}z$$

In this sense, the function $2\delta(\rho)$ here acts like the eikonal in geometrical optics. The approximation in scattering theory is therefore often called the *eikonal approximation*. It must be emphasized, however, that the scattering amplitude does not reduce to its quasi-classical value, since the conditions $\theta l \gg 1$, $\delta_l \gg 1$ are not in general satisfied.

cross-sections can be determined by considering "complex paths" in the classically inaccessible region of motion.†

In the classical scattering problem the relation between the angle of deviation θ of the particle in a field $U(r)$ and the impact parameter ρ is given by

$$\tfrac{1}{2}(\pi \mp \theta) = \int\limits_{r_0}^{\infty} \frac{\rho \, dr}{r^2\sqrt{(1-\rho^2/r^2-U/E)}}, \tag{131.11}$$

where r_0 is the minimum distance from the centre, a root of the equation

$$1-\rho^2/r^2-U/E = 0; \tag{131.12}$$

see (127.5). The case of interest to us corresponds to the range of angles which cannot occur in the scattering of a classical particle.‡ These angles therefore correspond to complex solutions $\rho(\theta)$ of equation (131.11) (with corresponding complex values of r_0). From the function $\rho(\theta)$ thus found and the classical orbital angular momentum $mv\rho$ of the particle we calculate the action

$$S(\theta) = mv \int \rho(\theta) \, d\theta, \tag{131.13}$$

where v is the velocity of the particle at infinity. The scattering amplitude is

$$f \sim \exp\left(-\frac{1}{\hbar} \operatorname{im} S(\theta)\right). \tag{131.14}$$

Equation (131.12) has in general more than one complex root. The value of r_0 in (131.11) must be taken as that root which gives the smallest positive imaginary part im S. In addition, if the function $U(r)$ has complex singularities, they must also be considered as possible values of r_0.‖

The region $r \sim r_0$ is the most important in the integral (131.11). For large energies E, the term U/E under the radical can be omitted. Carrying out the integration, we then have

$$\rho = r_0 \cos\tfrac{1}{2}\theta. \tag{131.15}$$

If r_0 is a singular point of the function $U(r)$, it depends only on the properties of the field, but not on ρ or E. Calculating S from (131.13), we find in this case that the scattering amplitude is

$$f \sim \exp\left(-\frac{2mv}{\hbar} \sin\tfrac{1}{2}\theta \operatorname{im} r_0\right). \tag{131.16}$$

---

† A discussion of the coefficient of the exponential is given by A. Z. Patashinskii, V. L. Pokrovskii and I. M. Khalatnikov, *Soviet Physics JETP* **18**, 683, 1964.

‡ The method described here is valid not only for large E but generally for all cases of exponentially small scattering.

‖ It may be recalled (see §126) that, if $U(r)$ has a singularity for real r, the decrease of the cross-section is not exponential.

If, however, r_0 has to be taken as a root of equation (131.12), the form of the exponent depends on the particular properties of the field. For example, with the function

$$U = U_0 e^{-(r/a)^2}$$

(which has no singularity at a finite distance) we obtain from the equation

$$U/E = 1-\rho^2/r^2 \approx \sin^2\tfrac{1}{2}\theta$$

the result

$$r_0 = ia\sqrt{\log([E/U_0]\sin^2\tfrac{1}{2}\theta)}. \tag{131.17}$$

Owing to the very slight dependence on θ, r_0 may be regarded as constant in the integration in (131.13), and we find for the scattering amplitude the formula (131.16) with r_0 given by (131.17).

PROBLEMS

PROBLEM 1. Determine the total scattering cross-section for a spherical square potential well of radius a and depth U_0 with the condition (131.1): $U_0 \ll \hbar^2k^2/m$.

SOLUTION. We have

$$\int_{-\infty}^{\infty} U\,dz = -2U_0\sqrt{(a^2-\rho^2)}.$$

According to (131.7), the forward-scattering amplitude ($\mathbf{q} = 0$) is

$$f(0) = -\frac{ik}{2\pi}\int_0^a \left[\exp\left(\frac{2iU_0}{\hbar v}\sqrt{(a^2-\rho^2)}\right)-1\right].\,2\pi\rho\,d\rho$$

$$= -ika^2\int_0^1 (e^{2i\nu x}-1)x\,dx$$

$$= \tfrac{1}{2}ka^2\left[i-\frac{e^{2i\nu}}{\nu}-\frac{i}{2\nu^2}(e^{2i\nu}-1)\right],$$

where $\nu = U_0a/\hbar v$ is the "Born parameter". By means of the optical theorem (125.9), we hence find the total cross-section:

$$\sigma = 2\pi a^2\left[1+\frac{1}{2\nu^2}-\frac{\sin 2\nu}{\nu}-\frac{\cos 2\nu}{2\nu^2}\right].$$

In the limiting (Born) case $\nu \ll 1$, this gives $\sigma = 2\pi a^2\nu^2$, in agreement with §126, Problem 1. In the opposite limiting case $\nu \gg 1$, we have simply $\sigma = 2\pi a^2$, i.e. twice the geometrical cross-section. The latter result has a simple significance. For $\nu \gg 1$, all particles with impact parameter $\rho < a$ are scattered, i.e. are removed from the incident beam. In this sense the well behaves as an "absorbing" sphere; and, according to Babinet's principle (see *Fields*, end of §61), the total cross-section is twice the "absorption" cross-section.

PROBLEM 2. The same as Problem 1, but for a field $U = U_0\exp(-r^2/a^2)$.

SOLUTION. In this case

$$\int_{-\infty}^{\infty} U\,dz = a\sqrt{\pi}U_0\exp(-\rho^2/a^2).$$

Substituting in (131.7) and making an obvious change of variable in the integral, we obtain for the scattering amplitude at zero angle

$$f(0) = -\tfrac{1}{2}ika^2 \int_0^{\nu\sqrt{\pi}} (e^{-iu}-1)\, du/u,$$

where again $\nu = U_0 a/\hbar v$. Hence the total cross-section is

$$\sigma = 2\pi a^2 \int_0^{\nu\sqrt{\pi}} (1-\cos u)\, du/u.$$

For $\nu \ll 1$ the integrand is $\frac{1}{2}u$ and the cross-section $\sigma = \frac{1}{2}\pi a^2\nu^2$, in accordance with the result of §126, Problem 2 (for $ka \gg 1$). For $\nu \gg 1$, we write the integrand as $(1-e^{-\lambda u}\cos u)/u$ with a small parameter λ, which afterwards is made to tend to zero. Integration by parts then gives

$$\int_0^{\nu\sqrt{\pi}} (1-\cos u)\, du/u \approx \log(\nu\sqrt{\pi}) - \int_0^{\infty} \log u \sin u\, du$$

$$= \log(\nu\sqrt{\pi}) + C,$$

where C is Euler's constant. Thus

$$\sigma = 2\pi a^2 \log(\nu\sqrt{\pi}e^C) \text{ for } \nu \gg 1.$$

PROBLEM 3. Determine the cross-section for small-angle scattering of electrons in a magnetic field concentrated in a cylindrical region of radius a (Y. Aharonov and D. Bohm 1959).

SOLUTION. Let the magnetic field be along the y-axis, which is also the axis of the cylindrical region, and let the direction of incidence of the electrons be taken as the z-axis. Then the scattering is independent of the coordinate y, and we can consider a two-dimensional problem in the xz-plane.

Outside the cylindrical region, the field $\mathbf{H} = 0$, but the vector potential is not zero:

$$\mathbf{A} = (\Phi/2\pi)\nabla\phi, \tag{1}$$

where ϕ is the polar angle in the xz-plane and Φ the magnetic flux; for, integrating over the area of a circle (with radius $r > a$) in this plane, we have

$$\int H\, dx\, dz = \oint \mathbf{A}\cdot d\mathbf{l} = [\Phi\phi/2\pi]_0^{2\pi} = \Phi.$$

The potential (1) changes the phase of the electron wave function (plane wave); according to (111.9), we have

$$\psi = e^{ikz}\exp(ie\Phi\phi/2\pi\hbar c). \tag{2}$$

This expression, however, is inapplicable in a narrow region (width $\sim a$) along the half-axis $z > 0$, since the motion of the particles that have passed through the field region is perturbed by the field. This explains the apparent non-uniqueness of the function (2) when the angle ϕ increases by 2π as we pass round the origin. In reality, there is a cut (of finite width) near the half-axis $z > 0$, resulting from the invalidity of formula (2); on the two sides of the cut, ϕ has values differing by 2π, for example $\mp\pi$.

For scattering at small angles θ with a small momentum transfer $q \approx k\theta$ ($qa \ll 1$, $\theta \ll 1$), transverse distances $x \sim 1/q \gg a$ are important, and the width of the cut may be neglected.

Considering the region $z \gg |x|$, we can also neglect the dependence of ψ on x on either side of the z-axis, obtaining†

$$\left.\begin{aligned} \psi = e^{ikz}F(x), \qquad F(x) &= \exp(-ie\Phi/2\hbar c),\ x>0, \\ &= \exp(ie\Phi/2\hbar c),\ x<0. \end{aligned}\right\} \quad (3)$$

The "two-dimensional" scattering amplitude is calculated from formula (131.7a).‡ For $q \neq 0$ we have

$$f = \sqrt{\frac{k}{2\pi i}}\left\{\exp\left(-\frac{ie\Phi}{2\hbar c}\right)\int_0^\infty e^{-iqx}\,dx + \text{complex conjugate}\right\}.$$

The integral is calculated by including a factor $e^{-\lambda x}$ and then taking the limit $\lambda \to 0$. The result is

$$f = -\frac{1}{q}\sqrt{\frac{2k}{\pi i}}\sin\frac{e\Phi}{2\hbar c}.$$

Hence the scattering cross-section is

$$d\sigma = |f|^2\,d\theta = \frac{2}{\pi k}\sin^2\frac{e\Phi}{2\hbar c}.\,d\theta/\theta^2. \quad (4)$$

For $e\Phi/\hbar c \ll 1$, we obtain

$$d\sigma = \frac{e^2\Phi^2}{2\pi k\hbar^2c^2}\frac{d\theta}{\theta^2},$$

which corresponds to the case where perturbation theory is applicable.

Note the periodic dependence of the cross-section (4) on the magnetic field strength, and the divergence of the total cross-section when $\theta \to 0$, although the field is concentrated in a finite region of space. These are both specifically quantum effects.

§132. The scattering of slow particles

Let us consider the properties of elastic scattering in the limiting case where the velocities of the particles undergoing scattering are so small that their wavelength is large compared with the radius of action a of the field $U(r)$ (i.e. $ka \ll 1$), and their energy is small compared with the field within that radius. The solution of this problem requires an elucidation of the limiting form of the dependence of the phases δ_l on the wave number k when the latter is small.

For $r \lesssim a$ we can neglect only the term in k^2 in the exact Schrödinger's equation (123.7):

$$R_l''+2R_l'/r-l(l+1)R_l/r^2 = 2mU(r)R_l/\hbar^2. \quad (132.1)$$

In the range $a \ll r \ll 1/k$, on the other hand, we can also omit the term in $U(r)$, leaving

$$R_l''+2R_l'/r-l(l+1)R_l/r^2 = 0. \quad (132.2)$$

† Formula (3), like (131.4), is not applicable for very large z, when diffraction effects become important.

‡ This formula (for $q \neq 0$) can, as already mentioned, also be derived without using Schrödinger's equation in the potential field.

The general solution of this equation is

$$R_l = c_1 r^l + c_2/r^{l+1}. \qquad (132.3)$$

The values of the constants c_1 and c_2 can in principle be determined only by solving equation (132.1) for a particular function $U(r)$; they are, of course different for different l. At still greater distances, $r \sim 1/k$, the term in $U(r)$ can be omitted from Schrödinger's equation, but the term in k^2 cannot be neglected, so that we have

$$R_l'' + \frac{2}{r}R_l' + \left[k^2 - \frac{l(l+1)}{r^2}\right]R_l = 0, \qquad (132.4)$$

i.e. the equation of free motion. The solution of this equation is (see §33)

$$R_l = c_1(-1)^l \frac{(2l+1)!!}{k^{2l+1}} r^l \left(\frac{\mathrm{d}}{r\,\mathrm{d}r}\right)^l \frac{\sin kr}{r} + c_2(-1)^l \frac{r^l}{(2l-1)!!}\left(\frac{\mathrm{d}}{r\,\mathrm{d}r}\right)^l \frac{\cos kr}{r}. \qquad (132.5)$$

The constant coefficients have been chosen so that, for $kr \ll 1$, this solution becomes (132.3); this ensures the "joining" of the solution (132.3) in the region $kr \ll 1$ to the solution (132.5) in the region $kr \sim 1$.

Finally, for $kr \gg 1$ the solution (132.5) takes the asymptotic form (§33)

$$R_l \approx c_1(2l+1)!! \frac{\sin(kr - \frac{1}{2}l\pi)}{rk^{l+1}} + \frac{c_2 k^l}{(2l-1)!!\,r}\cos(kr - \tfrac{1}{2}l\pi).$$

This expression can be put in the form

$$R_l \approx \text{constant} \times \frac{\sin(kr - \frac{1}{2}l\pi + \delta_l)}{r}, \qquad (132.6)$$

where the phase δ_l is given by the equation

$$\tan\delta_l \approx \delta_l = c_2 k^{2l+1}/c_1(2l-1)!!\,(2l+1)!! \qquad (132.7)$$

(since k is small, all the phases δ_l are small).

According to (123.15) the partial scattering amplitudes are

$$f_l = \frac{1}{2ik}(e^{2i\delta_l} - 1) \approx \delta_l/k,$$

and so we conclude that in the limit of low energies

$$f_l \sim k^{2l}. \qquad (132.8)$$

Thus all the partial amplitudes with $l \neq 0$ are small compared with the scattering amplitude with $l = 0$ (called *s-wave scattering*). Neglecting them, we obtain for the total amplitude

$$f(\theta) \approx f_0 = \delta_0/k = c_2/c_1 \equiv -\alpha, \tag{132.9}$$

so that $d\sigma = \alpha^2\, do$, and the total cross-section is

$$\sigma = 4\pi\alpha^2. \tag{132.10}$$

At low velocities the scattering is isotropic, and the cross-section is independent of the particle energy.† The constant α is called the *scattering length*; it may be either positive or negative.

In the above discussion it has been tacitly assumed that the field $U(r)$ decreases at large distances ($r \gg a$) sufficiently rapidly for the approximations made to be legitimate. It is easy to see how rapidly $U(r)$ must in fact decrease. For large r, the second term in the function R_l (132.3) is small in comparison with the first. In order for the retention of this term to be nevertheless legitimate, the small terms $\sim c_2/r^{l+1}r^2$ retained in equation (132.2) must be large compared with the term $UR_l \sim Uc_1 r^l$ omitted in going from (132.1) to (132.2). Hence it follows that $U(r)$ must decrease more rapidly than $1/r^{2l+3}$ if the result (132.8) is to be valid for the partial amplitude f_l. In particular, the calculation of f_0, and therefore the result (132.9) of isotropic scattering independent of energy, are valid only when $U(r)$ decreases at large distances more rapidly than $1/r^3$.

If the field $U(r)$ decreases exponentially at large distances, we can draw certain conclusions regarding the nature of the subsequent terms in the expansion of the amplitudes f_l in powers of k. We have seen in §128 that in this case the amplitude f_l, regarded as a function of the complex variable E, is real when E is real and negative.‡ The same is therefore true of the function $g_l(E)$ in the expression (125.15):

$$f_l = 1/(g_l - ik)$$

(ik is real for $E < 0$). The function $g_l(E)$ is also real (by definition) when $E > 0$. Thus this function is real for all real E, and can therefore be expanded in integral powers of E, i.e. in even powers of k. The amplitude $f_l(k)$ itself, therefore, can be expanded in integral powers of ik; all terms with even powers of k are real, while those with odd powers of k are imaginary. According to (132.8) the expression of $f_l(k)$ begins with the term $\sim\delta_l/k \sim k^{2l}$; accordingly, the expansion of $g_l(k)$ begins with a term proportional to k^{-2l}.

When the field decreases at large distances according to a power law $U \approx \beta r^{-n}$ with $n < 3$, the result (132.9) that the amplitude is constant is, as already stated, invalid.

---

† In the scattering of electrons by atoms, the length a with which $1/k$ must be compared (the condition $ka \ll 1$) is represented by the radius of the atom, which is several times the Bohr radius (several times $\hbar^2/me^2$) for complex atoms. Owing to the large value of this radius, the constancy of the effective cross-section actually applies here only up to energies of the order of fractions of an electron-volt; at greater electron energies there is a marked energy dependence of the cross-section (called the *Ramsauer effect*).

‡ For small E, the condition (128.6) is satisfied even when U decreases as $e^{-r/a}$.

Let us now consider the situations which occur for various values of n. For $n \leqslant 1$ and sufficiently small velocities, the condition

$$\rho|U(\rho)| \gg \hbar v \tag{132.11}$$

is satisfied for practically all values of the impact parameter ρ, and so the scattering is described by the classical formulae (cf. the condition (127.9)).

For $1 < n < 2$ the inequality (132.11) is satisfied over a considerable range of fairly small values of ρ; accordingly, the scattering is classical for angles which are not too small. There is also, however, a range of values of ρ for which

$$\rho|U(\rho)| \ll \hbar v, \tag{132.12}$$

i.e. the condition for perturbation theory to be valid is satisfied (cf. (126.2)).

For $n > 2$ the inequality

$$|U| \ll \hbar^2/mr^2 \tag{132.13}$$

holds at large distances, and therefore the contribution to the scattering which arises from interaction at these distances can be calculated by means of perturbation theory (whereas at smaller distances the condition for perturbation theory to be applicable may not be satisfied).† Let r_0 be a value of r such that for $r \gg r_0$ the inequality (132.13) holds, while $r_0 \ll 1/k$. The contribution to the scattering amplitude from the region $r \gg r_0$ is, according to (126.12), given by the integral

$$-\frac{2m\beta}{\hbar^2}\int_{r_0}^{\infty}\frac{1}{r^n}\frac{\sin qr}{qr}r^2\,dr = -\frac{2m\beta}{\hbar^2}q^{n-3}\int_{qr_0}^{\infty}\frac{\sin\xi}{\xi^{n-1}}\,d\xi. \tag{132.14}$$

For $2 < n < 3$ this integral converges at the lower limit, and for low velocities ($kr_0 \ll 1$) we can replace this limit by zero, so that the integral is proportional to $q^{-(3-n)}$, i.e. a negative power of the velocity. This contribution to the amplitude is therefore in this case the main one, so that

$$f \sim q^{-(3-n)}, \qquad 2 < n < 3. \tag{132.15}$$

This determines the dependence of the scattering cross-section on the velocity of the particles and on the angle of scattering.

For $n = 3$ the integral (132.14) diverges logarithmically at the lower limit. It is still the main part of the scattering amplitude, so that

$$f \sim \log(\text{constant}/q), \qquad n = 3. \tag{132.16}$$

For $n > 3$ the contribution from the region $r \gg r_0$ decreases as $k \to 0$, and the scattering is determined by the constant amplitude (132.9)). However, the contribution (132.14), despite its relative smallness, is still of some interest through being "anomalous". The "normal" situation when $U(r)$

† The scattering at low velocities is in this case nowhere quasi-classical, since the inequality (132.11) is incompatible with the simultaneous requirement that $|U(\rho)| \lesssim E$.

decreases sufficiently rapidly is that $f(k)$ can be expanded in integral powers of k, and all the real terms in the expansion are proportional to even powers of k. When the integral (132.14) is integrated several times by parts (lowering the power of ξ in the denominator), we can separate from it a part containing even powers of k and leave an integral convergent as $qr_0 \to 0$ and proportional to k^{n-3}, which is not in general an even power.†

PROBLEMS

PROBLEM 1. Determine the scattering cross-section for slow particles in a spherical square potential well of depth U_0 and radius a.

SOLUTION. The wave number of the particle is assumed to satisfy the conditions $ka \ll 1$ and $k \ll \kappa$, where $\kappa = \sqrt{(2mU_0)}/\hbar$. We are interested only in the phase δ_0. Hence we put $l = 0$ in equation (132.1), and obtain for the function $\chi(r) = rR_0(r)$ the equation

$$\chi'' + \kappa^2\chi = 0 \text{ for } r < a.$$

The solution which vanishes at $r = 0$ (χ/r must be finite at $r = 0$) is

$$\chi = A \sin \kappa r \qquad (r < a).$$

For $r > a$, the function χ satisfies the equation $\chi'' + k^2\chi = 0$ (i.e. equation (132.4) with $l = 0$), whence

$$\chi = B \sin(kr + \delta_0) \qquad (r > a).$$

From the continuity of χ'/χ at $r = a$, we obtain the equation

$$\kappa \cot \kappa a = k \cot(ka + \delta_0) \approx k/(ka + \delta_0),$$

from which we determine δ_0. As a result, we have for the scattering amplitude‡

$$f = \frac{\tan \kappa a - \kappa a}{\kappa}.$$

For $\kappa a \ll 1$ (i.e. $U_0 \ll \hbar^2/ma^2$) this formula gives $\sigma = (4\pi a^2/9)(\kappa a)^4$, in accordance with the result of the Born approximation (see §126, Problem 1).

PROBLEM 2. The same as Problem 1, but for scattering by a spherical "potential hump" of height U_0.

SOLUTION. The solution is obtained from that of Problem 1 if we change the sign of U_0 (which means replacing κ by $i\kappa$), and obtain for the scattering amplitude

$$f = \frac{\tanh \kappa a - \kappa a}{\kappa}.$$

In the limit $\kappa a \gg 1$ we have

$$f = -a, \quad \sigma = 4\pi a^2.$$

† If n is an odd integer $2p+1$, then $n-3 = 2p-2$ is an even number. In this case also, however, the integral (132.14) has an "anomalous" part, which gives a contribution to the scattering amplitude proportional to $q^{2p-2} \log q$.

‡ This formula becomes inapplicable if the width and depth of the well are such that κa is close to an odd multiple of $\frac{1}{2}\pi$. For such values of κa the discrete spectrum of negative energy levels includes one which is close to zero (see §33, Problem 1), and the scattering is described by formulae which we shall derive in the next section.

This corresponds to scattering from an impenetrable sphere of radius a; we note that classical mechanics would give a result four times smaller ($\sigma = \pi a^2$).

PROBLEM 3. Determine the scattering cross-section for particles of low energy in a field $U = \alpha/r^n$, $\alpha > 0$, $n > 3$.

SOLUTION. Equation (132.1) with $l = 0$ is

$$\chi'' - \gamma^2\chi/r^n = 0, \qquad \gamma = \sqrt{(2m\alpha)}/\hbar.$$

By the substitutions

$$\chi = \phi\sqrt{r}, \qquad r = [2\gamma/(n-2)x]^{2/(n-2)}$$

it can be brought to the form

$$\frac{d^2\phi}{dx^2} + \frac{1}{x}\frac{d\phi}{dx} - \left(1 + \frac{1}{(n-2)^2x^2}\right)\phi = 0,$$

i.e. Bessel's equation of order $1/(n-2)$, with imaginary argument ix. The solution which vanishes at $r = 0$ (i.e. $x = \infty$) is, apart from a constant factor,

$$\chi = \sqrt{r}H_{1/(n-2)}{}^{(1)}(2i\gamma r^{-(n-2)/2}/(n-2)).$$

Using the well-known formulae

$$H_p{}^{(1)}(z) = i[e^{-ip\pi}J_p(z) - J_{-p}(z)]/\sin p\pi,$$

$$J_p(z) \approx z^p/2^p\Gamma(p+1) \qquad (z \ll 1),$$

we obtain for the function χ at large distances ($\gamma \ll r \ll 1/k$) the expression $\chi = \text{constant} \times (c_1 r + c_2)$, and from the ratio c_2/c_1 we find the scattering amplitude

$$f = -\left(\frac{\gamma}{n-2}\right)^{2/(n-2)} \frac{\Gamma[(n-3)/(n-2)]}{\Gamma[(n-1)/(n-2)]}.$$

PROBLEM 4. Determine the scattering amplitude for slow particles in a field which decreases at large distances as $U \approx \beta r^{-n}$ with $2 < n \leqslant 3$.

SOLUTION. The principal term in the scattering amplitude is given by the expression (132.14), in which the lower limit in the integral can be replaced by zero. The calculation of the integral leads to the result

$$f = \frac{\pi m\beta}{\hbar^2} \frac{q^{n-3}}{\Gamma(n-1)\cos\frac{1}{2}\pi n}, \qquad 2 < n < 3, \tag{1}$$

and for $n = 3$

$$f = -\frac{2m\beta}{\hbar^2}\log\frac{\text{constant}}{q}. \tag{2}$$

Expanding (1) in Legendre polynomials, we obtain the partial scattering amplitudes (defined in accordance with (123.14)):

$$f_l = -\frac{\sqrt{\pi}m\beta}{2\hbar^2}\frac{\Gamma(\frac{1}{2}n-\frac{1}{2})\Gamma(l-\frac{1}{2}n+\frac{3}{2})}{\Gamma(\frac{1}{2}n)\Gamma(\frac{1}{2}n+\frac{1}{2}+l)}k^{n-3}. \tag{3}$$

For $n > 3$ the same formula (1) determines the "anomalous" part of the scattering amplitude. In the partial amplitudes the quantity (3) is always the principal part for values of l such that $2l > n-3$, and instead of (132.8) we then have $f_l \sim k^{n-3}$.

PROBLEM 5. Determine the scattering amplitude for slow particles in a field $U(r) = -U_0\exp(-r/a)$, $U_0 > 0$.

SOLUTION. After the change of variable

$$x = 2a\kappa e^{-r/2a}, \qquad \kappa = \sqrt{(2mU_0)}/\hbar,$$

equation (132.1) for the function $\chi = rR_0$ becomes

$$\frac{d^2\chi}{dx^2}+\frac{1}{x}\frac{d\chi}{dx}+\chi = 0.$$

The general solution of this equation is

$$\chi = AJ_0(x)+BN_0(x),$$

where J_0 and N_0 are the Bessel functions of the first and second kinds. The condition $\chi = 0$ for $r = 0$ gives

$$A/B = -N_0(2\kappa a)/J_0(2\kappa a).$$

The region $a \ll r \ll 1/k$ corresponds to $x \ll 1$ (it is assumed, of course, that $a\kappa \exp(-1/ak) \ll 1$; here

$$\chi \approx A+B\frac{2}{\pi}\log \tfrac{1}{2}\gamma x = A+\frac{2B}{\pi}\log \kappa a\gamma-\frac{Br}{\pi a},$$

where $\gamma = e^C = 1{\cdot}78 \ldots$ (C is Euler's constant). This expression corresponds to formula (132.3), and from the values of c_1 and c_2 thus obtained we find the scattering amplitude

$$f = -a\left(\frac{\pi A}{B}+2\log \kappa a\gamma\right)$$

$$= \frac{a\pi}{J_0(2\kappa a)}\left[N_0(2\kappa a)-\frac{2}{\pi}\log \kappa a\gamma\, J_0(2\kappa a)\right].$$

In the limit $\kappa a \ll 1$, $f = 2a^3\kappa^2$, in accordance with the Born approximation (126.14). For $\kappa a \gg 1$ we have $f = -2a \log \kappa a\gamma$.

PROBLEM 6. Determine, in the second approximation of perturbation theory, the scattering amplitude in the limit of low energies (I. Ya Pomeranchuk 1948).

SOLUTION. For $k \to 0$ the integral in the second term of (130.13) becomes

$$-\int \frac{U_{-\mathbf{k}''}U_{\mathbf{k}''}}{k''^2}\,d^3k'' = -\iint U(\mathbf{r})U(\mathbf{r}')e^{i\mathbf{k}''.(\mathbf{r}-\mathbf{r}')}\frac{d^3k''}{k''^2}\,dV\,dV'$$

$$= -2\pi^2\iint\frac{U(\mathbf{r})U(\mathbf{r}')}{|\mathbf{r}-\mathbf{r}'|}\,dV\,dV';$$

here we have used the formula

$$\int e^{i\mathbf{k}'.(\mathbf{r}-\mathbf{r}')}\frac{4\pi}{k^2}\frac{d^3k}{(2\pi)^3} = \frac{1}{|\mathbf{r}-\mathbf{r}'|};$$

see *Fields*, §51. Thus the scattering amplitude is

$$f = -\frac{m}{2\pi\hbar^2}\int U\,dV+\left(\frac{m}{2\pi\hbar}\right)^2\iint\frac{U(\mathbf{r})U(\mathbf{r}')}{|\mathbf{r}-\mathbf{r}'|}\,dV\,dV'. \qquad (1)$$

For a central field, this formula gives

$$f = -\frac{2m}{\hbar^2}\int Ur^2\,dr+\frac{8m^2}{\hbar^4}\iint U(r)U(r')r^2\,dr\,.\,r'\,dr'.$$

The second term in (1) is always positive (as is clear from the original form of the integral in **k**-space). In a repulsive field ($U > 0$), the first Born approximation therefore always gives too high a value, and in an attractive field ($U < 0$) too low a value, for the scattering cross-section at low energies.

§133. Resonance scattering at low energies

Particular consideration must be given to the scattering of slow particles ($ka \ll 1$) in an attractive field when the discrete spectrum of negative energy levels includes an s state whose energy is small compared with the value of the field U within its range of action a. We denote this level by ϵ ($\epsilon > 0$). The energy E of the particle undergoing scattering, being small, is close to ϵ, i.e. it is, as we say, almost in *resonance* with the level. This leads, as we shall see, to a considerable increase in the scattering cross-section.

The existence of the shallow level can be taken into account in scattering theory by means of a formal method based on the following arguments.

In the exact Schrödinger's equation for the function $\chi = rR_0$ (with $l = 0$),

$$\chi'' + (2m/\hbar^2)[E - U(r)]\chi = 0$$

in the "inner" region of the field ($r \lesssim a$) we can neglect E in comparison with U:

$$\chi'' - (2m/\hbar^2)U(r)\chi = 0, \qquad r \sim a. \tag{133.1}$$

In the "outer" region ($r \gg a$), on the other hand, we can neglect U:

$$\chi'' + (2m/\hbar^2)E\chi = 0, \qquad r \gg a. \tag{133.2}$$

The solution of equation (133.2) must be "joined" at some r_1 (such that $1/k \gg r_1 \gg a$) to the solution of equation (133.1) which satisfies the boundary condition $\chi(0) = 0$; the joining condition is that the ratio χ'/χ should be continuous. This ratio does not depend on the normalization factor in the wave function.

However, instead of considering the motion in the region $r \sim a$, we apply to the solution in the outer region a suitably chosen boundary condition on χ'/χ for small r; since the solution in the outer region varies only slowly as $r \to 0$, we can formally apply this condition at the point $r = 0$. The equation (133.1) for the region $r \sim a$ does not contain E; the boundary condition which replaces it must therefore also be independent of the energy of the particle. In other words, it must be of the form

$$[\chi'/\chi]_{r\to 0} = -\kappa, \tag{133.3}$$

where κ is some constant. But, κ being independent of E, the same condition (133.3) must also apply to the solution of Schrödinger's equation for small negative energy $E = -|\epsilon|$, i.e. to the wave function of the corresponding stationary state of the particle. For $E = -|\epsilon|$ we have from (133.2)

$$\chi = A_0 e^{-r\sqrt{(2m|\epsilon|)}/\hbar}, \tag{133.4}$$

where A_0 is a constant, and substitution of this function in (133.3) shows that κ is a positive quantity,

$$\kappa = \sqrt{(2m|\epsilon|)}/\hbar. \tag{133.5}$$

Let us now apply the boundary condition (133.3) to the wave equation for free motion,

$$\chi = \text{constant} \times \sin(kr + \delta_0),$$

which is the exact general solution of equation (133.2) for $E > 0$. Thus we have for the required phase δ_0

$$\cot\delta_0 = -\kappa/k$$
$$= -\sqrt{(|\epsilon|/E)}. \tag{133.6}$$

Since the energy E is here restricted only by the condition $ka \ll 1$, and need not be small compared with $|\epsilon|$, the phase δ_0 and the s-wave scattering amplitude may not be small.

The phases δ_l with $l > 0$ and the corresponding partial amplitudes are again small. Hence we can again regard the total amplitude as being equal to the s-wave scattering amplitude

$$f \approx \frac{1}{2ik}(e^{2i\delta_0}-1)$$
$$= 1/k(\cot\delta_0 - i),$$

Substituting (133.6) we obtain

$$f = -1/(\kappa+ik) \tag{133.7}$$

and for the total scattering cross-section

$$\sigma = \frac{4\pi}{\kappa^2+k^2} = \frac{2\pi\hbar^2}{m}\frac{1}{E+|\epsilon|}. \tag{133.8}$$

Thus the scattering is again isotropic, but the cross-section depends on the energy, and in the resonance region ($E \sim |\epsilon|$) is large compared with the squared range of action of the field a^2 (since $ka \ll 1$). The form of (133.8) is not affected by the details of the interaction of the particles at small distances, and depends only on the value of the resonance level.†

The above formula is somewhat more general than the assumption made in its derivation. Let the function $U(r)$ be slightly modified; this alters also the value of the constant κ in the boundary condition (133.3). By an appropriate change in $U(r)$, κ can be made to vanish, and then to become small and negative. This gives the same formulae (133.7) for the scattering amplitude and (133.8) for the cross-section. In the latter, however, the quantity $|\epsilon| = \hbar^2\kappa^2/2m$ is now simply a constant characteristic of the field $U(r)$, and not an energy level in that field. In such cases the field is said to have a *virtual level*, since, although there is no actual level close to zero, a slight change in the field would be sufficient to cause one to appear.

In the analytical continuation of the function (133.7) in the complex plane of E, ik becomes $-\sqrt{(-2mE)}/\hbar$ on the left half of the real axis (see §128), and we see that the scattering amplitude has a pole at $E = -|\epsilon|$, in accordance with the general results of §128. On the other hand, the virtual level corresponds, as we should expect, to no singularity of the scattering amplitude on

† Formula (133.8) was first derived by E. Wigner (1933); the idea of the derivation given here is due to H. A. Bethe and R. E. Peierls (1935).

the physical sheet. (The scattering amplitude has a pole at $E = -|\epsilon|$ on the non-physical sheet; see the first footnote to §128.)

Formally, the expression (133.7) corresponds to the case where in the expression (125.15),

$$f_0 = \frac{1}{g_0(k) - ik},$$

the first term in the expansion of the function $g_0(k)$ is negative and anomalously small. To refine the formula, we can take account of the second term in the expansion:

$$f_0 = \frac{1}{-\kappa_0 + \frac{1}{2}r_0 k^2 - ik} \tag{133.9}$$

(L. D. Landau and Ya. A. Smorodinskiĭ 1944); it may be recalled that, when the field decreases sufficiently rapidly, the functions $g_l(k)$ can be expanded in even powers of k (see §132). In (133.9) we have denoted by $-\kappa_0$ the value of $g_0(0)$, in order to retain the notation κ for the quantity (133.5), which is related to the energy level ϵ. According to the above discussion, κ is given by the value of $-ik$ which makes the denominator in (133.9) equal to zero, i.e. by the root of the equation

$$\kappa = \kappa_0 + \tfrac{1}{2}r_0\kappa^2. \tag{133.10}$$

The correction term $\frac{1}{2}r_0 k^2$ in the denominator in (133.9) is small compared with κ_0, since k is assumed small, but it is itself of "normal" order of magnitude: the coefficient $r_0 \sim a$ and is always positive (see Problem 1). It should be emphasized that the inclusion of this term is a legitimate refinement in the formula for the scattering amplitude when contributions from angular momenta $l \neq 0$ are neglected; it gives a correction to f of relative order ka, whereas the contribution from scattering with $l = 1$ is of relative order $(ka)^3$. When $k \to 0$, the amplitude $f_0 \to 1/\kappa_0$, i.e. $1/\kappa_0$ is equal to the scattering length α defined in §132. The coefficient r_0 in the formula

$$\begin{aligned} g_0(k) &\equiv k \cot\delta_0 \\ &= -1/\alpha + \tfrac{1}{2}r_0 k^2 \end{aligned} \tag{133.11}$$

is called the *effective range* of the interaction.†

For the cross-section we have, from (133.9),

$$\sigma = \frac{4\pi}{(\kappa_0 - \frac{1}{2}r_0 k^2)^2 + k^2}.$$

† The values of the constants α and r_0 may be mentioned for the important case of the interaction of two nucleons. For a neutron and a proton with parallel spins (isotopic state with $T = 0$), $\alpha = 5{\cdot}4 \times 10^{-13}$ cm, $r_0 = 1{\cdot}7 \times 10^{-13}$ cm; these correspond to a true level with energy $|\epsilon| = 2{\cdot}23$ MeV, the ground state of the deuteron. For a neutron and a proton with antiparallel spins (isotopic state with $T = 1$), $\alpha = -24 \times 10^{-13}$ cm, $r_0 = 2{\cdot}7 \times 10^{-13}$ cm; these values correspond to a virtual level with $|\epsilon| = 0{\cdot}067$ MeV. Owing to isotopic invariance, the latter values must apply also to a system of two neutrons with antiparallel spins; parallel spins of the *nn* system in the *s* state are prohibited by Pauli's principle.

If we neglect the term in k^4 in the denominator (though it may legitimately be included), this formula can be written (using (133.10)) in the form

$$\sigma = \frac{4\pi(1+r_0\kappa)}{k^2+\kappa^2} = \frac{4\pi\hbar^2}{m}\frac{1+r_0\kappa}{E+|\epsilon|}. \qquad (133.12)$$

Let us return to the expression (133.4) for the wave function of the bound state in the "outer" region, and relate the normalization coefficient to the parameters defined above. On calculating the residue of the function (133.9) at its pole $E = \epsilon$ and comparing with (128.11), we find

$$\frac{1}{A_0{}^2} = \frac{1}{2\kappa} - \tfrac{1}{2}r_0. \qquad (133.13)$$

The second term is a small correction to the first, since $\kappa r_0 \sim \kappa a \ll 1$. Without this correction, $A_0{}^2 = 2\kappa$, i.e.

$$\chi = \sqrt{(2\kappa)}e^{-\kappa r}, \qquad \psi = \frac{\chi}{\sqrt{(4\pi)}r} = \sqrt{\frac{\kappa}{2\pi}}\frac{e^{-\kappa r}}{r}, \qquad (133.14)$$

corresponding to the normalization that would occur if (133.14) were valid in all space.

We shall briefly discuss resonance in scattering with non-zero orbital angular momenta. The expansion of the function $g_l(k)$ begins with a term $\sim k^{-2l}$; retaining the first two terms in the expansion, we write the partial scattering amplitude as

$$f_l = -\frac{1}{bE^{-l}(-\epsilon+E)+ik}, \qquad (133.15)$$

where b and ϵ are two constants, with $b > 0$ (see below). The case of resonance corresponds to an anomalously low value of the coefficient of E^{-l}, i.e. an anomalously small ϵ. However, since E is small, the term $b\epsilon E^{-l}$ may still be large in comparison with k.

If $\epsilon < 0$, the denominator in the expression (133.15) has a real root $E \approx -|\epsilon|$, so that ϵ is a discrete energy level (with angular momentum l),† but in contrast to resonance in s-wave scattering the amplitude (133.15) is never large compared with a; the amplitude of resonance scattering with angular momentum $l+1$ is only of the same order of magnitude as that of non-resonance scattering with angular momentum l.

If $\epsilon > 0$, however, the amplitude (133.15) becomes of the order of $1/k$ in the region $E \sim \epsilon$, i.e. large compared with a. The relative width of this region is small: $\Delta E/\epsilon \sim (ka)^{2l-1}$. Thus in this case there is a sharp resonance.

† For $\epsilon < 0$, and E close to $|\epsilon|$,

$$f_l \approx (-1)^{l+1}|\epsilon|^l/b(E+|\epsilon|).$$

A comparison with (128.17) shows that $b > 0$.

This type of resonance scattering occurs because a positive level with $l \neq 0$, though not a true discrete level, is quasi-discrete: owing to the presence of the centrifugal potential barrier, the probability that a particle of low energy will escape from this state to infinity is small, so that the "lifetime" of the state is long (see §134). This is the reason why resonance scattering with $l \neq 0$ is different in nature from that in the s state, where there is no centrifugal barrier. The denominator in (133.15) with $\epsilon > 0$ vanishes when $E = E_0 - i \cdot \frac{1}{2}\Gamma$, where

$$E_0 \approx \epsilon, \qquad \Gamma = \frac{2\sqrt{(2m)}}{b\hbar}\epsilon^{l+1/2}. \tag{133.16}$$

This pole of the scattering amplitude is, however, on the non-physical sheet. The small quantity Γ is the width of the quasi-discrete level (see §134).

Finally, we may mention an interesting property of the phases δ_l which is easily derived from the above results. We shall regard the phases $\delta_l(E)$ as continuous functions of the energy, and not restrict them to the range from 0 to π (cf. the footnote following (33.20)). We shall show that the equation

$$\delta_l(0) - \delta_l(\infty) = n\pi \tag{133.17}$$

then holds, where n is the number of discrete levels with angular momentum l in the attractive field $U(r)$ (N. Levinson 1949).

To prove this, we note that, in a field which satisfies the condition $|U| \ll \hbar^2/ma^2$, the Born approximation is valid at all energies, so that $\delta_l(E) \ll 1$ for all E, and $\delta_l(\infty) = 0$, since for $E \to \infty$ the scattering amplitude tends to zero, while $\delta_l(0) = 0$ in accordance with the general results of §132. In such a field there are no discrete levels (see §45), and so $n = 0$. We now consider the variation of the difference $\delta_l(\Delta) - \delta_l(\infty)$, where Δ is some given small quantity, as the potential well $U(r)$ gradually becomes deeper. As this occurs, the first, second etc. levels successively appear at the top of the well, and the phases $\delta_l(\Delta)$ are increased by π each time.† On reaching the given $U(r)$ and then making $\Delta \to 0$, we obtain formula (133.17).

PROBLEMS

PROBLEM 1. Express the effective interaction range r_0 in terms of the wave function of the stationary state $E = \epsilon$ in the "inner" region $r \sim a$ (Ya. A. Smorodinskii 1948).

SOLUTION. Let χ_0 be the wave function in the region $r \sim a$, normalized by the condition that $\chi_0 \to 1$ as $r \to \infty$. Then the square of the wave function can be written in all space in the form $\chi^2 = A_0^2(e^{-2\kappa r} + \chi_0^2 - 1)$; this expression becomes $A_0^2 e^{-2\kappa r}$ for $\kappa r \gg 1$ and $A_0^2\chi_0^2$ for $\kappa r \ll 1$. It must be normalized by the condition

$$\int_0^\infty \chi^2\, dr = A_0^2\left(\frac{1}{2\kappa} - \int_0^\infty (1-\chi_0^2)\, dr\right) = 1,$$

† In formula (133.6) this corresponds to a change of δ_0 from 0 to π when, for a given small value of k, the quantity κ changes from a negative value ($-\kappa \gg k$) to a positive value $\kappa \gg k$. When $l \neq 0$, the same follows from the formula $k \cot \delta_l = -bE^{-l}(E-\epsilon)$ when, for a given $E = \Delta$, ϵ varies from $\epsilon \gg \Delta$ to $-\epsilon \gg \Delta$.

and a comparison with (133.13) gives

$$r_0 = 2 \int_0^\infty (1-\chi_0^2)\, dr.$$

From equation (133.1) with $U(r) < 0$, the solution of which is χ_0, it follows that $\chi_0(r) < \chi_0(\infty) = 1$. Hence we always have $r_0 > 0$.

PROBLEM 2. Determine the change in the phases δ_l when the field $U(r)$ is varied.

SOLUTION. Varying $U(r)$ in Schrödinger's equation

$$\chi_l'' + \frac{2m}{\hbar^2}\left[E - \frac{l(l+1)}{r^2} - U\right]\chi_l = 0,$$

we obtain

$$\delta\chi_l'' + \frac{2m}{\hbar^2}\left[E - \frac{l(l+1)}{r^2} - U\right]\delta\chi_l = \frac{2m}{\hbar^2}\chi_l \delta U.$$

Multiplying the first equation by $\delta\chi_l$, the second by χ_l, subtracting, and integrating with respect to r, we find

$$[\chi_l \delta\chi_l' - \chi_l' \delta\chi_l]_{r\to\infty} = \frac{2m}{\hbar^2}\int_0^\infty \chi_l^2 \delta U\, dr.$$

Substituting on the left-hand side the asymptotic expressions

$$\chi_l = \sin(kr - \tfrac{1}{2}l\pi - \delta_l),$$

$$\delta\chi_l = \delta(\delta_l)\cos(kr - \tfrac{1}{2}l\pi + \delta_l)$$

(the choice of the coefficient 1 in this expression determining the normalization used), we obtain

$$\delta(\delta_l) = -\frac{2m}{k\hbar^2}\int_0^\infty \chi_l^2 \delta U\, dr.$$

From this formula we can draw certain conclusions regarding the sign of the phases δ_l, considered as continuous functions of energy. To avoid the ambiguity in the definition of these functions (an additive multiple of π) we shall normalize them by the condition $\delta_l(\infty) = 0$.

Starting from $U = 0$, when all the δ_l are zero, and gradually increasing $|U|$, we find that in a repulsive field ($U > 0$) all the $\delta_l < 0$, and in an attractive field ($U < 0$) $\delta_l > 0$. In a repulsive field $\delta_l(0) = 0$ and therefore, for small energies, the δ_l are small; the scattering amplitude is therefore negative: $f \approx \delta_0/k < 0$. In an attractive field the corresponding deduction that f is positive can be made only if there are no discrete levels. Otherwise, when E is small, the phases δ_l are close to $n\pi$, not to zero (see (133.17)), and no conclusion can be drawn concerning the sign of f.

PROBLEM 3. Find the scattering length α and the effective range of interaction r_0 for a spherical square potential well of radius a and depth U_0 containing a single discrete energy level near zero.

SOLUTION. We proceed as in §132, Problem 1, except that in the region within the well we do not neglect the particle energy $E = \hbar^2 k^2/2m$ in comparison with U_0. The equation to determine the phase δ_0 is found to be

$$k \cot(\delta_0 + ak) = K \cot aK, \quad K = \frac{1}{\hbar}\sqrt{[2m(U_0+E)]}.$$

In order that the well should contain only one level, close to zero, it is necessary that

$$U_0 = (\pi^2\hbar^2/8ma^2)(1+\Delta)$$

with $\Delta \ll 1$; see §33, Problem 1. Expanding the above equation in powers of ka and Δ, we find that

$$k \cot \delta_0 \approx -\frac{\pi^2}{8a}\Delta + \tfrac{1}{2}ak^2,$$

whence $\alpha = 1/\kappa_0 = 8a/\pi^2\Delta$, $r_0 = a$. The value of κ_0 coincides, as it should, with that of $\sqrt{(2m|E_1|)}/\hbar$, where E_1 is the energy of the level in the well; see §33, Problem 1.

PROBLEM 4. Express the integral

$$\int_0^a \chi^2 \, dr$$

of the squared wave function of the s state in terms of the phase $\delta_0(k)$ for a field $U(r)$ that is zero outside a sphere of radius a (G. Lüders 1955).

SOLUTION. According to (128.10),

$$\int_0^a \chi^2 \, dr = \frac{1}{2k}\left[\chi' \frac{\partial \chi}{\partial k} - \chi \left(\frac{\partial \chi}{\partial k}\right)'\right]_{r=a},$$

where the prime denotes differentiation with respect to r (and the derivatives with respect to E in (128.10) are replaced by those with respect to $k = \sqrt{(2mE)}/\hbar$). Since, at $r = a$, there is no field, we can use on the right-hand side the wave function of free motion, $\chi = 2 \sin (kr + \delta_0)$ (normalized as in (33.20)). The result is

$$\int_0^a \chi^2 \, dr = 2\left(a + \frac{d\delta_0}{dk}\right) - \frac{1}{k} \sin 2(ka + \delta_0) > 0.$$

Since the integral of χ^2 is certainly positive, the expression on the right must also be positive.†

§134. Resonance at a quasi-discrete level

A system which can disintegrate does not, strictly speaking, have a discrete energy spectrum. The particle leaving it when it disintegrates recedes to infinity; in this sense, the motion of the system is infinite, and hence the energy spectrum is continuous.

It may happen, however, that the disintegration probability of the system is very small. The simplest example of this kind is given by a particle surrounded by a fairly high and wide potential barrier. Another possible reason for metastability of a state is that the spin of the system must change in a disintegration due to a weak spin-orbit interaction.

For such systems with a small disintegration probability, we can introduce the concept of *quasi-stationary* states, in which the particles move "inside the system" for a considerable period of time, leaving it only when a fairly long time interval τ has elapsed; τ may be called the *lifetime* of the almost stationary state concerned ($\tau \sim 1/w$, where w is the disintegration probability per unit time). The energy spectrum of these states will be *quasi-discrete*; it consists of a series of broadened levels, whose *width* is related to the lifetime by $\Gamma \sim \hbar/\tau$ (see (44.7)). The widths of the quasi-discrete levels are small compared with the distances between them.

In discussing the quasi-stationary states, we can use the following formal method. Until now we have always considered solutions of Schrödinger's equation with a boundary condition requiring the finiteness of the wave

† This inequality had previously been derived in a different manner by Wigner (1955).

function at infinity. Instead of this, we shall now look for solutions which represent an outgoing spherical wave at infinity; this corresponds to the particle finally leaving the system when it disintegrates. Since such a boundary condition is complex, we cannot assert that the eigenvalues of the energy must be real. On the contrary, by solving Schrödinger's equation, we obtain a set of complex values, which we write in the form

$$E = E_0 - \tfrac{1}{2}i\Gamma, \tag{134.1}$$

where E_0 and Γ are two constants, which are positive (see below).

It is easy to see the physical significance of the complex energy values. The time factor in the wave function of a quasi-stationary state is of the form

$$e^{-(i/\hbar)Et} = e^{-(i/\hbar)E_0 t} e^{-(\Gamma/\hbar)t/2}.$$

Hence all the probabilities given by the squared modulus of the wave function decrease with time as $e^{-(\Gamma/\hbar)t}$.† In particular, the probability of finding the particle "inside the system" decreases according to this law. Thus Γ determines the lifetime of the state; the disintegration probability per unit time is

$$w = \Gamma/\hbar. \tag{134.2}$$

At large distances the wave function of the quasi-stationary state (the outgoing wave) contains the factor

$$\exp[ir\sqrt{\{2m(E_0 - \tfrac{1}{2}i\Gamma)\}}/\hbar],$$

which increases exponentially as $r \to \infty$ (the imaginary part of the root is negative). Hence the normalization integral $\int |\psi|^2 \, dV$ for these functions diverges. It may be noted, incidentally, that this resolves the apparent contradiction between the decrease with time of $|\psi|^2$ and the fact that the normalization integral can be shown from the wave equation to be a constant.

Let us ascertain the form of the wave function which describes the motion of a particle with energy close to one of the quasi-discrete levels of the system.

As in §128, we write down the asymptotic form (at large distances) of the

---

† We may note that this shows the physical necessity for Γ to be positive, a condition which is automatically satisfied on account of the boundary condition imposed at infinity on the solution of the wave equation, or by the equivalent (see §130) rule of passage round poles in the formulae of perturbation theory. Let transitions from the discrete level n to the states ν of the continuous spectrum be caused by a constant perturbation V. Then the second-order correction to the energy level is

$$E_n^{(2)} = \int \frac{|V_{n\nu}|^2 \, d\nu}{E_n^{(0)} - E_\nu + i0};$$

cf. (33.10). The rule (43.10) gives

$$\Gamma = -2 \operatorname{im} E_n^{(2)} = 2\pi \int |V_{n\nu}|^2 \, \delta(E_n^{(0)} - E_\nu) \, d\nu,$$

in agreement with (43.1) for the transition probability.

radial part of the wave function in the form (128.1):

$$R_l = \frac{1}{r}\left[A_l(E)\exp\left(-\frac{\sqrt{(-2mE)}}{\hbar}r\right)+B_l(E)\exp\left(\frac{\sqrt{(-2mE)}}{\hbar}r\right)\right], \quad (134.3)$$

and regard E as a complex variable. For real positive E,

$$R_l = \frac{1}{r}[A_l(E)e^{ikr}+B_l(E)e^{-ikr}], \qquad k = \sqrt{(2mE)}/\hbar, \quad (134.4)$$

and $A_l(E) = B_l^*(E)$ (see (128.3), (128.4)); the function $B_l(E)$ is here taken on the upper edge of a cut along the right half of the real axis.

The condition which determines the complex eigenvalues of the energy consists in the absence of an ingoing wave from the asymptotic expression (134.3). This means that for $E = E_0 - \frac{1}{2}i\Gamma$ the coefficient $B_l(E)$ must vanish:

$$B_l(E_0-\tfrac{1}{2}i\Gamma) = 0. \quad (134.5)$$

Thus the quasi-discrete energy levels, like the true discrete levels, are zeros of the function $B_l(E)$. However, unlike the zeros which correspond to true levels, they do not lie on the physical sheet: in writing the condition (134.5) we have assumed that the required wave function of the quasi-stationary state arises from the same term in (134.3), which is an outgoing wave ($\sim e^{ikr}$) when $E > 0$ also (in (134.4)). But the point $E = E_0 - \frac{1}{2}i\Gamma$ lies below the positive real axis. This point can be reached from the upper edge of the cut (where the coefficients in (134.4) are defined), without leaving the physical sheet, only by passing round the point $E = 0$. Then $\sqrt{-E}$ changes sign, so that the outgoing wave becomes an ingoing one. Consequently, to preserve the outgoing wave the point must be reached by going directly down through the cut, on to another, non-physical, sheet.

Let us now consider real positive energy values close to the quasi-discrete level (assuming, of course, that Γ is small, since otherwise no such close values could exist). Expanding the function $B_l(E)$ in powers of the difference $E-(E_0-\frac{1}{2}i\Gamma)$ and taking only the first-order term, we have

$$B_l(E) = (E-E_0+\tfrac{1}{2}i\Gamma)b_l, \quad (134.6)$$

where b_l is a constant. Substituting in (134.4), we obtain the following expression for the wave function of a state close to the quasi-stationary state:

$$R_l = \frac{1}{r}[(E-E_0-\tfrac{1}{2}i\Gamma)b_l^*e^{ikr}+(E-E_0+\tfrac{1}{2}i\Gamma)b_l e^{-ikr}]. \quad (134.7)$$

The phase δ_l of this function is given by

$$\begin{aligned} e^{2i\delta_l} &= e^{2i\delta_l^{(0)}}\frac{E-E_0-\frac{1}{2}i\Gamma}{E-E_0+\frac{1}{2}i\Gamma} \\ &= e^{2i\delta_l^{(0)}}\left(1-\frac{i\Gamma}{E-E_0+\frac{1}{2}i\Gamma}\right), \end{aligned} \quad (134.8)$$

where

$$e^{2i\delta_l^{(0)}} = (-1)^{l+1} b_l^*/b_l. \tag{134.9}$$

For $|E-E_0| \gg \Gamma$, the phase δ_l is equal to $\delta_l^{(0)}$, so that $\delta_l^{(0)}$ is the value of the phase far from the resonance.

In the resonance region δ_l varies considerably with energy. If we rewrite formula (134.8), using the result

$$e^{2i\tan^{-1}\lambda} = \frac{e^{i\tan^{-1}\lambda}}{e^{-i\tan^{-1}\lambda}} = \frac{1+i\lambda}{1-i\lambda},$$

in the form

$$\delta_l = \delta_l^{(0)} + \tan^{-1}\frac{\Gamma}{2(E-E_0)}, \tag{134.10}$$

we see that the phase changes by π in a passage through the whole resonance region (from $E \ll E_0$ to $E \gg E_0$).

For $E = E_0 - \frac{1}{2}i\Gamma$, the function (134.7) becomes $R_l = -(1/r)i\Gamma b_l^* e^{ikr}$. If the wave function is normalized by the condition that the integral of $|\psi|^2$ over the region within the system is unity, the total current in this outgoing wave, equal to $v|i\Gamma b_l^*|^2$, must be equal to the distintegration probability (134.2). Hence we find

$$|b_l|^2 = 1/\hbar v\Gamma. \tag{134.11}$$

These results enable us to determine the amplitude of elastic scattering of a particle with energy E close to some quasi-discrete level E_0 of the *compound system* consisting of the scattering system together with the particle undergoing scattering. In the general formula (123.11) we must substitute the expression (134.8) in the term with the value of l which corresponds to the level E_0. This gives

$$f(\theta) = f^{(0)}(\theta) - \frac{2l+1}{k}\cdot\frac{\frac{1}{2}\Gamma}{E-E_0+\frac{1}{2}i\Gamma} e^{2i\delta_l^{(0)}} P_l(\cos\theta), \tag{134.12}$$

where $f^{(0)}(\theta)$ is the scattering amplitude far from the resonance, which is independent of the properties of the quasi-stationary state (it is given by formula (123.11) with $\delta_l = \delta_l^{(0)}$ in each term of the sum).† The amplitude $f^{(0)}(\theta)$ is called the *potential scattering* amplitude, and the second term in (134.12) the *resonance scattering* amplitude. The latter has a pole at $E = E_0 - \frac{1}{2}i\Gamma$, which, as shown above, is not on the physical sheet.‡

Formula (134.12) determines the elastic scattering near resonance at one of the quasi-discrete levels of the compound system. Its range of validity

---

† If scattering of a charged particle by a system of charged particles is considered, the expression (135.11) must be used for the phases $\delta_l^{(0)}$.

‡ It may be noted that formula (133.15) for resonance scattering of slow particles by a positive energy level ϵ with $l \neq 0$, with E close to ϵ, is in exact correspondence with the resonance term in (134.12). The values of E_0 and Γ are given by formulae (133.16), and since E is small the phase $\delta_l^{(0)}$ is small, so that $e^{2i\delta_l^{(0)}} \approx 1$.

is defined by the requirement that the difference $|E-E_0|$ should be small compared with the distance D to the adjoining quasi-discrete levels:

$$|E-E_0| \ll D. \tag{134.13}$$

This formula is somewhat simplified if the scattering of slow particles is being considered, i.e. if the wavelength of the particles in the resonance region is large compared with the dimensions of the scattering system. Here only s-wave scattering is important; we shall suppose that the level E_0 does in fact belong to motion with $l = 0$. The potential scattering amplitude then reduces to a real constant $-\alpha$ (see §132).† In the resonance scattering amplitude we put $l = 0$ and replace $e^{2i\delta_0^{(0)}}$ by unity, since $\delta_0^{(0)} = -\alpha k \ll 1$. Thus we find

$$f(\theta) = -\alpha - \frac{\frac{1}{2}\Gamma}{k(E-E_0+\frac{1}{2}i\Gamma)}. \tag{134.14}$$

In a narrow range $|E-E_0| \sim \Gamma$ the second term is large compared with the amplitude α, and the latter may be omitted. Farther from the resonance, however, the two terms may be comparable.

In the above derivations it has been tacitly assumed that the value E_0 of the level itself is not too small, and that the resonance region is not in the neighbourhood of the point $E = 0$. If resonance at the first quasi-discrete level of the compound system is considered, which lies at a distance from $E = 0$ small compared with the distance to the next level ($E_0 \ll D$), the expansion (134.6) may be no longer permissible. This is seen from the fact that the amplitude (134.14) does not tend to a constant limit as $E \to 0$, as would be necessary for s-wave scattering according to the general theory.

Let us consider the case of a quasi-discrete level close to zero, again assuming that in the resonance region the particles undergoing scattering are so slow that only s-wave scattering is of importance.

The expansion of the coefficients $B_l(E)$ in the wave function must now be made in powers of the energy E itself. The point $E = 0$ is a branch point of the functions $B_l(E)$, and a passage round this point from the upper to the lower edge of the cut changes $B_l(E)$ into $B_l^*(E)$. This means that the expansion is in powers of $\sqrt{-E}$, which changes sign on the above-mentioned passage. We write the first terms in the expansion of the function $B_0(E)$ for real positive E in the form

$$B_0(E) = (E-\epsilon_0+i\gamma\sqrt{E})b_0(E), \tag{134.15}$$

where ϵ_0 and γ are real constants, and $b_0(E)$ a function of energy, which can also be expanded in powers of $\sqrt{E}$ but has no zero near the point $E = 0$.‡ The quasi-discrete level $E = E_0-\frac{1}{2}i\Gamma$ corresponds to the vanishing of the

---

† It is assumed that the scattering field decreases sufficiently rapidly with increasing distance. In §145 the results given here will be applied to the scattering of slow neutrons by nuclei.

‡ The function $b_0(E)$ determines, according to (134.9), the phase of the potential scattering. In the scattering of slow particles, the first terms in its expansion are $b_0(E) = \text{constant} \times i(1+i\alpha k)$.

factor $E-\epsilon_0+i\gamma\sqrt{E}$, continued into the lower half-plane of the non-physical sheet; we therefore have for the determination of E_0 and Γ the equation

$$E_0-\tfrac{1}{2}i\Gamma-\epsilon_0+i\gamma\sqrt{(E_0-\tfrac{1}{2}i\Gamma)} = 0 \tag{134.16}$$

(the constants ϵ_0 and γ must be positive in order that E_0 and Γ should be positive). For example, a level with width $\Gamma \ll E_0$ corresponds to the relation $\epsilon_0 \gg \gamma^2$ between these constants, and from (134.16) we have $E_0 = \epsilon_0$, $\Gamma = 2\gamma\sqrt{\epsilon_0}$.

The expression (134.15) replaces in this case formula (134.6); the subsequent formulae must be correspondingly modified (everywhere replacing E_0 by ϵ_0 and Γ by $2\gamma\sqrt{E}$). Hence we obtain for the scattering amplitude, instead of (134.14), the expression

$$f = -\alpha - \frac{\hbar\gamma}{\sqrt{(2m)}(E-\epsilon_0+i\gamma\sqrt{E})} \tag{134.17}$$

(where we have put $k = \sqrt{(2mE)}/\hbar$, m being the reduced mass of the particle and the scattering system). For $E \to 0$ this amplitude tends to a constant limit, as it should, thus confirming the form of the expansion (134.15).

It may be noted that the expression (134.17) also covers the case of a true discrete level of the compound system close to zero, which is given by an appropriate relation between the constants ϵ_0 and γ. If $|\epsilon_0| \ll \gamma^2$, the first term E may be neglected in the denominator of the resonance term for energies $E \ll \gamma^2$.

Neglecting also the potential scattering amplitude α, we obtain the formula

$$f = -\frac{1}{ik-\sqrt{(2m)}\epsilon_0/\hbar\gamma},$$

which is the same as formula (133.7) (with $\kappa = -\sqrt{(2m)}\epsilon_0/\hbar\gamma$). This corresponds to resonance at the level $E = \epsilon_0^2/\gamma^2$, which is a true or virtual discrete level according as the constant κ is positive or negative.

§135. Rutherford's formula

Scattering in a Coulomb field is of interest from the point of view of physical applications. It is also of interest in that, for this case, the quantum-mechanical collision problem can be solved exactly.

When there is a direction (in this case, the direction of incidence of the particle) which can be distinguished from the remainder, Schrödinger's equation in the Coulomb field is conveniently solved in parabolic coordinates ξ, η, ϕ (§37). The problem of the scattering of a particle in a central field is axially symmetric. Hence the wave function ψ is independent of the angle ϕ. We write the particular solution of Schrödinger's equation (37.6) in the form

$$\psi = f_1(\xi)f_2(\eta); \tag{135.1}$$

this is (37.7) with $m = 0$. Accordingly, after separating the variables, we obtain† equations (37.8) with $m = 0$:

$$\left.\begin{aligned}\frac{\mathrm{d}}{\mathrm{d}\xi}\left(\xi\frac{\mathrm{d}f_1}{\mathrm{d}\xi}\right)+(\tfrac{1}{4}k^2\xi-\beta_1)f_1 = 0,\\ \frac{\mathrm{d}}{\mathrm{d}\eta}\left(\eta\frac{\mathrm{d}f_2}{\mathrm{d}\eta}\right)+(\tfrac{1}{4}k^2\eta-\beta_2)f_2 = 0, \quad \beta_1+\beta_2 = 1.\end{aligned}\right\} \tag{135.2}$$

The energy of the particle scattered is, of course, positive; we have put $E = \frac{1}{2}k^2$. The signs in equations (135.2) are for the case of a repulsive field; exactly the same final result is obtained for the scattering cross-section in an attractive field.

We have to find that solution of Schrödinger's equation which, for negative z and large r, has the form of a plane wave:

$$\psi \sim e^{ikz} \quad \text{for } -\infty < z < 0, \quad r \to \infty,$$

corresponding to a particle incident in the positive direction of the z-axis. We shall see from what follows that the condition imposed can be satisfied by a single particular integral (135.1); a sum of integrals with various values of β_1, β_2 is not needed.

In parabolic coordinates, this condition takes the form

$$\psi \sim e^{ik(\xi-\eta)/2} \quad \text{for } \eta \to \infty \text{ and all } \xi.$$

This can be satisfied only if

$$f_1(\xi) = e^{ik\xi/2} \tag{135.3}$$

and $f_2(\eta)$ is subject to the condition

$$f_2(\eta) \sim e^{-ik\eta/2} \quad \text{for } \eta \to \infty. \tag{135.4}$$

Substituting (135.3) in the first of equations (135.2), we see that this function does in fact satisfy the equation, provided that the constant $\beta_1 = \frac{1}{2}ik$. The second equation (135.2), with $\beta_2 = 1-\beta_1$, then takes the form

$$\frac{\mathrm{d}}{\mathrm{d}\eta}\left(\eta\frac{\mathrm{d}f_2}{\mathrm{d}\eta}\right)+(\tfrac{1}{4}k^2\eta-1+\tfrac{1}{2}ik)f_2 = 0.$$

Let us seek its solution in the form

$$f_2(\eta) = e^{-ik\eta/2}w(\eta), \tag{135.5}$$

where the function $w(\eta)$ tends to a constant as $\eta \to \infty$. For $w(\eta)$ we have the equation

$$\eta w''+(1-ik\eta)w'-w = 0, \tag{135.6}$$

† In this section we use Coulomb units (see §36).

which, by introducing the new variable $\eta_1 = ik\eta$, can be reduced to the equation for a confluent hypergeometric function with parameters $\alpha = -i/k$, $\gamma = 1$. We have to choose that solution of equation (135.6) which, on being multiplied by $f_1(\xi)$, contains only an outgoing (i.e. scattered) and not an ingoing spherical wave. This solution is the function

$$w = \text{constant} \times F(-i/k, 1, ik\eta).$$

Thus, on assembling the expressions obtained, we find the following exact solution of Schrödinger's equation, describing the scattering:

$$\psi = e^{-\pi/2k}\Gamma(1+i/k)e^{ik(\xi-\eta)/2}F(-i/k, 1, ik\eta). \tag{135.7}$$

We have chosen the normalization constant in ψ such that the incident plane wave has unit amplitude (see below).

In order to separate the incident and scattered waves in this function, we must consider its form at large distances from the scattering centre. Using the first two terms of the asymptotic expansion (formula (d.14)) for the confluent hypergeometric function, we have for large η

$$F(-i/k, 1, ik\eta) \approx \frac{(-ik\eta)^{i/k}}{\Gamma(1+i/k)}\left(1+\frac{1}{ik^3\eta}\right)+\frac{(ik\eta)^{-i/k}}{\Gamma(-i/k)}\frac{e^{ik\eta}}{ik\eta}$$

$$= \frac{e^{\pi/2k}}{\Gamma(1+i/k)}\left(1+\frac{1}{ik^3\eta}\right)e^{(i/k)\log(k\eta)}-\frac{(i/k)e^{\pi/2k}}{\Gamma(1-i/k)}\frac{e^{ik\eta}}{ik\eta}e^{-(i/k)\log(k\eta)}.$$

Substituting this in (135.7) and changing to spherical polar coordinates ($\xi-\eta = 2z$, $\eta = r-z = r(1-\cos\theta)$), we have the following final asymptotic expression for the wave function:

$$\psi = \left[1+\frac{1}{ik^3r(1-\cos\theta)}\right]e^{ikz+(i/k)\log(kr-kr\cos\theta)}+\frac{f(\theta)}{r}e^{ikr-(i/k)\log(2kr)}, \tag{135.8}$$

where

$$f(\theta) = -\frac{1}{2k^2\sin^2\frac{1}{2}\theta}e^{-(2i/k)\log\sin\theta/2}\frac{\Gamma(1+i/k)}{\Gamma(1-i/k)}. \tag{135.9}$$

The first term in (135.8) represents the incident wave. We see that, in consequence of the slow decrease of the Coulomb field, the plane wave is distorted even at large distances from the centre, as is shown by the presence of the logarithmic term in the phase and of the $1/r$ term in the amplitude.† The distorting logarithmic term in the phase is found also in the scattered

† The origin of this distortion may be elucidated classically. If we consider a family of classical Coulomb hyperbolic paths with the same direction of incidence (parallel to the z-axis), the equation of the surface normal to them at large distances from the scattering centre ($z \to -\infty$) is easily shown to tend to $z+k^{-2}\log k(r-z) = \text{constant}$, not $z = \text{constant}$. This is the surface of constant phase of the incident wave in (135.8).

spherical wave given by the second term in (135.8). These differences from the usual asymptotic form of the wave function (123.3) are unimportant, however, since they give a correction to the current density which tends to zero as $r \to \infty$.

Thus we obtain for the scattering cross-section $d\sigma = |f(\theta)|^2\, do$ the formula

$$d\sigma = do/4k^4 \sin^4\tfrac{1}{2}\theta,$$

or, in ordinary units,

$$d\sigma = (\alpha/2mv^2)^2\, do/\sin^4\tfrac{1}{2}\theta, \qquad (135.10)$$

where the velocity v of the particle $= k\hbar/m$. This is the familiar *Rutherford's formula* given by classical mechanics. Thus, for scattering in a Coulomb field, quantum and classical mechanics give the same result (N. Mott, and W. Gordon, 1928). Born's formula (126.12) naturally leads to the same expression (135.10) also.

We shall give for reference the expression for the scattering amplitude (135.9), written as a sum of spherical harmonics. This is obtained by substituting in (124.5) the phases from (36.28), i.e.†

$$\exp(2i\delta_{l\,\mathrm{Coul}}) = \Gamma(l+1+i/k)/\Gamma(l+1-i/k). \qquad (135.11)$$

Thus we find

$$f(\theta) = \frac{1}{2ik}\sum_l (2l+1)\frac{\Gamma(l+1+i/k)}{\Gamma(l+1-i/k)}P_l(\cos\theta). \qquad (135.12)$$

The signs in the scattering amplitude (135.9) correspond to a repulsive field. In an attractive Coulomb field, formula (135.9) is replaced by the complex conjugate expression. $f(\theta)$ then becomes infinite at the poles of the function $\Gamma(1-i/k)$, i.e. at points where the argument of the gamma function is a negative integer or zero (when im $k>0$ and the function $r\psi$ decreases at infinity). The corresponding energy values are $\frac{1}{2}k^2 = -1/2n^2$ ($n = 1, 2, 3, \ldots$), and coincide with the discrete energy levels in the Coulomb field (cf. §128).

§136. The system of wave functions of the continuous spectrum

In the analysis of motion in a centrally symmetric field (Chapter V) we have considered stationary states in which the particle has definite values of the energy, the orbital angular momentum l, and the component m of this angular momentum. The wave functions of such states of the discrete spectrum (ψ_{nlm}) and the continuous spectrum (ψ_{klm}, energy $\hbar^2k^2/2m$) together form a complete set in terms of which the wave function of any state

† The value of $\delta_{l,\mathrm{Coul}}$ in this formula differs from the true (divergent) Coulomb phase by a quantity which is the same for all l.

may be expanded. Such a set of functions is, however, not appropriate for problems in scattering theory. Here another set is convenient, in which the wave functions of the continuous spectrum are described by a particular asymptotic behaviour: at infinity there is a plane wave and an outgoing spherical wave. In these states the particle has a definite energy, but no definite angular momentum magnitude or component.

According to (123.6) and (123.7), such wave functions, here denoted by $\psi_{\mathbf{k}}^{(+)}$, are given by

$$\psi_{\mathbf{k}}^{(+)} = \frac{1}{2k}\sum_{l=0}^{\infty} i^l(2l+1)e^{i\delta_l} R_{kl}(r)P_l(\mathbf{k}\cdot\mathbf{r}/kr). \tag{136.1}$$

The argument of the Legendre polynomials is written as $\cos\theta = \mathbf{k}\cdot\mathbf{r}/kr$, and the expression therefore does not involve any particular choice of the coordinate axes as it did in (123.6) (where the z-axis was the direction of propagation of the plane wave). By giving the vector $\mathbf{k}$ all possible values, we obtain a set of wave functions, which, as we shall now show, are orthogonal and normalized by the usual rule for the continuous spectrum:

$$\int \psi_{\mathbf{k}'}^{(+)*}\psi_{\mathbf{k}}^{(+)}\,dV = (2\pi)^3\,\delta(\mathbf{k}'-\mathbf{k}). \tag{136.2}$$

To prove this,† we note that the product $\psi_{\mathbf{k}'}^{(+)*}\psi_{\mathbf{k}}^{(+)}$ is expressed by a double sum over l and l' of terms containing the products

$$P_l(\mathbf{k}\cdot\mathbf{r}/kr)P_{l'}(\mathbf{k}'\cdot\mathbf{r}/k'r).$$

The integration over the directions of $\mathbf{r}$ is effected by means of the formula

$$\int P_l(\mathbf{k}\cdot\mathbf{r}/kr)P_{l'}(\mathbf{k}'\cdot\mathbf{r}/k'r)do = \delta_{ll'}\frac{4\pi}{2l+1}P_l(\mathbf{k}\cdot\mathbf{k}'/kk'); \tag{136.3}$$

cf. (c.12) in the Mathematical Appendices. This leaves

$$\int \psi_{\mathbf{k}'}^{(+)*}\psi_{\mathbf{k}}^{(+)}\,dV = \frac{\pi}{kk'}\sum_{l=0}^{\infty}(2l+1)e^{i[\delta_l(k)-\delta_l(k')]}P_l(\cos\gamma)\int_0^{\infty} R_{k'l}(r)R_{kl}(r)r^2\,dr,$$

where γ is the angle between $\mathbf{k}$ and $\mathbf{k}'$. The radial functions R_{kl} are orthogonal, however, and are normalized by

$$\int_0^{\infty} R_{k'l}R_{kl}\,r^2\,dr = 2\pi\delta(k'-k).$$

† Essentially, only the orthogonality of the $\psi_{\mathbf{k}}^{(+)}$ needs to be proved separately; the normalization could be derived directly from the asymptotic form of the functions (cf. §21). In this sense, the validity of (136.2) is evident from the fact that, as $r\to\infty$, the only non-decreasing term in these functions is $\psi_{\mathbf{k}}^{(+)}\approx e^{i\mathbf{k}\cdot\mathbf{r}}$.

Hence we can put $k = k'$ in the coefficients in front of the integrals; using also the relation (124.3), we have

$$\int \psi_{\mathbf{k}'}^{(+)*}\psi_{\mathbf{k}}^{(+)}\,\mathrm{d}V = \frac{2\pi^2}{k^2}\delta(k'-k)\sum_{l=0}^{\infty}(2l+1)P_l(\cos\gamma)$$

$$= \frac{8\pi^2}{k^2}\delta(k'-k)\delta(1-\cos\gamma).$$

The expression on the right is zero for $\mathbf{k} \neq \mathbf{k}'$; on being multiplied by $2\pi k^2 \sin\gamma\, \mathrm{d}k\, \mathrm{d}\gamma/(2\pi)^3$ and integrated over all **k**-space it gives 1, and this proves formula (136.2).

Together with the system of functions $\psi_{\mathbf{k}}^{(+)}$, we can also introduce a system corresponding to states in which there are at infinity a plane wave and an ingoing spherical wave. These functions, which we denote by $\psi_{\mathbf{k}}^{(-)}$, are obtained directly from the $\psi_{\mathbf{k}}^{(+)}$:

$$\psi_{\mathbf{k}}^{(-)} = \psi_{-\mathbf{k}}^{(+)*}, \tag{136.4}$$

since the complex conjugate of e^{ikr}/r (outgoing wave) is e^{-ikr}/r (ingoing wave), and the plane wave becomes $e^{-i\mathbf{k}.\mathbf{r}}$, so that, in order to retain the previous definition of **k** (plane wave $e^{i\mathbf{k}.\mathbf{r}}$), we must replace **k** by $-\mathbf{k}$, as in (136.4). Noticing that $P_l(-\cos\theta) = (-1)^l P_l(\cos\theta)$, we obtain from (136.1)

$$\psi_{\mathbf{k}}^{(-)} = \frac{1}{2k}\sum_{l=0}^{\infty} i^l(2l+1)e^{-i\delta_l}R_{kl}(r)P_l(\mathbf{k}\,.\,\mathbf{r}/kr). \tag{136.5}$$

The case of a Coulomb field is of great importance. Here the functions $\psi_{\mathbf{k}}^{(+)}$ (and $\psi_{\mathbf{k}}^{(-)}$) can be written in a closed form, which is obtained directly from formula (135.7). We express the parabolic coordinates by

$$\tfrac{1}{2}k(\xi-\eta) = kz = \mathbf{k}\,.\,\mathbf{r}, \qquad k\eta = k(r-z) = kr-\mathbf{k}\,.\,\mathbf{r}.$$

Thus we obtain for a repulsive Coulomb field†

$$\psi_{\mathbf{k}}^{(+)} = e^{-\pi/2k}\Gamma(1+i/k)e^{i\mathbf{k}.\mathbf{r}}F(-i/k, 1, ikr-i\mathbf{k}\,.\,\mathbf{r}), \tag{136.6}$$

$$\psi_{\mathbf{k}}^{(-)} = e^{-\pi/2k}\Gamma(1-i/k)e^{i\mathbf{k}.\mathbf{r}}F(i/k, 1, -ikr-i\mathbf{k}\,.\,\mathbf{r}). \tag{136.7}$$

† Using Coulomb units.

The wave functions for an attractive Coulomb field are found by simultaneously changing the signs of k and r:

$$\psi_{\mathbf{k}}^{(+)} = e^{\pi/2k}\Gamma(1-i/k)e^{i\mathbf{k}.\mathbf{r}}F(i/k, 1, ikr-i\mathbf{k}.\mathbf{r}), \tag{136.8}$$

$$\psi_{\mathbf{k}}^{(-)} = e^{\pi/2k}\Gamma(1+i/k)e^{i\mathbf{k}.\mathbf{r}}F(-i/k, 1, -ikr-i\mathbf{k}.\mathbf{r}). \tag{136.9}$$

The action of the Coulomb field on the motion of the particle near the origin may be characterized by the ratio of the squared modulus of $\psi_{\mathbf{k}}^{(+)}$ or $\psi_{\mathbf{k}}^{(-)}$ at the point $r = 0$ to the squared modulus of the wave function $\psi_{\mathbf{k}} = e^{i\mathbf{k}.\mathbf{r}}$ for free motion. Using the formula

$$\begin{aligned}\Gamma(1+i/k)\,\Gamma(1-i/k) &= (i/k)\,\Gamma(i/k)\,\Gamma(1-i/k)\\ &= \pi/k\sinh(\pi/k),\end{aligned}$$

we easily find, for a repulsive field,

$$\frac{|\psi_{\mathbf{k}}^{(+)}(0)|^2}{|\psi_{\mathbf{k}}|^2} = \frac{|\psi_{\mathbf{k}}^{(-)}(0)|^2}{|\psi_{\mathbf{k}}|^2} = \frac{2\pi}{k(e^{2\pi/k}-1)}, \tag{136.10}$$

and for an attractive field,

$$\frac{|\psi_{\mathbf{k}}^{(+)}(0)|^2}{|\psi_{\mathbf{k}}|^2} = \frac{|\psi_{\mathbf{k}}^{(-)}(0)|^2}{|\psi_{\mathbf{k}}|^2} = \frac{2\pi}{k(1-e^{-2\pi/k})}. \tag{136.11}$$

The functions $\psi_{\mathbf{k}}^{(+)}$ and $\psi_{\mathbf{k}}^{(-)}$ play an important part in problems relating to the application of perturbation theory in the continuous spectrum. Let us suppose that, as a result of some perturbation $\hat{V}$, the particle makes a transition between states of the continuous spectrum. The transition probability is determined by the matrix element

$$\int \psi_f{}^* \hat{V} \psi_i \, dV. \tag{136.12}$$

The question arises of which solutions of the wave equation are to be taken as the initial (ψ_i) and final (ψ_f) wave functions, in order to obtain the amplitude for a transition of the particle from a state with momentum $\hbar\mathbf{k}$ to one with momentum $\hbar\mathbf{k}'$ at infinity.† We shall show that this requires that

$$\psi_i = \psi_{\mathbf{k}}^{(+)}, \quad \psi_f = \psi_{\mathbf{k}}^{(-)} \tag{136.13}$$

(A. Sommerfeld 1931).

† An example of such a process is an electron colliding with a heavy nucleus at rest and emitting a photon, thereby changing its energy and its direction of motion; the perturbation $\hat{V}$ is the interaction between the electron and the radiation field, and the Coulomb field of the nucleus is the field U for which the functions $\psi_{\mathbf{k}}^{(+)}$ and $\psi_{\mathbf{k}}^{(-)}$ are defined (see *RQT*, §§90 and 93). Another example is a collision of an electron with an atom, accompanied by ionization of the latter; see §148, Problem 4.

This becomes clear if we consider how the problem would be solved by perturbation theory applied not only as regards the perturbation $\hat{V}$ but also as regards the field $U(r)$ in which the particle is moving. In the zero-order approximation (with respect to U), the matrix element (136.12) is

$$V_{\mathbf{k}'\mathbf{k}} = \int e^{-i\mathbf{k}'.\mathbf{r}}\hat{V}e^{i\mathbf{k}.\mathbf{r}}\,dV.$$

In subsequent approximations with respect to U, this integral is replaced by a series of which each term is an integral

$$\int \frac{V_{\mathbf{k}'\mathbf{k}_1}U_{\mathbf{k}_1\mathbf{k}_2}\dots U_{\mathbf{k}_n\mathbf{k}}}{(E_{\mathbf{k}}-E_{\mathbf{k}_1}+i0)\dots(E_{\mathbf{k}}-E_{\mathbf{k}_n}+i0)}\,d^3k_1\dots d^3k_n;$$

cf. §§43 and 130. The numerator contains the matrix elements (in varying order) with respect to the unperturbed plane waves, and all poles are avoided in the integrations, according to one fixed rule. On the other hand, this series can be obtained as the matrix element (136.12) with the wave functions ψ_i and ψ_f as perturbation-theory series with respect to the field U. The fact that the result must be a sum of integrals in which all poles are avoided by the same rule means, therefore, that the poles in the terms of the series representing ψ_i and $\psi_f{}^*$ must be avoided by a similar rule. But if the wave equation is solved by perturbation theory with this avoidance rule, we necessarily obtain a solution whose asymptotic form includes an outgoing (as well as a plane) wave. In other words, the wave functions, which in the zero-order approximation (with respect to U) have the form

$$\psi_i = e^{i\mathbf{k}.\mathbf{r}}, \quad \psi_f{}^* = e^{-i\mathbf{k}'.\mathbf{r}},$$

must be replaced by exact solutions of the wave equation, respectively $\psi_{\mathbf{k}}^{(+)}$ and $\psi_{-\mathbf{k}'}^{(+)} + (\psi_{\mathbf{k}'}^{(-)})^*$. This proves the rule (136.13).

The choice of $\psi_{\mathbf{k}'}^{(-)}$ as the final wave function applies also to transitions from the discrete to the continuous spectrum; here there is, of course, no problem of choosing ψ_i.

§137. Collisions of like particles

The case where two identical particles collide requires special consideration. The identity of the particles leads in quantum mechanics to the appearance of a peculiar exchange interaction between them. This has an important effect on scattering also (N. F. Mott 1930).†

The orbital wave function of a system of two particles must be symmetric or antisymmetric with respect to the particles, according as their total spin is even or odd (see §62). The wave function which describes the scattering, and which is obtained by solving the usual Schrödinger's equation, must therefore be symmetrized or antisymmetrized with respect to the particles. An interchange of the particles is equivalent to reversing the direction of the

† Here the direct spin–orbit interaction is again ignored.

radius vector joining them. In the coordinate system in which the centre of mass is at rest, this means that r remains unchanged, while the angle θ is replaced by $\pi-\theta$ (and so $z = r\cos\theta$ becomes $-z$). Hence, instead of the asymptotic expression (123.3) for the wave function, we must write

$$\psi = e^{ikz}\pm e^{-ikz}+e^{ikr}[f(\theta)\pm f(\pi-\theta)]/r. \tag{137.1}$$

By virtue of the identity of the particles it is, of course, impossible to say which of them scatters and which is scattered. In the coordinate system in which the centre of mass is at rest, we have two equal incident plane waves, propagated in opposite directions (e^{ikz} and e^{-ikz}). The outgoing spherical wave in (137.1) takes into account the scattering of both particles, and the probability current calculated from it gives the probability that either of the particles will be scattered into the element do of solid angle considered. The scattering cross-section is the ratio of this current to the current density in either of the incident plane waves, i.e. it is given, as before, by the squared modulus of the coefficient of e^{ikr}/r in the wave function (137.1).

Thus, if the total spin of the colliding particles is even, the scattering cross-section is of the form

$$d\sigma_s = |f(\theta)+f(\pi-\theta)|^2\, do, \tag{137.2}$$

while if the total spin is odd, it is

$$d\sigma_a = |f(\theta)-f(\pi-\theta)|^2\, do. \tag{137.3}$$

The appearance of the interference term $f(\theta)f^*(\pi-\theta)+f^*(\theta)f(\pi-\theta)$ characterizes the exchange interaction. If the particles were different, as they are in classical mechanics, the probability that either of them would be scattered into a given element of solid angle do would simply be equal to the sum of the probabilities that one particle is deviated through an angle θ and the other through $\pi-\theta$; in other words, the cross-section would be

$$\{|f(\theta)|^2+|f(\pi-\theta)|^2\}\, do.$$

In the limiting case of low velocities, the scattering amplitude tends to a constant value independent of the angle (§132) if the interaction of the particles decreases sufficiently rapidly with increasing distance. It is seen from (137.3) that $d\sigma_a$ is then zero, i.e. only particles with even total spin scatter each other.

In formulae (137.2), (137.3) it is supposed that the total spin of the colliding particles has a definite value. If the particles are not in definite spin states, then to determine the cross-section it is necessary to average, assuming all possible spin states to be all equally probable. We have shown in §62 that, of the total number of $(2s+1)^2$ different spin states of a system of two particles with spin s, $s(2s+1)$ states correspond to an even total spin and $(s+1)(2s+1)$ to an odd total spin (if s is half-integral), or *vice versa* if s is integral. Let us first suppose that the spin s of the particles is half-integral. Then the probability that the system of two colliding particles will have even S is $s(2s+1)/(2s+1)^2 = s/(2s+1)$, while the probability of odd S is $(s+1)/(2s+1)$. Hence the cross-section is

$$d\sigma = \frac{s}{2s+1} d\sigma_s + \frac{s+1}{2s+1} d\sigma_a. \tag{137.4}$$

Substituting here (137.2), (137.3), we obtain

$$d\sigma = \{|f(\theta)|^2 + |f(\pi-\theta)|^2 - \frac{1}{2s+1}[f(\theta)f^*(\pi-\theta) + f^*(\theta)f(\pi-\theta)]\} \, do. \tag{137.5}$$

Similarly, we find for integral s

$$d\sigma = \{|f(\theta)|^2 + |f(\pi-\theta)|^2 + \frac{1}{2s+1}[f(\theta)f^*(\pi-\theta) + f^*(\theta)f(\pi-\theta)]\} \, do. \tag{137.6}$$

As an example, we shall write out the formulae for the collision of two electrons interacting by Coulomb's law ($U = e^2/r$). Substitution of the expression (135.9) in the formula (137.5) with $s = \frac{1}{2}$ gives (in ordinary units), after a simple calculation,

$$d\sigma = \left(\frac{e^2}{m_0 v^2}\right)^2 \left[\frac{1}{\sin^4 \frac{1}{2}\theta} + \frac{1}{\cos^4 \frac{1}{2}\theta} - \frac{1}{\sin^2 \frac{1}{2}\theta \cos^2 \frac{1}{2}\theta} \cos\left(\frac{e^2}{\hbar v} \log \tan^2 \tfrac{1}{2}\theta\right)\right] do, \tag{137.7}$$

where we have introduced the mass m_0 of the electron in place of the reduced mass $m = \frac{1}{2}m_0$. This formula is considerably simplified if the velocity is so large that $e^2 \ll v\hbar$; we notice that this is just the condition for perturbation theory to be applicable to a Coulomb field. Then the cosine in the third term can be replaced by unity, and we have

$$d\sigma = \left(\frac{2e^2}{m_0 v^2}\right)^2 \frac{4 - 3\sin^2\theta}{\sin^4\theta} \, do. \tag{137.8}$$

The opposite limiting case, $e^2 \gg v\hbar$, corresponds to the passage to the limit of classical mechanics (see the end of §127). In formula (137.7) this transition occurs in a very curious way. For $e^2 \gg v\hbar$, the cosine in the third term in the square brackets is a rapidly oscillating function. For any given θ, formula (137.7) gives for the scattering cross-section a value which in general differs considerably from the Rutherford value. However, on averaging over even a small range of values of θ, the oscillating term in (137.7) vanishes, and we obtain the classical formula.

All the above formulae for the cross-section refer to a system of co-ordinates in which the centre of mass is at rest. The transition to a system in which one of the particles is at rest before the collision is effected (according to (123.2)) simply by replacing θ by 2ϑ. Thus, for a collision of electrons we have from (137.7)

$$d\sigma = \left(\frac{2e^2}{m_0 v^2}\right)^2 \left[\frac{1}{\sin^4\vartheta} + \frac{1}{\cos^4\vartheta} - \frac{1}{\sin^2\vartheta\cos^2\vartheta}\cos\left(\frac{e^2}{\hbar v}\log\tan^2\vartheta\right)\right]\cos\vartheta\, do, \tag{137.9}$$

where do is the element of solid angle in the new system of coordinates. In replacing θ by 2ϑ, the element of solid angle do must be replaced by $4\cos\vartheta\, do$, since $\sin\theta\, d\theta d\phi = 4\cos\vartheta\sin\vartheta d\vartheta d\phi$.

PROBLEM

Determine the scattering cross-section for two identical particles of spin $\frac{1}{2}$, with given mean spin values $\bar{\mathbf{s}}_1$ and $\bar{\mathbf{s}}_2$.

SOLUTION. The dependence of the cross-section on the polarizations of the particles must be expressed by a term proportional to the scalar $\bar{\mathbf{s}}_1.\bar{\mathbf{s}}_2$. We look for $d\sigma$ in the form $a + b\bar{\mathbf{s}}_1.\bar{\mathbf{s}}_2$. For unpolarized particles ($\bar{\mathbf{s}}_1 = \bar{\mathbf{s}}_2 = 0$), the second term is absent, and according to (137.4) $d\sigma = a = \frac{1}{4}(d\sigma_s + 3d\sigma_a)$. If both particles are completely polarized in the same direction ($\bar{\mathbf{s}}_1.\bar{\mathbf{s}}_2 = \frac{1}{4}$), the system is certainly in a state with $S = 1$; in this case, therefore, $d\sigma = a + \frac{1}{4}b = d\sigma_a$. With a and b determined from these two equations, we have

$$d\sigma = \tfrac{1}{4}(d\sigma_s + 3d\sigma_a) + (d\sigma_a - d\sigma_s)\bar{\mathbf{s}}_1 \cdot \bar{\mathbf{s}}_2.$$

§138. Resonance scattering of charged particles

In the scattering of charged nuclear particles (e.g. of protons by protons), as well as the short-range nuclear forces there is the Coulomb interaction, which decreases only slowly. The theory of resonance scattering in this case is developed by the same method as that described in §133. The only difference is that the wave function in the region outside the range of action of the nuclear forces ($r \gg a$) must be, instead of the solution of the equation of free motion (133.2), the exact general solution of Schrödinger's equation in a Coulomb field. Here the velocity of the particles is again assumed only so small that $ka \ll 1$; the relation between $1/k$ and the Coulomb unit of length $a_c = \hbar^2/mZ_1Z_2e^2$ (where m is the reduced mass of the colliding particles) is left arbitrary.†

For motion with $l = 0$ in a repulsive Coulomb field, Schrödinger's equation for the radial function $\chi = rR_0$ is

$$\chi'' + \left(k^2 - \frac{2}{r}\right)\chi = 0; \tag{138.1}$$

here we use Coulomb units. In §36 the solution of this equation has been found, subject to the requirement that χ/r is finite at $r = 0$. This solution,

† The theory given below is due to L. D. Landau and Ya. A. Smorodinskiĭ (1944).

which we here denote by F_0, has the form (see (36.27) and (36.28))

$$\left.\begin{aligned} F_0 &= Ae^{ikr}krF(i/k+1,\, 2,\, -2ikr), \\ A^2 &= \frac{2\pi/k}{e^{2\pi/k}-1}. \end{aligned}\right\} \tag{138.2}$$

The asymptotic expression for this function at large distances is

$$\left.\begin{aligned} F_0 &\approx \sin\left(kr - \frac{1}{k}\log 2kr + \delta_{0,\mathrm{Coul}}\right), \\ \delta_{0,\mathrm{Coul}} &= \arg\Gamma(1+i/k), \end{aligned}\right\} \tag{138.3}$$

and the leading terms of the expansion for small r ($kr \ll 1$, $r \ll 1$) are

$$F_0 = Akr(1+r+\ldots). \tag{138.4}$$

Now, however, with the changed boundary condition, the behaviour of the function at the origin becomes unimportant, and we need the general solution of equation (138.1), which is a linear combination of two independent integrals.

The parameters of the confluent hypergeometric function in (138.2) are such (the value of $\gamma = 2$ being integral) that the case described at the end of §d of the Mathematical Appendices occurs. In accordance with the discussion given there, we obtain the second integral of equation (138.1) by replacing the function F in (138.2) by some other linear combination of two terms whose sum is, according to (d.14), the confluent hypergeometric function. Taking the difference of these terms as the combination in question, we find the second independent solution of equation (138.1) (denoted by G_0) in the form†

$$G_0 = 2\,\mathrm{im}\frac{Ae^{-ikr}kr}{\Gamma(1+i/k)}(-2ikr)^{-1+i/k}G(1-i/k,\, -i/k,\, -2ikr); \tag{138.5}$$

the function F_0 is the real part of the same expression. The asymptotic form at large distances is

$$G_0 \approx \cos\left(kr - \frac{1}{k}\log 2kr + \delta_{0,\mathrm{Coul}}\right), \tag{138.6}$$

and the leading terms of the expansion for small r are

$$G_0 = \frac{1}{A}\{1+2r[\log 2r + 2C - 1 + h(k)] + \ldots\}, \tag{138.7}$$

† The functions F_0 and G_0 (and the correspondingly defined functions F_l and G_l with $l \neq 0$) are called *regular* and *irregular Coulomb functions* respectively.

where $C = 0{\cdot}577\ldots$ is Euler's constant, and $h(k)$ denotes the function

$$h(k) = \mathrm{re}\,\psi(-i/k)+\log k, \tag{138.8}$$

$\psi(z) = \Gamma'(z)/\Gamma(z)$ being the logarithmic derivative of the Γ function.†

The general integral of equation (138.1) may be written as the sum

$$\chi = \text{constant}\times(F_0 \cot\delta_0 + G_0), \tag{138.9}$$

where $\cot\delta_0$ is a constant. The notation is chosen so that the asymptotic form of this solution is

$$\chi \sim \sin\left(kr - \frac{1}{k}\log 2kr + \delta_{0,\mathrm{Coul}} + \delta_0\right). \tag{138.10}$$

Thus δ_0 is the additional phase shift of the wave function due to the short-range forces. We have to relate it to the constant appearing in the boundary

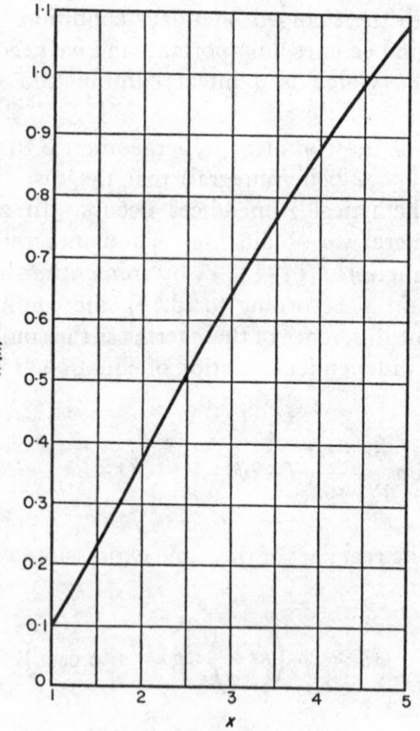

FIG. 49

† The expansion (138.7) is obtained from (138.5) by means of the expansion (d.17), using the well-known relation

$$\psi(1+z) = \psi(z)+1/z$$

(which is easily derived from $\Gamma(z+1) = z\Gamma(z)$) and the values $\psi(1) = -C$, $\psi(2) = -C+1$.

condition $[\chi'/\chi]_{r\to 0}$ = constant, which replaces the treatment of the wave function in the region where nuclear forces act. Owing to the logarithmic divergence of the logarithmic derivative χ'/χ as $r \to 0$, this condition must be applied at some arbitrarily small but finite value $r = \rho$, not at $r = 0$. Calculating by means of formulae (138.4) and (138.7) the derivative $\chi'(\rho)/\chi(\rho)$ and equating it to a constant, we obtain the boundary condition in the form

$$kA^2 \cot \delta_0 + 2[\log 2\rho + 2C + h(k)] = \text{constant}.$$

The expression on the left-hand side of the equation contains the constants $2 \log 2\rho$ and $4C$, which are independent of k; we include these in the constant on the right, and then denote it by $-\kappa$. The final expression for $\cot \delta_0$ is, in ordinary units,

$$\cot \delta_0 = -\frac{1}{\pi}(e^{2\pi/ka_c}-1)[h(ka_c)+\tfrac{1}{2}\kappa a_c]; \qquad (138.11)$$

in the limit $1/a_c \to 0$, i.e. for uncharged particles, formula (138.11) becomes the relation $\cot \delta_0 = -\kappa/k$, i.e. (133.6).

Figure 49 shows a graph of the function $h(x)$.†

Thus, when there is a Coulomb interaction, the "constant" is

$$\frac{2\pi \cot \delta_0}{a_c(e^{2\pi/ka_c}-1)} + \frac{2}{a_c}h(ka_c) = -\kappa. \qquad (138.12)$$

We have put the word "constant" in quotation marks, since κ is actually the first term in an expansion in powers of the small quantity ka of some function which depends on the properties of the short-range forces. As stated in §133, resonance at low energies corresponds to the case where the value of the constant κ is anomalously small. Consequently, in order to improve the accuracy, we must take account also of the next term ($\sim k^2$) in the expansion, which

---

† To calculate the function $h(k)$, we can use the formula

$$h(k) = k^{-2}\sum_{n=1}^{\infty}\frac{1}{n(n^2+k^{-2})} - C + \log k,$$

which is easily obtained by means of the formula

$$\psi(z) = -C - \frac{1}{z} + z\sum_{n=1}^{\infty}\frac{1}{n(n+z)};$$

see Whittaker and Watson, *Course of Modern Analysis*, Cambridge, 1944, §12.16. The limiting expressions for $h(k)$ are

$$h(k) \approx k^2/12 \quad \text{for} \quad k \ll 1,$$

$$h(k) = -C + \log k + 1{\cdot}2/k^2 \quad \text{for} \quad k \gg 1;$$

the latter formula gives values of $h(k)$ which are correct to within 4% even for $k > 2{\cdot}5$.

contains a coefficient of "normal" magnitude, i.e. in (138.12) $-\kappa$ must be replaced by $-\kappa_0+\frac{1}{2}r_0k^2$.†

The existence of resonance may, as stated in §133, be due to either a true or a virtual discrete bound state of the system. It can be shown‡ that the sign of the constant κ is again the criterion which determines whether the level is true or virtual.

The total phase shifts of the wave functions are, according to (138.10), the sums $\delta_{l,\text{Coul}}+\delta_l$. The scattering cross-section is therefore

$$f(\theta) = \frac{1}{2ik}\sum_{l=0}^{\infty}(2l+1)[e^{2i(\delta_{l,\text{Coul}}+\delta_l)}-1]P_l(\cos\theta). \tag{138.13}$$

The difference in the brackets may be written

$$e^{2i(\delta_{l,\text{Coul}}+\delta_l)}-1 = [e^{2i\delta_{l,\text{Coul}}}-1] + [e^{2i\delta_{l,\text{Coul}}}(e^{2i\delta_l}-1)]. \tag{138.14}$$

The Coulomb phases $\delta_{l,\text{Coul}}$ contribute equally, in order of magnitude, to the scattering amplitude for all l. The phases δ_l relating to the short-range forces are small for $l\neq 0$ at low energies. Hence, in substituting (138.14) in (138.13), we retain the first bracket in every term of the sum; the sum of these terms is the Coulomb scattering amplitude (135.9)

$$f_{\text{Coul}}(\theta) = -\frac{1}{2a_ck^2\sin^2\frac{1}{2}\theta}\exp\left(-\frac{2i}{ka_c}\log\sin\tfrac{1}{2}\theta+2i\delta_{0,\text{Coul}}\right). \tag{138.15}$$

The second bracket in (138.14) is retained only in the term with $l = 0$. Thus the total scattering amplitude is

$$f(\theta) = f_{\text{Coul}}(\theta)+\frac{1}{2ik}(e^{2i\delta_0}-1)e^{2i\delta_{0,\text{Coul}}}. \tag{138.16}$$

The second term in this expression may be called the *nuclear scattering amplitude*. It should be emphasized, however, that the division is arbitrary: in view of the definition of δ_0 in (138.11), the presence of the Coulomb interaction has a considerable effect on this term also, which is quite different from the corresponding term with the same short-range forces for uncharged particles. In particular, when $ka_c\to 0$ the phase δ_0, and therefore the whole of

---

† The values of the constants $\alpha = 1/\kappa_0$ and r_0 for proton–proton scattering are $\alpha = -7{\cdot}8\times 10^{-13}$ cm, $r_0 = 2{\cdot}8\times 10^{-13}$ cm (Coulomb unit of length $2\hbar^2/m_pe^2 = 57{\cdot}6\times 10^{-13}$ cm). These values relate to a pair of protons with antiparallel spins; when the spins are parallel a system of two protons cannot be in the s state, by Pauli's principle.

‡ See L. Landau and Ya. A. Smorodinskii, *Zhurnal éksperimental'noi i teoreticheskoi fiziki* **14**, 269, 1944.

the second term in (138.16), tend exponentially to zero as $e^{-2\pi/ka_c}$, i.e. the nuclear scattering is entirely masked by the Coulomb repulsion.

In the scattering cross-section the two parts of the amplitude interfere:

$$\frac{d\sigma}{do} = |f(\theta)|^2 = \left(\frac{Z_1 Z_2 e^2}{2mv^2}\right)^2 \left[\frac{1}{\sin^4 \frac{1}{2}\theta} - \frac{4ka_c}{\sin^2 \frac{1}{2}\theta} \sin\delta_0 \cos\left(\frac{2}{ka_c}\log\sin\tfrac{1}{2}\theta + \delta_0\right) + 4(ka_c)^2 \sin^2\delta_0\right]. \quad (138.17)$$

Here it is assumed that the colliding particles are different; for like particles, the scattering amplitude must be symmetrized before being squared (cf. §137).

§139. Elastic collisions between fast electrons and atoms

Elastic collisions between fast electrons and atoms can be treated by means of the Born approximation if the velocity of the incident electron is large compared with those of the atomic electrons.

Owing to the large difference in mass between the electron and the atom, the latter may be regarded as at rest during the collision, and the system of coordinates in which the centre of mass is fixed is the same as that in which the atom is fixed. Then $\mathbf{p}$ and $\mathbf{p}'$ in formula (126.7) denote the momenta of the electron before and after the collision, m the mass of the electron, and the angle θ is the same as the angle of deviation ϑ of the electron. The potential energy $U(r)$ in formula (126.7) must be defined appropriately.

In §126 we have calculated the matrix element $U_{\mathbf{p}'\mathbf{p}}$ of the interaction energy with respect to the wave functions of a free particle before and after the collision. In a collision with an atom it is necessary to take into account also the wave functions describing the internal state of the atom. In an elastic collision, the state of the atom is left unchanged. Hence $U_{\mathbf{p}'\mathbf{p}}$ must be determined as the matrix element with respect to the wave functions $\psi_{\mathbf{p}}$ and $\psi_{\mathbf{p}'}$ of the electron; it is diagonal with respect to the wave function of the atom. In other words, $U(r)$ in formula (126.7) must be taken to be the potential energy of the interaction of the electron with the atom, averaged with respect to the wave function of the latter. It is $e\phi(r)$, where $\phi(r)$ is the potential of the field at the point r due to the mean distribution of charges in the atom.

Denoting the density of the charge distribution in the atom by $\rho(r)$, we have, for the potential ϕ, Poisson's equation:

$$\triangle\phi = -4\pi\rho(r).$$

The required matrix element $U_{\mathbf{p}'\mathbf{p}}$ is essentially the Fourier component of U (i.e. of ϕ) corresponding to the wave vector $\mathbf{q} = \mathbf{k}' - \mathbf{k}$. Applying Poisson's equation to each Fourier component separately, we have

$$\triangle(\phi_{\mathbf{q}} e^{i\mathbf{q}\cdot\mathbf{r}}) = -q^2 \phi_{\mathbf{q}} e^{i\mathbf{q}\cdot\mathbf{r}} = -4\pi\rho_{\mathbf{q}} e^{i\mathbf{q}\cdot\mathbf{r}},$$

so that

$$\phi_{\mathbf{q}} = 4\pi\rho_{\mathbf{q}}/q^2,$$

i.e.

$$\int \phi e^{-i\mathbf{q}\cdot\mathbf{r}}\,dV = (4\pi/q^2)\int \rho e^{-i\mathbf{q}\cdot\mathbf{r}}\,dV. \tag{139.1}$$

The charge density $\rho(\mathbf{r})$ consists of the electron charges and the charge on the nucleus:

$$\rho = -en(r) + Ze\delta(\mathbf{r}),$$

where $en(r)$ is the electron charge density in the atom. Multiplying by $e^{-i\mathbf{q}\cdot\mathbf{r}}$ and integrating, we have

$$\int \rho e^{-i\mathbf{q}\cdot\mathbf{r}}\,dV = -e\int n e^{-i\mathbf{q}\cdot\mathbf{r}}\,dV + Ze.$$

Thus we obtain for the integral in question the expression

$$\int U e^{-i\mathbf{q}\cdot\mathbf{r}}\,dV = \frac{4\pi e^2}{q^2}[Z - F(q)], \tag{139.2}$$

where $F(q)$ is defined by the formula

$$F(q) = \int n e^{-i\mathbf{q}\cdot\mathbf{r}}\,dV \tag{139.3}$$

and is called the *atomic form factor*. It is a function of the scattering angle and of the velocity of the incident electron.

Finally, substituting (139.2) in (126.7), we obtain the following expression for the cross-section for the elastic scattering of fast electrons by an atom†:

$$d\sigma = \frac{4m^2e^4}{\hbar^4 q^4}[Z - F(q)]^2\,do, \qquad q = \frac{2mv}{\hbar}\sin\tfrac{1}{2}\vartheta. \tag{139.4}$$

Let us consider the limiting case $qa_0 \ll 1$, where a_0 is of the order of magnitude of the dimensions of the atom. Small scattering angles correspond to small q: $\vartheta \ll v_0/v$, where $v_0 \sim \hbar/ma_0$ is of the order of magnitude of the velocities of the atomic electrons.

Let us expand $F(q)$ as a series of powers of q. The zero-order term is $\int n\,dV$, which is the total number Z of electrons in the atom. The first-

† We are neglecting exchange effects between the fast electron which undergoes scattering and the atomic electrons, i.e. we do not symmetrize the wave function of the system. The legitimacy of this procedure is evident: the interference between the rapidly oscillating wave function of the free particle and the wave function of the atomic electrons in the "exchange integral" has the result that the corresponding contribution to the scattering amplitude is small.

order term is proportional to $\int \mathbf{r}n(r)\,dV$, i.e. to the mean value of the dipole moment of the atom; this vanishes identically (see §75). We must therefore continue the expansion up to the second-order term, obtaining

$$Z-F(q) = \frac{1}{6}q^2 \int nr^2\,dV;$$

substituting in (139.4), we obtain

$$d\sigma = \left|\frac{me^2}{3\hbar^2}\int nr^2\,dV\right|^2 do. \tag{139.5}$$

Thus, in the range of small angles, the cross-section is independent of the scattering angle, and is given by the mean square distance of the atomic electrons from the nucleus.

In the opposite limiting case of large q ($qa_0 \gg 1$, i.e. $\vartheta \gg v_0/v$), the factor $e^{-i\mathbf{q}.\mathbf{r}}$ in the integrand in (139.3) is a rapidly oscillating function, and therefore the whole integral is nearly zero. Consequently, we can neglect $F(q)$ in comparison with Z, so that

$$d\sigma = \left(\frac{Ze^2}{2mv^2}\right)^2 \frac{do}{\sin^4\frac{1}{2}\vartheta}. \tag{139.6}$$

i.e., we have Rutherford scattering at the nucleus of the atom.

We may also calculate the *transport cross-section*

$$\sigma_{tr} = \int (1-\cos\vartheta)\,d\sigma. \tag{139.7}$$

In the range of angles $\vartheta \ll v_0/v$ we have, according to (139.5), $d\sigma = \text{constant} \times \sin\vartheta\,d\vartheta = \text{constant} \times \vartheta\,d\vartheta$, where the constant is independent of ϑ. Hence, in this region, the integrand in the above integral is proportional to ϑ^3, so that the integral converges rapidly at the lower limit. In the region $1 \gg \vartheta \gg v_0/v$ we have $d\sigma \approx \text{constant} \times d\vartheta/\vartheta^3$; the integrand is proportional to $1/\vartheta$, and the integral (139.7) diverges logarithmically. Hence we see that this range of angles plays the chief part in the integral, and we need integrate only over this range. The lower limit of integration must be taken as of the order of v_0/v; we shall write it in the form $e^2/\gamma\hbar v$, where γ is a dimensionless constant. As a result we have the formula

$$\sigma_{tr} = 4\pi(Ze^2/mv^2)^2 \log(\gamma\hbar v/e^2). \tag{139.8}$$

An exact calculation of the constant γ requires a consideration of scattering through angles $\vartheta > v_0/v$, and cannot be carried out in a general form; σ_{tr} depends only slightly on the choice of this constant, since it enters only in a logarithm, and multiplied by the large quantity $\hbar v/e^2$.

For a numerical calculation of the atomic form factor for heavy atoms, we can use the Thomas–Fermi distribution of the density $n(r)$. We have seen that, in the Thomas–Fermi model, $n(r)$ has the form

$$n(r) = Z^2 f(rZ^{1/3}/b);$$

all quantities in this and the following formulae are measured in atomic units. It is easy to see that the integral (139.3), when calculated with such a function $n(r)$, will contain q only in the combination $qZ^{-1/3}$:

$$F(q) = Z\phi(bqZ^{-1/3}). \tag{139.9}$$

Table 11 gives, for reference, the values of the function $\phi(x)$, which holds for all atoms.†

TABLE 11

The atomic form factor on the Thomas–Fermi model

x	$\phi(x)$	x	$\phi(x)$	x	$\phi(x)$
0	1·000	1·08	0·422	2·17	0·224
0·15	0·922	1·24	0·378	2·32	0·205
0·31	0·796	1·39	0·342	2·48	0·189
0·46	0·684	1·55	0·309	2·64	0·175
0·62	0·589	1·70	0·284	2·79	0·167
0·77	0·522	1·86	0·264	2·94	0·156
0·93	0·469	2·02	0·240		

With the atomic form factor (139.9), the cross-section (139.4) will have the form

$$d\sigma = (4Z^2/q^4)[1-\phi(bqZ^{-1/3})]^2\, do = Z^{2/3}\Phi(Z^{-1/3}v \sin\tfrac{1}{2}\vartheta)\, do, \tag{139.10}$$

where $\Phi(x)$ is a new function holding for all atoms. The total cross-section may be obtained by integration. The chief part in the integral is played by the range of small ϑ. Hence we can write

$$d\sigma \approx Z^{2/3}\Phi(Z^{-1/3}v\vartheta/2)2\pi\vartheta\, d\vartheta,$$

and extend the integration over ϑ to infinity:

$$\sigma = 2\pi Z^{2/3}\int\limits_0^\infty \Phi(Z^{-1/3}v\vartheta/2)\vartheta\, d\vartheta = (8\pi/v^2)Z^{4/3}\int\limits_0^\infty x\Phi(x)\, dx.$$

Thus σ is of the form

$$\sigma = \text{constant}\times Z^{4/3}/v^2. \tag{139.11}$$

Similarly, it is easy to see that the constant γ in formula (139.8) will be proportional to $Z^{-1/3}$.

† It must be borne in mind that this formula is not applicable for small q, since the integral of nr^2 cannot in practice be calculated by the Thomas–Fermi method (see the third note to §113). It should also be mentioned that the Thomas–Fermi model does not represent the individual properties of atoms or their systematic variation with atomic number.

PROBLEM

Calculate the cross-section for the elastic scattering of fast electrons by a hydrogen atom in the ground state.

SOLUTION. The wave function of the normal state of the hydrogen atom is (in atomic units) $\psi = \pi^{-1/2}e^{-r}$, so that $n = e^{-2r}/\pi$. The integration over angles in (139.3) is effected as in the derivation of formula (126.12); we have

$$F = \frac{4\pi}{q}\int_0^\infty n(r)\sin qr\,.\,r\,dr = \frac{1}{(1+\frac{1}{4}q^2)^2}.$$

Substituting in (139.4), we obtain

$$d\sigma = \frac{4(8+q^2)^2}{(4+q^2)^4}\,do,$$

where $q = 2v\sin\frac{1}{2}\vartheta$. The total cross-section is calculated by putting $do = 2\pi\sin\vartheta\,d\vartheta = (2\pi/v^2)q\,dq$ and integrating over q from 0 to $2v$; since v is assumed large and the integral converges, the upper limit may be replaced by infinity. The result is

$$\sigma = 7\pi/3v^2.$$

The transport cross-section is calculated as

$$\sigma_{tr} = \frac{1}{2v^2}\int q^2\,d\sigma.$$

Changing the variable of integration by putting $u = 4+q^2$, and taking the upper limit as infinity everywhere except in the term du/u, we obtain

$$\sigma_{tr} = \frac{4\pi}{v^4}(\log v + \tfrac{1}{12}),$$

in accordance with (139.8).

§140. Scattering with spin–orbit interaction

Hitherto we have considered only collisions of particles whose interaction does not depend on their spins. Under these conditions the spins either do not affect the scattering process at all, or have an indirect influence due to exchange effects (§137).

Let us now examine the generalization of the theory of scattering given in §123 to the case where the interaction of the particles depends significantly on their spins, as occurs in collisions of nuclear particles.

We shall discuss in detail the simplest case, where one of the colliding particles (for definiteness taken to be the particle in the incident beam) has spin $\frac{1}{2}$, and the other (the target particle) has spin zero.

For a given (half-integral) angular momentum j of the system, the orbital angular momentum can have only the two values $l = j \pm \frac{1}{2}$, corresponding to states of different parities. In this case, therefore, the conservation of the absolute magnitude of the orbital angular momentum follows from that of $\mathbf{j}$ and the parity.

The operator $\hat{f}$ (§125) now acts not only on the orbital variables but also on the spin variables of the wave function of the system. It must commute with the operator of the conserved quantity $\mathbf{l}^2$. The most general form of such an operator is

$$\hat{f} = \hat{a} + \hat{b}\hat{\mathbf{l}} \cdot \hat{\mathbf{s}}, \tag{140.1}$$

where $\hat{a}$ and $\hat{b}$ are orbital operators depending only on $\mathbf{l}^2$.

The S-matrix, and therefore the matrix of the operator $\hat{f}$, are diagonal with respect to the wave functions of states with definite values of the conserved quantities l and j (and the component m of the total angular momentum), and the diagonal elements are expressed in terms of the phases δ of the wave functions by formula (123.15). For given l and given total angular momentum $j = l+\frac{1}{2}$ or $l-\frac{1}{2}$ the eigenvalues of $\mathbf{l} \cdot \mathbf{s}$ are $\frac{1}{2}l$ and $-\frac{1}{2}(l+1)$ respectively (see (118.5)). Hence, to determine the diagonal matrix elements of the operators $\hat{a}$ and $\hat{b}$ (denoted by a_l and b_l), we have the relations

$$\left.\begin{aligned} a_l+\tfrac{1}{2}lb_l &= \frac{1}{2ik}(e^{2i\delta_l^+}-1),\\ a_l-\tfrac{1}{2}(l+1)b_l &= \frac{1}{2ik}(e^{2i\delta_l^-}-1), \end{aligned}\right\} \tag{140.2}$$

where the phases δ_l^+ and δ_l^- correspond to states with $j = l+\frac{1}{2}$ and $j = l-\frac{1}{2}$ respectively.

We are interested, however, not in the diagonal elements themselves of the operator $\hat{f}$ with respect to the states with given l and j, but in the scattering amplitude as a function of the directions of the incident and scattered waves. This amplitude is still an operator, but only with respect to the spin variables—an operator which is non-diagonal with respect to the spin component σ. In the rest of this section $\hat{f}$ will denote this operator.

To derive this operator we must apply the operator (140.1) to the function (125.17) which corresponds to a plane wave incident along the z-axis. Thus

$$\hat{f} = \sum_{l=0}^{\infty} (2l+1)(a_l+b_l\hat{\mathbf{l}} \cdot \hat{\mathbf{s}})P_l(\cos\theta). \tag{140.3}$$

Here we must also calculate the result of the action of the operator $\hat{\mathbf{l}} \cdot \hat{\mathbf{s}}$ on the function $P_l(\cos\theta)$. This can be done by writing

$$\hat{\mathbf{l}} \cdot \hat{\mathbf{s}} = \tfrac{1}{2}(\hat{l}_+\hat{s}_- + \hat{l}_-\hat{s}_+) + \hat{l}_z\hat{s}_z$$

(see (29.11)) and using formulae (27.12) for the matrix elements of the operators $\hat{l}_\pm$, or still more simply by using the operator expressions (26.14), (26.15). The result is

$$\hat{\mathbf{l}} \cdot \hat{\mathbf{s}}P_l(\cos\theta) = i\boldsymbol{\nu} \cdot \hat{\mathbf{s}}P_l^1(\cos\theta),$$

where P_l^1 is the associated Legendre polynomial and $\boldsymbol{\nu}$ a unit vector in the direction $\mathbf{n}\times\mathbf{n}'$ which is perpendicular to the plane of scattering ($\mathbf{n}$ being the direction of incidence (the z-axis) and $\mathbf{n}'$ the direction of scattering, defined by the spherical polar angles θ, ϕ).

On determining a_l and b_l from (140.2) and substituting in (140.3), we have finally

$$\hat{f} = A + 2B\boldsymbol{\nu}\cdot\hat{\mathbf{s}}, \tag{140.4}$$

$$\left.\begin{aligned} A &= \frac{1}{2ik}\sum_{l=0}^{\infty}[(l+1)(e^{2i\delta_l^+}-1)+l(e^{2i\delta_l^-}-1)]P_l(\cos\theta), \\ B &= \frac{1}{2k}\sum_{l=1}^{\infty}(e^{2i\delta_l^+}-e^{2i\delta_l^-})P_l^1(\cos\theta). \end{aligned}\right\} \tag{140.5}$$

The matrix elements of this operator give the scattering amplitude for definite values of the spin component in the initial (σ) and final (σ') states. Let us consider the cross-section summed over all possible values of σ' and averaged with respect to the probabilities of various values σ in the initial state (in the incident beam). The cross-section is given by

$$d\sigma = \overline{(\hat{f}^+\hat{f})_{\sigma\sigma}}\, do; \tag{140.6}$$

by taking the diagonal matrix elements of the product $\hat{f}^+\hat{f}$ we effect the summation over final states, and the bar denotes the averaging with respect to initial states.† If all spin directions are equally probable in the initial state, this averaging reduces to taking the trace of the matrix, divided by the number of possible values of the spin component σ:

$$d\sigma = \tfrac{1}{2}\,\mathrm{tr}\,(\hat{f}^+\hat{f})\, do. \tag{140.7}$$

On substitution of (140.4) in (140.6) the mean value of the square $(\boldsymbol{\nu}.\mathbf{s})^2$ is calculated as $\frac{1}{3}\boldsymbol{\nu}^2\mathbf{s}^2 = \frac{1}{3}s(s+1) = \frac{1}{4}$. The result is

$$d\sigma/do = |A|^2+|B|^2+2\,\mathrm{re}(AB^*)\boldsymbol{\nu}\cdot\mathbf{P}, \tag{140.8}$$

where $\mathbf{P} = 2\bar{\mathbf{s}}$ is the initial polarization of the beam, defined as the ratio of the mean spin in the initial state to its maximum possible value ($\frac{1}{2}$). In the case of spin $\frac{1}{2}$ the vector $\bar{\mathbf{s}}$ completely describes the spin state (§59).

It may be pointed out that the polarization of the incident beam leads to an azimuthal asymmetry of the scattering: owing to the factor $\boldsymbol{\nu}.\mathbf{P}$ in the last term, the cross-section (140.8) depends not only on the polar angle θ

---

† If the squared modulus $|f_{0n}|^2$ of the matrix element of some operator for the transition $0 \to n$ is summed over final states n, we have

$$\sum_n |f_{0n}|^2 = \sum_n f_{0n}(f_{0n})^* = \sum_n f_{0n}(f^+)_{n0} = (ff^+)_{00}.$$

To avoid misunderstanding, it should be emphasized that the sign $^+$ denoting the conjugate refers in (140.6) and henceforward to $\hat{f}$ as a spin operator; in particular, the transposition of $\mathbf{n}$ and $\mathbf{n}'$ is not implied.

but also on the azimuth ϕ of the vector $\mathbf{n}'$ relative to $\mathbf{n}$ (if the polarization is not perpendicular to $\boldsymbol{\nu}$, so that $\boldsymbol{\nu}.\mathbf{P} \neq 0$).

The polarization of the scattered particles can be calculated from the formula

$$\mathbf{P}' = 2\overline{(f^+\mathbf{s}f)_{\sigma\sigma}}/\overline{(f^+f)_{\sigma\sigma}}. \tag{140.9}$$

For example, if the initial state is unpolarized ($\mathbf{P} = 0$), a simple calculation gives

$$\mathbf{P}' = \frac{2\,\mathrm{re}(AB^*)}{|A|^2+|B|^2}\boldsymbol{\nu}. \tag{140.10}$$

Thus scattering leads, in general, to the appearance of a polarization perpendicular to the plane of scattering. This effect is, however, absent in the Born approximation: if all the phases δ are small, the coefficient A is real in the first approximation with respect to the phases, and B is purely imaginary, so that $\mathrm{re}(AB^*) = 0$.

The fact that the polarization $\mathbf{P}'$ (140.10) is in the direction of $\boldsymbol{\nu}$ is obvious *a priori*. $\mathbf{P}'$ is an axial vector, and $\boldsymbol{\nu}$ is the only axial vector which can be constructed from the available polar vectors $\mathbf{n}$ and $\mathbf{n}'$. It is therefore evident that this property will also be possessed by the polarization resulting from the scattering of an unpolarized beam of particles with spin $\frac{1}{2}$ by an unpolarized target composed of nuclei with any spin (not necessarily zero).†

In formulating the reciprocity theorem for scattering in the presence of spins it must be borne in mind that time reversal changes the signs not only of the momenta but also of the angular momenta. Hence the symmetry of scattering with respect to time reversal must in this case be expressed by the equality of amplitudes for processes which differ not only in the interchange of the initial and final states and the reversal of the directions of motion but also in that the signs of the spin components of the particles are changed in both states. Here, however, the signs of these amplitudes may differ because, according to (60.3), time reversal introduces a factor $(-1)^{s-\sigma}$ in the spin wave function. This has the result that the reciprocity theorem must be formulated as follows:‡

$$f(\sigma_1, \sigma_2, \mathbf{n}; \sigma_1', \sigma_2', \mathbf{n}') = (-1)^{\Sigma(s-\sigma)}f(-\sigma_1', -\sigma_2', -\mathbf{n}'; -\sigma_1, -\sigma_2, -\mathbf{n}). \tag{140.11}$$

Here $f(\sigma_1, \sigma_2, \mathbf{n}; \sigma_1', \sigma_2', \mathbf{n}')$ is the amplitude of scattering with change in the spin components of the colliding particles from σ_1, σ_2 to σ_1', σ_2'. The sum in the exponent is taken over both particles before and after scattering.

† Here we have in mind a target with a completely random distribution of spin directions. For $s > \frac{1}{2}$, it will be recalled, the mean value of the spin vector does not fully determine the spin state, and if this mean value is zero there is not necessarily a complete absence of ordering of the spins.

‡ The derivation of this relation is similar to that of formula (125.12). The amplitudes of the ingoing and outgoing waves must contain spin factors, and instead of (125.10) we have the condition $\hat{K}^{-1}\tilde{\hat{S}}\hat{K} = \hat{S}$, where $\hat{K}$ is an operator which not only effects inversion but also changes the spin state in accordance with (60.3).

In the Born approximation, the scattering has a further symmetry; the probabilities of processes differing by the interchange of the initial and final states, without change in the signs of the momenta and spin components of the particles as in time reversal, are the same (see §126). Combining this property with the reciprocity theorem, we find that the scattering is symmetrical with respect to a change in sign of all the momenta and spin components, without interchange. Hence we easily conclude that in the Born approximation there can be no polarization in the scattering of any unpolarized beam by an unpolarized target. For, under the transformation mentioned, the polarization vector $\mathbf{P}$ changes sign, while the unit vector $\mathbf{k} \times \mathbf{k}'$, whose direction must be the same as that of $\mathbf{P}$, remains unaltered. Thus the property noted above for the scattering of particles with spin $\frac{1}{2}$ by particles with spin zero is actually a general one.

In the case of arbitrary spins of the colliding particles, the general formulae for the angular distributions are very complicated, and we shall not pause to derive them here, but merely calculate the number of parameters by which these distributions must be determined.

The case considered above of a collision between particles of spin $\frac{1}{2}$ and 0 has, in particular, the property that to given values of j and the parity there corresponds only one state of the system of two particles (apart from the unimportant orientation of the total angular momentum in space). Each such state leads to one real parameter (the phase δ) in the scattering amplitude. For other spins there are in general several different states with the same total angular momentum J and parity; these states differ in the values of the total spin S of the particles and the orbital angular momentum l of their relative motion. Let the number of such states be n. It is easy to see that each such group of states contributes $\frac{1}{2}n(n+1)$ real parameters in the scattering amplitude. For the S-matrix is, with respect to these states, a matrix having unitary symmetry (owing to the reciprocity theorem), with $n\,.\,n$ complex elements. The number of independent quantities in this matrix is conveniently calculated by noting that, if the operator $\hat{S}$ is written in the form $\hat{S} = \exp(i\hat{R})$, the unitarity condition is automatically satisfied when $\hat{R}$ is any Hermitian operator (see (12.13)). If the matrix $\hat{S}$ is symmetrical, so is the matrix $\hat{R}$, which, being Hermitian, is therefore real, and a real symmetrical matrix has $\frac{1}{2}n(n+1)$ independent components.

As an example, for two particles with spins $\frac{1}{2}$ the number $n = 2$: for given J there are in all four states, two with $l = J$ and total spin $S = 0$ or 1, and two with $l = J \pm 1$, $S = 1$. It is evident that two of these states are even (l is even) and two are odd (l is odd).

The general form of the scattering amplitude for particles with spin $\frac{1}{2}$, as an operator relating to the spin variables of the two particles, is easily written down from the necessary invariance conditions: it must be a scalar invariant under time reversal. To construct this expression we have the two axial vectors $\mathbf{s}_1$ and $\mathbf{s}_2$ of the particle spins and two ordinary (polar) vectors $\mathbf{n}$ and $\mathbf{n}'$. Each of the operators $\hat{\mathbf{s}}_1$ and $\hat{\mathbf{s}}_2$ must appear linearly in the amplitude, since any function of an operator of spin $\frac{1}{2}$ can be reduced to a linear function. The most general form of operator satisfying these conditions can be written as

$$f = A + B(\hat{\mathbf{s}}_1 \cdot \boldsymbol{\lambda})(\hat{\mathbf{s}}_2 \cdot \boldsymbol{\lambda}) + C(\hat{\mathbf{s}}_1 \cdot \boldsymbol{\mu})(\hat{\mathbf{s}}_2 \cdot \boldsymbol{\mu}) + \\ + D(\hat{\mathbf{s}}_1 \cdot \boldsymbol{\nu})(\hat{\mathbf{s}}_2 \cdot \boldsymbol{\nu}) + E(\hat{\mathbf{s}}_1 + \hat{\mathbf{s}}_2) \cdot \boldsymbol{\nu} + F(\hat{\mathbf{s}}_1 - \hat{\mathbf{s}}_2) \cdot \boldsymbol{\nu}. \qquad (140.12)$$

The coefficients $A, B, \ldots$ are scalar quantities, which can depend only on the scalar $\mathbf{n} \cdot \mathbf{n}'$, i.e. on the scattering angle θ (and on the energy); $\boldsymbol{\lambda}, \boldsymbol{\mu}, \boldsymbol{\nu}$ are three mutually perpendicular unit vectors along $\mathbf{n}+\mathbf{n}'$, $\mathbf{n}-\mathbf{n}'$ and $\mathbf{n}\times\mathbf{n}'$ respectively. The operations of time reversal correspond to the changes

$$\mathbf{s}_1 \to -\mathbf{s}_1, \qquad \mathbf{s}_2 \to -\mathbf{s}_2, \qquad \mathbf{n} \to -\mathbf{n}', \qquad \mathbf{n}' \to -\mathbf{n},$$

so that

$$\boldsymbol{\lambda} \to -\boldsymbol{\lambda}, \qquad \boldsymbol{\mu} \to \boldsymbol{\mu}, \qquad \boldsymbol{\nu} \to -\boldsymbol{\nu}$$

and the invariance of the operator (140.12) is obvious.

In the mutual scattering of nucleons (protons and neutrons) the last term in (140.12) does not appear. This is evident from the fact that the nuclear forces acting between nucleons conserve the absolute magnitude of the total spin S of the system; the operator $\hat{\mathbf{s}}_1 - \hat{\mathbf{s}}_2$, however, does not commute with the operator $\hat{\mathbf{S}}^2$. (The remaining terms in (140.12) are expressed, according to (117.4), in terms of the total spin operator $\hat{\mathbf{S}}$, and therefore commute with $\hat{\mathbf{S}}^2$.) In the scattering of like nucleons (*pp* or *nn*), the coefficients $A, B, \ldots$ as functions of the angle of scattering also satisfy certain symmetry relations as a result of the identity of the two particles (see Problem 2).

PROBLEMS

PROBLEM 1. Determine the polarization after the scattering of particles with spin $\frac{1}{2}$ by particles with spin zero when the polarization before scattering is non-zero.

SOLUTION. A calculation using formula (140.9) is conveniently effected in components, with the z-axis in the direction of ν. The result is

$$\mathbf{P}' = \frac{(|A|^2 - |B|^2)\mathbf{P} + 2|B|^2\boldsymbol{\nu}(\boldsymbol{\nu}\cdot\mathbf{P}) + 2\,\mathrm{im}(AB^*)\boldsymbol{\nu}\times\mathbf{P} + 2\boldsymbol{\nu}\,\mathrm{re}(AB^*)}{|A|^2 + |B|^2 + 2\,\mathrm{re}(AB^*)\boldsymbol{\nu}\cdot\mathbf{P}}.$$

PROBLEM 2. Find the symmetry conditions satisfied by the coefficients in the scattering amplitude for two like nucleons, as functions of the angle θ (R. Oehme 1955).

SOLUTION. We regroup the terms in (140.12) in such a way that each is non-zero only for singlet ($S = 0$) or triplet ($S = 1$) states of the system of two nucleons:

$$f = a(\hat{\mathbf{s}}_1 \cdot \hat{\mathbf{s}}_2 - \tfrac{1}{4}) + b(\hat{\mathbf{s}}_1 \cdot \hat{\mathbf{s}}_2 + \tfrac{3}{4}) + c[\tfrac{1}{4} + (\hat{\mathbf{s}}_1 \cdot \boldsymbol{\nu})(\hat{\mathbf{s}}_2 \cdot \boldsymbol{\nu})] + \\ + d[(\hat{\mathbf{s}}_1 \cdot \mathbf{n})(\hat{\mathbf{s}}_2 \cdot \mathbf{n}') + (\hat{\mathbf{s}}_1 \cdot \mathbf{n}')(\hat{\mathbf{s}}_2 \cdot \mathbf{n})] + e(\hat{\mathbf{s}}_1 + \hat{\mathbf{s}}_2) \cdot \boldsymbol{\nu}. \qquad (1)$$

Using formulae (117.4), we easily see that the first term is non-zero only for $S = 0$ and the remainder only for $S = 1$. Owing to the identity of the particles, the scattering amplitude must be symmetric with respect to interchange of the particle coordinates for $S = 0$, and antisymmetric for $S = 1$. This transformation is equivalent to $\theta \to \pi - \theta$, or to a change in sign of one of the vectors **n** and **n**′ (cf. §137). From these conditions we obtain the relations

$$\left.\begin{aligned} a(\pi-\theta) = a(\theta), \quad b(\pi-\theta) = -b(\theta), \quad c(\pi-\theta) = -c(\theta), \\ d(\pi-\theta) = d(\theta), \quad e(\pi-\theta) = e(\theta). \end{aligned}\right\} \quad (2)$$

Owing to isotopic invariance, the scattering amplitude is the same for *nn* and *pp* scattering and for *np* scattering in the isotopic state with $T = 1$. For the *np* system, however, the state with $T = 0$ is also possible, and the *np* scattering amplitude is therefore described by other coefficients $a, b, \ldots$ in (1), which do not possess the symmetry properties (2).

§141. Regge poles

In §128 we have considered the analytical properties of the scattering amplitude as a function of the complex variable E, the energy of the particle; the orbital angular momentum l acted as a parameter having real integral values. Further properties of the scattering amplitude that are of methodological importance appear if we now regard l as a continuous complex variable for real values of the energy E.†

As in §128, we shall take radial wave functions whose asymptotic form (as $r \to \infty$) is

$$\chi_l = rR_l = A(l, E)\exp\left(-\frac{\sqrt{(-2mE)}}{\hbar}r\right) + B(l, E)\exp\left(\frac{\sqrt{(-2mE)}}{\hbar}r\right). \quad (141.1)$$

These functions are solutions of Schrödinger's equation (32.8) (in which l is now regarded as a complex parameter); the choice from the two independent solutions is governed by the condition

$$R_l \approx \text{constant} \times r^l \quad \text{for } r \to 0. \quad (141.2)$$

It is immediately evident that this condition places a certain limitation on the permissible values of the parameter l: the general form of the solution of equation (32.8) for small r is

$$R_l \approx c_1 r^l + c_2 r^{-l-1}$$

(see the end of §32). In order for the second solution to be clearly distinguished from the first solution and eliminated, the term in r^{-l-1} must exceed that in r^l as $r \to 0$. For complex l, this leads to the condition re $l >$ re $(-l-1)$, or

$$\text{re}\,(l+\tfrac{1}{2}) > 0. \quad (141.3)$$

In the following, we shall consider only this half of the complex l-plane, to the right of the vertical line $l = -\frac{1}{2}$.

† These properties were first investigated by T. Regge (1958).

The wave function $R(r; l, E)$, being a solution of a differential equation with coefficients analytic in the parameter l, is an analytic function of l, having no singularities in the half-plane (141.3). This applies, in particular, to the asymptotic expression (141.1), and the functions $A(l, E)$ and $B(l, E)$ therefore have no singularities with respect to l. Here, however, it is assumed that the retention of both terms in (141.1) as $r \to \infty$ is in fact legitimate. When $E > 0$ this is always true; when $E < 0$ it is true if the field $U(r)$ satisfies the condition (128.6) or (128.13). In these arguments it is important that the form of the asymptotic behaviour (with respect to r) of the wave function depends on E but not on l. The approach to the asymptotic form is therefore unaffected by the fact that l is complex.

Comparing (141.1) with the asymptotic formula (128.15), we find the S-matrix element in the form

$$S(l, E) = \exp[2i\delta(l, E)] = e^{i\pi l} A(l, E)/B(l, E), \tag{141.4}$$

which is valid for complex l also (although the "phase shift" δ is of course then not real).

For real l, and $E > 0$, the functions A and B are related by (128.4): $A(l, E) = B^*(l, E)$. Hence it follows that, for complex l,

$$A(l^*, E) = B^*(l, E) \qquad \text{for } E > 0, \tag{141.5}$$

and $S(l, E)$ therefore satisfies the *complex unitarity* condition

$$S^*(l, E)S(l^*, E) = 1. \tag{141.6}$$

Since $A(l, E)$ and $B(l, E)$ have no singularities as functions of l, the function $S(l, E)$ and thus the partial scattering amplitude $f(l, E)$ have singularities (poles) only at the zeros of the function $B(l, E)$. The poles of the scattering amplitude in the complex l-plane are called *Regge poles*. Their position depends, of course, on the value of the real parameter E. The functions

$$l = \alpha_i(E),$$

which determine the positions of the poles, are called *Regge trajectories*; when E varies, the poles move along certain lines in the l-plane. The subscript i which labels the poles will be omitted henceforward.

Going on now to study the properties of the Regge trajectories, we shall show first of all that for $E < 0$ all the $\alpha(E)$ are real functions. To do so, let us consider the equation

$$\chi'' + \left[\frac{2m}{\hbar^2}(E - U(r)) - \frac{\alpha(\alpha+1)}{r^2}\right]\chi = 0, \tag{141.7}$$

which is satisfied by the wave function with $l = \alpha$. Multiplying this equation by χ^* and integrating with respect to r (with integration by parts in the first term), we obtain

$$-\int_0^\infty |\chi'|^2\,\mathrm{d}r + \frac{2m}{\hbar^2}\int_0^\infty (E-U)|\chi|^2\,\mathrm{d}r - \alpha(\alpha+1)\int_0^\infty \frac{|\chi|^2}{r^2}\,\mathrm{d}r = 0.$$

Here we have used the fact that for $B = 0$ (the condition determining the Regge poles) the wave function decreases exponentially as $r\to\infty$, so that all the integrals converge. The first two terms in the above equation are real, and so is the integral in the last term. Hence we must have

$$\operatorname{im}\alpha(\alpha+1) = \operatorname{im}(\alpha+\tfrac{1}{2})^2 = 2\operatorname{re}(\alpha+\tfrac{1}{2})\operatorname{im}\alpha = 0.$$

But, since we are considering only poles in the half-plane (141.3), we certainly have $\operatorname{re}(\alpha+\frac{1}{2})>0$, and this gives the desired result

$$\operatorname{im}\alpha(E) = 0 \quad \text{for } E<0. \tag{141.8}$$

Next, we proceed as follows with (141.7), in a similar manner to the derivation of equation (128.10): differentiate with respect to E, multiply by χ, and multiply (141.7) by $\partial\chi/\partial E$, and subtract. This gives the identity

$$\left[\chi'\frac{\partial\chi}{\partial E} - \chi\left(\frac{\partial\chi}{\partial E}\right)'\right]' - \frac{2m}{\hbar^2}\chi^2 + \frac{\chi^2}{r^2}\frac{\mathrm{d}\alpha(\alpha+1)}{\mathrm{d}E} = 0.$$

Integration with respect to r from 0 to ∞, again using the fact that $\chi\to 0$ as $r\to\infty$, shows that the integral of the first term is zero, and we have

$$\frac{\mathrm{d}\alpha(\alpha+1)}{\mathrm{d}E}\int_0^\infty \frac{\chi^2}{r^2}\,\mathrm{d}r = \frac{2m}{\hbar^2}\int_0^\infty \chi^2\,\mathrm{d}r. \tag{141.9}$$

Since we know that α is real, the wave function is also real, and both integrals in (141.9) are therefore positive. Hence

$$\frac{\mathrm{d}}{\mathrm{d}E}\alpha(\alpha+1) = 2(\alpha+\tfrac{1}{2})\frac{\mathrm{d}\alpha}{\mathrm{d}E} > 0,$$

and, since $\alpha+\frac{1}{2}>0$,

$$\mathrm{d}\alpha/\mathrm{d}E > 0 \text{ for } E<0.$$

Thus, for $E<0$, the functions $\alpha(E)$ increase monotonically with E.

The negative values of E for which the functions $\alpha(E)$ take "physical" values (i.e. are integers $l = 0, 1, 2, \ldots$) correspond to the discrete energy levels of the system. Note that this gives rise to a new principle of classification of bound states, according to the Regge trajectories on which they lie.

As an example, let us consider Regge trajectories for motion in an attractive Coulomb field. The scattering matrix elements are then given by†

$$S_l = \frac{\Gamma(l+1-i/k)}{\Gamma(l+1+i/k)}, \tag{141.10}$$

with k in Coulomb units. The poles of this expression are at points where the argument of $\Gamma(l+1-i/k)$ is a negative integer or zero. For $E<0$ we have $k = i\sqrt{(-2E)}$, so that

$$\alpha(E) = -n_r - 1 + 1/\sqrt{(-2E)}, \quad E<0, \tag{141.11}$$

where $n_r = 0, 1, 2, \ldots$ is the number of the Regge trajectory. Equating $\alpha(E)$ to an integer $l = 0, 1, 2, \ldots$, we obtain the familiar Bohr formula for discrete energy levels in a Coulomb field:

$$E = -\tfrac{1}{2}(n_r+1+l)^{-2}.$$

The number n_r here coincides with the radial quantum number which determines the number of nodes of the radial wave function. Each Regge trajectory (i.e. each given value of n_r) corresponds to an infinity of levels with different values of the orbital angular momentum.

Let us now consider the properties of the functions $\alpha(E)$ for $E>0$. The functions $A(l, E)$ and $B(l, E)$ of the complex variable E in (141.1) are defined on a plane with a cut along the right half of the real axis (see §128). Correspondingly, the functions $l = \alpha(E)$, for which $B(l, E) = 0$, have a similar cut. On the upper and lower edges of the cut, $\alpha(E)$ has complex conjugate values, with im $\alpha>0$ on the upper edge. Without pausing to give a formal proof of this, we shall present a more physical explanation of the reason.

When l is complex, so is the centrifugal energy and therefore the effective potential energy $U_l = U + l(l+1)/2mr^2$. Repeating the derivation in §19, we now have instead of (19.6)

$$\frac{\partial}{\partial t}|\Psi|^2 + \text{div}\,\mathbf{j} = 2|\Psi|^2 \,\text{im}\, U_l.$$

When $l = \alpha$, and im $\alpha>0$, we also have im $U_l>0$. Then the right-hand side of the equation is positive, signifying an emission of new particles in the field volume. Accordingly, the asymptotic expression for the wave function (which, when $B = 0$, contains only the first term in (141.1)) must represent

† Cf. (135.11), in which the sign of k must be changed to convert from repulsion to attraction.

an outgoing wave, and this occurs on the upper edge of the cut; cf. the derivation of (128.3) from (128.1).

Since, for $E > 0$, the functions $\alpha(E)$ are complex, they cannot here take their "physical" values $l = 0, 1, 2, \ldots$. They may, however, be close to these values in the complex l-plane. We shall show that there is then a resonance in the partial scattering amplitude (corresponding to the integral value of l in question).

Let l_0 be the integral value close to the function $\alpha(E)$, and let E_0 be the (real and positive) value of the energy for which re $\alpha(E_0) = l_0$. Then, near this value, we have

$$\alpha(E) \approx l_0 + i\eta + \beta(E - E_0), \tag{141.12}$$

where $\eta = \text{im}\ \alpha(E_0)$ is a real constant. We shall consider values $\alpha(E)$ on the upper edge of the cut; according to the preceding discussion, $\eta > 0$ in that case (and $\eta \ll 1$, from the assumption that α is close to l_0). It is easy to see that the constant β (i.e. the derivative $d\alpha/dE$ for $E = E_0$) may be regarded as real and positive: since $\alpha(E)$ is almost real, so is the wave function $\chi(r;\ \alpha,\ E)$. Neglecting quantities of a higher order of smallness with respect to η, we can neglect the imaginary part of χ, and then it follows that β is positive, since the integrals in (141.9) are positive.†

Since $l = \alpha(E)$ is a zero of $B(l, E)$, the latter is proportional to $\alpha - l$ near the point α, E_0. Using (141.12), we therefore have

$$B(l_0, E) \approx \text{constant} \times [a(E - E_0) + i\eta]. \tag{141.13}$$

The form of this expression is the same as that of (134.6), with E_0 the energy and $\Gamma = 2\eta/a > 0$ the width of the quasi-discrete level. Thus the closeness of the Regge trajectory (for $E > 0$) to integral values of l corresponds to quasi-stationary states of the system. For these states, therefore, there exists the same principle of classification as for strictly stationary states: each Regge trajectory can correspond to a family of discrete and quasi-discrete levels.

The treatment of l as a complex variable enables us to derive a useful

---

† To elucidate the structure of these integrals, we note that the asymptotic region $r \gg a$ (where a is the range of action of the field), in which the expression (141.1) for the wave function is valid, makes only a small contribution to the integrals if η is small. For, if $l = \alpha(E)$ is a zero of $B(l,E)$, then by (141.5) $l = \alpha^*$ is a zero of $A(l,E)$. Hence $A(\alpha,E)$ and therefore $\chi(r;\alpha,E)$ in the region $r \gg a$, are small quantities $\sim \eta^{1/2}$; see (134.11). In estimating the integrals, it is also important that, on the upper edge of the cut (in relation to E), the wave function contains a factor e^{ikr}: $\chi(r;\alpha,E) = A(\alpha,E)e^{ikr}$. On this edge, we can regard E as $E + i\delta$ ($\delta \to +0$); then k also has a small positive imaginary part, which ensures the convergence of the integrals in (141.9). Physically, the smallness of the contribution to the integrals from the region $r \gg a$ is due to the fact that the energy E_0 corresponds to a quasi-stationary state (see below); the particle therefore reaches this region only as a result of an improbable decay of the state. The principal contribution to the integrals comes from the region $r \sim a$, in which the wave function is almost real.

integral form of the total scattering amplitude (for $E>0$), given by the series (123.11):

$$f(\mu) = \frac{1}{2ik}\sum_{l=0}^{\infty}(2l+1)[S(l, E)-1]P_l(\mu), \quad \mu = \cos\theta. \qquad (141.14)$$

To obtain this, we must first define the functions $P_l(\mu)$ not only for integral $l \geqslant 0$ but also for complex l. This can be done by taking $P_l(\mu)$ as the solution of equation (c.2):

$$(1-\mu^2)P_l''(\mu) - 2\mu P_l'(\mu) + l(l+1)P_l(\mu) = 0 \qquad (141.15)$$

with the boundary condition $P_l(1) = 1$. The $P_l(\mu)$ thus defined as a function of l has no singularities for finite values of l.†

It is easily seen that the series (141.14) is equal to the integral

$$f(\mu) = \frac{1}{4k}\int_C \frac{2l+1}{\sin \pi l}[S(l, E)-1]P_l(-\mu)\mathrm{d}l, \qquad (141.16)$$

taken along a contour C that passes in a negative direction (clockwise) round all the points $l = 0, 1, 2, \ldots$ on the real axis and is closed at infinity:

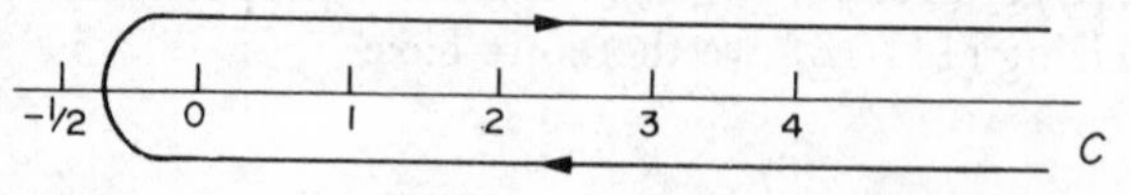

All the poles $l = \alpha_1, \alpha_2, \ldots$ of the function $S(l, E)$ (which are not on the real axis if $E>0$) must remain outside the contour C. The integral (141.16) reduces to $-2\pi i$ times the sum of the residues of the integrand at the points $l = 0, 1, 2, \ldots$, which are poles of the function $1/\sin \pi l$, and the residues of this function itself are $(-1)^l/\pi$. Since for integral l we have $P_l(-\mu) = (-1)^l P_l(\mu)$, we arrive at (141.14) from (141.16).‡

† By comparison of (141.15) with (e.2) we can express $P_l(\mu)$ as a hypergeometric function: $P_l(\mu) = F(-l, l+1, 1; \tfrac{1}{2}-\tfrac{1}{2}\mu)$.

‡ A more detailed account of the ideas discussed in this section (for non-relativistic theory) is given in the book by de Alfaro and Regge quoted in §123.

CHAPTER XVIII

INELASTIC COLLISIONS

§142. Elastic scattering in the presence of inelastic processes

COLLISIONS are said to be *inelastic* when they are accompanied by a change in the internal state of the colliding particles. Here we understand "a change in the internal state" in the widest sense; in particular, the very nature of the particles may be altered. For example, the change may consist in the excitation or ionization of atoms, the excitation or disintegration of nuclei, and so on. Where a collision (e.g. a nuclear reaction) may be accompanied by various physical processes, these are referred to as various *channels* of the reaction.

The existence of inelastic channels has a certain effect on the properties of elastic scattering also.

In the general case where various reaction channels exist, the asymptotic expression for the wave function of the system of colliding particles is a sum, with one term corresponding to each possible channel. Among these there is, in particular, a term describing the particles in the original unchanged state (the *input channel*). This is the product of the wave functions of the internal state of the particles and a function describing their relative motion (in a coordinate system where their centre of mass is at rest). The latter function is the one of interest here; we shall denote it by ψ, and seek its asymptotic form.

The wave function ψ in the input channel consists of an incident plane wave and an outgoing spherical wave corresponding to elastic scattering. It can also be represented as the sum of an ingoing and an outgoing wave, as in §123. The difference is that the asymptotic expression for the radial functions $R_l(r)$ cannot be taken in the form of the stationary wave. The stationary wave is the sum of ingoing and outgoing waves of equal amplitude. In purely elastic scattering this corresponds to the physical significance of the problem, but when there are inelastic channels the amplitude of the outgoing wave must be less than that of the ingoing wave. The asymptotic expression for ψ will therefore be given by formula (123.9):

$$\psi = \frac{1}{2ikr}\sum_{l=0}^{\infty}(2l+1)P_l(\cos\theta)[(-1)^{l+1}e^{-ikr}+S_l e^{ikr}], \qquad (142.1)$$

except that the S_l are no longer given by (123.10), but are certain quantities, in general complex, with moduli less than unity. The elastic scattering amplitude is given in terms of these quantities by formula (123.11):

$$f(\theta) = \frac{1}{2ik}\sum_{l=0}^{\infty}(2l+1)(S_l-1)P_l(\cos\theta). \tag{142.2}$$

For the total elastic scattering cross-section σ_e we have, instead of (123.12), the formula

$$\sigma_e = \frac{\pi}{k^2}\sum_{l=0}^{\infty}(2l+1)|1-S_l|^2. \tag{142.3}$$

The total inelastic scattering cross-section or *reaction cross-section* σ_r for all possible channels can also be expressed in terms of the S_l. To do so, we need only note that for each value of l the intensity of the outgoing wave is reduced in the ratio $|S_l|^2$ in comparison with that of the ingoing wave. This reduction must be ascribed entirely to inelastic scattering. It is therefore clear that

$$\sigma_r = \frac{\pi}{k^2}\sum_{l=0}^{\infty}(2l+1)(1-|S_l|^2), \tag{142.4}$$

and the total cross-section is

$$\sigma_t = \sigma_e + \sigma_r = \frac{2\pi}{k^2}\sum_{l=0}^{\infty}(2l+1)(1-\mathrm{re}\, S_l). \tag{142.5}$$

The partial amplitude for elastic scattering with angular momentum l, determined from (123.15), is

$$f_l = (S_l-1)/2ik, \tag{142.6}$$

and each of the terms in the sum in (142.3) and (142.4) is the partial cross-section for elastic or inelastic scattering of particles with angular momentum l:

$$\left.\begin{aligned} \sigma_e^{(l)} &= (\pi/k^2)(2l+1)|1-S_l|^2, \\ \sigma_r^{(l)} &= (\pi/k^2)(2l+1)(1-|S_l|^2), \\ \sigma_t^{(l)} &= (2\pi/k^2)(2l+1)(1-\mathrm{re}\, S_l). \end{aligned}\right\} \tag{142.7}$$

The value $S_l = 1$ corresponds to the complete absence of scattering (with a given l). The case $S_l = 0$ corresponds to total "absorption" of particles with angular momentum l (there is no outgoing partial wave with this l in (142.1)); the cross-sections for elastic and inelastic scattering are then equal:

$$\sigma_e^{(l)} = \sigma_r^{(l)} = (\pi/k^2)(2l+1), \tag{142.8}$$

Although elastic scattering can occur without inelastic scattering (when $|S_l| = 1$), the opposite situation is impossible: the presence of inelastic scattering necessarily implies the simultaneous presence of elastic scattering. For a given value of $\sigma_r^{(l)}$, the elastic scattering cross-section must be in the range

$$\sqrt{\sigma_0} - \sqrt{(\sigma_0 - \sigma_r^{(l)})} \leqslant \sqrt{\sigma_e^{(l)}} \leqslant \sqrt{\sigma_0} + \sqrt{(\sigma_0 - \sigma_r^{(l)})}, \tag{142.9}$$

where $\sigma_0 = (2l+1)\pi/k^2$.

Taking the value of $f(\theta)$ from (142.2) for $\theta = 0$ and comparing with (142.5), we find

$$\operatorname{im} f(0) = k\sigma_t/4\pi, \tag{142.10}$$

which is a generalization of the optical theorem (125.9). Here $f(0)$ is again the amplitude of elastic scattering through zero angle, but the total cross-section σ_t includes the inelastic component.

The imaginary parts of the partial amplitudes f_l are related to the partial cross-section $\sigma_t^{(l)}$ by

$$\operatorname{im} f_l = \frac{k}{4\pi} \frac{\sigma_t^{(l)}}{2l+1}, \tag{142.11}$$

which follows directly from (142.6) and (142.7).

The fact that the coefficients S_l in the asymptotic expression for the wave function are not of unit modulus does not affect the conclusions of §128 concerning the singular points of the elastic scattering amplitude as a function of complex E. These conclusions remain valid when inelastic processes occur. The analytical properties of the amplitude are, however, changed in that it is no longer real on the negative real axis ($E < 0$), and its values on the upper and lower sides of the cut for $E > 0$ are not complex conjugate quantities (and accordingly its values at all points in the upper and lower half-planes symmetrical about the real axis are not complex conjugate quantities).

When we go from the upper edge of the cut to the lower edge by passing round the point $E = 0$, the quantity $\sqrt{E}$ changes sign, i.e. this process changes the sign of the quantity k, which is real (for $E > 0$). The ingoing and outgoing waves in (142.1) are interchanged, and so the coefficient S_l is replaced by its reciprocal $1/S_l$ (which is not equal to S_l^*). The amplitudes f_l on the upper and lower edges of the cut may be denoted by $f_l(k)$ and $f_l(-k)$ (only $f_l(k)$ is a physical amplitude, of course). According to (142.6) we have

$$f_l(k) = \frac{S_l - 1}{2ik}, \quad f_l(-k) = -\frac{1/S_l - 1}{2ik}.$$

Eliminating S_l from these two equations gives

$$f_l(k) - f_l(-k) = 2ikf_l(k)f_l(-k); \tag{142.12}$$

in the absence of inelastic processes, $f(-k) = f^*(k)$, and the relations (142.11) and (142.12) are the same.

Writing (142.12) in the form

$$\frac{1}{f_l(k)} - \frac{1}{f_l(-k)} = -2ik,$$

we see that the sum $1/f_l(k)+ik$ must be an even function of k, and if this is denoted by $g_l(k^2)$, then

$$f_l(k) = \frac{1}{g_l(k^2)-ik}. \tag{142.13}$$

The even function $g_l(k^2)$, however, is not now real as it was in (125.15).†

When a beam of particles passes through a scattering medium consisting of a large number of scatterers, it is gradually attenuated owing to the removal of particles from it which undergo various collision processes. This attenuation is entirely determined by the amplitude of elastic scattering through zero angle and, under certain conditions (see below), can be described by the following formal method.‡

Let $f(0,E)$ be the amplitude of scattering through angle zero by each individual particle of the medium. We shall suppose that f is small in comparison with the mean distance $d \sim (V/N)^{1/3}$ between the particles. Then the scattering by each particle may be considered separately. We use as an auxiliary quantity an "effective field" U_{eff} of a fixed centre, so defined that the Born scattering amplitude for scattering through angle zero in this field is equal to the actual amplitude $f(0,E)$; this does not mean, of course, that the Born approximation can be used to calculate $f(0,E)$ from the actual interaction of the particles. Then, by definition, we have (see (126.4))

$$\int U_{\text{eff}}\,dV = -\frac{2\pi\hbar^2}{m}f(0,E), \tag{142.14}$$

where m is the mass of the scattered particle. The field thus defined is, like the amplitude f, complex. The relation between its range of action a and the quantity U_{eff} is obtained from an estimate of the two sides of equation (142.14):

$$a^3 U_{\text{eff}} \sim \hbar^2 f/m. \tag{142.15}$$

The definition (142.14) is, of course, not unique. We shall impose the further condition that the field U_{eff} satisfies the condition for perturbation theory to be applicable:

$$|U_{\text{eff}}| \ll \hbar^2/ma^2, \tag{142.16}$$

† The foregoing arguments, and the conclusion that the function g_l is even, assume that the interaction decreases sufficiently rapidly as $r \to \infty$, so that there are no cuts in the left half-plane of E and a complete circuit round the point $E = 0$ is possible.

‡ The following treatment can be used, in particular, for the description of scattering of fast neutrons (with energies of the order of hundreds of MeV) by nuclei, the wavelengths of such neutrons being so small that the nucleus may be regarded as an inhomogeneous macroscopic medium.

with $|f| \ll a$. It is easy to see that the attenuation of the scattered beam can then be described as the propagation of a plane wave in a homogeneous medium in which the particle has a constant potential energy given by

$$\overline{U_{\text{eff}}} = \frac{N}{V}\int U_{\text{eff}}\,dV$$

$$= -\frac{N}{V}\frac{2\pi\hbar^2}{m}f(0,E), \qquad (142.17)$$

which is obtained by averaging the effective fields of all N particles in the medium over its volume V. This becomes evident if we first consider scattering by a region of the medium which contains many scattering centres but for which the scattering effect is still small; the possibility of selecting such regions is ensured by the condition (142.16). The attenuation of the beam on passing through such a region is determined by the amplitude of scattering through angle zero, which in turn is determined, in the Born approximation, by the integral of the scattering field over the volume of the scattering region. This means that the scattering properties of interest here are entirely determined by the field (142.17) averaged over the volume of the medium.

Thus the beam of particles passing through the medium can be described by a plane wave $\sim e^{ikz}$ with wave number

$$k = \frac{1}{\hbar}\sqrt{[2m(E-\overline{U_{\text{eff}}})]}.$$

In terms of the wave number $k_0 = \sqrt{(2mE)}/\hbar$ of the incident particles, we can write k in the form nk_0, where the quantity

$$n = \sqrt{\left(1-\frac{\overline{U_{\text{eff}}}}{E}\right)}$$

$$= \sqrt{\left(1+\frac{N}{V}\frac{2\pi\hbar^2}{mE}f(0,E)\right)} \qquad (142.18)$$

plays the part of a "refractive index" of the medium with respect to the beam of particles passing through it. It is in general complex (the amplitude f being complex) and its imaginary part gives the attenuation of the beam intensity. If $E \gg |\overline{U_{\text{eff}}}|$, then (142.18) gives, as it should,

$$\operatorname{im} n = \frac{N}{V}\frac{\pi\hbar^2}{mE}\operatorname{im} f(0,E)$$

$$= \frac{N}{V}\frac{\sigma_t}{2k},$$

where σ_t is the total scattering cross-section, and we have used the optical theorem (142.10). This expression corresponds to the obvious result that the intensity of the wave is damped according to the law

$$|e^{ikz}|^2 \sim e^{-N\sigma_t z/V}.$$

As well as the absorption, the refractive index (142.18) also determines (by its real part) the law of refraction of the beam on entering and leaving the scattering medium.†

PROBLEM

Neutrons are scattered by a heavy nucleus whose radius a is large compared with the wavelength of the neutrons ($ka \gg 1$). It is assumed that all neutrons incident with orbital angular momentum $l < ka \equiv l_0$ (i.e. with impact parameter $\rho = \hbar l/mv = l/k < a$) are absorbed by the nucleus, while those with $l > l_0$ do not interact with it at all. Determine the cross-section for elastic scattering through small angles.

SOLUTION. Under the conditions stated, the motion of the neutrons is mainly quasi-classical, and elastic scattering results from a slight deflection entirely analogous to Fraunhofer diffraction of light by a black sphere. The required cross-section can therefore be written immediately from the known solution of the diffraction problem:‡

$$\mathrm{d}\sigma_e = \pi a^2 \frac{J_1^2(ka\theta)}{\pi\theta^2}\,\mathrm{d}o.$$

The same result can also be derived from (142.3). According to the conditions of the problem, we have $S_l = 0$ for $l < l_0$ and $S_l = 1$ for $l > l_0$. The elastic scattering amplitude is therefore

$$f(\theta) = -\frac{1}{2ik}\sum_{l=0}^{l_0}(2l+1)P_l(\cos\theta).$$

The chief part in the sum is played by the terms with large l. We therefore write $2l$ in place of $2l+1$, use the approximate expression (49.6) for $P_l(\cos\theta)$ with θ small, and change from summation to integration:

$$f(\theta) = \frac{i}{k}\int_0^{l_0} lJ_0(\theta l)\,\mathrm{d}l$$

$$= \frac{i}{k\theta} l_0 J_1(\theta l_0)$$

† An interesting example of the application of (142.17) is the displacement of the higher levels of an alkali-metal atom in a gas. In a highly excited state, the valency electron is at a mean distance $\bar{r}$ from the centre of the atom that is large compared with the dimensions a of both the rest of the atom and the neutral gas atoms. The latter atoms within a sphere of radius $\sim\bar{r}$ act as scattering centres for the valency electron and shift its energy level by an amount (142.17). Since the de Broglie wavelength of the excited valency electron is also large in comparison with a, the amplitude $f(0, E) \approx -\alpha$, where α is the scattering length; cf. (132.9). Thus this effect displaces the levels by a constant amount $2\pi\hbar^2 a\nu/m$, where m is the electron mass and ν the number density of the gas particles (E. Fermi 1934).

‡ See *Fields*, §61, Problem 3 (the problem of diffraction from a black sphere is equivalent to that of diffraction from a circular aperture cut in an opaque screen). The cross-section is obtained by dividing the intensity of the diffracted waves by the incident flux density.

$$= (ia/\theta)J_1(ka\theta),$$

as it should be.†

The total elastic scattering cross-section is

$$\sigma_e = \pi a^2 \int_0^\infty \frac{J_1^2(ka\theta)}{\pi\theta^2} 2\pi\theta \, d\theta$$

$$= \pi a^2;$$

the integration can be extended to infinity because of the rapid convergence. This is the result to be expected under the conditions stated (cf. (142.8)), and is the same as the absorption cross-section, simply the geometrical cross-section of the sphere. The total cross-section $\sigma_t = 2\pi a^2$.

§143. Inelastic scattering of slow particles

The derivation of the limiting law of elastic scattering at low energies given in §132 can easily be generalized to the case where inelastic processes are involved.

As before, the scattering with $l = 0$ is the most important at low energies. According to the results of §132, the corresponding element of the S-matrix is

$$S_0 = e^{2i\delta_0} \approx 1 + 2i\delta_0 = 1 - 2ik\alpha.$$

The properties of the wave function described in §132 are changed only in that the condition imposed on it at infinity (the asymptotic form (142.1)) is now complex, instead of the real stationary wave which occurs in the case of purely elastic scattering. The constant $\alpha = -c_2/c_1$ is therefore complex also. The modulus $|S_0|$ is no longer equal to unity; the condition $|S_0| < 1$ means that the imaginary part of $\alpha = \alpha' + i\alpha''$ must be negative ($\alpha'' < 0$).

Substituting S_0 in (142.7), we find the cross-sections for elastic and inelastic scattering:

$$\sigma_e = 4\pi|\alpha|^2, \tag{143.1}$$

$$\sigma_r = 4\pi|\alpha''|/k. \tag{143.2}$$

Thus the elastic scattering cross-section is again independent of velocity, but the inelastic cross-section is inversely proportional to the particle velocity—the $1/v$ *law* (H. A. Bethe 1935). Consequently, as the velocity diminishes, inelastic processes become more and more important in comparison with elastic scattering.‡

---

† A similar discussion can be given for the problem of diffraction scattering of fast charged particles by a "black" nucleus. The limiting value l_0 must here be determined from the condition that the shortest distance between the nucleus and a particle moving along a classical path in a Coulomb field is just equal to the radius of the nucleus. For $l < l_0$ we must again put $S_l = 0$, and for $l > l_0$ $S_l = e^{2i\delta_l}$, where δ_l are the Coulomb phases given by (135.11). See A. I. Akhiezer and I. Ya. Pomeranchuk, *Some Problems of Nuclear Theory* (*Nekotorye voprosy teorii yadra*), Gostekhizdat, Moscow 1950, §22; *Journal of Physics* **9**, 471, 1945.

‡ The velocity dependence of the partial reaction cross-sections for various non-zero orbital angular momenta l can be determined similarly. It is given by $\sigma_r^{(l)} \sim k^{2l-1}$. The elastic scattering cross-sections $\sigma^{(l)}$ are, as before, proportional to k^{4l}, i.e. they decrease more rapidly than $\sigma_r^{(l)}$ with the same l as $k \to 0$.

The limiting laws (143.1) and (143.2) are, of course, only the first terms of expansions of the cross-sections in powers of k. It is interesting to note that the next term in the expansion for each cross-section contains no constants other than those which appear in (143.1) and (143.2) (F. L. Shapiro 1958). This result is due to the fact that the function $g_0(k^2)$, in the expression (142.13)

$$f_0(k) = \frac{1}{g_0(k^2) - ik}$$

for the partial scattering amplitude ($l = 0$), is even. For small k this function can therefore be expanded in even powers of k, and the term following $g_0 \approx -1/\alpha$ is $\sim k^2$. If we neglect this term, we can still write two terms of the expansion in $f_0(k)$:

$$f_0(k) \approx -\alpha(1 - ik\alpha).$$

Correspondingly we can retain the next terms of the expansions in the cross-sections, for which the following expressions are easily obtained:

$$\sigma_e = 4\pi|\alpha|^2(1 - 2k|\alpha''|), \tag{143.3}$$

$$\sigma_r = 4\pi|\alpha''|(1 - 2k|\alpha''|)/k. \tag{143.4}$$

These results assume a sufficiently rapid decrease of the interaction at large distances. We have seen in §132 that the elastic scattering amplitude tends to a constant limit as $k \to 0$ if the field $U(r)$ decreases more rapidly than r^{-3}. This is a necessary condition also for the validity of the analogous result (143.1) when inelastic channels are present.†

The $1/v$ law for the reaction cross-section is subject to a weaker condition: the field must decrease more rapidly than r^{-2}, as is clear from the following intuitive derivation of the $1/v$ law.

The probability that a reaction will occur in a collision is proportional to the squared modulus of the wave function of the incident particle in the "reaction zone" (in the region $r \sim a$). Physically, this statement expresses the fact that, for example, a slow neutron colliding with a nucleus can bring about a reaction only if it "penetrates" into the nucleus. If the interaction decreases more rapidly than r^{-2}, it does not change the order of magnitude of the wave function between large r and $r \sim a$; in other words, the ratio $|\psi(a)/\psi(\infty)|^2$ tends to a finite limit as $k \to 0$ (this is seen from the fact that the term $U\psi$ in Schrödinger's equation is small compared with $\Delta\psi$). The reaction cross-section is obtained by dividing $|\psi|^2$ by the current density. If ψ is taken as a plane wave normalized to unit current density, we have $|\psi|^2 \sim 1/v$, the required result.

In collisions of charged nuclear particles, there is a slowly decreasing Coulomb field in addition to the short-range nuclear forces. The Coulomb field may considerably alter the magnitude of the incident wave in the reaction zone. The reaction cross-section is found by multiplying $1/v$ by

† The formula (143.3), which takes into account the next term in the expansion in powers of k, requires that U should decrease more rapidly than r^{-4}.

the ratio of the squared moduli of the Coulomb and free wave functions as $r \to 0$. This ratio is given by formulae (136.10), (136.11). The result is (in Coulomb units)

$$\sigma_r = \frac{2\pi A}{k^2|e^{\pm 2\pi/k}-1|}; \tag{143.5}$$

the plus sign in the exponent corresponds to repulsion and the minus sign to attraction.

The coefficient A is the constant in the $1/v$ law; if the velocity is large compared with the Coulomb unit ($k \gg 1$), the Coulomb interaction plays no part and we return to the law $\sigma_r = A/k$.

If the velocity is small compared with the Coulomb unit ($k \ll 1$, or in ordinary units $Z_1Z_2e^2/\hbar v \gg 1$, where Z_1e, Z_2e are the charges of the colliding particles), the Coulomb interaction is predominant in determining the magnitude of the wave function in the reaction zone. Then for a collision between attractive particles

$$\sigma_r = 2\pi A/k^2, \tag{143.6}$$

and for a collision between repulsive particles

$$\sigma_r = (2\pi A/k^2)e^{-2\pi/k}. \tag{143.7}$$

In the latter case, the cross-section tends to zero as $k \to 0$. The exponential factor by which (143.6) and (143.7) differ is the probability of passage through the Coulomb potential barrier; in ordinary units it is $\exp(-2\pi Z_1Z_2e^2/\hbar v)$.

Note that the limiting law (143.6) applies not only to the total cross-section but also to the partial cross-sections with each angular momentum l.† This is seen from the fact that in the expansion (136.1) of the functions $\psi_{\mathbf{k}}^{(+)}$ (which appear in the formulae (136.10) and (136.11) used above) the functions R_{kl} in every term of the sum have the same limiting dependence on k: in the limit $k \to 0$, the radial functions (for the case of attraction) are given by the expressions (36.25), and near the centre we have $R_{kl} \sim \sqrt{k}r^l$. The contributions of the individual angular momenta to the square of the wave function in the reaction zone are $\sim a^{2l}/k$, i.e. all depend on k in the same way, although they are reduced by the small factor $(a/a_c)^{2l}$, where $a_c = \hbar^2/mZ_1Z_2e^2$ is the Coulomb unit of length.

§144. The scattering matrix in the presence of reactions

The cross-section σ_r considered in §§142 and 143 was the total cross-section for all possible inelastic scattering channels. We shall now describe the derivation of the general theory of inelastic collisions, in which each channel can be considered separately.

We shall suppose that, as a result of the collision of two particles, two

† The same is true of (143.7).

particles (which may be the same or different ones) are formed. We number all possible reaction channels (for a given energy), and denote quantities pertaining to them by appropriate suffixes.

Let channel i be the input channel. The wave function of the relative motion of the colliding particles (in the centre-of-mass system) in this channel is given by the sum already mentioned of the incident plane wave and the elastically scattered outgoing wave:

$$\psi_i = e^{ik_i z} + f_{ii}(\theta)\frac{e^{ik_i r}}{r}. \tag{144.1}$$

The square of the amplitude f_{ii} gives the cross-section for elastic scattering in channel i:

$$d\sigma_{ii} = |f_{ii}|^2 do. \tag{144.2}$$

In other channels (suffix f) the wave functions of the relative motion of the particles represent outgoing waves. As explained above, these waves are conveniently represented in the form†

$$\psi_f = f_{fi}(\theta)\sqrt{\frac{m_f}{m_i}}\frac{e^{ik_f r}}{r}, \tag{144.3}$$

where $\mathbf{k}_f$ is the wave vector of the relative motion of the reaction products in channel f, θ the angle between it and the z-axis, and m_i, m_f the reduced masses of the two initial and two final particles. The scattered flux in the solid angle do is obtained by multiplying the square $|\psi_f|^2$ by $v_f r^2 do$, and the cross-section for the corresponding reaction is found by dividing this flux by the incident flux density, which is v_i. Thus

$$d\sigma_{fi} = |f_{fi}|^2 \frac{p_f}{p_i} do_f, \tag{144.4}$$

where the momenta $p_i = m_i v_i$, $p_f = m_f v_f$.

In §125 we have defined the scattering operator $\hat{S}$, which converts an ingoing wave into an outgoing one. When several channels are present, this operator has matrix elements for transitions between different channels. The elements which are "diagonal" with respect to the channels correspond to elastic scattering, and the non-diagonal elements correspond to various inelastic processes. All these elements remain operators with respect to the other variables. They are determined as follows.

Similarly to the method used in §125, we define operators $\hat{f}_{ii}$, $\hat{f}_{fi}$ related to the amplitudes f_{ii}, f_{fi}, by

$$\hat{S}_{fi} = \delta_{fi} + 2i\sqrt{(k_i k_f)}\hat{f}_{fi}. \tag{144.5}$$

It is easily seen that with this definition we obtain an S-matrix which must

† Here we again denote the initial state of the system by the suffix i and the final state by f (cf. the first footnote to §41). In the scattering amplitude, the suffix for the final state is written to the left of that for the initial state, in accordance with the placing of the suffixes in the matrix elements. For uniformity, the suffixes in the cross-sections will be put in the same order.

satisfy the unitarity condition. For we can write the wave function in the input channel as a set of ingoing and outgoing waves, as in §125:

$$\psi_i = F(-\mathbf{n}')\frac{e^{-ik_ir}}{r\sqrt{v_i}} - (1+2ik_i\hat{f}_{ii})F(\mathbf{n}')\frac{e^{ik_ir}}{r\sqrt{v_i}}$$

$$= F(-\mathbf{n}')\frac{e^{-ik_ir}}{r\sqrt{v_i}} - \hat{S}_{ii}F(\mathbf{n}')\frac{e^{ik_ir}}{r\sqrt{v_i}}. \tag{144.6}$$

Here, for convenience, we have introduced a further factor $1/\sqrt{v_i}$ in comparison with (125.3). Then, with the amplitudes as defined above, the wave function in channel f is

$$\psi_f = 2ik_i\sqrt{\frac{m_f}{m_i}}\hat{f}_{fi}F(\mathbf{n}')\frac{e^{ik_fr}}{r\sqrt{v_i}}$$

$$= \hat{S}_{fi}F(\mathbf{n}')\frac{e^{ik_fr}}{r\sqrt{v_f}}. \tag{144.7}$$

The flux in the ingoing waves must be equal to the sum of the fluxes in the outgoing waves in all channels. This requirement expresses the obvious condition that the sum of the probabilities of all processes (elastic and inelastic) which can occur in the collision must be unity. On account of the factor $\sqrt{v}$ in the denominators of the spherical waves, the velocity does not appear in the flux densities in these waves. The above condition therefore means simply that the normalizations of the ingoing wave and the assembly of outgoing waves must be the same. It is consequently again expressed by the condition of unitarity of the scattering operator, regarded as a matrix, with respect to (in particular) the channel numbers. For the operator $\hat{f}_{fi}$ this condition becomes

$$\hat{f}_{fi}-\hat{f}_{if}^{+} = 2i\sum_n k_n\hat{f}_{fn}\hat{f}_{in}^{+}, \tag{144.8}$$

which is analogous to (125.7). The index + denotes taking the complex conjugate and transposing with respect to all the matrix suffixes except the channel number.

The S-matrix is diagonal with respect to states having definite values of the orbital angular momentum l; the corresponding matrix elements are distinguished by the index (l). By applying the operators $\hat{f}_{ii}$ and $\hat{f}_{fi}$ to the function (125.17), we obtain the amplitudes for elastic and inelastic scattering processes in the form

$$\left.\begin{aligned} f_{ii} &= \frac{1}{2ik_i}\sum_{l=0}^{\infty}(2l+1)(S_{ii}^{(l)}-1)P_l(\cos\theta), \\ f_{fi} &= \frac{1}{2i\sqrt{(k_ik_f)}}\sum_{l=0}^{\infty}(2l+1)S_{fi}^{(l)}P_l(\cos\theta). \end{aligned}\right\} \tag{144.9}$$

The corresponding integral cross-sections are

$$\left.\begin{aligned}\sigma_{ii} &= \frac{\pi}{k_i^2}\sum_{l=0}^{\infty}(2l+1)|1-S_{ii}^{(l)}|^2,\\ \sigma_{fi} &= \frac{\pi}{k_i^2}\sum_{l=0}^{\infty}(2l+1)|S_{fi}^{(l)}|^2.\end{aligned}\right\}\tag{144.10}$$

The former is the same as (142.3). The total reaction cross-section σ_r (from input channel i) is

$$\sigma_r = \sum_f{}' \sigma_{fi}$$

taken over all $f \neq i$. Since the S-matrix is unitary, we have

$$\sum_f{}' |S_{fi}|^2 = 1-|S_{ii}|^2,$$

which gives formula (142.4) for σ_r.

The symmetry of the scattering process with respect to time reversal (the *reciprocity theorem*) is given by the equation

$$\hat{S}_{fi} = \hat{S}_{i^*f^*},\tag{144.11}$$

or, what is the same thing,

$$\hat{f}_{fi} = \hat{f}_{i^*f^*}.\tag{144.12}$$

The symbols i^* and f^* denote states which differ from i and f by a change in the signs of the momenta and spin components of the particles;† they are said to be *time-reversed* relative to the states i and f. The relations (144.11) and (144.12) generalize formulae (125.11) and (125.12) for elastic scattering.‡

Equation (144.12) leads to the following relation for the reaction cross-sections:

$$\mathrm{d}\sigma_{fi}/p_f^2\mathrm{d}o_f = \mathrm{d}\sigma_{i^*f^*}/p_i^2\mathrm{d}o_{i^*}.\tag{144.13}$$

This expresses the *principle of detailed balancing*.

† For complex particles (atoms and nuclei) the "spin" is here to be taken as the total intrinsic angular momentum, consisting of the spins and the orbital angular momenta of the internal motions of the constituent parts.

‡ Here we omit the factor −1 which may appear in collisions of particles having spin (cf. (140.11)). This, of course, does not affect formula (144.13) for the cross-sections.

It has been mentioned in §126 that, if perturbation theory is applicable, then in the first approximation we have not only the reciprocity theorem but also a further relation between the amplitudes of the direct and reverse processes (in the literal sense), $i \to f$ and $f \to i$. This property, expressed by the equation $f_{fi} = f_{if}^*$, holds good for inelastic processes (in the same approximation). The corresponding cross-sections are then related by

$$\frac{d\sigma_{fi}}{p_f^2 do_f} = \frac{d\sigma_{if}}{p_i^2 do_i}. \tag{144.14}$$

The difference between the transitions $i \to f$ and $i^* \to f^*$ no longer exists if we consider the cross-sections integrated over directions of $\mathbf{p}_f$ and summed over directions of the spins s_{1f}, s_{2f} of the resulting particles and averaged over the directions of the momentum $\mathbf{p}_i$ and spins s_{1i}, s_{2i} of the initial particles. Let this cross-section be

$$\overline{\sigma_{fi}} = \frac{1}{4\pi(2s_{1i}+1)(2s_{2i}+1)} \sum_{(m_s)} \int d\sigma_{fi} do_i;$$

the sum is taken over the spin components of all particles, and the factor before the sum and integral is due to the fact that we average, not sum, over quantities pertaining to the initial particles. Writing (144.13) in the form

$$p_i^2 d\sigma_{fi} do_{i*} = p_f^2 d\sigma_{i*f*} do_f$$

and effecting the integrations and summations, we obtain

$$g_i p_i^2 \overline{\sigma_{fi}} = g_f p_f^2 \overline{\sigma_{if}}. \tag{144.15}$$

Here

$$g_i = (2s_{1i}+1)(2s_{2i}+1), \quad g_f = (2s_{1f}+1)(2s_{2f}+1); \tag{144.16}$$

these determine the numbers of possible spin orientations of the initial pair or the final pair of the particles, and are called the *statistical weights* of the states i and f.

Finally, we may note the following property of the amplitudes f_{fi}. We have seen in §140 that the cross-section σ_{fi} varies as $1/p_i$ when $p_i \to 0$ (if the interaction decreases sufficiently rapidly at large distances). According to formula (144.4), this means that $f_{fi} \to$ constant as $p_i \to 0$. Hence it follows from the symmetry property (144.12) that f_{fi} tends to a constant limit as $p_f \to 0$ also. We shall return to this result in §147.

§145. Breit and Wigner's formulae

In §134 we have introduced the concept of quasi-stationary states as being those which have a finite but relatively long lifetime. A wide class of such states arises in the field of nuclear reactions at not too high energies which pass through the stage of formation of a *compound nucleus*.†

An intuitive physical picture of the processes occurring is that the particle incident on the nucleus interacts with the nucleons in the nucleus and "coalesces" with them, forming a compound system in which the energy contributed by the particle is distributed between many nucleons. The resonance energies correspond to the quasi-discrete levels of this compound system. The long lifetime of the quasi-stationary states (compared with the periods of the motion of the nucleons in the nucleus) is due to the fact that for the greater part of the time the energy is distributed between many particles, so that none of them has sufficient energy to overcome the attraction of the other particles and leave the nucleus. Sufficient energy for this purpose is only comparatively rarely concentrated on one particle. The disintegration of the compound nucleus can then take place in various ways corresponding to the various possible reaction channels.‡

This description of such collisions shows that the possibility of inelastic processes does not affect the potential part of the elastic scattering amplitude, which is not related to the properties of the compound nucleus (see §134); inelastic processes change only the resonance part of the elastic scattering amplitude. For the same reason the amplitudes of inelastic scattering processes which pass through the stage of formation of the compound nucleus are purely resonance in character. The resonance denominators of all amplitudes which relate to the vanishing of the coefficient of the ingoing wave for $E = E_0 - \frac{1}{2}i\Gamma$ retain their form $(E - E_0 + \frac{1}{2}i\Gamma)$, Γ being again the total probability of decay of any given quasi-stationary state of the compound nucleus.

These arguments, together with the unitarity condition which must be satisfied by the scattering amplitudes, are sufficient to establish the form of these amplitudes.

The calculations may conveniently be made in a symmetrical form by numbering all possible channels of disintegration of the compound nucleus and not specifying beforehand which of them is the input channel for the reaction concerned. The suffixes denoting the channel numbers will be represented by a, b, c, ... We shall also consider the partial scattering amplitudes corresponding to the value of l for the quasi-stationary state in question.‖ We accordingly seek these amplitudes in the form

$$f_{ab}^{(l)} = \frac{1}{2ik_a}(e^{2i\delta_a}-1)\delta_{ab} - \frac{1}{2\sqrt{(k_a k_b)}} e^{i(\delta_a+\delta_b)} \frac{\Gamma M_{ab}}{E - E_0 + \frac{1}{2}i\Gamma} \tag{145.1}$$

(the index (l) to the constants δ_a and M_{ab} is omitted for simplicity). The

† The concept of the compound nucleus is due to N. Bohr (1936).

‡ The competing reactions include also radiative capture of the incident particle, in which the compound nucleus goes from an excited state to its ground state with the emission of a γ-quantum. This process is also "slow", owing to the relatively low probability of the transition with emission.

‖ We shall at first ignore the complications which arise from the spins of the particles involved in the process.

first term appears only if $a = b$, and represents the amplitude of potential elastic scattering in channel a; the constants δ_a are the same as the phases $\delta_l^{(0)}$ which appear in (134.12). The second term in (145.1) corresponds to resonance processes. The form of the coefficient of the resonance factor in this term is chosen so as to simplify the result of applying the unitarity conditions (see below).

Since we are considering scattering for a given value of the absolute magnitude of the orbital angular momentum, a quantity which does not change sign under time reversal, the reciprocity theorem (symmetry with respect to time reversal) is expressed simply by the symmetry of the amplitudes $f_{ab}^{(l)}$ with respect to the suffixes a and b. Hence it follows that the coefficients M_{ab} must also be symmetrical ($M_{ab} = M_{ba}$).

The unitarity conditions for the amplitudes $f_{ab}^{(l)}$ are

$$\operatorname{im} f_{ab}^{(l)} = \sum_c k_c f_{ac}^{(l)} f_{bc}^{(l)*}; \tag{145.2}$$

cf. (144.8). Substituting the expressions (145.1), we find after a straightforward calculation

$$\frac{M_{ab}^*}{E-E_0-\frac{1}{2}i\Gamma} - \frac{M_{ab}}{E-E_0+\frac{1}{2}i\Gamma} = \frac{i\Gamma \sum_c M_{ac}M_{bc}^*}{(E-E_0)^2+\frac{1}{4}\Gamma^2}.$$

If this equation is satisfied identically for all energies E, we must have first of all $M_{ab} = M_{ab}^*$, i.e. the quantities M_{ab} are real. We then find

$$M_{ab} = \sum_c M_{ac}M_{bc}, \tag{145.3}$$

i.e. the matrix of coefficients M_{ab} must be equal to its own square.

The real symmetrical matrix M_{ab} can be brought to diagonal form by a suitable orthogonal linear transformation $\hat{U}$. Denoting the diagonal elements (eigenvalues) of the matrix by $M^{(\alpha)}$, we can write this transformation in the form

$$\sum_{a,b} U_{\alpha a} U_{\beta b} M_{ab} = M^{(\alpha)} \delta_{\alpha\beta},$$

where the transformation coefficients satisfy the orthogonality relations

$$\sum_c U_{\alpha c} U_{\beta c} = \delta_{\alpha\beta}. \tag{145.4}$$

Conversely

$$M_{ab} = \sum_\alpha U_{\alpha a} U_{\alpha b} M^{(\alpha)}. \tag{145.5}$$

The relations (145.3) give the conditions $M^{(\alpha)} = (M^{(\alpha)})^2$ for the eigenvalues $M^{(\alpha)}$, so that these must be zero or unity. If only one of the $M^{(\alpha)}$ is different from zero (say $M^{(1)} = 1$), then (145.5) gives

$$M_{ab} = U_{1a} U_{1b}, \tag{145.6}$$

i.e. all the matrix elements M_{ab} are expressed in terms of the set of quantities U_{1a}, $a = 1, 2, \dots$. If several of the $M^{(\alpha)}$ are non-zero, then the elements M_{ab} are sums expressed in terms of several sets of quantities $U_{1a}, U_{2a}, \dots$, these quantities being related only by the orthogonality relations and otherwise independent. This case would correspond to accidental degeneracy, where several different quasi-stationary states of the compound nucleus correspond to the same quasi-discrete energy level.† Ignoring these unimportant cases, i.e. considering non-degenerate levels, we therefore conclude that the matrix elements M_{ab} are products of quantities each depending on the number of only one channel.

With the notation

$$|U_{1a}| = \sqrt{(\Gamma_a/\Gamma)},$$

we can write formula (145.6) as

$$M_{ab} = \pm\sqrt{(\Gamma_a\Gamma_b)}/\Gamma; \tag{145.7}$$

the sign of M_{ab} depends on those of U_{1a} and U_{1b}, and remains indeterminate. On account of the equation $\Sigma U_{1c}U_{1c} = 1$, the quantities Γ_a thus defined satisfy the relation

$$\sum_a \Gamma_a = \Gamma. \tag{145.8}$$

They are called the *partial widths* of the various channels. Formulae (145.1), (145.7) and (145.8) give the required general form of the scattering amplitudes.

Let us now rewrite the final formulae, taking some definite channel as the input channel.‡ The partial width of this channel will be denoted by Γ_e (the *elastic width*) and the widths of channels corresponding to various reactions by $\Gamma_{r1}, \Gamma_{r2}, \dots$.

The total elastic scattering amplitude is

$$f_e(\theta) = f^{(0)}(\theta) - \frac{2l+1}{2k}\frac{\Gamma_e}{E-E_0+\frac{1}{2}i\Gamma}e^{2i\delta_l^{(0)}}P_l(\cos\theta), \tag{145.9}$$

where k is the wave number of the incident particle and $f^{(0)}$ the potential scattering amplitude. This formula differs from the expression (134.12) in that Γ in the numerator of the resonance term is replaced by the smaller quantity Γ_e.

The amplitudes of inelastic processes are, as already mentioned, of purely resonance type. The differential cross-sections are

$$d\sigma_{ra} = \frac{(2l+1)^2}{4k^2}\frac{\Gamma_e\Gamma_{ra}}{(E-E_0)^2+\frac{1}{4}\Gamma^2}[P_l(\cos\theta)]^2, \tag{145.10}$$

† This is particularly clear in the case where all the $M^{(\alpha)} = 1$. It follows from (145.4) and (145.5) that then $M_{ab} = \delta_{ab}$, i.e. there are no transitions between different channels. In other words, this case would correspond to a number of independent quasi-discrete states, each occurring in elastic scattering in one channel.

‡ These formulae were first obtained by G. Breit and E. Wigner (1936).

and the integral cross-sections are

$$\sigma_{ra} = (2l+1)\frac{\pi}{k^2}\frac{\Gamma_e \Gamma_{ra}}{(E-E_0)^2+\frac{1}{4}\Gamma^2}. \tag{145.11}$$

The total cross-section for all possible inelastic processes is

$$\sigma_r = (2l+1)\,\frac{\pi}{k^2}\,\frac{\Gamma_e \Gamma_r}{(E-E_0)^2+\frac{1}{4}\Gamma^2}, \tag{145.12}$$

where $\Gamma_r = \Gamma - \Gamma_e$ is the total *inelastic width* of the level.

It is also of interest to know the value of the reaction cross-section integrated over the range of energy near the resonance value $E = E_0$. Since σ_r decreases rapidly away from the resonance, the integration with respect to $E - E_0$ can be extended from $-\infty$ to $+\infty$, giving

$$\int \sigma_r \mathrm{d}E = (2l+1)\frac{2\pi^2}{k^2}\frac{\Gamma_e \Gamma_r}{\Gamma}. \tag{145.13}$$

In the scattering of slow neutrons (for which the wavelength is large compared with the dimensions of the nucleus), only *s*-wave scattering is important, and the potential scattering amplitude is a real constant $-\alpha$. Then (134.14) becomes

$$f_e = -\alpha - \frac{\Gamma_e}{2k(E-E_0+\frac{1}{2}i\Gamma)}. \tag{145.14}$$

The total elastic scattering cross-section is

$$\sigma_e = 4\pi\alpha^2 + \frac{\pi}{k^2}\frac{\Gamma_e^2+4\alpha k\Gamma_e(E-E_0)}{(E-E_0)^2+\frac{1}{4}\Gamma^2}. \tag{145.15}$$

The term $4\pi\alpha^2$ may be called the *potential scattering cross-section.* We see that in the resonance region there is interference between the potential scattering and the resonance scattering. The amplitude α can be negligible only in the immediate neighbourhood of the level ($E - E_0 \sim \Gamma$) (we recall that $|\alpha k| \ll 1$), and the formula for the slow neutron elastic scattering cross-section then becomes

$$\sigma_e = \frac{\pi}{k^2}\frac{\Gamma_e^2}{(E-E_0)^2+\frac{1}{4}\Gamma^2}. \tag{145.16}$$

The total cross-section for elastic and inelastic scattering is

$$\sigma_t = \sigma_e + \sigma_r = \frac{\pi}{k^2}\frac{\Gamma_e \Gamma}{(E-E_0)^2+\frac{1}{4}\Gamma^2}. \tag{145.17}$$

When potential scattering is negligible, the cross-sections σ_e, σ_{ra} can be put in the form

$$\sigma_e = \sigma_t \Gamma_e / \Gamma, \quad \sigma_{ra} = \sigma_t \Gamma_{ra} / \Gamma.$$

The quantity σ_t is the sum of the cross-sections for all possible resonance processes, and may be regarded as the cross-section for the formation of the compound nucleus. The cross-sections for the various elastic and inelastic processes are obtained by multiplying σ_t by the relative probabilities of particular types of disintegration of the compound nucleus, which are given by the ratios of the corresponding partial widths to the total width of the level. The possibility of this representation of the cross-sections is the result of the factorization of the coefficients M_{ab} in the numerators of the scattering amplitudes. It corresponds to the physical picture of the collision process as occurring in two stages: the formation of the compound nucleus in a certain quasi-stationary state, and its disintegration through one or another channel.†

As already mentioned in §134, the range of applicability of the formulae considered here is limited only by the requirement that the difference $|E-E_0|$ should be small compared with the distance D between neighbouring quasi-discrete levels of the compound nucleus (with equal values of the angular momentum). It was also mentioned, however, that the formulae as written do not allow the passage to the limit $E \to 0$, which is relevant if the value $E = 0$ lies in the resonance region. In this case the formulae must be modified by replacing the energy E_0 by some related constant ϵ_0, and the elastic width Γ_e by $\gamma_e \sqrt{E}$; the inelastic width Γ_r must again be regarded as constant (H. A. Bethe and G. Placzek 1937).‡ This change causes the inelastic cross-section (145.12) to increase as $1/\sqrt{E}$ when $E \to 0$, in accordance with the general theory of the inelastic scattering of slow particles (§143).

When the spins of the colliding particles are taken into account, the formulae are in general very complicated. We shall consider only the simplest, though important, case of the scattering of slow neutrons, when only orbital angular momenta $l = 0$ are involved in the scattering. The spin of the compound nucleus is obtained by adding the spin i of the target nucleus to the spin $s = \frac{1}{2}$ of the neutron, i.e. it can take the values $j = i \pm \frac{1}{2}$ (we assume that $i \neq 0$, since otherwise the formulae are unchanged). Each quasi-discrete level of the compound nucleus relates to a definite value of j. The reaction cross-section is therefore obtained by multiplying the expression (145.12) (with $l = 0$) by the probability $g(j)$ that the system of nucleus + neutron will have the necessary value of j for which there is a resonance level.

We shall suppose that the spins of the neutrons and of the target nuclei

---

† All the above calculations have been based on a reaction of the form $a+X = b+Y$, in which two initial particles (the nucleus and the incident particle) give rise to two particles. This assumption is not, however, of fundamental importance, as is clear from the physical nature of the results obtained. Formulae of the type (145.11) for the integral cross-sections are valid also for reactions where more than one particle leaves the nucleus.

‡ It is important to note that, for inelastic processes which are possible at small energies (for example, radiative capture), the value $E = 0$ is not a threshold value. A change in the partial widths Γ_{ra} similar to that specified for Γ_e would be necessary for energies close to the threshold of the reaction in question, below which it cannot occur at all.

are oriented at random. There are altogether $(2i+1)(2s+1) = 2(2i+1)$ possible orientations of the pair of spins **i** and **s**. Of these, $2j+1$ correspond to a given value j of the total angular momentum. Assuming that all orientations are equally probable, we find that the probability of a given value j is

$$g(j) = \frac{2j+1}{2(2i+1)}. \tag{145.18}$$

The formula for the elastic scattering cross-section must be modified similarly. Here it must be borne in mind that, in potential scattering, both values of j are involved. The factor $g(j)$ (with j corresponding to the resonance level) must therefore be included in the second term in (145.15), while the term $4\pi\alpha^2$ must be replaced by the sum

$$\sum_j g(j) \,.\, 4\pi[\alpha^{(j)}]^2.$$

The fact that resonance reactions go through the stage of formation of a compound nucleus in a definite quasi-stationary state leads to some general conclusions concerning the angular distribution of the products of these reactions. Each quasi-stationary state has a certain parity (in addition to its other characteristics). The system of particles $b+Y$ formed in the disintegration of the compound nucleus will therefore have the same parity. This means that the wave function of this system, and therefore the reaction amplitudes, can only be multiplied by ± 1 when the coordinate system is inverted; the squared amplitudes, i.e. the cross-sections, therefore remain unchanged. Inversion of the co-ordinates signifies (in the centre-of-mass system) the changes $\theta \to \pi-\theta$, $\phi \to \pi+\phi$ for the polar angle and the azimuth which determine the direction of scattering. The angular distribution of the reaction products must therefore be invariant under this change. In particular, after averaging with respect to the directions of the spins of all the particles participating in the reaction, the cross-section depends only on the scattering angle θ, and the distribution with respect to this angle must be symmetrical with respect to the change $\theta \to \pi-\theta$, i.e. the angular distribution (in the centre-of-mass system) is symmetrical about a plane perpendicular to the direction of collision of the particles.†

Owing to the very large number of closely-packed levels of the compound nucleus, the detailed variation with energy of the cross-sections for various scattering processes is extremely complex. This complexity makes difficult, in particular, the discovery of any systematic changes in the properties of the cross-sections from one nucleus to another. It is therefore reasonable to consider the behaviour of the cross-sections apart from the details of the resonance structure, i.e. averaged over energy ranges which are large compared with the distances between levels. With this treatment we also make no distinction between the various types of inelastic process, but

† For particles without spin, the differential reaction cross-section would be simply proportional to $[P_l(\cos\theta)]^2$, and the symmetry is obvious.

divide the scattering only into "elastic" and "inelastic" (in the sense defined below).†

To demonstrate the significance of the averaging processes, we again omit the complexities associated with spin, and consider the partial cross-sections for scattering with $l = 0$.

According to formulae (142.7),

$$\left.\begin{aligned}\sigma_e = \frac{\pi}{k^2}|S-1|^2, \quad \sigma_r = \frac{\pi}{k^2}(1-|S|^2),\\ \sigma_t = \frac{\pi}{k^2}.2(1-\operatorname{re} S),\end{aligned}\right\} \tag{145.19}$$

the elastic and inelastic scattering cross-sections, and therefore the total cross-section, are expressed in terms of the same quantity S (the index (0) is omitted for brevity). In averaging over the energy interval, the total cross-section, which depends linearly on S, is given in terms of the mean value of S by

$$\bar{\sigma}_t = (\pi/k^2).2(1-\operatorname{re}\bar{S}); \tag{145.20}$$

the factor $1/k^2$, which varies only slowly, is unaffected by the averaging. The averaged "elastic" cross-section is defined as

$$\bar{\sigma}_e{}^{\mathrm{opt}} = (\pi/k^2)|\bar{S}-1|^2, \tag{145.21}$$

which is not in general equal to the mean value $\bar{\sigma}_e$. In other words, we define the elastic scattering by first averaging the amplitude in the outgoing wave Se^{ikr}/r. With this definition the elastic scattering of a wave packet leaves it unchanged in form; we can say that the cross-section (145.21) relates to the "coherent" part of the scattering. This means that the part of the elastic scattering which occurs through the formation of a compound nucleus is excluded: when a long-lived compound nucleus is formed and then disintegrates, the specific features of the incident wave packet are, of course, lost. The "inelastic" scattering in the averaged model is now naturally defined as the difference $\bar{\sigma}_a{}^{\mathrm{opt}} = \bar{\sigma}_t - \bar{\sigma}_e{}^{\mathrm{opt}}$, i.e.

$$\bar{\sigma}_a{}^{\mathrm{opt}} = (\pi/k^2)(1-|\bar{S}|^2). \tag{145.22}$$

This includes, therefore, not only the various inelastic processes but also that part of the elastic scattering which occurs with the formation of an intermediate compound nucleus.

It is easy to see that this interpretation gives a correct account of the limiting cases, and therefore serves as a reasonable interpolation.

In the region of low energies, where the resonances are well resolved ($\Gamma \ll D$), S is given near each level by the formula

$$S = e^{2i\delta^{(0)}}\left(1-\frac{i\Gamma_e}{E-E_0+\frac{1}{2}i\Gamma}\right).$$

† The following method of averaging (for proceeding to what is called the *optical model* of nuclear scattering) was proposed by V. F. Weisskopf, C. E. Porter and H. Feshbach (1954).

Averaging gives

$$\bar{S} = e^{2i\delta^{(0)}}(1-\pi\bar{\Gamma}_e/D), \tag{145.23}$$

where $\bar{\Gamma}_e$ and D are the elastic width and the mean distance between the levels, averaged over the levels occurring in the energy range concerned; the slowly varying function $\delta^{(0)}(E)$ may be regarded as constant in the averaging. Hence we find

$$\bar{\sigma}_a{}^{\text{opt}} = \frac{\pi}{k^2}\frac{2\pi\bar{\Gamma}_e}{D}, \tag{145.24}$$

where small terms $\sim \Gamma/D$ have been omitted.† This expression in fact coincides with the mean value of the cross-section (145.17), which, as previously mentioned, corresponds to the formation of a compound nucleus.

As the excitation energy of the compound nucleus increases, the distances between its levels decrease, and the disintegration probabilities (and so also the total widths of the levels) increase, so that the levels begin to overlap (in which case the concept of quasi-discrete levels loses much of its significance). The irregularities of the function $S(E)$ are then smoothed out, so that the difference between the exact and the averaged functions becomes small, and the cross-section (145.22) is the same as σ_r given by (145.19). This is in accordance with the fact that at high energies the disintegration of the compound nucleus through the input channel is unimportant in comparison with the numerous other modes of disintegration possible at such energies. In this range, therefore, all processes which involve the formation of a compound nucleus may be regarded as inelastic.

Thus in the averaged model the scattering is again determined by a single quantity $\bar{S}$, which is now a smooth function of energy. In the *optical model*, in order to calculate this function, the scattering properties of the nucleus are approximated by a field of force with a complex potential. The imaginary part of the potential has the result that absorption of particles occurs as well as elastic scattering. This absorption, the cross-section for which is given by the expression (145.22), is identified with "inelastic" scattering in the averaged model.

§146. Interaction in the final state in reactions

The interaction between particles formed as a result of a reaction may have a considerable effect on their distribution in energy and angle. This effect will naturally be particularly marked when the relative velocity of the interacting particles is small. Such a phenomenon occurs, for example, in nuclear reactions accompanied by the emission of two or more nucleons, the effect here being due to the nuclear forces which act between free nucleons.‡

Let $\mathbf{p}_0$ be the momentum of the centre of mass of a pair of emergent

† Terms arising in the region of a level owing to the presence of other levels would be of the same order of magnitude.

‡ The results given below were obtained first by A. B. Migdal (1950) and, independently, by K. M. Watson (1952).

nucleons, and $\mathbf{p}$ the momentum of their relative motion. We shall suppose that $p \ll p_0$, and so the relative energy $E = p^2/m$ (m being the nucleon mass) is small compared with the energy $E_0 = p_0^2/4m$ of the motion of the centre of mass. We also suppose that the energy E_0 is large compared with the energy ϵ of the level (real or virtual) belonging to the system of two nucleons. That is, only the relative motion of the nucleons is assumed "slow", while the nucleons themselves are "fast".

The probability of reaction is proportional to the squared modulus of the wave function of the particles formed when they are in the "reaction zone", i.e. at a distance apart which is of the order of the range a of action of nuclear forces (cf. the similar discussion in §143 relating to primary particles). In the present case our object is to determine the dependence of the reaction probability only on the characteristics of the relative motion of one pair of nucleons. It is therefore sufficient to consider only the wave function $\psi_{\mathbf{p}}(\mathbf{r})$ of this motion, so that the probability of the formation of a pair of nucleons with relative momentum in the range d^3p is

$$dw_{\mathbf{p}} = \text{constant} \times |\psi_{\mathbf{p}}(a)|^2 d^3p. \tag{146.1}$$

It has been shown in §136 that, in order to find the probability that a system will enter, through scattering, a state with a definite direction of motion, we must take as the wave functions of the final state functions $\psi_{\mathbf{p}}^-$ which contain (at infinity) only an ingoing wave together with a plane wave; these functions must be normalized by the delta function of momentum. The functions $\psi_{\mathbf{p}}^{(-)}$ are also obtained directly (by taking the complex conjugate and changing the sign of $\mathbf{p}$) from the functions $\psi_{\mathbf{p}}^{(+)}$, which contain (at infinity) outgoing spherical waves, i.e. those which correspond to the mutual scattering of two particles. On substitution in (146.1) this difference is not significant, so that $\psi_{\mathbf{p}}$ in (146.1) may be taken to be the functions $\psi_{\mathbf{p}}^{(+)}$, and the problem is therefore reduced to that of the resonance scattering of slow particles, which has already been discussed.

Although the actual form of the function $\psi_{\mathbf{p}}$ in the region $r \sim a$ is unknown, in order to find the dependence of the probability on the energy E it is sufficient to consider this function at distances $r \lesssim 1/k \gg a$ (where $\mathbf{k} = \mathbf{p}/\hbar$; it is assumed that $ka \ll 1$), and then continue it in order of magnitude to distances $r \sim a$.† The main contribution to $\psi_{\mathbf{p}}$ comes from the spherical wave (containing the factor $1/r$). This wave is an assembly of partial waves with various values of l, whose amplitudes are the corresponding scattering amplitudes. To determine the square $|\psi_{\mathbf{p}}(a)|^2$ it is sufficient to consider the s-wave alone, since at low energies the scattering amplitudes with $l \neq 0$ are relatively small. According to formula (133.7) we therefore have

$$\psi_{\mathbf{p}} \sim \frac{1}{\kappa + ik} \frac{e^{ikr}}{r}, \tag{146.2}$$

† This procedure is permissible because in the region $r \ll 1/k$ the energy E may be neglected in Schrödinger's equation which determines the function $\psi_{\mathbf{p}}$. The dependence of $\psi_{\mathbf{p}}$ on E in this region is therefore entirely determined by the "joining" to the function in the region $r \sim 1/k$.

where $\kappa = \sqrt{(m|\epsilon|)}/\hbar$ and ϵ is the energy of the bound (or virtual) state of the two-nucleon system.† Substituting this expression in (146.1), we obtain

$$dw_{\mathbf{p}} = \text{constant} \times \frac{d^3p}{E+|\epsilon|}. \tag{146.3}$$

Thus the distribution with respect to direction of the momentum (in the centre-of-mass system of the two nucleons) is isotropic. The distribution with respect to energy of the relative motion is given by

$$dw_E = \text{constant} \times \frac{\sqrt{E}dE}{E+|\epsilon|}. \tag{146.4}$$

We see that the interaction of the nucleons leads to the appearance of a maximum in the distribution in the range of small E, at $E \sim |\epsilon|$.‡

In the laboratory system of coordinates, small angles θ between the momenta of the two nucleons correspond to small values of the relative momentum ($p \ll p_0$). Thus in this system an angular correlation between the directions of emission of the nucleons corresponds to the maximum in the distribution with respect to E, and leads to an increased probability of small values of θ.

Let $\mathbf{p}_1$ and $\mathbf{p}_2$ be the momenta of the nucleons in the laboratory system. Then

$$\mathbf{p}_0 = \mathbf{p}_1 + \mathbf{p}_2, \quad \mathbf{p} = \tfrac{1}{2}(\mathbf{p}_2 - \mathbf{p}_1)$$

(the reduced mass of two equal particles is $\frac{1}{2}m$). The vector product of these equations gives $\mathbf{p}_0 \times \mathbf{p} = \mathbf{p}_1 \times \mathbf{p}_2$, and so if $p \ll p_0$ we have

$$p_0 p_\perp = p_1 p_2 \sin\theta \approx \tfrac{1}{4} p_0^2 \theta,$$

or $\theta = 4p_\perp/p_0$, where $p_\perp$ is the component of the vector $\mathbf{p}$ transverse with respect to the direction of $\mathbf{p}_0$, and θ is the small angle between the directions of $\mathbf{p}_1$ and $\mathbf{p}_2$. Rewriting formula (146.3) in the form

$$dw_{\mathbf{p}} = \text{constant} \times \frac{2\pi p_\perp dp_\perp dp_\parallel}{(p_\perp^2 + p_\parallel^2)/m + |\epsilon|}$$

and integrating with respect to $p_\parallel$, we find the probability distribution as a function of the angle θ. Owing to the rapid convergence of the integral, the integration can be extended from $-\infty$ to $+\infty$, and the final result is

$$dw_\theta = \text{constant} \times \frac{\theta d\theta}{\sqrt{(\theta^2 + 4|\epsilon|/E_0)}}. \tag{146.5}$$

† We are here considering an *np* pair with parallel or antiparallel spins, or an *nn* pair with antiparallel spins. For a *pp* pair the situation is complicated by the Coulomb repulsion, and this case must be treated by means of the theory given in §138.

‡ Strictly speaking, the constant coefficients in formulae (146.3) and (146.4) may also depend on E through the remaining parts of the wave function for the whole system of reaction products. This dependence is only slight, however: the coefficient varies appreciably, as a function of E, only over the whole energy range ($\sim E_0$) available to the nucleon pair in the reaction considered. Thus this dependence may be neglected, as regards the distribution in the range $E \ll E_0$, in comparison with the strong dependence given by formula (146.4).

The angular distribution relative to the solid angle element do $\approx 2\pi\theta\, d\theta$ has a maximum at $\theta \sim \sqrt{(|\epsilon|/E_0)}$.

§147. Behaviour of cross-sections near the reaction threshold

If the sum of the internal energies of the reaction products exceeds the corresponding sum for the original particles, the reaction has a *threshold*: it can occur only when the kinetic energy E of the colliding particles (in the centre-of-mass system) is greater than a certain "threshold" value E_t. Let us examine the nature of the energy dependence of the reaction cross-section near the threshold. We shall assume that the reaction produces only two particles (type $A+B = A'+B'$).

Near the threshold, the relative velocity v' of the particles formed is small. Such a reaction is the opposite of one in which the velocity of the colliding particles is small. The dependence of its cross-section on v' is therefore easily found by means of the principle of detailed balancing (144.13) and the known energy dependence of the reaction where v' is the velocity in the input channel (§143). In a wide class of reactions where there is no Coulomb interaction between the particles A' and B' (such as nuclear reactions in which a slow neutron is formed), we therefore find that the reaction cross-section is proportional to $v'^2\,(1/v')$, i.e.†

$$\sigma_r \sim v'. \tag{147.1}$$

Similarly we find the dependence of the cross-section on the energy of the colliding particles: the velocity v', and therefore the reaction cross-section, are proportional to the square root of the difference $E-E_t$:

$$\sigma_r = A\sqrt{(E-E_t)}. \tag{147.2}$$

The scattering amplitudes in different channels are related by the unitarity conditions. The opening of a new channel therefore leads to the appearance of certain singularities in the energy dependence of the cross-sections for other processes also, including the elastic scattering cross-section (E. P. Wigner 1948; A. I. Baz' 1957; G. Breit 1957). To elucidate the origin and nature of this phenomenon, let us consider the simple case where only elastic scattering is possible below the reaction threshold.

Near the threshold, the particles A' and B' arc formed in a state with orbital angular momentum $l = 0$ (corresponding to (147.2)). If the reacting particles have no spin, the orbital angular momentum is conserved, and the system of particles $A+B$ is also in the s-state. According to (142.7), the partial reaction cross-section for $l = 0$ is related to the S-matrix element for elastic scattering by

$$\sigma_r^{(0)} = \frac{\pi}{k^2}(1-|S_0|^2), \tag{147.3}$$

† This result corresponds to the constant limit of the amplitude f_{fi} as $p_f \to 0$ derived at the end of §144. The cross-section (144.4) is proportional to p_f.

where k is the wave number of the colliding particles. Equating (147.2) and (147.3), we find that just above the reaction threshold the modulus $|S_0|$ is given, to within quantities of higher order than $\sqrt{(E-E_t)}$, by

$$|S_0| = 1 - \frac{k_t^2}{2\pi} A\sqrt{(E-E_t)} \quad (E > E_t), \tag{147.4}$$

where $k_t = \sqrt{(2mE_t)}/\hbar$, and m is the reduced mass of the particles A and B. Below the threshold we have only elastic scattering, so that

$$|S_0| = 1 \quad (E < E_t). \tag{147.5}$$

The scattering amplitude, and therefore S_0, must be analytic functions for all values of the energy. The function concerned, which takes the values (147.4) and (147.5) above and below the threshold, is given to the same accuracy by the formula

$$S_0 = e^{2i\delta_0}\left[1 - \frac{k_t^2}{2\pi} A\sqrt{(E-E_t)}\right], \tag{147.6}$$

where δ_0 is constant; for $E < E_t$ the root becomes imaginary, and the modulus of the expression in the brackets differs from unity only by a quantity of a higher order of smallness.

For all $l \neq 0$ there is no inelastic scattering, so that

$$S_l = e^{2i\delta_l} \quad (l \neq 0), \tag{147.7}$$

and in the region near the threshold the phases δ_l must be taken equal to their values for $E = E_t$.†

Substituting the values obtained for S_l in (142.2), we find the following expression for the scattering amplitude near the reaction threshold:

$$f(\theta, E) = f_t(\theta) - \frac{k_t}{4\pi i} A\sqrt{(E-E_t)} e^{2i\delta_0}, \tag{147.8}$$

where $f_t(\theta)$ is the scattering amplitude for $E = E_t$. The differential scattering cross-section is therefore

$$\frac{d\sigma}{do} = |f_t(\theta)|^2 + \frac{k_t}{2\pi} A\sqrt{(E-E_t)} \text{ im } \{f_t(\theta)e^{-2i\delta_0}\} \quad \text{for } E > E_t,$$

$$= |f_t(\theta)|^2 - \frac{k_t}{2\pi} A\sqrt{(E-E_t)} \text{ re } \{f_t(\theta)e^{-2i\delta_0}\} \quad \text{for } E < E_t.$$

Writing the amplitude f_t in the form $|f_t|e^{i\alpha(\theta)}$, we can finally put this result in the form

$$\frac{d\sigma}{do} = |f_t(\theta)|^2 - \frac{k_t}{2\pi} A|f_t(\theta)|\sqrt{(|E-E_t|)} \times \left.\begin{matrix} \sin(2\delta_0 - \alpha), & E > E_t, \\ \cos(2\delta_0 - \alpha), & E < E_t. \end{matrix}\right\} \tag{147.9}$$

† Since the functions $\delta_l(E)$ are real both for $E > E_t$ and for $E < E_t$, they can be expanded as a series of integral powers of the difference $E - E_t$.

Depending on whether the angle $2\delta_0 - \alpha$ is in the first, second, third or fourth quadrant, the energy dependence of the cross-section described by this formula has the forms shown in Fig. 50a, b, c, d. In every case there are two branches lying on either side of a common vertical tangent.

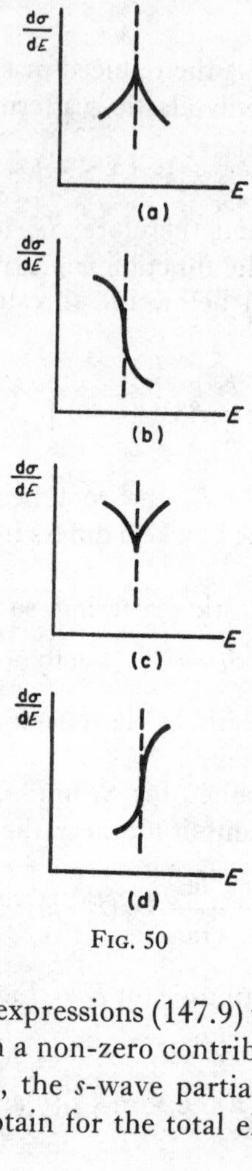

FIG. 50

In the integration of the expressions (147.9) with respect to o, the integrals of the second terms contain a non-zero contribution only from the isotropic part of the amplitude $f_t(\theta)$, the *s*-wave partial elastic scattering amplitude $(e^{2i\delta_0}-1)/2ik_t$. We thus obtain for the total elastic scattering cross-section near the threshold

$$\sigma = \sigma_t - 2A\sqrt{(|E-E_t|)} \times \left.\begin{matrix} \sin^2\delta_0 & \text{for } E > E_t, \\ \sin\delta_0\cos\delta_0 & \text{for } E > E_t. \end{matrix}\right\} \qquad (147.10)$$

This has the form (a) or (b) in Fig. 50, for positive and negative $\sin\delta_0\cos\delta_0$ respectively.

Thus the existence of a reaction threshold leads to a characteristic singularity in the energy dependence of the elastic scattering cross-section. If the particles have spin, of course, the formulae are quantitatively different, but the general nature of the effect remains the same.† If other reactions as well as elastic scattering are possible below the threshold, then corresponding singularities will appear in the cross-sections for such reactions. They all have a singularity at $E = E_t$ near which they are linear functions of the root $\sqrt{(|E-E_t|)}$ with different slopes above and below the threshold.

In nuclear reactions with emission of a positively charged particle, we have a case where Coulomb repulsion forces act between the reaction products (the particles A' and B'). In this case the reaction cross-section, together with all its derivatives with respect to energy, tends exponentially to zero as $v' \to 0$ (i.e. as $E \to E_t$), and there is no singularity in the cross-sections for other processes.

Finally, let us consider reactions in which two oppositely charged slow particles are formed, so that Coulomb attraction forces act between them. The cross-section for such a reaction is related by the principle of detailed balancing to the cross-section (143.6) for the opposite reaction between two slow attracting particles. Thus we find that as $v' \to 0$ the cross-section tends to a constant limit:

$$\sigma_r = \text{constant as } v' \to 0, \tag{147.11}$$

i.e. the reaction begins suddenly with a finite cross-section as the threshold is passed.

We may elucidate the nature of the singularity of the elastic scattering cross-section near the threshold for such a reaction (A. I. Baz' 1959). This cannot, however, be done directly from the known law (147.11) above the threshold by the simple method used previously for uncharged particles. In comparison with the latter case the situation is now complicated by the fact that the system of particles $A'+B'$ has bound states in the region near the threshold (with $E < E_t$), corresponding to discrete energy levels in the Coulomb attraction field. These states can be formed, so far as energy is concerned, in a collision of particles A and B, but owing to the possibility of elastic scattering they are only quasi-stationary states. Their existence must nevertheless cause resonance effects in the elastic scattering below the threshold, analogous to the Breit–Wigner resonances.

To solve the foregoing problem, let us consider the structure of the wave functions which describe the collision process. In accordance with the presence of two channels, Schrödinger's equation for the system of interacting particles has two independent solutions finite in all configuration space. Let two such solutions, arbitrarily selected and arbitrarily normalized, be denoted by ψ_1 and ψ_2. From these functions we can construct linear combinations which describe the scattering in the case where one or other of the channels is the input channel. Let the channels corresponding to the pairs of particles A, B, A', B', be denoted by a and b, and let the sum $\psi = \alpha_1\psi_1 + \alpha_2\psi_2$ correspond to the case of input channel a; it describes elastic scattering

† For non-zero spins, the system of particles $A'+B'$ in the s state may have a non-zero total angular momentum, and there can therefore be different orbital states of the system $A+B$.

of particles A and B and the reaction $A+B \to A'+B'$. Near the reaction threshold, the coefficients α_1 and α_2 depend considerably on the small momentum k_b, while the arbitrarily chosen functions ψ_1 and ψ_2 do not have singularities at $k_b = 0$.

At large distances, the function ψ must represent the sum of two terms corresponding to the motion of pairs of particles in the channels a and b. These terms are the products of the "internal" functions of the particles and the wave function of their relative motion.† In channel a the latter function has the form $R_a^- - S_{aa}R_a^+$, and in channel b it is $-S_{ab}R_b^+$, where R^+ and R^- are the outgoing and ingoing waves in the corresponding channels. At distances r_0 which are large compared with the range of action of the short-range forces and small in comparison with $1/k_b$, these functions (and their derivatives) must join on to the values calculated from the wave function ψ in the "reaction zone". These conditions are expressed by equations of the form

$$\alpha_1 a_1 + \alpha_2 a_2 = [R_a^- - S_{aa}R_a^+]_{r_0}, \quad \alpha_1 b_1 + \alpha_2 b_2 = [-S_{ab}R_b^+]_{r_0},$$

$$\alpha_1 a_1' + \alpha_2 a_2' = [R_a^- - S_{aa}R_a^+]'_{r_0}, \quad \alpha_1 b_1' + \alpha_2 b_2' = [-S_{ab}R_b^+]'_{r_0},$$

where $a_1, a_1', b_1, b_1', \ldots$ are quantities calculated from the functions ψ_1 and ψ_2; according to the above discussion, they may be regarded as constants independent of k_b near the threshold. Dividing these two pairs of equations, we obtain two linear equations for two unknowns (α_1/α_2 and S_{aa}), the coefficients in these equations involving only one quantity which depends "critically" on k_b, namely the logarithmic derivative of the outgoing wave in channel b. We define this as

$$\lambda = \frac{1}{2\pi}\left[\frac{(rR_b^+)'}{rR_b^+}\right]_{r=r_0}.$$

There is no need to derive the actual solution of these equations; it is sufficient to note that the quantity S_{aa} of interest here (which determines the elastic scattering amplitude) is a fractional-linear function of λ. Below the threshold the quantity λ is real, since the wave function R_b^+ is real, being the solution of a real Schrödinger's equation with a real condition at infinity (decrease as $e^{-\kappa_b r}$, where $\kappa_b = \sqrt{[2m_b(E_t - E)]}/\hbar$). Below the threshold we must have $|S_{aa}| = 1$, whence it follows that the fractional-linear function $S_{aa}(\lambda)$ must have the form

$$S_{aa} = \frac{1+\beta\lambda}{1+\beta^*\lambda} e^{2i\eta}, \tag{147.12}$$

where η is a real constant and β a complex constant.

Let us determine the value of λ as a function of the momentum k_b. Since

† The law (147.11) holds not only for the total cross-section but also for the partial cross-sections with various values of l; cf. the end of §143. The singularity discussed below therefore occurs also for all the partial scattering cross-sections. Its nature is entirely evident from the treatment for the case $l = 0$ given below. The index 0 will be omitted, for simplicity, from the corresponding partial amplitudes.

Coulomb attraction forces act between the particles A and B, rR_b^+ is the Coulomb wave function which is asymptotically proportional to e^{ik_br} at infinity. In a Coulomb repulsion field this function is given by the sum G_0+iF_0, with G_0 and F_0 as in (138.4) and (138.7). The change to an attractive field is effected by simultaneously reversing the signs of k and r.† Making this change and calculating the logarithmic derivative (see §138), we have‡

$$\lambda = \frac{i}{1-e^{-2\pi/k_b}} - \frac{1}{\pi}\left\{\log k_b + \tfrac{1}{2}\left[\psi\left(\frac{i}{k_b}\right)+\psi\left(-\frac{i}{k_b}\right)\right]\right\}. \tag{147.13}$$

Here k_b is assumed real, so that the formula pertains to the region above the threshold. For $k_b \to 0$, the first term in (147.13) tends to i, and the second tends to zero (see the fourth footnote to §138). Thus we have above the threshold

$$\lambda = i \quad (E > E_t). \tag{147.14}$$

The passage to the region below the threshold is achieved by replacing k by $i\kappa$, which gives from (147.13) with $\kappa \to 0$‖

$$\lambda = -\cot(\pi/\kappa_b) \quad (E < E_t). \tag{147.15}$$

These formulae solve the problem under consideration. The elastic scattering cross-section is

$$\sigma_e = \pi k_a^{-2}|S_{aa}-1|^2.$$

Above the threshold we have

$$S_{aa} = \frac{1+i\beta}{1+i\beta^*}e^{2i\eta} \quad (E > E_t); \tag{147.16}$$

like the reaction cross-section, the scattering cross-section is constant in this region. The condition $|S_{aa}| < 1$ signifies that im $\beta > 0$.

Below the threshold

$$S_{aa} = e^{2i\eta}\frac{\beta-\tan(\pi/\kappa_b)}{\beta^*-\tan(\pi/\kappa_b)} \tag{147.17}$$

This expression has an infinite number of resonances whose density increases towards the point $E = E_t$. The resonance energies are the roots of the expression

$$S_{aa} = -1, \text{ i.e. re } e^{i\eta}[\beta-\tan(\pi/\kappa_b)] = 0;$$

† In what follows we use Coulomb units. The change in sign of k and r corresponds formally to a change in sign of the Coulomb unit of length.

‡ To simplify the subsequent formulae, we omit from the expression in the braces the real constant $-\log 2r_0 - 2C$, which is independent of k_b; this amounts to an unimportant redefinition of the complex quantity β and the real quantity η in (147.12).

‖ The first term in (147.13) gives $-\frac{1}{2}\cot(\pi/\kappa_b)+\frac{1}{2}i$, and the expression in the braces tends to $\frac{1}{2}\pi\cot(\pi/\kappa_b)+\frac{1}{2}i\pi$. Here we have used the formula $\psi(x)-\psi(-x) = -\pi\cot\pi x - 1/x$, which can be obtained by logarithmic differentiation of the well-known relation $\Gamma(x)\Gamma(-x) = -\pi/x\sin\pi x$, and the limiting expression $\psi(x) \approx \log x - 1/2x$ as $x \to \infty$.

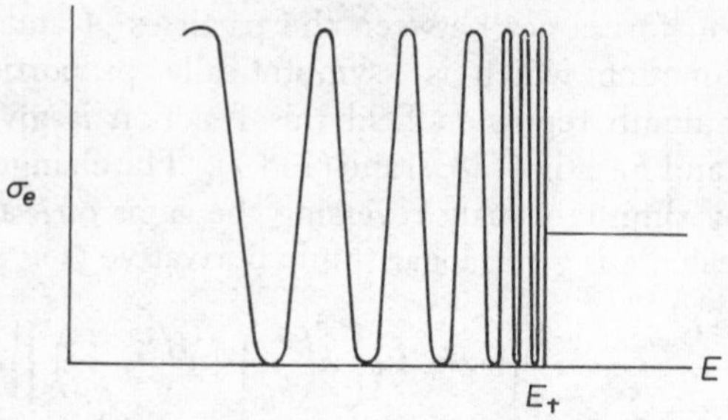

FIG. 51

they are somewhat displaced relative to the purely Coulomb levels (the roots of $\tan(\pi/\kappa_b) = 0$) owing to the short-range forces. As the energy E approaches the threshold, the elastic scattering cross-section oscillates between zero and $4\pi/k_a{}^2$, as shown diagrammatically in Fig. 51. The width of the entire sub-threshold region in which the resonance structure occurs is determined by the energy of the first Coulomb level.†

§148. Inelastic collisions between fast electrons and atoms

Inelastic collisions between fast electrons and atoms can be considered by means of the Born approximation in the same way as elastic collisions in §139.‡ The condition for the Born approximation to be applicable is, as before, that the velocity of the incident electron should be large compared with those of the atomic electrons. The energy loss in the collision may have any value. If the electron loses a considerable part of its energy, the atom is ionized, the energy being transferred to one of its electrons. However, we can always regard as the scattered electron that which has the greater velocity after the collision; thus, if the velocity of the incident electron is large, that of the scattered electron is large also.

In a collision between an electron and an atom, the coordinate system in which their centre of mass is at rest may, as already remarked, be identified with that in which the atom is at rest; this latter system will in fact be used below.

An inelastic collision is accompanied by a change in the internal state of the atom. The atom may go from the normal state into an excited state of the discrete or continuous spectrum; the latter case signifies an ionization of the

† Another interesting case of reactions near the threshold is the ionization of an atom by an electron whose energy is only slightly greater than the first ionization energy of the atom. In these conditions the collision process may be regarded as quasi-classical, but the problem is greatly complicated by the presence of three charged particles in the final state. The general solution of this difficult problem has been given by G. H. Wannier (*Physical Review* **90**, 817, 1953). The probability of ionization of a neutral atom is found to be proportional to $(E-I)^\alpha$, where $\alpha = \frac{1}{4}(\sqrt{(91/3)}-1) = 1{\cdot}13$ and $E-I$ is the amount by which the electron energy exceeds the ionization threshold.

‡ Most of the results given in §§148–150 were obtained by H. A. Bethe (1930).

atom. In deriving the general formulae, we can consider these two cases together.

We start (as in §126) from the general formula for the transition probability between states of the continuous spectrum, and apply it to the system consisting of the incident electron and the atom. Let **p**, **p**′ be the momenta of the incident electron before and after the collision, and E_0, E_n the corresponding energies of the atom. For the transition probability, we have instead of (126.9) the expression

$$dw_n = \frac{2\pi}{\hbar}|\langle n, \mathbf{p}'|U|0, \mathbf{p}\rangle|^2 \,\delta\left(\frac{p'^2-p^2}{2m}+E_n-E_0\right)\frac{d^3p'}{(2\pi\hbar)^3}, \qquad (148.1)$$

where the matrix element is that of the energy of interaction between the incident electron and the atom,

$$U = Ze^2/r - \sum_{a=1}^{Z} e^2/|\mathbf{r}-\mathbf{r}_a|;$$

here **r** is the radius vector of the incident electron, $\mathbf{r}_a$ those of the atomic electrons; the origin is at the nucleus of the atom, and m is the mass of the electron.

The wave functions $\psi_{\mathbf{p}}$, $\psi_{\mathbf{p}'}$ of the electron are determined by the previous formulae (126.10), (126.11); then dw is the cross-section $d\sigma$ for the collision. The wave functions of the atom in the initial and final states we denote by ψ_0, ψ_n. If the final state of the atom belongs to the discrete spectrum, then ψ_n (like ψ_0) is normalized to unity in the usual manner. If, on the other hand, the atom enters a state of the continuous spectrum, the wave function is normalized by the delta function of the parameters ν which determine these states (these parameters may be, for instance, the energy of the atom, and the momentum components of the electron which leaves the atom in the ionization), The cross-sections thus obtained give the probability of a collision in which the atom enters states of the continuous spectrum lying in the range of parameters between ν and $\nu+d\nu$.

Integration of (148.1) over the absolute magnitude p' gives

$$d\sigma_n = \frac{mp'}{4\pi^2\hbar^4}|\langle n\mathbf{p}'|U|0\mathbf{p}\rangle|^2 do',$$

where p' is determined from the law of conservation of energy:

$$(p^2-p'^2)/2m = E_n - E_0. \qquad (148.2)$$

Substituting in the matrix element the wave functions of the electron from (126.10), (126.11), we obtain

$$d\sigma_n = \frac{m^2}{4\pi^2\hbar^4}\frac{p'}{p}\left|\iint Ue^{-i\mathbf{q}.\mathbf{r}}\psi_n{}^*\psi_0\, d\tau\, dV\right|^2 do, \qquad (148.3)$$

where $d\tau = dV_1\, dV_2 \ldots dV_Z$ is the element of configuration space of the Z electrons in the atom, and we omit the prime to do.† For $n = 0$ and $p = p'$, (148.3) becomes the formula for the elastic scattering cross-section.

Since the functions ψ_n and ψ_0 are orthogonal, the term in U which contains the interaction Ze^2/r with the nucleus vanishes on integration over τ, and so we have for inelastic collisions

$$d\sigma_n = \frac{m^2}{4\pi^2\hbar^4}\frac{p'}{p}\left|\sum_a \iint \frac{e^2}{|\mathbf{r}-\mathbf{r}_a|} e^{-i\mathbf{q}\cdot\mathbf{r}}\psi_n{}^*\psi_0\, d\tau dV\right|^2 do. \qquad (148.4)$$

The integration over V can be effected as in §139. The integral

$$\phi_{\mathbf{q}}(\mathbf{r}_a) = \int e^{-i\mathbf{q}\cdot\mathbf{r}}\, dV/|\mathbf{r}-\mathbf{r}_a|$$

is formally the same as the Fourier component of the potential at the point $\mathbf{r}$ due to charges distributed in space with density $\rho = \delta(\mathbf{r}-\mathbf{r}_a)$. Formula (139.1) therefore gives

$$\phi_{\mathbf{q}}(\mathbf{r}_a) = (4\pi/q^2)e^{-i\mathbf{q}\cdot\mathbf{r}_a}. \qquad (148.5)$$

Substituting this expression in (148.4), we finally obtain the following general expression for the inelastic collision cross-section:

$$d\sigma_n = \left(\frac{e^2m}{\hbar^2}\right)^2 \frac{4k'}{kq^4}|\langle n|\sum_a e^{-i\mathbf{q}\cdot\mathbf{r}_a}|0\rangle|^2\, do; \qquad (148.6)$$

here the matrix element is taken with respect to the wave functions of the atom, and we have introduced, in place of the momenta, the wave vectors $\mathbf{k}' = \mathbf{p}'/\hbar$, $\mathbf{k} = \mathbf{p}/\hbar$. This formula gives the probability of a collision in which the electron is scattered into an element of solid angle do and the atom enters the nth excited state. The vector $-\hbar\mathbf{q}$ is the momentum given to the atom by the electron in the collision.

In effecting the calculations, it is more convenient to refer the cross-section, not to the element of solid angle, but to the element dq of the absolute magnitudes of the vector $\mathbf{q}$. The vector $\mathbf{q}$ is defined by $\mathbf{q} = \mathbf{k}'-\mathbf{k}$; for its absolute magnitude we have

$$q^2 = k^2+k'^2-2kk'\cos\vartheta. \qquad (148.7)$$

Hence, for given k, k', i.e. for a given loss of energy by the electron,

$$q\,dq = kk'\sin\vartheta\, d\vartheta = (kk'/2\pi)\, do. \qquad (148.8)$$

† In this form, it is a general result of perturbation theory, applicable not only to collisions of electrons with an atom but also to any inelastic collisions of two particles, and determines the scattering cross-section in a system of coordinates where the centre of mass of the particles is at rest (m being then the reduced mass of the two particles).

Formula (148.6) may therefore be written

$$d\sigma_n = 8\pi\left(\frac{e^2}{\hbar v}\right)^2 \frac{dq}{q^3}|\langle n|\sum_a e^{-i\mathbf{q}.\mathbf{r}_a}|0\rangle|^2. \tag{148.9}$$

The vector $\mathbf{q}$ plays an important part in the following calculations. Let us examine more closely its relation to the scattering angle ϑ and to the energy $E_n - E_0$ transferred in the collision. We shall see below that the most important collisions are those which cause scattering through small angles ($\vartheta \ll 1$), with a transfer of energy which is small in comparison with the energy $E = \frac{1}{2}mv^2$ of the incident electron: $E_n - E_0 \ll E$. The difference $k - k'$ is in this case also small ($k - k' \ll k$), and

$$E_n - E_0 = \hbar^2(k^2 - k'^2)/2m \approx \hbar^2 k(k-k')/m = \hbar v(k-k').$$

Since ϑ is small, we have from (148.7)

$$q^2 \approx (k-k')^2 + (k\vartheta)^2,$$

and finally

$$q = \sqrt{[\{(E_n - E_0)/\hbar v\}^2 + (k\vartheta)^2]}. \tag{148.10}$$

The minimum value of q is

$$q_{\min} = (E_n - E_0)/\hbar v. \tag{148.11}$$

In the region of small angles we can further distinguish between different regions depending on the relation between the small quantities ϑ and v_0/v, where v_0 is of the order of the velocity of the atomic electrons. If we consider energy transfers of the order of the energy ϵ_0 of the atomic electrons ($E_n - E_0 \sim \epsilon_0 \sim mv_0^2$), then for $(v_0/v)^2 \ll \vartheta \ll 1$ we have

$$q = k\vartheta = (mv/\hbar)\vartheta; \tag{148.12}$$

the first term under the radical in (148.10) can be neglected in comparison with the second. In this range of angles, therefore, q is independent of the energy transfer. For $\vartheta \ll 1$, q may be either large or small in comparison with $1/a_0$ (where a_0 is a quantity of the order of atomic dimensions). On the same assumption regarding the energy transfer we have

$$qa_0 \sim 1 \text{ for } \vartheta \sim v_0/v. \tag{148.13}$$

Let us now apply the general formula (148.9) for small q ($qa_0 \ll 1$, i.e. $\vartheta \ll v_0/v$). In this case we can expand the exponential factors as series of powers of $\mathbf{q}$:

$$e^{-i\mathbf{q}.\mathbf{r}_a} \approx 1 - i\mathbf{q}.\mathbf{r}_a = 1 - iqx_a;$$

we choose a coordinate system with the x-axis along the vector $\mathbf{q}$. On substituting this expansion in (148.9), the terms containing 1 give zero, by the orthogonality of the wave functions ψ_0 and ψ_n, and we obtain

$$d\sigma_n = 8\pi\left(\frac{e}{\hbar v}\right)^2 \frac{dq}{q}|\langle n|d_x|0\rangle|^2 = \left(\frac{2e}{\hbar v}\right)^2 |\langle n|d_x|0\rangle|^2 \frac{do}{\vartheta^2}, \tag{148.14}$$

where $d_x = e\sum_a x_a$ is the component of the dipole moment of the atom. We see that the cross-section (for small q) is given by the squared modulus of the matrix element of the dipole moment for the transition which corresponds to the change in state of the atom.†

It may happen, however, that the matrix element of the dipole moment vanishes identically for the transition considered, on account of the selection rules (a *forbidden transition*). Then the expansion of $e^{-i\mathbf{q}.\mathbf{r}_a}$ must be continued to the next term, and we obtain

$$d\sigma_n = 2\pi\left(\frac{e^2}{\hbar v}\right)^2 |\langle n|(\sum_a x_a^2)|0\rangle|^2 q\, dq. \qquad (148.15)$$

Let us now consider the opposite limiting case of large q ($qa_0 \gg 1$). If q is large, this means that the atom receives a momentum which is large compared with the original intrinsic momentum of the atomic electrons. It is evident from physical considerations that, in this case, we can regard the atomic electrons as free, and the collision with the atom as an elastic collision between the incident electron and the atomic electrons, the latter being originally at rest. This can also be seen from the general formula (148.9). For large q, the integrand in the matrix element contains rapidly oscillating factors $e^{-i\mathbf{q}.\mathbf{r}_a}$, and the integral is almost zero if ψ_n does not contain a similar factor. Such a function ψ_n corresponds to an ionized atom, with the electron emitted from it with momentum $-\hbar\mathbf{q} = \mathbf{p}-\mathbf{p}'$ given by the law of conservation of momentum, as it would be in a collision of two free electrons.

In a collision with a large transfer of momentum, the incident electron and the atomic electron may have final velocities that are comparable in magnitude. The exchange effect arising from the identity of the colliding particles therefore becomes important, although it was not taken into account in the general formula (148.9). The scattering cross-section for fast electrons when exchange is allowed for is given by formula (137.9); this formula relates to a coordinate system in which one of the electrons is at rest before the collision. For a fast electron the cosine in the last term in (137.9) may be put equal to unity. Multiplying by the number of electrons in the atom, Z, we obtain the cross-section for the collision of an electron with an atom, in the form

$$d\sigma = 4Z\left(\frac{e^2}{mv^2}\right)^2\left[\frac{1}{\sin^4\vartheta}+\frac{1}{\cos^4\vartheta}-\frac{1}{\sin^2\vartheta\cos^2\vartheta}\right]\cos\vartheta\, do. \qquad (148.16)$$

In this formula it is convenient to express the scattering angle in terms of the energy which the electrons have after the collision. As is well known, when a particle of energy $E = \frac{1}{2}mv^2$ collides with one of the same mass at

† The cross-section $d\sigma_n$, summed over all directions of the angular momentum of the atom in the final state and averaged over the directions of the angular momentum in the initial state, is what is usually of physical interest. After this summation and averaging, the square $|\langle n|d_x|0\rangle|^2$ is independent of the direction of the x-axis.

rest, the energy of the particles after the collision is

$$\epsilon = E\sin^2\vartheta, \qquad E-\epsilon = E\cos^2\vartheta.$$

In order to find the cross-section referred to the interval $d\epsilon$, we express do in terms of $d\epsilon$ by the relation $\cos\vartheta\, do = 2\pi\sin\vartheta\cos\vartheta\, d\vartheta = (\pi/E)\, d\epsilon$. Substituting in (148.16), we obtain the final formula

$$d\sigma_\epsilon = \pi Ze^4\left[\frac{1}{\epsilon^2}+\frac{1}{(E-\epsilon)^2}-\frac{1}{\epsilon(E-\epsilon)}\right]\frac{d\epsilon}{E}. \qquad (148.17)$$

If one of the energies ϵ and $E-\epsilon$ is small compared with the other, only one of the three terms in this formula (the first or the second) is important. This is as it should be, since, for a great difference between the energies of the two electrons, the exchange effect becomes insignificant, and we then return to the familiar Rutherford's formula.†

The integration of the differential cross-section over all angles (or, what is the same thing, over q) gives the total cross-section σ_n for a collision in which the atom is excited to the state in question. The dependence of σ_n on the velocity of the incident electron is closely related to the existence or otherwise of the matrix element, for the corresponding transition, of the dipole moment of the atom. Let us first suppose that this matrix element is not zero. Then, for small q, $d\sigma_n$ is given by formula (148.14), and we see that, as q diminishes, the integral over q diverges logarithmically. In the region of large q, on the other hand, the cross-section (for a given energy transfer E_n-E_0) decreases exponentially as q increases, because of the presence (already pointed out) of a rapidly oscillating factor in the integrand of the matrix element in (148.9). Thus the region of small q plays the principal part in the integral over q, and we can restrict ourselves to an integration from the minimum value (148.11) to some value of the order of $1/a_0$.

As a result we obtain

$$\sigma_n = 8\pi(e/\hbar v)^2|\langle n|d_x|0\rangle|^2\log(\beta_n v\hbar/e^2), \qquad (148.18)$$

where β_n is a dimensionless constant, which cannot be calculated in a general form.‡

If, on the other hand, the matrix element of the dipole moment vanishes for the transition in question, the integral over q converges rapidly both for small q (as we see from (148.15)) and for large q. The most important range in the integral is in this case $q \sim 1/a_0$. No general quantitative formula

---

† For a collision of a positron with an atom there is no exchange effect, and Rutherford's formula

$$d\sigma_\epsilon = (\pi Z_e^4/E)\, d\epsilon/\epsilon^2$$

holds for all $q \gg 1/a_0$.

‡ We suppose that E_n-E_0 is of the order of the energy ϵ_0 of the atomic electrons. For larger energy transfers ($E_n-E_0 \sim E \gg \epsilon_0$), the formulae (148.14), (148.18) are still inapplicable, since the matrix element of the dipole moment becomes very small, and it is not possible to take only the first term of the expansion in powers of q.

can be obtained, and we can deduce only that σ_n is inversely proportional to the square of the velocity:

$$\sigma_n = \text{constant}/v^2. \tag{148.19}$$

This follows at once from the general formula (148.9), according to which $d\sigma_n$ is proportional to $1/v^2$ for $q \sim 1/a_0$.

Let us determine the cross-section $d\sigma_r$ for inelastic scattering into a given element of solid angle regardless of the state entered by the atom. To do this, we have to sum the expression (148.9) for all $n \neq 0$, i.e. over all the states of the atom (of both the discrete and the continuous spectrum) except the normal state. We omit from consideration the ranges of large and small angles, and suppose that $1 \gg \vartheta \gg (v_0/v)^2$. Then, by (148.12), q is independent of the amount of energy transferred.†

The latter circumstance makes it easy to calculate the total inelastic collision cross-section, i.e. the sum

$$d\sigma_r = \sum_{n\neq 0} d\sigma_n = 8\pi\left(\frac{e^2}{\hbar v}\right)^2 \sum_{n\neq 0} |\langle n|\sum_a e^{-i\mathbf{q}\cdot\mathbf{r}_a}|0\rangle|^2 \frac{dq}{q^3}$$

$$= \left(\frac{2e^2}{mv^2}\right)^2 \sum_{n\neq 0} |\langle n|\sum_a e^{-i\mathbf{q}\cdot\mathbf{r}_a}|0\rangle|^2 \frac{do}{\vartheta^4}. \tag{148.20}$$

To do so, we note that, for any quantity f, we have by the multiplication rule for matrices

$$\sum_n |f_{0n}|^2 = \sum_n f_{0n} f_{0n}^* = \sum_n f_{0n}(f^+)_{n0} = (ff^+)_{00}.$$

The summation here is over all n, including $n = 0$. Hence

$$\sum_{n\neq 0} |f_{0n}|^2 = \sum_n |f_{0n}|^2 - |f_{00}|^2 = (ff^+)_{00} - |f_{00}|^2. \tag{148.21}$$

Applying this relation for $f = \sum e^{-i\mathbf{q}\cdot\mathbf{r}_a}$, we have

$$d\sigma_r = \left(\frac{2e^2}{mv^2}\right)^2 \{\langle|\sum_a e^{-i\mathbf{q}\cdot\mathbf{r}_a}|^2\rangle - |\langle\sum_a e^{-i\mathbf{q}\cdot\mathbf{r}_a}\rangle|^2\}\frac{do}{\vartheta^4} \tag{148.22}$$

where $\langle\ldots\rangle$ denotes averaging with respect to the normal state of the atom (i.e. taking the diagonal matrix element 00). The mean value $\langle\sum e^{-i\mathbf{q}\cdot\mathbf{r}_a}\rangle$ is, by definition, the atomic form factor $F(q)$ for the atom in the normal state. In the first term in the braces we can write

$$\left|\sum_{a=1}^{Z} e^{-i\mathbf{q}\cdot\mathbf{r}_a}\right|^2 = Z + \sum_{a\neq b} e^{i\mathbf{q}\cdot(\mathbf{r}_a-\mathbf{r}_b)}.$$

† The summation in (148.9) is taken over states with $E_n - E_0 \gg \epsilon_0$ also, for which (148.12) does not hold. However, the effective cross-section for transitions with a large energy transfer is relatively small, and these terms in the sum are unimportant. The condition $\vartheta \ll 1$ is imposed so that the exchange effects need not be taken into account.

Thus we find the general formula

$$d\sigma_r = \left(\frac{2e^2}{mv^2}\right)^2 \{Z - F^2(q) + \langle \sum_{a \neq b} e^{i\mathbf{q}\cdot(\mathbf{r}_a - \mathbf{r}_b)} \rangle\} \frac{do}{\vartheta^4}. \quad (148.23)$$

This formula is much simplified for small q, when we can expand in powers of q ($v_0/v \ll qa_0 \ll 1$, corresponding to angles $(v_0/v)^2 \ll \vartheta \ll v_0/v$). Instead of effecting the expansion from formula (148.23), it is more convenient to sum again over n, using for $d\sigma_n$ the expression (148.14). Summing with the aid of the relation (148.21) with $f = d_x$, and recalling that $\langle d_x \rangle = 0$, we have

$$d\sigma_r = (2e/\hbar v)^2 \langle d_x^2 \rangle do/\vartheta^2. \quad (148.24)$$

It is of interest to compare this expression with the cross-section (139.5) for elastic scattering through small angles; whereas the latter is independent of ϑ, the cross-section for inelastic scattering into the solid angle element do increases as $1/\vartheta^2$ when ϑ decreases.

For angles ϑ such that $1 \gg \vartheta \gg v_0/v$ (so that $qa_0 \gg 1$), the second and third terms in the braces in (148.23) are small, and we have simply

$$d\sigma_r = Z(2e^2/mv^2)^2 do/\vartheta^4, \quad (148.25)$$

i.e. Rutherford scattering from the Z atomic electrons (without allowance for exchange). We recall that, for elastic scattering, we had the result (139.6), which is proportional to Z^2 and not to Z.

Finally, integrating over angles, we have the total cross-section σ_r for inelastic scattering at all angles and with any excitation of the atom. In an exactly similar manner to the calculation of σ_n (148.18), we obtain

$$\sigma_r = 8\pi(e/\hbar v)^2 \langle d_x^2 \rangle \log(\beta v\hbar/e^2). \quad (148.26)$$

PROBLEMS†

PROBLEM 1. Determine the angular distribution for $1 \gg \vartheta \gg v^{-2}$ from the inelastic scattering of fast electrons by a hydrogen atom (in the normal state).

SOLUTION. For the hydrogen atom, the third term in the braces in (148.23) vanishes, while the atomic form factor $F(q)$ has been calculated in §139, Problem. Substituting, we find

$$d\sigma_r = \frac{4}{v^4\vartheta^4}\frac{(1+v^2\vartheta^2/4)^4-1}{(1+v^2\vartheta^2/4)^4} do.$$

PROBLEM 2. Determine the differential cross-section for collisions of electrons with a hydrogen atom in the normal state, the latter being excited to the nth level of the discrete spectrum (where n is the principal quantum number).

SOLUTION. The matrix elements are conveniently calculated in parabolic coordinates. We take the z-axis in the direction of the vector $\mathbf{q}$; then

$$e^{i\mathbf{q}\cdot\mathbf{r}} = e^{iqz} = e^{\frac{1}{2}iq(\xi-\eta)}.$$

† We use atomic units in all the Problems.

The wave function of the normal state is

$$\psi_{000} = \pi^{-\frac{1}{2}} e^{-\frac{1}{2}(\xi+\eta)}.$$

The matrix elements are non-zero only for transitions to states with $m = 0$. The wave functions of these states are the functions

$$\psi_{n_1 n_2 0} = (1/\sqrt{\pi n^2}) e^{-\frac{1}{2}(\xi+\eta)/n} F(-n_1, 1, \xi/n) F(-n_2, 1, \eta/n)$$

$(n = n_1+n_2+1)$. The required matrix elements are the integrals

$$\langle n_1 n_2 0|e^{i\mathbf{q}\cdot\mathbf{r}}|000\rangle = \int_0^\infty\int_0^\infty e^{iq(\xi-\eta)/2}\psi_{000}\psi_{n_1 n_2 0}\frac{\xi+\eta}{4}2\pi\, d\xi d\eta.$$

The integration is effected by means of the formulae of §f in the Mathematical Appendices. The result is

$$|\langle n_1 n_2 0|e^{i\mathbf{q}\cdot\mathbf{r}}|000\rangle|^2 = 2^8 n^6 q^2 \frac{[(n-1)^2+(qn)^2]^{n-3}}{[(n+1)^2+(qn)^2]^{n+3}}[(n_1-n_2)^2+(qn)^2].$$

All states with the same $n_1+n_2 = n-1$ have the same energy. Summing over all possible values of n_1-n_2 for the given n, and substituting the result in (148.9), we obtain the required cross-section:

$$d\sigma_n = 2^{11}\pi\frac{1}{v^2}n^7[\tfrac{1}{3}(n^2-1)+(qn)^2]\frac{[(n-1)^2+(qn)^2]^{n-3}}{[(n+1)^2+(qn)^2]^{n+3}}\frac{dq}{q}.$$

PROBLEM 3. Determine the total cross-section for the excitation of the first excited state of the hydrogen atom.

SOLUTION. We have to integrate

$$d\sigma_2 = 2^8\pi\frac{1}{v^2}\frac{dq}{q(q^2+9/4)^5}$$

over all q from $q_{\min} = (E_2-E_1)/v = 3/8v$ to $q_{\max} = 2v$, only the terms of the highest degree in v being retained. The integration is elementary, and the result is†

$$\sigma_2 = \frac{2^{18}\pi}{3^{10}v^2}\left[\log(4v) - \frac{25}{24}\right] = \frac{4\pi}{v^2}0{\cdot}555\log\frac{v^2}{0{\cdot}50}.$$

PROBLEM 4. Determine the cross-section for the ionization of a hydrogen atom (in the normal state), with the emission of a secondary electron in a given direction; the energy of the secondary electron is small in comparison with that of the primary, and so exchange effects are unimportant (H. S. W. Massey and C. B. O. Mohr 1933).

SOLUTION. The wave function of the atom in the initial state is $\psi_0 = e^{-r}/\sqrt{\pi}$. In the final state, the atom is ionized, and the secondary electron emitted from it has a wave vector which we denote by κ (and energy $\epsilon = \frac{1}{2}\kappa^2$). This state is described by a function $\psi^{(-)}_{\kappa}$.

† The cross-section can also be calculated for arbitrary n. By numerical calculation, we can obtain also the total cross-section for inelastic scattering by a hydrogen atom:

$$\sigma_r = \frac{4\pi}{v^2}\log(v^2/0{\cdot}160).$$

This includes the following contributions from collisions in which states of the discrete spectrum are excited, and from those in which the atom is ionized:

$$\sigma_r = \frac{4\pi}{v^2}\log(v^2/0{\cdot}160).$$

$$\sigma_{io} = \frac{4\pi}{v^2}\times 0{\cdot}285\log(v^2/0{\cdot}012).$$

(136.9), in which the outgoing part consists (at infinity) only of a plane wave propagated in the direction of κ. The function $\psi_{\kappa}^{(-)}$ is normalized by the delta function in $\kappa/2\pi$-space; hence the cross-section calculated from it will relate to $d^3\kappa/(2\pi)^3$ or to $\kappa^2\, d\kappa\, do_\kappa/(2\pi)^3$, where do_κ is an element of solid angle about the direction of the secondary electron. Thus

$$d\sigma = \frac{4k'\kappa^2}{(2\pi)^3 kq^4}|\langle\kappa|e^{-i\mathbf{q}\cdot\mathbf{r}}|0\rangle|^2\, do\, do_\kappa\, d\kappa,$$

where do is an element of solid angle about the direction of the scattered electron, and

$$\langle\kappa|e^{-i\mathbf{q}\cdot\mathbf{r}}|0\rangle = \int \psi_\kappa^{(-)*} e^{-i\mathbf{q}\cdot\mathbf{r}}\psi_0\, dV = \frac{e^{-\pi/2\kappa}\Gamma(1-i/\kappa)}{\pi^{1/2}} I,$$

$$I = \left[-\frac{\partial}{\partial\lambda}\int e^{-i\mathbf{q}\cdot\mathbf{r}-i\boldsymbol{\kappa}\cdot\mathbf{r}-\lambda r}F(i/\kappa, 1, i(\kappa r+\boldsymbol{\kappa}\cdot\mathbf{r}))\frac{dV}{r}\right]_{\lambda=1}.$$

We effect the integration in parabolic coordinates, with the z-axis in the direction of κ and the angle ϕ measured from the $(\mathbf{q}, \boldsymbol{\kappa})$ plane:

$$I = \left[-\tfrac{1}{2}\frac{\partial}{\partial\lambda}\int_0^\infty\int_0^\infty\int_0^{2\pi}\exp\{-\tfrac{1}{2}iq(\xi-\eta)\cos\gamma+iq\sqrt{(\xi\eta)}\sin\gamma\cos\phi-\tfrac{1}{2}\lambda(\xi+\eta)-\tfrac{1}{2}i\kappa(\xi-\eta)\}\times \right.$$
$$\left. \times F(i/\kappa, 1, i\kappa\xi)\, d\phi\, d\xi\, d\eta\right]_{\lambda=1},$$

where γ is the angle between $\boldsymbol{\kappa}$ and $\mathbf{q}$. The integration over ϕ and η is easily performed by substituting $\sqrt{\eta}\cos\phi = u$, $\sqrt{\eta}\sin\phi = v$, which gives

$$\frac{I}{2\pi} = \left[\frac{\partial}{\partial\lambda}\int_0^\infty \exp\left\{\frac{-q^2\sin^2\gamma+\lambda^2+(\kappa+q\cos\gamma)^2}{2[i(\kappa+q\cos\gamma)-\lambda]}\xi\right\}\times\frac{F(i/\kappa, 1, i\kappa\xi)\, d\xi}{i(\kappa+q\cos\gamma)-\lambda}\right]_{\lambda=1}.$$

The integral here is found from the formula (f.3) with $\gamma = 1$, $n = 0$. The subsequent calculations, though lengthy, are elementary, and give as a result the following expression for the cross-section:

$$d\sigma = \frac{2^8 k'\kappa[q^2+2q\kappa\cos\gamma+(\kappa^2+1)\cos^2\gamma]}{\pi kq^2[q^2+2q\kappa\cos\gamma+1+\kappa^2]^4[(q+\kappa)^2+1][(q-\kappa)^2+1](1-e^{-2\pi/\kappa})}\times$$
$$\times e^{-(2/\kappa)\tan^{-1}[2\kappa/(q^2-\kappa^2+1)]}\, do\, do_\kappa\, d\kappa.$$

The integration over all angles of emission of the secondary electron is elementary, and gives the distribution of scattering over directions, for a given energy $\tfrac{1}{2}\kappa^2$ of the emitted electron:

$$d\sigma = \frac{2^{10}k'\kappa\,[q^2+\tfrac{1}{3}(1+\kappa^2)]e^{-(2/\kappa)\tan^{-1}[2\kappa/(q^2-\kappa^2+1)]}}{kq^2\quad[(q+\kappa)^2+1]^3[(q-\kappa)^2+1]^3(1-e^{-2\pi/\kappa})}\, do\, d\kappa.$$

For $q \gg 1$, this expression has a sharp maximum at $\kappa \approx q$; near the maximum,

$$d\sigma = \frac{2^5}{3\pi\kappa^4}\frac{d\kappa\, do}{[1+(q-\kappa)^2]^3}.$$

Integrating over o, with $do = 2\pi q\, dq/k^2 \approx (2\pi\kappa/k^2)d(q-\kappa)$, we obtain the expression $8\pi\, d\kappa/k^2\kappa^3$; this is the same as the first term in formula (148.17), as it should be.

§149. The effective retardation

In applications of collision theory, the calculation of the mean energy lost by a colliding particle is of great importance. This energy loss is conveniently characterized by the quantity

$$d\kappa = \sum_n (E_n - E_0)\, d\sigma_n, \tag{149.1}$$

which we shall call the (differential) *effective retardation*; the summation is taken, of course, over states of both the discrete and the continuous spectrum. $d\kappa$ relates to scattering into a given element of solid angle.†

The general formula for the effective retardation of fast electrons is

$$d\kappa = 8\pi\left(\frac{e^2}{\hbar v}\right)^2 \sum_n (E_n - E_0)|\langle n|\sum_a e^{-i\mathbf{q}\cdot\mathbf{r}_a}|0\rangle|^2 \frac{dq}{q^3}, \tag{149.2}$$

where $d\sigma_n$ has been taken from (148.9). As in the derivation of (148.23), we exclude from consideration the region of very small angles, and suppose that $1 \gg \vartheta \gg (v_0/v)^2$. Then q is independent of the amount of energy transferred, and the sum over n can therefore be calculated in a general form.

This is done by means of a *summation theorem* derived as follows. The matrix elements of some quantity f, a function of the coordinates, and of its derivative $\dot{f}$ with respect to time are related by

$$(\dot{f})_{0n} = -(i/\hbar)(E_n - E_0) f_{0n}. \tag{149.3}$$

Hence we have

$$\sum_n (E_n - E_0)|f_{0n}|^2 = \sum_n (E_n - E_0) f_{0n} f_{0n}^*$$

$$= \sum_n (E_n - E_0) f_{0n} (f^+)_{n0} = i\hbar \sum_n (\dot{f})_{0n} (f^+)_{n0} = i\hbar (\dot{f} f^+)_{00}.$$

The wave functions of the stationary states of the atom can be taken real. Then the matrix elements of the function f of the coordinates are related by $f_{0n} = f_{n0}$, and for the matrix elements (149.3) we accordingly have $(\dot{f})_{0n} = -(\dot{f})_{n0}$. Thus we can also write the sum in question as

$$-i\hbar \sum_n (f^+)_{0n} (\dot{f})_{n0} = -i\hbar (f^+ \dot{f})_{00}.$$

Taking half the sum of these two equations, we have the required theorem:

$$\sum_n (E_n - E_0)|f_{0n}|^2 = \tfrac{1}{2} i\hbar (\dot{f} f^+ - f^+ \dot{f})_{00}. \tag{149.4}$$

We apply it to the quantity

$$f = \sum_a e^{-i\mathbf{q}\cdot\mathbf{r}_a}.$$

According to (19.2), its derivative with respect to time is represented by the operator

$$\dot{f} = -(\hbar/2m) \sum_a [e^{-i\mathbf{q}\cdot\mathbf{r}_a}(\mathbf{q}\cdot\nabla_a) + (\mathbf{q}\cdot\nabla_a) e^{-i\mathbf{q}\cdot\mathbf{r}_a}].$$

A direct calculation gives

$$\dot{f} f^+ - f^+ \dot{f} = -(i\hbar/m) q^2 Z.$$

† If an electron is passing through a gas, the scattering at various atoms is independent, and $N d\kappa$ (where N is the number of gas atoms per unit volume) is the energy lost by the electron in unit path by collisions in which it deviates into the given element of solid angle.

Substituting in (149.4), we obtain the formula

$$\sum_n \frac{2m}{\hbar^2 q^2}(E_n-E_0)|\langle n|\sum_a e^{-i\mathbf{q}\cdot\mathbf{r}_a}|0\rangle|^2 = Z, \tag{149.5}$$

which effects the summation required.†

Thus we find for the differential effective retardation the formula

$$d\kappa = 4\pi\frac{Ze^4}{mv^2}\frac{dq}{q} = \frac{2Ze^4}{mv^2}\frac{do}{\vartheta^2}. \tag{149.6}$$

The range of applicability of this formula is given by the inequality

$$(v_0/v)^2 \ll \vartheta \ll 1, \text{ i.e. } v_0/v \ll a_0 q \ll v/v_0.$$

Next, let us determine the total effective retardation $\kappa(q_1)$ for all collisions in which the transfer of momentum does not exceed some value q_1 such that $v_0/v \ll a_0 q_1 \ll v/v_0$:

$$\kappa(q_1) = \sum_n \int_{q_{\min}}^{q_1} (E_n-E_0)\, d\sigma_n; \tag{149.7}$$

$q_{\min}$ is given by (148.11). The integration and summation signs cannot be transposed, since $q_{\min}$ depends on n.

We divide the range of integration into two parts, from $q_{\min}$ to q_0 and from q_0 to q_1, where q_0 is some value of q such that $v_0/v \ll q_0 a_0 \ll 1$. Then, over the whole range of integration from $q_{\min}$ to q_0, we can use for $d\sigma_n$ the expression (148.14):

$$\kappa(q_0) = 8\pi\left(\frac{e}{\hbar v}\right)^2 \sum_n |\langle n|d_x|0\rangle|^2 (E_n-E_0)\int_{q_{\min}}^{q_0}\frac{dq}{q},$$

whence

$$\kappa(q_0) = 8\pi\left(\frac{e}{\hbar v}\right)^2 \sum_n |\langle n|d_x|0\rangle|^2 (E_n-E_0)\log\frac{q_0\hbar v}{E_n-E_0}. \tag{149.8}$$

In the range from q_0 to q_1, on the other hand, we can first sum over n, which gives the expression (149.6) for $d\kappa$, and then on integrating over q we have

$$\kappa(q_1)-\kappa(q_0) = 4\pi(Ze^4/mv^2)\log(q_1/q_0). \tag{149.9}$$

To transform the above expressions, we use the summation theorem obtained from formula (149.4) by putting there

$$f = d_x/e = \sum_a x_a, \qquad \dot{f} = (1/m)\sum_a \hat{p}_{xa}.$$

† In deriving this relation we have nowhere used the fact that the state denoted by the suffix 0 is the normal state of the atom. The relation therefore holds for any initial state.

Commuting $\hat{f}^+$ and $\hat{f}$ gives (in the present case, $\hat{f}^+$ is the same as $\hat{f}$) $\hat{f}\hat{f}^+ - \hat{f}^+\hat{f} = -i\hbar Z/m$, so that†

$$\sum_n N_{0n} \equiv \sum_n (2m/e^2\hbar^2)(E_n - E_0)|\langle n|d_x|0\rangle|^2 = Z. \qquad (149.10)$$

The quantities N_{0n} are called the *oscillator strengths* for the corresponding transitions.

We introduce some mean energy I of the atom, defined by the formula

$$\log I = \sum_n N_{0n} \log(E_n - E_0)/\sum_n N_{0n}$$

$$= (1/Z) \sum_n N_{0n} \log(E_n - E_0). \qquad (149.11)$$

Then, using (149.10), we can rewrite formula (149.8) in the form $\kappa(q_0) = (4\pi Ze^4/mv^2)\log(q_0\hbar v/I)$. Adding this to (149.9), we have finally

$$\kappa(q_1) = (4\pi Ze^4/mv^2)\log(q_1\hbar v/I). \qquad (149.12)$$

Only one constant characterizing the atom concerned appears in this formula.‡

Expressing q_1 in terms of the scattering angle ϑ_1 by means of $q_1 = mv\vartheta_1/\hbar$, we obtain the effective retardation in scattering through all angles $\vartheta \leqslant \vartheta_1$:

$$\kappa(\vartheta_1) = (4\pi Ze^4/mv^2)\log(mv^2\vartheta_1/I). \qquad (149.13)$$

If $q_1 a_0 \gg 1$ (i.e. $\vartheta_1 \gg v_0/v$), we can express κ as a function of the greatest amount of energy that can be transferred from the incident electron to the atom. We have shown in the previous section that, for $qa_0 \gg 1$, the atom is ionized, almost all the momentum $\hbar\mathbf{q}$ and energy being given to one atomic electron. Hence $\hbar\mathbf{q}$ and ϵ are related by being the momentum and energy of an electron, i.e. by $\epsilon = \hbar^2q^2/2m$. Substituting in (149.12) $q_1^2 = 2m\epsilon_1/\hbar^2$, we obtain the effective retardation in collisions where the energy transfer is $\epsilon \leqslant \epsilon_1$:

$$\kappa(\epsilon_1) = (2\pi Ze^4/mv^2)\log(2m\epsilon_1 v^2/I^2). \qquad (149.14)$$

In conclusion, we may make the following remark. The energy levels of the discrete spectrum of an atom mainly involve excitations of a single (outer) electron; the excitation of even two electrons usually requires an

† The remark made concerning (149.5) applies to this relation also.

‡ For hydrogen, $I = 0{\cdot}55me^4/\hbar^2 = 14{\cdot}9$ eV. For heavy atoms we should expect to get good accuracy on calculating the constant I by the Thomas–Fermi method. It is easy to establish how the values of I thus calculated will depend on Z. In the quasi-classical case, the eigenfrequencies of the system of particles correspond to the differences of the energy levels. The mean eigenfrequency of the atom is of the order of v_0/a_0; hence we can deduce that $I \sim \hbar v/a_0$. The velocities of the atomic electrons in the Thomas–Fermi model depend on Z as $Z^{2/3}$, while the dimensions of the atom vary as $Z^{-1/3}$. Thus we find that I should be proportional to Z: $I = \text{constant} \times Z$. From experimental results it can be found that the constant is of the order of magnitude of 10 eV.

energy sufficient to ionize the atom. Hence, in the sum of oscillator strengths, the transitions to states of the discrete spectrum form only a part, of the order of unity, while those which involve ionization form a part of the order of Z. Hence it follows that the main part in retardation (by heavy atoms) is played by those collisions which are accompanied by ionization.

PROBLEM

Determine the total effective retardation of an electron by a hydrogen atom ($I = 0{\cdot}55$ atomic units); for large energy transfers, the faster of the two colliding electrons is taken to be the primary.

SOLUTION. When the primary and secondary electrons have comparable energies after the collision, the exchange effect must be taken into account. Hence, for retardation with an energy transfer between some value ϵ_1 ($1 \ll \epsilon_1 \ll v^2$) and the maximum value $\epsilon_{max} = \frac{1}{2}E = \frac{1}{4}v^2$ (by our definition of the primary electron), we must use the effective cross-section (148.17):

$$\kappa(\epsilon_{max}) - \kappa(\epsilon_1) = \frac{\pi}{E}\int_{\epsilon_1}^{\frac{1}{2}E} \epsilon\left[\frac{1}{\epsilon^2} + \frac{1}{(E-\epsilon)^2} - \frac{1}{\epsilon(E-\epsilon)}\right] d\epsilon$$

$$= \frac{\pi}{E}[\log(E/8\epsilon_1)+1].$$

Adding this to (149.14), we obtain†

$$\kappa = \frac{4\pi}{v^2}\log\left[\frac{v^2}{2I}\sqrt{(\tfrac{1}{2}e)}\right] = \frac{4\pi}{v^2}\log\frac{v^2}{1{\cdot}3}$$

in atomic units.

§150. Inelastic collisions between heavy particles and atoms

The condition for the Born approximation to be applicable to collisions between heavy particles and atoms, expressed in terms of the velocity of a particle, remains the same as for electrons:

$$v \gg v_0.$$

This follows immediately from the general condition (126.2) for perturbation theory to be applicable ($Ua_0/\hbar v \ll 1$), if we notice that the mass of the particle does not appear there, while $Ua_0/\hbar$ is of the order of magnitude of the velocity of the atomic electrons.

In a system of coordinates in which the centre of mass of the atom and the particle is at rest, the cross-section is given by the general formula (148.3), in which m is now the reduced mass of the particle and the atom. It is more convenient, however, to consider the collision in a system of co-ordinates in which the scattering atom is at rest before the collision. To do

† For collisions between a positron and a hydrogen atom there is no exchange effect, and the total retardation is obtained by simply substituting $\epsilon_{max} = E = \frac{1}{2}v^2$ in place of ϵ_1 in (149.14):

$$\kappa = (4\pi/v^2)\log(v^2/0{\cdot}55).$$

this, we start from formula (148.1); in a system of coordinates in which the atom is at rest before the collision, the argument of the delta function which expresses the law of conservation of energy is of the form

$$\tfrac{1}{2}p'^2/M-\tfrac{1}{2}p^2/M+\tfrac{1}{2}(\mathbf{p}'-\mathbf{p})^2/M_a+E_n-E_0, \qquad (150.1)$$

where M is the mass of the incident particle and M_a that of the atom. The third term is the kinetic recoil energy of the atom (and could be entirely neglected when considering a collision between an atom and an electron).

For a collision of a fast heavy particle with an atom, the change in the momentum of the particle is almost always small in comparison with its original momentum. If this condition holds, we can neglect the recoil energy of the atom in the argument of the delta function, and we then arrive at exactly the same formula (148.3), except that m in the latter must be replaced by the mass M of the incident particle (and not by the reduced mass of the particle and the atom). Bearing in mind that the transfer of momentum is supposed small in comparison with the original momentum, we put $p \approx p'$; then the cross-section in a system of coordinates in which the atom is at rest before the collision is

$$d\sigma_n = (M^2/4\pi^2\hbar^4)\left|\int\int Ue^{-i\mathbf{q}\cdot\mathbf{r}}\psi_n^*\psi_0\, d\tau dV\right|^2 do. \qquad (150.2)$$

Taking into account the fact that the charge on the particle may differ from that on the electron, we write ze^2 in place of e^2, where ze is the charge on the incident particle. The general formula for inelastic scattering, written in the form (148.9):

$$d\sigma_n = 8\pi\left(\frac{ze^2}{\hbar v}\right)^2 |\langle n|\sum_a e^{-i\mathbf{q}\cdot\mathbf{r}_a}|0\rangle|^2\frac{dq}{q^3}, \qquad (150.3)$$

does not contain the mass of the particle. Hence it follows that all the formulae derived from it remain applicable to collisions with heavy particles, provided that these formulae are expressed in terms of v and q.

It is easy to see how the formulae must be modified when they are expressed in terms of the scattering angle ϑ (the angle of deviation of the heavy particle on colliding with the atom). To see this, we notice first of all that the angle ϑ is always small in an inelastic collision with a heavy particle. For, when the momentum transfer is large (compared with the momenta of the atomic electrons), we can regard the inelastic collision with the atom as an elastic collision with free electrons; when a heavy particle collides with a light one (the electron), however, the heavy particle hardly deviates at all. In other words, the transfer of momentum from the heavy particle to the atom is small in comparison with the original momentum of the particle; an exception is formed by elastic scattering through large angles, but this is extremely improbable.

Thus, over the whole range of angles, we can put

$$q = \sqrt{\{[(E_n-E_0)/v]^2+(Mv\vartheta)^2\}}/\hbar, \qquad (150.4)$$

which in practice reduces to

$$q\hbar \approx Mv\vartheta \tag{150.5}$$

everywhere except for very small angles. On the other hand, when considering the collisions of electrons with an atom, we had (for small angles)

$$q = \sqrt{\{[(E_n - E_0)/v]^2 + (mv\vartheta)^2\}}/\hbar.$$

Hence we can deduce that the formulae which we obtained for collisions between electrons and atoms, if expressed in terms of the velocity and the angle of deviation, become formulae for the collision of heavy particles if we everywhere make the substitution

$$\vartheta \to M\vartheta/m \tag{150.6}$$

(including the element of solid angle $do = 2\pi \sin\vartheta\, d\vartheta \approx 2\pi\vartheta\, d\vartheta$), the velocity of the incident particle remaining unchanged. Qualitatively, this means that the whole picture of small-angle scattering is (for a given velocity) compressed in the ratio m/M.

The rules obtained above relate also to elastic scattering through small angles. Making the transformation (150.6) in formula (139.4) with $\vartheta \ll 1$, we have the cross-section

$$d\sigma_e = 8\pi(ze^2/Mv^2)^2[Z - F(Mv\vartheta/\hbar)]^2\, d\vartheta/\vartheta^3. \tag{150.7}$$

The elastic scattering of heavy particles through angles $\vartheta \sim 1$ reduces to Rutherford scattering at the nucleus of the atom.

Inelastic scattering in which the atom is ionized with a large transfer of momentum requires special consideration. Unlike the situation for ionization by an electron, there are of course no exchange effects. For heavy particles it is characteristic that a large momentum transfer ($qa_0 \gg 1$) does not mean a deviation through a large angle; ϑ always remains small. The cross-section for ionization with the emission of an electron of energy between ϵ and $\epsilon + d\epsilon$ is found immediately from formula (148.25), which we write in the form

$$d\sigma_r = 8\pi(ze^2/\hbar v)^2 Z\, dq/q^3,$$

putting $\hbar^2 q^2/2m = \epsilon$ (the whole of the momentum $\hbar\mathbf{q}$ is given to a single atomic electron). This gives

$$d\sigma_\epsilon = (2\pi Z z^2 e^4/mv^2)\, d\epsilon/\epsilon^2. \tag{150.8}$$

In collisions of heavy particles with atoms, the total cross-section and effective retardation are of particular interest. The total inelastic scattering cross-section is given by the previous formula (148.26). The total effective retardation is obtained by substituting the maximum possible momentum transfer q_{max} in place of q_1 in (149.12). q_{max} is easily expressed in terms of the velocity of the particle as follows. Since even $\hbar q_{max}$ is small compared with the original momentum Mv of the particle, the change in its energy is related to the change in momentum by $\Delta E = \mathbf{v} \cdot \hbar\mathbf{q}$. On the other hand, for a large momentum transfer nearly all this energy is given to one atomic electron, so that we can write

$$\epsilon = \hbar^2 q^2/2m = \hbar \mathbf{v}\cdot\mathbf{q} \leqslant \hbar v q.$$

Hence we have $\hbar q \leqslant 2mv$, i.e.

$$\hbar q_{\max} = 2mv, \qquad \epsilon_{\max} = 2mv^2. \tag{150.9}$$

We may notice that the maximum angle of deviation of the particle in an inelastic collision is

$$\vartheta_{\max} = \hbar q_{\max}/Mv = 2m/M.$$

Substituting (150.9) in (149.12), we obtain the total effective retardation of a heavy particle:

$$\kappa = (4\pi Z z^2 e^4/mv^2)\log(2mv^2/I). \tag{150.10}$$

§151. Scattering of neutrons

In various physical problems of collision theory it is necessary to consider how the scattering process is affected by the motion of the scattering centres. Under certain conditions we can solve such problems by means of a type of perturbation theory devised by E. Fermi (1936), even if perturbation theory is not applicable to scattering by each centre individually. These problems include, in particular, that of the scattering of slow neutrons by a system of atoms, such as a molecule. We shall refer to this specific problem.

Neutrons are scarcely scattered by electrons, so that practically all the scattering takes place at the nuclei.† We shall suppose that the amplitude for scattering by an individual nucleus is small in comparison with the distances between the atoms. Then the amplitude of the wave scattered by each nucleus in the molecule is small even at the positions of the other nuclei. Under these conditions the amplitude for scattering by the molecule is the sum of those for scattering by the individual nuclei.

Perturbation theory is not in general applicable to a collision between a neutron and a nucleus: although the range of nuclear forces is small, they are very strong within that range. It is important, however, that the amplitude for scattering of a slow neutron (whose wavelength is large compared with the dimensions of the nucleus) is a constant independent of the velocity. Let f_a be the amplitude for scattering by the ath nucleus; $|f_a|^2\, do$ is the differential cross-section for elastic scattering of the neutron by a free nucleus (in their centre-of-mass system).

The constant amplitude can be formally obtained from perturbation theory

† It is also assumed that the molecule has no magnetic moment. Otherwise, there is a further specific scattering effect due to the interaction of the magnetic moments of the molecule and the neutron.

if we describe the interaction of the neutron with the nucleus by a "point" potential energy

$$U(\mathbf{r}) = -\frac{2\pi\hbar^2}{M}f\delta(\mathbf{r}), \tag{151.1}$$

where M is the reduced mass of the neutron and the nucleus; when this expression is substituted in Born's formula (126.4), the delta function makes the integral a constant independent of $\mathbf{q}$. The "field" $U(\mathbf{r})$ thus defined is called a *pseudo-potential*. It should be emphasized that the possibility of defining this is due to the fact that f is constant. In the general case of an arbitrary neutron energy, the scattering amplitude depends on the initial and final momenta $\mathbf{p}$ and $\mathbf{p}'$ separately, and not only on their difference $\mathbf{q}$, whereas the amplitude given by the Born approximation can depend only on $\mathbf{q}$.†

If the scattering nucleus executes a given motion (for example, vibrations in a molecule), and we average over this motion, then the interaction (151.1) is "smeared" over a region of dimensions in general large compared with the scattering amplitude f. For such a "smeared" interaction the condition (126.1) for the Born approximation to be valid is satisfied.

Thus we can describe the neutron–molecule interaction by the pseudo-potential

$$U(\mathbf{r}) = -2\pi\hbar^2 \sum_a \frac{1}{M_a} f_a \delta(\mathbf{r}-\mathbf{R}_a), \tag{151.2}$$

where the summation is over all the nuclei in the molecule, $\mathbf{R}_a$ are their radius vectors, and $\mathbf{r}$ that of the neutron. Substituting this expression in formula (148.3), with M_m in place of m, M_m being the reduced mass of the molecule and the neutron, we obtain the following formula for the cross-section for scattering of a neutron by a molecule, in the centre-of-mass system:

$$d\sigma_n = M_m^2 \frac{p'}{p} \left| \sum_a \frac{1}{M_a} f_a \langle n | e^{-i\mathbf{q}\cdot\mathbf{R}_a} | 0 \rangle \right|^2 do; \tag{151.3}$$

the matrix elements are taken with respect to the wave functions of the stationary states of motion of the nuclei with energies E_0 and E_n, and the momenta p and p' are related by the law of conservation of energy:

$$(p^2 - p'^2)/2M_m = E_n - E_0.$$

Formula (151.3) describes an inelastic collision with a definite change of

† Although the pseudo-potential gives the correct value of the scattering amplitude when perturbation theory is formally applied, this does not mean that perturbation theory is actually applicable to this field. For a potential well of depth U_0 which tends to infinity in such a way that $U_0 a^3$ = constant (where a is the radius of the well, tending to zero), the conditions (126.1), (126.2) are certainly not satisfied.

state of the motion of the nuclei in the molecule (transition $0 \to n$), and is the solution of the problem stated: from the amplitudes (assumed known) for scattering of neutrons by free nuclei, it determines the cross-section for scattering by the molecule, taking account of the intrinsic motion of the nuclei and interference effects due to scattering by different nuclei.

If the nuclei have non-zero spin, the fact must also be taken into account that the scattering amplitudes f_a depend on the total spin of the scattering nucleus and the neutron. This can be done as follows.

The total spin of the nucleus and the neutron can take two values, $j_a = i_a \pm \frac{1}{2}$, where i_a is the spin of the nucleus; we denote the corresponding scattering amplitudes by f_a^+ and f_a^-. We form a spin operator whose eigenvalues for definite values of j_a are f_a^+ and f_a^- respectively. This operator is

$$\hat{f}_a = a_a + b_a \hat{\mathbf{s}} \cdot \hat{\mathbf{i}}_a, \tag{151.4}$$

where $\hat{\mathbf{i}}_a$ and $\hat{\mathbf{s}}$ are the spin operators of the nucleus and the neutron, and the coefficients a_a and b_a are given by the formulae

$$\left.\begin{aligned} a_a &= \frac{1}{2i_a+1}\left[(i_a+1)f_a^+ + i_a f_a^-\right], \\ b_a &= \frac{2}{2i_a+1}(f_a^+ - f_a^-). \end{aligned}\right\} \tag{151.5}$$

This is easily seen if we note that, for a given value of j, the eigenvalue of the operator $\hat{\mathbf{s}} \cdot \hat{\mathbf{i}}$ is

$$\mathbf{s} \cdot \mathbf{i} = \tfrac{1}{2}[j(j+1) - i(i+1) - \tfrac{3}{4}].$$

The operators (151.4) must replace f_a in formula (151.3), and their matrix elements corresponding to the transition considered must be taken. If the incident neutrons and the target nuclei are unpolarized, then the scattering cross-section must be appropriately averaged.

PROBLEMS

PROBLEM 1. Average formula (151.3), assuming the directions of the spins of the neutrons and the nuclei to be distributed entirely at random, and all the nuclei in the molecule to be different.

SOLUTION. The averagings with respect to the directions of the spins of the neutrons and of the nuclei are independent, and each spin gives zero on averaging; hence $\overline{\mathbf{s} \cdot \mathbf{i}_a} = 0$. If the molecule contains no two like atoms, there is no exchange interaction of the nuclear spins, and, since their direct interaction is negligible, the directions of the spins of the various nuclei in the molecule may be regarded as independent; the products of the form $(\mathbf{s} \cdot \mathbf{i}_1)(\mathbf{s} \cdot \mathbf{i}_2)$ therefore also give zero on averaging. For the square $(\mathbf{s} \cdot \mathbf{i})^2$ we have

$$\overline{(\mathbf{s} \cdot \mathbf{i})^2} = \tfrac{1}{3}\mathbf{s}^2\mathbf{i}^2 = \tfrac{1}{3}s(s+1)i(i+1) = \tfrac{1}{4}i(i+1).$$

This gives the following expression for the averaged cross-section:

$$d\sigma_n = M_m{}^2\frac{p'}{p}\Big[|\Sigma\frac{1}{M_a}a_a\langle n|e^{-i\mathbf{q}.\mathbf{R}_a}|0\rangle|^2+$$

$$+\tfrac{1}{4}\sum_a\frac{i_a(i_a+1)}{M_a{}^2}b_a{}^2|\langle n|e^{-i\mathbf{q}.\mathbf{R}_a}|0\rangle|^2\Big]\,do.$$

PROBLEM 2. Apply formula (151.3) to the scattering of slow neutrons by parahydrogen and orthohydrogen (J. Schwinger and E. Teller 1937).

SOLUTION. Before the matrix elements of the spin operators are taken, the expression (151.3) for the scattering by a hydrogen molecule is

$$d\sigma_n = \frac{16}{9}\frac{p'}{p}|a\langle n|e^{-i\mathbf{q}.\mathbf{r}/2}+e^{i\mathbf{q}.\mathbf{r}/2}|0\rangle]+$$

$$+b\hat{\mathbf{s}}\,.\,\langle n|\hat{\mathbf{i}}_1e^{-i\mathbf{q}.\mathbf{r}/2}+\hat{\mathbf{i}}_2e^{i\mathbf{q}.\mathbf{r}/2}|0\rangle|^2\,do, \qquad (1)$$

$$a = \tfrac{1}{4}(3f^++f^-), \qquad b = f^+-f^-,$$

$\pm\mathbf{r}/2$ being the radius vectors of the two nuclei in the molecule relative to their centre of mass.

The rotational and vibrational states of the molecule are defined by the quantum numbers K, M_K, v (which are together represented by n in (1)). In the electron ground state of the H_2 molecule, even values of K are possible only for a total nuclear spin $I = 0$ (parahydrogen), and odd values of K only for $I = 1$ (orthohydrogen) (see §86). We must therefore distinguish two cases: (1) transitions between rotational states with values of K of the same parity, which are possible only for unchanged I (ortho-ortho and para-para transitions), (2) transitions between states with values of K of different parity, which are possible only when I changes (ortho-para and para-ortho transitions). In the first case we have

$$\langle n|e^{-i\mathbf{q}.\mathbf{r}/2}|0\rangle = \langle n|e^{i\mathbf{q}.\mathbf{r}/2}|0\rangle = \langle n|\cos\tfrac{1}{2}\mathbf{q}\,.\,\mathbf{r}|0\rangle;$$

it should be remembered that the rotational wave function is multiplied by $(-1)^K$ when the sign of $\mathbf{r}$ is changed. The spin operator in (1) then becomes $2a+b\hat{\mathbf{s}}.\hat{\mathbf{I}}$, where $\hat{\mathbf{I}} = \hat{\mathbf{i}}_1+\hat{\mathbf{i}}_2$. This operator is diagonal with respect to I, in accordance with the above discussion. The square $(2a+b\mathbf{s}.\mathbf{I})^2$ is averaged, as in Problem 1, giving

$$4a^2+\tfrac{1}{4}b^2I(I+1).$$

The result is

$$d\sigma_n = \frac{4}{9}\frac{p'}{p}|\langle n|\cos\tfrac{1}{2}\mathbf{q}\,.\,\mathbf{r}|0\rangle|^2[(3f^++f^-)^2+I(I+1)(f^++f^-)^2]\,do. \qquad (2)$$

In the second case

$$\langle n|e^{i\mathbf{q}.\mathbf{r}/2}|0\rangle = -\langle n|e^{-i\mathbf{q}.\mathbf{r}/2}|0\rangle = i\langle n|\sin\tfrac{1}{2}\mathbf{q}\,.\,\mathbf{r}|0\rangle,$$

and the spin operator in (1) becomes $\hat{\mathbf{s}}.(\hat{\mathbf{i}}_1-\hat{\mathbf{i}}_2)$; it has only matrix elements which are non-diagonal with respect to I. The squared moduli of these elements, summed over all possible values of the component of the total spin $\mathbf{I}'$ in the final state, are calculated as the mean values (diagonal elements) of the square $[\mathbf{s}.(\mathbf{i}_1-\mathbf{i}_2)]^2$ (see the first footnote to §140):

$$\overline{[\mathbf{s}.(\mathbf{i}_1-\mathbf{i}_2)]^2} = \tfrac{1}{3}.\tfrac{3}{4}\overline{(\mathbf{i}_1-\mathbf{i}_2)^2}$$

$$= \tfrac{1}{4}(2\mathbf{i}_1{}^2+2\mathbf{i}_2{}^2-\mathbf{I}^2)$$

$$= \tfrac{1}{4}[3-I(I+1)].$$

The result is

$$d\sigma_n = (1)(3)\frac{4}{9}\frac{p'}{p}|\langle n|\sin\tfrac{1}{2}\mathbf{q}\,.\,\mathbf{r}|0\rangle|^2(f^+-f^-)^2\,do, \qquad (3)$$

where the coefficient 1 appears for ortho-para transitions and the coefficient 3 for para-ortho transitions.

If the neutrons are so slow that their wavelength is large even compared with the size of the molecule, then we can put $\cos(\tfrac{1}{2}\mathbf{q}.\mathbf{r}) = 1$, $\sin(\tfrac{1}{2}\mathbf{q}.\mathbf{r}) = 0$ in the matrix elements in (2) and (3), so that they are all zero except the diagonal element 00; in these conditions, of course, only elastic scattering is possible. The elastic scattering cross-section in this case is

$$d\sigma_e = \frac{4}{9}[(3f^+ + f^-)^2 + I(I+1)(f^+ - f^-)^2]do.$$

PROBLEM 3. Determine the cross-section for the scattering of neutrons by a bound proton, regarded as an isotropic three-dimensional oscillator of frequency ω (E. Fermi 1936).

SOLUTION. Considering the proton as oscillating about a point fixed in space, we must put in formula (151.3), from its derivation, $M_m = M$ and $M_a = \frac{1}{2}M$ (M being the mass of the proton). Then

$$d\sigma_n = \frac{p'}{p}\frac{\sigma_0}{\pi}\Sigma\left|\int e^{-i\mathbf{q}\cdot\mathbf{r}}\psi_{000}(\mathbf{r})\psi_{n_1n_2n_3}(\mathbf{r})dV\right|^2 do,$$

where $\sigma_0 = 4\pi|f|^2$ is the cross-section for scattering by a free proton, and $\psi_{n_1n_2n_3}$ are the eigenfunctions of the three-dimensional oscillator corresponding to the energy levels $E_n = \hbar\omega(n+3/2)$; the summation is over all values of n_1, n_2 and n_3 whose sum has a given value n. The functions $\psi_{n_1n_2n_3}$ are products of the wave functions of three linear oscillators (see §33, Problem 4). The required integral therefore falls into the product of three integrals of the form

$$\int_{-\infty}^{\infty} e^{-iq_x x/2}e^{-\alpha^2x^2/2}e^{-\alpha^2x^2/2}H_{n_1}(\alpha x)dx$$

($\alpha = \sqrt{(M\omega/\hbar)}$), which are found by substituting $H_{n_1}(x)$ in the form (a.4) and integrating n_1 times by parts. The result is

$$d\sigma_n = \frac{1}{\pi}\frac{v'}{v}\frac{\sigma_0}{2^n\alpha^{2n}}\sum\frac{q_x^{2n_1}q_y^{2n_2}q_z^{2n_3}}{n_1!n_2!n_3!}e^{-q^2/2\alpha^2}do.$$

The summation is effected by the binomial theorem, and the final result is

$$d\sigma_n = \frac{\sigma_0}{\pi n!}\sqrt{\frac{E'}{E}}\left(\frac{q^2}{2\alpha^2}\right)^n e^{-q^2/2\alpha^2}do.$$

In particular, the elastic scattering cross-section ($n = 0$, $E = E'$) is

$$d\sigma_e = \frac{\sigma_0}{\pi}e^{-q^2/2\alpha^2}do, \qquad \sigma_e = \sigma_0\frac{\hbar\omega}{E}(1 - e^{-4E/\hbar\omega});$$

as $E/\hbar\omega \to 0$, $\sigma_e \to 4\sigma_0$.

§152. Inelastic scattering at high energies

The eikonal approximation used in §131 for the problem of mutual scattering of two particles can be generalized to cover also processes (including inelastic ones) in the collision of a fast particle with a system of particles, or "target" (R. J. Glauber 1958).

In this generalization, the principal assumptions made are as before. The energy E of the incident particle is assumed so large that $E \gg |U|$ and $ka \gg 1$, where U is the energy of interaction between this particle and the target particles, and a the range of this interaction. We consider scattering with a relatively small momentum transfer: the change $\hbar\mathbf{q}$ in the momentum of the incident particle is small in comparison with its original value $\hbar\mathbf{k}$ ($q \ll k$). This condition, however, now implies not only that the angle of scattering is small but also that the energy transferred is relatively small.

We shall also suppose that the velocity v of the incident particle is large compared with the velocities v_0 of the particles within the target:

$$v \gg v_0. \tag{152.1}$$

For the scattering of charged particles by atoms, this condition is equivalent to the validity of the Born approximation (cf. §§148 and 150): if $v \gg v_0$, it necessarily follows that $|U|a/\hbar v \ll 1$. There is therefore no need for the present theory in that case. The situation is different, however, for nuclear targets, where the particles are held by nuclear and not Coulomb forces. In the following we shall discuss the particular case of the scattering of a fast particle by a nucleus.†

The condition (152.1) enables us to consider the motion of the incident particle for fixed positions of the nucleons in the nucleus.‡ That is, the wave function of the particle–target system may be written

$$\psi(\mathbf{r}, \mathbf{R}_1, \mathbf{R}_2, \ldots) = \phi(\mathbf{r}; \mathbf{R}_1, \mathbf{R}_2, \ldots)\Phi_i(\mathbf{R}_1, \mathbf{R}_2, \ldots). \tag{152.2}$$

Here $\Phi_i(\mathbf{R}_1, \mathbf{R}_2, \ldots)$ is the wave function of the ith internal state of the nucleus; $\mathbf{R}_1, \mathbf{R}_2, \ldots$ are the radius vectors of the nucleons in it. The factor $\phi(\mathbf{r}; \mathbf{R}_1, \mathbf{R}_2, \ldots)$ is the wave function of the particle undergoing scattering ($\mathbf{r}$ being its radius vector) for given values of $\mathbf{R}_1, \mathbf{R}_2, \ldots$, which act as parameters in Schrödinger's equation

$$\left[-\frac{\hbar^2}{2m}\triangle + \sum_a U_a(\mathbf{r}-\mathbf{R}_a)\right]\phi = \frac{\hbar^2 k^2}{2m}\phi, \tag{152.3}$$

where $U_a(\mathbf{r}-\mathbf{R}_a)$ is the energy of interaction of the particle with the ath nucleon, and $\hbar\mathbf{k}$ the momentum of the particle at infinity.||

If we find a solution of (152.3) with the asymptotic form

$$\phi = e^{i\mathbf{k}.\mathbf{r}} + F(\mathbf{n}', \mathbf{n}; \mathbf{R}_1, \mathbf{R}_2, \ldots)e^{ikr}/r, \tag{152.4}$$

where $\mathbf{n}' = \mathbf{r}/r$, $\mathbf{n} = \mathbf{k}/k$, then the wave function (152.2)

$$\psi = e^{i\mathbf{k}.\mathbf{r}}\Phi_i + F\Phi_i e^{ikr}/r \tag{152.5}$$

will describe scattering by a nucleus that is in its ith state before the collision: the incident wave $e^{i\mathbf{k}.\mathbf{r}}$ appears in (152.5) as a product with Φ_i. The second

---

† The condition (152.1) gives relativistic velocities v for any heavy nuclei. In presenting here the formalism in non-relativistic theory, we ignore the question of its practical applicability to any particular scattering problems.

‡ This approximation is analogous to the one on which the theory of molecules is based, in which the electron state is considered for fixed positions of the nuclei.

|| In (152.3) it is assumed that the interaction of the particle with the nucleus is equal to a sum of binary interactions with the individual nucleons.

term in (152.5) represents the scattered wave. This expression, however, is appropriate for determining the scattering amplitude only if the change in the energy of the incident particle is sufficiently small, i.e. if the change in the internal energy of the nucleus is sufficiently small; thus, by considering the motion of the particle in the constant field of "rigidly fixed" nucleons (corresponding to equation (152.3)), we neglect a possible change in the energy of this motion.

To separate the scattering amplitude with a definite change of the internal state of the nucleus, we must put ψ in the form

$$\psi = e^{i\mathbf{k}.\mathbf{r}}\Phi_i + \sum_f f_{fi}(\mathbf{n}', \mathbf{n})\Phi_f e^{ikr}/r, \tag{152.6}$$

where the summation is over various states of the nucleus; $f_{fi}(\mathbf{n}', \mathbf{n})$ then gives the required scattering amplitude with a particular transition $i \rightarrow f$ of the nucleus, as a function of the scattering angle (the angle between $\mathbf{n}$ and $\mathbf{n}'$). Comparison of (152.6) and (152.5) shows that

$$f_{fi}(\mathbf{n}', \mathbf{n}) = \int \Phi_f^* F \Phi_i \, d\tau, \tag{152.7}$$

where $d\tau = d^3R_1 d^3R_2 \ldots$ is the volume element in the configuration space of the nucleus. We must again emphasize that this formula is valid only when the energies of the states i and f have a relatively slight difference.

The solution (152.4) itself of equation (152.3) is found by the method described in §131.† Similarly to (131.7) we have

$$F(\mathbf{n}', \mathbf{n}; \mathbf{R}_1, \mathbf{R}_2, \ldots) = \frac{k}{2\pi i} \int [S(\boldsymbol{\rho}, \mathbf{R}_1, \mathbf{R}_2, \ldots) - 1] e^{-i\mathbf{q}\cdot\boldsymbol{\rho}} \, d^2\rho, \tag{152.8}$$

where

$$\left.\begin{aligned} S(\boldsymbol{\rho}, \mathbf{R}_1, \mathbf{R}_2, \ldots) &= \exp[2i\delta(\boldsymbol{\rho}, \mathbf{R}_1, \mathbf{R}_2, \ldots)], \\ \delta(\boldsymbol{\rho}, \mathbf{R}_1, \mathbf{R}_2, \ldots) &= \sum_a \delta_a(\boldsymbol{\rho} - \mathbf{R}_{a\perp}), \\ \delta_a(\boldsymbol{\rho} - \mathbf{R}_{a\perp}) &= -\frac{1}{2\hbar v} \int_{-\infty}^{\infty} U_a(\mathbf{r} - \mathbf{R}_a) dz. \end{aligned}\right\} \tag{152.9}$$

It will be recalled that $\boldsymbol{\rho}$ is the component of the radius vector $\mathbf{r}$ in the xy-plane (perpendicular to $\mathbf{k}$), and $\mathbf{R}_{a\perp}$ is the corresponding component of the radius vector $\mathbf{R}_a$; $\hbar\mathbf{q} = \mathbf{p}' - \mathbf{p}$ is the change in momentum of the

† It has been noted in §131 that the initial expression (131.4) for the wave function is valid only at distances $z \ll ka^2$. This was not important in the further derivation in §131, but for scattering by a system of particles (such as a nucleus) it leads to another limitation: the expression (131.4) must be valid throughout the volume occupied by the scattering system, i.e. we must have $R_0 \ll ka^2$, where R_0 is the radius of the nucleus and a the range of the potentials U.

scattered particle, and only its transverse components appear in (152.8). The functions δ_a determine the amplitudes for elastic scattering of the particle by the individual free nucleons:

$$f^{(a)} = \frac{k}{2\pi i}\int \{\exp[2i\delta_a(\boldsymbol{\rho})]-1\}\, e^{-i\mathbf{q}\cdot\boldsymbol{\rho}}\, d^2\rho. \qquad (152.10)$$

When $i = f$, (152.7) and (152.8) give the amplitude for elastic scattering by the nucleus:

$$f_{ii}(\mathbf{n}', \mathbf{n}) = \frac{k}{2\pi i}\int [\bar{S}(\boldsymbol{\rho})-1]e^{-i\mathbf{q}\cdot\boldsymbol{\rho}}\, d^2\rho, \qquad (152.11)$$

where the bar denotes averaging with respect to the internal state of the nucleus:

$$\bar{S}(\boldsymbol{\rho}) = \int S(\boldsymbol{\rho}, \mathbf{R}_1, \mathbf{R}_2, \ldots)|\Phi_i(\mathbf{R}_1, \mathbf{R}_2, \ldots)|^2\, d\tau. \qquad (152.12)$$

This formula generalizes the previous formula (131.7).

Putting in (152.11) $\mathbf{n}' = \mathbf{n}$ and using the optical theorem (142.10), we obtain the total scattering cross-section:

$$\sigma_t = 2\int (1-\operatorname{re}\bar{S})\, d^2\rho. \qquad (152.13)$$

The integrated elastic scattering cross-section σ_e is obtained by integrating $|f_{ii}|^2$ over the directions of $\mathbf{n}'$. For small scattering angles θ, we have $q \approx k\theta$ and the solid-angle element $do \approx d^2q/k^2$. Hence

$$\sigma_e = \int |f_{ii}|^2\, d^2q/k^2.$$

Writing $f^*_{ii}f_{ii}$, with f_{ii} from (152.11), as a double integral over $d^2\rho d^2\rho'$, we can carry out the integration with respect to d^2q by means of the formula

$$\int e^{-i\mathbf{q}\cdot(\boldsymbol{\rho}-\boldsymbol{\rho}')}\, d^2q = (2\pi)^2\delta(\boldsymbol{\rho}-\boldsymbol{\rho}'),$$

and the delta function is then eliminated by integration over $d^2\rho'$. The result is

$$\sigma_e = \int |\bar{S}-1|^2\, d^2\rho. \qquad (152.14)$$

The total reaction cross-section is

$$\sigma_r = \sigma_t - \sigma_e = \int (1-|\bar{S}|^2)\, d^2\rho. \qquad (152.15)$$

Note that (152.13)–(152.15) are in agreement with the general formulae (142.3)–(142.5): on changing, in the latter, from summation (over large l) to integration over $d^2\rho$ with $\rho = l/k$, and replacing S_l by the function $\bar{S}(\rho)$, we obtain (152.13)–(152.15).

PROBLEMS

PROBLEM 1. Express the amplitude for elastic scattering of a fast particle by a deuteron in terms of those for the proton and the neutron (R. J. Glauber 1955).

SOLUTION. According to (152.11), the amplitude for elastic scattering by a deuteron is

$$f^{(d)}(\mathbf{q}) = \frac{k}{2\pi i}\int |\psi_d(\mathbf{R})|^2\{\exp[2i\delta_n(\boldsymbol{\rho}-\tfrac{1}{2}\mathbf{R}_\perp)+2i\delta_p(\boldsymbol{\rho}+\tfrac{1}{2}\mathbf{R}_\perp)]-1\}\times e^{-i\mathbf{q}\cdot\boldsymbol{\rho}}\, d^3R\, d^2\rho. \tag{1}$$

Here $\psi_d(\mathbf{R})$ is the wave function of the relative motion of the neutron (n) and the proton (p) in the deuteron; $\mathbf{R} = \mathbf{R}_n - \mathbf{R}_p$, and $\mathbf{R}_\perp$ is the component of $\mathbf{R}$ in the plane perpendicular to the wave vector $\mathbf{k}$ of the incident particle. The difference in the braces in (1) may be written

$$\exp(2i\delta_n+2i\delta_p)-1 = (e^{2i\delta_n}-1)+(e^{2i\delta_p}-1)+(e^{2i\delta_n}-1)(e^{2i\delta_p}-1).$$

The integrals can then be transformed, using the definition of the scattering amplitudes for the neutron $f^{(n)}$ and the proton $f^{(p)}$, according to (152.10) and the converse formulae

$$\exp[2i\delta_a(\boldsymbol{\rho})]-1 = \frac{2\pi i}{k}\int f^{(a)}(\mathbf{q})e^{i\mathbf{q}\cdot\boldsymbol{\rho}}\, d^2q/(2\pi)^2.$$

The result is

$$f^{(d)}(\mathbf{q}) = f^{(n)}(\mathbf{q})F(\mathbf{q})+f^{(p)}(\mathbf{q})F(-\mathbf{q})$$
$$-\frac{1}{2\pi ik}\int F(2\mathbf{q}')f^{(n)}(\tfrac{1}{2}\mathbf{q}+\mathbf{q}')f^{(p)}(\tfrac{1}{2}\mathbf{q}-\mathbf{q}')\, d^2q', \tag{2}$$

where

$$F(\mathbf{q}) = \int |\psi_d(\mathbf{R})|^2\, e^{-i\mathbf{q}\cdot\mathbf{R}/2}\, d^3R$$

is the deuteron form factor.

Putting in (2) $\mathbf{q} = 0$ (with $F(0) = 1$) and using the optical theorem (142.10), we find the total deuteron scattering cross-section:

$$\sigma_t^{(d)} = \sigma_t^{(n)}+\sigma_t^{(p)}+\frac{2}{k^2}\,\mathrm{re}\int F(2\mathbf{q})f^{(n)}(\mathbf{q})f^{(p)}(-\mathbf{q})\, d^2q. \tag{3}$$

PROBLEM 2. Determine the cross-section for disintegration of a fast deuteron into a separate neutron and proton in scattering by a heavy absorbing nucleus whose radius R_0 is large compared with the deuteron wavelength ($kR_0 \gg 1$, where $\hbar\mathbf{k}$ is the deuteron momentum) and the deuteron radius (E. L. Feinberg 1954; R. J. Glauber 1955; A. I. Akhiezer and A. G. Sitenko 1955).

SOLUTION. With regard to the incident deuteron plane wave, the large ($kR_0 \gg 1$) absorbing nucleus acts as an opaque screen at which the wave is diffracted. The wave function of the incident deuterons is $e^{i\mathbf{k}\cdot\mathbf{r}}\psi_d(R)$, where $\psi_d(R)$ is the internal wave function of the deuteron ($\mathbf{R} = \mathbf{R}_n - \mathbf{R}_p$ is the radius vector between the neutron and the proton in it, $\mathbf{r} = \frac{1}{2}(\mathbf{R}_n+\mathbf{R}_p)$ that of their centre of mass). The presence of the absorbing nucleus "removes" a part of this function corresponding to transverse coordinates of the neutron and the proton ($\boldsymbol{\rho}_n$ and $\boldsymbol{\rho}_p$) that lie in the "shadow" of the nucleus, i.e. within a circle of radius R_0. Thus the wave function becomes

$$\psi = e^{i\mathbf{k}\cdot\mathbf{r}}S(\boldsymbol{\rho}_n, \boldsymbol{\rho}_p)\psi_d(R),$$

where $S = 1$ for $\rho_n, \rho_p \geqslant R_0$ and $S = 0$ if either ρ_n or ρ_p is less than R_0.† This function,

† The Coulomb interaction of the deuteron with the nucleus is neglected.

without the factor ψ_d, corresponds to the expression for the incident wave in the form (131.5) (it neglects the curvature of the rays by diffraction), and the factor S therefore has the same significance as in §§131 and 152.

Analogously to (152.13) and (152.14), the total deuteron scattering cross-section σ_t (including all inelastic processes) and the elastic scattering cross-section σ_e are

$$\sigma_t = 2\int (1-\bar{S})\, d^2\rho, \qquad \sigma_e = \int (\bar{S}-1)^2\, d^2\rho,$$

where $\boldsymbol{\rho} = \frac{1}{2}(\boldsymbol{\rho}_n + \boldsymbol{\rho}_p)$ and we have used the fact that S is real; the averaging of S is with respect to the ground state of the deuteron,

$$\bar{S}(\rho) = \int S\psi_d^2\, d^3R.$$

For ψ_d, it is sufficient to use the function

$$\psi_d = \sqrt{\frac{\kappa}{2\pi}}\frac{e^{-\kappa R}}{R},$$

which is valid for distances R beyond the range of the nuclear forces between the neutron and the proton (cf. (133.14); $\kappa = \sqrt{(m|\epsilon|)}/\hbar$, where $|\epsilon|$ is the deuteron binding energy and m the nucleon mass). From the definition of S, $1-S$ is not zero if one or both of the nucleons come within the circle of radius R_0 and are absorbed by the nucleus; hence

$$\sigma_{\text{capt}} = \int (1-\bar{S})\, d^2\rho = \tfrac{1}{2}\sigma_t \tag{1}$$

is the cross-section for the capture of one or both nucleons. On the other hand, $\sigma_t = \sigma_{\text{capt}} + \sigma_e + \sigma_{\text{dis}}$, where σ_{dis} is the required cross-section for "diffractive" disintegration of the deuteron. Hence

$$\sigma_{\text{dis}} = \tfrac{1}{2}\sigma_t - \sigma_e = \int \bar{S}(1-\bar{S})\, d^2\rho. \tag{2}$$

When $R_0\kappa \gg 1$, the important distances from the edge of the nucleus are those which are small and $\sim 1/\kappa$; then the integration along the edge gives a factor $2\pi R_0$, and the integration in the perpendicular direction can be carried out as if the shadow region were bounded by a straight line. Taking the latter as the y-axis, and the x-axis outwards from the shadow, we have

$$\sigma_{\text{dis}} = 2\pi R_0 \int_0^\infty \bar{S}(x)[1-\bar{S}(x)]\, dx,$$

with the integral

$$\bar{S}(x) = \int_{-\infty}^{\infty}\int_{-\infty}^{\infty}\int_{-2x}^{2x} \psi_d^2(R)\, dX\, dY\, dZ, \qquad R = \sqrt{(X^2+Y^2+Z^2)},$$

taken over the region X_n, $X_p \geqslant 0$ for a given $x = \frac{1}{2}(X_n + X_p)$ or, what is the same thing, over the region $|X| = |X_n - X_p| \leqslant 2x$. The integral is transformed by changing to the variables X and R and to the polar angle in the YZ-plane (with $dY\, dZ \to 2\pi R\, dR$), becoming

$$\bar{S}(x) = 1 - e^{-4\kappa x} + 4\kappa x \int_{4\kappa x}^{\infty} \frac{e^{-\xi}}{\xi}\, d\xi. \tag{3}$$

The integral (2) with this function $\bar{S}(x)$ is calculated by repeated integration by parts, using the formula

$$\int_0^\infty (e^{-\xi} - e^{-2\xi})\frac{d\xi}{\xi} = \log 2.$$

The result is†

$$\sigma_{\text{dis}} = \frac{\pi}{3\kappa} R_0 (\log 2 - \tfrac{1}{4}).$$

---

† With the same condition $\kappa R_0 \gg 1$, the capture cross-section is

$$\sigma_{\text{capt}} = \pi R_0^2 + \pi R_0/4\kappa;$$

the integral (1) over the region $\rho > R_0$ is calculated by means of (3), and that over the region $\rho < R_0$ gives πR_0^2. This cross-section includes both capture of the whole deuteron and capture of one nucleon with release of the other (stripping reaction). The cross-section for the latter reaction is calculated as the impact area (averaged with respect to ψ_d^2) corresponding to the incidence of only one nucleon on the shadow region, and is

$$\sigma_{\text{capt},\, n} = \sigma_{\text{capt},\, p} = \pi R_0/4\kappa$$

(R. Serber 1947).

MATHEMATICAL APPENDICES

§a. Hermite polynomials

The equation

$$y'' - 2xy' + 2ny = 0 \tag{a.1}$$

belongs to a class which can be solved by what is called *Laplace's method.*†

This method is applicable to any linear equation of the form

$$\sum_{m=0}^{n} (a_m + b_m x) \frac{d^m y}{dx^m} = 0,$$

whose coefficients are of degree in x not higher than the first, and consists in the following procedure. We form the polynomials

$$P(t) = \sum_{m=0}^{n} a_m t^m, \qquad Q(t) = \sum_{m=0}^{n} b_m t^m,$$

and from them the function

$$Z(t) = (1/Q) e^{\int (P/Q)\,dt},$$

which is determined to within a constant factor. Then the solution of the equation under consideration can be expressed as a complex integral:

$$y = \int_C Z(t) e^{xt}\, dt,$$

where the path of integration C is taken so that the integral is finite and non-zero, and the function

$$V = e^{xt} Q Z$$

returns to its original value when t describes the contour C (which may be either closed or open).

In the case of equation (a.1) we have

$$P = t^2 + 2n, \qquad Q = -2t, \qquad Z = -\frac{1}{2t^{n+1}} e^{-t^2/4}, \qquad V = \frac{1}{t^n} e^{xt - t^2/4},$$

so that its solution is

$$y = \int e^{xt - t^2/4}\, dt/t^{n+1}. \tag{a.2}$$

† See, for instance, E. Goursat, *Cours d'Analyse Mathématique*, Vol. II, Gauthier-Villars, Paris; V. I. Smirnov, *Course of Higher Mathematics*, Vol. III, Part 2, Pergamon, Oxford, 1964.

For physical applications we need only consider values $n > -\frac{1}{2}$. For these values the contour of integration can be taken as C_1 or C_2 (Fig. 52); these satisfy the required conditions†, since the function V vanishes at their ends ($t = +\infty$ or $t = -\infty$).

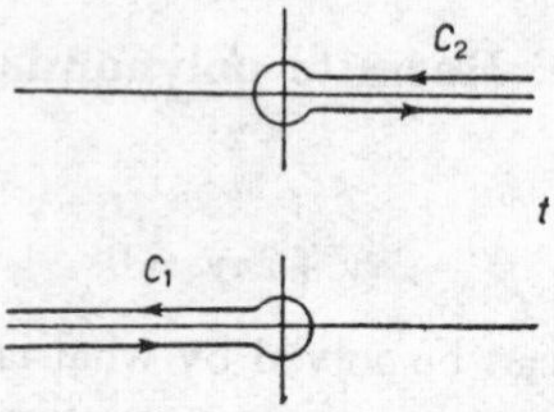

FIG. 52

Let us find the values of the parameter n for which equation (a.1) has solutions finite for all finite x, which tend to infinity, as $x \to \pm\infty$, not more rapidly than every finite power of x. First, we consider non-integral values of n. The integrals (a.2) along C_1 and C_2 then give two independent solutions of equation (a.1). We transform the integral along C_1 by introducing the variable u such that $t = 2(x-u)$. Omitting a constant factor, we find

$$y = e^{x^2} \int_{C_1'} e^{-u^2} du/(u-x)^{n+1}, \tag{a.3}$$

where the integration is taken over the contour C_1' in the complex plane of u, as shown in Fig. 53.

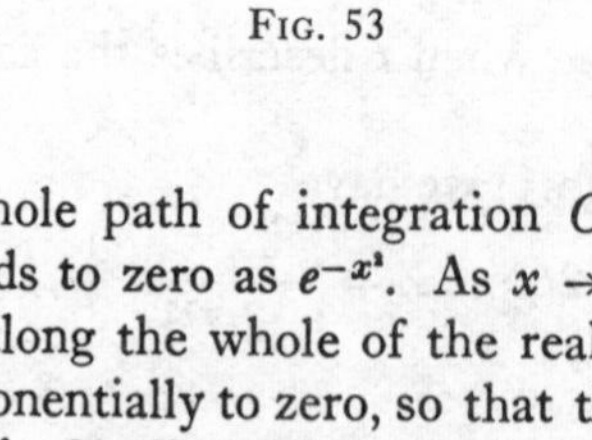

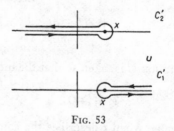

FIG. 53

As $x \to +\infty$, the whole path of integration C_1' moves to infinity, and the integral in (a.3) tends to zero as e^{-x^2}. As $x \to -\infty$, however, the path of integration extends along the whole of the real axis, and the integral in (a.3) does not tend exponentially to zero, so that the function $y(x)$ becomes infinite essentially as e^{x^2}. Similarly, it is easy to see that the integral (a.2) along the contour C_2' diverges exponentially as $x \to +\infty$.

† These paths will not serve for negative integral n, since the integral (a.2) along them then vanishes identically.

For positive integral n (including zero), on the other hand, the integrals along the straight parts of the path of integration cancel, and the two integrals (a.3), along C_1' and C_2', reduce to an integral along a closed path round the point $u = x$. Thus we have the solution

$$y(x) = e^{x^2} \oint e^{-u^2} \, \mathrm{d}u/(u-x)^{n+1},$$

which satisfies the conditions stated. According to Cauchy's well-known formula for the derivatives of an analytic function,

$$f^{(n)}(x) = \frac{n!}{2\pi i} \oint \frac{f(t)}{(t-x)^{n+1}} \, \mathrm{d}t,$$

$y(x)$ is, apart from a constant factor, an *Hermite polynomial*:

$$H_n(x) = (-1)^n e^{x^2} \frac{\mathrm{d}^n}{\mathrm{d}x^n} e^{-x^2}. \tag{a.4}$$

The polynomial H_n, expanded in decreasing powers of x, has the open form

$$H_n(x) = (2x)^n - \frac{n(n-1)}{1}(2x)^{n-2} + \frac{n(n-1)(n-2)(n-3)}{1.2}(2x)^{n-4} - \dots. \tag{a.5}$$

It contains only powers of x which are of the same parity as n. We may write out here the first few Hermite polynomials:

$$H_0 = 1, \quad H_1 = 2x, \quad H_2 = 4x^2-2, \quad H_3 = 8x^3-12x, \quad H_4 = 16x^4-48x^2+12. \tag{a.6}$$

To calculate the normalization integral, we replace $e^{-x^2} H_n$ by its expression in (a.4) and integrate n times by parts:

$$\int_{-\infty}^{\infty} e^{-x^2} H_n^2(x) \, \mathrm{d}x = \int_{-\infty}^{\infty} (-1)^n H_n(x) \frac{\mathrm{d}^n}{\mathrm{d}x^n} e^{-x^2} \, \mathrm{d}x$$

$$= \int_{-\infty}^{\infty} e^{-x^2} \frac{\mathrm{d}^n}{\mathrm{d}x^n} H_n \, \mathrm{d}x.$$

But $\mathrm{d}^n H_n/\mathrm{d}x^n$ is a constant, $2^n n!$. Thus

$$\int_{-\infty}^{\infty} e^{-x^2} H_n^2(x) \, \mathrm{d}x = 2^n n! \sqrt{\pi}. \tag{a.7}$$

§b. The Airy function

The equation

$$y'' - xy = 0 \qquad \text{(b.1)}$$

is of Laplace's type (see §a). Following the general method, we form the functions

$$P = t^2, \qquad Q = -1, \qquad Z = -e^{-t^3/3}, \qquad V = e^{xt-t^3/3},$$

so that the solution can be represented in the form

$$y(x) = \text{constant} \times \int_C e^{xt-t^3/3}\, dt. \qquad \text{(b.2)}$$

The path of integration C must be chosen so that the function V vanishes at both ends of it. These ends must therefore go to infinity in the regions of the complex plane of t in which re $t^3 > 0$ (the shaded regions in Fig. 54).

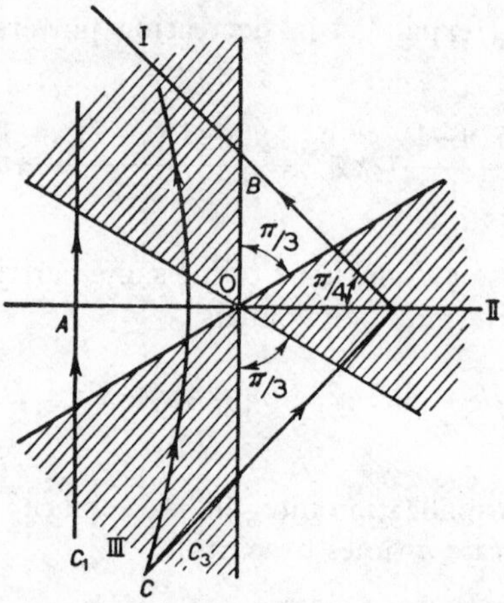

FIG. 54

A solution finite for all x is obtained by taking the path C as shown in the figure. It can be displaced in any manner provided that the ends of it go to infinity in the same two shaded sectors (I and III in Fig. 54). We notice that, by taking a path which lay in sectors III and II (say), we should obtain a solution which becomes infinite as $x \to \infty$.

Deforming the path C so that it goes along the imaginary axis, we obtain the function (b.2) in the form (substituting $t = iu$)

$$\Phi(x) = \frac{1}{\sqrt{\pi}} \int_0^\infty \cos(ux + \tfrac{1}{3}u^3)\, du. \qquad \text{(b.3)}$$

The constant in (b.2) has been put equal to $-i/2\sqrt{\pi}$, and we have denoted the function thus obtained by $\Phi(x)$; it is called the *Airy function*.†

The asymptotic expression for $\Phi(x)$ for large values of x is obtained by calculating the integral (b.2) by the *saddle-point method*. For $x > 0$, the exponent in the integrand has an extremum for $t = \pm\sqrt{x}$, and the "direction of steepest descent" of the integrand is parallel to the imaginary axis. Accordingly, to obtain the asymptotic expression for large positive x, we expand the exponent in powers of $t+\sqrt{x}$ and integrate along the line C_1 (Fig. 54), which is parallel to the imaginary axis; the distance $OA = \sqrt{x}$. Making the substitution $t = -\sqrt{x}+iu$, we have

$$\Phi(x) \approx -\frac{i}{2\sqrt{\pi}}\int_{-\infty}^{\infty} e^{-(2/3)x^{3/2}-u^2\sqrt{x}}\,du,$$

whence

$$\Phi(x) \approx \tfrac{1}{2}x^{-1/4}e^{-(2/3)x^{3/2}}. \tag{b.4}$$

Thus, for large positive x, the function $\Phi(x)$ diminishes exponentially.

To obtain the asymptotic expression for large negative values of x, we notice that, for $x < 0$, the exponent has an extremum for $t = i\sqrt{|x|}$ and $t = -i\sqrt{|x|}$, and the direction of steepest descent at these points is along lines at angles $-\frac{1}{4}\pi$ and $\frac{1}{4}\pi$ respectively to the real axis. Taking as the path of integration the broken line C_3 (the distance $OB = \sqrt{|x|}$), we have, after some simple transformations,

$$\Phi(x) = |x|^{-1/4}\sin(\tfrac{2}{3}|x|^{3/2}+\tfrac{1}{4}\pi). \tag{b.5}$$

Thus, in the region of large negative x, the function $\Phi(x)$ is oscillatory. We may mention that the first (and highest) maximum of the function $\Phi(x)$ is $\Phi(-1{\cdot}02) = 0{\cdot}95$.

The Airy function can be expressed in terms of Bessel functions of order $\frac{1}{3}$. The equation (b.1), as can easily be seen, has the solution

$$\sqrt{x}\,Z_{1/3}(\tfrac{2}{3}x^{3/2}),$$

where $Z_{1/3}(x)$ is any solution of Bessel's equation of order $\frac{1}{3}$. The solution which is the same as (b.3) is

$$\left.\begin{aligned}\Phi(x) &= \tfrac{1}{3}\sqrt{(\pi x)}[I_{-1/3}(\tfrac{2}{3}x^{3/2})-I_{1/3}(\tfrac{2}{3}x^{3/2})]\\ &\equiv \sqrt{(x/3\pi)}K_{1/3}(\tfrac{2}{3}x^{3/2}) \text{ for } x>0,\\ \Phi(x) &= \tfrac{1}{3}\sqrt{(\pi|x|)}[J_{-1/3}(\tfrac{2}{3}|x|^{3/2})+J_{1/3}(\tfrac{2}{3}|x|^{3/2})] \text{ for } x<0,\end{aligned}\right\} \tag{b.6}$$

where

$$I_\nu(x) = i^{-\nu}J_\nu(ix),\ K_\nu(x) = \frac{\pi}{2\sin\nu\pi}[I_{-\nu}(x)-I_\nu(x)].$$

† We follow the definition proposed by V. A. Fok; see G. D. Yakovleva, *Tablitsy funktsiĭ Éĭri* (*Tables of Airy Functions*), Nauka, Moscow, 1969. The function $\Phi(x)$ is one of two defined by Fok, who denotes it by $V(x)$. In the literature, another definition of the Airy function is also found, which differs from (b.3) by a constant factor: $\text{Ai}(x) = \Phi(x)/\sqrt{\pi}$.

Using the recurrence relations

$$K_{\nu-1}(x) - K_{\nu+1}(x) = -\frac{2\nu}{x}K_\nu(x),$$

$$2K_\nu'(x) = -K_{\nu-1}(x) - K_{\nu+1}(x),$$

we easily find for the derivative of the Airy function

$$\Phi'(x) = -\frac{x}{\sqrt{(3\pi)}}K_{2/3}(\tfrac{2}{3}x^{3/2}) \text{ for } x > 0. \tag{b.7}$$

When $x = 0$,

$$\Phi(0) = \frac{\sqrt{\pi}}{3^{2/3}\Gamma(\frac{2}{3})} = 0{\cdot}629,$$

$$\Phi'(0) = -\frac{3^{1/6}\Gamma(\frac{2}{3})}{2\sqrt{\pi}} = -0{\cdot}459.$$

Figure 55 shows a graph of the Airy function.

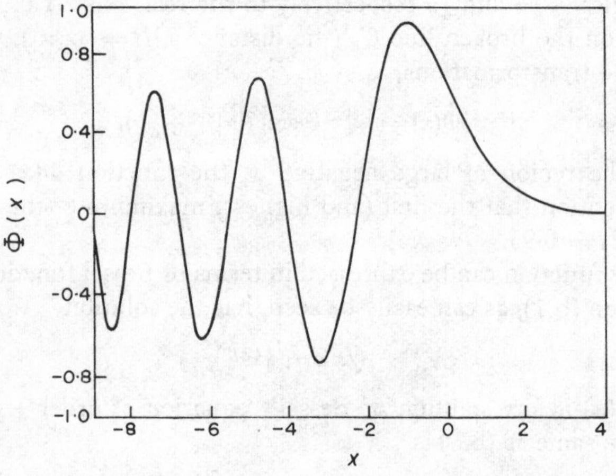

FIG. 55

§c. Legendre polynomials†

The *Legendre polynomials* $P_l(\cos\theta)$ are defined by the formula

$$P_l(\cos\theta) = \frac{1}{2^l l!}\frac{d^l}{(d\cos\theta)^l}(\cos^2\theta - 1)^l. \tag{c.1}$$

† There are in the mathematical literature many good accounts of the theory of spherical harmonics. Here we shall give, for reference, only a few basic relations, and make no attempt at a systematic discussion of the theory of these functions.

They satisfy the differential equation

$$\frac{1}{\sin\theta}\frac{\mathrm{d}}{\mathrm{d}\theta}\left(\sin\theta\frac{\mathrm{d}P_l}{\mathrm{d}\theta}\right)+l(l+1)P_l = 0. \tag{c.2}$$

The *associated Legendre polynomials* are defined by

$$P_l^m(\cos\theta) = \sin^m\theta\,\frac{\mathrm{d}^m P_l(\cos\theta)}{(\mathrm{d}\cos\theta)^m}$$

$$= \frac{1}{2^l l!}\sin^m\theta\,\frac{\mathrm{d}^{l+m}}{(\mathrm{d}\cos\theta)^{l+m}}(\cos^2\theta-1)^l, \tag{c.3}$$

or, equivalently,

$$P_l^m(\cos\theta) = (-1)^m\frac{(l+m)!}{(l-m)!2^l l!}\sin^{-m}\theta\,\frac{\mathrm{d}^{l-m}}{(\mathrm{d}\cos\theta)^{l-m}}(\cos^2\theta-1)^l, \tag{c.4}$$

with $m = 0, 1, \ldots, l$. The associated polynomials satisfy the equation

$$\frac{1}{\sin\theta}\frac{\mathrm{d}}{\mathrm{d}\theta}\left(\sin\theta\frac{\mathrm{d}P_l^m}{\mathrm{d}\theta}\right)+\left[l(l+1)-\frac{m^2}{\sin^2\theta}\right]P_l^m = 0. \tag{c.5}$$

The normalization integral

$$\int_{-1}^{1}[P_l(\mu)]^2\,\mathrm{d}\mu$$

($\mu = \cos\theta$) for the Legendre polynomials is calculated by substituting (c.1) and integrating l times by parts, which gives

$$\frac{(-1)^l}{2^{2l}(l!)^2}\int_{-1}^{1}(\mu^2-1)^l\frac{\mathrm{d}^{2l}}{\mathrm{d}\mu^{2l}}(\mu^2-1)^l\,\mathrm{d}\mu = \frac{(2l)!}{2^{2l}(l!)^2}\int_{-1}^{1}(1-\mu^2)^l\,\mathrm{d}\mu.$$

Substitution of $u = \frac{1}{2}(1-\mu)$ reduces this integral to Euler's beta function, and the result is

$$\int_{-1}^{1}[P_l(\mu)]^2\,\mathrm{d}\mu = \frac{2}{2l+1}. \tag{c.6}$$

Similarly, it is easily seen that the functions $P_l(\mu)$ with different l are orthogonal:

$$\int_{-1}^{1}P_l(\mu)P_{l'}(\mu)\,\mathrm{d}\mu = 0, \quad l\neq l'. \tag{c.7}$$

The calculation of the normalization integral for the associated Legendre polynomials is easily effected by a similar method. We write $[P_l^m(\mu)]^2$ as a product of the expressions (c.3) and (c.4), and integrate $l-m$ times by parts; the result is

$$\int_{-1}^{1} [P_l^m(\mu)]^2 \, d\mu = \frac{2}{2l+1} \frac{(l+m)!}{(l-m)!}. \tag{c.8}$$

It is also easily seen that the functions P_l^m with different l (and the same m) are orthogonal:

$$\int_{-1}^{1} P_l^m(\mu) P_{l'}^m(\mu) \, d\mu = 0, \qquad l \neq l'. \tag{c.9}$$

The calculation of the integrals of products of three Legendre polynomials is discussed in §107.

The following *addition theorem* holds for Legendre polynomials. Let γ be the angle between two directions defined by the spherical angles θ, ϕ and θ', ϕ': $\cos\gamma = \cos\theta \cos\theta' + \sin\theta \sin\theta' \cos(\phi-\phi')$. Then

$$P_l(\cos\gamma) = P_l(\cos\theta) P_l(\cos\theta') + \sum_{m=1}^{l} 2\frac{(l-m)!}{(l+m)!} P_l^m(\cos\theta) P_l^m(\cos\theta') \cos m(\phi-\phi'). \tag{c.10}$$

This theorem can also be written in terms of the spherical harmonic functions defined by (28.7):

$$P_l(\mathbf{n}\,.\,\mathbf{n}') = \frac{4\pi}{2l+1} \sum_{m=-l}^{l} Y_{lm}^*(\mathbf{n}') Y_{lm}(\mathbf{n}). \tag{c.11}$$

Here $\mathbf{n}$ and $\mathbf{n}'$ are two unit vectors, and $Y_{lm}(\mathbf{n})$ denotes the spherical harmonic function of the polar angle and azimuth of the direction of $\mathbf{n}$ relative to a fixed system of coordinates.

If equation (c. 10) is multiplied by $P_l(\cos\theta)$ and integrated over $do = \sin\theta \, d\theta \, d\phi$, the integration with respect to ϕ gives zero for all terms on the right that contain factors $\cos m(\phi-\phi')$; using (c.6) and (c.7), we obtain

$$\int P_l(\cos\gamma) P_{l'}(\cos\theta) do = \delta_{ll'} \frac{4\pi}{2l+1} P_l(\cos\theta').$$

This result may be written in the symmetrical form

$$\int P_l(\mathbf{n}_1 . \mathbf{n}_2) P_{l'}(\mathbf{n}_1 . \mathbf{n}_3) do_1 = \delta_{ll'} \frac{4\pi}{2l+1} P_l(\mathbf{n}_2 . \mathbf{n}_3), \qquad \text{(c.12)}$$

where $\mathbf{n}_1$, $\mathbf{n}_2$, $\mathbf{n}_3$ are three unit vectors and the integration is with respect to the direction of $\mathbf{n}_1$.

Finally, we give the first few normalized spherical harmonics Y_{lm}:

$$\begin{aligned}
Y_{00} &= 1/\sqrt{(4\pi)};\\
Y_{10} &= i\sqrt{(3/4\pi)}\cos\theta, \qquad Y_{1,\pm 1} = \mp i\sqrt{(3/8\pi)}\sin\theta \,.\, e^{\pm i\phi};\\
Y_{20} &= \sqrt{(5/16\pi)}(1-3\cos^2\theta),\\
Y_{2,\pm 1} &= \pm\sqrt{(15/8\pi)}\cos\theta\sin\theta \,.\, e^{\pm i\phi},\\
Y_{2,\pm 2} &= -\sqrt{(15/32\pi)}\sin^2\theta \,.\, e^{\pm 2i\phi};\\
Y_{30} &= -i\sqrt{(7/16\pi)}\cos\theta(5\cos^2\theta-3),\\
Y_{3,\pm 1} &= \pm i\sqrt{(21/64\pi)}\sin\theta(5\cos^2\theta-1)e^{\pm i\phi},\\
Y_{3,\pm 2} &= -i\sqrt{(105/32\pi)}\cos\theta\sin^2\theta \,.\, e^{\pm 2i\phi},\\
Y_{3,\pm 3} &= \pm i\sqrt{(35/64\pi)}\sin^3\theta \,.\, e^{\pm 3i\phi}.
\end{aligned}$$

§d. The confluent hypergeometric function

The *confluent hypergeometric function* is defined by the series

$$F(\alpha, \gamma, z) = 1 + \frac{\alpha}{\gamma}\frac{z}{1!} + \frac{\alpha(\alpha+1)}{\gamma(\gamma+1)}\frac{z^2}{2!} + \ldots, \qquad \text{(d.1)}$$

which converges for all finite z; the parameter α is arbitrary, while the parameter γ is supposed not zero or a negative integer. If α is a negative integer (or zero), $F(\alpha, \gamma, z)$ reduces to a polynomial of degree $|\alpha|$.

The function $F(\alpha, \gamma, z)$ satisfies the differential equation

$$zu'' + (\gamma - z)u' - \alpha u = 0, \qquad \text{(d.2)}$$

as is easily seen by direct verification.† By the substitution $u = z^{1-\gamma}u_1$, this equation is transformed into another of the same form,

$$zu_1'' + (2-\gamma-z)u_1' - (\alpha-\gamma+1)u_1 = 0. \qquad \text{(d.3)}$$

Hence we see that, for non-integral γ, equation (d.2) has also the particular integral $z^{1-\gamma} F(\alpha-\gamma+1,\ 2-\gamma,\ z)$, which is linearly independent of (d.1), so that the general solution of equation (d.2) is of the form

$$u = c_1 F(\alpha, \gamma, z) + c_2 z^{1-\gamma} F(\alpha-\gamma+1, 2-\gamma, z). \qquad \text{(d.4)}$$

† The equation (d.2) with a negative integral γ does not require special discussion, since it can be reduced to a case of positive integral γ by the transformation which gives equation (d.3).

The second term, unlike the first, has a singular point at $z = 0$.

Equation (d.2) is of Laplace's type, and its solutions can be represented as contour integrals. Following the general method, we form the functions

$$P(t) = \gamma t-\alpha, \quad Q(t) = t(t-1), \quad Z(t) = t^{\alpha-1}(t-1)^{\gamma-\alpha-1},$$

so that

$$u = \int e^{tz}t^{\alpha-1}(t-1)^{\gamma-\alpha-1}\,dt. \tag{d.5}$$

The path of integration must be chosen so that the function $V(t) = e^{tz}t^{\alpha}(t-1)^{\gamma-\alpha}$ returns to its original value on traversing the path. Applying the same method to equation (d.3), we can obtain for u a contour integral of another form:

$$u = z^{1-\gamma}\int e^{tz}t^{\alpha-\gamma}(t-1)^{-\alpha}\,dt.$$

The substitution $tz \to t$ reduces this integral to the convenient form

$$u(z) = \int e^{t}(t-z)^{-\alpha}t^{\alpha-\gamma}\,dt, \tag{d.6}$$

and the function V to

$$V(t) = e^{t}t^{\alpha-\gamma+1}(t-1)^{1-\alpha}.$$

The integrand in (d.6) has in general two singular points, at $t = z$ and $t = 0$. We take a contour of integration C which passes from infinity (re $t \to -\infty$) round the two singular points in the positive direction and back to infinity (Fig. 56). This contour satisfies the required conditions, since $V(t)$ vanishes at its ends. The integral (d.6), taken along the contour C, has no singular point for $z = 0$; hence it must be the same, apart from a

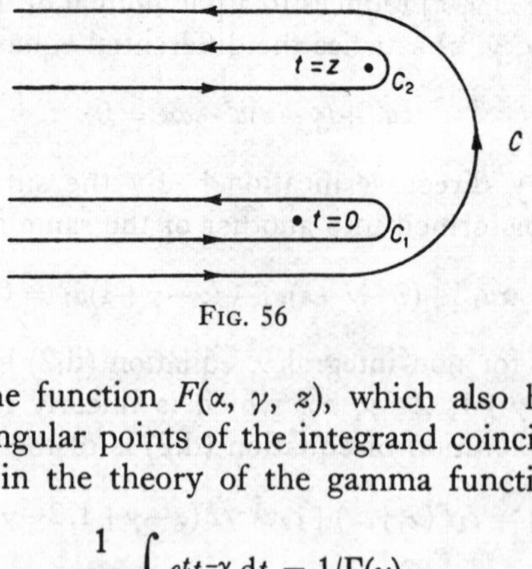

FIG. 56

constant factor, as the function $F(\alpha, \gamma, z)$, which also has no singularity. For $z = 0$ the two singular points of the integrand coincide; according to a well-known formula in the theory of the gamma function,

$$\frac{1}{2\pi i}\int_C e^{t}t^{-\gamma}\,dt = 1/\Gamma(\gamma). \tag{d.7}$$

Since $F(\alpha, \gamma, 0) = 1$, it is evident that

$$F(\alpha,\gamma,z) = \frac{\Gamma(\gamma)}{2\pi i}\int_C e^t(t-z)^{-\alpha}t^{\alpha-\gamma}\,dt. \tag{d.8}$$

The integrand in (d.5) has singular points at $t = 0$ and $t = 1$. If $\mathrm{re}(\gamma-\alpha) > 0$, and γ is not a positive integer, the path of integration can be taken as a contour C' starting from the point $t = 1$, passing round the point $t = 0$ in the positive direction, and returning to $t = 1$ (Fig. 57); for $\mathrm{re}(\gamma-\alpha) > 0$,

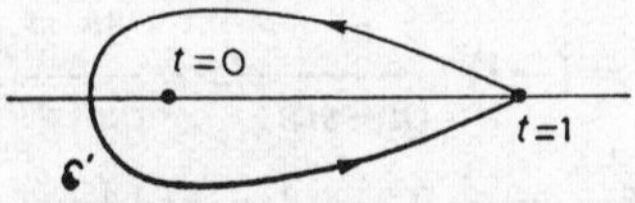

FIG. 57

the function $V(t)$ returns to its original value of zero on passing round such a contour.† The integral thus defined again has no singularity for $z = 0$, and is related to $F(\alpha, \gamma, z)$ by

$$F(\alpha,\gamma,z) = -\frac{1}{2\pi i}\frac{\Gamma(1-\alpha)\Gamma(\gamma)}{\Gamma(\gamma-\alpha)}\oint_{C'} e^{tz}(-t)^{\alpha-1}(1-t)^{\gamma-\alpha-1}\,dt. \tag{d.9}$$

The following remark should be made concerning the integrals (d.8), (d.9). For non-integral α and γ, the integrands are not one-valued functions. Their values at each point are supposed chosen in accordance with the condition that the complex quantity which is raised to a power is taken with the argument whose absolute value is least.

We may notice the useful relation

$$F(\alpha,\gamma,z) = e^z F(\gamma-\alpha,\gamma,-z), \tag{d.10}$$

which is obtained at once by substituting $t \to t+z$ in the integral (d.8).

We have already remarked that, if $\alpha = -n$, where n is a positive integer, the function $F(\alpha, \gamma, z)$ reduces to a polynomial. A concise formula can be obtained for these polynomials. Making in the integral (d.9) the substitution $t \to 1-t/z$ and applying Cauchy's formula to the resulting integral, we find

$$F(-n,\gamma,z) = \frac{1}{\gamma(\gamma+1)\dots(\gamma+n-1)}z^{1-\gamma}e^z\frac{d^n}{dz^n}(e^{-z}z^{\gamma+n-1}). \tag{d.11}$$

If also $\gamma =$ a positive integer m, we have the formula

$$F(-n,m,z) = \frac{(-1)^{m-1}}{m(m+1)\dots(m+n-1)}e^z\frac{d^{m+n-1}}{dz^{m+n-1}}(e^{-z}z^n). \tag{d.12}$$

† If γ is a positive integer, C' can be any contour which passes round both the points $t = 0$ and $t = 1$.

This formula is obtained by applying Cauchy's formula to the integral derived from (d.8) by the substitution $t \to z-t$.

The polynomials $F(-n, m, z)$, $0 \leqslant m \leqslant n$, are (apart from a constant factor) the *generalized Laguerre polynomials*

$$\begin{aligned} L_n^m(z) &= (-1)^m \frac{(n!)^2}{m!\,(n-m)!} F(-[n-m], m+1, z) \\ &= \frac{n!}{(n-m)!} e^z \frac{\mathrm{d}^n}{\mathrm{d}z^n}(e^{-z} z^{n-m}) \\ &= (-1)^m \frac{n!}{(n-m)!} e^z z^{-m} \frac{\mathrm{d}^{n-m}}{\mathrm{d}z^{n-m}}(e^{-z} z^n). \end{aligned} \tag{d.13}$$

The polynomials L_n^m for $m = 0$ are denoted by $L_n(z)$ and are called simply *Laguerre polynomials*; from (d.13) we have

$$L_n(z) = e^z \frac{\mathrm{d}^n}{\mathrm{d}z^n}(e^{-z} z^n).$$

The integral representation (d.8) is convenient for obtaining the asymptotic expansion of the confluent hypergeometric function for large z. We deform the contour into two contours C_1 and C_2 (Fig. 56), which pass round the points $t = 0$ and $t = z$ respectively; the lower branch of C_2 and the upper branch of C_1 are supposed to join at infinity. To obtain an expansion in inverse powers of z, we take $(-z)^{-\alpha}$ outside the parenthesis in the integrand. In the integral along the contour C_2, we make the substitution $t \to t+z$; the contour C_2 is thereby transformed into C_1. We thus represent the formula (d.8) as

$$\begin{aligned} F(\alpha, \gamma, z) &= \frac{\Gamma(\gamma)}{\Gamma(\gamma-\alpha)}(-z)^{-\alpha} G(\alpha, \alpha-\gamma+1, -z) + \\ &\quad + \frac{\Gamma(\gamma)}{\Gamma(\alpha)} e^z z^{\alpha-\gamma} G(\gamma-\alpha, 1-\alpha, z), \end{aligned} \tag{d.14}$$

where

$$G(\alpha, \beta, z) = \frac{\Gamma(1-\beta)}{2\pi i} \int_{C_1} \left(1+\frac{t}{z}\right)^{-\alpha} t^{\beta-1} e^t \, \mathrm{d}t. \tag{d.15}$$

In raising $-z$ and z to powers in the formula (d.14) we must take the arguments which have the smallest absolute value. Finally, expanding $(1+t/z)^{-\alpha}$ in the integrand in powers of t/z and applying formula (d.7), we have for $G(\alpha, \beta, z)$ the asymptotic series

$$G(\alpha, \beta, z) = 1 + \frac{\alpha\beta}{1!\,z} + \frac{\alpha(\alpha+1)\beta(\beta+1)}{2!\,z^2} + \ldots. \tag{d.16}$$

Formulae (d.14) and (d.16) give the asymptotic expansion of the function $F(\alpha, \gamma, z)$.

For positive integral γ, the second term in the general solution (d.4) of equation (d.2) is either the same as the first term (if $\gamma = 1$) or meaningless (if $\gamma > 1$). In this case we can take, as a set of two linearly independent solutions, the two terms in formula (d.14), i.e. the integrals (d.8) taken along the contours C_1 and C_2 (these contours, like C, satisfy the required conditions, so that the integrals along them are solutions of equation (d.2)). The asymptotic form of these solutions is given by the formulae already obtained; it remains for us to find their expansion in ascending powers of z. To do this, we start from equation (d.14) and the analogous equation for the function $z^{1-\gamma}\ F(\alpha-\gamma+1,\ 2-\gamma,\ z)$. From these two equations we express $G(\alpha, \alpha-\gamma+1, -z)$ in terms of $F(\alpha, \gamma, z)$ and $F(\alpha-\gamma+1, 2-\gamma, z)$; we then put $\gamma = p+\epsilon$ (p being a positive integer), and pass to the limit $\epsilon \to 0$, resolving the indeterminacy by L'Hospital's rule. A fairly lengthy calculation gives the following expansion:

$$G(\alpha, \alpha-p+1, -z) = -\frac{\sin \pi\alpha\,.\,\Gamma(p-\alpha)}{\pi\Gamma(p)} z^{\alpha}\times$$

$$\times\Big\{\log z\,.\,F(\alpha, p, z)+\sum_{s=0}^{\infty}\frac{\Gamma(p)\Gamma(\alpha+s)[\psi(\alpha+s)-\psi(p+s)-\psi(s+1)]}{\Gamma(\alpha)\Gamma(s+p)\Gamma(s+1)}z^{s}+$$

$$+\sum_{s=1}^{p-1}(-1)^{s+1}\frac{\Gamma(s)\Gamma(\alpha-s)\Gamma(p)}{\Gamma(\alpha)\Gamma(p-s)}z^{-s}\Big\}, \tag{d.17}$$

where ψ denotes the logarithmic derivative of the gamma function: $\psi(\alpha) = \Gamma'(\alpha)/\Gamma(\alpha)$.

§e. The hypergeometric function

The *hypergeometric function* is defined in the circle $|z| < 1$ by the series

$$F(\alpha, \beta, \gamma, z) = 1+\frac{\alpha\beta}{\gamma}\frac{z}{1!}+\frac{\alpha(\alpha+1)\beta(\beta+1)}{\gamma(\gamma+1)}\frac{z^2}{2!}+\ldots, \tag{e.1}$$

and for $|z| > 1$ it is obtained by analytical continuation of this series (see (e.6)). The hypergeometric function is a particular integral of the differential equation

$$z(1-z)u''+[\gamma-(\alpha+\beta+1)z]u'-\alpha\beta u = 0. \tag{e.2}$$

The parameters α and β are arbitrary, while $\gamma \neq 0, -1, -2, \ldots$. The function $F(\alpha, \beta, \gamma, z)$ is evidently symmetrical with respect to the parameters

α and β.† The second independent solution of equation (e.2) is

$$z^{1-\gamma}F(\beta-\gamma+1, \alpha-\gamma+1, 2-\gamma, z);$$

it has a singular point at $z = 0$.

We shall give here for reference a number of relations obeyed by the hypergeometric function.

The function $F(\alpha, \beta, \gamma, z)$ can be represented for all z, if $\mathrm{re}(\gamma-\alpha) > 0$, as an integral:

$$F(\alpha,\beta,\gamma,z) = -\frac{1}{2\pi i}\frac{\Gamma(1-\alpha)\Gamma(\gamma)}{\Gamma(\gamma-\alpha)}\oint_{C'}(-t)^{\alpha-1}(1-t)^{\gamma-\alpha-1}(1-tz)^{-\beta}\,dt, \tag{e.3}$$

taken along the contour C' shown in Fig. 57. That this integral in fact satisfies equation (e.2) is easily seen by direct substitution; the constant factor is chosen so as to give unity for $z = 0$.

The substitution $u = (1-z)^{\gamma-\alpha-\beta}u_1$ in equation (e.2) leads to an equation of the same form, with parameters $\gamma-\alpha$, $\gamma-\beta$, γ in place of α, β, γ respectively. Hence we have

$$F(\alpha,\beta,\gamma,z) = (1-z)^{\gamma-\alpha-\beta}F(\gamma-\alpha,\gamma-\beta,\gamma,z); \tag{e.4}$$

both sides of this equation satisfy the same equation, and they have the same value for $z = 0$.

The substitution $t \to t/(1-z+zt)$ in the integral (e.3) leads to the following relation between hypergeometric functions with variables z and $z/(z-1)$:

$$F(\alpha,\beta,\gamma,z) = (1-z)^{-\alpha}F(\alpha,\gamma-\beta,\gamma,z/(z-1)). \tag{e.5}$$

The value of the many-valued expression $(1-z)^{-\alpha}$ in this formula (and of similar expressions in all the following formulae) is determined by the condition that the complex quantity which is raised to a power is taken with the argument whose absolute value is least.

Next we shall give, without proof, an important formula relating hypergeometric functions with variables z and $1/z$:

$$F(\alpha,\beta,\gamma,z) = \frac{\Gamma(\gamma)\Gamma(\beta-\alpha)}{\Gamma(\beta)\Gamma(\gamma-\alpha)}(-z)^{-\alpha}F(\alpha,\alpha+1-\gamma,\alpha+1-\beta,1/z)+$$
$$+\frac{\Gamma(\gamma)\Gamma(\alpha-\beta)}{\Gamma(\alpha)\Gamma(\gamma-\beta)}(-z)^{-\beta}F(\beta,\beta+1-\gamma,\beta+1-\alpha,1/z). \tag{e.6}$$

† The confluent hypergeometric function is obtained from $F(\alpha, \beta, \gamma, z)$ by taking the limit

$$F(\alpha,\gamma,z) = \lim_{\beta\to\infty} F(\alpha,\beta,\gamma,z/\beta).$$

The notation ${}_2F_1(\alpha, \beta, \gamma, z)$ for the hypergeometric function and ${}_1F_1(\alpha, \gamma, z)$ for the confluent hypergeometric function is also used in the literature. The subscripts to the left and right of F show the numbers of parameters in the numerators and denominators respectively of the terms in the series.

This formula expresses $F(\alpha, \beta, \gamma, z)$ as a series which converges for $|z| > 1$, i.e. it is the analytical continuation of the original series (e.1).

The formula

$$F(\alpha,\beta,\gamma,z) = \frac{\Gamma(\gamma)\Gamma(\gamma-\alpha-\beta)}{\Gamma(\gamma-\alpha)\Gamma(\gamma-\beta)}F(\alpha,\beta,\alpha+\beta+1-\gamma,1-z)+$$
$$+\frac{\Gamma(\gamma)\Gamma(\alpha+\beta-\gamma)}{\Gamma(\alpha)\Gamma(\beta)}(1-z)^{\gamma-\alpha-\beta}F(\gamma-\alpha,\gamma-\beta,\gamma+1-\alpha-\beta,1-z) \tag{e.7}$$

relates hypergeometric functions of z and $1-z$; it is derived similarly to formula (e.6). Combining (e.7) with (e.6), we obtain the relations

$$F(\alpha,\beta,\gamma,z) = \frac{\Gamma(\gamma)\Gamma(\beta-\alpha)}{\Gamma(\beta)\Gamma(\gamma-\alpha)}(1-z)^{-\alpha}F(\alpha,\gamma-\beta,\alpha+1-\beta,1/(1-z))+$$
$$+\frac{\Gamma(\gamma)\Gamma(\alpha-\beta)}{\Gamma(\alpha)\Gamma(\gamma-\beta)}(1-z)^{-\beta}F(\beta,\gamma-\alpha,\beta+1-\alpha,1/(1-z)), \tag{e.8}$$

$$F(\alpha,\beta,\gamma,z) = \frac{\Gamma(\gamma)\Gamma(\gamma-\alpha-\beta)}{\Gamma(\gamma-\beta)\Gamma(\gamma-\alpha)}z^{-\alpha}F\left(\alpha,\alpha+1-\gamma,\alpha+\beta+1-\gamma,\frac{z-1}{z}\right)+$$
$$+\frac{\Gamma(\gamma)\Gamma(\alpha+\beta-\gamma)}{\Gamma(\alpha)\Gamma(\beta)}(1-z)^{\gamma-\alpha-\beta}z^{\alpha-\gamma}F\left(1-\beta,\gamma-\beta,\gamma+1-\alpha-\beta,\frac{z-1}{z}\right). \tag{e.9}$$

Each of the terms in the sums on the right of equations (e.6)–(e.9) is itself a solution of the hypergeometric equation.

If α (or β) is a negative integer or zero, $\alpha = -n$, the hypergeometric function reduces to a polynomial of the nth degree, and can be represented in the form

$$F(-n,\beta,\gamma,z) = \frac{z^{1-\gamma}(1-z)^{\gamma+n-\beta}}{\gamma(\gamma+1)\dots(\gamma+n-1)}\frac{\mathrm{d}^n}{\mathrm{d}z^n}[z^{\gamma+n-1}(1-z)^{\beta-\gamma}]. \tag{e.10}$$

These polynomials are the same, apart from a constant factor, as the *Jacobi polynomials*, defined by

$$P_n^{(a,b)}(z) = \frac{(a+1)(a+2)\dots(a+n)}{n!}F(-n,\, a+b+n+1,\, a+1,\, \tfrac{1}{2}-\tfrac{1}{2}z)$$
$$= \frac{(-1)^n}{2^n n!}(1-z)^{-a}(1+z)^{-b}\frac{\mathrm{d}^n}{\mathrm{d}z^n}[(1-z)^{a+n}(1+z)^{b+n}]. \tag{e.11}$$

For $a = b = 0$, the Jacobi polynomials are the Legendre polynomials. For $n = 0$, $P_0^{(a,b)} = 1$.

§f. The calculation of integrals containing confluent hypergeometric functions

Let us consider an integral of the form

$$J_{\alpha\gamma}{}^{\nu} = \int_0^\infty e^{-\lambda z} z^{\nu} F(\alpha, \gamma, kz)\, dz. \tag{f.1}$$

We assume that it converges. If this is so we must have re $\nu > -1$ and re $\lambda > |\text{re } k|$; if α is a negative integer, the latter condition can be replaced by re $\lambda > 0$. Using for $F(\alpha, \gamma, kz)$ the integral representation (d.9) and effecting the integration over z under the contour integral, we have

$$J_{\alpha\gamma}{}^{\nu} = -\frac{1}{2\pi i}\frac{\Gamma(1-\alpha)\Gamma(\gamma)}{\Gamma(\gamma-\alpha)}\oint_{C'}\int_0^\infty e^{-(\lambda-kt)z} z^{\nu}(-t)^{\alpha-1}(1-t)^{\gamma-\alpha-1}\, dt dz$$

$$= -\frac{1}{2\pi i}\frac{\Gamma(1-\alpha)\Gamma(\gamma)}{\Gamma(\gamma-\alpha)}\lambda^{-\nu-1}\Gamma(\nu+1)\times$$

$$\times \oint_{C'} (-t)^{\alpha-1}(1-t)^{\gamma-\alpha-1}(1-kt/\lambda)^{-\nu-1}\, dt.$$

Using (e.3), we have finally

$$J_{\alpha\gamma}{}^{\nu} = \Gamma(\nu+1)\lambda^{-\nu-1}F(\alpha, \nu+1, \gamma, k/\lambda). \tag{f.2}$$

In the cases where the function $F(\alpha, \nu+1, \gamma, k/\lambda)$ reduces to a polynomial, we have for the integral $J_{\alpha\gamma}{}^{\nu}$ an expression in terms of elementary functions:

$$J_{\alpha\gamma}{}^{\gamma+n-1} = (-1)^n\Gamma(\gamma)\frac{d^n}{d\lambda^n}[\lambda^{\alpha-\gamma}(\lambda-k)^{-\alpha}], \tag{f.3}$$

$$J_{-n,\gamma}{}^{\nu} = (-1)^n\frac{\Gamma(\nu+1)(\lambda-k)^{\gamma+n-\nu-1}}{\gamma(\gamma+1)\dots(\gamma+n-1)}\frac{d^n}{d\lambda^n}[\lambda^{-\nu-1}(\lambda-k)^{\nu-\gamma+1}], \tag{f.4}$$

$$J_{\alpha m}{}^{n} = \frac{(-1)^{m-n}}{k^{m-1}(1-\alpha)(2-\alpha)\dots(m-1-\alpha)}\Bigg\{-(m-1)!\frac{d^n}{d\lambda^n}[\lambda^{\alpha-1}(\lambda-k)^{m-\alpha-1}]+$$

$$+n!(m-n-1)\dots(m-1)\lambda^{\alpha-n-1}(\lambda-k)^{-1+m-n-\alpha}\frac{d^{m-n-2}}{d\lambda^{m-n-2}}[\lambda^{m-\alpha-1}(\lambda-k)^{\alpha-1}]\Bigg\}; \tag{f.5}$$

here m, n are integers, with $0 \leqslant n \leqslant m-2$.

Next, let us calculate the integral

$$J_{\nu} = \int_0^\infty e^{-kz} z^{\nu-1}[F(-n, \gamma, kz)]^2\, dz, \tag{f.6}$$

where n is an integer and re $\nu > 0$. To calculate this, we begin with a more general integral having $e^{-\lambda z}$ instead of e^{-kz} in the integrand. We write one of the functions $F(-n, \gamma, kz)$ as a contour integral (d.9), and then integrate over z, using formula (f.3):

$$\int_0^\infty e^{-\lambda z} z^{\nu-1}[F(-n,\gamma,kz)]^2\,dz = -\frac{1}{2\pi i}\frac{\Gamma(1+n)\Gamma(\gamma)}{\Gamma(\gamma+n)}\times$$

$$\times\int_0^\infty\oint_{C'}(-t)^{-n-1}(1-t)^{\gamma+n-1}e^{-(\lambda-kt)z}z^{\nu-1}F(-n,\gamma,kz)\,dt\,dz$$

$$= -\frac{1}{2\pi i}(-1)^n\frac{\Gamma(1+n)\Gamma^2(\gamma)\Gamma(\nu)}{\Gamma^2(\gamma+n)}\times$$

$$\times\oint_{C'}(\lambda-kt-k)^{\gamma+n-\nu}(-t)^{-n-1}(1-t)^{\gamma+n-1}\frac{d^n}{d\lambda^n}[(\lambda-kt)^{-\nu}(\lambda-kt-k)^{\nu-\gamma}]\,dt.$$

The nth derivative with respect to λ can evidently be replaced by a derivative of the same order with respect to t; we then put $\lambda = k$, and thereby return to the integral J_ν:

$$J_\nu = -\frac{1}{2\pi i}\frac{\Gamma(n+1)\Gamma(\nu)\Gamma^2(\gamma)}{\Gamma^2(\gamma+n)k^\nu}\times\oint_{C'}(-t)^{\gamma-\nu-1}(1-t)^{\gamma+n-1}\frac{d^n}{dt^n}[(1-t)^{-\nu}(-t)^{\nu-\gamma}]\,dt.$$

By integrating n times by parts, we transfer the operator d^n/dt^n to the expression $(-t)^{\gamma-\nu-1}(1-t)^{\gamma+n-1}$, and then expand the derivative by Leibniz' formula. As a result, we obtain a sum of integrals, each of which reduces to Euler's well-known integral. We finally have the following expression for the integral required:

$$J_\nu = \frac{\Gamma(\nu)n!}{k^\nu\gamma(\gamma+1)\ldots(\gamma+n-1)}\times$$

$$\times\left\{1+\sum_{s=0}^{n-1}\frac{n(n-1)\ldots(n-s)(\gamma-\nu-s-1)(\gamma-\nu-s)\ldots(\gamma-\nu+s)}{[(s+1)!]^2\gamma(\gamma+1)\ldots(\gamma+s)}\right\}. \tag{f.7}$$

It is easy to see that the integrals J_ν are related by

$$J_{\gamma+p} = \frac{(\gamma-p-1)(\gamma-p)\ldots(\gamma+p-1)}{k^{2p+1}}J_{\gamma-1-p}, \tag{f.8}$$

where p is any integer.

We similarly calculate the integral

$$J = \int_0^\infty e^{-\lambda z}z^{\gamma-1}F(\alpha,\gamma,kz)F(\alpha',\gamma,k'z)\,dz. \tag{f.9}$$

We represent the function $F(\alpha', \gamma, k'z)$ as a contour integral (d.9), and integrate over z, using formula (f.3) with $n = 0$:

$$J = -\frac{1}{2\pi i}\frac{\Gamma(1-\alpha')\Gamma(\gamma)}{\Gamma(\gamma-\alpha')}\oint_{C'}\int_0^\infty (-t)^{\alpha'-1}(1-t)^{\gamma-\alpha'-1}z^{\gamma-1}e^{-z(\lambda-k't)}F(\alpha,\gamma,kz)\,dz\,dt$$

$$= -\frac{1}{2\pi i}\frac{\Gamma(1-\alpha')\Gamma^2(\gamma)}{\Gamma(\gamma-\alpha')}\oint_{C'}(-t)^{\alpha'-1}(1-t)^{\gamma-\alpha'-1}(\lambda-k't)^{\alpha-\gamma}(\lambda-k't-k)^{-\alpha}\,dt.$$

By the substitution $t \to \lambda t/(k't+\lambda-k')$, this integral is brought to the form (e.3), giving

$$J = \Gamma(\gamma)\lambda^{\alpha+\alpha'-\gamma}(\lambda-k)^{-\alpha}(\lambda-k')^{-\alpha'}F\left(\alpha,\alpha',\gamma,\frac{kk'}{(\lambda-k)(\lambda-k')}\right). \qquad \text{(f.10)}$$

If α (or α') is a negative integer, $\alpha = -n$, this expression can be rewritten, using (e.7), as

$$J = \frac{\Gamma^2(\gamma)\Gamma(\gamma+n-\alpha')}{\Gamma(\gamma+n)\Gamma(\gamma-\alpha')}\lambda^{-n+\alpha'-\gamma}(\lambda-k)^n(\lambda-k')^{-\alpha'}\times$$

$$\times F\left(-n,\alpha',-n+\alpha'+1-\gamma,\frac{\lambda(\lambda-k-k')}{(\lambda-k)(\lambda-k')}\right). \qquad \text{(f.11)}$$

Finally, let us consider integrals of the form

$$J_\gamma^{sp}(\alpha,\alpha') = \int_0^\infty e^{-(k+k')z/2}z^{\gamma-1+s}F(\alpha,\gamma,kz)F(\alpha',\gamma-p,k'z)\,dz. \qquad \text{(f.12)}$$

The values of the parameters are supposed such that the integral converges absolutely; s and p are positive integers. The simplest of these integrals, $J_\gamma^{00}(\alpha, \alpha')$, is, by (f.10),

$$J_\gamma^{00}(\alpha,\alpha') = 2^\gamma\Gamma(\gamma)(k+k')^{\alpha+\alpha'-\gamma}(k'-k)^{-\alpha}(k-k')^{-\alpha'}F\left(\alpha,\alpha',\gamma,-\frac{4kk'}{(k'-k)^2}\right); \qquad \text{(f.13)}$$

if α (or α') is a negative integer, $\alpha = -n$, we can also write, by (f.11),

$$J_\gamma^{00}(-n,\alpha') = 2^\gamma\frac{\Gamma(\gamma)(\gamma-\alpha')(\gamma-\alpha'+1)\dots(\gamma-\alpha'+n-1)}{\gamma(\gamma+1)\dots(\gamma+n-1)}\times$$

$$\times(-1)^n(k+k')^{-n+\alpha'-\gamma}(k-k')^{n-\alpha'}F\left[-n,\alpha',\alpha'+1-n-\gamma,\left(\frac{k+k'}{k-k'}\right)^2\right]. \qquad \text{(f.14)}$$

The general formula for $J_\gamma^{sp}(\alpha, \alpha')$ can be derived, but it is so complex that it cannot be used conveniently. It is more convenient to use recurrence formulae, which enable us to reduce the integrals $J_\gamma^{sp}(\alpha, \alpha')$ to the integral with $s = p = 0$. The formula

$$J_\gamma^{sp}(\alpha, \alpha') = \frac{\gamma-1}{k}\{J_{\gamma-1}^{s,p-1}(\alpha, \alpha') - J_{\gamma-1}^{s,p-1}(\alpha-1, \alpha')\} \tag{f.15}$$

enables us to reduce $J_\gamma^{sp}(\alpha, \alpha')$ to the integral with $p = 0$. The formula

$$J_\gamma^{s+1,0}(\alpha, \alpha') = \frac{4}{k^2-k'^2}\{[\tfrac{1}{2}\gamma(k-k') - k\alpha + k'\alpha' - k's]J_\gamma^{s0}(\alpha, \alpha') + \\ + s(\gamma-1+s-2\alpha')J_\gamma^{s-1,0}(\alpha, \alpha') + 2\alpha' s J_\gamma^{s-1,0}(\alpha, \alpha'+1)\} \tag{f.16}$$

then makes possible the final reduction to the integral with $s = p = 0$.†

† See W. Gordon, *Annalen der Physik* [5] **2**, 1031, 1929.

INDEX

Acceleration, in quantum mechanics 56
Accidental degeneracy 119, 124, 254
Action 20, 165 n.
Adiabatic perturbations 148, 194–6
Airy function 75, 650–2
Alternation of functions 233
Angular momentum (IV) 82 ff.
  addition of 99–101, (XIV) 431 ff.
  commutation relations 83–5, 93, 97
  eigenfunctions 89–92
  eigenvalues 86–9
  orbital 198
    total 200
  spin 198
  vibrational 420
Annihilation operator 241
Anticommuting operators 98
Anti-Hermitian operator 15
Antisymmetric term 330
Antisymmetric unit tensor 84
Apparatus 2–3, 21–4
Atom (X) 249 ff.
  in magnetic field 461–8
  interaction at large distances 339–42
Atomic energy levels 249–50
Atomic units 117 n.
Auger effect 278–9
Axial vector 98
Azimuthal quantum number 104

Barrier, potential 79–81
  in quasi-classical case 178–85, 189–93
Bilateral axis 361
Binary transformations 205 n.
Bohr and Sommerfeld's quantization rule 171
Bohr magneton 453
Bohr radius 117 n.
Born approximation 513–18
Bose (–Einstein) statistics 226
  second quantization 240–5
Bosons 226
Bound state 29
Breit and Wigner's formula 603–11

Centrally symmetric field, motion in (V) 102 ff.
  quasi-classical 175–8
Centrifugal energy 104
Channel of reaction 591
  input 591
  widths 606–7

Charge symmetry 472
Class of group 359
Classically inaccessible 166
Clebsch–Gordan coefficients 434
Closed set of functions 8
Closed shell 251
Coherent states 71 n.
Collision of the second kind 343, 351
Collisions, *see* Elastic collisions; Inelastic collisions
Commutation relations 43, 46, 83–5, 93, 97, 198, 411, 455
Commutative operators 14
Commutator of operators 15
Commuting operators 14
Complete description 4, 5
Complete set
  of functions 8
  of quantities 5
Complex compounds 312
Compound nucleus 604
Compound system 558
Configuration space 6
Conservation of
  angular momentum 82
  energy 27
  momentum 42
  parity 98
Conserved quantity 27
  in Coulomb field 124
Continuous spectrum 8, 15–19, 563–7
Contravariant spinor components 205
Coordinate representation 44
Coriolis interaction 422
Coset 359
Coulomb degeneracy 119, 124–6
Coulomb field, motion in 117 ff.
  scattering in 560–3
Coulomb functions 571 n.
Coulomb units 117
Covariant spinor components 206
Creation operator 242
Cross-section
  partial 505
  reaction 592
  scattering 503, 607
  transport 516, 577
Current density 57
  in magnetic field 470–1

De Broglie wavelength 51

Degeneracy of energy levels 28, 88, 249
accidental 119, 124, 254
Coulomb 119, 124–6
permutational 235
removal of 139
Delta function 17, 43 n., 152, 507 n.
Density matrix 38–41
Derivative, in quantum mechanics 26, 31
Detailed balancing, principle of 602
Deuteron disintegration 192, 644–6
Diatomic molecule (XI) 298 ff.
Dimension of representation 197 n., 369
Dipole moment 279
Discrete spectrum 8–15
Dispersion relation 529–32
Doublets (*see also* Multiplet)
levels 250 n.
relativistic 277
screening 278
Dummy indices 206

Eigenfunctions 8
Eigenvalues 8
complex 556
Eikonal approximation 538 n., 640
Elastic collisions (XVII) 502 ff.
with inelastic processes 591–7
Electron
configuration 251, 272–5
diffraction experiment 1–2
states in the atom 250 ff.
terms in the diatomic molecule 298 ff.
intersection of 300–3
Element of group, *see* Group, element of
Energy 27
levels 28
atomic 249–50
complex 556
degenerate 28, 88, 249
hydrogen-like 254–5
of linear oscillator 68
anharmonic 136
in magnetic field 457
in potential well 65
in quasi-classical case 171–4
vibrational 403–5
virtual 550
width of 159, 555, 606
representation 33 n.
Equivalent
axes 361
planes 361
states 251
Eulerian angles 214
Even-even nuclei 484
Even states 97
of molecules 299
Exchange
integral 231
interaction 230, 568

"Fall" to the centre 54, 114–17
Fermi(–Dirac) statistics 226
second quantization 245–8
Fermions 226
Fine structure 250, 320
of atomic levels 265–9
Finite-displacement operator 45
Finite motion 29
Finite-rotation matrix 215
Forbidden transition 101
Form factor, atomic 576
Franck and Condon's principle 344
Free motion 50–2
in centrally symmetric field 105–14

Galileo's relativity principle 50
Generators of rotation group 388 n.
Ground state 28
Group
Abelian 358
class of 359
conjugate 360
continuous 368, 387–91
cubic 367 n.
cyclic 358
direct product of 360
double 392
element of 358
conjugate 359
generating 379
inverse of 358
multiplication of 358
order of 358
period of 358
product of 358
unit 358
finite 358
icosahedron 368
isomorphous 360
normal divisor of 360
octahedron 367
order of 359
point 360–8, 376–80, 387–95
representation of 368–83; *see also* Representation of group
rotation 387
sub- 358
conjugate 360
tetrahedron 365
theory 358 ff.
unit element of 358
Gyromagnetic factor 463, 485

Hamiltonian (operator) 26
of freely moving particle 50
of interacting particles 51
of linear oscillator 67
Heisenberg representation 37–8
Helium atom 255–8

Hermite polynomials 69, 647–9
Hermitian
conjugate operator 11
matrix 32
operator 11
Heteropolar binding 310
High energies
elastic scattering at 535–42, 575–9
inelastic scattering at 620–9, 640–6
Hole 252, 277
Homogeneous field, motion in 74–6
Homopolar binding 311
Hund's rule 251
Hydrogen atom 254–5, 287–97, 460–1, 579, 627–9
Hydrogen-like energy levels 254–5
Hyperfine structure
of atomic levels 496–9
of molecular levels 499–501
Hypergeometric functions 655–65

Identical particles (IX) 225 ff.
collisions of 567–70
Index of sub-group 359
Indistinguishability of similar particles 225
Inelastic collisions (XVIII) 591 ff.
Inert gases 273
Intermediate groups 272, 273–5, 311–12
Intermediate states 156
Intersection of electron terms 300–3
Invariant integration 388
Invariant sub-group 360
Inverse operator 14
Inversion transformation 97–9
Ionization of atoms
in α decay 150–1
in β decay 150
by electric field 292
by electrons 620, 628–9
by heavy particles 635
near threshold 620 n.
Irreducible tensor 212
Isobaric invariance 472
Isospin 473
Isotopic
invariance 472–6
shift 494–6
spin 473

Jacobi polynomials 217, 661
Jahn-Teller theorem 407
jj coupling 269, 483

Kernel of operator 10
Kramers' theorem 223
Kronecker product 374

Laboratory system of coordinates 502
Laguerre polynomials 658
generalized 119, 658
Lambda (Λ)-doubling 336–9
Landau levels 457
Landé *g*-factor 463
Landé's interval rule 267
Langevin's formula 464
Large distances, atoms at 339–42
Legendre polynomials 652–5
associated 90–1, 653–5
Linear operator 10 n.
Linear oscillator 67–74
anharmonic 136
Line of nodes 214
Logarithmic accuracy 460
Low energies
elastic scattering at 542–55
inelastic scattering at 597–9
LS coupling 268
Luminescence 523 n.

Magic numbers 485
Magnetic field, motion in (XV) 453 ff.
Magnetic quantum number 104, 130
Magnetons 453
Mass of particle 50
reduced 102
Matrix 30 ff.
density 38–41
diagonal form of 33
elements 31
for addition of angular momenta 448–50
of angular momentum 88–9
for axially symmetric systems 450–2
in classical limit 173–4
of derivative 31
diagonal 32
for diatomic molecule 332–6
quasi-classical 185–9
reduced 94, 440
selection rules for 383–7
of tensors 439–42
of vectors 92–6
finite-rotation 215
Hermitian 32
multiplication of 32
Pauli 202
trace of 36
Mean value 10
Measurements 3–5, 21–4, 157–9
predictable 5
Mirror nuclei 472 n.
Mixed states 39 n.
Molecular terms
and atomic terms 303–7
classification of 423–30
Molecule, *see* Diatomic molecule; Polyatomic molecules

Momentum 42–5
  commutation relations 43, 46
  representation 44
Multiplet
  inverted 267
  normal 267
  splitting 250, 320
  terms in diatomic molecule 319–29
Multiplicity 250, 299
  of frequency 397
Multipole moments 279–82

Negative terms 329, 423
Neutrons, scattering of 636–40
9*j*-symbol 447–8
Nodes 59
  line of 214
Normal coordinates 397
Normal state 28
Normalization 7, 16–17
Notation xiv, 104, 250, 277, 298
Nuclear forces 472 ff.
  charge symmetry of 472
  saturation of 479
  spin dependence of 476–8
Nuclear magneton 453
Nuclear scattering 574
Nuclear structure (XVI) 472 ff.
Nucleons 472 ff.
Nuclei (*see also* Nuclear forces; Nuclear structure)
  compound 604
  non-spherical 489–94

Occupation numbers 240
Odd-odd nuclei 484
Odd states 97
  of molecules 299
One-dimensional motion 60–81
  quasi-classical case 164–74
Operators 10 ff.
  addition of 13
  annihilation 241
  anticommuting 98
  anti-Hermitian 15
  commutative 14
  commutator of 15
  commuting 14
  creation 242
  differentiation of 26–7
  of finite displacement 45
  of finite rotation 213–19
  Hermitian 11
  Hermitian conjugate 11
  inverse 14
  linear 10 n.
  momentum 42–5
  multiplication of 14
  particle density 245
Operators (*cont.*)
  self-conjugate 12
  spin 202
  symmetrized products of 15
  transposed 11
  unitary 36
Optical model 610 n., 611
Optical theorem for scattering 509, 593
Orbital angular momentum 198
  total 200
Orthogonal functions 10
Orthohelium 256 n.
Orthohydrogen 332, 639–40
Orthonormal functions 10
Oscillation theorem 60–1
Oscillator
  in external field 149
  linear 67–74
    anharmonic 136
  strength 632
  three-dimensional 111

Parabolic
  coordinates 128
  quantum numbers 130
Parahelium 256 n.
Parahydrogen 332, 639–40
Parity 97
  addition rule for 101
  selection rules for 98
Particle density operator 245
Paschen–Back effect 465, 466
Pauli matrices 202
Pauli's principle 228
Periodic system 269–77
Permutations 232–9
Perturbation theory (VI) 133 ff.
Perturbations
  adiabatic 148, 194–6
  in diatomic molecule 350–1
Phase
  factor 7
  shift 109
  space 172, 259, 479
Physical quantity 5
Physical sheet 524
Planck's constant 20
Plane wave, resolution of 112–14
Point groups 360–8, 376–80, 387–95
  continuous 368, 387–91
Poisson bracket 27 n., 36 n.
Polar vector 98
Polarizability 283
Polarization of particles, partial 219–21
Polyatomic molecules (XIII) 396 ff.
Positive terms 329, 423
Potential barrier 79–81
  in quasi-classical case 178–85, 189–93
Potential scattering 558, 607

Potential wall 76–8, 170
Potential well 63–7, 72–4, 110–11, 127–8, 137–8, 160, 162–3, 546
Predictable measurements 5
Pre-dissociation 342–53
Principal groups 272–3, 309–11
Principal quantum number 119, 250, 483
Principle of least action 20
Probability current density 57
Product
  of elements of group 358
  of groups 360
  of matrices 32
  of operators 14
  of quantities 13
  of representations of groups 374
  of spinors 207
Pseudo-potential 637
Pseudoscalar 98
ψ-operators 244
Pure states 39 n.

Quadrupole moment 279–82, 494
Quantum mechanics 2
  basic concepts of (I) 1 ff.
  and classical mechanics 2–3, 20–1, 51–2, 54, 97, 160, (VII) 164 ff., 197 f., 225, 454, 507, 518–23
Quantum number
  azimuthal 104
  magnetic 104, 130
  parabolic 130
  principal 119, 250, 483
  radial 104, 119
  vibrational 317
Quasi-classicality condition 165
Quasi-classical systems 20, 48, 114, (VII) 164 ff., 518–23
Quasi-discrete spectrum 555
Quasi-stationary states 159, 555, 589, 604

Racah coefficients 443
Radial
  quantum number 104, 119
  wave function 103–5
    in Coulomb field 117–26
    in free motion 106–9
Rainbow scattering 523 n.
Ramsauer effect 544 n.
Reciprocity theorem for scattering 510, 582, 602
Reflection above the barrier 190–1
Reflection coefficient 77
  in quasi-classical case 189–93
Regge poles 586
Regge trajectories 586
Representation
  coordinate 44
  energy 33 n.
Representation (*cont.*)
  momentum 44, 532–5
  of group 368–83
    antisymmetric product of 375
    basis of 369
    character of 370
    dimension of 197 n., 369
    direct product of 374
    equivalent 370
    irreducible 197 n., 370–83
    Kronecker product of 374
    physically irreducible 381
    reducible 370
    regular 373
    symmetric product of 375
    total vibrational 397
    two-valued 391–5
    unit 372
  of matrices 33
  of operators 37–8, 44
  of wave functions 18
  Schrödinger 37
Resonance 152
  scattering 548–60, 570–5
Retardation, effective 629–33
Rigid body, rotation of 410–19
Rotary-reflection axis 355
Rotation of molecules 314–19, 321–9, 419–23
Russell–Saunders coupling 268
Rutherford's formula 563
Rydberg's correction 254

Saddle-point method 651
Scattering (*see also* Elastic collisions; Inelastic collisions)
  amplitude 503, 523–35
    nuclear 574
    partial 505
  general theory of 502 ff.
  length 544
  in magnetic field 541–2
  matrix 509 ff., 599–603
  operator 509
  potential 558, 607
  quasi-classical 518–23
  rainbow 523 n.
  resonance 548–60, 570–5
Schrödinger representation 37
Schrödinger's equation 51 ff.
  in central field 102–3
  for free particle 51
  in homogeneous field 74
  for linear oscillator 69
  in magnetic field 453–6
  in one dimension 60, 64
  in quasi-classical case 164
Second quantization 239–48
Secular equation 139
Selection rules 93–4, 98, 383–7, 440

Self-adjoint operator, *see* Self-conjugate operator
Self-conjugate operator 12
Self-consistent field 250, 255–8, 481
Shell model of the nucleus 480–9
Sign of terms 329–30, 423
6*j*-symbols 442–7
Slow particles
  elastic scattering of 542–55
  inelastic scattering of 597–9
S-matrix 509; *see also* Scattering matrix
Spectral terms 250, 298
Spectrum
  continuous 8, 15–19, 563–7
  discrete 8–15
  of eigenvalues 8
Spherical Bessel functions 107
Spherical harmonics 91, 652–5
Spherical tensor 439
Spherical top 411
Spin (VIII) 197 ff.
  commutation relations 199
  components 198
  in magnetic field 468–70
  nuclear 483
  operator 202
  total 200
  variable 198
Spin–axis interaction 319
Spin–orbit interaction 265–8, 319, 482, 579–85
Spin–spin interaction 265, 268, 324
Spinors 204 ff.
  contraction of 207
  metric 207
  multiplication of 207
  symmetrical 207
  and tensors 208–13
  unit 207
Spur of matrix 36
Stability of molecules 405–10
Stark effect 282–97
  in diatomic molecule 335–6
  linear 287
Stationary states 28
Statistical weight 331 n., 425 n., 603
Sum of quantities 13
Summation theorem 630
Superposition principle 7
s-wave scattering 543
Symmetric term 330
Symmetrization
  of functions 233
  of spinors 207–10
Symmetry
  axis of 354
    bilateral 361
    equivalent 361
  centre of 355
  charge 472
  groups 357 ff.
Symmetry (*cont.*)
  plane of 354
    equivalent 361
  of terms
    in diatomic molecule 298–9, 329–32
    in polyatomic molecules 396 ff.
  theory of (XII) 354 ff.
  transformations 354–7

Tensor
  antisymmetric unit 84
  forces 477
  irreducible 212
  matrix elements of 439–42
  spherical 439
Thomas–Fermi method 259–64, 275–7
3*j*-symbols 431–8
Threshold of reactions 614
Time reversal 24, 55, 205 n., 221–4, 434 n., 453 n., 510, 602
Time-reversed states 602
Top 410–19
  asymmetrical 413–16
  spherical 411
  symmetrical 412–13
Trace of matrix 36
Transition
  frequency 31
  probability, in quasi-classical case 189–93
Transmission coefficient 77
  in quasi-classical case 180–5
Transport cross-section 516, 577
Transposed operator 11
Tunnel effect 171
Turning points 165

Uncertainty
  principle 2
  relations 47
    for energy 157
Unitarity condition 509
  complex 586
Unitary operator 36

Valency 307–14
Van der Waals forces 340
Variational principle 58–60
Vector addition coefficients 434
Vector model 100
Velocity, in quantum mechanics 4, 55–6, 455
Vibrational
  angular momentum 420
  coordinates 397
  energy levels 403–5
  quantum number 317
  states and rotational states in diatomic molecule 314–19, 321–9

Vibrations, molecular
anharmonic 403
classification of 396–403
interaction with rotations 419–23
Virtual level 550

Wall, potential 76–8, 170
Wave equation 26
Wave function 6, 21–4, 57 ff.
antisymmetrical 226
of boson system 227
coordinate 228
of fermion system 227
in magnetic field 455, 456–7
near nucleus 264–5
orbital 228
quasi-classical 164 ff.
radial 103–5, 106–9, 117–26
spin 228
for arbitrary spin 208–10
symmetrical 226
Wave mechanics 2
Wave number 61, 106
Wave packet 21, 47
Well, potential, *see* Potential well
Width
of channel 606–7
of level 159, 555
Wigner–Eckart theorem 440
Wigner 3*j*-symbols 433

X-ray terms 277–9

Young diagrams 234–9

Zeeman effect 462–8, 499
anomalous 463 n.